REVISITING THE FOUNDATIONS OF RELATIVISTIC PHYSICS

BOSTON STUDIES IN THE PHILOSOPHY OF SCIENCE

Editors

ROBERT S. COHEN, *Boston University*
JÜRGEN RENN, *Max-Planck-Institute for the History of Science*
KOSTAS GAVROGLU, *University of Athens*

Editorial Advisory Board

THOMAS F. GLICK, *Boston University*
ADOLF GRÜNBAUM, *University of Pittsburgh*
SYLVAN S. SCHWEBER, *Brandeis University*
JOHN J. STACHEL, *Boston University*
MARX W. WARTOFSKY†, *(Editor 1960–1997)*

VOLUME 234

REVISITING THE FOUNDATIONS OF RELATIVISTIC PHYSICS

Festschrift in Honor of John Stachel

Editorial Team
(Max Planck Institute for the History of Science)

JÜRGEN RENN
LINDY DIVARCI
PETRA SCHRÖTER

Editorial Committee

ABHAY ASHTEKAR
ROBERT S. COHEN
DON HOWARD
JÜRGEN RENN
SAHOTRA SARKAR
ABNER SHIMONY

KLUWER ACADEMIC PUBLISHERS
DORDRECHT / BOSTON / LONDON

A C.I.P. Catalogue record for this book is available from the Library of Congress.

ISBN 1-4020-1284-5

Published by Kluwer Academic Publishers,
P.O. Box 17, 3300 AA Dordrecht, The Netherlands.

Sold and distributed in North, Central and South America
by Kluwer Academic Publishers,
101 Philip Drive, Norwell, MA 02061, U.S.A.

In all other countries, sold and distributed
by Kluwer Academic Publishers,
P.O. Box 322, 3300 AH Dordrecht, The Netherlands.

Printed on acid-free paper

All Rights Reserved
© 2003 Kluwer Academic Publishers
No part of this work may be reproduced, stored in a retrieval system, or transmitted
in any form or by any means, electronic, mechanical, photocopying, microfilming, recording
or otherwise, without written permission from the Publisher, with the exception
of any material supplied specifically for the purpose of being entered
and executed on a computer system, for exclusive use by the purchaser of the work.

Printed in the Netherlands.

TABLE OF CONTENTS

Autobiographical Reflections
John Stachel xi

Introduction
Abhay Ashtekar, Jürgen Renn and Abner Shimony xv

I: Historical and Philosophical Roots of Relativity

The Prehistory of Relativity
Jean Eisenstaedt 3

Interpretations and Equations of the Michelson Experiment
and its Variations
Horst Melcher 13

The Trouton Experiment, $E = mc^2$, and a Slice of Minkowski Space-Time
Michel Janssen 27

The N-Stein Family
John D. Norton 55

Eclipses of the Stars:
Mandl, Einstein, and the Early History of Gravitational Lensing
Jürgen Renn and Tilman Sauer 69

The Varieties of Unity: Sounding Unified Theories 1920-1930
Catherine Goldstein and Jim Ritter 93

Indiscernibles, General Covariance, and Other Symmetries:
The Case for Non-Reductive Relationalism
Simon Saunders 151

On Relative Orbital Rotation in Relativity Theory
David B. Malament 175

II: Foundational Issues in Relativity and their Advancement

The Unique Nature of Cosmology
George F. R. Ellis 193

Time, Structure, and Evolution in Cosmology
Lee Smolin 221

Timekeeping in an Expanding Universe
James L. Anderson 275

Gravitational Lensing from a Space-Time Perspective
Jürgen Ehlers, Simonetta Frittelli, and Ezra T. Newman .281

Rigidly Rotating Disk Revisted
C. V. Vishveshwara 305

DSS 2+2
Ray A. d'Inverno 317

Geometry, Null Hypersurfaces and New Variables
David C. Robinson 349

On Vacuum Twisting Type-N Again
Jerzy F. Plebański and Maciej Przanowski 361

Quasi-Local Energy
Joshua N. Goldberg 375

Space-Time Defects: Open and Closed Shells Revisited
Reinaldo J. Gleiser and Patricio S. Letelier 383

Dimensionally Challenged Gravities
S. Deser 397

A Note on Holonomic Constraints
Wlodzimierz M. Tulczyjew 403

Towards an Action-at-a-Distance Concept of Spacetime
Daniel H. Wesley and John A. Wheeler 421

III: Foundational Issues in Quantum Physics and their Advancement

Inevitability, Inseparability and *Gedanken* Measurement
Mara Beller 439

The Concept of Quantum State: New Views on Old Phenomena
Michel Paty 451

Elementary Processes
David Ritz Finkelstein 479

On Quantum Non-Locality, Special Relativity, and Counterfactual Reasoning
Abner Shimony and Howard Stein 499

Coherence, Entanglement, and Reductionist Explanation in Quantum Physics
Greg Jaeger and Sahotra Sarkar 523

IV: Science, History, and the Challenges of Progress

Physics and Science Fiction
Allen I. Janis 545

Can We Learn From History? Do We Want To?
Lazlo Tisza. 555

Patterns of Appropriation in the Greek Intellectual Life of the 18th Century:
Case Study on the Notion of Time
Kostas Gavroglu and Manolis Patiniotis 569

Darwin, Marx, and Warranted Progress:
Materialism and Views of Development in Nineteenth-Century Germany
Wolfgang Lefèvre 593

Albert Einstein and the Founding of Brandeis University
Silvan S. Schweber 615

Appendix

John Stachel's Publications 641

AUTOBIOGRAPHICAL REFLECTIONS

A talk given on June 5, 1998 by John Stachel at the Max Planck Institute for the History of Science in Berlin on the occasion of a workshop honoring his 70th birthday "Space-Time, Quantum Entanglement and Critical Epistemology"

I have found from experience—in Boston and Berlin—that when one turns seventy, one's friends forget all one's sins of omission and commission (I hope mine have been mostly those of omission)—at least for the moment—and remember only the good deeds which they recall, often with an exaggeration that I can only ascribe to fondness.

In the course of my life, I have made many friends and not a few enemies. It seems to me that a life worth living needs both. There have been many gratifications in my life, but also many disappointments and still unresolved problems. My experiences have been such as to make me critical of many current shibboleths, including many that once were my own. As Iago said, "I am nothing if not critical."[1] I hope my experiences have made me much less dogmatic now than I used to be. Let me recall just a few of these problems.

I grew up a Jew in a predominantly Christian society, yet a non-religious Jew with no links to the organized Jewish community—what Isaac Deutscher called a non-Jewish Jew. My family roots in East-European Yiddish culture, on the one hand, on the other, the "civilized" anti-Semitism of my homeland, and above all the barbaric forms which anti-Semitism took in Germany while I was growing up, culminating in the holocaust, left me in no doubt of my Jewish identity. Indeed, it is a good example of the "List der Vernunft," the cunning of reason, that a celebration of my 70th birthday should be taking place in Berlin. If someone had foretold this fifty years ago, I would have been incredulous, to say the least.

Yet my pride as a Jew in the accomplishments of my people in surviving and overcoming the murderous assault on their very existence is tempered by the experience of seeing how the victims of oppression have themselves become oppressors of another people. Until the Palestinian people have fully achieved their liberation, I cannot think about the state of Israel with unmixed emotions.

I was brought up in a Communist family, in the heartland of capitalism. This was during the 1930s, a time when American capitalism was suffering its most profound crisis—so far—and the American Communist Party was at its strongest. The Soviet Union then still embodied the hopes of many people around the world, including quite a few in the United States, for a socialist transformation of society. My experiences inculcated in me a spirit of internationalism and inoculated me against an uncritical worship of the "American Way of Life." Of course, a healthy sense of pride in one's national heritage is entirely compatible with an internationalist orientation—and in that sense I am proud to be an American; but an all-too-common jingoistic nationalism is not. Certainly, I could not take pride in the America of the Vietnam War, nor in its current role as would-be policeman of the world. For me, shame over and struggle

A. Ashtekar et al. (eds.). Revisiting the Foundations of Relativistic Physics, ix-xiv.
© 2003 Kluwer Academic Publishers: Printed in the Netherlands.

against the wrongs done in the name of one's country are part of true patriotism.

My father was a leader of the American Communist Party, imprisoned for his beliefs during the McCarthy era. For many years I was a dogmatic Communist. But, as happened to so many in the Communist movement, at a certain point—in my case the Soviet invasion of Hungary in 1956—I had to reexamine my views. I came to see, as someone well put it, that the problem with the Communist Party was not that it was too far to the left, but that it was too far to the East. That is, notwithstanding many good causes it championed in the United States, it became a servant of Soviet Russian interests that had nothing to do with the cause of socialism.

But unlike many who left the Communist Party, I turned left rather than right, and returned—or rather turned for the first time—to a critical examination of Marx's work. I found—and still find—that his analysis of capitalism, which for me is the heart of his work, provides the best starting point, the best critical tools, with which—suitably developed—to understand contemporary capitalism. I remind you that this year is also the sesquicentennial of the Communist Manifesto, a document that still haunts the capitalist world.

This understanding of capitalism inoculated me against the current wave of worship of the not-so-free "free market" that followed the collapse of the Soviet bloc. This collapse affected the core of Marx's analysis of capitalism as little as the collapse of the Holy Roman Empire affected Jesus of Nazareth's teachings in the Sermon on the Mount.

After over fifty years of grappling with this problem, and over forty of trying to free myself from dogmatic approaches to it, this is how I see the problem of capitalism and socialism. In order to survive, capitalism must continually deal with a series of antagonisms inherent to it, starting with the antagonistic relation between capital and labor that lies at its core. Any temporary solution to one or another form that these antagonisms take ultimately generates new forms. Thus, the system changes through a series of crises, partial or global. No one of these is "the final crisis," in the sense that the antagonisms provoking it cannot be solved within the capitalist framework. But these moments of crisis do offer an opportunity—and no more—for interventions by those forces within the society—primarily the labor movement and its allies—that are motivated by their position in society to move beyond capitalism toward a system of non-antagonistic relations of production that we may call socialism.

I no longer believe in the inevitability of socialism. But I do believe in the non-inevitability of capitalism. History shows that, at least in the past, there have been societies based on non-capitalistic social relations. If I am correct in asserting that capitalism cannot ever surmount its basic antagonisms, then there is always the possibility that the labor movement will succeed in an attempt to transcend capitalist social relations.

I see two main problems facing the current labor movement:

- 1) the concept of labor is too narrowly defined, even by the labor movement itself. What Marx called the collective laborer includes many intellectual workers who today neither define themselves nor are accepted by the labor movement as part of it. Until this split between labor of head and labor of hand is overcome in theory (which is easy) and in practice (which is not), the potential power of the labor movement will remain badly crippled.

- 2) the globalization of capital has far outstripped the ability of current labor movements, organized at best on a national level, to conduct an effective defense of the interests of labor within capitalism, let alone to seriously challenge the capitalist system. To develop some form—or forms—of international organization of labor, long an ideological challenge ("Workers of the World Unite") has now become an urgent matter of survival for the labor movements of the world.

Here is a challenge, on which I think broad agreement is possible: Even those who think capitalism is capable of indefinite survival must agree that it has functioned best in the past—for example, during the long period of post-World War II expansion—when the power of capital has been effectively limited by the countervailing power of labor. Effective exercise of that power has always depended on overcoming the segmentation of labor due to such factors as locality, race, gender, occupation, etc., which still remain important. Above, I have singled out the two factors that today seem key to me: the split between mental and manual labor, and segmentation by nationality. Let all concerned about the current state of capitalism work to build up the countervailing power of labor, and let time show whether this results in nothing more than the better functioning of capitalism, or whether a new challenge to the system ultimately emerges.

One thing seems clear to me even now: if we proceed along the path we are currently on, with the power of capital exerted almost without challenge on a global scale, we are heading for economic, social and ecological disasters that will make our descendants look back nostalgically on the horrors of the twentieth century![2]

In science too, my experience has made me critical. I became a physicist, which put me in a small minority; then a theoretical physicist, another minority within a minority; then a general relativist at a time (the late 1950s) when we formed a tiny and often despised minority within that minority. Here again, my experience as a general relativist inoculated me against the prevailing view among theoretical physicists that quantum theory—and in particular quantum field theory—provided the answers to all the problems of physics, such problems always boiling down to computation of a number. I saw the dynamization and relativizing of the space-time structures as the unique contribution of general relativity to fundamental physics, a contribution that was not—and is not—easy to fit together with the quantization schemes so successful in the rest of physics, based as they are on fixed background space-time structures (usually taken as those of Minkowski space).

I reacted against what I saw as the two physics imperialisms: that of unified field theories, which (following Einstein) hoped to somehow annex all quantum phenomena by finding the right classical field-theoretical generalization of general relativity; and that of special-relativistic quantum field theory, which hoped to annex gravitational phenomena by treating the field equations of general relativity as no more than a non-linear special-relativistic theory with a particularly nasty gauge group, but otherwise to be treated by standard quantization techniques.

Forty years ago such skeptical views were heretical. Roger Penrose, who shared similar views, ironically dubbed a session on possible alternative approaches (at which he spoke) "the crackpot session." Today, after sixty-odd years of failure of conventional attempts at quantization, alternative approaches receive a much more respectful hearing. The problem of the relation between quantum theory and general relativity—for short the problem of quantum gravity—remains in my opinion the outstanding challenge to the fundamental physics of the next century.

As one grows older, it is tempting to dwell on the past; in particular, for me to try to hold on to a beautiful moment like this. Yet I must resist this temptation, and say with Faust:[3]

> Werd' ich zum Augenblicke sagen
> "Verweile doch! du bist so schön!"
> Dann magst du mich in Fesseln schlagen
> Dann will ich gern zugrunde gehn.

Life means change, and to resist change is death. "Nur wer sich wandelt, bleibt mit mir verwandt" says Nietzsche.[4] One must stay open to experience. One of my favorite graphic images, which I have already described to many, is a self portrait by the aged Goya: a very old man with long white hair and beard, bent over and supported by two canes, it bears the caption: "Aun aprendo," I am still learning.[5]

But one must not only continue to learn, to guard against all rigidity of belief, all dogmatism. One must continue to act in the world, not be paralyzed by the knowledge that all our opinions are fallible. We must act to change the world, our personal world, our social world, our intellectual world, guided by our best current beliefs, but always ready to change these in the face of new information. Our knowledge may fallible, but it is corrigible!

One of the greatest gratifications that one can have in life comes from the family ties that one forms This is true of the biological family, of course. I cannot begin to tell you how much I owe to my dearest friend, lover, and companion in life for over forty-five years, my wife Evelyn, and to our children and grandchildren. To one who has not been so blessed, these joys are simply indescribable.

But for those of us in the scholarly world, there is another kind of kinship that supplements the biological: we are privileged to form intellectual families. We have our intellectual parents, siblings, and offspring. It has been and continues to be a source of deep satisfaction to me, both in Boston and Berlin, to see in person and to hear by letter from so many members of my intellectual family. Of course, like all parents; I take special pride in the younger people who think of me as their colleague, and some of whom I like to think of (to myself) as my intellectual children. I take the often exaggerated words of appreciation uttered by members of this intellectual family in the same spirit in which I take such words from members of my biological family: less as an indication of my own merits, more as an expression of the depth of their feelings and good wishes. Thank you all! As long as body and mind permit, I shall try to keep up our intellectual family contacts, to continue to give a little to you, in return for so much that over the years you have given to me.

NOTES

1. William Shakespeare, *Othello*, Act II, Scene 1.
2. When I spoke these words almost five years ago, I did not expect my gloomy forecast to be realized so quickly; but, for Americans, 9/11/2001 seems to have begun its fulfillment with a vengeance.
3. J.W. Goethe, *Faust*, Part I, Scene 6.
4. Friedrich Nietzsche, Nachgesang, "Aus hohen Bergen," in *Jenseits von Gut und Böse*.
5. It is in the Prado, Madrid.

ABHAY ASHTEKAR, JÜRGEN RENN, ABNER SHIMONY

INTRODUCTION

This Festschrift is dedicated to John Stachel in honor of his seventieth birthday. The broad range of authors and themes represented in this volume testify to John Stachel's own wide intellectual horizons as well as to the large area covered by the intellectual circle around him. The contributions assembled here represent a gift not only to an eminent scholar but also to a warm and inspiring human being. Written by prominent authors in physics as well as in the history and philosophy of physics, these contributions may also constitute a useful compendium for any reader interested in the present discussion about the foundations of relativistic physics. They demonstrate that this discussion can only be fully appreciated if normally disparate strands are brought together and, in particular, if historians of science are willing to learn from active scientists, if philosophers of science are prepared to learn from historians not only about case studies but also about contexts, and if physicists are patient enough to listen to both historians and philosophers. The outstanding scholarly achievements of the physicist, philosopher, and historian John Stachel amply document the benefits that can be drawn from such lessons. He has not only succeeded in crossing disciplinary boundaries in his own work but, perhaps even more importantly, has created a interdisciplinary intellectual community, stimulated by his work and example. The contributions to this volume represent a small selection of the production of this community which has found, for many years now, their *Zentralorgan* in the *Einstein Studies*, co-edited by John Stachel. Many of the volumes that have appeared in this series of books are based on conferences on the history and philosophy of general relativity, another initiative of Stachel.

In accordance with the main themes relevant to Stachel's own oeuvre, the present volume has been divided into four parts, dealing with the "Historical and Philosophical Roots of Relativity," "Foundational Issues in Relativity and their Advancement," "Foundational Issues in Quantum Physics and their Advancement," and with "Science, History, and the Challenges of Progress," respectively. The volume opens with an autobiographical essay by John Stachel, written on the occasion of a workshop held in his honor in 1998 at the Max Planck Institute for the History of Science in Berlin. As one of this institute's first visiting scholars, he has in fact contributed much to shaping its scholarly profile from the very beginning. The volume closes with an appendix displaying the impressive list of Stachel's publications.

The first part is dedicated to the historical roots of the special and general theories of relativity, as well as the philosophical implications of relativity theory. Each contribution to this part is in one way or another related to John Stachel's own seminal work in the history of relativity which has profoundly shaped the entire field. As a matter of fact, it is hardly conceivable that *any* scholarly contribution to the understanding of Einstein's work today would not be indebted to *The Collected Papers of Albert Einstein* and thus to their founding editor John Stachel. With his editorial team he not only gathered, identified, and analyzed primary sources, publishing them together with enlightening annotations and masterful surveys that have since been reprinted and translated into numerous languages, he also used the institution of the Collected Papers in an exemplary way making Einstein materials openly available and thus actively shaping the intellectual community surrounding the project. He has, in particular, stimulated many colleagues, and in particular younger scholars, to take up specific themes or materials coming up in the editorial process, thus promoting their careers in a decisive way.

Some of Stachel's key insights into the emergence of special and general relativity figure prominently in the background of the studies presented here. According to him, the tendency to oversimplify special relativity resulted from Einstein's embracing first the relativity principle and then the principle of the constancy of the speed of light, each of them understood as embodying the essence of the classical knowledge to be preserved in the new theory, mechanics and electrodynamics respectively. General relativity, on the other hand, is described by Stachel as the result of a drama in three acts, the first of which is represented by the conception of the Equivalence Principle in 1907, the second by Einstein's recognition of the non-Euclidean nature of the metric in 1912, and the third by the formulation of the field equations of gravitation in 1915. Among his many contributions to the analysis of this drama, three stand out as having attracted special attention not only from historians but also from physicists and philosophers of science. The first was his identification of the rigidly rotating disk as the "missing link" in the history of relativity, which was crucial to the second act of the drama—the identification of the metric tensor as the representation of the gravitational potential. Secondly, Stachel influentially interpreted Einstein's hole argument, originally formulated in 1913, as not only providing a stumbling block on the road to general relativity but as also representing the corner stone for a deep conceptual insight into the nature of spacetime. Thirdly, in a paper co-authored by Stachel, the myth of the anticipation of the discovery of the gravitational field equations by Hilbert was unraveled. This achievement was accomplished, as were several of his other influential contributions to the history of relativity, in the context of a major research project on the genesis of general relativity pursued at the Max Planck Institute for the History of Science.

Some of these issues are taken up by the contributors to the first part of this volume. A beautiful illustration of a conceptual analysis of the problem of rotation in relativity is provided by Malament's contribution, which deals with a definition of relative orbital rotation and its consequences. The contribution of Saunders focuses on Stachel's interpretation of the hole argument, using it to make the case for a non-reductive relationalism in an argument that reaches from Leibniz to Stachel. Stachel has in fact given what most relativists regard as the clearest expositions of Einstein's hole argument. His writings on this subject continue to have a major impact, especially on younger researchers. More generally, he has been a forceful spokesman for the deep lessons Einstein taught us. In numerous conferences and workshops bringing gravity theorists and particle physicists together, he has argued passionately for the theme "gravity is geometry," offering sharp and well-thought-out criticisms of the tendency to treat gravity just as another force in Minkowskian physics. Although largely unrecorded, these are also very significant contributions to the physics community, deeply appreciated by many.

The contributions in the first part of the volume offer a fair impression of some of the main aspects of the present discussion about the history and philosophy of relativity. This discussion is, in fact, characterized, on the one hand, by an ingenious use of the full technical apparatus of mathematical relativity for addressing philosophical questions about space and time, as exemplified by the papers by Malament and Saunders. On the other hand, it also brings out the long-range character of historical developments and of the role ordinary practice plays in the progress of science, beyond the supposedly isolated contributions of a few great discoverers. The importance of long-range developments is particularly evident in Eisenstaedt's paper on the effect of gravity on light in the framework of Newtonian theory. It is also evident in Norton's pursuit of counterfactual history of relativity as pioneered by John Stachel. Actually, Norton's paper illustrates both the continuity of problems between classical and relativistic gravitation theory and the possibility to use methods of counterfactual history in order to address philosophical issues such as the relation between the extension of the relativity of motion to acceleration and the representation of gravitational free fall by a curved affine structure. The collective practice of science is at the focus of Goldstein's and Ritter's analysis of the contributions to unified field theories between 1920 and 1930. The role of the ordinary practice of science—in the sense of practice by ordinary people—for the progress of science may also be illustrated by Einstein's interaction with an amateur scientist and the curious role this interaction played in the early history of gravitational lensing as described in the contribution by Renn and Sauer. Two further contributions to the first part deal with experiments relevant to the history of special relativity, the essay by Melcher on the Michelson experiment and the study by Janssen on the Trouton experiment of 1901. Janssen's painstaking analysis not only resurrects an ingenious but almost forgotten experiment to the glory it warrants. It also illustrates the extent to which even the establishment of special relativity was not a sudden breakthrough but a laborious process. In fact, a full explanation of the negative result of the experiment involved several fundamental conceptual insights of relativity theory introduced only in the decade following the experiment.

The second part, dedicated to "Foundational Issues in Relativity and their Advancement," circles around John Stachel's contributions to relativistic physics but also takes up stimuli produced by his historical work as is illustrated by Vishveshwara's paper on the rigidly rotating disk. Stachel's physics contributions are in fact primarily in the realm of general relativity and related areas. As historians and philosophers would expect of John Stachel, they carry his hallmark of conceptual depth. He has thus significantly contributed to preserving the style so characteristic of Einstein's own achievements, combining technical work with a conceptual analysis informed by philosophical and historical awareness. In the present volume, this style is particularly evident in the contributions by Ellis on the unique nature of cosmology, by Smolin on time, structure, and evolution in cosmology, and in the contribution by Anderson on timekeeping in an expanding universe.

Stachel's contributions to relativistic physics also testify to his remarkable mathematical abilities. In many cases, it is striking to see how far ahead of his time he was—for example in his very first work, his Master's thesis, which dealt with cylindrical gravitational waves. In it, Stachel investigated the global properties of these waves using the then newly discovered framework of null infinity and showed that, contrary to what one might have expected from the presence of the symmetry, these space-times are in fact asymptotically flat except along a single generator of null infinity. Although the mathematical ideas underlying this global analysis were very recent, his work was remarkably complete. Indeed, it could be further developed only in the early nineties. Stachel made another early contribution to the theory of gravitational radiation in exact general relativity by generalizing the Bondi-Sachs framework to include pure radiation matter fields. A second interesting set of contributions is in the area of the Cauchy problem and the issue of identifying the "true degrees of freedom" of the non-linear gravitational field. In particular, he laid the foundations of what is now known as a 2+2 formulation in which space-time is split by a family of two 2-dimensional surfaces, rather than the more familiar 3+1 decomposition in terms of a family of space-like 3-surfaces.

Accordingly, the second part includes a number of technical papers related to Stachel's interest in the foundations and further development of general relativity. Some of them challenge standard convictions or discuss exotic gravity models, in the spirit of both Einstein's and Stachel's unconventionalism, such as the paper by Wesley and Wheeler, proposing the vision of an action-at-a-distance concept of spacetime or the paper by Deser on "dimensionally challenged gravities." Some touch issues dear to Stachel's heart as they are, in one way or another, related to his own work, for example the 2 + 2 formalism exposed by d'Inverno or the masterful exposition of gravitational lensing from a spacetime perspective by Ehlers, Fritelli, and Newman. The subjects of other papers in this part may be more indirectly related to Stachel's own work, such as the issue of new variables approaches to the canonical formalism for general relativity discussed in Robinson's contribution. The canonical formalism is also used in the contribution by Goldberg suggesting an expression for a quasi-local energy in general relativity.

A further area of Stachel's research in relativistic physics concerns the equations of motion and conservation laws for particles with speeds small compared to the speed of light. The thrust of the early work in the subject was on deriving the conservation laws directly from equations of motion. This procedure can turn out to be cumbersome and is often conceptually obscure. In collaboration with Peter Havas, Stachel traced the origin of these conservation laws to symmetries of the appropriately defined Lagrangians and Hamiltonians, thereby providing a deeper understanding of the known laws and obtaining their generalizations to new situations. In his research he was also concerned with strings. In the early eighties, Stachel developed an elegant geometric description of time-like and null strings in general curved space-times. He then considered dust and fluids of strings (rather than point particles) and studied their equations of motion and conservation laws. It is remarkable that this mathematical and technically subtle work was completed during the time he was working on *The Collected Papers of Albert Einstein* in Princeton.

Stachel has enriched the research on relativistic physics with several other insightful papers. For example, he realized that the difference between exact and closed forms can be exploited to construct space-times which are locally static but globally only stationary and these lead to a gravitational analog of the Aharanov-Bohm effect. In another paper, he carried out a careful analysis of the structure of the Curzon space-time and discovered a number of interesting global properties. Yet another paper, devoted to geometry and carried out with Hubert Goenner, provides an elegant treatment of 3-dimensional isometry groups with 2-dimensional orbits in space-time. Further contributions to the second part represent an indirect reflection of this versatility of themes and methods of his physical research, in particular the paper by Plebanski and Przanowski on Cartan's structure equations for the vacuum twisting type-N and the proposal of a new description of autonomous mechanical systems by Tulczyjew.

The third part of this volume is dedicated to "Foundational Issues in Quantum Physics and their Advancement." Again, the contributions reflect themes central to Stachel's own work. Some general remarks on the character and achievements of this work may therefore not be out of place. In view of Stachel's integrity and harmoniousness of character, it is not surprising to find a continuity between features of his worldview and his ruminations on quantum mechanics. His Marxism is manifest in a pervasively critical attitude, a constant awareness of the reality of the material world and the constraints that it imposes on human activities, and an equal awareness of the reality of the social world, with its constraints on language, judgments, and ideals. His lifelong study of Einstein has implanted—or strengthened his native inclination towards—deep curiosity, independence of judgment, search for deep principles, exploratory attitude, love for mathematical beauty, and sense of wonder towards nature (which is almost a religious element in a non-religious temperament).

Though perhaps less known than his contributions to the history of relativity, Stachel has also significantly widened our views on the history of the quantum, focussing on Einstein's adumbrations, contributions, critical evaluations, and philosophical assessments of quantum mechanics. There are some novel discoveries and psycho-historical conjectures in Stachel's history, and new light is thrown even upon the well-known episodes. Stachel marshals evidence, for instance, to support the astonishing thesis that in 1904-5 Einstein had ideas about using the quantum of action to explain the structure of the atom and the emission spectrum.

In 1925 Einstein applied Bose's idea of the statistics of a gas of photons to a gas of identical atoms (later recognized to be of integral spin and named "bosons"). Stachel notes that in the studies of black-body radiation statistics and the statistics of massive bosons Einstein had intimations of the idea of entanglement before Schrödinger exhibited this feature in 1926 in many-particle wave functions. Another surprise revealed by Stachel's compiling of evidence is that Einstein anticipated Born's statistical interpretation of the wave function. He has also pointed out that, in spite of Einstein's life-long attachment to the concept of field, Einstein also explored the possibility that non-continuum concepts are appropriate for the foundations of physics—if only adequate mathematical methods were invented. The contribution by Finkelstein to this volume on elementary processes outlines such a radical non-continuum reformulation of fundamental physics (e.g., 'the chronon replaces infinitesimal locality by finite locality'). Although he does not mention Einstein's tentative remarks about the possibility of a non-continuum physical theory, he attempts to develop and apply the mathematical tools whose non-availability Einstein lamented. The historical dimension of Stachel's work on the quantum is, in this volume, reflected by Beller's discussion of inevitability, inseparability and "Gedanken measurement." In this paper, which is based on her recent book, she summarizes different ways of understanding the Copenhagen interpretation of quantum mechanics and strongly criticizes those variants that make the most radical claims, such as the final overthrow of the law of causality.

In his own contributions to the understanding of quantum physics, such as his penetrating discussion of Feynman's "sum over paths" formulation of quantum mechanics, Stachel has benefitted both from his historical and from his broad philosophical background, making him suspicious of what he sees as uncritical positivism or uncritical idealism. For Stachel, certain approaches to quantum logic, for instance, offer the superficial consolation that "reality is logical." He sees the danger that such consolations may divert us from confronting tensions within the existing theoretical and experimental structure of quantum physics, or between that structure and other, unassimilated elements—tensions that could lead to a deeper comprehension, modifications, or even the complete overthrow of that structure. It is therefore only fitting that the third part contains a discussion of quantum logic, the paper by Shimony and Stein on quantum non-locality, special relativity, and counterfactual reasoning. They apply in fact logic to a specific quantum mechanical situation in order to address an argument by Henry Stapp. For several decades Stapp has maintained not only that there is a conflict between quantum mechanics and relativistic locality, but also that this conflict can be exhibited without any assumption about "elements of physical reality" like that made by Einstein, Podolsky, and Rosen. Shimony and Stein study the logic of counterfactuals and make some emendations in Stapp's application of this logic. They find flaws in Stapp's demonstration—which relies upon counterfactual reasoning—of the inconsistency of special relativistic locality with quantum mechanical predictions and conclude that Stapp has not succeeded in demonstrating such an inconsistency without making a supplementary assumption about elements of physical reality—which he intended to avoid by resorting to counterfactual reasoning.

- In his analysis of Feynman's "sum over paths" formulation of quantum mechanics, Stachel has reexamined also the meaning of fundamental concepts of quantum mechanics such as the concept of quantum states and the concept of entanglement, which is Schroedinger's term for non-factorizable quantum states of several-particle systems. Stachel maintains that the latter concept can be extended in a natural way to a single particle in Feynman's formulation, by thinking of the entanglement of paths rather than of states of distinct particles. These concepts are also central to two further contributions; Paty's exposition of the new light thrown by recent experiments on such concepts; and Jaeger's and Sarkar's discussion of coherence, entanglement, and reductionist explanation in quantum physics. Paty celebrates the extraordinary achievements of experimentalists in the last few decades and argues from the phenomena they demonstrated that the quantum level is just as physical as the classical level, even though there are epistemological differences between these levels. He discusses, for instance, how improved interference experiments produce an interference pattern even though particles in a dilute beam pass through the interferometer one at a time (with rare exceptions) and argues that the wave function should therefore be ascribed to each individual particle in the beam. Jaeger and Sarkar discuss various types of reductionist explanation in physics and argue that the criteria for strong reduction are violated by explanations in quantum mechanics involving entangled states, a conclusion with potential impact also on the issue of reductionism in biological explanations.

Various contributions revolving around the themes of science, history, and the challenges of progress are assembled in the fourth part. It is a tribute to Stachel, *homme des lettres,* political thinker, and *philosophe* in the Enlightenment sense of the word. It is a tribute, more specifically, to his lifelong struggle for a humane science which includes a responsibility to mankind and its great struggles, for example the workers' movement for social progress or the Jewish community's fight for its rights. Questions such as whether one can learn from history, raised in Tisza's contribution, are therefore central to Stachel's life and work, well beyond the domain of science on which Tisza's paper focuses. In his grand review of the development of physics and chemistry since Newton, Tisza argues for a two-stage model of scientific progress, according to which a heuristic stage must be followed by a stage of consolidation or rational reconstruction. According to Tisza, it is now finally time for a consolidation phase in the case of quantum physics.

The subjects of other contributions to this part, ranging from interstellar travel to the logic of economic development according to Marx, can hardly be fitted into a simple scheme of progress. Janis in his paper, for example, discusses the physics of science fiction literature, conceived as material for teaching, and demonstrates in this way how scientific knowledge can become more attractive to students when it is embedded in a larger, if only speculative vision. Lefèvre analyzes views of development in 19th century Germany. He traces the impact of these views on the reception of Darwin's and Marx's theories by their contemporaries who found it difficult to grasp the novel ideas about the nature of development inherent in these theories. Read from the perspective of Stachel's unflinching commitment to a liberal and open-minded Marxism, Lefèvre's subtle analysis can be read as a demand for *engagement* and as an implicit appeal against resignation and cynicism since, as the title and conclusion of his paper suggest, progress is indeed not warranted and requires that our struggle for it must not be abandoned. How difficult, attritional, and sometimes nevertheless successful such a struggle can actually be is illustrated by two further contributions to this part: Gavroglu's and Patiniotis' study of patterns of the appropriation of Enlightenment ideas in the Greek intellectual life of the 18th century, and Schweber's concise account of Einstein's involvement in the founding history of Brandeis University. Gavroglu and Patiniotis argue convincingly that the spread of scientific knowledge from a "center" to a "periphery" should not be conceived as a simple transmission leaving this knowledge essentially unaffected. According to their analysis, this process should rather be understood as an active appropriation of new knowledge, largely governed by internal patterns of reception and constraints of the appropriating community. The paper introduces its reader to the fascinating world of Greek intellectual life after the fall of Constantinople and illustrates the consequences of local social and intellectual conditions on the reception of Newton's concept of time in Greece at that time. How much labor may be required in order to put great ideas to work, and how much power struggle and plotting may accompany the pursuit of humanistic ideals is illustrated by Schweber's account of the founding of Brandeis University, in some ways reminiscent of the founding of the Einstein Papers Project. Schweber's case study follows the history of Brandeis University from the first ideas of a secular university sponsored by the American Jewish community, via its rather tense birth phase, to its splendid establishment as a first-class university, and hence also demonstrates that such labors may well turn out to be worthwhile in the end.

To all of us who have contributed to this volume it has been a great pleasure to give back to John Stachel, in this modest way, some of what he has given to us over the years, not only in a scholarly but also in a human sense, and to pay tribute to an extraordinary individual to whose intellectual family we proudly belong.

I

Historical and Philosophical Roots of Relativity

JEAN EISENSTAEDT

THE PREHISTORY OF RELATIVITY*

INTRODUCTION

Years ago, I wrote a paper on the prehistory of black holes (Eisenstaedt 1991). It dealt essentially with the history of the concept of dark bodies, so named by Simon de Laplace (1796, 2:304-306) who most probably took the idea from the work of John Michell. Actually, I showed how John Michell constructed, essentially in his 1784 article (Michell 1784), the Newtonian theory of the action of gravitation on light. Elsewhere I showed how Michell's ideas are deeply rooted in Newton's theories: not only his gravitation of course but also his corpuscular theory of light. At the beginning of the nineteenth century, Arago was to use Michell's ideas in order to think over and perform his well-known experiment on the velocity of light. He was to show then that the velocity of light was constant and his experiment, which was a predecessor of the Michelson experiment, drove him to support Fresnel's ideas. In this article, I will come back to these trains of thought which concern light and gravitation from Newton to Arago.[1]

1. EINSTEIN ON THE BENDING OF LIGHT

How Einstein came to the conclusion that light could be subject to gravitation is fairly well known. Two of his 1905 articles were important in this context: his "electrodynamics of moving bodies" of course, but also the 1905 idea of a corpuscular view of light which was important in order to think over the action of a gravitational field on light. In 1907, Einstein was already working on what were to become his two favorite concepts: light and gravitation. In 1911, his paper published in *Annalen der Physik* was chiefly concerned with the question of the influence of gravitation on the propagation of light. During the summer 1913 he wrote to Erwin Finlay-Freundlich who aimed at observing the bending of light by different techniques,

> that the idea of the bending of light rays appeared at the time of the theory of emission [was] rather natural ... (Eisenstaedt 1991, 378).

Actually it happens that the bending of light in a Newtonian context is explicitly discussed in Bernstein, a popular handbook that Einstein read in his youth.

Thus to Einstein, such a train of thought was logical and coherent. At the end of 1915, he predicted the precise formula of the relativistic deflection of light in a field of gravitation. In May 1919 the English astronomical expedition showed for the very first time that light was actually influenced by gravitation, an essential confirmation of general relativity.

One year later, Sir Joseph Larmor published a paper entitled "Gravitation and Light" where he quoted Newton's well-known *Query* on the bending of rays by bodies. But in the same line he also quoted "the physically-minded John Michell" who had

> insisted that the Newtonian corpuscles of light must be subject to gravitation like other bodies, [and] that the velocities of the corpuscles shot out from one of the more massive stars would be sensibly diminished by the backward pull of its gravitation (Larmor 1920, 324).

In 1921 Lenard partially republished an almost unknown paper on the gravitational bending of light rays, an article that had been first published in 1801 by a German astronomer, Johann Georg von Soldner.[2] When general relativity was at a low ebb (Eisenstaedt 1989), Laplace's (actually Michell's!) dark bodies were mentioned by none other than Eddington in 1926.

In order to understand the emergence of this corpus and to put things more clearly, we must first come back to Newton's corpuscular theory of light.

2. NEWTON'S BALLISTIC THEORY OF LIGHT

Newton's corpuscular theory of light is essentially a dynamical theory of light, a ballistic theory of a light-corpuscle.[3] In Section XIV, Book I of Newton's *Principia*, refraction is described as an attractive force—the refringent force—acting perpendicularly to the surface separating the two media. Thus the incident corpuscles of light are accelerated and bent by this attractive force.

Fifty years after *Principia*, Newton's ballistic theory has been precisely expounded—in an algebraic way—by Clairaut (1741). It shows up a mathematized optics which implies two different laws:

- first of course Descartes' law of refraction that comes out with the conservation of the 'impulsion' parallel to the plane separating the two media:

$$\sin i = \frac{v_r}{v_i} \sin r = n \sin r$$

(where n is the relative index, i the angle of incidence, r that of refraction, v_i the velocity of the incident corpuscle of light and v_r the velocity of the refracted corpuscle),

- and second, something like a conservation law of the energy of a light corpuscle, implicit in Newton's demonstration:

$$v_r^2 - v_i^2 = -2\int_0^b f(x)dx = \varepsilon v_0^2$$

(where f is the refringent force per unity of mass and where v_0^2 is the refringent energy per unity of mass, ε being equal to + or -1 depending on the path of the corpuscle).

From these equations it is clear that in Newton's corpuscular theory of light refraction was linked to the velocity of the incident light. As we will see later on, such a dependence was to be used by Michell and Arago in order to try to show a difference in the velocities of incident corpuscles of light. But Newton also used this dependence as an attempt for a theory of chromatic dispersion.

3. FROM CHROMATIC DISPERSION TO THE THEORY OF EMISSION

As we know, for Newton, white light consisted in a stream of corpuscles of different colors. In the year 1690, it seems that he hoped to explain chromatic dispersion by establishing a connection between color and velocity, each corpuscle of different color being endowed with a different velocity. Actually, velocity would be the parameter of color.

Thus a particle with a greater velocity would be less deflected by the attractive force than a particle with a smaller velocity. The faster a corpuscle is, the less bent and less refracted it will be. The slower it is, the more bent and more refracted it will be; just like a cannonball in the Earth's gravitation field. As a consequence, a less deflected particle (like a "red-making" particle) is supposed to have a greater velocity than the more deflected one (like the "violet-making" particle). At hand, we have a possible model for chromatic dispersion and an explanation of the spectrum.

Also, such a mechanical model of dispersion would imply that a moon of Jupiter would have its color modified as it appeared (or disappeared) behind the planet: at emersion, over a short period of time, the colors of the spectrum would appear in turn beginning with the fastest (red) rays.

On August 10th 1691, Newton wrote to John Flamsteed, Royal Astronomer and long-time correspondent, to ask if he had observed any change of color in Jupiter's satellites before they disappeared. As Flamsteed had not observed any change of color in the light of the appearing or disappearing satellites of Jupiter, Newton as a consequence abandoned this hypothesis.

Actually, Newton was never really involved in either his mechanical model of dispersion nor in his corpuscular theory; after publishing *Principia* he never alluded to it in public again. Nevertheless, in 1694 as Biot later showed, Newton used his corpuscular theory as a model to compute his table of astronomical refraction.

Some fifty years later, in France and in Britain, Newton's corpuscular theory still dominated the field. Alexis-Claude Clairaut was certainly one of the philosophers most interested in it. The question of the color of Jupiter's satellites at emersion was taken very seriously by Jean-Jacques D'Ortous de Mairan, by the Marquis de Courtrivon and Clairaut in France, by Thomas Melvill in Scotland, and finally by James Short, the well-known London optician. After many interesting discussions and a tentative observation by Short it became clear—as Newton had understood long ago—that it was no longer possible to deal with such a model.

In any case, Newton's corpuscular—ballistic—theory was to be used for quite a long time, but without any reference to colors; the angle of refraction was still related to the velocity of the incoming corpuscle but no longer to its color.

At the time, this reduced interpretation of Newton's corpuscular theory was the most valued theory in optics, in France it was called "La Théorie de l'Émission." Simon de Laplace, Jean-Baptiste Biot, Étienne-Louis Malus, Siméon-Denis Poisson, François Arago, all of the "Société d'Arcueil" were all working in that way. Meanwhile, it was still possible to use the emission theory in order to explain or predict physical effects, for example, double refraction, astronomical refraction, or the Boscovich effect.

4. JOHN MICHELL AND THE ACTION OF GRAVITATION ON LIGHT

For many a Newtonian, light was composed of corpuscles whose velocity was finite, and as James Bradley had shown through aberration, it seemed to be a constant. But from a theoretical point of view there was at the time absolutely no reason for the velocity of light to be a constant. More precisely, Galileo's principle impeded the velocity of light being shown as a constant. Here was one of the most important contradictions of the century.

Thus for John Michell[4] (1724-1793), a friend of Henry Cavendish and a most convinced Newtonian philosopher, there was no contradiction to suppose that light was subject to gravitation. In a context closely related to Newton's corpuscular theory he was to apply Newton's theory of gravitation to light. To him light was supposed to be subject to gravitation in the very same way as an ordinary material corpuscle, but it was endowed with a greater (emission) velocity:

> Let us now suppose the particle of light to be attracted in the same manner as all other bodies with which we are acquainted; that is, by forces bearing the same proportion to their vis inertiae, of which there can be no reasonable doubt, gravitation being, as far as we know, or have any reason to believe, an universal law of nature (Eisenstaedt 1991, 329).

Thus light could be slowed down or accelerated by gravitation. In his 1784 article[5] Michell calculated in a geometical way the gravitational slowing down of a corpuscle of light. Fifteen years later the very same calculation was to be performed by Laplace (1799) but in an algebraic way: $c(r)$, the velocity of light at a distance r of a star was to take the form:

$$c^2(r) = c_0^2 - \frac{2GM}{r_0} + \frac{2GM}{r}$$

(where M is the mass of the star, r_0 its radius; c_0 is the emission velocity: actually the velocity of light at emission at r_0).

From his (geometrical) formulation Michell, followed by Laplace, inferred the possible existence of "dark bodies"; simply that $c^2(r)$ vanishes at infinity if

$$c_0^2 \leq \frac{2GM}{r_0} :$$

> Hence [...] if the semi-diameter of a sphaere of the same density with the sun were to exceed that of the sun in the proportion of 500 to 1, a body falling from an infinite height towards it, would have acquired at its surface a greater velocity than that of light, and consequently, supposing light to be attracted by the same force in proportion to its vis inertiae, with other bodies, all light emitted from such a body would be made to return towards it, by its own proper gravity (*ibid.*, 332).

Of course Michell's effect also allows for a diminution of the velocity of light, a slowing down of light by gravity:

> But if the semi-diameter of a sphaere, of the same density with the sun, was of any other size less than 497 times that of the sun, though the velocity of the light emitted from such a body, whould never be wholly destroyed, yet would it always suffer some diminution, more or less, according to the magnitude of the said sphaere (*ibid.*, 338).

Even if Michell foresaw this, such an effect allows for an acceleration of light by gravity. In the same way, Michell stuck to the radial case and didn't consider that a grazing ray of light could be deviated or bent by gravitation. It was Soldner's aim (referring to Laplace but not to Michell) to work out this idea. His calculation can be found in his 1801 essay:

> Thus when a ray of light passes by a celestial body, it will, instead of going on in a straight direction, be forced by its attraction to describe a hyperbola whose concave side is directed against the attracted body.[6]

This calculation was also independently performed (but never published) by Henry Cavendish at the beginning of the nineteenth century (Jungnickel and McCormmach 1996, note 33, 303).

5. MICHELL'S EXPERIMENTS OF 1783

Michell was essentially interested in astronomy. From statistical considerations, applied to the distribution of stars in the sky, he predicted in 1767 the existence of groups of physically connected stars, of "double stars," long before William Herschel could observe their movements. Michell's aim was not so much to focus on the retardation of light; he was primarily interested in measuring the distance of the stars.

Michell thought that in some double star systems, the mass of the central star had to be much more important than that of its companion. If he could observe such a system, he would (try to) collect at the same time the two beams of light, the first one emitted by the central mass and the second one by its companion. He would then analyze their lights with the help of a prism. The light of the massive star had to be gravitationally retarded in relation to that of the smaller star. Thus, due to Newton's corpuscular theory of light it had to be differentially refracted on the prism: the analysis of the observation would show a difference in the refraction of the beams due to the difference in the velocities of the corpuscle of lights coming from the two stars. Such a measurement would have provided one more piece of data to help determine the distance of the star system:

> Now the means by which we may find what this diminution amounts to seems to be supplied by the difference which would be occasioned in consequence of it, in the refrangibility of the light, whose velocity should be so diminished (Eisenstaedt 1991, 343).

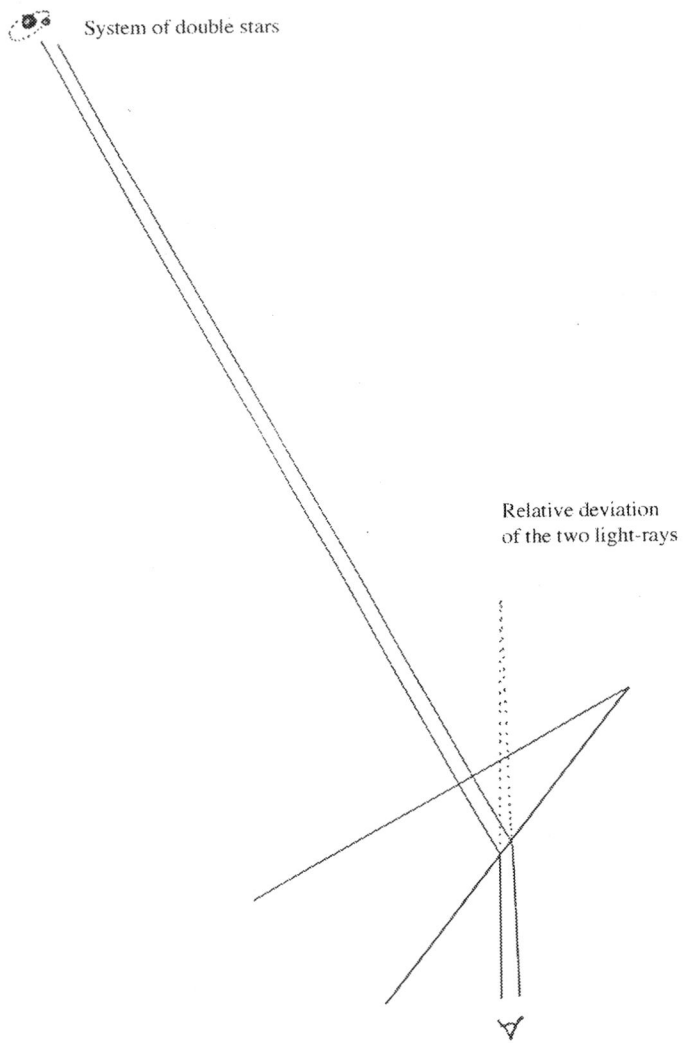

*Figure 1. John Michell's Experiment, 1783.
Light coming from the central star will be more deviated by the prism
than that coming from its less massive companion.*

But what kind of precision did Michell think he could obtain?

> As a prism might be made use for this purpose, which should have a much larger refracting angle than that we have proposed, especially if it was constructed in the achromatic way, according to Mr. Dollond's principles, not only such a diminution, as one part in twenty, might be made still more distinguishable; but we might be able to discover considerably less diminutions in the velocity of light, as perhaps a hundredth, a two-hundredth, a five-hundredth, or even a thousandth part of the whole (*ibid.*, 345).

The experiment was jointly performed by the Royal Astronomer Nevil Maskelyne and by William Herschel during the summer of 1783 in the direction of the Pleiades but without success. Herschel even ground a prism for the occasion. Henry Cavendish was also involved in the experiment and had predicted that a reduction in the focal length of an achromatic lens would be a further consequence of Michell's theory; Maskelyne performed the new observation with the lens but he too failed to detect the effect.

6. ARAGO ON THE VELOCITY OF LIGHT

Arago began his experiments on the velocity of light as early as 1806. But it was only on his return from Africa in 1809-1810 that these experiments were resumed. The results were made public at the time but were not published before 1853. To Arago, who took these ideas from Michell, a prism was a perfect tool for comparing the velocities of light rays.

For Arago, there were many reasons for the velocity of light not to be constant, and consequently differences had to exist in the velocities of lights. He even thought that the velocity of light could double ($2c_0$!) in certain circumstances. Such an effect might be due to different causes, of course linked to the law of addition of velocities: the velocity of the Earth (rotation or translation) or the probable proper velocities of the stars themselves. But, of course, it might also be due to refraction (light being accelerated in a denser medium due to the emission theory), or even as he put it

> one of the most powerful cause of change in the velocity of light seems to be the unequal size between the diameters of the stars (Arago 1853, 47)

which is due to Michell's effect of gravitation on light. Actually, Arago is well aware of Michell's theory and even explains properly the dark body idea:

> One finds actually, through calculations, that a star of the same density as the Sun and whose diameter would be a little hundred times greater than that of this star, would, through its attraction, annihilate totally the velocity of its rays, which consequently, could not reach us.[7]

Arago was to compare the prismatic deviation of a light-ray coming from a definite star at twelve hours difference. The velocity of the Earth in relation to this star, v, will be $-v$ twelve hours later; and as a consequence the velocity of a light corpuscle coming from such a star will be respectively $c + v$ and $c - v$ in relation to a terrestrial observer. The angles of refraction at the prism will have to be different. Actually, Arago is seeking for a Doppler-shift with Michell's prism. A simple calculation made in the context of Newton's emission theory relates the difference in the velocities as a function of the deviation of the ray. Arago was to perform this experiment a number of times from 1806 to 1810. At last he was convinced that no effect was detectable.

But how can this very strange result be explained? Of course it was not possible to question Galileo's principle of relativity. Was Newton's emission theory wrong? None of these conclusions were possible. At the time Arago assumed an *ad-hoc* hypothesis:

> among the infinity of rays with every velocities that emanate from a luminous body, only those of a determined velocity are visible.[8]

The reason is that the rays are seen and measured at last in the "vitreous humour of the eye" of the observer, where they will display—due to Newton's emission theory—the same velocity in the same medium; at least, if all the observers' vitreous humour has the same refractive index. This is Arago's ultimate explanation that he most probably picked up from de Mairan (Eisenstaedt 1996, 148).

As Arago put it in his paper:

> this work [has] been the point of departure of experimental and theoretical researches (Arago 1853, 38).

A conclusion also made by Tetu Hirosige:

> ...in the first half of the century [...] Arago's experiment and the aberration of light were considered the touchstone of [a legitimate theory of light] (Hirosige 1976, 12).

In the year 1815 Arago came to support Fresnel's ideas and the wave theory of light. Afterwards and due to the success of this line of thought it was no longer possible to think in terms of Michell's influence of gravitation on light, probably because it proved to be quite complicated for gravitation to act on light waves. For a long time Michell's theory was totally forgotten.

7. THE PREHISTORY OF RELATIVITY

To conclude, I will address the following question: is it justified to talk of a prehistory of relativity, as my title implies? First, the two main ingredients of general relativity, gravity and light, are also at the very basis of Michell's theory of the action of *gravitation* on the propagation of *light*. And second, also at least two important predictions: that of the bending of light by Soldner and that of the dark bodies. As is well known, these predictions actually differ qualitatively and quantitatively, which shows how different the two theories of gravitation are. Light is not trapped in dark bodies as it is in black holes, and the intensity of the bending of light rays, in the context of Newton's theory, is only half of what it is in general relativity. But what is really important is that the effect of deviation is actually present.

I want to state clearly that this prediction is coherent with the vision that one has then of a corpuscle of light but also with Newton's theory of gravitation and also with Galilean kinematics. A grain of light, as it was often called then, might be accelerated or slowed down without question. Actually, the idea of the constancy of the velocity of light came to the fore at the end of the nineteenth century from numerous experiments and observations, but it was not ideologically coherent with the philosophy of the time.

Moreover there is one additional link between Michell's theory and Einstein's theory of general relativity: more precisely, between Michell's 1783 experiment and the 1925 experiment by Adams (1925) on the displacement of the spectral lines by the Companion of Sirius. Both experiments aimed at the same physical prospect; the influence of gravity on light. Both observations relied on the same material, a binary system and on the same equipment, a telescope and a prism. Of course both observations relied on completely different theories. Moreover, Michell was not waiting for a spectral shift but for a differential refraction—spectroscopy was not yet born. But the idea was almost the same. Michell's effect, the slowing down of light due to a gravitational field, is nearly parallel to the "Einstein effect," the displacement of the spectral lines of an atom due to a field of gravitation. But is that so astonishing? We have long known that general relativity is not the only theory predicting such effects, at least from a qualitative point of view. The theory of the influence of gravitation on light, as it developed at the end of the eighteenth century, is without doubt an important part of the prehistory of Einstein's theory of gravitation.

Observatoire de Paris

ACKNOWLEDGEMENT

I would like to thank Raymond Fredette for revising the English of this text.

NOTES

* In homage to John Stachel for his 70th birthday.

1. Much of this matter, references and quotations are to be found in (Eisenstaedt 1991, 1996, 1997). This paper is essentially a summary of these articles, forthcoming in an article on Arago and the velocity of light.
2. Concerning this point see (Jaki 1978).
3. Concerning these sections see (Eisenstaedt 1996).
4. There are few historical works on John Michell as an astronomer and a physicist; see (McCormmach 1968; Eisenstaedt 1991; Jungnickel and McCormmach 1996; Vignolles 2000).
5. Actually one finds early developments on this topic in Michell's notes published in Priestley's History of Optic: see (Eisenstaedt 1991, 322).
6. Soldner (1801) translated in (Jaki 1978, 945).
7. "On trouve en effet, par le calcul, qu'une étoile de même densité que le Soleil, et dont le diamètre serait un petit nombre de centaines de fois plus considérable que celui de cet astre, anéantirait totalement par son attraction la vitesse de ses rayons, qui n'arriveraient par conséquent pas jusqu'à nous" (Arago 1853, 47).
8. "dans l'infinité des rayons de toutes les vitesses qui émanent d'un corps lumineux, il n'y a que ceux d'une vitesse déterminée qui soient visibles" (Arago 1853, 47).

REFERENCES

Adams, W.S. 1925. "The Relativity Displacement of the Spectral Lines in the Companion of Sirius." *National Academy of Sciences. Proceedings* 11:382-387.

Arago, Dominique François Jean. 1853. "Mémoire sur la vitesse de la lumière, lu à la première Classe de l'Institut le 10 décembre 1810." *Académie des Sciences Paris. Comptes Rendus* 36:38-49; 53.

Clairaut, Alexis-Claude. 1741. "Sur les explications Cartésiennes et Newtoniennes de la Réfraction de la Lumière." *Académie Royale des Sciences Paris. Mémoires pour 1739*:259-275.

Eisenstaedt, Jean. 1989. "The Low Water-Mark of General Relativity, 1925-1955." Pp. 277-292 in *Einstein and the History of General Relativity*, eds. John Stachel and Don Howard. Proceedings of the 1986 Osgood Hill Conference. [*Einstein Studies*, Vol. 1.] Boston: Birkhäuser.

———. 1991."De l'influence de la gravitation sur la propagation de la lumière en théorie newtonienne. L'archéologie des trous noirs." *Archive for History of Exact Sciences* 42:315-386.

———. 1996. "L'optique balistique newtonienne à l'épreuve des satellites de Jupiter." *Archive for History of Exact Sciences* 50:117-156.

———. 1997. "Laplace: l'ambition unitaire ou les lumières de l'astronomie." *Académie des Sciences Paris. Comptes Rendus Série* IIb 324:565-574.

Hirosige, Tetu. 1976. "The Ether Problem, the Mechanistic Worldview, and the Origins of the Theory of Relativity." *Historical Studies in the Physical Sciences* 7:3-82.

Jaki, Stanley L. 1978. "Johann Georg von Soldner and the Gravitational Bending of Light, with an English Translation of His Essay on It Published in 1801." *Foundations of Physics* 8:927-950.

Jungnickel, Christa, and Russel McCormmach. 1996. *Cavendish*. Philadelphia: American Philosophical Society.

Laplace, Pierre-Simon. 1796. *Exposition du systême du monde*, Ed. originale. 2 vols. Paris: Imprimerie du Cercle-Social.

———. 1799. "Beweis des Satzes, dass die anziehende Kraft bey einem Weltkörper so gross seyn könne, dass das Licht davon nicht ausströmen kann." *Allgemeine Geographische Ephemeriden* 4:1-6. F. von Zach: Weinar.

Larmor, Sir Joseph. 1920. "Gravitation and Light." *Cambridge Philosophical Society. Proceedings* 19:324-344.

McCormmach, Russell. 1968. "John Michell and Henry Cavendish: Weighing the Stars." *The British Journal for the History of Science* 4:126-155.

Michell, John. 1784. "On the Means of Discovering the Distance, Magnitude, &c. of the Fixed Stars, in consequence of the Diminution of the Velocity of their Light, in case such a Diminution should be found to take place in any of them, and such other Data should be procured from Observations, as would be farther necessary for that Purpose." By the Rev. John Michell, B. D. F. R. S. in a letter to Henry Cavendish, Esq. F. R. S. and A. S. Royal Society of London. *Philosophical Transactions* 74:35-57.

Vignolles, Hélène. 2000. "La distance des étoiles au dix-huitième siècle: l'échelle des magnitudes de John Michell." *Archive for History of Exact Sciences* 55:77-101.

HORST MELCHER

INTERPRETATIONS AND EQUATIONS OF THE MICHELSON EXPERIMENT AND ITS VARIATIONS*

INTRODUCTION

The null results of Michelson's experiment (1881 in Potsdam) and of its repetition in improved form by Michelson and Morley (1887 in Cleveland) are to be expected and even predicted on the basis of the relativity principle. Einstein had mentioned this in his work *"Über das Relativitätsprinzip und die aus demselben gezogenen Folgerungen"* (Einstein 1907,1908), without, however, giving a detailed explanation. The interpretation of the MM (Michelson and Morley) experiment with the help of the relativity principle remains widely ignored in physics textbooks. It was John Stachel (1982) who first emphasized this. As an inspection of the extensive textbook and secondary literature shows, without exception it is the equality (constancy) of the speed of light in any spatial direction that is mentioned rather than the relativity principle. Not infrequently, the experiment's result is even (erroneously) cited or suggested as evidence for the second principle of special relativity theory (SRT) concerning the invariance of the vacuum speed of light (VSL). In no case however, with the exception of Einstein, is the relativity principle (RP) mentioned in connection with the MM experiment and its result interpreted with its help.

Just as in thermodynamics—except in historical discussions, processes and phenomena are no longer explained on the basis of the hypothetical substance *phlogiston*, so about a hundred years after the end of ether-physics one should proceed similarly; especially since the many erroneous, circuitous and by-roads in ether-physics only burden and surely do not aid physical thinking. Therefore, arguments from the pre-relativistic era play no role in the following discussion; the ether was eliminated or rather recognized as superfluous by Einstein (1905).

In addition to three experiments by Michelson before 1905, a considerable number of variants of experiments of this kind were performed after 1905, although according to Einstein (1907) the results were "predictable" because of the RP. Because of faulty as well as incorrect interpretations of them, in what follows the results of the most essential experiments *à la* Michelson are interpreted in more detail with the help of the RP, and in addition their correct equations are also given.

Analysis of the MM experiment from the standpoint of a moving inertial frame of reference shows immediately, because of the absolute simultaneity of coincidences, that an extensively-used equation, which still appears without criticism in the textbook literature, contradicts the physical facts.

Using relativistic addition of velocities, which plays a role in some of the equations relevant to the experiment, the unity of mechanics and electrodynamics, established by Einstein in his famous paper *"Zur Elektrodynamik bewegter Körper"* (1905) is presented here in the form of a tabular survey.

1. THE NULL RESULTS OF MM EXPERIMENTS INTERPRETED ON THE BASIS OF THE RELATIVITY PRINCIPLE (RP)

As shown in figure 1, we consider emitted and reflected electromagnetic signals, which are registered by a receiver at the same position as the source. Reflectors B and C are kept at the same distance l' from the emitter. In this set-up, there is no velocity difference between the source and the receiver; the source is rigidly (immovably) connected to the rest of the apparatus. The system S' represents the whole experimental set-up, which is situated in a closed cabin (lab), e.g., on board a ship, plane, or railroad carriage, or on a celestial body. An observer inside the closed lab would not notice whether he was at rest or if his location (e.g. the earth) were moving at a constant velocity, i.e., without acceleration. Observers in such a system would believe (in all cases of rest or motion with constant velocity) that they are at rest. This is the proper content of the (special) RP, and the simplest to formulate: All physical processes and phenomena are completely, whether they occur in a system at rest one moving with v = const. Einstein recognized that this principle of relativity, originally restricted to mechanics by Galilei and Newton, is valid in all fields of physics.

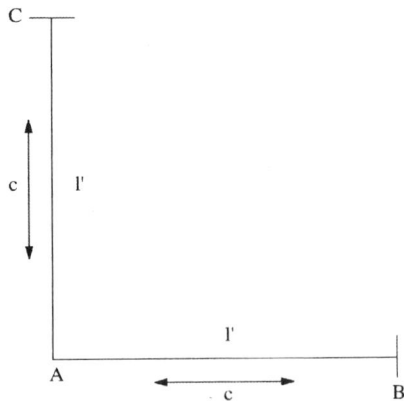

Figure 1. The principle of the Michelson experiment:
Two equally long travel distances l' for light pulses that are reflected in B and C, respectively.
The difference in travel times yields $\Delta t' = 0$; coincidence in A but also in the system.

What is measured with such an experimental set-up, which basically represents the MM experiment? Given the conditions indicated, for the equal distances between A and B, and A and C, respectively, equal travel times are recorded for the electromagnetic signals i.e. $\Delta t'_{AB} = 2l'/c$ and $\Delta t'_{AC} = 2l'/c$. The results of the experiment is a null one: $\Delta t'_{AB} - \Delta t'_{AC} = \Delta t' = 0$ in accord with the theoretical expectation, namely with the principle of relativity (fig. 1).

It is the unchanging (invariant) proper time in system S' that is measured. The difference between two equal proper time intervals in this system is zero. Is this trivial? If the RP is trivial, then this result is trivial—and vice versa. The principle of conservation of energy, the range of applicability of which is comparable, similarly seems to be obvious or trivial in many examples. (Obviously, it is easier to see that a perpetuum mobile cannot work because of the energy principle than the assertion that an experiment *à la* Michelson must give a null result because of the RP.)

Not only textbook authors—famous physicists among them do not form an exception—ignore the RP in connection with the MM experiment, while only referring to the constancy of the vacuum speed of light (VSL). Robert Shankland, one of Michelson's successors at Case Western Reserve University in Cleveland, Ohio, visited Einstein shortly before his death and discussed questions concerning the MM experiment. Shankland reports on his talks with Einstein (Shankland 1973, 896) and writes:

> it is evident that the importance of the Michelson-Morley experiment for Einstein was that it gave positive confirmation to his belief that the speed of light is invariant in all inertial frames, independent of the motion of the source, apparatus or observer.

This statement is obviously not only inaccurate, but incorrect: There is no relative velocity at all between the light source, the apparatus, and the observer. Measurements are made in their rest frame. The reference to the second principle of SRT—for short "invariance of VSL"—is completely inappropriate. It is actually only a question of the RP, evidence for which is provided by the MM experiment or, conversely: The null result of the MM experiment is understood most simply in terms of the RP, which is just what Einstein indicated. Therefore the words transmitted by Shankland are unacceptable or at least a misunderstanding.

Why should the same RP known from mechanics not underlie experiments with reflected light signals and their travel times?: Experiments with projectiles or simple ball games take the same course in a uniformly moving train or plane as in the rest state. The fact that the velocity of earth in its orbit around the sun cannot be measured by means of the MM experiment is known from analogous mechanical experiments in isolated railroad or aircraft compartments. And in the end this is exactly the RP, formulated as a principle of impossibility. If then, in countless books, the constancy of the VSL is stated as the result of the MM experiment—with a source of light and recording apparatus at rest in the lab—then that is nothing remarkable, since one knows that the constant velocity of a train or airplane has no influence at all on, e.g., falling or projected bodies or processes of any sort in any kind of compartment. These facts form the content of the RP, in the light of which the result of the MM experiment appears as quite a natural consequence. In this case as well, something that Einstein once told Heisenberg is appropriate:

> It is theory that decides what to measure.

An experimental apparatus of any kind, in particular the Michelson interferometer, and any physical experiments are not affected in any way if they are situated in systems that are moving with constant velocity. Experiments and measurements always correspond to those performed in the "rest state." Physical laws remain unaltered in their forms. So in the MM experiment, no shift in the interference pattern as a measure of different travel times is to be expected. The travel times do not change—because of the RP.

2. THE INCORRECT EQUATIONS OF MICHELSON AND LORENTZ

It is known that the equation set up by Michelson for his first experiment (Potsdam in 1881) was incorrect. Alfred Potier and independently Hendrik Antoon Lorentz came up with the (presumably) correct equation, which to this day can still be found in many textbooks. The equation for the first Michelson experiment, incorrect by the factor 2, had no influence on the interpretation of the result.

Was the equation set up by Lorentz for the second experiment, done together with Morley in Cleveland, correct? It was based on an equation from mechanics that works for a swimmer or boat if the motion is once directed downstream and then upstream, and next back and forth across the stream. In both cases, equal distances have to be covered for the passages to and fro. Let v be the speed of the current and u the speed of the swimmer or the boat, respectively, when the water is at rest. In any case the time for the movements to and fro across the stream will be smaller than the sum of times of the movements with and against the current; so that a finite time difference $\Delta t \neq 0$ results:

$$\Delta t = \frac{l}{u+v} + \frac{l}{u-v} - \left(\frac{l}{u^2-v^2} + \frac{l}{\sqrt{u^2-v^2}}\right). \qquad (1)$$

This equation is correct if the following restrictions are effective: $u > v$ and $u \ll c$ as well as $v \ll c$. In any case, for $v \neq 0$ a finite time interval $\Delta t \neq 0$ follows.

This equation of classical mechanics is now transformed, with VSL $u = c$, into the following form, which still appears in many books, without any indication of its invalidity:

$$\Delta t = \frac{2l}{c}\left(\frac{1}{1-\frac{v^2}{c^2}} - \frac{1}{\sqrt{1-\frac{v^2}{c^2}}}\right). \qquad (2)$$

The validity of this equation had not been tested when it was set up, its validity was assumed without justification and mistakenly. The invalidity of the Galilean law of addition of velocities when the speed of light is involved has been known since 1905—quite apart from Fizeau's experiment of 1851, which had been repeated with increased precision by Michelson and Morley in 1886. On this basis, such incorrect equations should no longer be discussed, certainly not if their range of validity is not indicated.

For $v \neq 0$ equation (2) gives the finite time interval $\Delta t \neq 0$. This stands in contradiction to the experimental result, which always yields $\Delta t = 0$. An incorrect equation can never be in accordance with an experimental result; in other words, equation (2) corresponds to no real physical fact. Hence it seems strange that, even after 1905, so many MM experiments—indeed all possible variations—were carried out. As generally known, Einstein said explicitly that, on account of the RP, the null results were "predictable."

Once, however, a positive result was announced, which Miller believed to have found through his own repetition of a Michelson experiment. Einstein was in the USA at the time and, when he received the news, he said—being completely convinced of the RP—the famous words:

> The Lord God is subtle, but he is not malicious.

These words are engraved on the mantelpiece of Fine Hall, the Mathematics Institute of Princeton University. Einstein's suspicion of Miller's result was later confirmed: Miller's measurements were distorted by temperature influences.

In pre-relativistic times, when one still believed in the existence of the ether, H.A. Lorentz brought equation (2) into accord with the experimental result by an arbitrary act (so-called ad hoc hypothesis). In this connection, it will never be quite clear whether the mathematical trick came first, or the physical interpretation by postulation of a shortening of the length, that has been named the Lorentz contraction. In equation (2) one multiplies the first term in parentheses by $(1 - v^2/c^2)^{1/2}$, and says that the term $l(1 - v^2/c^2)^{1/2}$ is a real contraction. In this way one obtains $\Delta t = 0$ in agreement with the experimental result. As the basis for this artificial procedure, it is supposed that, in forward motion through the ether, a real contraction takes place, caused by interatomic forces and is independently equal in magnitude for all materials. By assuming such compression without any (transverse) bulging, Lorentz saved the stationary ether, the existence of which Michelson had abolished previously by virtue of his experiment. How artificial and unrealistic such hypothesis is perhaps becomes manifest if one returns from equation (2) to the classically valid equation (1) and asks, for example, if with a speed ratio of $u/v = 3/5 = 60\%$ a contraction (reduction in length) of $l/l_o = 4/5 = 80\%$ would occur. What remarkable properties the ether would have to possess!

The terminus technicus "Lorentz contraction," since it arose in ether physics as the model of a real contraction, should, therefore, be conceptually distinguished from the relativistic contraction. In SRT, one is concerned with inertial reference systems in relative motion, and the contraction of lengths occurring there have a universal, different conceptual basis.

One would expect that, after SRT was established, further MM experiments would have become superfluous. Yet, in 1930 Georg Joos still carried out another similar, high precision experiment at Zeiss Industries in Jena, using an automated apparatus.

Strangely enough, one can still today read in textbooks that the Michelson experiment could only be explained with the help of the Lorentz contraction. (See, for example, Macke 1958; Jansen 1984; Janossy 1971; Döring 1973, 83-93.)

3. THE EXPERIMENT OF KENNEDY AND THORNDIKE (1932)

In contrast to the MM experiment Kennedy and Thorndike chose the two perpendicular arms l' of the interferometer to be unequally long, so that now $l_1' \neq l_2'$ holds (fig. 2). The interferometer was not swivel-mounted, but rigidly attached to the laboratory. A displacement of interference fringes was expected in consequence of the change in the earth's velocity along its annual orbit around the sun. However, as previously in the original Michelson experiments, no shift in the interference pattern occurred. On the basis of the RP this is actually again completely clear. But in A.P. French (1971), it is merely said that this experiment cannot be explained with the help of the Lorentz contraction. It is not indicated how the experimental result is to be interpreted.

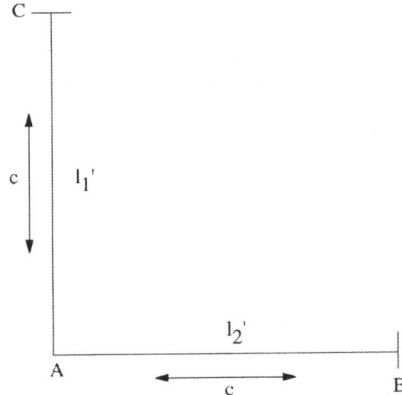

Figure 2. The principle of the experiment of Kennedy and Thorndike:
The two travel distances are unequally long: $l_1 \neq l_2$. The permanently constant difference in travel times is $\Delta t' \neq 0$. In the system S, $t\Delta \neq \Delta t'$ would be measured.

K.R. Atkins (1974, 473) writes about the Kennedy-Thorndike experiment:

> We are not able to explain this experiment by the Lorentz contraction alone. The decisive point is that RT predicts the Lorentz contraction *plus other effects.*

The reader is left in obscurity, the author obviously as well. But with the help of the RP the interpretation of the experimental result is almost painfully simple. The apparatus is placed in the rest system S' and a constant uniform velocity has no influence on physical processes. While in the MM experiment, the two equally long interferometer arms represent two clocks that remain in time with each other, there are, in the KT experiment, two clocks each with its own time interval. The difference of these two unequal intervals is constant, since the interference fringes do not shift, but are not equal to zero:

$$\Delta t'_{AB} = \frac{2l'_1}{c}, \Delta t'_{AC} = \frac{2l'_2}{c}, \text{ hence } \Delta t'_{AB} - \Delta t'_{AC} = \Delta t' \neq 0.$$

As a measure of the proper time, this interval of course stays constant. The interference pattern of the interferometer does not change. Experiment and theoretical prediction are in accord. Guesswork after reading French (1971) becomes unnecessary. Another question would be, does the time interval as determined from a different inertial reference frame S, which has the velocity v relative to the system S', have the same value? This is not the case, $\Delta t'$ and Δt are different, the transformation formula $\Delta t = \Delta t'/(1 - v^2/c^2)^{1/2}$ holds between them.

4. EXPERIMENTS OF TOMASCHEK (1924) AND MILLER (1925) WITH LIGHT FROM EXTRATERRESTRIAL SOURCES

In this case, the light source is no longer rigidly fixed to the measuring apparatus, but is external to the system S' (fig. 3). Depending on the season of the year, there is a distinguishably large relative speed v between light source (sun, stars, planets) and receiver. Nevertheless, an unchanging interference pattern is recorded as a measure of the permanently constant travel time differences.

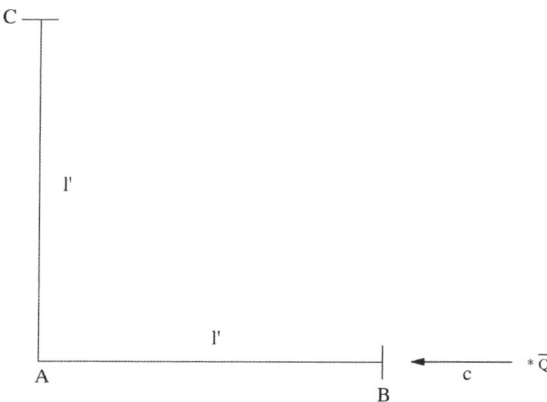

Figure 3. The principle of the experiments of Tomaschek and Miller: The light source is positioned outside the measuring apparatus. There is a relative velocity v between source and receiver. The light caught by the apparatus always propagates at the constant speed c because of the relativistic addition of velocities and the invariance principle respectively.

In this case, of course, the interpretation of the result is not done using the RP, but the second Einstein principle, the principle of the "invariance of the VSL." This short form of the principle easily leads to misunderstandings, as probably did not happen rarely in the interpretation of Michelson's experiment. The principle actually states that, when combining the VSL with another velocity v, the value of the VSL always results. The VSL remains invariant no matter how fast the source and/or the observer are moving. In particular, it is even true that the sum or difference of two VSLs again yields the VSL.

Relative velocities between source and receiver have no effect on the interference pattern. These experiments of Tomaschek and Miller confirm the validity of the second Einstein principle of SRT and the relativistic addition theorem for velocities.

If the often-misinterpreted, shortened denotation constancy-principle is replaced by "independence principle", since c is always obtained independent of how the VSL is combined with a source speed v and/or an observer's speed, it should be noted that, essentially, the term "independence principle" has already been allocated: In Galilei-Newtonian mechanics, motions superpose independently of each other— without any mutual interference. In this case, there is a linear principle of superposition. The independent composition of velocities is confined to velocities much smaller than the VSL. Then and only then does the parallelogram rule hold for the superposition. Because of the relativistic composition of velocities, the parallelogram rule, i.e. the independence principle, no longer holds.

5. FIZEAU'S EXPERIMENT (1851)

Fizeau's experiment (1851) provided a first indication that the speed of light in water flowing with velocity v is not composed in accordance with the independence principle known from classical mechanics. If c_o stands for the velocity of light in water at rest, then the "usual" or familiar relation $c_w = c_o \pm v$ does not hold for the composition of velocities; but rather

$$c_w = c_o \pm \left(1 - \frac{1}{n^2}\right)v \qquad (3)$$

The application of this formula is limited to $v \ll c$. The expression $f = (1 - 1/n^2)$ is called Fresnel's coefficient, where n is the index of refraction of the optical medium in question; for water $n = 4/3$, $c_o = c/n$.

Max Laue (later ennobled to von Laue) showed in 1907 that equation (3), and hence also Fresnel's coefficient, can be derived as an approximation from the relativistic addition of velocities by inserting $u_x = c/n$ into the relativistic relation of velocities

$$u'_x = \frac{u_x + v}{1 + \frac{u_x v}{c^2}} \qquad (4)$$

and taking into account that $v \ll c$.

Einstein repeatedly called the Fizeau experiment "highly significant," without giving an argument for this statement. Apparently, Einstein's judgement is based on the fact that, here, the limitation of the Galilean law of addition of velocities became obvious. Did Einstein make his statement before or after Laue's publication? I was unable to find any notes about this in the Einstein Archive in Boston in 1987, nor any notes of Einstein on incorrect equations. But there exists—written down on a sheet of paper by Einstein himself—the formula, well known from many textbooks,

$$\Delta t = \frac{2l}{c} \cdot \frac{1}{2} \cdot \frac{v^2}{c^2}, \tag{5}$$

yet without any comment. This relation results from equation (2), recognized above to be incorrect, by approximation for $v \ll c$. Hence, equation (5) does not reflect any real physical fact either. Taking into account the relativistic law of addition, one finds—as already given in (Melcher 1988, 96)—instead of (5), the correct equation

$$\Delta t = \frac{2l'}{c} \cdot n'^3 \cdot \frac{1}{2} \cdot \frac{v^2}{c^2}\left(1 - \frac{1}{n'^2}\right). \tag{6}$$

It is obvious that, for Fizeau's experiment, a finite time interval $\Delta t \neq 0$ is obtained for $n' \neq 1$ and $v \neq 0$. And thus for the MM experiment, it is only for the VSL that one obtains $\Delta t = 0$ both for $v = 0$ as well as for $v \neq 0$. Thus the incorrectness of equation (5) is once again made clear.

6. ON THE UNITY OF MECHANICS AND ELECTRODYNAMICS

The results given above for $\Delta t = 0$ and $\Delta t \neq 0$ are collected in the following table. The classical approximation, i.e., the velocity addition according to the Galilean law, was incorrectly applied—see equations (1) and (2)—to the case of the VSL by Michelson and Lorentz. It is only for equal distances $l'_1 \neq l'_2$ that $\Delta t' = 0$ is obtained. In this case of coincidence, a coincidence is also observed in any other inertial system S (absolute coincidence): $\Delta t = 0$. The null result of the Michelson experiment is, of course, obvious because of RP.

If the light source (broadcasting station) is external to the measurement chamber and moves relative to the receiver with a relative velocity v, so that the VSL is superposed with v (experiments of Tomaschek and Miller), for $l'_1 = l'_2$ equal travel times for the light signals or radar echoes are measured in all directions.

This result is interpreted with the help of the principle of the constancy of the VSL, according to which the speed of light in vacuo is not changed, and thus remains constant, by superposition with any other speed (of source or receiver). If the length intervals travelled are different i.e. $l'_1 \neq l'_2$ then the finite time interval $\Delta t' \neq 0$ remains constant for all directions of the measuring arms; the interference pattern will not vary with interferometer orientation.

In the experiment of Kennedy and Thorndike, the light source remains at rest with respect to the apparatus with unequally long length intervals travelled by the electromagnetic signals. The null result is, once again, predictable because of the RP: The interference pattern does not change, it is independent of spatial orientation. A finite travel time difference $\Delta t' > 0$ remains constant but will have a different value $\Delta t' \neq \Delta t$ in a different inertial system S.

$\Delta t = 0$	Michelson 1881, 1887 Piccard 1927 RP Joos 1930	$t'_1 - t'_2 = 0$ $\Delta t = 0$	Photons	Non-linear
$\Delta t' = 0$	Tomaschek 1924 Miller 1925 c=const	$t'_1 - t'_2 = 0$ $\Delta t' = \Delta t = 0$	$m_o = 0$	Range:
$\Delta t' \neq 0$	Kennedy and Thorndike 1932 $l'_1 \neq l'_2$	$t'_1 - t'_2 \neq 0$ $\Delta t' = \Delta t = 0$	$n' = 1$	Lorentz
	Fizeau 1851 $c_w = \dfrac{c}{n'} + v\left(1 - \dfrac{1}{n'^2}\right)$	$c_w < c$	Photons $n' > 1$	
$\Delta(t' \neq 0)$	Elementary Particle and High-Energy Physics $u_x = \dfrac{u'_x + v}{1 + \dfrac{u'_x v}{c^2}}, v \leq c, u' \leq c, v \leq c,$ $u' \leq c$	$m_o \neq 0$		Transformation
$\Delta(t' \neq 0)$	Vehicles, Vessels, Aircrafts: $m_o \gg 0$ Classical Approximation: $v \ll c$ $u' \ll c, u_x = u'_x + v$			Linear Range: Galilean Transformation

In Fizeau's experiment, a finite travel time difference is measured in spite of equal travel distances for the electromagnetic signals. In this case, it is not a question of the VSL, but of the lower speed of light in water. Agreement between experiment and theory is only achieved with this experiment if the superposition of the light velocity with the water velocity is no longer calculated with the Galilean law of velocity addition. For small current velocities $v \ll (c/n)$, Fresnel's relation (neglecting dispersion) is sufficiently precise. This equation is derived from the theorem of relativistic addition of velocities by approximation.

In high-energy and elementary-particle physics the classical addition law fails. Equation (4) is valid for the entire velocity range of masses (photons included). The applications reviewed here of the composition of velocities, in which both electromagnetic signals and moving bodies appear, show the unity of mechanics and electrodynamics brought about by Einstein. The examples also explain the title of Einstein's famous work, which serves as the birth certificate of SRT: *On the Electrodynamics of Moving Bodies*.

In the present work, the inadequacy discussed above of a classical equation for the composition of velocities to properly describe the coincidences occurring in system S ($\Delta t' = \Delta t = 0$) provides the motivation to seek the correct equation and find the adequate transformation. In today's ether-free physics, radar echo experiments and radio navigation of satellites are a contemporary starting point, experiences from *the electrodynamics of moving bodies* encountered in daily life, which were anticipated by Einstein.

7. ON THE IMPORTANCE OF THE MICHELSON EXPERIMENT: DEVALUED OR OVERVALUED?

Einstein clearly showed only a limited interest in the Michelson experiment although he passed beyond the experiment's result, according to which "the hypothesis of a stationary ether is false," by explaining that a light-propagating ether of any sort, even one that is dragged along, is "superfluous."

On the other hand, it is possible that the almost incalculable profusion of literature on the Michelson experiment—although often misjudging its actual purpose and significance—contributes to the extreme schematization of this ether drift experiment, to the fact that here and there it has attained the rank of the legendary and mythical. In so doing, as precisely the textbook and secondary literature proves, the real essence of the Michelson experiment, as presented here in Einstein's interpretation—is not seen; so that many attempt without success to grasp the meaning of the experiment and to understand why it is—often at great length—described at all. Overemphasis on the Michelson experiment is not conducive to the understanding of the SRT. The impossibility of measuring the velocity of the earth in its orbit around the sun by means of this experiment is just what is decreed by the RP—just as, analogously, the principle of energy conservation accounts for the impossibility of a perpetuum mobile. Michelson's ingenious interferometer is an instrument for the ages, which initiated a unique technique of measurement in many fields of physics and which will be used as a detector for gravitational waves in the century to come. Michelson's experimental mastery and proficiency remain completely unaffected by the unexpected and disappointing result—probably for that reason the null result is also often called a "negative result."

Incorrect interpretations—that the experiment can be explained only by lengthy contraction (Macke 1958)—or unfounded raptures:

> Unfortunately, the Michelson-Morely experiment cannot be mastered by means of high school physics (Jansen 1984),

as well as the nebulous and pompous phrases of Skinner, drive beginners to despair by parading before their eyes an underdeveloped ability to understand. In (Skinner 1982, 26) one can read:

> The Michelson-Morley experiment extends our range of experience and shows that our usual concepts of space and time are not valid in that extended range.

The beginner hardly sees an "extended range," but rather asks himself why the Michelson experiment was repeated so frequently. Were there more doubts about the experimental results than about the correctness of the underlying theoretical and mathematical assumptions? One cannot outwit a natural law by continual increase of experimental precision. In a different context Einstein once said:

> By pharmacist's methods one cannot find one of God's secrets.

Probably many people were struck with astonishment, which they concealed out of politeness, by several sentences in the telegram that Ronald Reagan addressed to the participants in the Michelson-Morley Centennial Celebration in Cleveland, Ohio in 1987. The Celebration extended over eight months. The telegram, dated March 16, 1987, reads as follows:

> I am pleased to send greetings to everyone celebrating the centennial of the Michelson-Morley experiment. This astonishing achievement, which found that Earth's motion does not affect the speed of light, was and is a milestone in man's quest for mastery over nature. Einstein himself credited it with opening the door to his own revolutionary theory of relativity. The experiment also led to quantum mechanics and profoundly affected not only physics but other sciences and technology as well. Therefore, it is hard to exaggerate the debt we owe to Albert Michelson and Edward Morley. I know their brilliant work will continue to inspire other Americans to strive and to succeed as they did. God bless you. Ronald Reagan.

Once again, not the RP but the speed of light is discussed and, astonishingly, the MM experiment is also supposed to have led to quantum mechanics. Did Einstein misjudge the experiment and its significance, so that the impression now emerges that he devalued or underrated it? Anyway, the null result of the Michelson experiment is evident because of the RP and needs no discussion. Just as the perpetuum mobile is a "non- problem" today—because of the principle of energy conservation, so the null result of the MM experiment is a "non-problem"—because of the RP.

Michelson's (1881) historical ether drift experiment and Michelson and Morley (1887) are neither the starting point of SRT nor a component of this theory. There is no sort of genetic connection between this experiment and Einstein's theory. The experiment is neither necessary nor sufficient for the foundation of the SRT nor for its presentation. Superfluously, even after the foundation of SRT, additional Michelson experiments were performed. For Einstein, the result of the experiment provided further evidence for his principle of (special) relativity.[1]

Potsdam, Germany

NOTES

* To Professor John Stachel in honor of his seventieth birthday

1. Even in books published in the nineties, the Michelson experiment is related to the "constancy of the speed of light" instead of to the "relativity principle" (Ruder 1993, 48).

REFERENCES

Atkins, K. R. 1974. *Physics*, 2nd ed. New York, 1970. W. de Gruyter & Co.
Döring, W. 1973. *Physik und Didaktik* 1.
Einstein, A. 1905. "On the Electrodynamics of Moving Bodies." *Annalen der Physik.* 17:891-921.
———. 1907. *Jahrbuch der Radioaktivität und Elektronik* 4, 411-462.
———. 1908. *Jahrbuch der Radioaktivität und Elektronik* 5, 98-99.
French, A. P. 1971. *Special Relativity* (M.I.T. Introductory Physics Series). Cambridge.
Janossy, L. 1971. *Theory of Relativity Based on Physical Reality.* Budapest.
Jansen, L. 1984. *Physik und Didaktik* 12, Heft 4, 273-278.
Macke, W. *Wellen.* 1958. *Ein Lehrbuch der Theoretischen Physik.* Leipzig, 334.
Melcher, H. 1988. *The Michelson Era in American Science* 1870-1930. Conference Proceedings 179. American Institute of Physics, New York.
Ruder, H. and M. 1993. *Die Spezielle Relativitätstheorie.* Braunschweig, Wiesbaden.
Shankland, R. S. 1973. *Am. J. Phys.* 41.
Skinner, R. 1982. *Relativity for Scientists and Engineers.* New York.
Stachel, J. 1982. *Astronomische Nachrichten* 303, Heft 1, 47-53.

MICHEL JANSSEN

THE TROUTON EXPERIMENT, $E = MC^2$, AND A SLICE OF MINKOWSKI SPACE-TIME

1. THE FORGOTTEN PRECURSOR TO THE TROUTON-NOBLE EXPERIMENT

In the Fall of 1900, Frederick T. Trouton started work on an ingenious experiment in his laboratory at Trinity College in Dublin. The purpose of the experiment was to detect the earth's presumed motion through the ether, the 19th-century medium thought to carry light waves and electric and magnetic fields. The experiment was unusual in that, unlike most of these so-called ether drift experiments, it was not an experiment in optics.[1] Trouton tried to detect ether drift by charging and discharging a capacitor in a torsion pendulum at its resonance frequency, which he hoped would set the system oscillating. The basic idea behind the experiment came from George Francis FitzGerald, whose assistant Trouton was at the time. According to FitzGerald, a capacitor moving through the ether should experience an impulse, a jolt, upon being charged or discharged. Trouton's torsion pendulum was designed to detect these jolts. Not surprisingly from a modern relativistic point of view, Trouton found no such effect. FitzGerald died in February 1901 before the experiment was concluded. It was thus left to others to try and reconcile Trouton's result with then current electromagnetic theory.

The first to do so was Joseph Larmor, who not only got closely involved with Trouton's experiment after FitzGerald's death, but who also became the editor of a volume of FitzGerald's scientific papers published the following year. Trouton's paper on the experiment suggested by FitzGerald was reprinted in this volume accompanied by an interesting four-page editorial note. Larmor, however, devoted only one short paragraph of his note to Trouton's original experiment, confidently asserting that no effect should have been expected in the first place. He was far more interested in a new and promising variant of the experiment that Trouton, most likely with input from Larmor himself, had proposed in his paper. The idea behind this new experiment was to detect ether drift not through linear impulses upon charging or discharging a capacitor in a torsion pendulum, but through a turning couple on a carefully insulated charged capacitor in a torsion pendulum of a slightly different design. Trouton eventually carried out this experiment at University College in London, where he was appointed professor of physics in 1903. He was assisted by one of his research students, Henry R. Noble.

Figure 1. FitzGerald was given to flights of fancy.

This Trouton-Noble experiment continues to intrigue theorists and experimentalists, in- and outside the physics mainstream, to this day (see for example, Teukolsky 1996; Hayden 1994; Cornille, Naudin, and Szames 1998, 1999). Its precursor, the Trouton experiment, on the other hand, has been all but forgotten. This is unfortunate, for there is as much to be learned from the original experiment as there is from its better-known sequel. In this paper, I will therefore focus on the Trouton experiment. I hope to present my findings concerning the Trouton-Noble experiment on some other occasion (for a preliminary version, see Janssen 1995). For a more historically oriented discussion of the experiments, including a careful analysis of the role Larmor played in Trouton's work, I refer to Warwick 1995.[2] The emphasis in this paper will be on conceptual issues.

The full explanation of the Trouton experiment requires that one take into account the inertia of energy expressed in Einstein's famous equation $E = mc^2$. The Trouton experiment can actually be seen as a practical version of a thought experiment with which Einstein tried to show that $E = mc^2$ is both necessary and sufficient to ensure that the center-of-mass theorem, according to which no process in an isolated system can change the state of motion of the system's center of mass, holds for systems involving both electromagnetic fields and ordinary matter (Einstein 1906).[3] As these observations suggest, the analysis of the Trouton experiment provides valuable insights into the transition from classical to relativistic mechanics.[4] On this score, the Trouton and Trouton-Noble experiments have much more to tell us than such famous optical ether drift experiments as the Michelson-Morley experiment.

That Larmor was so quick to dismiss FitzGerald's original proposal was probably because he realized that the predicted effect would violate the center-of-mass theorem.[5] The problem with Larmor's *reductio* is that, at the time, it was at best unclear whether current electromagnetic theory, based on the notion of a stationary or immobile ether, was at all compatible with the center-of-mass theorem or with such closely related laws as Newton's third law and the law of momentum conservation. In 1895,

H. A. Lorentz, the leading proponent of the immobile ether theory, had in fact already renounced Newton's third law as a universal law of nature. In Lorentz's theory, matter can never set the ether in motion, yet ether can set matter in motion via the Lorentz forces that fields in the ether exert on charged particles in matter. This is clearly in violation of Newton's principle that action equals reaction.

In 1903, Max Abraham introduced the concept of electromagnetic momentum, with the help of which the notion of an immobile ether could at least be reconciled with the law of momentum conservation for processes involving both fields and matter. In his well-known paper on the electrodynamics of moving bodies of 1904, Lorentz availed himself of this new concept of electromagnetic momentum in his analysis of both the Trouton and the Trouton-Noble experiment. In the case of the Trouton experiment, Lorentz's analysis vindicated FitzGerald's prediction that a moving capacitor should experience a jolt upon being charged or discharged. His calculations, however, indicated that Trouton's torsion pendulum had not been sensitive enough to detect the effect.

If we put Larmor's and Lorentz's accounts of the Trouton experiment side by side, we arrive at the following dilemma. If the effect predicted by FitzGerald does occur, the center-of-mass theorem is violated (as is the relativity principle, one may add). That is what Larmor tells us. If, however, the effect does *not* occur, momentum conservation appears to be violated. That is what Lorentz tells us. It seems that we have to choose between the center-of-mass theorem and momentum conservation, two laws that are essentially equivalent in Newtonian mechanics. $E = mc^2$ allows us to escape this dilemma. Once the inertia of energy is properly taken into account in Lorentz's analysis of the Trouton experiment, a negative result in the experiment, even if the sensitivity of the apparatus were greatly improved, is seen to be compatible both with the center-of-mass theorem and with momentum conservation.

It is tempting to speculate that the Trouton experiment would have lived on in physics textbooks as an elegant illustration of $E = mc^2$, if only the connection between the experiment and the equation had been recognized in the early years of special relativity. This in turn raises the question why nobody in fact made this connection, which with hindsight appears to be so obvious. In general, such questions are not very fruitful, but in this case it will direct us to another complication we are faced with in working out a detailed explanation of the Trouton experiment (and, for that matter, the Trouton-Noble experiment). I already referred obliquely to this complication: how does one *properly* take into account $E = mc^2$ in Lorentz's analysis of the Trouton experiment? The third part of the title of my paper refers to this problem.

Mathematically, the definition of electromagnetic momentum used by Lorentz in his analysis of the Trouton and Trouton-Noble experiments is an instance of the standard relativistic definition of the momentum of spatially extended systems, a definition that is not Lorentz invariant. Lorentz defined the electromagnetic momentum of a charged capacitor moving through the ether as a space integral of the momentum density of the capacitor's electromagnetic field. Such space integrals—to borrow some modern relativistic terminology—are integrals over hyperplanes of simultaneity in a frame at rest in the ether, regardless of the velocity of the capacitor with respect to the ether. It follows that the electromagnetic momentum of a capacitor at rest in the ether and the electromagnetic momentum of the same Lorentz contracted capacitor carrying the same charge in uniform motion through the ether are *not* related to one another via a Lorentz transformation. Under the standard definition—to once again put it in relativistic terms—electromagnetic momentum does not behave as the spatial part of a four-vector under Lorentz transformations. And, as we shall see, for the explanation of the Trouton experiment based on the inertia of energy to work out in detail, it is crucial that it does.

In special relativity, the easiest way out is to use an alternative Lorentz-invariant definition of the four-momentum of spatially extended systems. This alternative definition was first proposed by Fermi (1922) and made popular by Rohrlich (1960, 1965).[6] Under this new definition, the electromagnetic momentum of a capacitor is the integral of the electromagnetic energy-momentum density over a hyperplane of simultaneity in the capacitor's rest frame, regardless of the frame of reference in which the momentum is calculated. Defined in this way, the electromagnetic momentum of the capacitor does transform as the spatial part of a four-vector under Lorentz transformations. In special relativity, it is ultimately a matter of convention whether one chooses the standard definition or the definition of Fermi, Rohrlich, *et al.*[7] In Lorentz's ether theory, however, the hyperplanes of simultaneity in the frames at rest in the ether are privileged, which commits the proponents of the theory to a definition that mathematically is a special case of the standard definition. Hence, the easy way out is not available to the ether theorist.

Under the standard definition, only the four-momentum of a *closed* system transforms as a four-vector under Lorentz transformations, not the four-momentum of its open sub-systems. This result was first announced in 1911 by Max Laue for the special case of *static* closed systems.[8] The electromagnetic field of a capacitor is not a closed system and its four-momentum (under the standard definition) does not transform as a four-vector. A satisfactory account of the Trouton experiment in Lorentz's theory therefore can not be given without considering the material part of the capacitor as well.

From the point of view of classical mechanics, this is a completely unexpected complication. Classical mechanics, of course, already tells us that there will be stresses in the material part of a charged capacitor that prevent the capacitor from collapsing under the Coulomb attraction between its plates. But relativistic mechanics (at least under the standard definition of the four-momentum of spatially extended systems) predicts the entirely new effect that stresses in a system's rest frame give rise to momentum in a frame in which the system is moving. It is only the sum of this momentum and the electromagnetic momentum that transforms as the spatial part of a four-vector under Lorentz transformations as is required for the detailed explanation of the Trouton experiment. Laue, unaware of the conventional element in his definition of momentum, put great emphasis on this relation between stresses and momentum. He presented it as on par with the relation between energy current and momentum density, which, as was first noted by Planck (1908), can be seen as an expression of the inertia of energy. For a modern relativist Laue's effect is just an artifact of a definition that the relativist is under no obligation to adopt. The ether theory does not allow such leeway, and Laue's peculiar effect is another element that has to be taken over from relativistic mechanics, along with the inertia of energy, to arrive at a satisfactory account of the Trouton experiment.

So, the full explanation of the negative result of the Trouton experiment, be it in special relativity or in Lorentz's ether theory, involves three elements completely foreign to classical mechanics that were introduced in the decade following the experiment: electromagnetic momentum (Abraham 1903), the inertia of energy (Einstein 1905b, 1906), and the effect that stresses give rise to momentum (Laue 1911a).[9] In special relativity, the third element can be replaced by the later insight that there is a certain freedom in defining the four-momentum of spatially extended systems. In view of all this, I would say that the Trouton experiment fully deserves to be rescued from oblivion.

2. LARMOR, THE TROUTON EXPERIMENT, AND THE CENTER OF MASS THEOREM

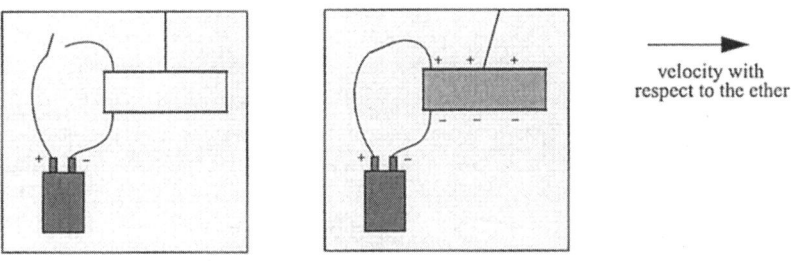

Figure 2. The basic idea behind the Trouton experiment.

Figure 2 illustrates the basic idea behind the Trouton experiment. A battery is used to charge a capacitor. If the power is switched on, an electromagnetic field is produced largely confined to the volume between the plates of the capacitor. If the system is at rest in the ether, the charges will only produce an electric field; if the system is moving, the charges will also produce a magnetic field. As Trouton wrote in his paper on the experiment:

> The question then naturally arises as to the source supplying the energy required to produce this magnetic field. If we attribute it to the electric generator, say a battery, there is no difficulty [...] FitzGerald's view, however, was that it would be found to be supplied through there being a mechanical drag on the condenser itself at the moment of charging (Trouton 1902, 557–558)

In other words, FitzGerald thought that the energy for the magnetic field would come from the capacitor's kinetic energy. Elementary Newtonian mechanics tells us that in that case a moving capacitor upon being charged should experience a jolt in the direction opposite to its direction of motion. In figure 2, the effect is illustrated for a capacitor suspended on a wire from the ceiling of the laboratory with its plates parallel to the direction of motion.

The actual arrangement, shown in figure 3 taken from Trouton's paper, was a little more subtle. At FitzGerald's suggestion, Trouton made the capacitor part of a torsion pendulum. The capacitor was "charged and discharged continuously by means of a clock-work, at the intervals corresponding to the free period of swing of the apparatus. In this way any effect produced would cumulate and be made easier of observation" (Trouton 1902, 560).[10]

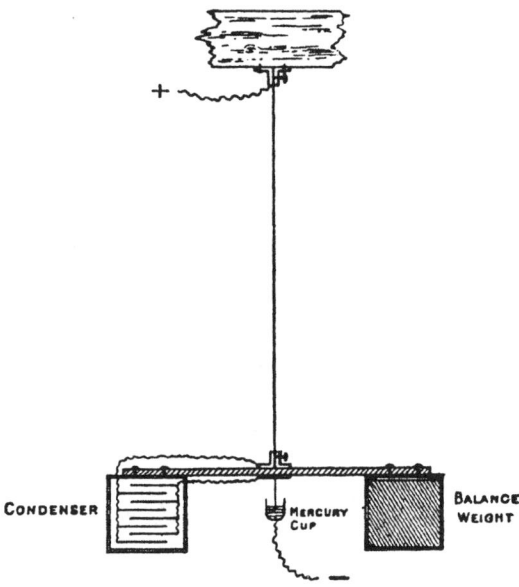

Figure 3. Trouton's torsion pendulum with capacitor (from: Trouton 1902, 560).

Trouton did not find any effect. He concluded: "it is evident that some other source for the energy or some countervailing effect must clearly be looked for" (Trouton 1902, 562). The passage quoted at the beginning of this section suggests that Trouton, probably under Larmor's influence, preferred the first option. FitzGerald, as Trouton reports, was leaning toward the second:

> On the last opportunity I had of discussing the matter with Professor FitzGerald, preliminary experiments had been made, giving as far as they went negative results: the final results not being completed till after Science had to deplore the grievous loss it sustained at his death. FitzGerald, on that occasion, made a remark which, as well as I remember, was to the effect that should the negative results then obtained be sustained by further work, he would attribute the non-occurrence of any observable effect to the same general cause as produced the negative results in Michelson and Morley's experiments on the relative motion of the Earth and the ether by means of the interference of light. (Trouton 1902, 562)

As Trouton goes on to explain, "the same general cause" refers to the Lorentz-FitzGerald contraction, or, as Trouton and Larmor preferred, the FitzGerald-Lorentz contraction. In the next paragraph, Trouton elaborates on FitzGerald's suggestion: "From some such cause a diminution of the electrostatic energy might be brought about [...] just sufficient in amount to provide the energy required for the magnetic field" (Trouton 1902, 562–563). It is not clear whether this elaboration is Trouton's or FitzGerald's. However, no matter whose idea it was, it is hard to see how it could be made to work.

Trouton acknowledged the help of Larmor in his paper. And Larmor included Trouton's paper in his edition of FitzGerald's scientific papers, adding an interesting note himself. For the most part Larmor's note deals with the proposal in Trouton's paper to look for a turning couple on a charged capacitor rather than for impulses upon charging and discharging a capacitor. Anticipating the negative result of the Trouton-Noble experiment aimed at detecting this turning couple, Larmor argued that a negative result could be accounted for on the basis of the Lorentz-FitzGerald contraction.[11] The original Trouton experiment is disposed of in the final short paragraph of Larmor's note:

> If the condenser [...] is held *absolutely fixed* while it is being charged, any impulsive torque there might be could do no work; yet the condenser gets its energy. This seems by itself sufficient to negative the suggestion that the energies of charge and discharge [...] have to do directly with mechanical forces (Larmor 1902, 569; italics in the original).

I take it that by 'absolutely fixed' Larmor meant 'fixed with respect to the laboratory rather than suspended on a torsion wire as in the actual experiment.' The alternative reading 'fixed with respect to the ether' does not seem to make sense, since the problem only arises for a moving capacitor. At first glance, Larmor's argument is a gross *non-sequitur*. It is perfectly consistent to maintain that if the capacitor is held fixed with respect to the laboratory the energy of the magnetic field comes from an ever so slight decrease in the kinetic energy of the earth as a whole. It is hard to believe that Larmor could have overlooked such an obvious point. It is more likely that his *reductio* is, in fact, to the absurdity of the notion that the earth's kinetic energy could decrease in this experiment. But why would Larmor have thought this to be absurd?

As Andrew Warwick (1995, 317) has emphasized, Larmor believed that it was impossible to extract energy from an object's motion through the ether, except maybe a very small amount, proportional to some higher power of v/c, the ratio of the object's velocity with respect to the ether and the velocity of light. If it were possible to extract more, the whole universe, or so Larmor believed, would have come to rest in the ether a long time ago. If FitzGerald were right, the energy extracted from the motion of the capacitor and the earth through the ether in the Trouton experiment would be of order (v^2/c^2). Perhaps this then is the absurdity Larmor sensed in FitzGerald's proposal. His attitude toward the new experiment Trouton proposed would seem to support this interpretation. Referring to the latter experiment, Larmor writes: "Thus the energy of motion of the Earth through the æther is available for mechanical work to an unlimited extent, unless [...] the FitzGerald-Lorentz shrinkage of moving bodies is a fact" (Larmor 1902, 568). Larmor appears to have been quite confident that the result of the experiment would be negative. The importance of the experiment for him was that it would provide further evidence for the contraction hypothesis. There is an important difference though between Larmor's attitude toward the original experiment and his attitude toward Trouton's new proposal. In the latter case, Larmor offered a detailed explanation as to why it would be impossible to extract energy from the earth's motion through the ether in this fashion. In the case of the original Trouton experiment, he simply dismissed FitzGerald's proposal out of hand. This suggests that Larmor had some other reason for believing FitzGerald's idea to be absurd.

John Stachel (see note 5) has suggested such a reason. If the fully isolated system of earth and capacitor were to change its velocity upon charging or discharging the capacitor, we would have a blatant violation of the center-of-mass theorem. One might object that the earth-capacitor system is not fully isolated since it interacts with the ether. However, since Larmor assumed the ether to be immobile, the center-of-mass theorem is also violated if we consider the earth-capacitor-ether system. It seems very plausible to me that this violation of the center-of-mass theorem is indeed the absurdity that Larmor sensed in FitzGerald's proposal. In that case, it has to be said that Larmor's intuition was right on the mark. The connection between the Trouton experiment and the center-of-mass theorem, however, turns out to be much more complicated than Larmor, given the extreme brevity of his remarks, can possibly have realized at the time.

3. THE ACTION-EQUALS-REACTION PRINCIPLE AND THE INTRODUCTION OF ELECTROMAGNETIC MOMENTUM

The center-of-mass theorem is closely related to Newton's third law, the principle that action equals reaction, which, in turn, is closely related to the law of momentum conservation. When Larmor wrote his comment on the Trouton experiment, the status of momentum conservation in its various guises in theories positing an immobile ether had been the subject of some serious debate, notably between Lorentz and Poincaré.[12] In 1902, the situation was unclear at best.

Newton's principle of the equality of action and reaction is hard to reconcile with the notion of an ether that can set matter in motion (through the Lorentz forces of electromagnetic fields on charged particles), yet can itself never be set in motion by matter. Lorentz clearly stated this obvious difficulty in his widely read monograph of 1895. After discussing the problem of how to make sense of forces acting on an immobile ether and concluding that the easiest way to solve the problem would be never to apply the notion of force to the ether at all, Lorentz wrote, in an often quoted passage:

> It is true that this conception would violate the principle of the equality of action and reaction—because we do have grounds for saying that the ether *exerts* forces on ponderable matter—but nothing, as far as I can see, forces us to elevate that principle to the rank of a fundamental law of unlimited validity. (Lorentz 1895, 28; italics in the original.)

Poincaré strongly objected to this aspect of Lorentz's theory, especially to the violations of the center-of-mass theorem it entails. In fact, he made this the topic of his contribution to a *Festschrift* on the occasion of the 25th anniversary of Lorentz's doctorate (Poincaré 1900b).

Poincaré illustrated his objection with the example of a mirror recoiling upon the reflection of light (Poincaré 1900b, 273). He used this same example in an important lecture during the International Congress of Arts and Sciences in St. Louis in 1904:

> Imagine, for example, a Hertzian oscillator, like those used in wireless telegraphy; it sends out energy in every direction; but we can provide it with a parabolic mirror, as Hertz did with his smallest oscillators, so as to send all the energy produced in a single direction. What happens then according to the theory? The apparatus recoils, as if it were a cannon and the projected energy a ball; and that is contrary to the [action equals reaction] principle of Newton, *since our projectile here has no mass, it is not matter, it is energy*" (Poincaré 1904, 101; my italics).

The italicized final remark, which is not to be found in Poincaré's more detailed discussion of the example in 1900, shows how tantalizingly close he came to the resolution of the problem through $E = mc^2$.[13]

In a letter to Poincaré in response to the latter's contribution to his *Festschrift*, Lorentz reiterated that any theory based on an immobile ether—or, as Lorentz put it, "any theory that can explain Fizeau's experiment," which was generally understood to provide strong evidence for the notion of an immobile ether (see Janssen and Stachel 1999)—will violate the action-equals-reaction principle and thereby the center-of-mass theorem. He made it clear that he did not see this as a serious problem for his theory.[14]

From the point of view of classical mechanics, Poincaré's recoiling mirror example also violates momentum conservation. This can be avoided by ascribing momentum to the electromagnetic field. The concept of electromagnetic momentum was introduced by Abraham in 1903. Today we are so accustomed to a concept of momentum that is broader than mechanical momentum that it is easy to forget that this was by no means obvious at the beginning of the century.[15] It may not be inappropriate therefore to illustrate this point with the following extensive quotation from a paper by Planck. The paper is based on a lecture delivered during the annual meeting of the *Versammlung Deutscher Naturforscher und Ärzte* in Cologne on September 23, 1908, two days after Minkowski's famous lecture on "Space and Time" (Minkowski 1909). Planck's paper, entitled "Comments on the Principle of Action and Reaction in General Dynamics," contains a vivid description of the difficulties surrounding the action-equals-reaction principle around the turn of the century:

> As is well-known, the real content of the Newtonian principle of the equality of action and reaction is the theorem of the constancy of the quantity of motion or of the momentum of motion; I therefore want to talk about this principle only in the sense of that theorem, and, more specifically, about its relevance for general dynamics, which not only includes mechanics in a more restricted sense, but also electrodynamics and thermodynamics.
>
> Many of us will still recall the stir it caused, when H. A. Lorentz, in laying the foundations of an atomistic electrodynamics on the basis of a stationary ether, denied Newton's third axiom absolute validity, and inevitably this circumstance was turned into a serious objection against Lorentz's theory, as was done, for instance, by H. Poincaré. A calmness of sorts [*eine Art Beruhigung*] only returned when it became clear, especially through the investigations of M. Abraham, that the reaction principle could be saved after all, in its full generality at that, if only one introduces, besides the mechanical quantity of motion, the only kind known at that point, a new quantity of motion, the electromagnetic kind.

> Abraham made this notion even more plausible by a comparison between the conservation of the quantity of motion and the conservation of energy. Just as the energy principle is violated if one does not take electromagnetic energy into account and satisfied if one does introduce this form of energy, so is the reaction principle violated if one only considers the mechanical quantity of motion but satisfied as soon as one also takes into account the electromagnetic quantity of motion.
>
> However, this comparison, incontestable in and of itself, leaves one essential difference untouched. In the case of energy, we already knew a whole series of different kinds—kinetic energy, gravitation [sic], elastic energy of deformation, heat, chemical energy—so it does not constitute a fundamental innovation if one adds electromagnetic energy to these different forms of energy as yet another form. In the case of the quantity of motion, however, we only knew one kind so far: the mechanical kind. Whereas energy was already a universal physical concept, the quantity of motion had so far been a typically mechanical concept and the reaction principle had been a typically mechanical theorem. Consequently, its generalization, while recognized to be necessary, was bound to be experienced as a revolution of a fundamental nature, through which the up to that point relatively simple and uniform concept of the quantity of motion acquired a considerably more complicated character (Planck 1908, 828–829).

Planck may have exaggerated the difficulties physicists were experiencing with the notion of electromagnetic momentum somewhat for rhetorical purposes (he goes on to show that the idea of putting energy and momentum on equal footing is a very natural one in relativity theory), but this passage does make it clear that the introduction of electromagnetic momentum was indeed, as Planck says, a "fundamental innovation."

4. LORENTZ, THE TROUTON EXPERIMENT, AND MOMENTUM CONSERVATION

Both the Trouton and the Trouton-Noble experiment are discussed in the paper that forms the crowning achievement of Lorentz's work on the electrodynamics of moving bodies before the advent of special relativity (Lorentz 1904a). Lorentz discussed both experiments in terms of electromagnetic momentum. This does not mean that he accepted the interpretation of this quantity as a new kind of momentum. In the case of the Trouton-Noble experiment, Lorentz made it clear that he only used the phrase electromagnetic momentum to describe the result of a derivation that is completely independent of the quantity's interpretation as a form of momentum.[16] Lorentz's caution illustrates the physics community's somewhat reluctant acceptance of electromagnetic momentum as momentum *sui generis* in the early years of this century (cf. the quotation from Planck 1908 above).[17] In his discussion of the Trouton experiment, as we shall see shortly, Lorentz uncharacteristically did invoke the interpretation of electromagnetic momentum as a form of momentum. It cannot be ruled out, however, that he simply suppressed a derivation that he felt would justify his argument without relying on this interpretation. It is important to keep in mind in this context that the Trouton experiment plays a rather modest role in Lorentz's paper. The Trouton-Noble experiment is prominently discussed in the introduction of the paper, where it is presented as one of two new ether drift experiments that partly motivated the paper. Discussion of the Trouton experiment is relegated to the final section of the paper, a section that has the character of an appendix.

After these cautionary remarks, let us see what Lorentz actually had to say about the Trouton experiment:

> I take this opportunity for mentioning an experiment that has been made by Trouton at the suggestion of FitzGerald, and in which it was tried to observe the existence of a sudden impulse acting on a condenser at the moment of charging or discharging; for this purpose the condenser was suspended by a torsion-balance, with its plates parallel to the earth's motion. For forming an estimate of the effect that may be expected, it will suffice to consider a condenser with aether as dielectricum. Now if the apparatus is charged there will be [...] an electromagnetic momentum
>
> $$\mathfrak{G} = \frac{2U}{c^2} \mathbf{w} \qquad 18$$
>
> (Terms of the third and higher orders are here neglected). This momentum being produced at the moment of charging and disappearing at that of discharging, the condenser must experience in the first case an impulse $-\mathfrak{G}$ and at the second an impulse $+\mathfrak{G}$. However Trouton has not been able to observe these jerks.
> I believe it may be shown (though his calculations have led him to a different conclusion) that the sensibility of the apparatus was far from sufficient for the object Trouton had in view (Lorentz 1904a, 829–830).

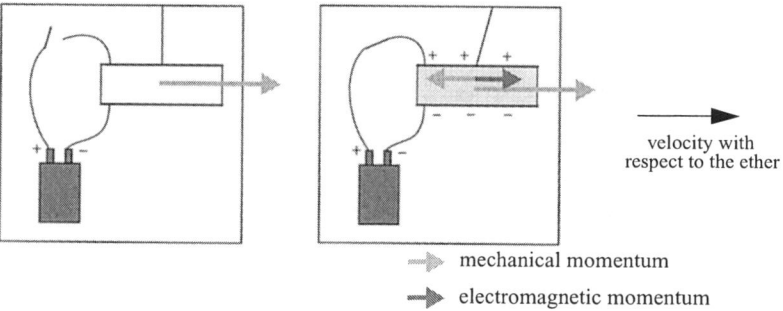

Figure 4. *Lorentz's analysis of the Trouton experiment.*

Figure 4 illustrates Lorentz's analysis in this passage. As the moving capacitor is charged, a certain amount of electromagnetic momentum is created in the direction of motion (or rather, to first order in (v/c), in the direction parallel to the plates of the capacitor,[19] which Lorentz, following Trouton, assumed to coincide with the direction of motion). Implicitly invoking momentum conservation, Lorentz concludes that the capacitor's gain in electromagnetic momentum must be compensated by a loss of ordinary mechanical momentum. Hence, the capacitor should experience a jerk backwards upon being charged, just as FitzGerald had originally predicted.

Here we have the dilemma mentioned in the introduction. Momentum conversation seems to require a positive result if the Trouton experiment were repeated with increased accuracy, whereas the center-of-mass theorem seems to require a strictly negative result. As we saw in the preceding section, Lorentz was quite willing to give up the center-of-mass theorem, so for him there was no real dilemma.

Of course, a positive result would violate not only the center-of-mass theorem but also the relativity principle. In ether-theoretic terms, a positive result in the Trouton experiment would have provided evidence of the earth's motion through the ether. In 1904, this would have been no more troubling for Lorentz than a possible violation of the center-of-mass theorem. Citing Poincaré's scathing criticism of the way in which he had introduced the contraction hypothesis to explain the negative result of the Michelson-Morley experiment, Lorentz described the task he had set himself in his 1904 paper as showing "by means of certain fundamental assumptions, and without neglecting terms of one order of magnitude or another, that *many* electromagnetic actions are entirely independent of the motion of the system" (Lorentz 1904a, 811; my emphasis). In fact, the reason that the Trouton-Noble experiment played such an important role in the paper was that the theory developed in the paper provided an elegant explanation of the experiment's negative result. However, as the statement just quoted shows, Lorentz did not claim that no experiment could ever detect ether drift.

After 1905 the situation changed. Borrowing some insights from special relativity, in particular the notion that moving observers will actually measure the "fictive" Lorentz transformed quantities of Lorentz's theorem of corresponding states instead of the "real" Galilean transformed ones, Lorentz perfected his theory in such a way that it predicted in full generality that no experiment could ever detect ether drift. Lorentz put great emphasis on the fact that his theory was therefore empirically indistinguishable from special relativity.[20] Yet, as far as I know, he never returned to the Trouton experiment to explain what was wrong with his analysis of 1904.

5. $E = MC^2$: HOW THE CENTER-OF-MASS THEOREM AND MOMENTUM CONSERVATION CAN BOTH HOLD IN THE TROUTON EXPERIMENT

The dilemma that we arrived at in the preceding section is easily resolved once we realize that energy has mass. Qualitatively, the argument runs as follows. If energy has mass, a transfer of energy from the battery to the capacitor means a transfer of mass, and, in a frame of reference in which battery and capacitor are moving, a transfer of momentum. So, figure 4 showing the momentum of the capacitor in the Trouton experiment before and after it is charged should be replaced by figure 5 below showing the momentum of both the capacitor and the battery before and after the capacitor is charged. When the moving capacitor is charged, it gains a certain amount of energy, mass, and momentum, while the moving battery loses that same amount of energy, mass, and momentum. The total amount of momentum is conserved. Contrary to what Lorentz thought in 1904, this does not require the capacitor to change its velocity. The increase in the capacitor's momentum corresponds to a change in the capacitor's mass, not to a change in its velocity. Hence, there is no violation of the center-of-mass theorem. Once the inertia of energy is taken into account, a strictly negative result of the Trouton experiment is thus seen to be compatible both with momentum conservation and with the center-of-mass theorem.

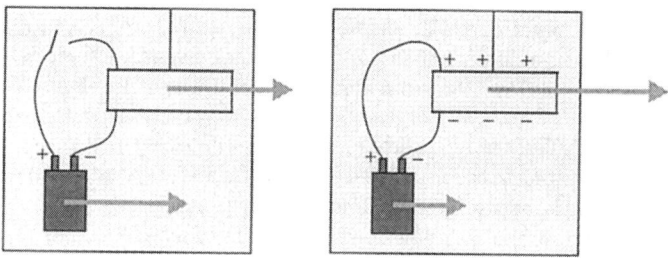

Figure 5. Transfer of momentum in the Trouton experiment.

All that is needed for this simple argument to work out quantitatively as well, is that energy and momentum transform as the components of a four-vector under Lorentz transformations. Let the transfer of four-momentum from the battery to the capacitor in a frame with coordinates $x'^\mu = (ct', x', y', z')$ in which battery and capacitor are at rest, be:

$$\Delta p'^\mu_{\text{Capacitor}} = -\Delta p'^\mu_{\text{Battery}} = \left(\frac{\Delta E'}{c}, 0, 0, 0\right). \tag{1}$$

In a frame of reference in which the laboratory is moving with a velocity v in the x-direction—a frame with coordinates $x^\mu = \Lambda^\mu{}_\nu x'^\nu$ [21]—the transfer of four-momentum is:[22]

$$\Delta p^\mu = \Lambda^\mu{}_\nu \Delta p'^\nu = \Lambda^\mu{}_0\left(\frac{\Delta E'}{c}\right) = \left(\gamma\frac{\Delta E'}{c}, \gamma\left(\frac{\Delta E'}{c^2}\right)v, 0, 0\right). \tag{2}$$

The spatial components of this four-vector give the transfer of ordinary three-momentum:

$$\Delta \mathbf{p} = \gamma\left(\frac{\Delta E'}{c^2}\right)\mathbf{v}. \tag{3}$$

Since $\Delta \mathbf{p}_{\text{Capacitor}} = -\Delta \mathbf{p}_{\text{Battery}}$, momentum is conserved. Yet, there is no violation of the center-of-mass theorem: $\Delta \mathbf{p}$ corresponds to a change in mass ($\Delta m = \gamma \Delta E'/c^2$), not to a change in velocity ($\Delta \mathbf{v} = 0$).

About a year after he first introduced the inertia of energy (Einstein 1905b), Einstein published a paper, entitled "The Principle of the Conservation of Motion of the Center of Gravity and the Inertia of Energy," in which he showed that $E = mc^2$ is necessary and sufficient to ensure that the center-of-mass theorem holds for systems in which "not only mechanical, but also electromagnetic processes take place"

(Einstein 1906, 627). As Einstein acknowledges, his paper is similar to Poincaré's contribution to the Lorentz *Festschrift* (Poincaré 1900b). Einstein showed that in order to avoid the kind of violations of the center-of-mass theorem discussed by Poincaré, one has to assume that energy has inertia. Instead of Poincaré's recoiling mirror, Einstein considered the thought experiment illustrated in figure 6.

Consider a box of mass M and length L. Suppose some energy E is stored on the inside of the left wall of the box, and suppose that at time $t = 0$ this energy is somehow converted into electromagnetic radiation travelling to the other side of the box. The radiation is absorbed at the other end of the box, where the energy is converted back to its original form. According to standard electromagnetic theory, the box will recoil upon emission of the radiation, and it will recoil again upon re-absorption of the radiation, bringing the box back to rest. Standard electromagnetic theory tells us that the radiation will have momentum (E/c). Momentum conservation requires that the box will recoil with that same momentum in the opposite direction. So, what this thought experiment shows is that by moving energy from one side of the box to the other, the completely isolated system of box plus energy E can move itself. If the energy E has no mass, this is in blatant violation of the center-of-mass theorem. With the help of figure 6, it can easily be shown that the only way to avoid this consequence is to ascribe mass $m = E/c^2$ to the energy E.

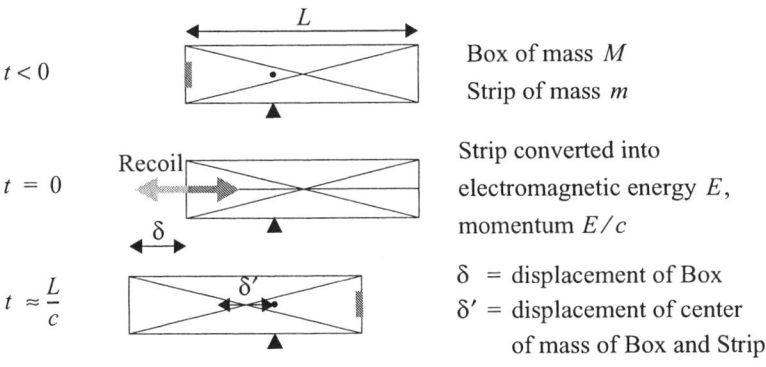

Figure 6: *Einstein's thought experiment to show that $E = mc^2$ is necessary and sufficient for the center-of-mass theorem.*

Let the energy E initially be contained in a strip of as yet unknown mass $m \ll M$ stuck against the inside of the left wall of the box. This means that the center of mass of box plus strip will be slightly to the left of the middle of the box. The energy is then converted into electromagnetic radiation and, a short time later, reconverted into a strip of mass m stuck against the inside of the right wall of the box. The center of mass of box plus strip is now slightly to the right of the middle of the box. The displacement δ' of the center of mass can be calculated from the following condition that determines where a wedge supporting the system should be placed so that the system is perfectly balanced:

$$M\left(\frac{\delta'}{2}\right) = m\left(\frac{L-\delta'}{2}\right). \tag{4}$$

It follows that, to a very good approximation, the displacement of the center of mass is given by:

$$\delta' \approx \left(\frac{m}{M}\right)L. \tag{5}$$

The center-of-mass theorem is satisfied if and only if the displacement δ' of the center of mass to the right is equal to the distance δ that the box travels to the left during the time it takes for the radiation to move from one end of the box to the other. To a very good approximation, the time that the box is in flight can be set equal to (L/c), and the velocity of the box can be set equal to its momentum E/c divided by its mass M. Hence, to a very good approximation, the distance travelled by the box is given by:

$$\delta \approx \left(\frac{E/c^2}{M}\right)L. \tag{6}$$

Comparing eqs. (5) and (6), one sees that indeed

$$\delta = \delta' \Leftrightarrow E = mc^2. \tag{7}$$

The conclusion is that $E = mc^2$ is the necessary and sufficient condition for the center-of-mass theorem to hold in systems in which processes involving both electromagnetic fields and ordinary matter occur. Strictly speaking, there should of course be approximately-equal signs rather than equal signs in eq. (7), just as in eqs. (5) and (6). In other words, the thought experiment only yields the conclusion to a very good approximation. Einstein was happy to leave it at that (Einstein 1906, 629).[23]

The Trouton experiment can be seen as a practical version of Einstein's thought experiment. In the case of the Trouton experiment it is the conversion of chemical energy of the battery into the energy of the electromagnetic field between the plates of the capacitor that would lead to a violation of the center-of-mass theorem were it not for the inertia of energy expressed in $E = mc^2$.[24]

6. THE TROUTON EXPERIMENT FORGOTTEN

Why did the Trouton experiment not live on in textbooks on relativity as a practical version of Einstein's thought experiment of 1906? Something along these lines, after all, did happen to the Trouton-Noble experiment. Ever since Max Laue (1911b) first made the connection between the Trouton-Noble experiment and a thought experiment involving a turning couple on a moving right-angled lever (Lewis and Tolman 1909), the Trouton-Noble experiment has frequently been cited as a physical instantiation of this thought experiment (see for example, Pauli 1921, 128). The Trouton experiment, however, seems to have been forgotten before anybody could make the connection between the experiment and $E = mc^2$.

The main reason for this sad fate is undoubtedly that the Trouton-Noble experiment completely overshadowed the Trouton experiment. We have seen this in the work of Larmor and Lorentz. Larmor was interested in the Trouton-Noble experiment as a test of FitzGerald's contraction hypothesis, but saw no merit in the reasoning behind FitzGerald's prediction of the effect sought after in the Trouton experiment. And FitzGerald was no longer there to defend or elucidate his argument. Lorentz took the Trouton experiment more seriously, but the fact remains that the experiment does not play the prominent role that the Trouton-Noble experiment plays in the exposition of the 1904 version of his theory for the electrodynamics of moving bodies. To make matters worse, the section on the Trouton experiment was omitted when Lorentz's paper was reprinted as part of *The Principle of Relativity*, an anthology still popular today (Lorentz et al. 1913, 1922). The omission was probably not just to save space. The editors may have found it more than a little awkward for a collection of papers on the relativity principle to include a passage implying that a more accurate repetition of some experiment should produce a violation of this principle.[25]

The far greater importance that both Larmor and Lorentz attached to the Trouton-Noble experiment appears to have sealed the fate of the Trouton experiment. The only subsequent mention of the experiment in the physics literature that I am aware of occurs in a review article on the experimental evidence in support of the relativity principle written by Jakob Laub, one of Einstein's early collaborators (Laub 1910, 428–430).[26] Laub listed the experiment as one of four first-order electrodynamical (as opposed to optical) ether drift experiments with a negative result. The Trouton-Noble experiment is also listed, as one of two second-order electrodynamical ether drift experiments with a negative result. The classifications 'first order' and 'second order' refer to the fact that the experiments aimed at detecting effects of first and second order in (v/c), respectively.

The fact that the Trouton-Noble experiment was a second-order experiment was an important factor favoring it over the first-order Trouton experiment. In the years before the advent of special relativity, it was widely felt that Lorentz's theory gave a satisfactory account of almost all first-order experiments and that the main problem was how to extend this theory to cover second-order experiments as well. This is reflected both in the statement of purpose in Lorentz 1904a, quoted in section 4, and in Poincaré's criticism of the theory presented in Lorentz 1895 to which Lorentz was in part responding in 1904. And it is precisely for this reason that the Trouton-Noble experiment was so important for Lorentz in 1904.

Laub's discussion of the Trouton experiment nicely illustrates another factor that may have been partly responsible for the neglect of the Trouton experiment after 1905. Recall that Einstein, following the example of thermodynamics, presented special relativity in his 1905 paper in the form of what he would later call a "theory of principle" (Einstein 1919).[27] One can thus take the attitude that in special relativity the negative result of a particular ether drift experiment needs as little explanation as the impossibility of some alleged perpetuum mobile in thermodynamics. This indeed appears to have been Laub's attitude. He reported that according to Lorentz's theory there should have been a small effect in the Trouton experiment but that the apparatus had not been sensitive enough to detect it. He then noted that the negative result of the experiment is in accordance with the relativity principle and left it at that. He made no attempt to pinpoint what was wrong with Lorentz's analysis.

The analogy between special relativity and thermodynamics obviously cannot fully explain the lack of interest in producing a detailed account of the Trouton experiment, for in the case of the Trouton-Noble experiment there was no such lack of interest. It also does not explain why after 1905 Lorentz did not work out a detailed account of the Trouton experiment in his own "constructive theory" (to use the terminology of Einstein 1919 again), which was now supposed to predict null results for all conceivable ether drift experiments. Lorentz was in an ideal position, it seems, to produce such an account. He was familiar with all relevant papers. Trouton 1902 and Larmor 1902 are cited in Lorentz 1904a. And he must have read Einstein 1906 with special interest given its relation to Poincaré 1900b, a contribution to a *Festschrift* in his honor. In fact, in his lectures on relativity at the University of Leiden in 1910–12, Lorentz used the thought experiment of Einstein 1906 in explaining $E = mc^2$ (Lorentz 1922, 242–243). Yet, despite all this, he apparently never made the connection between the Trouton experiment and Einstein's thought experiment. It is tempting to ask why not.

There is a plausible answer to this question. As the reader may have noticed, there appears to be a discrepancy between Lorentz's expression for the momentum generated upon charging a moving capacitor and the expression given in the explanation of the Trouton experiment on the basis of $E = mc^2$ in section 5. According to Lorentz, the momentum gained by the capacitor (to first order in v/c) is $2(U/c^2)\mathbf{v}$, whereas our relativistic analysis suggests that the gain is only half that amount (cf. eq. 3). As we shall see in the next section, it is a non-trivial task to find out where the extra factor 2 is coming from. In fact, this was not fully clarified until Laue's work on relativistic mechanics in 1911. Suppose that Lorentz had realized sometime between 1905 and 1911 that the transfer of energy from battery to capacitor in a laboratory moving through the ether is accompanied by a transfer of momentum. This would still only account for half the momentum gained by the capacitor. Hence, momentum conservation would still require the capacitor to experience a jolt upon being charged.

By the time that Laue had provided the tools with which Lorentz's analysis of the Trouton experiment can fully be reconciled with the simple relativistic analysis given in section 5, the experiment seems to have been thoroughly forgotten. Years later, Laue—by now Max *von* Laue—chose the inertia of energy as the topic for his contribution to the volume in *The Library of Living Philosophers* devoted to Einstein (Laue 1949).[28] Von Laue discussed Einstein's thought experiment of 1906 and his own analysis of the Trouton-Noble experiment in the same section of this essay. He did not mention the Trouton experiment.

7. THE TROUTON EXPERIMENT AND DEFINING THE FOUR-MOMENTUM OF SPATIALLY EXTENDED SYSTEMS.

The account of the Trouton experiment in section 5 was based on the assumption that energy and momentum transform as the components of a four-vector under Lorentz transformations. As a matter of fact, the energy and momentum of the electromagnetic field of a charged capacitor do *not* transform in this manner, at least not under the definitions of these quantities that were used by Lorentz, nor for that matter under the closely related definitions still standard in special relativity today. If we want to retain these definitions, we have to revise the simple account of the Trouton experiment. If we want to retain this simple account, we have to adopt new definitions.

In modern notation,[29] Lorentz, following Abraham, defined the electromagnetic momentum **G** of an electric field **E** and a magnetic field **B** produced by a charge distribution moving through the ether as

$$\mathbf{G} \equiv \int \varepsilon_0 (\mathbf{E} \times \mathbf{B}) \, d^3 x, \qquad (8)$$

where ε_0 is the dielectric constant of the ether. In Lorentz's theory, the integral is to be taken in a coordinate system moving with the charge distribution. For Lorentz this coordinate system is related to a coordinate system at rest in the ether through a Galilean transformation. So, in relativistic terms, the integral is to be taken over a hyperplane of simultaneity in a frame that, from Lorentz's point of view, would be at rest in the ether. I will call such frames 'ether frames' for short.

To first order in (v/c), the field of a charged capacitor moving at velocity **v** with respect to the ether, its plates tilted at an angle ϑ with **v**, has momentum

$$G = 2 \left(\frac{U'}{c^2} \right) v \cos \vartheta \qquad (9)$$

pointing in the direction of the plates.[30] In this equation, U' is the energy of the electromagnetic field of the same capacitor at rest in the ether, or—what in special relativity comes to the same thing—the energy in the moving capacitor's rest frame. For the special case that the plates are parallel to the direction of motion ($\vartheta = 0$), we recover the expression for the electromagnetic momentum given in Lorentz's 1904 discussion of the Trouton experiment (see the quotation from Lorentz 1904a in section 4 above). As I mentioned at the end of section 6, even in this special case and to first order in (v/c),

$$\mathbf{G} \neq \left(\frac{U'}{c^2}\right)\mathbf{v}, \quad (10)$$

as would be required for the simple explanation of the Trouton experiment based on $E = mc^2$ given in section 5 (cf. eqs. 1–3). In other words, the electromagnetic momentum of the capacitor does not transform as the spatial part of a four-vector under Lorentz transformations. In general, it is not even in the direction of motion.

With the standard definition in special relativity of the four-momentum of spatially extended systems,

$$P^\mu \equiv \frac{1}{c}\int T^{\mu 0} d^3x, \quad (11)$$

where $T^{\mu\nu}$ is the system's energy-momentum tensor, we run into the exact same problem. In fact, the quantity $\mathbf{G}$ as defined in eq. (8) is just the spatial part of the quantity P^μ as defined in eq. (11), if for $T^{\mu\nu}$ one takes the energy-momentum tensor of the electromagnetic field of a moving charge distribution in an ether frame.[31]

How is the four-momentum of the electromagnetic field of the moving capacitor related to the four-momentum of the electromagnetic field of that same capacitor at rest, or—what in special relativity comes to the same thing—to the four-momentum of the electromagnetic field of the moving capacitor in its rest frame? I will answer this question along the lines of a seminal paper by Max Laue (1911a), in which the energy-momentum tensor was for the first time put at the center of relativistic mechanics.[32]

With the help of the transformation law for the energy-momentum tensor, eq. (11) can be rewritten as

$$P^\mu = \frac{1}{c}\Lambda^\mu{}_\rho \Lambda^0{}_\sigma \int T'^{\rho\sigma} d^3x \quad (12)$$

where $T'^{\mu\nu}$ the energy-momentum tensor for the field of the capacitor at rest. Since a charged capacitor is a static system, the $i0$- and $0i$-components ($i = 1, 2, 3$) of $T'^{\mu\nu}$ vanish, and the volume integral in the x^μ-frame can be set equal to $1/\gamma$ times the volume integral in the x'^μ-frame.[33] Using in addition that $\Lambda^0{}_0 = \gamma$, we can rewrite eq. (12) for this special case as

$$P^\mu = \frac{1}{c}\Lambda^\mu{}_0 \int T'^{00} d^3x' + \frac{1}{\gamma c}\Lambda^\mu{}_i \Lambda^0{}_j \int T'^{ij} d^3x'. \quad (13)$$

The first term on the right-hand side is equal to $\Lambda^\mu{}_0(U'/c)$. This is just the Lorentz transform of the four-momentum $P'^\mu = (U'/c, 0, 0, 0)$ of the system at rest (cf. eqs. 1–2). It is the second term that is responsible for the fact that P^μ does not transform as a four-vector.

This term depends on the electromagnetic stresses T'^{ij}. Relativistic mechanics thus seems to predict a rather peculiar effect. Stresses in a system's rest frame give rise to energy and momentum in a frame in which the system is in motion. In general, such momentum will not be in the direction of motion. This means that it will give rise to a turning couple.[34] This is just the type of effect sought after in the Trouton-Noble experiment. The electromagnetic stresses in a charged capacitor, however, will be exactly balanced by stresses in the material part of the capacitor that prevent the plates from collapsing onto one another under the influence of their mutual Coulomb attraction. These material stresses will also give rise to energy and momentum in a frame in which the system is in motion. The energy and momentum coming from the material stresses will be equal and opposite to the energy and momentum coming from the electromagnetic stresses. The material stresses will thus give rise to a turning couple compensating the turning couple coming from the electromagnetic stresses. This, essentially, is Laue's explanation of the negative result of the Trouton-Noble experiment, which has become the standard explanation of the experiment in special relativity (see for example, Pauli 1921, 129).

Laue was able to show that

$$\int T'^{ij} d^3 x' = 0 \qquad (14)$$

for the total energy-momentum tensor of any so-called "complete static system" (*Vollständiges statisches System*, Laue 1911a, 539), where 'complete' means 'closed' (i.e., $\partial_\nu T^{\mu\nu} = 0$).[35] So, for a complete static system, the second term on the right-hand side of eq. (13) vanishes. Therefore, the total four-momentum of a complete static system, unlike the four-momentum of its various constituents, transforms as a four-vector under Lorentz transformations.[36] One of the examples Laue gives of a complete static system is that of a charged capacitor and its electromagnetic field.

The upshot of these considerations based on Laue's work of 1911 is that the simple explanation of the Trouton experiment of section 5 based on $E = mc^2$ works if we consider *both* the electromagnetic field of the capacitor *and* the stresses in the material part of the capacitor, but that it does *not* work if we *only* consider the electromagnetic field, as Lorentz did in 1904. Nothing in classical mechanics could have prepared Lorentz for Laue's strange effect of stresses producing momentum. So, even if Lorentz had immediately recognized the connection between the thought experiment used to derive $E = mc^2$ in Einstein's 1906 paper and the Trouton experiment, he would have been in no position to correct his 1904 analysis of the Trouton experiment before Laue's work of 1911.

From a modern point of view, Laue's analysis of the conditions under which P^μ transforms as a four-vector is rather clumsy. Most importantly, the restriction to static systems that Laue was forced to impose is completely unnecessary. The total four-momentum of any closed system, static or not, transforms as a four-vector under Lorentz transformations. This can easily be shown with the help of the obvious generalization from 3-dimensional space to 4-dimensional space-time of the standard theorem of Gauss that says that the integral of any vector field over a closed surface is equal to the integral of the divergence of that vector field over the volume enclosed by that surface. It will still be worth our while to take a somewhat closer look at this derivation, for it will reveal that special relativity allows us a certain freedom in defining the four-momentum of spatially extended systems. Taking advantage of this freedom, we can simplify the explanation of the Trouton and Trouton-Noble experiments considerably.

We begin by defining the manifestly Lorentz invariant quantity

$$P^\mu_\Sigma \equiv \int_\Sigma T^{\mu\nu} n_\nu d\Sigma, \tag{15}$$

where Σ is a space-like hyperplane in Minkowski space-time defined in some arbitrary frame with coordinates x^μ through $n_\mu x^\mu = c\tau$ (n^μ is the normal to the hyperplane, $c\tau$ the hyperplane's distance to the origin of the x^μ-frame). For arbitrary but fixed hyperplanes Σ, P^μ_Σ transforms as a four-vector under Lorentz transformations. It is an example of what Felix Klein called a "free vector" (Klein 1918, 398–399).[37]

The standard definition of the four-momentum of spatially extended systems is obtained from eq. (15) by stipulating that in every frame we choose Σ to be a hyperplane of simultaneity in that frame. In that case, $n^\mu = (1, 0, 0, 0)$ and the right-hand side of eq. (15) reduces to the right-hand side of eq. (11). With this convention, four-momentum will, in general, not transform as a four-vector. Let $P'^\mu_{\Sigma(x'^\mu)}$ be the four-momentum in the x'^μ-frame. So, $\Sigma(x'^\mu)$ is a hyperplane of simultaneity in the x'^μ-frame. The Lorentz transform $P^\mu_{\Sigma(x'^\mu)} = \Lambda^\mu{}_\nu P'^\nu_{\Sigma(x'^\mu)}$ in the x^μ-frame will still be an integral over $\Sigma(x'^\mu)$, which is *not* a hyperplane of simultaneity in the x^μ-frame. So, $P^\mu_{\Sigma(x'^\mu)}$ will in general not give the four-momentum in the x^μ-frame. It will only give the four-momentum in the x^μ-frame if $P^\mu_{\Sigma(x'^\mu)} = P^\mu_{\Sigma(x^\mu)}$, i.e., if the values of the integrals do not depend on which hyperplane of simultaneity is being integrated over. The generalization of Gauss's theorem referred to above tells us that under fairly general conditions (e.g., that $T^{\mu\nu}$ falls off sufficiently rapidly as we go to infinity):

$$(P^\mu_\Sigma \text{ independent of } \Sigma) \Leftrightarrow (\partial_\nu T^{\mu\nu} = 0 \text{ everywhere}). \tag{16}$$

So, under the standard definition, the four-momentum of a spatially extended system transforms as a four-vector under arbitrary Lorentz transformations if and only if the system is closed.[38]

In 1922, Enrico Fermi proposed an alternative Lorentz invariant definition of the four-momentum of spatially extended systems. His proposal was forgotten, then independently rediscovered several times, until Fritz Rohrlich (1960, 1965) made sure that it would not be forgotten again. I will refer to this alternative definition as the Fermi-Rohrlich definition. It differs from the standard definition in the convention that is adopted for choosing the hyperplane Σ in eq. (15). The standard convention, as we just saw, is to pick different families of hyperplanes in different frames of reference, viz. the hyperplanes of simultaneity in whatever frame we happen to be using. This results in a definition that is not Lorentz invariant. The Fermi-Rohrlich definition stipulates that we pick the same family of hyperplanes in all frames of reference, viz. the hyperplanes of simultaneity in the rest frame of the system under consideration.[39]

Given this definition, all four-momentum transforms as a four-vector under Lorentz transformations. It does not matter whether we are dealing with closed systems or with their open subsystems.

The relativist is free to adopt either the standard definition or the Fermi-Rohrlich definition. After all, the only difference between the two is a convention about how to slice Minkowski space-time. A proponent of Lorentz's theory, however, is committed to a definition that mathematically is a special case of the standard definition. To convince the reader that the proponent of Lorentz's theory is indeed thus committed and does not enjoy the same freedom as his relativistic counterpart in this matter, I will once again resort to describing the state of affairs in Lorentz's theory with the help of some relativistic terminology.

After 1905, as I mentioned at the end of section 4, Lorentz came to realize that a system in motion through the ether will always *appear* to a co-moving observer as if it were at rest in the ether (in relativistic terms, the co-moving observer will always measure the quantities in the system's rest frame). Even after 1905, however, the *true* description of the system continues to be the description in terms of the quantities of a coordinate system either at rest in the ether or related to such a coordinate system by a Galilean transformation.

Consider the specific example of a charged capacitor moving through the ether. To a co-moving observer the capacitor will appear as if it is at rest in the ether.[40] The co-moving observer will thus measure the four-momentum of the capacitor's electromagnetic field in the capacitor's rest frame. There is no need to specify under which definition of four-momentum because in the capacitor's rest frame the standard definition and the Fermi-Rohrlich definition coincide. It is also no problem, of course, to introduce the notion of four-momentum into Lorentz's theory for the combination of energy and momentum.

Now consider the true description of the system, which, up to a Galilean transformation, is a description in an ether frame. Hence, the true four-momentum of the electromagnetic field of a moving capacitor will be an integral over a hyperplane of simultaneity in an ether frame. Lorentz's definition of the four-momentum of electromagnetic fields or other spatially extended systems is thus obtained from eq. (15) by stipulating that P_Σ^μ always be evaluated in an ether frame and that Σ be a hyperplane of simultaneity in such a frame. From a purely mathematical point of view, this makes Lorentz's definition a special case of the standard definition in special relativity.[41]

The explanations of the Trouton and Trouton-Noble experiments are much simpler under the Fermi-Rohrlich definition of the four-momentum of spatially extended systems than under the standard definition. In the case of the Trouton experiment, the Fermi-Rohrlich definition allows us to use the simple explanation of section 5 (1–3) without ever having to worry about stresses in the material part of the capacitor. This is because under the Fermi-Rohrlich definition the four-momentum of the capacitor's electromagnetic field taken by itself will transform as a four-vector under Lorentz transformations. So, for the energy $\Delta E'$ transferred from the battery to the capacitor (see eq. 1) we can simply substitute the energy U' of the electric field in the capacitor's rest frame. In the case of the Trouton-Noble experiment, the simplification brought about by adopting the Fermi-Rohrlich definition is even greater. Instead of the delicately balanced turning couples that we find under the standard definition, there will be no turning couples whatsoever under the Fermi-Rohrlich definition![42] As I explained above, the Fermi-Rohrlich definition is unacceptable to proponents of Lorentz's theory. In Lorentz's theory one is therefore stuck with the more cumbersome explanations of these experiments on the basis of the standard definition. I think that this circumstance can be turned into a strong argument for preferring special relativity over an empirically equivalent version of Lorentz's theory. But that is a story for another paper.

ACKNOWLEDGMENTS

I am greatly indebted to John Stachel for his perceptive comments following a talk I gave on the experiments of Trouton and Noble in a colloquium of the Max-Planck-Institut für Bildungsforschung in Berlin in the Summer of 1994 (cf. notes 3 and 5). I also want to thank Jon Dorling, Tony Duncan, Gordon Fleming, David Hillman, A. J. Kox, John Norton, Jürgen Renn, and again John Stachel for helpful discussions on many other occasions.

NOTES

1. For a concise survey of 19th-century ether drift experiments and their theoretical ramifications, with selected references to further literature, see Janssen and Stachel 1999.
2. Warwick does not always distinguish carefully, as I think one should, between the effect sought after in the Trouton experiment and the effect sought after in the Trouton-Noble experiment (see for example, Warwick 1995, 318).
3. John Stachel first drew my attention to the connection between the Trouton experiment and this paper by Einstein.
4. This focus on developments at the level of mechanics in my analysis of the history of special relativity was partly inspired by Damerow et al. 1992 and by several discussions over the years with Jürgen Renn.
5. John Stachel first suggested this interpretation of Larmor's rather cryptic comments on the Trouton experiment to me.
6. Both Fermi and Rohrlich proposed this alternative definition to give a Lorentz-invariant definition of the electromagnetic four-momentum of the classical electron model of Lorentz and Poincaré. In this model, the electron is a physical system very similar to a charged capacitor.
7. One can avoid such arbitrary conventions altogether by accepting that the four-momentum of a spatially extended system is an example of a hyperplane-dependent quantity. This position, to which I am very sympathetic, has been championed by Gordon Fleming (1998).
8. Laue's proof was not entirely satisfactory even for this special case (see note 33 below). A satisfactory proof, without the unnecessary restriction to static systems, was first given by Felix Klein in the context of general rather than special relativity (Klein 1918). Klein's proof was inspired by correspondence with Einstein in 1918 (see Schulmann et al. 1998, in particular doc. 581, note 9). Klein coined the name "free vector" for such non-local quantities as the four-momentum of spatially extended systems obtained through integration over some hypersurface in space-time.
9. Readers who think of Einstein and Lorentz as proponents of competing paradigms, research programs, or what have you, will probably feel uneasy about this identification of elements common to the ether-theoretic and the relativistic explanations of the Trouton experiment. This is not the place to argue the point (see Janssen 1995, 1997), but I think the difference between the positions of Lorentz and Einstein is much more naturally understood as a difference of opinion over the interpretation of the Lorentz invariance of a formalism they agreed upon. Although Lorentz formally retained remnants of

Newtonian mechanics in his theory, he accepted that physical systems are in fact governed by relativistic mechanics (see for example, Lorentz 1922, based on lectures of 1910–12).

10. There is an obvious improvement of Trouton's design. As Trouton explains in his paper: "It was originally intended to have two condensers, one at each end of the cross arm, the one to be charged at the moment the other was discharged, not only to double the effect, but also to secure a pure torque acting on the wire. This idea had to be abandoned in the final experiment, owing to all the condensers available breaking down under the excessive voltage employed save only one" (Trouton 1902, 559).
11. See Janssen 1995, section 1.3, for my reconstruction and analysis of Larmor's argument.
12. For an insightful discussion of this issue, see Darrigol 1995.
13. In accordance with his infamous attribution of special relativity to Lorentz and Poincaré, Sir Edmund T. Whittaker has Poincaré proclaim the inertia of energy in 1900: "In 1900 Poincaré suggested that electromagnetic energy might possess mass density [...] that is to say $E = mc^2$ [...] and he remarked that if this were so, then a Hertz oscillator, which sends out electromagnetic energy preponderantly in one direction, should recoil as a gun does when it is fired" (Whittaker 1951–53, II, 51). The passage quoted above from Poincaré's lecture four years later clearly refutes Whittaker's claim. I am grateful to Tony Duncan for alerting me to Whittaker's claim.
14. Lorentz to Poincaré, January 20, 1901. The letter is quoted in full in Miller 1986, 6–7. Lorentz also discussed these issues in his lectures at Columbia University in New York in 1906 (Lorentz 1915, 30–33).
15. Given the initial resistance in the 1920s to the notion of spin as a non-mechanical form of angular momentum it should perhaps not surprise us that there was some resistance to the notion of electromagnetic momentum as non-mechanical momentum two decades earlier.
16. Lorentz only sketched this derivation in his paper and referred to one of his contributions to the *Encyclopädie der mathematischen Wissenschaften* (Lorentz 1904b) for further details. Using nothing but Newtonian mechanics, the Maxwell-Lorentz equations, and the expression for the Lorentz force, Lorentz showed that the force **F** and the turning couple **T** that a charged system, that is static except for an overall velocity **v** with respect to the ether, experiences from its self-field are given by $\mathbf{F} = -d\mathbf{G}/dt$ and $\mathbf{T} = -\mathbf{v} \times \mathbf{G}$, respectively (where **G** is the self-field's electromagnetic momentum). These equations are just what one would expect on the basis of the interpretation of **G** as a form of momentum. They express conservation of momentum and angular momentum, respectively. (See Janssen 1995, Secs. 1.2, 1.4.2, and 3.4.2, for reconstructions of Lorentz's derivations in modern notation.)
17. In his lectures at Columbia University two years later, Lorentz was still careful not to commit himself to the interpretation of electromagnetic momentum as a new form of momentum (see Lorentz 1915, 32).
18. In this equation, U is "the energy of the charged condenser in the state of rest" (Lorentz 1904a, 830-831), c is the velocity of light, and w is the velocity of the capacitor with respect to the ether.
19. See section 7, eq. (9), below.
20. For a more detailed account of the development of Lorentz's theory, see (Janssen 1995—which also contains a detailed critique of older accounts of this development; Janssen 1997; and Janssen and Stachel 1999).
21. Where $\Lambda^\mu{}_\nu = \begin{pmatrix} \gamma & \gamma\beta & 0 & 0 \\ \gamma\beta & \gamma & 0 & 0 \\ 0 & 0 & 1 & 0 \\ 0 & 0 & 0 & 1 \end{pmatrix}$, with $\beta \equiv \frac{v}{c}$ and $\gamma \equiv \frac{1}{\sqrt{1-\beta^2}}$.
22. Notice that it does not matter what the angle between the plates of the capacitor and its velocity is.
23. The argument as it stands cannot be made exact, because it tacitly involves the assumption that the box in the thought experiment can be treated as a rigid body, a notion incompatible with special relativity. In fact, the disturbance at the left end of the box upon emission of the radiation will not even have reached the right side of the box at the time of reabsorption of the radiation at the right end! The problem can be circumvented by modifying the thought experiment. One option (suggested to me by John Norton) is to have the energy E transmitted from one end of the box to the other in small parcels over

a long period of time. Another option is to only retain the walls at the two ends of the box without anything connecting them.

24. Thinking of the Trouton experiment in terms of Einstein's thought experiment, we see that there should still be a minuscule effect in the former due to the slight displacement of the center of mass of the battery-capacitor system that results from transferring energy from the battery to the capacitor.
25. In terms of distorting the history of special relativity, the omission of this section of Lorentz 1904a pales in comparison, of course, to the decision to include only the section on the Michelson-Morley experiment of Lorentz 1895 (cf. Miller 1981, 391–392).
26. I am grateful to John Stachel for bringing this paper to my attention. For discussion of the collaboration between Einstein and Laub, see the editorial note, "Einstein and Laub on the Electrodynamics in Moving Media," in Stachel et al. 1989, 503–507.
27. Einstein contrasted theories of principle with constructive theories. Roughly, the distinction is that a constructive theory provides a detailed model of (features of) the physical world, whereas a theory of principle only provides constraints on such modeling, constraints based on empirically well confirmed regularities.
28. In his response to the papers brought together in this volume, Einstein characterized Von Laue's essay as "An historical investigation of the development of the conservation postulates, which, in my opinion, is of lasting value. I think it would be worth while to make this essay easily accessible to students by way of independent publication" (Einstein 1949, 686).
29. I use SI or MKSA units. For conversion to other units, see for example, Jackson 1975, 817–819.
30. See Janssen 1995, section 1.4, for a reconstruction in modern notation of Lorentz's derivation of this result. That **G** will be in the direction of the plates of the capacitor to first order in v/c can be seen upon inspection of the integrand in eq. (8): the **E**-field will be perpendicular to the plates, while the **B**-field will be parallel to the plates and perpendicular to the velocity **v**.
31. Use $T^{\mu\nu} \equiv \frac{1}{\mu_0}\left(\eta_{\alpha\beta}F^{\mu\alpha}F^{\beta\nu} + \frac{1}{4}\eta^{\mu\nu}\eta_{\alpha\rho}\eta_{\beta\sigma}F^{\rho\sigma}F^{\alpha\beta}\right)$, where

$$F^{\mu\nu} \equiv \begin{pmatrix} 0 & -E_x/c & -E_y/c & -E_z/c \\ E_x/c & 0 & -B_z & B_y \\ E_y/c & B_z & 0 & -B_x \\ E_z/c & -B_y & B_x & 0 \end{pmatrix},$$

$\eta^{\mu\nu} = \eta_{\mu\nu} \equiv \text{diag}(1,-1,-1,-1)$, and $c = 1/\sqrt{\varepsilon_0\mu_0}$ (with μ_0 the permeability *in vacuo*).

32. For further discussion of this work by Laue, see Norton 1992, section 9.
33. In general, this will not be true: a volume integral in the x^μ-frame and a volume integral in the x'^μ-frame are integrals over different hyperplanes of simultaneity. For static systems, however, the results of the two integrals differ only by a factor γ. It is unclear whether Laue, when he wrote his 1911 paper, realized that the assumption that the system is static is crucial at this juncture. For further discussion, see Janssen 1995, section 2.1.4.
34. Cf. the equation $\mathbf{T} = -\mathbf{v} \times \mathbf{G}$ in note 16.
35. This result is sometimes called "Laue's theorem" (Miller 1981, 373; unfortunately, Miller uses the phrase "perfectly static system" instead of "complete static system").
36. It follows that there will never be a net turning couple acting on a complete static system.
37. Cf. note 8.
38. For a more detailed version of this proof, see Rohrlich 1965, 89–90, 279–281.
39. As Gordon Fleming has pointed out (private communication), it is not clear exactly how to define the rest frame of an arbitrary spatially extended system. For *static* systems, however, such as a charged capacitor and its electromagnetic field, this problem does not arise.
40. This observer, for instance, will not notice the Lorentz-FitzGerald contraction that the capacitor experiences as a result of its motion through the ether.
41. It follows from Lorentz's definition (cf. the discussion following eq. (15) above) that the four-momentum of the electromagnetic field of a capacitor moving through the ether is not the Lorentz transform

of the four-momentum of the electromagnetic field of the same capacitor at rest in the ether. The Lorentz transform of the four-momentum of the field of the capacitor at rest in the ether will be an integral over a hyperplane of simultaneity in the moving system's rest frame, whereas the (true) four-momentum of the field of the capacitor in motion through the ether will be an integral over a hyperplane of simultaneity in the ether frame. Since the electromagnetic field of a capacitor is not a closed system, these integrals over different hyperplanes will give different results.

42. See Butler 1968, Janssen 1995, Teukolsky 1996.

REFERENCES

Abraham, Max. 1903. "Prinzipien der Dynamik des Elektrons." *Annalen der Physik* 10:105–179.
Butler, J. W. 1968. "On the Trouton-Noble experiment." *American Journal of Physics* 36:936–941.
Cornille, Patrick, Jean-Louis Naudin, and Alexandre Szames. 1998. *Way Back to the Future: Why Did the Trouton-Noble Experiment Fail and How to Make it Succeed*. Paper presented at the Sixth Conference on the Physical Interpretations of the Relativity Theory (PIRT), London, UK, September 11–14. Forthcoming
———. 1999. *Stimulated Forces Demonstrated: Why the Trouton-Noble Experiment Failed and How to Make it Succeed*. Paper presented at the Space Technology and Applications International Forum (STAIF), Albuquerque, NM, USA, January 31–February 4, 1999. Forthcoming.
Damerow, Peter, Gideon Freudenthal, Peter McLaughlin, and Jürgen Renn. 1992. *Exploring the Limits of Preclassical Mechanics*. New York: Springer.
Darrigol, Olivier. 1995. "Henri Poincaré's Criticism of Fin de Siècle Electrodynamics." *Studies in History and Philosophy of Modern Physics* 26:1–44.
Einstein, Albert. 1905a. "Zur Elektrodynamik bewegter Körper." *Annalen der Physik* 17:891-921. Reprinted in facsimile as Doc. 23 in Stachel et al. 1989.
———. 1905b. "Ist die Trägheit eines Körpers von seinem Energieinhalt abhängig?" *Annalen der Physik* 18:639–641. Reprinted in facsimile as Doc. 24 in Stachel et al. 1989.
———. 1906. "Das Prinzip von der Erhaltung der Schwerpunktsbewegung und die Trägheit der Energie." *Annalen der Physik* 20:627–633. Reprinted in facsimile as Doc. 35 in Stachel et al. 1989.
———. 1919. "What is the Theory of Relativity?" *The London Times*, November 28, 1919. Reprinted in Albert Einstein. *Ideas and Opinions*. New York: Bonanza, 227–232.
———. 1949. "Remarks Concerning the Essays Brought Together in this Co-operative Volume." Pp. 665–688 in *Albert Einstein: Philosopher-Scientist*, ed. Paul Arthur Schilpp. Evanston, IL: Library of Living Philosophers.
Fermi, Enrico. 1922. "Über einen Widerspruch zwischen der elektrodynamischen und der relativistischen Theorie der elektromagnetischen Masse." *Physikalische Zeitschrift* 23:340–344.
Fleming, Gordon. 1998. "Reeh–Schlieder meets Newton-Wigner." *Proceedings of the 1998 Meeting of the Philosophy of Science Association*, Kansas City, Missouri, October 22–25, 1998. Forthcoming.
Hayden, Howard C. 1994. "High Sensitivity Trouton-Noble Experiment." *Review of Scientific Instruments* 65:788–792.
Jackson, John D. 1975. *Classical Electrodynamics*. 2nd edition, New York: John Wiley & Sons.
Janssen, Michel. 1995. *A Comparison Between Lorentz's Ether Theory and Special Relativity in the Light of the Experiments of Trouton and Noble*. Ph.D. Thesis. University of Pittsburgh.
———. 1997. *Reconsidering a Scientific Revolution: the Case of Einstein versus Lorentz*. Unpublished manuscript.
Janssen, Michel, and John Stachel. 1999. "The Optics and Electrodynamics of Moving Bodies." In *Storia della scienza*. Instituto della Enciclopedia Italiana, forthcoming.
Klein, Felix. 1918. "Über die Integralform der Erhaltungssätze und die Theorie der räumlich-geschlossenen Welt." *Königliche Gesellschaft der Wissenschaften zu Göttingen. Nachrichten*:394–423.
Larmor, Joseph. 1902. "Can Convection Through the Æther Be Detected Electrically? Note on the Foregoing Paper." Pp. 566–569 in *The Scientific Writings of the Late George Francis FitzGerald*, ed. J. Larmor. Dublin: Hodges, Figgis, & Co.; London: Longmans, Green, & Co.

Laub, Jakob. 1910. "Über die experimentellen Grundlagen des Relativitätsprinzips." *Jahrbuch der Radioaktivität und Elektronik* 7:405–463.

Laue, Max. 1911a. "Zur Dynamik der Relativitätstheorie." *Annalen der Physik* 35:524–542.

———. 1911b. "Ein Beispiel zur Dynamik der Relativitätstheorie." *Verhandlungen der Deutschen Physikalischen Gesellschaft* 13:513–518.

Laue, Max von. 1949. "Inertia and Energy." Pp. 501–533 in *Albert Einstein: Philosopher-Scientist*, ed. Paul Arthur Schilpp. Evanston, IL: Library of Living Philosophers.

Lewis, Gilbert N., and Richard C. Tolman. 1909. "The Principle of Relativity, and Non–Newtonian Mechanics." *Philosophical Magazine* 18:510–523.

Lorentz, Hendrik Antoon. 1895. *Versuch einer Theorie der electrischen und optischen Erscheinungen in bewegten Körpern.* Leiden: Brill.

———. 1904a. "Electromagnetische verschijnselen in een stelsel dat zich met willekeurige snelheid, kleiner dan die van het licht, beweegt." *Koninklijke Akademie van Wetenschappen te Amsterdam. Wisen Natuurkundige Afdeeling. Verslagen van de Gewone Vergaderingen* 12 (1903–04):986–1009. Reprinted in translation as "Electromagnetic Phenomena in a System Moving with Any Velocity Smaller Than That of Light." *Koninklijke Akademie van Wetenschappen te Amsterdam. Section of Sciences. Proceedings* 6 (1903–04):809–831.

———. 1904b. "Weiterbildung der Maxwellschen Theorie. Elektronentheorie." Vol. 5, *Physik*, part 2, 145–280 in *Encyklopädie der mathematischen Wissenschaften, mit Einschluß ihrer Anwendungen*, ed. Arnold Sommerfeld. Leipzig: Teubner, 1904–1922. Issued 16 June 1904.

———. 1915. *The Theory of Electrons and Its Applications to the Phenomena of Light and Radiant Heat. A Course of Lectures Delivered in Columbia University, New York, in March and April 1906.* 2d ed. Leipzig: Teubner.

———. 1922. *Lessen over theoretische natuurkunde aan de Rijksuniversiteit te Leiden gegeven.* Vol. 6. *Het relativiteitsbeginsel voor eenparige translaties (1910–1912).* Adriaan D. Fokker, ed. Leiden: Brill. English translation: *Lectures on Theoretical Physics.* Vol. 3. London: Macmillan and Co, 1931. Page references are to this translation.

Lorentz, Hendrik A., Albert Einstein, and Hermann Minkowski. 1913. *Das Relativitätsprinzip. Eine Sammlung von Abhandlungen.* Leipzig: Teubner.

Lorentz, Hendrik A., Albert Einstein, Hermann Minkowski, and Hermann Weyl. 1922. *Das Relativitätsprinzip. Eine Sammlung von Abhandlungen.* 4th ed. Leipzig: Teubner. English translation: *The Principle of Relativity.* New York: Dover, 1952.

Miller, Arthur I. 1981. *Albert Einstein's Special Theory of Relativity. Emergence (1905) and Early Interpretation (1905–1911).* Reading, MA: Addison–Wesley.

———. 1986. *Frontiers of Physics: 1900–1911.* Boston: Birkhäuser.

Minkowski, Hermann. 1909. "Raum und Zeit." *Physikalische Zeitschrift* 10:104–111. Reprinted in *Lorentz et al. 1913, Lorentz et al, 1922.*

Norton, John D. 1992. "Einstein, Nordström and the Early Demise of Scalar, Lorentz Covariant Theories of Gravitation." *Archive for the History of Exact Sciences* 45:17-94.

Pauli, Wolfgang. 1921. "Relativitätstheorie." Vol. 5, *Physik*, part 2, 539–775 in *Encyklopädie der mathematischen Wissenschaften, mit Einschluß ihrer Anwendungen*, ed. Arnold Sommerfeld. Leipzig: Teubner, 1904–1922. Issued 15 November 1921. Reprinted in translation, with supplementary notes, as *Theory of Relativity.* G. Field, trans. London: Pergamon, 1958. Page references to this reprint.

Planck, Max. 1908. "Bemerkungen zum Prinzip der Aktion und Reaktion in der allgemeinen Dynamik." *Deutsche Physikalische Gesellschaft. Verhandlungen* 6:728–732.

Poincaré, Henri. 1900a. "Sur les rapports de la physique expérimentale et de la physique mathématique." Vol. 1, 1–29 in *Rapports présentés au Congrès international de Physique réuni à Paris en 1900.* Paris: Gauthier–Villars.

———. 1900b. "La théorie de Lorentz et le principe de réaction." *Archives Néerlandaises des Sciences Exactes et Naturelles* 2:252–278.

———. 1904. "L'état actuel et l'avenir de la physique mathématique." *Bulletin des Sciences Mathématiques* 28:302–324. Reprinted in translation as Chs. 7–9 of Henri Poincaré, *The Value of Science.* New York: Dover, 1952. Page references to this reprint.

Rohrlich, Fritz. 1960. "Self-Energy and the Stability of the Classical Electron." *American Journal of Physics* 28:639–643.

———. 1965. *Classical Charged Particles: Foundations of Their Theory*. Reading, MA: Addison-Wesley.

Schulmann, Robert, A. J. Kox, Michel Janssen, and József Illy, eds. 1998. *The Collected Papers of Albert Einstein. Vol. 8. The Berlin Years: Correspondence, 1914–1918*. Princeton: Princeton University Press.

Stachel, John, David C. Cassidy, Jürgen Renn, and Robert Schulmann, eds. 1989. *The Collected Papers of Albert Einstein. Vol. 2. The Swiss Years: Writings, 1900–1909*. Princeton: Princeton University Press.

Teukolsky, Saul A. 1996. "The Explanation of the Trouton-Noble Experiment Revisited." *American Journal of Physics* 64:1104–1106.

Trouton, Frederick T. 1902. "The Results of an Electrical Experiment, Involving the Relative Motion of the Earth and Ether, Suggested by the Late Professor FitzGerald," *Transactions of the Royal Dublin Society* 7:379–384. Reprinted in: J. Larmor, ed., *The Scientific Writings of the Late George Francis FitzGerald*. Dublin: Hodges, Figgis, & Co.; London: Longmans, Green, & Co, 1902. Pp. 557–565. Page references are to this reprint.

Trouton, Frederick T., and Henry R. Noble. 1903. "The Mechanical Forces Acting On a Charged Electric Condenser Moving Through Space." *Philosophical Transactions of the Royal Society*, London 202:165–181.

Warwick, Andrew. 1995. "The Sturdy Protestants of Science: Larmor, Trouton, and the Earth's Motion Through the Ether." Pp. 300–343 in *Scientific Practice. Theories and Stories of Doing Physics*, ed. Jed Buchwald. Chicago: University of Chicago Press.

Whittaker, Edmund T. 1951–53. *A History of the Theories of Aether and Electricity*. 2 Vols. London: Thomas Nelson & Sons, Ltd.

JOHN D. NORTON

THE N-STEIN FAMILY*

1. THE STORY OF NEWSTEIN

The work of Newstein is now so familiar to us, thanks to Professor Stachel's efforts, that it bears only the briefest recapitulation. Sometime after 1880 but before the advent of general relativity, Newstein brooded on the equality of inertial and gravitational mass. Through an ingenious thought experiment—the Newstein elevator—he hit upon the idea of an essential unity of gravitation and inertia. This was expressed in the indistinguishability of the effects of acceleration in a uniformly accelerated frame of reference from a homogeneous gravitational field in an inertial frame of reference. Now having to consider the behavior of the gravitational force as it is transformed from unaccelerated to accelerated frames of reference, Newstein found it no longer behaved like the familiar vector. Puzzled, he turned to his mathematician friend Weylmann, another neglected figure in history of mathematics. His extraordinary achievement, as revealed by Professor Stachel, was to formulate the notion of affine connection around 1880, decades before the much better known formulation of Levi-Civita of 1917. Weylmann recognized that the puzzling transformation behavior of gravitational force was simply that of the components of a four-dimensional affine connection.

This provided the insight needed to write the now famous Newstein-Weylmann paper. It developed a formulation of Newton's theory of gravitation akin to Cartan and Friedrich's later proposals of the 1920s. In it, the chronogeometrical structure of spacetime remained absolute, but inertia and gravitation are combined in an affine structure. The Poisson equation for the gravitational potential is absorbed into an equation relating the Ricci tensor of the connection with the gravitational field's sources. With the association of gravitation with a curved, four-dimensional affine structure, the scene was now set for an Einstein to merge this viewpoint with the chronogeometry of special relativity, as captured in the spacetime metric of Minkowski, to yield the general theory of relativity.

2. THE FATAL OBJECTION?

Sadly, however, the Newstein-Weylmann proposal was neglected. As Professor Stachel tells it:

> Their work was regarded by contemporaries, in so far as they took any notice of it at all, as an ingenious mathematical tour-de-force; but since it had no new physical consequences, it did not much impress Newstein's positivistically-inclined physics colleagues.

There is no doubt that this diagnosis reveals part of reasons for the hesitation over the Newstein-Weylmann proposal. But there is more to say. We are inclined now to draw an analogy between special relativity and the Newstein-Weylmann proposal. Special relativity proceeds from the recognition that classical theories proposed the existence of an aether state of rest. What was objectionable in that proposal was that the aether state of rest was itself unobservable. That in turn resulted from its indeterminate nature. Any inertial state of motion proved to be an equally viable candidate for the aether state of rest. Both observation and theory were powerless to decide between them. Here we accord fully with the positivist sentiments of Newstein's physics colleagues in so far as they regarded the unverifiable aether state of rest as something to be purged from our physical theories.

The Newstein-Weylmann proposal seems very similar. The classical theory portrays free fall motions as the resultant of inertial motion and a gravitational deflection. But which of all possible motions are we to choose as the true inertial motion? All we observe are the resultant free fall motions. It would seem that the background inertial structure that fixes these inertial motions is as indeterminate as an aether state of rest. We eschew this aether state of rest in special relativity and build our theory of inertial motions alone. Should we do the same in gravitation theory: eschew the background inertial motions and build our theory on what is observed, the free fall motions, to which Newstein-Weylmann directly adapt their affine structure?

Compelling as this consideration may seem to us now, Newstein's colleagues were unconvinced. There was a telling disanalogy between the two cases. While the true inertial motions were not directly observable, they could be picked out uniquely by very natural conditions in the standard examples used in gravitation theory. Take the case of the gravitational field of the sun. We make the standard and natural presumptions of classical theory: the background inertial structure can be represented by a flat affine structure and the gravitational field of the sun is spherically symmetric in the space about the sun. This now provides a unique decomposition of the free fall motions around the sun into a background inertial structure and a gravitational deflection. The background inertial structure is perfectly determinate. Not even a strong dose of positivistic skepticism can undo that and repeal the sense that this determinate split into inertial motion and gravitational deflection reflects reality.

While this objection seems fatal, there was an answer. Natural conditions may pick out a unique inertial structure in some cases, but there are others in which demonstrably no such conditions can succeed. The realm of possibility is large and we may well wonder whether someone hit upon these examples and their import in the history of physics.

3. THE N-STEIN FAMILY: EINUNDZWANZIGSTEIN

My primary purpose in this paper is to announce the discovery not just of a single unnoticed 'stein in the history of science, but of a family of such figures:[1] Einstein, Newstein, Zweistein, ... The first two of these family members now enjoy the celebrity that their work warrants, thanks to the efforts of Professor Stachel. The mathematically inclined reader will immediately see that they form not just a family but an n-parameter family, where n takes suitable values: Ein, New,... For our purposes what is important is that one family member did hit upon the response that defeats the objection sketched above to the Newstein-Weylmann proposal. The work of this hitherto unrecognized figure, Albert Einundzwanzigstein, was revealed using techniques of historical research pioneered by Professor Stachel.[2] The content of the 1905 volume number 17 of *Annalen der Physik* is widely known; it contains the five papers of "Einstein's Miraculous Year."[3] What has remained unrecognized until now is the existence of a supplementary volume (see fig. 1) in which Einundzwanzigstein's "On the Cosmology of Free Falling Bodies" was published (see fig. 2). There Einundzwanzigstein showed that there is one case in Newtonian gravitation theory in which no natural conditions on the inertial structure and gravitational field can enforce a unique split of free fall motions into a background inertial motion and a gravitational deflection.

Einundzwanzigstein's result was expressed as the recognition that Newtonian cosmology is covariant under transformations between inertial frames and accelerated frames and that this covariance reflects the equivalence of observation for inertial and accelerated observers. It follows immediately that there are no unique background inertial motions identifiable, for these inertial motions cannot be invariant under a transformation to an accelerated frame. Einundzwanzigstein's argument is closely analogous to that of Einstein's 1905 "On the Electrodynamics of Moving Bodies." In Einstein's theory, an absolute state of rest is purged from the laws of physics by the principle of relativity since that state fails to remain invariant under a transformation between inertial frames of reference. We shall see that this similarity of strategy is reflected by closer analogies in the two papers.

ANNALEN
DER
PHYSIK.

BEGRÜNDET UND FORTGEFÜHRT DURCH
F. A. C. GREN, L. W. GILBERT, J. C. POGGENDORFF, G. UND E. WIEDEMANN.

VIERTE FOLGE.

BAND 17.

DER GANZEN REIHE 322. BAND.

BEILAGE
JOHANNES STACHEL
ZU SEINEM SIEBZIGSTEN GEBURTSTAG

KURATORIUM:
F. KOHLRAUSCH, M. PLANCK, G. QUINCKE,
W. C. RÖNTGEN, E. WARBURG.

UNTER MITWIRKUNG
DER DEUTSCHEN PHYSIKALISCHEN GESELLSCHAFT
UND INSBESONDERE VON
M. PLANCK

HERAUSGEGEBEN VON

PAUL DRUDE.

MIT FÜNF FIGURENTAFELN.

LEIPZIG, 1905.
VERLAG VON JOHANN AMBROSIUS BARTH.

Figure 1.

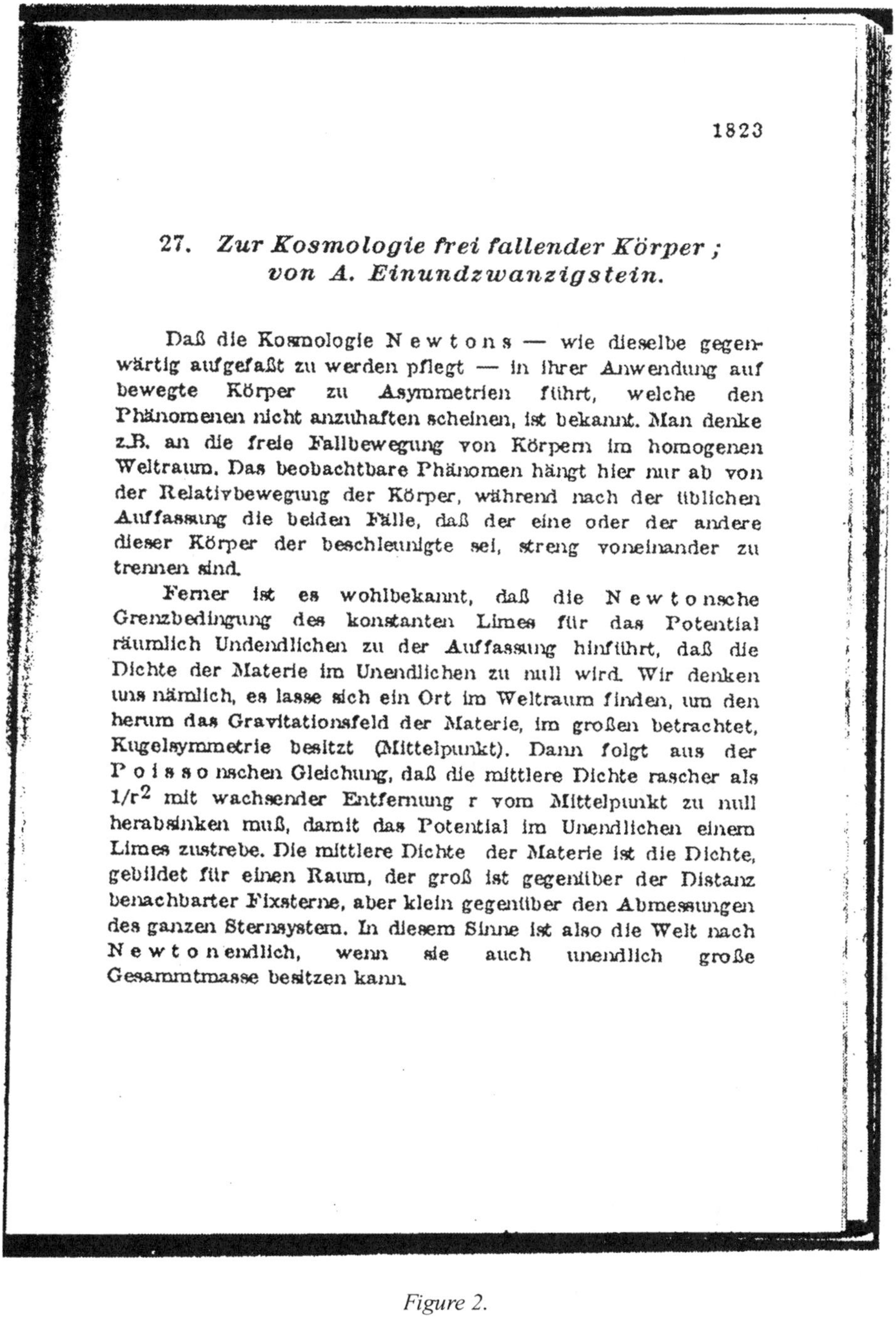

1823

27. *Zur Kosmologie frei fallender Körper;*
von A. Einundzwanzigstein.

Daß die Kosmologie Newtons — wie dieselbe gegenwärtig aufgefaßt zu werden pflegt — in ihrer Anwendung auf bewegte Körper zu Asymmetrien führt, welche den Phänomenen nicht anzuhaften scheinen, ist bekannt. Man denke z.B. an die freie Fallbewegung von Körpern im homogenen Weltraum. Das beobachtbare Phänomen hängt hier nur ab von der Relativbewegung der Körper, während nach der üblichen Auffassung die beiden Fälle, daß der eine oder der andere dieser Körper der beschleunigte sei, streng voneinander zu trennen sind.

Ferner ist es wohlbekannt, daß die Newtonsche Grenzbedingung des konstanten Limes für das Potential räumlich Unendlichen zu der Auffassung hinführt, daß die Dichte der Materie im Unendlichen zu null wird. Wir denken uns nämlich, es lasse sich ein Ort im Weltraum finden, um den herum das Gravitationsfeld der Materie, im großen betrachtet, Kugelsymmetrie besitzt (Mittelpunkt). Dann folgt aus der Poissonschen Gleichung, daß die mittlere Dichte rascher als $1/r^2$ mit wachsender Entfernung r vom Mittelpunkt zu null herabsinken muß, damit das Potential im Unendlichen einem Limes zustrebe. Die mittlere Dichte der Materie ist die Dichte, gebildet für einen Raum, der groß ist gegenüber der Distanz benachbarter Fixsterne, aber klein gegenüber den Abmessungen des ganzen Sternsystem. In diesem Sinne ist also die Welt nach Newton endlich, wenn sie auch unendlich große Gesammtmasse besitzen kann.

Figure 2.

4. NEWTONIAN COSMOLOGY

Einundzwanzigstein's paper addressed a natural formulation of the cosmology of a homogeneous universe as afforded by Newton's theory of gravitation. Space is assumed to be infinite and Euclidean and filled with a uniform matter distribution of density $\rho(t)$, which will vary as a function of time. The gravitational potential φ is governed by the Poisson equation

$$\nabla^2 \varphi = 4\pi G \rho \qquad (1)$$

where G is the constant of universal gravitation. These assumptions combined provide the framework of Newtonian cosmology. One might expect that, these assumptions are sufficient to fix the gravitational potential uniquely. But that is not so. Any of the class of solutions

$$\varphi(r) = \left(\frac{2}{3}\right)\pi G \rho(t)(r - r_0)^2 \qquad (2)$$

satisfies the condition, where the vector position $r = (x, y, z)$, for Cartesian spatial coordinates x, y and z and r_0 is any arbitrarily chosen position in space.[4] It follows directly from (2) that the force on a unit test mass is

$$f = -\left(\frac{4}{3}\right)\pi G \rho (r - r_0). \qquad (3)$$

This in turn enables a very simple expression for the gravitational tidal force. The differential force Δf on two unit masses separated by a distance[5] Δr is given by

$$\Delta f = -\left(\frac{4}{3}\right)\pi G \rho \Delta r. \qquad (4)$$

Since no other forces are presumed to prevail on the bodies forming the matter distribution ρ, these cosmic masses are in free fall with accelerations and relative accelerations given by (3) and (4) respectively.

5. THE ASYMMETRY OF NEWTONIAN COSMOLOGY

In addressing this simple system, Einundzwanzigstein commenced his "On the Cosmology of Free Falling Bodies" by noticing the existence of an asymmetry between theory and observation in the system that was strongly reminiscent of the asymmetry Einstein used to launch his "On the Electrodynamics of Moving Bodies" Einundzwanzigstein wrote:

> It is known that Newton's cosmology—as usually understood at the present time—when applied to moving bodies, leads to asymmetries which do not appear to be inherent in the phenomena. Take, for example, the motion of bodies in free fall in a homogeneous space. The observable phenomenon here depends only on the relative motion of the bodies, whereas the customary view draws a sharp distinction between the two cases in which one or the other of the bodies is accelerated.

Einundzwanzigstein's point is recoverable immediately from equation (4).

The observables sustain a perfect equivalence of all bodies in the cosmology. What is observable is the relative motion of the bodies. That observable is the same for any of the cosmic bodies. Each is in free fall and, according to (4), each sees neighboring masses accelerating towards it with an acceleration proportional to distance. As far as the observables are concerned, every body is fully equivalent to every other. If we find ourselves on one of them, no observation of motions can decide which that is. Inertial forces can supply no guide; since every body is in free fall, none of them experience inertial forces.

Newtonian gravitation theory, however, is unable to sustain this equivalence. According to it, at most *one* of all the cosmic masses of the distribution ρ can be unaccelerated, that is, in inertial motion. All the rest are truly accelerated. That body has the role of a unique center of the universe. All the other bodies accelerate towards it. It is designated by the position vector r_0. While that position vector appears in the expression for the many different fields φ of (2) and f of (3), it does not appear in the equation (4) that governs the observable of motion, the tidal force.

Einundzwanzigstein's response was analogous to Einstein's response to the corresponding problem in the electrodynamics of moving bodies.[6] The aether state of rest Einstein observed in 1905, was superfluous for the treatment of electrodynamics. All inertial motions are equivalent. The designation of any reference system as "at rest" is purely a matter of convenience. Electrodynamics embodies a relativity of inertial motion. Correspondingly Einundzwanzigstein declared the notion of a preferred class of inertial motions as superfluous to the cosmology. All inertial and uniformly accelerated motions are equivalent. The designation of any reference system as "inertial" is purely a matter of convenience. Newtonian cosmology embodies a relativity of uniform acceleration.

6. COVARIANCE OF NEWTONIAN COSMOLOGY UNDER ACCELERATION TRANSFORMATIONS

In 1905, Einstein gave formal expression to this relativity of inertial motion by demonstrating that electrodynamics is covariant under the transformations that connect inertial systems of reference, the Lorentz transformation. Correspondingly, Einundzwanzigstein demonstrated that Newtonian cosmology is covariant under an acceleration transformation. To display this covariance, he chose a reference system (x, y, z, t) as "inertial." In it, there is just one cosmic body whose motion is inertial (i.e. its position coordinates are linear functions of the time coordinate). The origin of the reference system is so selected that this body remains at position $x = y = z = 0$. The gravitational potential and the acceleration of cosmic bodies are given as

$$\varphi = \left(\frac{2}{3}\right)\pi G r^2 \qquad \frac{d^2 r(t)}{dt^2} = -\left(\frac{4}{3}\right)\pi G r \qquad (5)$$

where $r = (x, y, z)$ and $r^2 = |r|^2$. Einundzwanzigstein now selected arbitrarily another comic body at position $R(t)$. Its trajectory over time is governed by

$$\frac{d^2 R(t)}{dt^2} = -\left(\frac{4}{3}\right)\pi G \rho R(t). \qquad (6)$$

This arbitrarily chosen body in turn can be used to define an acceleration transformation[7] from the original reference system to the new system (x', y', z', t')

$$r' = r - R(t) \qquad t' = t. \qquad (7)$$

If we write $R(t) = (X(t), Y(t), Z(t))$, this transformation can also be written as

$$x' = x - X(t) \quad y' = y - Y(t) \quad z' = z - Z(t) \quad t' = t.$$

Under this transformation, the gravitational potential and the acceleration of cosmic bodies is now given as

$$\varphi' = \left(\frac{2}{3}\right)\pi G r'^2 \qquad \frac{d^2 r'(t)}{dt'^2} = -\left(\frac{4}{3}\right)\pi G r' \qquad (5')$$

expressions identical in form to (5). The Lorentz covariance of Maxwell's theory expresses the relativity of inertial motion; the elimination of the aether state of rest lies just in the failure of that state to be invariant under Lorentz transformation. The covariance of Newtonian cosmology under transformation (7) expresses a relativity of acceleration. The selection of one class of motions as inertial corresponds to a choice of one subclass of the reference systems of the theory. That choice is not invariant under the transformation (7); motions that are inertial in (x, y, z, t) are accelerated in (x', y', z', t') and *vice versa*. Further, the distinction between the different potential and force fields of (2) and (3) loses physical significance. That is, the designation of which body occupies the preferred position r_0 of the unique inertial moving body is not invariant under the transformation (7). By suitable choice of $R(t)$, any body can be brought to the origin of coordinates and thus to this preferred position.

The transformation of (5) to (5') requires that the gravitational potential φ *not* transform as a scalar. Rather it must transform as[8]

$$\varphi' = \varphi + r \cdot \left(\frac{d^2 R}{dt^2}\right) + \varphi(R). \tag{8}$$

That φ does not transform as a scalar has no effect on observables. There are two additional terms in the transformation law (8). The second, $\varphi(R)$, is just the adding of a constant to the potential; such a constant does not affect the observables, since it has no effect on the motions. The first term added, $r \cdot (d^2 R/dt^2)$, corresponds to the addition of a homogeneous field to the force field associated with φ. That force field is given by the negative gradient of φ and is $-\nabla\varphi' = -\nabla\varphi - (d^2 R/dt^2)$. It is augmented by a vector $d^2 R/dt^2$, which is a constant over space at any instant. Such a homogeneous field does affect accelerations, but it does not affect the observable, relative accelerations, since it accelerates all bodies alike.

This new transformation law for the gravitational potential corresponds to the Lorentz transformation law for electric and magnetic fields in special relativity.

7. THE GEOMETRIC FORMULATION

Einundzwanzigstein's point is that there was a relativity of acceleration built into Newtonian gravitation theory that is closely analogous to the relativity of inertial motion of special relativity. That relativity of acceleration is hard to see in the context of the usual examples. In the case of the gravitational field of the sun, for example, the *observable* inhomogeneity of the field picked out a preferred trajectory in space (that of the sun) and this in turn defined a preferred inertial motion. The case of Newtonian cosmology allowed no such selection. In terms of observables, the motion of all bodies were fully equivalent, even though they were in relative acceleration.

The methods and formalism of Einundzwanzigstein's paper was that of Einstein's 1905 paper on special relativity. Just as the ideas of Einstein's paper were soon translated by Minkowski into a geometrical language, the same translation was possible for Einundzwanzigstein's paper. It could be expressed in the language of the Newstein-Weylmann proposal, in which the free falls of Newtonian cosmology are just the geodesics of the affine spacetime structure. Now the Lorentz covariance of special relativity embodies a relativity of inertial motion because the Lorentz transformation is a symmetry of the Minkowski metric. Correspondingly covariance of Newtonian cosmology under (7) is expressed geometrically as the symmetry of the geometric structures of the spacetime, including the affine structure, under the transformation (7), now read as an active point transformation.[9] In each case, the relativity of a motion is expressed as a symmetry of the geometric structure.

In this context, Einundzwanzigstein's point can be given it sharpest expression. The attempt to preserve some absoluteness of inertial motion corresponds to the attempt to find some way to split the affine connection into a connection defining true inertial motions and a gravitational deflection. *No invariant condition can effect this split in a way that privileges the motion of any one cosmic body over any other.* For it follows immediately from the symmetry of the geometry that any property of one such motion must be shared equally by any other.[10] It is not even sufficient to require that the inertial affine structure be flat—this condition is met by each of the different, natural splits that render one or other body's motion inertial.

The transition to the geometric formulation can be made very quickly on the basis of the equations (1), (3), and (4). If we introduce an index notation so that $r = (x, y, z) = (x^1, x^2, x^3)$ and the time coordinate $t = x^0$, then, according to (3), the trajectories of masses in free fall are governed by

$$\frac{d^2 x^i}{dt^2} + \left(\frac{4}{3}\right)\pi G \rho x^i = 0 \tag{3'}$$

where $i = 1, 2, 3$. These motions are just the geodesics of the affine connection with symbols Γ^i_{km}, so that this condition (3') can be rewritten as

$$\frac{d^2 x^i}{dt^2} + \Gamma^i_{00} = 0 \tag{3''}$$

where t is an affine parameter and the only non-zero symbols are

$$\Gamma^i_{00} = \left(\frac{4}{3}\right)\pi G \rho x^i \tag{9}$$

which fixes the affine structure. Further, since Γ^i_{00} represents the gravitational force on a unit mass in the reference systems used by Einundzwanzigstein, we see that this force must transform like the coefficients of the connection.

From (4) we read off an expression for the relative acceleration of neighboring bodies in free fall

$$\frac{d^2 \Delta x^i}{dt^2} + \left(\frac{4}{3}\right)\pi G \rho \Delta x^i = 0. \tag{4'}$$

This corresponds to the equation of geodesic deviation

$$\frac{d^2 \Delta x^\alpha}{dt^2} + R^\alpha_{\gamma\delta\beta}\Delta x^\beta \left(\frac{dx^\gamma}{dt}\right)\left(\frac{dx^\delta}{dt}\right) = 0 \tag{4''}$$

where $\alpha, \beta, \gamma, \delta = 0, 1, 2, 3, 4$.

The comparison of (4') and (4'') is very fruitful. To begin we can see that the coefficients of the affine curvature tensor represented in (4'') must be constant. This suggests, but does not prove, the uniformity of the affine structure expressed in its symmetry under transformation (7). We can read sufficient of the coefficients of the curvature tensor to allow recovery of the Ricci tensor

$$R_{00k^i} = \left(\frac{4}{3}\right)\pi G\rho \delta^i_k \tag{10}$$

where $i, k = 1, 2, 3$. Contraction over the indices i and k allows us to recover[11] the R_{00} component of the Ricci tensor as

$$R_{00} = 4\pi G\rho. \tag{1'}$$

This is the analog of the Poisson equation (1) in the geometric formulation.

8. CONCLUSIONS, REFLECTIONS AND ADMISSIONS

Lest any readers be in doubt, Newstein, Weylmann, and Einundzwanzigstein are all fictitious and the history reported a fable—inspired by Professor Stachel's own creative endeavors. I have tried to ensure however that all footnoted material in the above fable is historically correct. The fable is intended to convey a serious moral and one that I have laid out in (Norton 1995), in response to David Malament's demonstration (Malament 1995) that the paradoxes of Newtonian cosmology are eradicated by the geometric approach. The usual decision to represent gravitational free falls by a curved affine structure in Newtonian theory is akin to extending the relativity of motion to acceleration, but there are significant disanalogies between it and Einstein's original introduction of the relativity of inertial motion in special relativity. Einstein introduced the relativity of inertial motion to express the indistinguishability of inertial motions that was itself revealed in the failure of experiments that would have picked out the aether state of rest. In general, the representation of gravitational free falls by a curved affine structure does not express a corresponding indistinguishability and the case for it is correspondingly weaker. Newtonian cosmology supplies a clear instance in which it does express such an indistinguishability and is hard to resist. But once it has been admitted in this case, the attempt to avoid it elsewhere becomes all the more contrived.

University of Pittsburgh

NOTES

* With great pleasure, I join the contributors to this volume in honoring Professor Stachel and celebrating his many achievements. My debt to him is great. I learned the real craft of history of science at his elbow when he generously allowed me to visit the Einstein Papers Project in 1982 and 1983 in Princeton and my career owes a great deal to his generosity and kindness. We have all learned so much from Professor Stachel's researches. However, when he revealed the hitherto unknown figure in history of physics, Newstein, in his (Stachel forthcoming) we may have learned somewhat more from him than even he intended, as this paper will demonstrate.

1. I am grateful to Don Howard for pointing out another 'stein that truly belongs to the family: Howard Stein for his (1967). We might also adopt Wolfgang Pauli as an honorary family member on the strength of his nickname, recalled for me by Professor Stachel: "Zweistein."

2. The long standing debate over whether Einstein knew of the Michelson-Morley experiment prior to his work on special relativity of 1905 was settled by the discovery of a letter from Einstein to Mileva Maric of September 1899) in which he recalls reading a paper by Wien (1898) that includes a report on the experiment. That paper was located in an 1898 supplement to the volume of *Annalen der Physik und Chemie*. See (Stachel 1987, 233-34, 407).
3. So named in (Stachel 1998).
4. That the presumptions of this cosmology did not force a unique solution for φ produced great confusion at this time that is not reflected in the above exposition. It was widely expected that any potential φ in the cosmology ought to respect the homogeneity and isotropy of the spatial geometry and matter distribution so that a constant φ was sought. The indeterminacy of φ, as expressed by the admissibility of any member of (2), surfaced in the result that the integral expressions for the gravitational potential, gravitational force and tidal force were not uniformly convergent; they could be integrated to give many conflicting results. A common response was the conclusion that the result was fatal to Newton's law of gravitation, which must be supplemented by other terms to eradicate this indeterminacy. Einstein (1917) used a related argument to motivate the cosmological constant in general relativity, for example. He noticed that the solutions (2) require the density of lines of force to grow without limit at r increases. For a detailed survey of the problem up to 1930, see (Norton 1999).
5. Δr need not be infinitesimally small because of the linearity of f in r according to (3).
6. Correspondingly, Einstein in 1905 argued that the observable phenomena of electrodynamics depend only the relative motions of bodies, whereas Maxwell's electrodynamics distinguished the cases according to which body was at rest in the aether. His example was the electric current induced by the relative motion of a magnet and conductor. The observable, the current, depended only on the relative motion of the magnet and conductor, but Maxwell's electrodynamics gave a very different account of the process according to which of the conductor or magnet was deemed at rest in the aether. If the conductor was at rest, the motion of the magnet led to the induction of a new entity, an electric field, which was not present in the case in which the magnet was at rest in the aether. This example, Einstein suggested, was typical.
7. This transformation (7) corresponds to a uniform acceleration in this sense. Let the trajectory of some body be $S(t)$. At some instant t, its acceleration will be $d^2S(t)/dt^2$. Under transformation (7), that acceleration becomes $d^2S'(t)/dt^2 = d^2S(t)/dt^2 - d^2R(t)/dt^2$. The acceleration has been reduced by the term $d^2R(t)/dt^2$, which is a constant over all space at time t, but will vary with t. That is, at a fixed instant, all accelerations in space are altered by the same amount, but that amount will vary from time to time.
8. Then we have $\varphi + r \cdot \left(\dfrac{d^2R}{dt^2}\right) + \varphi(R) = \left(\dfrac{2}{3}\right)\pi G\rho(r^2 - 2r \cdot R + R^2) = \left(\dfrac{2}{3}\right)\pi G\rho(r - R)^2 = \varphi'$.
9. This symmetry is set up and proved in (Malament 1995).
10. This result is the analog of the result in special relativity that no invariant condition can pick out a preferred state of rest from the inertial motions.
11. Recall that $R_{000}{}^0$ vanishes identically.

REFERENCES

Einstein, Albert. 1917. "Kosmologische Betrachtungen zur allgemeinen Relativitätstheorie." *Preussische Akademie der Wissenschaften, Sitzungsberichte*:142-152.

Malament, David. 1995. "Is Newtonian Cosmology Really Inconsistent?" *Philosophy of Science* 62:489-510.

Norton, John D. 1995. "The Force of Newtonian Cosmology: Acceleration is Relative." *Philosophy of Science* 62:511-22.

———. 1999. "The Cosmological Woes of Newtonian Gravitation Theory." Pp. 271-322 in *The Expanding Worlds of General Relativity* (Einstein Studies, Volume 7), eds. H. Goenner, J. Renn, J. Ritter, and T. Sauer. Boston: Birkhäser.

Stachel, John. (Forthcoming.) "The Story of Newstein or: Is Gravity Just Another Pretty Force?" In *Alter-

native Approaches to General Relativity: The Genesis of General Relativity, eds. J. Renn et al. Dordrecht: Kluwer.

———. 1998. *Einstein's Miraculous Year: Five Papers that Changed the Face of Physics.* Princeton University Press: Princeton.

Stachel, John et al. 1987. *The Collected Papers of Albert Einstein* (Volume 1: The Early Years, 1879-1901). Princeton University Press: Princeton.

Stein, Howard. 1967. "Newtonian Space-Time." *Texas Quarterly* 10:174-200.

Wien, Wilhelm. 1898. "Über die Fragen, welche die translatorisch Bewegung des Lichtäthers betreffen." *Annalen der Physik und Chemie* 65:3, Beilage:i-xvii.

JÜRGEN RENN AND TILMAN SAUER

ECLIPSES OF THE STARS*

Mandl, Einstein, and the Early History of Gravitational Lensing

> Aber rühmen wir nicht nur den Weisen
> Dessen Namen auf dem Buche prangt!
> Denn man muß dem Weisen seine Weisheit erst entreißen.
> Darum sei der Zöllner auch bedankt:
> Er hat sie ihm abverlangt.
>
> But the honour should not be restricted
> To the sage whose name is clearly writ.
> For a wise man's wisdom needs to be extracted.
> So the customs man deserves his bit.
> It was he who called for it.
>
> Bertold Brecht, Legende von der Entstehung des Buches Taoteking auf dem Weg des Laotse in die Emigration
> (Legend of the origin of the book Tao-te-ching on Lao-tsu's road into exile)

INTRODUCTION

This paper is about an odd but characteristic episode in Einstein's life, presenting him as an egalitarian intellectual, supportive of an outsider to the scientific establishment, unprejudiced and open to good ideas however humble their source may be, some ambivalence notwithstanding. It shows how one such humble idea eventually became a great scientific achievement—after much resistance and reluctance due to elitist attitudes towards science. It is a story about imagination and individual generosity, but also about science as a social enterprise and the role of contingency in its development.

1. AN AMATEUR'S IDEA

One day in spring 1936, Rudi W. Mandl, a Czech amateur scientist, walked into the building of the National Academy of Sciences in Washington and asked for the offices of the Science Service, an institution devoted to the popularization of science. He came with a new idea of his concerning a "proposed test for the relativity theory based on observations during eclipses of the stars."[1]

```
                    2101 CONSTITUTION AVENUE
                    WASHINGTON, D.C.

                                              Sept. 16, 1936

Prof. Albert Einstein
Institute for Advanced Study
Princeton, N.J.



Dear Prof. Einstein:

        Last spring an apparently sincere layman in science, Rudi
Mandl, came into our offices here in the building of the National
Academy of Sciences and discussed a proposed test for the relativity
theory based on observations during eclipses of the stars.

        We supplied Mr. Mandl with a small sum of money to enable
him to visit you at Princeton and discuss it with you. On his return
he showed us what were apparently authentic letters from you to him
regarding his suggestion.

        Mr. Mandl has since moved to New York City (108-11 Roosevelt
Ave., Corona, L.I.) but before he left he told us that you had agreed
to publish his ideas, or at least incorporate some of them in a tech-
nical paper to be prepared by you for some scientific journal.

        A letter has today come from Mr. Mandl asking us if this paper
has yet been published.

        Could you tell us what is the status of the Mandl proposal
from your point of view, with the promise that anything you would write
would be completely confidential?

                                        Sincerely yours,

          beantwortet.                  Robert D. Potter
                                        Robert D. Potter
                                        Science Service
```

Figure 1. R. D. Potter to A. Einstein, 16 Sept. 1936, EA 17039
© Einstein Archives, The Hebrew University Jerusalem.

He was looking for someone to help him publish his ideas and to persuade professional astronomers to take up investigations along his proposal. What to do with him? His intentions seemed sincere, his ideas not so easily refuted, and the man himself not so easy to dismiss either. After a while, somebody from the staff of the Science

Service suggested that Mandl discuss his ideas with an undisputed expert on relativity theory. Princeton was not too far away, so he might go there for a day and talk to Professor Einstein himself about his proposed test. They would finance the trip, and if Einstein found his ideas worthwhile he might come back and they would see what they could do for him.

What Mandl was to present to Einstein was, as we know from correspondence and manuscripts surviving in the Einstein Archives, a queer combination of ideas from general relativity, optics, astrophysics, and evolutionary biology. There is evidence that Mandl rather obsessively attempted to persuade professional scientists of his ideas, among them William Francis Gray Swann, director of a center of cosmic ray studies, the Nobel prize winners Arthur Holly Compton and Robert Andrews Millikan, and V. K. Zworykin, research scientist at the Radio Corporation of America (RCA) and inventor of the first all-electronic television system.[2] Some of them reacted with interest and gave Mandl's ideas some brief consideration, others excused themselves with lack of time or understanding. None of them, in any case, pursued the matter seriously.

When Mandl visited Einstein in Princeton on April 17, 1936, he found the professor friendly and willing to listen to his ideas in spite of their oddity. The core of Mandl's suggestion was in fact simple, it essentially amounts to the combination of an elementary insight from general relativity, the deflection of light rays by a gravitational field, with the lensing effect familiar from ray optics. Mandl proposed a simple model according to which one star focalizes the light of another star if both are aligned with the earth, thus constituting a gravitational lens and its object. He speculated that the effects of such a focalization might already have been observed, though their origin had remained undisclosed. Among the possible effects that Mandl took into consideration were the recently discovered annular shaped nebulae which he interpreted as gravitational images of distant stars, cosmic radiation which he conjectured to be an effect of the gravitational amplification of the radiation emitted by a distant galaxy, and the sudden extinction of biological species such as dinosaurs, which he attempted to relate to the momentary intensification of such radiation due to what he described as a stellar eclipse.

Though the range of Mandl's ideas was daring, at its core was an insight that would eventually—several decades later—indeed turn into an astrophysical confirmation of relativity theory. From a letter written a day after Mandl's visit to Einstein, we can gather what that idea was. (Consider the "old formula" and the corresponding sketch in the diagram that Mandl included in his letter, depicted in fig. 2.)

Light coming from an infinitely distant star, located exactly behind a massive, gravitating star of spherical radius R_0, located at distance D from a terrestrial observer, passes at a distance R from the center of that gravitating object. The light ray is bent inwards, i.e. towards the line between the observer and the gravitating star by gravitational deflection and is seen under an angle ε by the terrestrial observer. According to general relativity, the angle of deflection ε is inversely proportional to the offset R, i.e. $\varepsilon \sim (1/R)$. Let ε_0 be the angle of deflection for light rays just grazing the edge of the deflecting star, i.e. for $R = R_0$.[3] Since $\varepsilon_0 \sim 1/R_0$, the angle of deflection in general would be $\varepsilon = \varepsilon_0(R_o/R)$. Light rays visible on the earth, on the other hand, can be seen for angles $\varepsilon = (R/D)$. And from the latter two formulae the relation in eq. (1) follows.

$$\varepsilon = \sqrt{\varepsilon_0 \frac{R_0}{D}} \qquad (1)$$

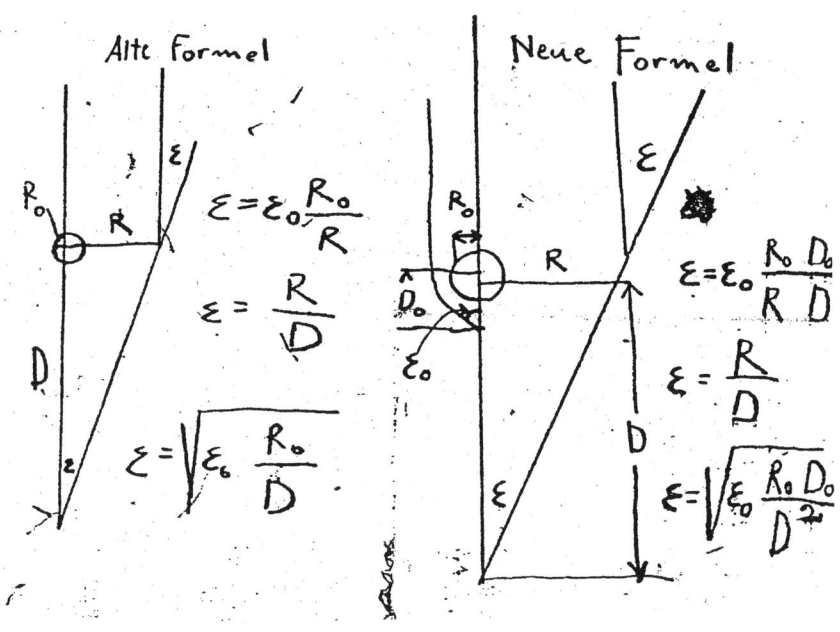

Figure 2. Sketch by Mandl. EA 17-028
© Einstein Archives, The Hebrew University Jerusalem

Clearly, what Mandl and Einstein had been discussing was some analogue of an optical lens, effected by the light deflection of the gravitating star. The angle ε then would be the apparent magnitude of the distant star, as seen from the observer. But in contrast to the lensing properties of common optical lenses, gravitational light deflection by a massive star does not collect parallel light rays at one single focal point but rather smeared out along a "focal line" as it were. For an observer at the distance D the condition holds only for light rays passing at a particular distance R, resulting in what we would now call an "Einstein ring." In any case, Mandl's idea seems to have been that this lensing effect would result in a considerable magnification of the star's light at the point of the terrestrial observer.

To Mandl the opportunity of a conversation with Einstein must have meant a lot. "Inexcusably" he had forgotten to express his thankfulness for the "friendly reception" and hastened to make up for this mistake by a first letter written a day after his visit. In this letter he also pointed out what he believed was an error in their considerations. He proposed a correction of the usual light deflection formula, now assuming without any further explanation that the angle of deflection would have to be taken in general as $\varepsilon = \varepsilon_0(R_0D_0)/(RD)$ with D_0 denoting some fixed distance of the lensing star.[4]

In the course of his meeting with Mandl Einstein let himself apparently be drawn into the matter by Mandl's enthusiasm and even entertained the possibility of publishing something about the idea. After some days of reflection, however, skepticism and caution prevailed so that Einstein changed his mind and wrote to Mandl:

> The first formula stays. The second one is based on an erroneous consideration. I have come to the conclusion that the phenomenon in question will, after all, not be observable so that I am no longer in favor of publishing anything about it.[5]

Mandl, on the other hand, relentlessly pursued his idea and, in a letter written the same day and crossing that of Einstein, reported further progress, addressing precisely the issue of Einstein's concern, the difficulty of empirically checking the phenomenon:

> Meanwhile I have found a method to measure the intensity increase in the domain of the focal line of a star and to confirm it experimentally. It would be, according to my view, in the interest of science to begin with these experiments as soon as possible.[6]

Mandl also asked for another meeting with Einstein, insisting that his idea would provide a very simple explanation of the origin of cosmic radiation. Ten days later, Mandl wrote once more. He had received Einstein's letter in the meantime and agreed that his second formula was wrong. Mandl emphasized that he did not want to persuade Einstein to publish something against his will but was, at the same time, well aware that he needed the latter's authority to gain recognition for his idea. He attempted to win Einstein for his publication plan by appealing to the need of defeating pseudo-science:

> The main reason why I bother you in order to gain your e[steemed] collaboration is that immediately after the publication of my investigations all astrologists and similar parasites of science will take possession of the results of my considerations, and it is my conviction that it would be of use for the "average citizen" if, from the very beginning, a man of your rank and fame would emphasize the nonsense of the pseudo-science of these charlatans.[7]

Ironically some of the ideas Mandl expounded in a four-page typescript enclosed with his letter must have sounded to Einstein very much like the pseudo-science Mandl pretended to fight. It contains a sober assessment of the observability of a stellar eclipse but also daring speculations about its possible effects on life on earth. He asked that Einstein read this manuscript and kindly point out fallacies if he found any.

Mandl's typescript[8] was never published. It started with the computation of some numerical values of the distance D of the "focal point" and of the apparent diameter ε of the ring image of the distant star as a function of the offset R both for the case of a lensing star of the mass of the sun and for a lensing star of 100 solar masses. In the following three pages, Mandl explained his idea of an "Einstein focal line" ("E. F. L.", Mandl also asked Einstein for permission to give the phenomenon his name) in somewhat more detail. Regarding its observability, he remarked:

> Given the extraordinarily small surface which the fixed stars cover in comparison with the surface of the sky, the transition of the earth through an E. F. L. [of a fixed star] is a very rare phenomenon and would be best compared, regarding both time and perhaps also extension, with a solar eclipse, a transition through an E. F. L. [of a galaxy], however, would be something ordinary and should even last over millenia.[9]

Mandl added in parentheses:

> (The relatively large quantity of cosmic radiation would be most easily explained by a transition through an E. F. L. of the milky way or other nebulae.)[10]

With the astronomical details explained, the remainder of the typescript discusses the surmised spectacular terrestrial consequences of this focal effect:

> In the sequel some general observations on Darwinism. The greatest unclearness and uncertainty in the Darwinian doctrine of evolution consists in the insufficiency of the explanation of the cause why evolution proceeds not gradually but by leaps and bounds and that after periods of intensive evolution and mutation a long period of stagnation followed and why all-of-a-sudden entire classes of highly-developed animals went extinct.[11]

With reference to H.J. Muller's 1927 experiments[12] on drosophila mutation after exposure to X-ray radiation, Mandl goes on to conjecture that the supposedly unexplained origin of discontinuous evolutionary steps in Darwin's theory might be due just to the gravitational focussing of distant cosmic light. He concluded:

> The theory of relativity thus delivers a key, as absurd as it may appear in the first moment, to the hitherto dark parts of the evolutionary doctrine.[13]

At the end of his typescript, Mandl disputed any vicinity of his considerations to a pseudo-science such as astrology. As a matter of fact, his final remarks represent, however, just the kind of disclaimer which raises rather than refutes suspicions:

> While the above consideration teaches us that the stars though being far away "are responsible" for the evolution or have brought it about, it has to be emphasized at this point that the "science"?? of astrology has nothing in common with the above considerations and that we are unable, even with our astronomical instruments of today, to reconstruct even merely by calculation which stars come into consideration for having influenced our earth in the sense pointed out above.[14]

Reading Mandl's memoir, it does not come as a surprise that, as Mandl reported in his covering letter, other physicists whom he had meanwhile approached about his theories reacted with caution or even disdain. While Swann and Compton excused themselves for not entering into the matter for lack of time, and asked that Mandl send them more details, Millikan had just given him short shrift with three brief "I don't understand."[15]

Although Einstein clearly shared the aversion of his established colleagues to the over-ambitious projects of an amateur-scientist with the air of a crackpot, he did, as we shall see, take the matter less lightly, not the least because Mandl would not let him easily "off the hook." Indeed, before Einstein even had a chance of responding to Mandl's memoir, he received yet another letter by Mandl, dated May 9. In it he argued against Einstein's claim that the phenomenon would be unobservable, pointed out that there would be a numerical error in his table, and expressed his hope to hear from Einstein soon. Three days later Einstein gave in to Mandl's bugging and finally wrote him. He had actually sat down to work on Mandl's ideas:

> Dear Mr. Mandl: I have calculated your intensification effect more precisely. The result is the following.[16]

Denoting the distance between an observer and a light-bending star ("ablenkender Stern") D, the radius of the bending star r_0, the amount of deflection of a light ray grazing the surface of the bending star α_0, the (vertical) distance of the observer from the line passing through the centers of the emitting and the light-bending star x, and introducing the quantity $l^2 = D\alpha_0 r_0$, Einstein reported the following expression for the intensification G:

$$G = \frac{l}{x} \frac{1 + \frac{x^2}{2l^2}}{\sqrt{1 + \frac{x^2}{4l^2}}}. \tag{2}$$

He remarked that a noteworthy intensification would take place if x would be small against l, in which case one would have approximately $G = l/x$ and hence infinite intensity on the central line itself. For a nearby fixed star l would be roughly 10 light seconds, and hence the intensification would be restricted to a very small zone. For an intensification of factor 10, that zone would be one light second, or roughly equal to the diameter of the sun. Einstein also pointed out that, strangely, l would grow like $\sqrt{D}$, and hence the effect of a distant star would be larger than that of one close by.

These dazzling results must have been a great satisfaction to Mandl. Regarding the observability of the phenomenon, Einstein now showed himself less pessimistic than in his first letter. Would he agree to some kind of joint publication? But although his calculations had opened up more promising perspectives for the observability of the phenomenon, Einstein, in the end, nevertheless came to a negative assessment and concluded:

> In any case there may well be more chance to occasionally observe this intensification effect than the "halo effect" with which we have dealt earlier. But the probability that we can get so precisely into the connecting line of the centers of two stars at very different distances is rather low, even lower the probability that the phenomenon, lasting in general only a few hours, happens to be observed.[17]

Even more than the problems of observability Einstein must have felt uneasy about the wild speculations Mandl had associated with his idea. But in spite of these hesitations and in contrast to the reactions of other colleagues who had dealt with Mandl's idea, he finally faced the possibility of a "modest publication":

> Your fantastic speculations associated with the phenomenon would only make you the laughing-stock of the reasonable astronomers. I warn you in your own interest against such a publication. On the other hand, one cannot object against a modest publication of a derivation of the two characteristic formulas for the "halo effect" and the "intensification effect."[18]

Einstein's letter triggered a series of enthusiastic epistles by Mandl. On May 17, he wrote a lengthy letter, thanked Einstein for his response, and repeated his request "to publish the results of the effects as well as your formulas as I am really lacking any possibility to do so."[19] Two days later, he apologized for having written in a rather confused way but he was sure:

> that it [the letter] will bring you great joy, a joy that will only be smaller by a bit than the joy which our common friend Hitler will have once he finds out that it was again one of those damned Jews who turned the Einstein theory into the Einstein law.[20]

While, for all we know, Hitler could not have cared less, an empirical confirmation of general relativity achieved by an unknown Jewish amateur scientist would not only have surprised German scientists of the period who, like Max Planck on the occasion of his meeting with Hitler in May 1933, distinguished between valuable and less valuable Jews.[21] That a non-professional could contribute to a physical theory of the complexity of general relativity was hardly conceivable also to the American academic establishment. In fact, Mandl's efforts to find a channel of publication for his views remained in vain. Two days after the letter quoted above, on 21 May, he turned once more to Einstein, this time in a rather discouraged vein:

> In the last days I undertook desperate attempts to publish the results of my researches together with your formulas and I heard, without exception, the question: 'Well, if Mr. Einstein appreciates your results why doesn't he publish them himself?' Hence it seems to depend on you to make the results accessible to the scientific world.[22]

In an environment in which participation in the dissemination of information depended heavily on academic status, an outsider had indeed little chance to place his ideas and needed, like an intellectual in the early modern period, the grace, benevolence, and authority of some patron to be admitted to the official world of learning. The patron in turn, inevitably took some responsibility for the actions of his protégé. When Mandl's last letter confronted Einstein with this choice, he finally gave in and agreed to published a short note of the lensing idea. In a letter to Mandl, dated June 2, 1936, he recapitulated the calculations.[23] He also mentioned that he had asked the text of the note to be translated into English but had not yet received the translation since his assistant was away on vacation. For the time being, Einstein seems to have laid the matter to rest with that.

Three months later, Mandl sent a letter to the Science Service asking what had become of Einstein's promise, a question that Robert D. Potter from Science Service duly passed on to Einstein without delay:

> he told us that you had agreed to publish his ideas, or at least incorporate some of them in a technical paper to be prepared by you for some scientific journal. [...]
> Could you tell us what is the status of the Mandl proposal from your point of view, with the promise that anything you would write would be completely confidential?[24]

Whether such a confidential answer by Einstein to this question was ever written is unclear. Mandl's appeal had, in any case, the desired effect: Einstein complied with his wish and submitted a short note to *Science*. The note is entitled "Lens-like Action of a Star by the Deviation of Light in the Gravitational Field" and was published in the December 4 issue of the journal (Einstein 1936), cp. fig. 3. It has since become the classical starting point for the officially recorded history of gravitational lensing research.[25]

How much self-conquest it must have taken Einstein to overcome his reluctance and publish this note is evident from its introductory sentence, as well as from a letter Einstein wrote to Professor J. McKeen Cattell, Editor of *Science*, on December 18, two weeks after the publication of Einstein's paper. The opening statement of the published note reads:

> Some time ago, R. W. Mandl paid me a visit and asked me to publish the results of a little calculation which I had made at his request. This note complies with his wish (ibid., 507).

This mention of Mandl reads less like giving due credit to the proprietor of a good idea than like a general disclaimer. In his letter to the editor of *Science,* Einstein distanced himself even more definitively from his publication, stressing that it was only written to appease Mandl:

> Let me also thank you for your cooperation with the little publication, which Mr. Mandl squeezed out of me. It is of little value, but it makes the poor guy happy.[26]

Einstein could apparently recognize no value in the theoretical analysis of what must have appeared to him as a "science-fiction effect" because gravitational lensing seemed so far out of the reach of any observational verification. Thus, his note ends with the remark that:

> there is no great chance of observing this phenomenon, even if dazzling by the light of the much nearer star B is disregarded (ibid., 508).

Einstein concludes with what reads like an attempt at a justification for publishing his "little calculation:"

> This apparent amplification of q by the lens-like action of the star B is a most curious effect, not so much for its becoming infinite, with x vanishing, but since with increasing distance D of the observer not only does it not decrease, but even increases proportionally to $\sqrt{D}$, (ibid., 508).

Evidently, it was, at least in Einstein's understanding, hardly legitimate for theoretical physics to decouple itself from experimental physics or observational astronomy to this extent and pursue theoretical consequences in a merely speculative way, even if they pointed to what he described as "a most curious effect." In fact, such consequences might be empirically substantiated only in some indefinite future, while there was, on the other hand, the plenitude of known phenomena still awaiting a satisfactory interpretation. Nevertheless, in spite of such hesitations, Einstein did in the end decide to publish his note.

2. A DEJA VU

Why did Einstein not simply dismiss Mandl as Professor Millikan had done? What was it that eventually tipped the scales? Do we have to assume that it was, after all, Einstein's "physical instinct," some subconscious capability of foreseeing the eventual success of Mandl's idea? One element of a more plausible answer is found in Einstein's above-mentioned letter to the editor of *Science,* James McKeen Cattell, formerly professor of psychology at Columbia University. This letter shows a political awareness on the part of both correspondents that provides a relevant context also to their support of Mandl:

> I know quite well that you had to leave Columbia. Researchers are here treated as are elsewhere waiters or salesmen in the retail business (The latter also merit, of course, to

work under better secured conditions, this expression hides no class arrogance); presently it is particularly bad if somebody is considered to be a 'radical.' The public is much too indifferent with regard to those violations of the freedom of teaching and teachers.[27]

DISCUSSION

LENS-LIKE ACTION OF A STAR BY THE DEVIATION OF LIGHT IN THE GRAVITATIONAL FIELD

Some time ago, R. W. Mandl paid me a visit and asked me to publish the results of a little calculation, which I had made at his request. This note complies with his wish.

The light coming from a star A traverses the gravitational field of another star B, whose radius is R_o. Let there be an observer at a distance D from B and at a distance x, small compared with D, from the extended central line $\overline{AB}$. According to the general theory of relativity, let α_o be the deviation of the light ray passing the star B at a distance R_o from its center.

For the sake of simplicity, let us assume that $\overline{AB}$ is large, compared with the distance D of the observer from the deviating star B. We also neglect the eclipse (geometrical obscuration) by the star B, which indeed is negligible in all practically important cases. To permit this, D has to be very large compared to the radius R_o of the deviating star.

It follows from the law of deviation that an observer situated exactly on the extension of the central line $\overline{AB}$ will perceive, instead of a point-like star A, a luminous circle of the angular radius β around the center of B, where

$$\beta = \sqrt{\alpha_o \frac{R_o}{D}}.$$

It should be noted that this angular diameter β does not decrease like $1/D$, but like $1/\sqrt{D}$, as the distance D increases.

Of course, there is no hope of observing this phenomenon directly. First, we shall scarcely ever approach closely enough to such a central line. Second, the angle β will defy the resolving power of our instruments. For, α_o being of the order of magnitude of one second of arc, the angle R_o/D, under which the deviating star B is seen, is much smaller. Therefore, the light coming from the luminous circle can not be distinguished by an observer as geometrically different from that coming from the star B, but simply will manifest itself as increased apparent brightness of B.

The same will happen, if the observer is situated at a small distance x from the extended central line $\overline{AB}$. But then the observer will see A as two point-like light-sources, which are deviated from the true geometrical position of A by the angle β, approximately.

The apparent brightness of A will be increased by the lens-like action of the gravitational field of B in the ratio q. This q will be considerably larger than unity only if x is so small that the observed positions of A and B coincide, within the resolving power of our instruments. Simple geometric considerations lead to the expression

$$q = \frac{l}{x} \cdot \frac{1 + \frac{x^2}{2l^2}}{\sqrt{1 + \frac{x^2}{4l^2}}},$$

where

$$l = \sqrt{\alpha_o D R_o}.$$

DECEMBER 4, 1936

If we are interested mainly in the case $q \gg 1$, the formula

$$q = \frac{l}{x}$$

is a sufficient approximation, since $\frac{x^2}{l^2}$ may be neglected.

Even in the most favorable cases the length l is only a few light-seconds, and x must be small compared with this, if an appreciable increase of the apparent brightness of A is to be produced by the lens-like action of B.

Therefore, there is no great chance of observing this phenomenon, even if dazzling by the light of the much nearer star B is disregarded. This apparent amplification of q by the lens-like action of the star B is a most curious effect, not so much for its becoming infinite, with x vanishing, but since with increasing distance D of the observer not only does it not decrease, but even increases proportionally to $\sqrt{D}$.

ALBERT EINSTEIN

INSTITUTE FOR ADVANCED STUDY,
PRINCETON, N. J.

Figure 3. Einstein's note in Science, presenting "the results of a little calculation" he had made at Mandl's request.

Cattell had also asked Einstein to participate in a political meeting informing the public about the "terrible state of affairs" ("furchtbaren Zustände") in Nazi Germany. Einstein declined for personal reasons; in fact, his wife, Elsa Einstein, was fatally ill and died only two days later. He also wrote that to avoid the suspicion of partiality it would be better if somebody who was not directly suffering from the current "criminal business" ("Verbrecherwirtschaft") in Germany would do the job.

Although in his letter to Cattell Einstein spoke of the "poor guy" Mandl with a somewhat condescending air, he did not just look upon him as a crackpot at the margins of science. In fact, Einstein was well aware that the Czech immigrant was one of the many victims of Nazi imperialism which needed support, a support that he was willing to extend without much ado also to those not in the limelight as outstanding scientists.

But Einstein's tortuous and hesitant decision to give in to Mandl's pestering, as well as the ambivalent character of his support, have deeper roots than his political awareness. This is evident from a circumstance that has so far played no role in our story, although it is intimately related to it. With a portion of luck, this circumstance, documented by an early Einstein notebook, was revealed in the course of joint work with John Stachel. Without this finding, the encounter between Mandl and Einstein would have entered the history of gravitational lensing merely as a fortunate coincidence of two biographies which in fact were worlds apart, that of a lonely amateur and that of a famous scientist. It has turned out, however, that Einstein himself had thought of gravitational lensing even earlier than Mandl, at a time when he was in a very similar position.

In fact, as an analysis of Einstein's scribblings in a notebook used during his years as a professor in Prague has shown, he had not only considered the very same idea in 1912 but also performed exactly the same calculations he made in 1936 at the request of Rudi Mandl (Renn, Sauer, and Stachel 1997). When Einstein first conceived the idea of gravitational lensing, he too was at the periphery of the academic establishment and desperately searched support for what to many of his colleagues appeared to be an outlandish idea, that gravitation might affect the course of light rays. In a paper published in 1911, Einstein discussed consequences of the *Influence of Gravitation on the Propagation of Light* (Einstein 1911), an idea quite unheard of since Newton's corpuscular theory of light had been discarded in favour of wave theoretic concepts. And just as Mandl was trying to do in 1936, Einstein also encouraged astronomers to pursue investigations along his ideas:

> It would be urgently desirable that astronomers take up the question broached here, even if the considerations presented above may appear insufficiently founded or even adventurous.[28]

But Einstein's attempts to contact established astronomers remained as unsuccessful as those of Mandl many years later.[29] Due to the mediation of one of his Prague students, Leo Pollak, Einstein had the opportunity of enlisting the help of a younger astronomer, Erwin Finlay Freundlich, who was willing to undertake an exploration of the observational consequences of his theory. Freundlich later remembered:

> 25 years ago Einstein, then professor in Prague, gave one of his first lectures on the gen. rel. theory and concluded with the words that he now needed an astronomical collaborator

> but that the astronomers were too much behind the time to follow the physicists. At that point a young student stood up and declared that, in Berlin, at the observatory which he had visited, he had become acquainted with a young astronomer for which this characterization did not fit.
> Einstein immediately wrote to me. From that our joint work and, de facto, my entire scientific life came into being. The student was Pollak, now ordinary professor in Prague.[30]

Freundlich, in turn, encountered similar problems as Einstein had encountered at the beginning of his career and as Mandl did much later. At the time assistant at the Prussian Royal Observatory, Freundlich could only pursue his plan to support Einstein against the resistance of his superior, the well known astronomer Struve, who regarded the endeavor with skepticism.[31] It may well have been therefore in an atmosphere of clandestineness that, on the occasion of a visit to Berlin in April 1912, Einstein met Freundlich and evidently discussed his daring ideas, including the possibility of a gravitational lensing effect. It is, in any case, among notes from this period that his early calculations on gravitational lensing are found.[32]

Einstein was similarly isolated in his academic context, in spite of his rapid academic career. In fact, his project to develop a relativistic theory of gravitation met in the beginning, and certainly during his years in Prague, that is, 1911 and 1912, with the disinterest if not disapproval of most of his established colleagues. The interest, encouragement, and support which he did find came from friends of his student years, in particular from Marcel Grossmann and Michele Besso. Taking into account their contributions, Einstein's unbending search for a relativistic theory of gravitation emerges as the direct continuation of his bohemian rebellion against established academic science for which the mock Olympia Academy, founded in 1902 in Berne together with friends, stands as a symbol. Einstein was then an employee of the Swiss patent office, where some years later, still far from climbing the academic ladder, he conceived the equivalence principle, the first step towards a theory of gravitation based on a generalization of the relativity principle.

After Einstein had become part of the academic establishment, as a member of the Prussian Academy and Director at the Kaiser-Wilhelm-Institute for Physics, he continued to battle for his project. His struggle to secure Freundlich a position which would allow him to work on the astronomical implications of the new theory of gravitation lasted for years. As late as 1918, Einstein intervened on behalf of Freundlich to Hugo Andres Krüss, ministerial director for Academy matters in the Prussian ministry of Education, complaining that:

> The disapproving behavior of his director has made it, for seven years, impossible for him to achieve the realization of his work plans directed towards the checking of the theory.[33]

Einstein's interventions for Freundlich are remarkably similar to his later dealings with Mandl, also in their ambivalence. In fact, in the same letter to Krüss, he acknowledged Freundlich's lack of qualification as a professional astronomer and emphasized, at the same time, his role as a pioneer in recognizing the astronomical significance of general relativity:

> Incomparably less gifted than Schwarzschild, he has nevertheless recognized several years before the latter the importance of the new gravitation theories for astronomy and

> has engaged himself with glowing zeal for the checking of the theory along an astronomical or rather astrophysical way. [34]

In contemporary letters to colleagues he was even more explicit about Freundlich's weaknesses[35] but, in the end, regularly came down on his side, not least because the latter essentially remained his most devoted support among German astronomers.

Together with Freundlich, Einstein had considered a number of possibilities to detect the light-deflecting effect of gravitation predicted by general relativity. In his first letter to Freundlich he lamented that nature did not provide him with a better environment for testing the theory, for instance by way of a planet sufficiently large to make the effect noticeable. In his letter he also stressed the crucial significance of such a test which, in fact, would be capable of distinguishing between his theory and alternative gravitation theories:

> But at least one thing can be stated with certainty: If such a deflection does not exist, then the premises of the theory are not adequate. For one has to keep in mind that, although these premises are plausible, they are nevertheless rather daring. If only we had a truly larger planet than Jupiter! But Nature did not deem it her business to make the identification of her laws comfortable for us.[36]

On the basis of Einstein's knowledge about the dimensions of the universe in the 1910s, which for him and his contemporaries essentially consisted of our own galaxy, the conditions under which gravitational lensing might be detected must have struck him as even more far-fetched than speculations about the observability of light deflection by Jupiter. Indeed, as we know from a letter to his friend Zangger of 15 October 1915, Einstein had given up the idea of finding an observational confirmation of general relativity on the basis of gravitational lensing even before he completed the theory of general relativity little more than a month later. In this letter he rejected an earlier speculative interpretation of nova stars as lensing phenomena:

> Now it has unfortunately dawned on me that the "new stars" have nothing to do with the "lensing effect," that furthermore the latter has to be, in view of the star densities occurring in the sky, such an enormously rare phenomenon that one would probably expect it in vain.[37]

When the deflection of light by the gravitational field of the sun was eventually confirmed by an English expedition during the total solar eclipse of 1919, it was a stroke of luck that the effect was just within the margins of observability.[38] This confirmation of the predictions of general relativity by Eddington and his collaborators made Einstein all of a sudden world-famous and put general relativity at the foundations of modern physics. But in spite of this breakthrough success, Einstein may well have remembered that general relativity had emerged from what to his colleagues once appeared as a crackpot idea. His support for outsiders such as Freundlich and Mandl becomes fully understandable, it seems to us, only on the background of this experience.

3. MANDL'S SUCCESS

The significance of gravitational lensing for the history of general relativity and of cosmology makes it natural to ask who had first suggested it and when. But the peculiar story of Einstein's double encounter with the idea of gravitational lensing—in 1912 and in 1936—and of Mandl's role in the second episode show that this question cannot easily be answered. Evidently, Einstein himself neither in 1912 nor in 1936 considered gravitational lensing to be a great idea, let alone a discovery. But the story is even more complicated. Almost immediately after Einstein's "little calculation"— made to comply with Mandl's wish—was eventually published in the "Discussion" section of the December 4 issue of *Science,* a number of other papers were published which further developed the idea, taking it much more seriously than Einstein himself had done. But what is more, various authors now also claimed fatherhood to what had evidently become, all of a sudden, a respectable child.

Thus Tikhov, in a publication triggered by Einstein's note, dated 25 June 1937 and entitled "Sur la déviation des rayons lumineux dans le champ de gravitation des étoiles" (Tikhov 1937), claimed in the introductory paragraph that he had had the idea as early as summer 1935 and that he had sent a first communication to the Poulkovo observatory by January 1936. In his paper Tikhov then gives a deduction of the lensing formulae both for what he calls the classical and the relativistic case.

Zwicky, in the second of two notes on gravitational lensing triggered by Einstein's publication, pointed out:

> Dr. G. Strömberg of the Mt Wilson Observatory kindly informs me that the idea of stars as gravitational lenses is really an old one. Among others, E. B. Frost, late director of the Yerkes Observatory, as early as 1923 outlined a program for the search of such lens effects among stars (Zwicky 1937b).

To our knowledge, however, neither Strömberg nor Frost ever published anything about their ideas, and whatever research they did was not given away in publications.

However, there were indeed also precursors who discussed the idea in published work—but strangely without leaving any mark on the history of the idea. Tikhov in his paper pointed to a publication from the year 1924 by O. Chwolson (Chwolson 1924). As Tikhov observed, it was the only reference he found on consulting the literature, that discussed the idea of a gravitational lens. Chwolson's note discussed both the possibility of observing double stars as well as the possible effect of a ring-shaped image for perfect alignment.

It is not unlikely that even Einstein was familiar with Chwolson's note and yet simply chose to ignore it. It was published in the prestigious *Astronomische Nachrichten*, at the time perhaps the most important astronomical journal in Europe. Einstein himself also published in the same journal. Indeed, a brief response by Einstein to Anderson on the electron gas came to be printed in the same issue, in fact just below Chwolson's note on the very same page, cp. fig. 4.

Über eine mögliche Form fiktiver Doppelsterne. Von *O. Chwolson*.

Es ist gegenwärtig wohl als höchst wahrscheinlich anzunehmen, daß ein Lichtstrahl, der in der Nähe der Oberfläche eines Sternes vorbeigeht, eine Ablenkung erfährt. Ist γ diese Ablenkung und γ_0 der Maximumwert an der Oberfläche, so ist $\gamma_0 \gtreqless \gamma \gtreqless 0$. Die Größe des Winkels ist bei der Sonne $\gamma_0 = 1.''7$; es dürften aber wohl Sterne existieren, bei denen γ_0 gleich mehreren Bogensekunden ist; vielleicht auch noch mehr. Es sei A ein großer Stern (Gigant), T die Erde, B ein entfernter Stern; die Winkeldistanz zwischen A und B, von T aus gesehen, sei α, und der Winkel zwischen A und T, von B aus gesehen, sei β. Es ist dann

$$\gamma = \alpha + \beta.$$

Ist B sehr weit entfernt, so ist annähernd $\gamma = \alpha$. Es kann also α gleich mehreren Bogensekunden sein, und der Maximumwert von α wäre etwa gleich γ_0. Man sieht den Stern B von der Erde aus an zwei Stellen: direkt in der Richtung TB und außerdem nahe der Oberfläche von A, analog einem Spiegelbild. Haben wir mehrere Sterne B, C, D, so würden die Spiegelbilder umgekehrt gelegen sein wie in einem gewöhnlichen Spiegel, nämlich in der Reihenfolge D, C, B, wenn von A aus gerechnet wird (D wäre am nächsten zu A).

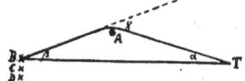

Der Stern A würde als fiktiver Doppelstern erscheinen. Teleskopisch wäre er selbstverständlich nicht zu trennen. Sein Spektrum bestände aus der Übereinanderlagerung zweier, vielleicht total verschiedenartiger Spektren. Nach der Interferenzmethode müßte er als Doppelstern erscheinen. Alle Sterne, die von der Erde aus gesehen rings um A in der Entfernung $\gamma_0 - \beta$ liegen, würden von dem Stern A gleichsam eingefangen werden. Sollte zufällig TAB eine gerade Linie sein, so würde, von der Erde aus gesehen, der Stern A von einem Ring umgeben erscheinen.

Ob der hier angegebene Fall eines fiktiven Doppelsternes auch wirklich vorkommt, kann ich nicht beurteilen.

Petrograd, 1924 Jan. 28.

O. Chwolson.

Antwort auf eine Bemerkung von *W. Anderson*.

Daß ein Elektronengas einer Substanz mit negativem Brechungsvermögen optisch äquivalent sein müßte, kann bei dem heutigen Stand unserer Kenntnisse nicht zweifelhaft sein, da dasselbe einer Substanz von verschwindend kleiner Eigenfrequenz äquivalent ist.

Aus der Bewegungsgleichung

$$\varepsilon X = \mu\, d^2x/dt^2$$

eines Elektrons von der elektrischen Masse ε und der ponderabeln Masse μ folgt nämlich für einen sinusartig pendelnden Prozeß von der Frequenz ν die Gleichung

$$\varepsilon X = -(2\pi\nu)^2 \mu x.$$

Berücksichtigt man, daß εx das »Moment« eines schwingenden Elektrons ist, so erhält man für die Polarisation $p = n\varepsilon x$ eines Elektronengases mit n Elektronen pro Volumeinheit

$$p = -\varepsilon^2 n/[\mu\,(2\pi\nu)^2] \cdot X.$$

Hieraus folgt, daß die scheinbare Dielektrizitätskonstante

$$D = 1 + 4\pi p/X = 1 - \varepsilon^2 n/(\pi\mu\,\nu^2)$$

ist. $\sqrt{D}$ ist in diesem Falle der Brechungsexponent, also jedenfalls kleiner als 1. Es erübrigt sich bei dieser Sachlage, auf das Quantitative einzugehen.

Es sei noch bemerkt, daß ein Vergleich des Elektronengases mit einem Metall unstatthaft ist, weil die bei der elementaren Theorie der Metalle zugrundegelegte »Reibungskraft« bei freien Elektronen fehlt; das Verhalten der letzteren ist allein durch die Einwirkung des elektrischen Feldes und durch die Trägheit bedingt.

Berlin, 1924 April 15.

A. Einstein.

Figure 4. A brief note by Einstein published on the same page of the Astronomische Nachrichten as Chwolson's 1924 note on the gravitational lensing phenomenon.

Clearly, Chwolson's note, though written by an established physicist in a well-known journal, did not have the least effect on Einstein and, it seems, did not have any effect at all.

Ironically, the more time passed by and the more gravitational lensing became a productive field of research, the more precursors were identified. In 1964, at a time of renewed interest in the effect of gravitational lenses,[39] Liebes went to the efforts of compiling

> the references which have been found in the literature to gravitational lens phenomena, apologizing for those oversights which undoubtedly have been made (Liebes 1964, B 835).

The first reference in Liebes' list is a half-page note by Oliver Lodge on "Gravitation and Light" published in *Nature* in the December 4 issue of 1919. While Lodge qualitatively conceived of the idea of a gravitational lens, he emphasized that

> it is not permissible to say that the solar gravitational field acts like a lens, for it has no focal length (Lodge 1919, 354).

Liebes then cited Eddington's book *Space, Time and Gravitation*, published in 1920. In his book, Eddington mentioned the possibility of observing a double image due to gravitational lensing in a section on observational tests of general relativity. He also considered the expected intensity of the deflected light rays and concluded that

> it is easily calculated that the increased divergence would so weaken the light as to make it impossible to detect it when it reached us (Eddington 1920, 134).

Einstein may well have seen Lodge's note, and was certainly aware of Eddington's book. Nevertheless, he did not mention these authors in his 1936 paper and for all we know, he responded as little to any of them as he did to Chwolson's note. It was evidently only Mandl's initiative that forced gravitational lensing to enter the historical stage. What was so special about this intervention of an amateur scientist?

The root of Mandl's idea was not a technical problem within a highly-specialized scientific discipline but a simple model of gravitational light deflection conceived in analogy to the action of a lens in geometrical optics. What made Mandl pursue this idea so vigorously until he finally turned Einstein's mind was the combination of this mental model with a grand vision of its implications for the understanding of nature on a cosmological scale. However, the simplicity of the mental model and the grandness of his vision did not fit into the grid of contemporary professional science.

Mandl's characterization as an "amateur scientist" is certainly justified in view of his lack of adequate professional training and academic status. But it does distract from a crucially important dimension of science, its foundation in a shared knowledge of nature that is not the exclusive property of outstanding individuals or of the institutions of professional science. On the contrary, we believe that it is hardly possible to understand the development of science, and in particular the scientific revolution represented by Einstein's theories of relativity, without taking into account this shared knowledge which, we believe, comprises both specialized theories but also elementary ideas e.g. of space, time, gravity, and light. It is these elementary ideas which by their very nature do not fall under the domain of any specialized domain of physics but cover areas of knowledge ranging from psychology via technological practices to cosmology. Mandl's role in the history of gravitational lensing illustrates that the exploration of this vast territory of knowledge is not necessarily the privilege of a few outstanding scientists such as Einstein. After all, even the most daring and seemingly ridiculous aspects of Mandl's vision turn out to be, under closer inspection, not as "unprofessional" as they might appear at first glance. His expectations about the promise of gravitational lensing to become an observational confirmation of general relativity have obviously been amply confirmed. But also Mandl's, at first sight, far-fetched speculations about a cosmic cause of the disappearance of dinosaurs are strikingly close to Alvarez's theory about a meteorite impact as the possible origin of an ecological catastrophe leading to the extinction of species, a theory that is today widely accepted.

What Mandl achieved, in the end, was to introduce a simple idea into the canon of accepted scientific knowledge, an idea which before was rejected only because it was not deemed observable. Why should astronomers care about an effect that seemed to be inaccessible to observations? Mandl's initiative, together with the fact that Einstein gave prominence to it with his 1936 publication, however, stimulated a broad discussion among astronomers and astrophysicists, even in spite of the absence of any immediate prospects of the observational verification of the lensing effect. And this discussion lasted until the effect was eventually confirmed by observations. Mandl's success was thus, as a matter of fact, also a victory of fantasy. How would a world look in which gravitational lensing would not, as Einstein and Eddington had originally surmised, be a minor, essentially unobservable effect? Einstein's publication stimulated his contemporaries to imagine such a world and thus to take the effect seriously and explore the conditions under which it might be observable after all.

One such publication was written by Henry Norris Russell, entitled "A Relativistic Eclipse," and appeared in the February 1937 issue of the *Scientific American* (Russell 1937). As the author line tells the reader, Russell was "PhD. Chairman of the Department of Astronomy and Directory of the Observatory at Princeton University. Research Associate of the Mount Wilson Observatory of the Carnegie Institution of Washington. President of the American Astronomical Society." Since both lived in Princeton, Russell and Einstein had probably been talking the matter over personally. The paper, in any case, dated December 2, 1936, acknowledges Einstein's help:

> My hearty thanks are due to Professor Einstein, who permitted me to see the manuscript of his note before its publication.[40]

Russell focused on the issue of observability and agreed that the lensing effect would not be verifiable for terrestrial observers. But he was not disheartened by this and further pursued the idea. The subject of his paper is:

> What Might be Seen from a Planet Conveniently Placed Near the Companion of Sirius. Perfect Tests of General Relativity that are Unavailable (ibid. 1937).

Discussing the orders of magnitude of the lensing effect with a white dwarf, Russell considers taking Sirius's companion as a gravitational lens, and Sirius itself as the light source. He imagines a small planet orbiting around Sirius's companion with just the right distance and considers how Sirius would appear to observers on this planet if distorted by the lensing effect of its companion. The paper gives some sketches of the distorted forms of Sirus's image as seen from the imaginary planet when its lensing companion passes through the line of sight. Paraphrasing the sketches, Russell compares the event to an ordinary solar eclipse (without gravitational deflection). He describes an intermediate state of the eclipse and the appearance of the lensing effect like this:

> a bright crescent has appeared on the *opposite* side of the eclipsing disk. This is produced by light coming from the part of the geometrical disk of Sirius nearest the center of the companion and deflected around the far side of the latter.

The final drawing shows the case of perfect alignment, or the case of the later so-called "Einstein Ring":

> ... for central eclipse, it looks like an annular eclipse of a large disk by a small one, instead of the actual total phase. From this point, all the previous phases occur in reverse order.

Russell concluded:

> Our hypothetical space-tourist, therefore, could settle down with his planet in such a place that general relativity would no longer be a matter of the utmost refinement of theory and observation. It would instead be needed to account for the most bizarre and spectacular phenomena of the heavens, as he saw them.

While Russell's world was purely imaginary, his paper contributed much to keep the interest alive and was often cited in the sequel. It also contributed to give a realistic twist to the abstract question of how a world would have to look like to make gravitational lensing an important effect.

The challenge of using gravitational lensing in order to probe cosmic dimensions was taken up in another immediate response to Einstein's paper, which was sent as a "Letter to the Editor" of *Physical Review* and which appeared in the February 15 issue of that journal (Zwicky 1937a). It was entitled "Nebulae as Gravitational Lenses" and was written by the Swiss astronomer Fritz Zwicky who then worked at the Norman Bridge Laboratory[41] at the California Institute of Technology in Pasadena. Zwicky's concern as well was the observability of the phenomenon. His brief note also started with a reference to Mandl's idea:

> Einstein recently published some calculations concerning a suggestion made by R. W. Mandl, namely, that a star B may act as a "gravitational lens" for light coming from another star A which lies closely enough on the line of sight behind B. As Einstein remarks the chance to observe this effect for stars is extremely small.

The next passage makes it clear that also Zwicky had first encountered the problem of gravitational lensing as an indirect consequence of Mandl's persistence:

> Last summer Dr. V. K. Zworykin (to whom the same idea had been suggested by Mr. Mandl) mentioned to me the possibility of an image formation through the action of gravitational fields. As a consequence I made some calculations which show that extragalactic *nebulae* offer a much better chance than *stars* for the observation of gravitational lens effects.

It was in fact the aim of Zwicky's communication to point out that extragalactic nebulae, as a consequence of their masses and apparent dimensions, were much more likely candidates for the observation of gravitational lenses. He argued that the discovery of lensing images "would be of considerable interest" not only since it would provide a test of relativity theory but also because one might find nebulae at greater distance through the lensing effect and also because one might get further information about the masses of those nebulae that act as gravitational lenses. In the final sentence, Zwicky optimistically announced the publication of a "detailed account of the problems sketched here."

Two months later, Zwicky, instead of a detailed account, submitted another letter to the editor of *Physical Review* (Zwicky 1937b). It was entitled "On the Probability of Detecting Nebulae Which Act as Gravitational Lenses." Zwicky now argued that:

> the probability that nebulae which act as gravitational lenses will be found becomes practically a certainty. [...]
> Present estimates of masses and diameters of cluster nebulae are such that the observability of gravitational lens effects among the nebulae would seem ensured.

But in spite of his optimism, Zwicky was aware that the search for gravitational lenses would actually be laborious:

> In searching through actual photographs, a number of nebular objects arouse our suspicion. It will, however, be necessary to investigate certain composite objects spectroscopically, since differences in the red shift of the different components will immediately betray the presence of gravitational effects. Until such tests have been made, further discussion of the problem in question may be postponed.

It seems to have been postponed for quite a while, for Zwicky's detailed account, announced for the *Helvetica Physica Acta,* did not appear in that journal for some years.

As the papers by Russell, Zwicky, and others testify, gravitational lensing as a subject in its own right had been born with Einstein's 1936 *Science* note. Mandl's role in establishing this subject was crucial since he had helped turn gravitational lensing into a theoretical reality long before it became an observational reality. Only after Einstein's publication of the calculations he had made at Mandl's request, did other scientists like Russell, Zwicky, and Tikhov take up the idea and dare to publish their findings. From this point on, the idea of gravitational lensing was kept alive and became part of the theoretical program of general relativity. Henceforth, it was again and again tentatively applied to explain curious astronomical phenomena.[42] Whenever a new phenomenon appeared on the sky, it became routine to ask whether it could be related to gravitational lensing, until one of these curious phenomena turned out to be a perfect embodiment of the idea. But it was almost half a century after Einstein's publication before, in the sequel of the discovery of quasars, the cosmos known to us finally reached the dimensions and astrophysics the technical sophistication to make gravitational lensing a reality.[43]

ACKNOWLEDGEMENTS

We would like to thank Hubert Goenner for helpful suggestions on a draft of this paper and Michel Janssen for his advice in an earlier phase of our research. Without John Stachel, this paper would never have come into being—in many senses.

We are also grateful to Ze'ev Rosenkrantz, Bern Dibner Curator of the Albert Einstein Archives of the Hebrew University of Jerusalem, for granting us the permission to quote from Einstein sources as well as to the publishers for granting permission to reproduce their material.

Finally we thank Simone Rieger for her assistance in preparing the documentary appendix to the preprint version (MPI preprint 160) of this paper.

Max Planck Institute for the History of Science

NOTES

* The paper continues an investigation begun jointly with John Stachel and is dedicated to him in gratitude for all he has taught us about these issues. It has appeared earlier for a substantial collection of documents relevant to the early history of gravitational lensing. The reader is referred to the document appendix of the preprint version of this paper (MPI preprint 160).

1. R. D. Potter to A.Einstein, 16 September 1936, EA17039. Cp. fig. 1.
2. See Mandl to Einstein, 3 May 1936, EA17031, and Zwicky 1937a, 290.
3. Strictly speaking, only a determination of the numerical value of ε_0 would provide a test of General Relativity since the $1/R$ dependence follows already from the equivalence hypothesis alone and was, in fact, already derived in Einstein 1911.
4. Mandl to Einstein, 18 April 1936, EA 17027/28.
5. "Es bleibt bei der ersten Formel. Die zweite beruht auf einer falschen Überlegung. Ich habe mir überlegt, dass das fragliche Phänomen doch nicht beobachtbar sein wird, sodass ich nicht mehr dafür bin, etwas darüber zu publizieren." Einstein to Mandl, 23 April 1936, EA 17030.
6. "Ich habe in der Zwischenzeit eine Methode gefunden die Intensitaetssteigerung im Bereich der Focuslinie eines Sternes zu messen und experimentell bestaetigen zu koennen.
   Es waere meiner Ansicht nach im Interesse der Wissenschaft mit diesen Versuchen alsbald als moeglich zu beginnen." Mandl to Einstein, 23 April 1936, EA 17029.
7. "Der Hauptgrund warum ich Sie um Ihre w. Mitarbeit zu gewinnen belaestige ist dass unmittelbar nach Veroeffentlichung meiner Untersuchungen alle Astrologen und aehnliche Parasiten der Wissenschaft sich der Resultate meiner Ueberlegungen bemaechtigen werden und es ist meine Ueberzeugung dass es von Nutzen fuer den "Average citizen" waere wenn von vornherein ein Mann Ihren Ranges und Rufes denn Unnsinn der Pseudowissenschaft dieser Charlatane unterstreichen wuerde." Mandl to Einstein, 3 May 1936, EA 17031.
8. EA 17032.
9. "Bei der auserordentlich geringen Oberfläche den die Fixsterne im Vergleich zu der Himmelsoberflaeche einnehmen ist der Durchgang der Erde durch eine E. F. L. [eines Fixsterns] ein sehr seltenes Phaenomen und waere im Bezug auf Zeit und vielleicht auch in Ausdehnung am besten mit einer Sonnenfinsternis zu vergleichen, ein Durchgang jedoch durch ein E. F. L. [einer Galaxie] waere etwas alltaegliches und duerfte sogar Jahrtausende andauern." EA 17032.
10. "(Die verhaeltnismaessig grosse Quantitaet der Cosmischen Strahlung waere am leichtesten durch einen Durchgang durch eine E. F. L. der Milchstrasse und anderer Nebel zu erklaeren.)" EA 17032.
11. "Nachstehend einige allgemeine Bemerkungen ueber den Darvinismus.
    Die groesste Unklarheit und Ungewissheit in der Darvinischer Evolutionslehre besteht in der Unzulaenglichkeit der Erklaerung der Ursach warum die Evolution nich graduell sondern sprunghaft vor sich ging und dass nach Perioden einer intensiven Evolution und Mutation eine lange Periode von Stillstand kam und warum ploetzlich ganze Klassen von hoch entwickelten Tieren ausstarben." EA 17032.
12. Mandl's memorandum erroneously refers to "Professor J. P.Mueller von der University of Texas" but this is clearly a slip.
13. "Die Relativitaets-Theorie liefert also so wiedersinnig es im ersten Mommente erscheinen mag einen Schluessel zu den bisher dunklen Teilen der Evolutionslehre." EA 17032.
14. "Wenn auch obige Ueberlegung uns lehrt dass die Sterne moegen selbe noch so ferne stehen fuer die Evolution 'verantwortlich sind' oder sie zustande gebracht haben muss an dieser Stelle hervorgehoben werden das die 'Wissenschaft'?? der Astrologie mit obigen Ueberlegungen nichts gemein hat und dass wir selbst mit unseren heutigen Astronomischen Instr nicht im stande sind auch nur zurueckzuberechnen welche Sterne fuer unsere Erde in Betracht kommen diese im oben dargelegten Sinne beeinflust zu haben." EA 17032.
15. "I found, during the last days, occasion to discuss the consideration with Dr. Compton, Dr. Swann and Dr. Millikan, and the first two were interested in the matter and asked me (because of lack of time) for further information. Dr. Millikan, however, gave me short shrift with three brief 'I don't understand.' ("Ich fand Gelegenheit waehrend der letzten Tage mit Dr. Compton, Dr. Swann und Dr. Millikan die

Ueberlegung zu besprechen und die beiden erstgenannten interessierten sich fuer die Sache und ersuchten mich um (des Zeitmangels wegen) naehere Angaben Dr. Millikan jedoch fertigte mich mit drei kurzen 'I don't understand' ab." (Mandl to Einstein, 3 May 1936, EA 17031). The reference to Compton is probably to Arthur Holly Compton, then professor of physics at the University of Chicago, winner of the 1927 Nobel prize for the discovery of the "Compton-effect," who had changed his main research interests from X-rays to cosmic rays in the early 1930s which is probably why Mandl approached him. Mandl may, however, also refer to Arthur's brother Karl Taylor Compton who had been chair of the physics department at Princeton University for many years and was, since 1930, president of the Massachussetts Institute of Technology. In 1935-36 Karl Taylor Compton presided the American Association for the Advancement of Science. William Francis Gray Swann (1884-1962) since 1927 was director of the Bartol Research Foundation of the Franklin Institute, a center of cosmic ray studies, and author of the highly successful book The Architecture of the Universe (1934). Robert Andrews Millikan (1868-1953) was director of the Norman Bridge Laboratory at the California Institute of Technology in Pasadena, and had done extensive experimental research on cosmic rays. Believing for many years that cosmic rays were photons originating in processes of nuclear fusion (the "birth cries" of atoms), Millikan had to concede in the early 30's that some percentage of cosmic radiation consisted of charged particles, and by 1935 he also rejected his atom-building hypothesis. Millikan had won the Nobel prize in 1923.

16. "Sehr geehrter Herr Mandl: Ich habe Ihren Verstaerkungseffekt genauer ausgerechnet. Folgendes ist das Resultat." Einstein to Mandl, 12 May 1936, EA 17034/35.
17. "Immerhin ist wohl mehr Chance vorhanden, diesen Verstärkungseffekt gelegentlich einmal zu beobachten als den "Hof-Effekt", von dem wir früher gehandelt haben. Aber die Wahrscheinlichkeit, dass wir so genau in die Verbindungslinie der Mittelpunkte zweier sehr verschieden entfernten Sterne hineinkommen ist recht gering, noch geringer die Wahrscheinlichkeit, dass das im Allgemeinen nur wenige Stunden währende Phänomen zur Beobachtung gelangt." Einstein to Mandl, 12 May 1936, EA 17034, 35.
18. "Ihre an das Phänomen geknüpften phantastischen Spekulationen würden Ihnen nur den Spott der vernünftigen Astronomen eintragen. Ich warne Sie in ihrem eigenen Interesse vor einer derartigen Veröffentlichung. Dagegen ist gegen eine bescheidene Publikation einer Ableitung der beiden charakteristischen Formeln für den 'Hof-Effekt' und den 'Verstärkungs-Effekt' nichts einzuwenden." Einstein to Mandl, 12 May 1936, EA 17034/35.
19. "die Resultate der Effecte sowie Ihre Formeln zu veroeffentlichen, da mir ja jede Moeglichkeit dazu fehlt." Mandl to Einstein, 17 May 1936, EA 17036.
20. "das selber [Brief] Ihnen eine grosse Freude machen wird eine Freude die nur um ein klein wenig kleiner wird als die Freude die unser gemeinsamer Freund Hitler haben wird wenn er herausfindet dass es wieder einer der verdammten Juden war der die Einstein Theorie in das Einstein Gesetz verwandelte." Mandl to Einstein, 19 May 1936, EA 17037.
21. See Planck 1947, 143, and, for historical discussion, Albrecht 1993.
22. "Ich machte in den letzten Tagen verzweifelte Versuche, die Resultate meiner Forschungen mit Ihren Formeln zu veroeffentlichen und hoerte ausnahmslos die Frage 'Ja wenn Herr Einstein ihre Resultate gut befindet warum veoeffentlicht er sie nicht selber?' So es scheint dass an Ihnen liegt die Resultate der wissenschaftlichen Welt zugaenglich zu machen." Mandl to Einstein, 21 May 1936, EA 17038.
23. The letter was auctioned in 1995 and is facsimilized in Sotheby's Auction Catalogue "Fine Books and Manuscripts - Sale 6791," lot 73. Draft notes for this letter are extant in the Einstein Archives, EA 3-011-55, see (Renn, Sauer, Stachel 1997, 186) for a facsimile. Similar and related, but undated calculations are also found among Einstein's Princeton manuscripts, see EA 62-225, 62-275, 62-349, 62-368.
24. R. D. Potter to A. Einstein, 16 Sept. 1936, EA 17-039, cp. fig. 1.
25. For more explicit historical accounts of gravitational lensing, see, e.g., Barnothy 1989, Schneider, Ehlers, and Falco 1992, pp. sec. 1.1.
26. "Ich danke Ihnen noch sehr für das Entgegenkommen bei der kleinen Publikation, die Herr Mandl aus mir herauspresste. Sie ist wenig wert, aber diese arme Kerl hat seine Freude davon." Einstein to J. McKeen Cattell, 18 December 1936, EA65-603.
27. "Ich weiss sehr wohl, dass Sie Columbia verlassen mussten. Forscher werden hier behandelt wie anderwärts Kellner oder Verkäufer in Detailgeschäften (Letztere verdienten es natürlich auch, unter

besser gesicherten Verhältnissen zu arbeiten, es steckt kein Klassenhochmut hinter diesem Ausdruck.); gegenwärtig ist es besonders arg, wenn man von einem findet, er sei ein "Radical". Die Öffentlichkeit ist gegen jene Verstösse gegen die Freiheit der Lehre und der Lehrer viel zu gleichgültig." Einstein to Cattell, 18 December 1936, EA 65-603, the sentence in parenthesis was added as a footnote to the original letter.

28. "Es wäre dringend zu wünschen, daß sich Astronomen der hier aufgerollten Frage annähmen, auch wenn die im vorigen gegebenen Überlegungen ungenügend fundiert oder gar abenteuerlich erscheinen sollten." Einstein 1911, 908.

29. See, e.g., Einstein to Hale, 14 October 1913, and Hale to Einstein, 8 November 1913, The Collected Papers of Albert Einstein, Vol. 5, Docs. 477, 483. Even though Einstein by that time already was a well-established ETH professor, he still found it useful to have his ETH colleague Julius Maurer add a postscript to his letter thanking Hale in advance for a friendly reply.

30. "Vor 25 Jahren hielt Einstein, damals Professor in Prag, einen seiner ersten Vorträge über die allg. Rel. Theorie und schloss mit den Worten, dass er nunmehr einen astronomischen Mitarbeiter bedürfe, dass die Astronomen aber zu rückständig seien, um den Physikern zu folgen. Da erhob sich ein junger Student und teilte mit, er habe in Berlin an der Sternwarte, die er besucht hatte, einen jungen Astronomen kennen gelernt, auf den diese Charakterisierung nicht stimme. Einstein schrieb sofort an mich. Daraus ist unsere gemeinsame Arbeit und de facto mein wissenschaftliches Leben hervorgegangen. Der Student war Pollak, jetzt Ordinarius in Prag." Freundlich to Bosch, 4. 12. 193?, (from Unternehmensarchiv BASF, Ludwigshafen, Personalarchiv Carl Bosch W 1/folder 9/2; this letter was found by Dieter Hoffmann who generously made it available to us). See also the reference to Pollack's role as a mediator in note 6 to Doc. 278 of The Collected Papers of Albert Einstein, 5:313.

31. See Hentschel 1994, Hentschel 1997, and Renn, Castagnetti, and Damerow 1999.

32. See Renn, Sauer, and Stachel 1997, 185.

33. "Das ablehnende Verhalten seines Direktors hat es ihm jedoch sieben Jahre lang unmöglich gemacht seine auf die Prüfung der Theorie gerichteten Arbeitspläne zur Ausführung zu bringen." Einstein to Hugo A. Krüss, 10 January 1918, The Collected Papers of Albert Einstein, Vol. 8, Doc. 435.

34. "Ungleich weniger begabt als Schwarzschild hat er doch mehrere Jahre vor diesem die Wichtigkeit der neuen Gravitationstheorien für die Astronomie erkannt und sich mit glühendem Eifer für die Prüfung der Theorie auf astronomischem beziehungsweise astrophysikalischem Wege eingesetzt." Einstein to Hugo A. Krüss, 10 January 1918, The Collected Papers of Albert Einstein, Vol. 8, Doc. 435.

35. See, e.g. Einstein to Schwarzschild, 9 January 1916, The Collected Papers of Albert Einstein, Vol. 8, Doc. 181, Einstein to Sommerfeld, 2 February 1916, The Collected Papers of Albert Einstein, Vol. 8, Doc. 186, Einstein to Hilbert, 30 March 1916, The Collected Papers of Albert Einstein, Vol. 8, Doc. 207.

36. "Aber eines kann immerhin mit Sicherheit gesagt werden: Existiert keine solche Ablenkung, so sind die Voraussetzungen der Theorie nicht zutreffend. Man muss nämlich im Auge behalten, dass diese Voraussetzungen, wenn sie schon naheliegen, doch recht kühn sind. Wenn wir nur einen ordentlich grösseren Planeten als Jupiter hätten! Aber die Natur hat es sich nicht angelegen sein lassen, uns die Auffindung ihrer Gesetze bequem zu machen." Einstein to Freundlich, 1 September 1911, The Collected Papers of Albert Einstein, Vol. 5, Doc. 281.

37. "Es ist mir nun leider klar geworden, dass die "neuen Sterne" nichts mit der "Linsenwirkung" zu thun haben, dass ferner letztere mit Rücksicht auf die am Himmel vorhandenen Sterndichten ein so ungeheuer seltenes Phänomen sein muss, dass man wohl vergebens ein solches erwarten würde." Einstein to Zangger, 15 October 1915, The Collected Papers of Albert Einstein, Vol. 8, Doc. 130.

38. For a historical discussion, see Earman and Glymour 1980.

39. See, e.g., Darwin 1959, Mikhailov 1959, Metzner 1963, Klimov 1963, Refsdal 1964, and Schneider, Ehlers, and Falco 1992, sec. 1.1.

40. Einstein in turn had also seen Russell's paper prior to publication and had made comments on a previous draft, see H. N. Russell to A. Einstein, dated 27 November 1936, EA 20-067.

41. The Norman Bridge Laboratory was directed by Millikan, see note 15.

42. For a historical review, see Barnothy 1989.

43. See Walsh, Carswell, and Weymann 1979 and Young et al. 1980. For reviews, see Refsdahl and Surdej 1994 and Wambsgans 1998.

REFERENCES

Albrecht, H. 1993. *Max Planck:* "Mein Besuch bei Adolf Hitler – Anmerkungen zum Wert einer historischen Quelle." Pp. 41-63 in *Naturwissenschaft und Technik in der Geschichte,* ed. H. Albrecht. Stuttgart: Verlag für Geschichte der Naturwissenschaft und Technik.

Barnothy, J. M. 1989. "History of gravitational lenses and the phenomena they produce." Pp. 23-27 in *Gravitational Lenses,* eds. J. M. Moran, J. N. Hewitt, and K. Y. Lo. Berlin: Springer.

Chwolson, O. 1924. "Über eine mögliche Form fiktiver Doppelsterne." *Astronomische Nachrichten* 221:329-330.

Darwin, C. 1959. "The gravity field of a particle." *Proceedings of the Royal Society* A249, 180. London.

Eddington, A. S. 1928?check?. *Space, Time and Gravitation.* Cambridge: Cambridge University Press.

Einstein, A. 1911. "Über den Einfluß der Schwerkraft auf die Ausbreitung des Lichtes." *Annalen der Physik* 35:898-908.

———. 1924. "Antwort auf eine Bemerkung von W. Anderson." *Astronomische Nachrichten* 221:330.

Einstein, A. 1936. "Lens-like Action of a Star by the Deviation of Light in the Gravitational Field." *Science* [N. S. WB], 84 (2188):506-507.

Hentschel, K. 1994. "Erwin Finlay Freundlich and Testing Einstein's Theory of Relativity." *Archive for History of Exact Sciences* 47/2:143-201.

———. 1997. *The Einstein Tower. An Intertexture of Architecture, Astronomy, and Relativity Theory.* Stanford: Stanford University Press.

Idlis, G. M., and S. A. Gridneva. 1960. *Izv. Astrofiz.* Inst. Acad. Nauk. Kaz. SSR, 9:78.

Klein et al. eds. 1993. *The Collected Papers of Albert Einstein. Vol. 5 The Swiss Years: Correspondence 1902-1914.* Princeton: Princeton University Press.

Klimov, Y. G. 1963. "The Deflection of Light Rays in the Gravitational Fields of Galaxies." *Soviet Physics - Doklady* 8/2:119-122.

Liebes, S. 1964. "Gravitational Lenses". *Physical Review* 133:B835-B844.

Lodge, O. J. 1919. "Gravitation and Light". *Nature* 104:354.

Metzner, A. W. K. 1963. "Observable Properties of Large Relativistic Masses." *J. Math. Phys.* 4:1194-1205.

Mikhailov, A. A. 1959. "The Deflection of Light by the Gravitational Field of the Sun." *Monthly Notices Roy. Astron. Society* 119:593.

Moran, J. M., J. N. Hewitt, and K. Y. Lo, eds. 1989. *Gravitational Lenses.* Berlin: Springer.

Planck, M. 1947. "Mein Besuch bei Adolf Hitler." *Physikalische Blätter* 3:143.

Refsdahl, S. 1964. "The gravitational lens effect." *Mon. Not. R. Astron. Soc.* 128:295-308.

Refsdahl, S. and J. Surdej. 1994. "Gravitational Lenses." *Rep. Prog. Phys,* 56:117-185.

Renn, J., G. Castagnetti, and P. Damerow. 1999. "Albert Einstein: alte und neue Kontexte in Berlin." Pp. 333-354 in *Die Königlich Preußische Akademie der Wissenschaften zu Berlin im Kaiserreich,* ed. J. Kocka. Berlin: Akademie Verlag.

Renn, J., T. Sauer, and J. Stachel. 1997. "The Origin of Gravitational Lensing: A Postscript to Einstein's 1936 *Science* Paper." *Science* 275:184-186.

Russell, H. N. 1937. "A Relativistic Eclipse." *Scientific American* 156:76-77.

Schneider, P., J. Ehlers, and E. E. Falco. 1992. *Gravitational Lenses.* Berlin: Springer.

Schulmann et al. 1998. *The Collected Papers of Albert Einstein. Vol. 8 The Berlin Years: Correspondence 1914-1918.* Princeton: Princeton University Press.

Stockton, A. 1980. "The lens galaxy of the twin QSO 0957+561." *Astrophysical Journal* 242:L141.

Tikhov, G. A. 1937. "Sur la déviation des rayons lumineux dans le champ de gravitation des étoiles." *Dokl. Akad. Nauk S. S. R.* 16:199-204.

Walsh, D., R. F. Carswell, and R. J. Weymann. 1979. "0957+561 A, B: twin quasistellar objects or gravitational lens?" *Nature* 279:381-384.

Wambsgans, J. 1998. http://www.livingreviews.org/Articles/volume1/1998-12wamb/

Young, P., J. E. Gunn, J. Kristian, J. B. Oke, and J. A. Westphal. 1980. "The double quasar Q0957+561 A,B: A gravitational lens formed by a galaxy at z=0.39." *Astrophysical Journal* 241:507-520.

Zwicky, F. 1937a. "Nebulae as Gravitational Lenses." *Physical Review* 51:290.
———. 1937b. "On the Probability of Detecting Nebulae Which Act as Gravitational Lenses." *Physical Review* 51:679.
———. 1957. *Morphological Astronomy*. Berlin: Springer.

CATHERINE GOLDSTEIN AND JIM RITTER

THE VARIETIES OF UNITY: SOUNDING UNIFIED THEORIES 1920–1930*

> *Die Voraussetzungen, mit denen wir beginnen, sind keine willkürlichen, keine Dogmen, es sind wirkliche Voraussetzungen, von denen man nur in der Einbildung abstrahieren kann. Es sind die wirklichen Individuen, ihre Aktion und ihre materiellen Lebensbedingungen, sowohl die vorgefundenen wie die durch ihre eigne Aktion erzeugten. Die Voraussetzungen sind also auf rein empirischem Wege konstatierbar.*
>
> Karl Marx and Friedrich Engels
> *Deutsche Ideologie*

The "goal of the ultimate"[1] — that is, the unification of all fundamental physical phenomena in a single explanatory scheme — had perhaps never seemed so close at hand for many physicists and mathematicians as in the third decade of the twentieth century. And if there were those who saw a great promise in this, there were equally those who opposed it.[2] Even among its partisans, just what such an 'ultimate' might resemble was not clear, its scope and its formulation seemed infinitely extendible, varying by author and even in the same author, by period. Hermann Weyl's trajectory provides an object lesson on this theme. In 1919, the preface to the third edition of his celebrated *Raum, Zeit, Materie*, devoted to an exposition of Einstein's general theory of relativity, hopefully announced:

> A new theory by the author has been added, which ... represents an attempt to derive from world-geometry not only gravitational but also electromagnetic phenomena. Even if this theory is still only in its infant stage, I feel convinced that it contains no less truth than Einstein's Theory of Gravitation[3] (Weyl 1919, vi).

Ten years later, the unity in sight at the beginning of the decade had been blurred for Weyl in the failure of his own and many similar attempts; moreover, his idea of what a unified theory ought to take into account, and how, had changed. To an American journalist at the Science Service in Washington D. C., who had publicized a recent unified theory of Einstein, much in the classical mold of Weyl's first, the latter wrote:

> Einstein's work is a new contribution to a search which he first undertook some years ago — one among many, many others which have been tried in the last ten years. ... I believe that the development of quantum theory in recent years has so displaced the status of the problem that we cannot expect to find the sought-for unity without involving matter waves, by which wave mechanics replaces moving material particles, in the framework.[4]

When, in 1950, he was asked for a new preface to the first American printing of *Raum, Zeit, Materie*, his judgement was final:

> My book describes an attempt to attain this goal [of unification]. This attempt has failed. ... Quite a number of unified field theories have sprung up in the meantime. ... None has had a conspicuous success (Weyl 1952, v–vi).

This story has been told a number of times in recent years, sometimes stressing the growing isolation of Einstein, shackled to a dying program, sometimes underlining the later revivals of some of the theories and their impact on the development of contemporary differential geometry and physics.[5] The appeal for us, however, lies essentially elsewhere, in just the absence of an obvious winner or — *pace* Weyl — of an obvious loser, and the puzzling historiographical issues this raises.

Traditional history of science has aimed at retracing the path, however tortuous, leading to the establishment of new truths. Since the sixties, at least, this aim, indeed the very notion of truth and its connection to the scientific enterprise, has been much contested; debates among scientists, their choices of values or paradigms, their more or less efficient uses of arguments and rallying of allies from different spheres of activities, political as well as technical, have become the focus of the historian's attention. Even if the work of a loser is handled with the same historical tools and the same respect as that of a winner, the difference between them still often provides the incentive of the narrative. How then to deal with the numerous scientific situations where no theory has lost or won, no general agreement, at any level, has been reached?

Most of the unified theories proposed in the twenties have been rediscovered and buried again; in a few cases, repetitively. They have been commented upon, expanded, compared, tested. In short, they are part and parcel of professional scientific activities. But which part? Did these theories represent marginal forays? Or the speculative forefront of some trends in physics? Or a recognized branch of research ordinarily practiced by competing groups? How to understand the dynamics of such a topic, and its role? How did it relate and interact with the most famous innovations of the early twentieth century, relativity theory and quantum theory? Indeed to what extent did such attempts towards unification ever become a recognized scientific discipline at all?

These questions deal with *collective* processes. The investigation of a few landmarks, however detailed it may be, will not be appropriate to answer them. Our ultimate goal is to understand how the whole body of work devoted to unified theories is organized, to analyze the possible alternatives, not only concerning the path to follow, but even as to what such a path might look like, and to trace the links (or lack thereof) between the different approaches and debates.

Our point of departure has been as concrete as possible — "real individuals, their actions and the material conditions of their existence." The material life of scientists in our century is punctuated by the writing and the publication of papers and books; these are, for us, the marks of production as a collective process and so have constituted the basis of our study. Since we wish to examine *professional* responses, we need to remain within the limits of professional acceptability; we have thus chosen to construct our corpus among the articles summarized in the main professional review journals of the twenties, *Physikalische Berichte*, widely read by physicists, and *Jahrbuch über die Fortschritte der Mathematik*, which played the same role for mathematicians. And here a first difficulty presents itself; in this period, there was no specific section devoted to unified theories (in itself, this is of course significant). Most attempts in 1920 deal with Einstein's general relativity, either to integrate it or to replace it, and are reviewed in the sections devoted to relativity theory and gravitation. These sections are also the unique obvious choice which is common to the two review journals during the twenties. Thus, while there are of course relevant papers in other sections, we have provisionally here chosen to concentrate on the papers reviewed in the section "relativity theory and gravitation." Within them, we have selected every article in which the author expresses an ambition to relate classes of natural phenomena seen as distinct. As we shall have more than one occasion to repeat in what follows, however, the conceptions of unification thus expressed vary widely.

Unification can mean the more or less complete merging of two fields into a single object, a metric for instance, or a unique action functional. It can mean the creation of a single englobing framework where different, 'natural', components (like the two fundamental forms of a hypersurface in a general space) take charge of the various phenomena or where the same mathematical object houses in turn each of them, according to the need or the interest of the physicist. We shall also find reductionist programs, in which one class of phenomena is shown to be an apparent instance of another, as well as schemes that coordinate different theories by having one replace the phenomenological aspects of the other. As this (non-exhaustive) list suggests, a deep ontological commitment is not necessarily considered as essential to a unification project, and indeed is totally lacking in some of our papers.

In a preliminary section, we shall first use the data generated by our selection to display some general trends of evolution during the 1920s, for relativity and for quantum theory in general, as well as for unified theories compared to those of relativity.

Such a rough-grained quantitative analysis, however, cannot give us access to the collective *practice* of unified theories. To do this we must study the papers themselves, locate and trace elements which relate them to others; either positively, by integrating these elements into the work itself, or negatively, by airing criticisms or stating alternatives. In contemporary scientific texts, *references* provide precisely such a means of capturing linking elements and we shall use them as our main guidelines within the limited space of this paper. Citation analysis has been, of course, a standard routine for some time in bibliometric studies and 'network analysis'.[6] But our technique is different; in particular, it is not derived from counting or automatic indexation, and we shall understand the word "references" in a larger sense than that of explicit bibliographical citations. We shall, of course, examine explicit footnotes and in-text citations, but we shall also take into account vaguer allusions to an idea or a rallying cry, as well as the use of a specific mathematical technique, in so far as such indicators appear to signify a collective practice.

Since, again, our emphasis is not on the communication of knowledge, but on its (collective) production, we shall need to take into account how the citations are used and precisely what kind of relation each reveals. Some of the links we shall examine have an obvious role in the papers they relate and the configurations[7] that they delineate are clearly-cut, for instance, the set of articles linked by the fact that they develop one particular affine theory. Others, however, are more subtle — for instance, the fact of taking for granted Einstein's general relativity as the correct theory of gravitation — and their concrete implementation in each paper authorizes a larger range of possibilities, which can be illustrated only through a more detailed analysis of the texts. In short, a close reading of the papers will here often ground a social analysis — and vice versa.

To fit all this into a finite space, we shall present here only a *sondage*,[8] the analysis of the articles dealing with unified theories and reviewed in one of our abstracting journals for three years only: 1920, 1925, 1930. These dates are not *anni mirabiles*.[9] We shall try in each case to outline most of the papers in sufficient detail to give a flavor of their variety. But, like its archeological counterpart, the result of our *sondage* will mainly consist of snapshots of the global organization of our topic at these different dates, allowing us to locate the main axes of its production and some of its characters. Their comparison, then, will give access to a more precise sense of the normality of this, a priori, very abnormal subject, and of its transformation during our decade.

1. UNIFIED THEORIES: SOME QUANTITATIVE DATA

Let us first present briefly the two abstracting journals on which we have relied for the initial selection of our corpus. The *Physikalische Berichte* was founded in 1920 as the amalgamation, under the auspices of the Deutsche Physikalische Gesellschaft and the Deutsche Gesellschaft für technische Physik, of a number of pre-war German physics abstracting journals: *Fortschritte der Physik*, *Halbmonatliches Literaturverzeichnis*, and *Beiblätter zu den Annalen der Physik*. The rate of publication was biweekly, but, fortunately for us, the review was indexed thematically at the end of each year. That part of *Physikalische Berichte* which interests us here is the second, "Allgemeine Grundlagen der Physik" [General foundations of physics], including several sections: "Prinzipien der älteren Physik" [Principles of the older physics], "Relativitätsprinzip" [Relativity principle], "Quantenlehre" [Quantum theory],"Wahrscheinlichkeit und Statistik" [Probability and statistics], "Erkenntnistheorie" [Epistemology].[10] The last year of our study, 1930, saw a change in the name of the second section, which hereafter became "Relativitätstheorie."

In 1920, on the other hand, the *Jahrbuch über die Fortschritte der Mathematik* was already a long-established mathematical review organ (it had been founded in 1869). The rhythm of publication, however, was not stable — biannual for the years 1919–1922 and annual thereafter — and the publication date was quite irregular, delayed generally three or four years after the nominal date.[11] Starting precisely with volume 47 ('1919–1920'; published in 1924–1926), a new Section VII was added, intercalated between "Mechanik" [Mechanics] and "Astronomie, Geodäsie und Geophysik" [Astronomy, geodesy and geophysics]; it was entitled "Relativitätstheorie und Theorie der Gravitation" [Relativity theory and theory of gravitation]. From the reorganisation of the '1925' volume on,[12] Section VII becomes "Mathematische Physik" with a subsection "2. Relativitätstheorie" and a new addition, "3. Quantentheorie."

Our first idea had been to capture how mathematicians and physicists received and reviewed different unified theories through a comparison of the titles present in one or the other of the two journals, in their unique common section, relativity. We discovered that, in fact, most of the articles were reviewed in both journals, although not always in the same year. We have thus simply aggregated their information in the selection of our corpus. The following quantitative analysis will, however, be based only on *Physikalische Berichte*, because it will allow us to draw some comparisons with quantum theory during the whole decade.[13]

Fig. 1 below displays the number of publications reviewed per year in the section on relativity and in the section on quantum theory. We note immediately that the number of relativity articles oscillates, with peaks around 1921–1923 and 1927, and that relativity dominates quantum theory up to 1925, after which it lags far behind.

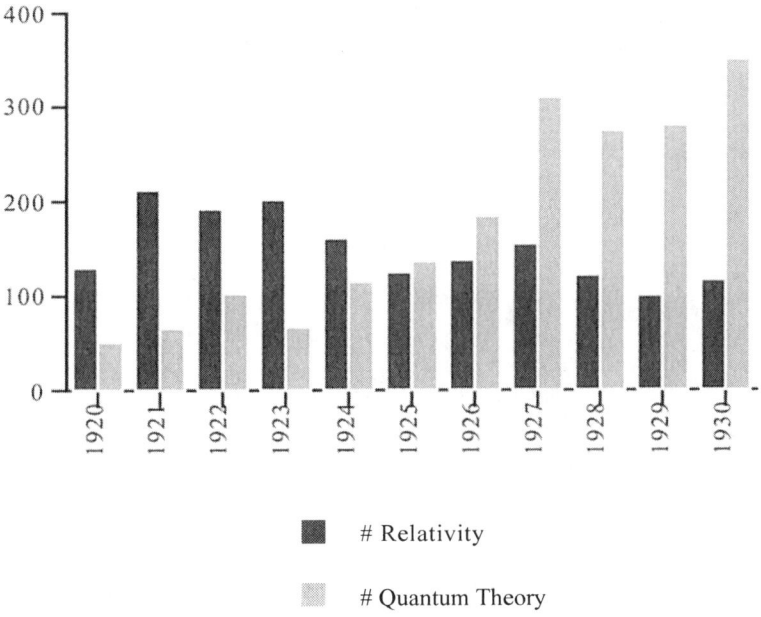

Figure 1. Number of relativity and quantum physics articles reviewed between 1920 and 1930. From Physikalische Berichte.

To gain more perspective, it is interesting to look at these publications as a percentage of *all* the physics articles and books published in the corresponding years and reviewed in *Physikalische Berichte* (fig. 2). Note that the combined production of the two sections, relativity and quantum theory (theoretical and experimental articles together), make up a nearly fixed percentage of total output, which rarely exceeds 5% and is never as much as 6%. The usual historiography of this 'golden age of physics' thus concentrates on a minute part of the activity in physics. Within this, relativity theory declines rather smoothly from 1922 on, while quantum theory takes up the slack.

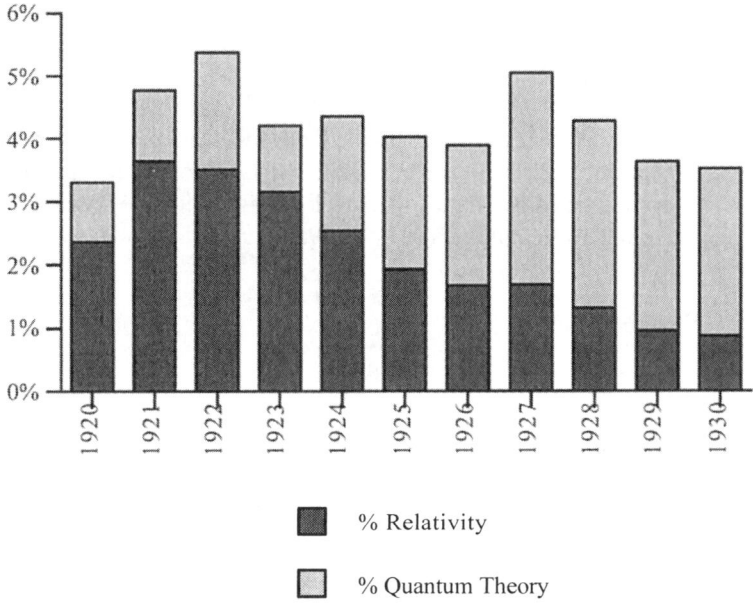

Figure 2. Relativity and quantum physics articles as a percentage of all physics articles: 1920–1930. From Physikalische Berichte.

Thus the view that spectacular advances in the 'new' quantum theory of Heisenberg and Schrödinger in 1925–1926 created a rival to relativity that drew interest away from the latter, may be, as we see, correct as a rough, global, picture, but does not capture the dynamics of the process:[14] the decline in the number of relativity publications had begun in 1924 in absolute numbers, but already in 1923 from a relative point of view, i.e., *before* the introduction of the 'new' quantum theory in the years immediately following. Indeed the real increase in quantum articles occurred only in 1927.

How to locate the unified field theories within this corpus? For neither of the two journals did they constitute a subcategory at this time. As mentioned before, we have restricted ourselves here to the papers in the section on "Relativity theory" (or "principle") in both journals.[15] Among them, we have selected, through the reading of both the reviews and the individual papers, those which attempt to unify two or more phenomena seen as fundamental and distinct: this means, for most cases, gravitation, electromagnetism or matter.[16] The type of integration, as we have discussed in the introduction, can vary widely, and we have imposed no restrictions on it in selecting our corpus.

Some 200 articles met these criteria.[17] Their number per year increases during our decade, which, coupled with the slow diminution in the total number of relativity papers, implies that articles dealing with unified theories constitute a non-negligible part of these last; to fix our ideas, they make up some 11% in 1920, 17% in 1925, and 36% in 1930, for articles reviewed in *Physikalische Berichte*.[18] Furthermore, between 15 and 25 authors each year among the eighty or so relativists devote one or several papers to unification.

These figures are large enough to suggest that working on unified theories was far from marginal in the twenties — at least inside the quite marginal area, in terms of production, that was relativity theory. They are also small enough to hold out the hope that a more systematic analysis is within reach. We shall begin it here, as announced above, through a detailed survey of the articles reviewed during three years: 1920, 1925, 1930, considered in turn. But we will take into account the two abstracting journals, which, in view of the relative shift in their dates of publication and in their handling of the material, means a covering of greater time intervals; in fact, the landscape we shall sketch includes almost half of the articles reviewed in the decade.

Note:

1. We have tried to distinguish carefully between papers in our corpus and those which fall outside. Papers which are not in the corpus, whether primary or secondary sources, are listed at the end, in the bibliography; references to them inside the text are given in the author–date system. The list of papers selected in our corpus for each year is given at the end of the part discussing this year; references to these papers are indicated by name alone (if the only paper by the author in that year's corpus) or by author–Roman numeral (if there are several).

2. In the list of titles for each year's corpus, the reference in brackets at the end of each entry refers to the review journal (P = *Physikalische Berichte*, J = *Jahrbuch*) and page number of the review.

3. The 69 authors selected use very different symbolic conventions — a given author will often even change them from one article to another. The choice of Greek or Latin letters for indices is non-systematic, the same Christoffel symbol is sometimes denoted $\left\{ {j \atop ik} \right\}$ and sometimes $\left\{ {ik \atop j} \right\}$, etc. For the sake of space and to facilitate reference to the original papers, we have nonetheless retained their notation unless otherwise indicated.

4. Foreign, especially Slavic, proper names have been generally transcribed following the international linguistic system except when used in a reference, where the printed version, often using another transcription system, has been retained.

2. 1920

It is in this first year[19] of our *sondage* that the question of the definition of a unified theory is posed in its most acute form. Applying the criteria discussed above, we have retained 34 papers, written by 18 authors.[20] Among them, 26 are original research papers, 5 summarize other articles in our corpus, and 3 are general, non-technical, discussions of the state of the art.[21]

The most striking feature of the 1920 unification theories is their heterogeneity; a heterogeneity which concerns equally the nature of the unification, the choice of phenomena to unify, and the technical means mobilized to this end. Indeed, their only common features are an awareness of Einstein's general relativity and the presence of electromagnetism among the fundamental phenomena — electromagnetism, not gravitation, despite our restriction to the sections on relativity.

To illustrate how large is the spectrum of unified theories in 1920, consider the case of the well-known mathematician, Harry Bateman: in a letter to the editor of the *Philosophical Magazine* he draws aggrieved attention to his own priority in the creation of a theory of general relativity. This priority is based on an article of his, published in 1910, now generally described as giving a proof of the conformal invariance of the Maxwell equations, but interpreted by him as part of a unification program.

> My work on the subject of General Relativity was published before that of Einstein and Kottler, and appears to have been overlooked by recent writers. In 1909 I proposed a scheme of electromagnetic equations... which are covariant for all transformations of co-ordinates which are biuniform in the domain we are interested in. These equations were similar to Maxwell's equations, except that the familiar relations $B = \mu H, D = kE$ of Maxwell's theory were replaced by more general equations, which implied that two fundamental integral forms were reciprocals with regard to a quadratic differential form
>
> $$\sum \sum g_{m,n} \, dx_m \, dx_n,$$
>
> which was assumed to be invariant for all transformations of co-ordinates (pp. 219–220).
>
> The idea that the coefficients of the quadratic form might be considered as characteristics of the mind interpreting the phenomena was also entertained,... and it was suggested that a correspondence or transformation of co-ordinates might be employed as a crude mathematical symbol for a mind. ... If we assume that the nature of an electromagnetic field depends on the type of fundamental quadratic form, which determine the constitutive relations, and thus depends indirectly on a transformation which alters the coefficients of this quadratic form, this dependence may be a symbol for the relation between physical and mental phenomena instead of giving the influence of gravitation on light as in Einstein's theory.
>
> Einstein and the others have attempted to formulate a set of equations of motion which will cover all physical phenomena; but... the true equations of motion should be capable of accounting for the phenomena of life (pp. 220–221).

In this limiting case we have a theory where electromagnetism and mental phenomena, seen as basic forces, are linked together through a quadratic form, the mind operating through coordinate transformations; moreover, Einstein's theory itself is reinterpreted not as theory of gravitation alone, but as a unified theory, encapsulating light and gravitation.

Like Bateman, 11 of the 18 authors present autonomous theories and refer mainly to their own previous work. In some cases these approaches appear to be recent, launched within the previous few years, like that of Hermann Weyl alluded to in the introduction, or the theories proposed by Théophile De Donder and Henri Vanderlinden, or by Ernst Reichenbächer; in others, it consists of an attempt to reactivate older material, produced in another period, and refurbished in new, primarily geometrical, clothing. Such a case is Joseph Larmor, for whom his currently proposed five-dimensional unified field theory is an elaboration of ideas contained in his classic book of 1900, *Æther and Matter*. Again, Emil Wiechert's major references are to his own work of the late 1890s. One obvious reason for this rejuvenation is the 1919 eclipse report on the deviation of light in the Sun's gravitational field, which confirmed Einstein's previsions and thus motivated more than a few sceptics of general relativity and defenders of the old electromagnetic world-view[22] (Larmor and Wiechert, in particular) to readjust their own projects, and integrate these phenomena within them; for, in Wiechert's words, "A discovery has been made!"[23] (Wiechert I, 301) and this had to be accounted for.

In most cases, then, references serving to designate the approach used are of an idiosyncratic nature. The two main exceptions are De Donder's work, which is discussed and developed by three authors, and, above all, Weyl's theory, which succeeds in gathering around it a small group of papers by several authors. Besides this type of quotation, and the references to general relativity which we will explore more closely later, there are also some mutual citations of a vaguer nature, mention, for instance, of work considered as related in some way.

These allusions rather neatly divide our corpus into two main sets, which essentially correspond to the language(s) (and place) of publication: German on the one hand and French or English on the other. German-language papers largely dominate, with 20 articles by 10 authors, among which 10 (by 5 authors) appear in *Annalen der Physik* and 3 (by 3 authors) in *Physikalische Zeitschrift*;[24] on the other side, we find 5 articles in English (by 3 authors) and 7 in French (by 4 authors). A single paper in Dutch completes the scene.

The meaning of such a dichotomy needs to be elucidated in detail.[25] It is, at least partly, a consequence of the First World War and we will see the situation change in the later years of our *sondage*. But how does it operate? What divides the two sets is not a question of approach; the variety mentioned above runs through both groups. Electromagnetic programs are common to Wiechert and Lodge, variational principles are important for Weyl and De Donder, but not for Reichenbächer. The division mainly indicates the limits of *reciprocal visibility*; the authors on one side seem almost not to see those on the other; not only do they not explore their proposals, they do not even engage in discussions or debates on the same issues. First established by the examination of mutual references, this observation can be reinforced by a number of further details, within or outside our corpus. When, for example, Einstein corresponds with De Donder, or mentions the latter or Bateman in his correspondence with third parties, he is either vague or critical.[26] And while Larmor publishes in 1921 a eulogistic review of De Donder's first book (De Donder 1921) in the London *Times* (Larmor 1921), he makes no mention of Wiechert or Weyl. Similarly, in the later editions of his *Raum-Zeit-Materie*, Weyl, in a note concerning projects analogous to his own, refers to Wiechert and even to the evidently quite marginal Reichenbächer[27] — but never to De Donder, who might be considered to be scientifically closer. Leading to the same grouping are the brief historical introductions to some of the papers; the Mie, Hilbert, Einstein trilogy (in varying order) of founding fathers, which appears in German-speaking authors like Reichenbächer, Humm, or Einstein himself, is never mentioned in articles written in other languages.

An obvious counterargument seems to be offered by Einstein himself; his name is cited by every author. Furthermore, in a 1920 article in the *Berliner Tageblatt* (Einstein 1920), he himself writes: "The greatest names in theoretical physics, H. A. Lorentz, M. Planck, Sommerfeld, Laue, Born, Larmor, Eddington, Debye, Langevin, Levi-Civita, based their work on the theory [of general relativity] and have, in general, made valuable contributions to it," thus drawing the implicit contour of an international community of one heart and mind. However, Einstein's visibility concerns his theory of general relativity, not his 1919 paper: that is, Einstein is no exception *as an author in our corpus*. His 1920 article in the *Berliner Tageblatt* was written in the context of the conflict with the "Anti-Relativity Inc." group inside Germany; even if Einstein's sincerity about his internationalism is not in doubt, the fact is that he only very rarely uses or cites at this date non-Germanophone contributions (other than Eddington) to his theory; moreover the (essentially critical) views of general relativity and its extensions of someone like Larmor, for instance, makes his inclusion seem more a move to extend the list than a reflection of serious scientific interaction.[28]

Let us then examine first the German-speaking landscape. The circle of authors, it should be pointed out, is surprisingly narrow. All have some close connection with Göttingen, with the exception of Reichenbächer — and, to some extent, of Mie and Einstein, whose works however were adopted and promoted by the Göttingen mathematicians (Corry 1999). Wiechert, of course, was professor of geophysics there and co-directed the Göttingen electron seminar of 1905 with Hilbert and Minkowski (Pyenson 1979), Weyl and Humm were students of Hilbert, Dällenbach in turn was a student of Weyl's in Zürich, Arthur Haas had studied in Göttingen in the first decade of the century. Wiechert, Roland Weitzenböck and Wolfgang Pauli collaborated in the *Encyclopädie der mathematischen Wissenschaften*, an enterprise tightly connected with Göttingen mathematician Felix Klein's perspective (Tobies 1994). Pauli, furthermore, was directly linked to Weyl through Sommerfeld and Einstein — a connection which permitted Weyl (I) and Pauli to make reference to each other, their papers having been exchanged as preprints. We have further testimony of direct, professional or personal encounters between some of our authors; for instance, the discussion of Weyl's communication at the Bad Nauheim conference on relativity (Weyl II), showcases Einstein, Pauli, Reichenbächer and Mie.

A closer look at the articles suggests however a finer delineation. A first set of papers, to which we have already alluded, cluster around the exploration of Weyl's theory. It includes in particular 2 articles by Weyl himself, 2 articles by Weitzenböck, 1 by Pauli, as well as the 2 expository papers of Haas; to these may be added 2 of the 5 papers by Reichenbächer (IV–V), which compare Weyl's theory to the author's own. Weyl's theory, the usual prototype for a unified field theory, has been well studied.[29] Let us here only recall that Weyl, in an attempt to develop a true "geometry of proximity," that would be adapted to a physics excluding action at a distance, extends the (pseudo-)Riemanian geometry of general relativity; in Weyl's geometry, not only can the direction of vectors not be transferred from one point to another independently of the path taken, but neither can their lengths. On the other hand, Weyl retains an extra condition of "gauge"-invariance for the laws of nature, besides the general covariance of Einstein's theory: the line-element $ds^2 = g_{ij}dx^i dx^j$ at a point $P(x^i)$ becomes $ds^2(1 + d\phi)$ at $P(x^i + dx^i)$, with $d\phi = \phi_i dx^i$. While the g_{ij}, as in Einstein's theory, are to be associated with the gravitational field, Weyl proposes identifying the linear metric element ϕ_i with the electromagnetic four-potential.

> Then not only gravitational but also electromagnetic forces would spring out of the world metric; and since no other truly fundamental forces other than these two are known to us, through the resulting theory, in a strange, unforeseen way, Descartes' dream of a purely geometrical physics would be fulfilled. In this it appears that physics, with its conceptual content, does not at all extend beyond geometry; *in matter and natural forces only the metric field reveals itself.* Gravitation and electricity would thus be accounted for in terms of a unified source[30] (Weyl 1, 112).

Weyl presented his theory in 1918, but criticisms on physical grounds by Einstein, among others, moved him to modify this initial version in two ways, both explained in his 1919 articles. The first is to separate the actual procedure of measurement by rods and clocks from the ideal procedure associated with parallel transport and basic to his geometry.[31] The other consists of exploring more precisely the field equations; Weyl derives them from a variational principle applied to the invariant integral $\int \mathfrak{W} \, dx$ for a specific Lagrangian $\mathfrak{W}$ ("the action function"), selected on formal and philosophical grounds, and he obtains both Maxwell's equations and new gravitational equations, different from Einstein's, and allowing a closed world without recourse to a cosmological term.

Both Pauli and Weitzenböck, as well as Weyl, explore in 1919 specific choices of the action function leading to the field equations. But they do so in a fashion which illustrates perfectly their differences in perspective. Weitzenböck's work is a search for the *complete* list of invariants of the theory.[32] Pauli, on the contrary, makes his choice among the possible action functions by examining their *physical* consequences, in particular, in the static case, stating independently some of Weyl's 1919 results. He shows that the field equations are symmetric with respect to the two kinds of electricity, positive and negative — a circumstance he sees at the time as an important drawback to the theory since no positively charged particle of electronic mass was then known — and that Weyl's equations can lead to a correct value for the advance of Mercury's perihelion. He also studies the problem of the electron, that is the possibility of static, spherically symmetric solutions.

On a more popular level, the two essays by Haas plead, from the tribune offered by Weyl's theory, for the reduction of physics to geometry. Haas acts as an unofficial spokesman for Hilbert's approach, including the advocacy of an axiomatic program for physics; this, he claims, is the essence of the Einsteinian revolution.[33] Haas concludes his interventions with the comment that now that gravitation and electromagnetism have already been taken care of in this approach, the next step should be the introduction of a discontinuous geometry in order to integrate matter.

> Thus all physical laws are reduced to the single problem of the metric of a four-dimensional space-time manifold. ... One of the most important tasks for the future in this respect ... is certainly the introduction of quantum theory into general relativity.

> To handle this problem, physical axiomatics must clearly enter into a thought that ... Riemann... had expressed: that the object of geometry could also be a *discontinuous manifold*. ... But if the ... manifold itself were taken as discontinuous, then it would be understandable why the quantity of action that appears in given physical processes necessarily must be an integer multiple of an *elementary quantum of action*[34] (Haas I, 749).

Haas is not isolated in his concern with quanta. Indeed, Weyl himself, in the preface to *Raum, Zeit, Materie*, evokes quantum physics as a possible boundary for his theory. At Bad Nauheim, in a question following Weyl's talk (Pauli *apud* Weyl II, 651), but aimed at Einstein as well, Pauli begins to challenge the capacity of a purely classical field approach (i.e., a continuous theory) to deal with the situation inside the electron, and as a consequence, to doubt the validity of Einstein's and Weyl's unification programs; from this time on, Pauli, as is well known, turned his efforts in quite another direction.[35]

Besides these explorations of a single theory, we find, within the German-speaking group in our corpus, several papers which share among themselves, and in common with those just discussed, one specific feature: they refer to Einstein's theory of gravitation in a positive way, integrating it as a part of their program or at least as an horizon bounding it. They are distinguished, however, by quite different approaches to matter: its nature, its physical properties and representations, and the way it interacts with gravitation.

The nearest to Weyl's program, in terms of objectives and of the role played in a communication network, is that of Einstein himself, in his first published attempt at tying gravitation and electromagnetism together more efficiently than in his 1915 theory, by proposing the second modification of his field equations in three years. Two problems have led him to doubt his original field equations. The cosmological question had obliged him (Einstein 1917) to add an extra cosmological term in order to have a closed, static model of the universe; now the problem of matter, the need to derive a solution corresponding to an electron, motivates his new approach. Leaving the (pseudo-)Riemannian geometry without change, he puts forward as his new equations for the case of gravitational and electromagnetic fields,

$$R_{ik} - \frac{1}{4} g_{ik} R = -\kappa T_{ik}^{\text{EM}},$$

R_{ik} being the Ricci tensor, R the scalar curvature and T_{ik}^{EM} the Maxwell energy tensor of the electromagnetic field. The new coefficient 1/4 is intended to kill two birds with one stone: to provide the cosmological term more naturally, as a constant of integration, and to yield a regular, static, spherically symmetric solution that will represent the electron. The indeterminacy of the solution was to push Einstein to explore other theories in the following years (Ritter 1993), though keeping always the same central ambition; the recuperation of matter from the interlinkage between gravitation and electromagnetism.

A quite different approach to the problem of matter is to be seen in the papers which constitute Dällenbach's thesis (Dällenbach I and II). He operates within the framework of a flat Minkowski space, but, he incorrectly believes, only as a technical simplification which "easily" generalizes to a generally covariant theory. Dällenbach essentially recasts Lorentz's electron-theoretic approach to electromagnetism (Lorentz 1904) in an Einsteinian mold, which allows him to derive the (phenomenological) constitutive equations of Maxwell as spatial averages integrated over properties of the electronic constituents of matter.

Gustav Mie, too, explores the interface between Einstein's theory of gravitation and classical electromagnetic questions.[36] He first finds the key to the "wonderful and consummately beautiful mathematical structure"[37] (Mie, 653) of Einstein's theory, and the clarification of its role in any future unified theory, in what he calls the "generalized principle of the relativity of gravitational action." This he understands as the possibility of transforming, through an appropriate choice of coordinates, the study of a moving test body in a gravitational field into one of a body at rest in a space with a non-Minkowskian geometry. Mie uses this principle to study the following paradox: Einstein's theory predicts that there can be no electromagnetic radiation from a charged particle moving in a gravitational orbit in empty space, in agreement with what is expected from Bohr's quantum theory; Maxwell's equations assure us of the contrary. Using the Schwarzschild metric to provide the geometry of space in which the transformed particle is at rest, a generalized Laplace electrostatic equation for the resulting Coulomb potential and a particular cylindrical coordinate system borrowed from the work of Reichenbächer (1917 and I–III) to transform the result back into the physical rotating situation, Mie exhibits a solution in Einstein's framework which indeed does not radiate. He is then able to identify it, in the far field, with a particular Maxwell equation solution, but one which represents a moving particle surrounded by a spherical standing wave, such as would be provided by a purely reflecting shell. Conversely, a radiating solution can be created for empty space in the Einstein theory, but only at the price of adding to the original potential one representing a sourceless electric rotation field; happily this turns out to be necessary in any case to provide conservation of energy.

Finally, two of our German-speaking authors — Reichenbächer and Wiechert — do not accept general relativity. They prefer to develop their own approach, in each case an electromagnetic reductionism, to take into account the new results brought in by Einstein's theory.[38] Their similarities stop here; there is no unified front in these alternative attempts, and their conception of matter, as well as the techniques they use, link them more directly to other authors previously discussed than to each other.

Ernst Reichenbächer situates his work within the Einsteinian geometric tradition, but seeks to base it on a direct expression of electronic properties, thus avoiding the phenomenological features he reads in both Einstein's and Weyl's proposals for the determination of the metric:

> Contrary to ordinary intuition, the mass density appears [in general relativity] not as a scalar, but as the 44-component of a sixteen-component tensor .... This and the fact that, because of their dependence on the choice of coordinate system, the $g_{\mu\nu}$ are nevertheless subjected to a restricted arbitrariness, did not please me in Einstein's theory. Therefore, in my [1917] article "Characteristics of a Theory of Electricity and Gravitation," I attempted to set up the theory of a scalar gravitational potential, which I identified with the speed of light and where I introduced certain conditions on the gravitational perturbation by electrons — positive and negative — which I saw as the only kind of matter. I completed

the simplest case ... of a single electron and set up the general equation

$$\mathfrak{K} = 2 \operatorname{Div} \operatorname{Grad} \lg l,$$

following the analogy of this case[39] (Reichenbächer I, 1).

More precisely, for the case of one electron, Reichenbächer builds his metric by gluing: far from the electron, the metric is supposed to be Minkowskian; inside the electron, the deviation of the metric from its Minkowskian values is interpreted as a rotation, of which the angles, associated with the 6 pairs of coordinate lines, are given by the components of the electromagnetic 6-vector, the first set of Maxwell equations providing exactly the required compatibility conditions. The second set of Maxwell equations is used to derive the fundamental equation quoted above. The gravitational force is then associated with the variable time-component of the metric, and appears as a (variable) velocity of light.[40] In his 1919 and 1920 papers, Reichenbächer generalizes his construction by recurrence for a finite number of electrons and derives from his theory an advance for the perihelion of Mercury and a deflection of light in a gravitational field, both one-half the general relativistic value. He also adapts his theory to the Weyl approach, in order to obtain covariant and gauge-invariant laws of gravitation and electromagnetism:

> It is then possible ... to arrive at a ... solution of the world problem ... by taking a realistic point of view instead of the more phenomenological one of the relativist[41] (Reichenbächer IV, 113).

Another dissident is Emil Wiechert, for whom, as for Reichenbächer, matter is of electromagnetic origin, here comprised of ether and electrons. Wiechert, with titles like "Gravitation as an Electrodynamic Phenomenon" (I) and "Remarks on an Electrodynamic Theory of Gravitation" (II) clearly announcing his program, is quite explicit:

> The foundation of the theory should be the acceptance that molecular matter is built up out of electrical particles. This then explains electrification as a basic property of all the building blocks of matter. The assumption appears as the natural consequence of the results of the molecular-physical research of the last three decades. Once having recognized through electrodynamics that electrification is an essential cause of *inertia*, the proof should now be sought that electrification is also an essential cause of *gravitation*[42] (Wiechert I, 331).

In contrast to Reichenbächer, his theory does not use any metric; he had already, in 1916, suggested an alternative Lagrangian to that of general relativity in order to compute the advance of the perihelion of Mercury, interpreting it also as a variable speed of light. The purpose of his 1919 papers is to obtain an electromagnetic theory of gravitation which could explain all the experimental results obtained by the Einstein theory. Starting from the Lagrangian $L = -2/3 \, (e^2/a)\sqrt{1 - v^2/c^2}$, an idea borrowed from Abraham (1902), he derives new field equations with two free parameters, which he then tries to evaluate on the basis of experimental data. He points out that his hope is to ultimately derive these values *ab initio* from a theory of the structure of ether, though this he has not yet been able to obtain.

If we now turn to the non-German language articles, further groupings are recognizable, though of a very different nature. The Cambridge trio, Lodge, Larmor and

Bateman, whom we have already met at the beginning of this section, share not only an interpretation of general relativity as a unified theory of gravitation and light, but also a number of mathematical techniques, in particular a stress on changes of coordinate systems, coming out of long-established ways of working in physics. But they do not share a well-defined project nor do they cite each other on the issue of unification.[43] On the contrary, De Donder, Vanderlinden and the Toulousan mathematician Adolphe Buhl, among the authors of papers in French, do participate in such a project. All explore De Donder's theory, putting the emphasis on analytical, not geometrical, techniques; they are convinced these will provide a common formal machinery which could uniformly accommodate different natural phenomena. For the sake of space, we shall briefly present only one example of each group.

A geometry with five dimensions is the framework suggested by Larmor (I) to englobe electromagnetism and gravitation.[44] Commenting on the work of a young mathematician who had rediscovered Clifford algebra, Larmor shows, in an article submitted in the summer of 1919, how such a mathematical framework might be used to give a new geometrical interpretation to the special theory of relativity, in which the electromagnetic field would be a four-dimensional flat surface embedded in a five-dimensional space. And then the 6 November news of the results of the eclipse expedition, and its verification of the general relativistic deflection of light, arrives prior to publication. Two weeks later, Larmor has written a generalization which allows gravitation an entry into this picture:

> *Note added 20 November 1919* — ... The phenomena of gravitation have been included by Einstein in this Minkowski scheme by altering slightly the expression for $\delta\sigma^2$. ... This generalisation can be still be brought within the range of the elements of the Clifford geometry ... by introducing into the analysis a new dimension (ξ), preferably of space; so that
> $$\delta\sigma^2 = \delta x^2 + \delta y^2 + \delta z^2 + \delta\xi^2 + \delta w^2, \qquad w = ict.$$
> ... Now any continuum of four dimensions, having a quadratic line-element, however complex, is expressible as a hypersurface in this homaloid [flat] continuum of five dimensions. If these considerations are correct, the Einstein generalization, made with a view to include gravitation within his four dimensions, must be interpretable as the geometry of some type of hypersurface constructed in this extended homaloid of five dimensions.
>
> ... Thus we postulate a fivefold electrodynamic potential ... in the Euclidean auxiliary space (x, y, z, ξ, ict). Then any section of this space and its vector-system is a hypersurface of four dimensions ... and represents a possible electrodynamic world process; including implicitly its gravitation, which would become apparent only when the hypersurface, actually already nearly flat, is forced into representation on a hyperplane (Larmor I, 353–354, 362).

In other words, electromagnetism was to provide the metric — the first fundamental form — of the four-dimensional embedded surface, while gravitation was to describe this embedding, i.e., determine the second fundamental form.

Théophile De Donder had, in 1914, proposed a theory analogous, as he claimed on various occasions, to Einstein's general relativity. In the following years he will devote numerous books and papers to the presentation and development of a "théorie de la gravifique," either his own, Einstein's or, later, Weyl's. He defines this theory as the study of relationships between a "twisted space-time" (which for some applications can be a flat Minkowski space) and different fields, either electromagnetic or "material"; this last category including, in opposition to Einstein, gravitational properties of matter. In particular, the geometrical setting, that is, the very definition of the spacetime, is a given, to be exhibited at the beginning of each paper, the work itself bearing on essentially analytical aspects. Here, De Donder, with the help of Vanderlinden, lays special emphasis on a modified variational principle formalism, used to obtain Einstein's field equations by means of ordinary calculus without recourse to tensor analysis; the aim is to develop this principle with as much generality as possible, and then specialize it for the treatment of various phenomena, one example being the inclusion of Poincaré stresses in the construction of atoms.[45]

1920 Corpus

Bateman, Harry
"On General Relativity" [Letter to the editor, 10 August 1918]. *Phil. Mag.* (6) 37 (1919) 219–223 [P70; J812]

Bloch, Léon
"Remarque sur la théorie de Lorentz comparée à celle de Mie." *CRASP* 171 (1920) 1379–1380 [J805]

Buhl, Adolf
I   "Sur les symétries du champ électromagnétique et gravifique." *CRASP* 171 (1920) 345–348 [J804]
II "Sur la formule de Stokes dans l'espace-temps." *CRASP* 171 (1920) 547–549 [J804]
III "Sur les symétries du champ gravifique et l'extension lorentzienne du principe d'Hamilton." *CRASP* 171 (1920) 786–788 [J804]

Dällenbach, Walter
I   "Die allgemein kovarianten Grundgleichungen des elektromagnetischen Feldes im Innern ponderabler Materie vom Standpunkt der Elektronentheorie." *Ann. Phys.* (4) 58 (1919) 523–548 [P126; J792]
II "Hamiltonsches Prinzip der elektromagnetischen Grundgleichungen im Innern ponderabler Materie." *Ann. Phys.* (4) 59 (1919) 28–32 [P127; J793]

De Donder, Théophile
"Le Tenseur gravifique." *Versl. Kon. Akad. A'dam* 27 (1918/19) 432–440 [J803]

De Donder, Théophile and Henri Vanderlinden
I   "Théorie nouvelle de la gravifique." *Bull. Acad. Roy. Belgique* (5) 6 (1920) 232–245 [J803]
II "Les Nouvelles Équations fondamentales de la gravifique." *CRASP* 170 (1920) 1107–1109 [P1178; J804]

Einstein, Albert
"Spielen Gravitationsfelder im Aufbau der materiellen Elementarteilchen eine wesentliche Rolle?" *SPAW* (1919) 349–356 [P193]

Haas, Arthur
I   "Die Axiomatik der modernen Physik." *Naturwiss.* 7 (1919) 744–750 [P520]
II "Die Physik als geometrische Notwendigkeit." *Naturwiss.* 8 (1920) 121–127 [P518]

Humm, Rudolf Jakob
  "Über die Energiegleichungen der allgemeinen Relativitätstheorie." *Ann. Phys.* (4) 58 (1919) 474–486 [J792]
Larmor, Joseph
I "On Generalized Relativity in Connection with Mr. W. J. Johnston's Symbolic Calculus." *Proc. Roy. Soc. London* 96 (1919) 334–363 [P123]
II [Report on Meeting of the Royal Society, 20 November 1919.] *Nature* 104 (1919) 365 [P456]
Lodge, Oliver
I "Connexion between Light and Gravitation." *Phil. Mag.* (6) 38 (1919) 737 [P791; J811]
II "Gravitation and Light." [Letter to the editor, 30 November 1919]. *Nature* 104 (1919) 354 [J812]
Mie, Gustav
  "Das elektrische Feld eines um ein Gravitationszentrum rotierenden geladenen Partikelchens." *Phys. Z.* 21 (1920) 651–659 [J786]
Nordström, Gunnar
  "Opmerking over het niet Uitstralen van een overeenkomstig kwantenvoorwaarden bewegende elektrische Lading." *Versl. Kon. Akad. A'dam* 28 (1919/20) 67–72 [J806]
Pauli, Wolfgang, jr.
  "Zur Theorie der Gravitation und der Elektrizität von Hermann WEYL." *Phys. Z.* 20 (1919) 457–467 [J791]
Reichenbächer, Ernst
I "Das skalare Gravitationspotential."*Ann. Phys.* (4) 61 (1920) 1–20 [P457; J794]
II "Die Krummung des Lichtstrahls infolge der Gravitation." *Ann. Phys.* (4) 61 (1920) 21–24 [P457; J794]
III "Die Punktbewegung im allgemeinen Gravitationsfelde." *Ann. Phys.*(4) 61 (1920) 25–31 [P457; J794]
IV "Über die Nichtintegrabilität der Streckenübertragung und die Weltfunktion in der Weylschen verallgemeinerten Relativitätstheorie." *Ann. Phys.* (4) 63 (1920) 93–114 [J794]
V "Die Feldgleichungen der Gravitation und der Elektrizität innerhalb der Materie." *Ann. Phys.* (4) 63 (1920) 115–144 [J795]
Weitzenböck, Roland
I "Über die Wirkungsfunktion in der Weyl'schen Physik. I." *SAW Wien* (2) 129 (1920) 683–696 [J784]
II "Über die Wirkungsfunktion in der Weyl'schen Physik. II." *SAW Wien* (2) 129 (1920)697–708 [J784]
Weyl, Hermann
I "Eine neue Erweiterung der Relativitätstheorie." *Ann. Phys.* (4) 59 (1919) 101–133 [P257; J782]
II "Elektrizität und Gravitation." *Phys. Z.* 21 (1920) 649–651 [J784]

Wiechert, J. Emil

I   "Die Gravitation als elektrodynamische Erscheinung." *Ann. Phys.* (4) 63 (1920) 301–381 [J789]

II   "Die Gravitation als elektrodynamische Erscheinung." *Nachr. Göttingen* (1920) 101–108 [J799]

III   "Bemerkungen zu einer elektrodynamischen Theorie der Gravitation." *Astronomische Nachrichten* 211 (1920) 275–284, 287–288 [J790]

3. 1925

The selection for this second year of our *sondage*, 1925, consists of 34 articles, written by 21 authors. The corpus, thus, is quite similar quantitatively to that obtained for 1920, but this apparent resemblance only underlines the limits of a purely quantitative approach; the situations in 1920 and in 1925 are very different indeed.

The first striking difference is the internationalization of the topic. Besides the languages and countries represented in 1920, we now find authors publishing in Japanese, American, Hungarian and Italian journals; moreover, the impressive domination in 1920 of a single outlet, the *Annalen der Physik*, has disappeared. Indeed, not a single paper from this journal appears in our list for 1925. In its place we find in particular 4 papers (2 authors in each case) in the *Physikalische Zeitschrift* and the *Notes aux Comptes rendus de l'Académie des sciences de Paris*, 3 (by 3 authors) in *Nature*, *Zeitschrift für Physik* and *Physical Review*, this last in the context of abstracts from meetings of the American Physical Society. A further element of this internationalization is the translation into a number of languages of two reference books, Weyl's *Raum-Zeit-Materie* and Eddington's *The Mathematical Theory of Relativity* of 1923; a number of our papers refer to them for their basic notation and an introduction to their main tools. This new distribution, however, does not mean a homogenous visibility and the place of publication is still a good marker for the readership and use of an article.[46] Yet some of the newcomers have a broad perspective on the various trends; significant in this respect, even if on quite an exceptional scale, is the paper of Manuel Sandoval Vallarta, a Mexican physicist working at the time at MIT, who quotes, with equal ease, De Donder and Vanderlinden as well as Weyl and Einstein, Bateman as well as the German quantum theorists Max Born, Werner Heisenberg and Pauli.

A second crucial difference with 1920 is that there are fewer authors (6) claiming that they pursue their own, personal theory. Most of the authors place themselves quite consciously in one of a few major traditions (or some combination of them). The principal one is now standardly referred to as the (Einstein-)Weyl-Eddington theory, in one of its variants; in 1925 we find, for example, papers following Weyl's theory (Vallarta, Eyraud I and Reichenbächer), Eddington's (Rice I, II), or Jan Schouten's and Élie Cartan's work (Eyraud II, III).[47] Even De Donder now presents his work as an extension of the Weyl-Eddington-Einstein trend, while it also constitutes an extension of his own work of 1920. What renders this difference with the situation in 1920 ambiguous is that there are almost as many traditions as papers; the explanation is that the kernel of papers is not so closed in 1925 as in 1920. The 1925 articles more frequently refer either to earlier (1921–1924) authors — who thus do not appear directly in our *sondage* — or to works which were reviewed in other sections of *Physikalische Berichte* and *Jahrbuch* ("differential geometry" for Cartan's papers or "quantum theory" for Bateman's, for example). The regrouping of the intellectual traditions at work, and the more elaborately structured organization of the programs, are thus counterbalanced by an enlargement of the possible sources for the theories that are to be adapted, mixed, or developed.

Two main technical innovations, to which we have already alluded, appear in these papers. The first concerns Eddington's generalization of Weyl's ideas (Eddington 1921), to initially posit a connection rather than a metric as the fundamental element of the theory. The connection Γ^i_{jk}, a concept borrowed by Weyl and then by Eddington from Levi-Civita's work (1917),[48] describes the parallel displacement of a contravariant vector A^i along a curve x^s, that is $dA^i = -\Gamma^i_{rs} A^r dx^s$. The Riemann and Ricci tensors could then be defined directly in terms of it. In the classical case of general relativity, the connection is given by the (symmetric) Christoffel symbol, defined by the metric and its derivatives; the Ricci tensor is then symmetric. Starting with a general (though still symmetric) connection allowed Eddington — and Einstein following him in 1923 (see Einstein 1923a) — to obtain a non-symmetric Ricci tensor; its antisymmetric part could then be taken as a representation of the (antisymmetric) electromagnetic field tensor.

This work led to the idea of relaxing in turn all the constraints on symmetry and on the relationship between the connection and the metric in order to find room to house electromagnetism; some of these further possibilities are explored in the 1925 corpus. For instance, the young Japanese physicist Bunsaku Arakatsu considers two symmetric connections, one of which he associates with the metric, and thus with the "geometry" of the space, the other with the "physics" of the space.[49] He then sets the Riemann curvature tensor built from the second connection to zero, interpreting this as saying that the space-time is "physically flat." This, in turn, implies that the curl of a certain vector, linked to the difference between the two connections, is also zero and thus that the vector can be chosen to encapsulate the electromagnetic field. Henri Eyraud, to whose work we shall return, following Schouten 1923, uses an asymmetric connection, whose antisymmetric part incarnates the electromagnetic tensor. Einstein introduces, in I, both a general metric and an independent general connection, the antisymmetric part of the metric serving to represent electromagnetism; his field equations are then obtained by *separate* applications of a variational principle to the metric and the connection.

The second new feature is the appearance of quantum theory as an integral part of some unification schemes. Though quantum matter had already appeared as a programmatic goal for a few authors (Haas, Mie, De Donder) at the beginning of the decade, there are in 1925 — and have been since 1922— more technical attempts to combine gravitation with the 'old' quantum theory of Bohr (1913) and Sommerfeld (see Sommerfeld 1919), specifically using their rules for determining the orbital parameters of electrons in hydrogenic atoms. While 8 of our papers take quantum theory as a main component of the unification, the compatibility with quantum effects, or the possibility of including quanta in a theory unifying gravitation and electromagnetic phenomena, is alluded to in more than a dozen others.

Let us now return to the relationships between these articles. As already indicated, a first kind of reference, that used to situate the article's immediate antecedents, structures our corpus into a number of short-term traditions; their choice can in some cases be linked to a direct, personal or institutional, relation, but there is no systematicity in the process. Another kind of citation is used, for instance, in dealing with a technical point, but, unlike the 1920 corpus, the 1925 papers do not permit an evaluation of their importance for the formation of social configurations. Finally, a third type of reference, more discursive and prominent in the review articles and letters to the editor of journals (6 such in 1925), as well as in the introductions to some of the other articles, allows us to scrutinize how the authors themselves represent the organization of the topic, how they envision the main choices offered to them. From this last, we obtain not one classification, but two: one based on the role that matter plays and the nature of its representation, the other on the degree of geometrization.

Structuring unified theories according to this double criteria may seem familiar; but we would like to clarify its specific historicity. First of all, while such classifications, of course, could be applied retrospectively, for example to 1920, our point is that they were put forward by only a minority of authors at this moment and that they were not operational in configuring the links among the articles. For instance, in 1920, the opponents of relativity theory were not necessarily explicit opponents of a geometrical approach, nor the contrary.[50]

Secondly, these two classifications are today often rigidly articulated: geometry associated with a preference for a field approach and a continuous conception of matter, quantum program seen basically as non-geometric.[51] If this identification may be relevant for later periods and will be explicitly promoted by both partisans of quantum theory and of general relativity, we would like to stress emphatically that it does not fit our 1925 corpus. We find, on the contrary, all combinations: the most analytical, non-geometric aspect of general relativity associated either with quantum theory (Kudar I, II,III) or with a concern for purely classical matter (Reißner I, II); geometrization of gravitation and electromagnetism without any reference to matter (Arakatsu) or linked to an ambitious quantization program (Vallarta). We will thus discuss and illustrate separately both axes.

The first alternative, then, is articulated around the opposition between those programs that include quanta as an integral part of the unification scheme and those based on a unification of gravitational and electromagnetic fields from which one hopes to derive a theory of matter. The two positions are well described at the beginning of a paper by Hans Reißner:

> Either one holds that time has yet not come and defers the solution [of the problem of electron and nucleus] until the perhaps identical sources of the still more mysterious quantum laws attached to the names of Planck, Einstein and Bohr are uncovered.
> Or one holds that a solution created on the foundations of Maxwell, Lorentz, Mie, Einstein, Hilbert, Weyl, etc. is possible[52] (Reißner I, 925).

Reißner, who situates himself in the second camp, tries to derive some essential properties of the electron and proton within the Einstein-Weyl framework by introducing into the electromagnetic source term an auxiliary non-Maxwellian tensor. Another attempt that had been made by Einstein in 1923 (Einstein 1923a) is reported on in an expository paper by Rolin Wavre. As Wavre explains it, Einstein had tried to derive the behavior of matter from the field, by providing extra partial differential equations in order to overdetermine the initial state, that is to be able to "define quantum physics inside relativistic physics like a species in a genus by the adjunction of a specific character" (Wavre, 300).[53] Einstein's paper in our corpus (Einstein 1), that we have mentioned above — as well as Einstein's 1919 attempt discussed in the preceding section — belongs to the same trend.[54] After his derivation of the field equations for the metric and the connection, he presents, as a future question to be settled, the possible appearance in this theory of an electrically charged mass with spherical symmetry and without singularity, a solution that would represent an electron.[55]

An example of the opposite position — that of treating quantum theory as a domain entering into the unification from the outset — is given by the series of three articles by Johann Kudar. Their aim is to show how the Sommerfeld quantum relations for conditionally periodic systems and the Bohr frequency relation can be used in the context of general relativity to derive the gravitational redshift law from first principles, without any use of the Einstein assumption that the proper frequency rate of atomic clocks is independent of the gravitational field. The result is thus established for those areas in which the "old" quantum theory gives reasonable results — hydrogenlike atomic spectral series and Deslandres-Schwarzschild band spectra — and Kudar clearly feels the way is open towards a program of replacing the phenomenological aspects of relativity theory with exact quantum mechanical principles.

An even more explicit promoter of this position is Vallarta, who announces a program to integrate Einstein's theory of gravitation and Bohr's approach. Vallarta's aim is to counter the objection that Sommerfeld's treatment of the fine structure of matter is not derivable in a unique way from special relativity by offering a well-defined approach from the standpoint of general relativity, which reduces to Sommerfeld's results in a weak gravitational field. He uses Nordström's solution for a static charged particle and the Weyl-Eddington equation of electronic motion to show that the curvature of the material field associated with the nucleus almost vanishes and that its field is nearly static, allowing him to treat an atom as a relativistic one-body problem. The paper, presented at the Annual Meeting of the American Physical Society, ends with a promise of a future, more thorough unification of quantum theory and general relativity, one which would resolve in particular the then pressing quantum problem of "unmechanical orbits" in atoms of more than one electron.[56]

We come now to the second alternative, that which concerns the relationship between geometry (and, more widely, mathematics) and physics. Though we have met this question first in Haas' 1920 paper and in Freundlich's answers, it seems no longer to be an important point of controversy for German-speaking authors in 1925. On the other hand, the issue is now quite acute among the English specialists, the two main camps being incarnated, to caricature only slightly, by Eddington on one side and Lodge on the other.[57] For example, a letter from Lodge to the editor of *Nature* comments, with his typical irony, on a 'pro-geometric' and 'anti-ether' lecture by James Jeans on the "present position in physics."

> Dr Jeans makes it clear that in his view the terms ether and force are unnecessary since all that they connote can be represented equally well by pure geometry ... It is marvellous

what hyper-geometry can be made to express, and what high reasoning about reality can be thus carried on (Lodge, 419).

This distrust of geometry is also expressed by a mathematician, Alfred North Whitehead. He comments favorably on George Temple's lecture at the Physical Society of London, in these terms:

> In investigating the laws of nature what really concerns us is our own experiences and the uniformities which they exhibit, and the extreme generalizations of the Einstein method are only of value in so far as they suggest lines along which these experiences may be investigated. There is a danger in taking such generalizations as our essential realities, and in particular the metaphorical "warp" in space-time is liable to cramp the imagination of the physicist, by turning physics into geometry (Whitehead *apud* Temple, 193).

An opposing position is defended by James Rice, whose paper (I) begins with a resume of some of Eddington's theses, in particular his principle of identification:

> If any further advance be made in physical science, conforming presumably to the Principle of Relativity and therefore involving the introduction of fresh tensors in its mathematical formulation, it may be possible to discover geometrical tensors possessing by virtues of identities just the same properties as the newly introduced physical tensors possess by reason of experimental facts (Rice I, 457).

This principle suggests then to Rice a system of natural units,[58] about which he states in his introduction that

> such an application finds but little favour in certain quarters where it is described as a geometrization of physics. ... The radius of curvature of [the Einstein-Eddington world] should be the natural unit of length on which the units employed in the geometrical tensors should be based. But if there be an underlying connexion between geometrical and physical tensors, such a fundamental unit of length might conceivably lead to the discovery of natural units in which to measure physical quantities (Rice I, 457–458).

To accord the preeminent place to geometry — or, inversely, to physics — is not only a theme for philosophical debate, it is put into action in the scientific work itself. But the ways in which this comes about vary and it is difficult to delineate possible corresponding solidarities; both physicists and mathematicians appear on each side, neither position seems to be linked to a specific group or place. Still, we may detect subtler links between the role accorded to geometry and the task of unification itself. To illustrate this, we shall present four cases, two in which the privileged place falls to geometry and two to physics.[59] We have chosen them to be as close as possible to each other in order to bring out the exact point at which this hierarchizing intervenes;[60] all four use the basic principles of Riemannian geometry, set themselves the task of integrating gravitational and electromagnetical fields and ignore the problem of the constitution of matter.

An extreme position is held by the mathematician George Rainich (I–III), who presents a new departure for unified field theory with his claim that general relativity already contains a unification of gravitation and electromagnetism.[61] After having underlined that, in standard general relativity,

> gravitation may be said to have been 'geometricized' — when the space is given, all the gravitational features are determined; on the contrary it seemed that the electromagnetic

> tensor is superposed on the space, that it is something external with respect to the space, that after space is given the electromagnetic tensor can be given in different ways (Rainich III, 106).

and pointed out that the attempts by Weyl and others to remedy this situation have failed as physical theories, Rainich remarks that the tensors of gravitation (the Einstein tensor G^i_j) and electromagnetism (the electromagnetic tensor F^i_j and its dual $^*F^i_j$) are connected through the "energy relation,"

$$G^i_j = F^i_s F^s_j - {}^*F^i_s {}^*F^s_j$$

and proceeds to study this relation more closely in the framework of classical Riemannian geometry, making essential use of the algebraic classification of tensors.

> The result of this study is quite unexpected; it is that under certain assumptions, the electromagnetic field is entirely determined by the curvature of space-time; ... without any modifications it takes care of the electromagnetic field, as far as "classical electrodynamics" is concerned (Rainich III, 107).

Our second example, Eyraud (II, III), is drawn from within the Weyl-Eddington affine tradition and shows how a maximal form of geometrization can be displayed by means of term-by-term identification between geometrical and physical magnitudes — that is, through the Eddington principle.[62] Eyraud's point of departure, as we have pointed out earlier, is an asymmetric connection, Γ^j_{ik}, whose antisymmetric part, Λ^j_{ik}, is the torsion, in Cartan's sense, of the four-dimensional space-time; R_{ik} as usual denotes the Ricci tensor. Eyraud *defines* the electromagnetic field as

$$E_{ik} = \frac{1}{2}\left(\frac{\Gamma^s_{is}}{x^k} - \frac{\Gamma^s_{ks}}{x^i}\right)$$

and the gravitational field as

$$K_{ik} = \frac{1}{2}(R_{ik} + R_{ki}).$$

Then the field equations are obtained from a variational principle applied to a *unique* Lagrangian, constructed as a function of E_{ik} and K_{ik}. Eyraud shows in particular that there exists a covariant vector Λ_k, such that the torsion can be written

$$\Lambda^j_{ik} = -\frac{1}{3}\left(\delta^j_i \Lambda_k - \delta^j_k \Lambda_i\right),$$

and that the symmetric part of the connection is given by

$$\frac{1}{2}\left(\Gamma^j_{ik} + \Gamma^j_{ki}\right) = \left\{{j \atop ik}\right\} + \frac{1}{6}\left(\delta^j_i \Lambda_k + \delta^j_k \Lambda_i\right).$$

In other terms, he proves that the connection is semisymmetric (an *hypothesis* in Schouten 1923) and also that the space-time has the same geodesics as a Riemannian space. Eyraud also extracts from his results some complementary geometrical interpretations of physical terms, for instance that the "potential vector finds its geometrical expression in the torsion"[63] (II, 129). Here too the geometrization of the physics is complete.

What do the cases where physics takes the lead look like? George Temple, our first example, follows a path opened up by Ludwig Silberstein (1918) and Alfred North Whitehead (1922), who both rejected the dominant role accorded to gravitation in Einstein's theory and its variants, a role based on its privileged relation to the space-time metric.[64] Temple distinguishes between two geometries: one, the "true" geometry, corresponds to the geometry of the real world, assumed to be a space with a metric dG^2 of constant curvature;[65] the other, the "fictitious" geometry, is a convenient tool which allows one to deal with physical dynamics. This dynamical manifold is represented, again strictly following Whitehead (Whitehead 1922, 79–82), by the "potential mass impetus," dJ, and the electromagnetic potential dF created by a particle of mass M and charge E, both taking the form of a metric,

$$dJ^2 = dG_M^2 - \frac{2}{c^2} \sum \psi_m \cdot dG_m^2,$$

$$dF = \sum_\mu F_\mu \, dx_\mu, \qquad \mu = 1, 2, 3, 4.$$

The dG_M^2 and dG_m^2 are the line elements in the *true* manifold along the path described by the corresponding particles of mass M or m, the sum on m being extended to all particles in the causal future of the particle of mass M; ψ_m is a retarded potential with a moving singularity due to the particle of mass m, which is intended to express the "law of the diminishing intensity of the perturbing influence of other particles" (Temple, 177) on the particle of mass M. Temple then exhibits an explicit expression for the potential ψ_m, by solving a differential equation which plays the role of a field equation; this equation is an empirical compound, borrowed from Silberstein.

The electromagnetic tensor $F_{\mu\nu}$, on the other hand, defined as the curl of the electromagnetic potential F_ν, satisfies the (non-covariant) Maxwell-Lorentz equations:

$$\frac{F_{\mu\nu}}{x_\lambda} + \frac{F_{\nu\lambda}}{x_\mu} + \frac{F_{\lambda\mu}}{x_\nu} = 0 \qquad \lambda, \mu, \nu \text{ different}$$

$$F^{\mu\nu}_{(\nu)} =: \frac{F^{\mu\nu}}{x_\nu} + \left\{ \begin{matrix} \nu\alpha \\ \mu \end{matrix} \right\} \cdot F^{\alpha\nu} + \left\{ \begin{matrix} \nu\alpha \\ \nu \end{matrix} \right\} \cdot F^{\mu\alpha} = \frac{4\pi\rho}{c} \frac{dx_\mu}{dx_4},$$

where ρ is the electric charge density, the x_μ are coordinates in the true manifold and indices are raised and lowered using the metric associated with dJ^2.

The path of a particle is then assumed to minimize the integral $\int (M dJ + c^{-1} E \cdot dF)$ along the path. Explicit calculations are made among others for planetary motions, leading to an expression for the perihelion advance equal to the general relativistic value plus a correction term involving the radius of curvature of space, R; supposing that this adds less than 1% to the Einstein value for Mercury gives a lower bound for R ($R > 2.5 \times 10^{16}$ km). Using this value to evaluate the corresponding correction to the Einsteinian value for the bending of light rays grazing the sun's surface gives a modification of 2×10^{-8}, "therefore wholly inappreciable" (Temple, 191).

The geometrical setting appears here as a mere framework, whose shape helps to treat the physical material, but not to produce it. Physics plays the central role: equations defining the metric are chosen a priori, as encapsulating experimental observations; the mathematics is then used to transform these initial data into explicit computations, with which in turn to confront experience. The dissociation of the dynamical space from the true space-time robs the two forces of any privileged status among possible phenomena and their coordination takes its legitimacy from observation, not geometry. As Whitehead puts it:

> A further advantage of distinguishing between space-time relations as universally valid and physical relations as contingent is that a wider choice of possible laws of nature (e.g., of gravity) thereby becomes available, and while the one actual law of gravity must ultimately be selected from these by experiment, it is advantageous to choose that outlook on Nature which gives the greater freedom to experimental enquiry (Whitehead *apud* Temple, 193).

In view of the representation of Einstein's work, both by his contemporaries and by some historians, our final example may appear strange: it is Einstein's own second article. Though of course Einstein I, as we have said, is a typical example of a theory in the Weyl-Eddington tradition, a pair of paragraphs at the end raises a point of another order: they purport to settle the question of the physical identification of parts of the electromagnetic tensor by considering the behavior of a solution under time reflection. In the sequel, II, however, Einstein reveals that the real question prompting the examination of this behavior is the difficulty that every negatively-charged solution of a given mass admits an equal-mass, positively-charged solution under the action of time reversal — and in 1925 the very unequally massive electron and proton are the only known charged elementary particles. He meanwhile has obtained an elementary proof that, in any set of covariant equations, no relabeling of the electromagnetic tensor will suffice to resolve the problem since the addition of a space-reflection reinstates the result. Even the requirement of a positive determinant for the transformation cannot help since a combined space- and time-reflection reproduces the difficulty.[66]

Finally, a note added in proof raises the point that, since the electric density ρ is equal to a square root, $\sqrt{g_{ik} \cdot F^{ir}/x^r \cdot F^{ks}/x^s}$, with its inherent ambiguity of sign, such unwanted symmetry is inescapable — unless the sign of ρ can be fixed within the field tensor. And this is possible if electromagnetism can be constructed with an attached "arrow of time." But this cannot be done in the gravitational case and Einstein surprisingly ends with the following:

> The conclusion seems to me to be essentially that an explanation of the disparity of the two electricities is only possible if a directional arrow is ascribed to time and this is then used in the definition of the principal physical quantities. In this respect, electromagnetism is fundamentally different from gravitation; thus it appears to me that the attempt to fuse electrodynamics and the laws of gravitation into a unity is no longer justified[67] (Einstein II, 334).

Confronted by a contradiction between geometry — for Einstein, as often, incarnated by tensors and symmetry conditions — and physics, Einstein does not consider new experiments to detect the particles predicted by the theory; on the contrary, he abandons (for a short moment as it turns out) the search for a unified field theory. In this classical setting, the retreat of geometry before physical considerations leads to a disruption of the unification program.

1925 Corpus

Arakatsu Bunsaku
    "The Theory of General Relativity in a Physically Flat Space." *Mem. Coll. Sci. Kyoto A* 8 (1925) 263–272 [J708]

Buhl, Adolf
    "Sur les formules fondamentales de l'électromagnétisme et de la gravifique. IV." *Ann. fac. sci. Toulouse* (3) 16 (1925) 1–28 [J705]

De Donder, Théophile
I   "Synthèse de la gravifique." *CRASP* 177 (1923) 106–108 [P165]
II   "La Gravifique de Weyl-Eddington-Einstein. I." *Bull. Acad. Roy. Belgique* (5) 10 (1924) 297–324 [P591]

Della Noce, G.
    "Confronti tra la teoria della relatività e la teoria del quanta." *Boll. Un. mat. Ital.* 4 (1925) 125–131 [J718]

Einstein, Albert
I   'Einheitliche Feldtheorie von Gravitation und Elektrizität." *SPAW* (1925) 414–419 [J704]
II   "Elektron und allgemeine Relativitätstheorie." *Physica* 5 (1925) 330–334 [J703]

Eyraud, Henri
I   "Sur le principe d'action et les lois de la dynamique de l'éther." *CRASP* 178 (1924) 761–763 [P591]
II   "Sur le caractère riemannien projectif du champ gravifique électromagnétique." *CRASP* 180 (1925) 127–129 [P1341; J716]
III   "La Théorie affine asymétrique du champ électromagnétique et le rayonnement atomique." *CRASP* 180 1245–1248 [P1341; J716]

Kudar, Johann
I   "Die Quantentheorie und die Rotverschiebung der Spektrallinien." *Phys. Z.* 26 (1925) 207–211 [P1344; J722]
II   "Über die atomdynamische Deutung der Uhrenhypothese." *Phys. Z.* 26 (1925) 331–334 [P1344; J722]
III   "Allgemeiner Beweis der 'Atomuhr' mit Hilfe der Hamilton-Jacobischen Theorie." *Phys. Z.* 26 (1925) 334 [P1344; J722]

Lanczos, Cornelius
I   "Zum Wirkungsprinzip der allgemeinen Relativitätstheorie." *Z. Phys.* 32 (1925) 163–172 [J710]
II   "Zur Anwendung des Variationsprinzips in der allgemeinen Relativitätstheorie." *Acta Litt. Sci. Szeged* 2 (1924/6) 182–192 [J710]

Larmor, Joseph
    "The Exploded Ether." [Letter to the editor, 7 March 1925]. *Nature* 115 (1925) 419 [P1342; J724]

Lodge, Oliver
    "Ether and Matter and Relativity." [Letter to the editor, 10 March 1925]. *Nature* 115 (1925) 419 [P1342; J724]

Lunn, Arthur C.
    "Relativity, the Quantum Phenomenon, and a Kinematic Geometry of Matter and Radiation." [Abstract, APS Meeting, Washington, 29–31 December 1924]. *Phys. Rev.* 25 (1925) 244 [P1249]

Mie, Gustav
    "Das Problem der Materie und die Relativitätstheorie." *Scientia* 37 (1925) 149–156, 225–234 (2 articles) [P1340; J703]

Rainich, George Yuri
  I "Electrodynamics in the General Relativity Theory." *Proc. Nat. Acad. Sci.* 10 (1924) 124–127 [P167]
  II "Second Note: Electrodynamics in the General Relativity Theory." *Proc. Nat. Acad. Sci.* 10 (1924) 294–298 [P167]
  III "Electrodynamics in the General Relativity Theory." *Trans. Am. Math. Soc.* 27 (1925) 106–136 [J713]

Reichenbächer, Ernst
    "Die mechanischen Gleichungen im elektromagnetischen Felde." *Z. Phys.* 33 (1925) 916–932 [J715]

Reissner, Hans
  I "Über eine Möglichkeit die wesentlichen Eigenschaften von Kern und Elektron aus dem metrisch-elektromagnetischen Felde abzuleiten." *Phys. Z.* 26 (1925) 925–932 [J717]
  II "Beitrag zur Theorie des Elektrons." *Z. Phys.* 31 (1925) 844–865 [J723]
  III "Elektron und Kern als Punktladungen." *SBMG* 24 (1925) 21–28 [J716]

Rice, James
  I "On Eddington's Natural Unit of the Field, and Possible Relations between It and the Universal Constants of Physics." *Phil. Mag.* (6) 49 (1925) 457–463 [P945; J712]
  II "On Eddington's Natural Unit of the Field." [Letter to the editor, 25 March 1925] *Phil. Mag.* (6) 49 (1925): 1056–1057 [J712]

Row, C. K. Venkata
    "Relativity Referred to a Flat Space-Time." [Letter to the editor, 1 January 1925]. *Nature* 115 (1925) 261–262 [P1342; J712]

Swann, William Francis Gray
    "A Generalization of Electrodynamics Correlating the Primary Features of Terrestrial Magnetism, Atmospheric Electricity, and Gravitation, under a Scheme Consistent with Restricted Relativity." [Abstract, APS Meeting, Washington, 29–31 December 1924]. *Phys. Rev.* 25 (1925) 253 [P944]

Temple, George F. J.
    "A Generalization of Professor Whitehead's Theory of Relativity." *Proc. Phys. Soc. Lond.* 36 (1924) 176–192 [P594]

Vallarta, Manuel Sandoval
    "Bohr's Atomic Model from the Standpoint of General Relativity." [Abstract, APS Meeting, New York, 27–28 February 1925]. *Phys. Rev.* 25 (1925) 582 [P1256]

Wavre, Rolin
    "A propos d'une tentative de conciliation de la théorie des quanta et de la théorie de la relativité." *Arch. Sci. Phys. Nat.* (5) 6 (1924) 294–301 [P592]

4. 1930

The last year of our *sondage* shows a massive increase in the number of unified theory publications: 57 articles written by 30 authors (or couples of authors).[68] Among these articles, 15 (by 9 authors) appear in *Zeitschrift für Physik*, 10 (5 authors) in *Proceedings of the National Academy of Science*, 5 (4 authors) in *Physical Review* and 5 (5 authors) in *Physikalische Zeitschrift*; other journals are represented by one or two authors each. This distribution manifests the new importance of the US scene for this subject. We would also point out that all the articles published in *Physikalische Zeitschrift* are summaries, written in German, of contributions to a conference held at Kharkov, in the Soviet Union, during the week from 19 to 25 May 1929.[69] Besides the participants in this conference, two other Soviet physicists make contributions to this year's corpus.[70] The other authors work in various European countries (France, Germany, Great Britain, Italy, Romania, Bulgaria, Hungary), as well as in China and India. But, as we shall see — and in opposition to the situation in 1920 — geographical location, either of publication or of residence, though in 1930 still significant insofar as they facilitate teamwork, are far less so for questions of visibility or citation of the articles of others.

Though the mean percentage of papers per author seems to have remained stable for the three years of our *sondage*, in this year we find more frequent traces of long series of papers and notes: 5 in our corpus authored by Einstein alone or with his assistant of this period, Walther Mayer, following 4 others by him directly connected to the same theme in the preceding months; 5 by Igor Tamm and his coauthors; 4 by Manuel Sandoval Vallarta and the MIT group, out of a series of 7; 6 by Tracy Yerkes Thomas; 5 by Gavrilov Raško Zaikov, the earliest in our corpus being, in fact, the last in a series of 6. This quick succession of articles is, in part, due to the reactions of readers to the first papers of the series and to the objections that are raised. Interactions, at least in an important subgroup in our corpus, are in 1930 effective and prompt, even sometimes hasty.

Moreover, in clear opposition to the two other years we have studied, a single theory dominates the scene. In 1928 Einstein had launched the unified field theory approach that was to attract the widest attention of any he was to put forward until his final attempt of 1945–1955, the theory of distant parallelism (*Fernparallelismus*). Newspapers as well as scientific journals welcomed articles on the question; the *New York Times* of 3 February 1929 put Einstein's photograph, together with a long article on "his new discoveries," on the first page of their Sunday supplement; it was followed by the London *Times* of 4 and 5 February and numerous other newspapers.[71] The theory also triggered a large number of more technical articles; witness the fact that 42 of our 57 papers — 21 out of our 30 authors — explain, comment, modify, combine, or complete the theory of distant parallelism.

In this theory, Einstein uses a kind of metric space in which a notion of parallelism between vectors (in the tangent spaces) at two distant points can be defined. He also introduces a new formalism, that of the *Vierbein* h_a^ν, which represents a local orthonormal coordinate system (Latin letters index the different vectors of the system, Greek letters the components of each).[72] Parallel transport is defined by the exact differential

$$dA^\nu = -h_a^\nu \frac{h_{\mu a}}{x^\sigma} A^\mu dx^\sigma,$$

with $h_{\mu a}$ such that $h_{\mu a} h_a^\nu = \delta_\mu^\nu$.[73] The torsion $\Lambda_{\alpha\beta}^\nu$, the antisymmetric part of the associated connection, is non-zero, while the Riemann curvature tensor is shown to vanish identically. The metric is given by $g_{\mu\nu} = h_{\mu a} h_{\nu a}$, and thus it is determined by the *Vierbein*, though the converse is not true. In particular, the 16 components of the *Vierbein*, compared to the 10 of the metric, offer a latitude which nourishes Einstein's hopes to fit electromagnetic phenomena into the theory; e.g., in his early papers on the subject, by viewing the trace of the torsion, $\Lambda_{\alpha\nu}^\nu$, as the electromagnetic potential vector.

Such a proposal, of course, seems quite close to some attempts already mentioned in the Weyl-Eddington lineage, and enters naturally into the general framework of affine spaces proposed by Élie Cartan as early as 1922 (Cartan 1922, 1923–25).[74] It seems at first sight only one of many possibilities to explore. Why, then, its particular attraction?

First of all, the classification of affine spaces is not considered to be obvious, nor is it common knowledge; Hans Reichenbach devotes the first half of his epistemologically oriented paper in our corpus to explain the relationship between Riemannian geometry and the geometry used in *Fernparallelismus*. Again, the introductory paper to the Kharkov conference by Vsevelod Frederiks and A. Isakson is precisely devoted to such a classification of metric spaces, partly inspired by Reichenbach's paper, partly by the classification given by Schouten in his 1924 *Der Ricci-Kalkül*. Einstein himself, in his third paper of our corpus, presents his theory, not as a kind of affine space, but as an *intermediate* case, situated between Riemannian and Euclidean geometry. Whereas Weyl's geometry provides no possibility of comparing either lengths or directions of vectors at a finite distance, and Riemann's only permits a comparison of lengths, the new geometry, like the Euclidean case, allows both, and Einstein proudly announces that he has found "a metric structure for the continuum which lies between the Riemannian and the Euclidean."[75] (Einstein 1929a, 130).

But there is more than a question of geometry in play. The field equations proposed by Einstein yield classical equations of gravitation and of electromagnetism only to first order.[76] For some indeed, this novelty constitutes a drawback of the theory, the experimental confirmations of general relativity being apparently lost with no compensatory gain.[77] But others express the opposite opinion. In Reichenbach's view, for instance, *Fernparallelismus* appears not only as a formally satisfying unification, but as a real cognitive advance over previous attempts, precisely because it is not reducible to Einstein's earlier theory of gravitation. Many physicists and mathematicians are in agreement; the theory is not seen simply as one of many, but as a creation of Einstein on a par with his earlier work, the next step beyond the special and the general relativity theories. In more than one instance, it is alluded to by others as "the" unified field theory, with at most a vague mention of a few analogous projects in the past — hence Weyl's annoyance mentioned in the introduction. Even those well-versed in some of the preceding attempts, like Zaikov, distinguish it from the others:

> But a return to the old theory of relativity (four-dimensional ... as well as five-dimensional ...) appears to be excluded once and for all. With the idea of *Fernparallelismus* a real step forward in our knowledge has been taken![78] (Zaikov II, 835).

Paradoxical as it might seem, a last feature favoring this overwhelming interest lies in the new Dirac quantum theory. Paul A. M. Dirac's (special-) relativistic theory of the electron was published in 1928 (Dirac 1928) and changed the topography of unified theories in important ways. Physically, it showed how one could couple a quantum charged particle to the electromagnetic field. Mathematically, it introduced *spinors* (*semi-vectors* for most of our authors at this time) as the mathematico-physical objects necessary to do this, in the context of a linear first-order differential equation. Different lines of inquiry were then explored to integrate a theory of gravitation with it.[79] From the start, *Fernparallelismus* appeared as a very promising candidate; this combination is indeed the very theme of the Kharkov conference and, among the 42 papers of our corpus devoted to distant parallelism, 9 (involving 3 authors) concern just such a combination. The hope is perfectly expressed by Norbert Wiener and Vallarta in a letter to the editor of *Nature* dated 7 February 1929.[80]

> May we be permitted to direct attention to a certain aspect of Einstein's three recent papers ... on distant parallelism which came to light in a discussion with Prof. D. J. Struik? The avowed aim of these papers is to develop an improved unified field theory of electricity and gravitation. A much more pressing need of general relativity theory is a harmonisation with quantum theory, particularly with Dirac's theory of the spinning electron. On the basis of Levi-Civita's parallelism the task seems hopeless, inasmuch as we have no adequate means of comparing spins at different points. On the other hand, the notion of a parallelism valid for the whole of space and of Einstein's n-uples enables us to carry over the Dirac theory into general relativity almost without alteration ... .
>
> ... The quantities ${}^s h_\lambda$ of Einstein['s *Fernparallelismus* theory] seem to have one foot in the macro-mechanical world formally described by Einstein's gravitational potentials and characterised by the index λ, and the other foot in a Minkowskian world of micromechanics characterised by the index s. ... This seems to us the most important aspect of Einstein's recent work, and by far the most hopeful portent for a unification of the divergent theories of quanta and gravitational relativity (Wiener and Vallarta 1929b).

This obvious center of interest, corroborated by the references cited in the papers, imposes a treatment of the papers related to *Fernparallelismus* as our first group.[81] A second smaller group of papers centers around five-dimensional theories, though even here showing important links with the articles on distant parallelism. The remaining papers are much more isolated, in some cases completely marginal, attempts to develop a unified theory, and for the sake of space we shall restrict ourselves to a few comments on them at the end.

What happens in the 42 papers devoted to distant parallelism? The difficulty in describing them is due to their organization; the resulting configuration is roughly star-shaped, with the core constituted by Einstein's series of papers. But a detailed chronology is necessary to follow the quick responses of some authors to Einstein's papers, and to understand the unfolding of events.[82] We will thus begin with the hero of the year, then examine some characteristic answers by others and finally discuss more globally the links between the papers.

Chronologically, the first of Einstein's papers in our corpus is dated 19 August 1929 and published in *Mathematischen Annalen* (Einstein I); his purpose is to establish the mathematical foundations of his *Fernparallelismus* theory and to present it in a way which makes it accessible to specialists of general relativity. While acknowledging that the manifolds used are not new, Einstein underlines the importance and originality of his work, "the discovery of the simplest field equations to which a Riemannian manifold with *Fernparallelismus* can be subjected"[83] (Einstein I, 685). Einstein posits directly his 22 field equations, without using a variational principle:

$$G^{\mu\alpha} = 0, \qquad F^{\mu\alpha} = 0,$$

where

$$G^{\mu\alpha} = (\Lambda^{\alpha}_{\beta\gamma} g^{\mu\beta} g^{\nu\gamma})_{;\nu} - (\Lambda^{\sigma}_{\beta\gamma} g^{\mu\beta} g^{\tau\gamma}) \Lambda^{\alpha}_{\sigma\tau}$$

and

$$F^{\mu\nu} = (\Lambda^{\alpha}_{\beta\gamma} g^{\mu\beta} g^{\nu\gamma})_{;\alpha}.$$

Here, as before, $\Lambda^{\alpha}_{\beta\gamma}$ is the antisymmetric part of the connection and the semicolon notation indicates a covariant derivative with respect to the affine connection associated with the *Vierbein*. In this case, the field equations, in first approximation, reduce to the classical Poisson (sourceless) gravitational equation and to the vacuum Maxwell equations. In his talk given at the Institut Henri Poincaré in Paris in November 1929, transcribed by Alexandre Proca (Einstein II), Einstein describes at length a heuristic path to obtain these equations, by first annihilating the various divergences built from the covariant derivatives of the torsion, then correcting the equations thus obtained by a careful examination of identities among the $\Lambda^{\alpha}_{\beta\gamma}$; in particular, he succeeds in giving a sufficient number of relations between these equations in 12 variables to ensure compatibility. Indeed, the overdetermination of the field equations appeals to him as much as it did in 1923: "The great charm of the theory for me lies in its unitarity and in the high (authorized) degree of its overdetermination"[84] (Einstein I, 697).

The determination of the relations among the field equations becomes the focus of Einstein's interest during the following months, and the core of his correspondence with Cartan, using the mathematician's suggestions to simplify and correct his previous publications (Debever 1979, Biezunski 1989). In the first of two notes to the Berlin Academy of Science, in January and July 1930 (Einstein III, V), he stresses once more the interest of overdetermination: "The higher the number of equations (and consequently also of the identities among them), the more the theory makes definite assertions beyond the requirement of mere determinism; and thus the more valuable the theory is, provided it is compatible with experimental facts"[85] (Einstein III, 21).

Once the field equations are obtained, the path of investigation is clearly drawn, as is explained repetitively by Einstein and his followers in their research papers and in the more popular expositions of the theory: first, to discover suitable solutions of the field equations that will represent elementary particles; secondly, to determine the laws of motion in these spaces; finally, to test these last through experiment.

The first problem is studied by Einstein in his joint paper with Mayer in February 1930 — and as we shall see, in the work of quite a number of other authors as well — where exact solutions are found in two specific cases: that of spherical and mirror symmetry, and that of a static, purely gravitational, field. In the first, which corresponds to the physical situation of the external field of a charged massive sphere, there appear exactly two constants, prompting, in a satisfying manner, a natural interpretation of them as mass and charge. The second case leads to a solution which is, at first sight, rather discouraging; any arbitrary distribution of non-charged masses will remain at rest! The authors, undismayed, point out that since the laws of motion of singularities in this version of the theory cannot be derived from the field equations, such a solution provides no argument against its applicability: "One should surely recognize that, in the new theory, solutions which will represent the elementary particles of matter must be required to be free from singularities"[86] (Einstein and Mayer, 120).

But the laws of motion remain elusive, and, with them, questions of experimental verification. On 11 April 1930, William Francis Gray Swann reports in *Science*:

> It now appears that Einstein has succeeded in working out the consequences of his general law of gravity and electromagnetism for two special cases.... It is hoped that the present solutions obtained by Einstein, or if not these, then others which may later evolve, will suggest some experiments by which the theory may be tested (Swann, 390).

We are still awaiting them.

But even viewed from 1930, the quasi-success story provided by Einstein's rational reconstruction of his own work and reinforced by the nature of our selection process seems already overtidy; between 7 June 1928, the date of the Berlin Academy of Science gathering which includes Einstein's first article on *Fernparallelismus* (Einstein 1928b), and 19 August 1929, the date of the first paper of his in our corpus, Einstein proposes in fact three different *Fernparallelismus* theories, with varying sets of field equations (Einstein 1928a, 1929b, c and I).[87] And despite his pride and pleasure, cited above, in the August 1929 equations, Einstein does not stop there: in June 1931, after a long silence, he writes to Cartan: "Meanwhile I have been working a great deal with Dr. Mayer on the subject and I have abandoned those field equations."[88] (Einstein 1931). This comedy of errors has its impact on the work of other authors: some try to work by sticking close to Einstein's heels, following his change of equations as soon as they are produced; some study specific problems associated with one set of field equations; some, finally, try to reformulate the theory and obtain their own field equations.

A first example of such a reframing of *Fernparallelismus* theory is presented at the Kharkov conference by the Leningrad physicist Heinrich Mandel', and then developed in a longer article in June 1929 (Mandel' I, II). He treats the relationship between distant parallelism and his own theory of 1927, a five-dimensional theory of Kaluza-Klein type with a cylindricality condition:

> According to the fundamental idea of the five-dimensional theory, the set of the ∞^4 world-points is not to be considered as the set of ∞^4 points of a hypersurface in [the five-dimensional space] R_5, but as the set of the ∞^4 lines (of the congruence X_5^i) in the cylindrical R_5 [89] (Mandel' II, 240).

As at every point, the local *Vierbein*, $X_1^i, X_2^i, X_3^i, X_4^i$, may be oriented orthogonally to X_5^i, it is possible to choose as the fundamental tensor $g_{ik} = \sum_{\rho=1}^{4} X_\rho^i X_\rho^k$. The geometry of the four-dimensional space-time is thus to be understood as the geometry of the projection of the *Vierbein* on the axes $X_1, \ldots, X_4$. Mandel' describes, for a special case, parallel displacement in such a geometry. He then expresses the fact that the Riemann curvature tensor is zero in terms of the five-dimensional curvature tensor and derives field equations, easily interpretable as gravitational equations. Morever, Mandel' suggests a geometrical interpretation, not of the potential vector, as Einstein has proposed up to this point, but of the electromagnetic field itself. Defining $F_{\mu k}^i$ as the difference between the connection of the space with distant parallelism and the classical Christoffel symbols, Mandel' shows that the equation of the geodesics (here the straightest lines), if interpreted as equations of motion of a charged body of charge e and mass m, leads to $F_{\mu k}^i \dot{x}^\mu = 2(e/m) M_k^i$, with $2M_k^i$ the tensor of the electromagnetic field.

Another interesting example is the series of articles (I–VI) by the mathematician Tracy Y. Thomas, communicated to the *Proceedings of the National Academy of Sciences* between 30 September 1930 and 2 April 1931. Thomas, a former student of Oswald Veblen, is a member of the Princeton group of differential geometers. Their trademark has been, for the previous decade, the development of a differential geometry starting from *paths*, i.e., autoparallel curves, rather than the more usual connection, as the fundamental geometric objects.[90] In his first note, Thomas axiomatically reformulates the fundamentals of the theory of spaces with distant parallelism. He requires the existence at each point of a local system of coordinates (z^i) ($i = 1, \ldots, 4$) such that the coordinate axes z^i are tangent to the vectors h_i^α of the *Vierbein*; the local metric is Minkowskian, with z^1 the time-coordinate, the paths of the space are straight lines in the local coordinates. He then expresses the covariant derivative in terms of the local system of coordinates, thus privileging another choice for the differentiation of tensors than that of Einstein. Thomas then *postulates* a system of 16 field equations analogous in local coordinates to a 4-dimensional wave equation:[91]

$$h_{j,11}^i - h_{j,22}^i - h_{j,33}^i - h_{j,44}^i = 0,$$

where the covariant derivatives $h_{j,kl}^i$ are given by

$$h_{j,kl}^i = \left(\frac{\partial^2 \left(h_\alpha^i \frac{\partial x^\alpha}{\partial z^j} \right)}{\partial z^k \partial z^l} \right)_{z=0}.$$

Thomas then suggests possible interpretations for the electromagnetic potentials in terms of the covariant components h_α^i of the *Vierbein* and for the gravitational potentials in terms of the coefficients $g_{\alpha\beta}$ of the metric. As Thomas comments in an endnote:

> There is a certain psychological influence exerted by the method itself upon the investigator ... . So, for example, the field equations proposed by Einstein have a very simple analytical form in terms of the covariant derivative used by him ... also the simple form of the field equations assumed in the above investigation is peculiar to the method of absolute differentiation which I have adopted (Thomas I, 776).

In his second note, however, Thomas himself modifies his field equations, displacing his emphasis from the obtaining of exact Maxwellian equations in the local system to that of a law of conservation. More precisely, he now sets to zero the divergence — a notion that Thomas has to redefine in his framework — of what corresponds to the electromagnetic forces, $\nabla_k h^i_{j,k} = 0$, introducing changes of second order in the $h^i_{j,k}$ in his former field equations. The following notes are then devoted to a standard, analytic, study of these new field equations. Thomas establishes, in particular, a general existence theorem of the Cauchy-Kovalevskaya type; examines, in the Hadamard tradition, the characteristic surfaces, "which appear to an observer in the local system as a spherical wave propagated with unit velocity." (Thomas IV, 112); and, finally, shows that the null geodesics are the light paths in this unified field theory, as in general relativity.

The range of activities around one set of field equations is exemplified by R. N. Sen and George McVittie. The first takes up some remarks made by Edmund Whittaker during his presidential adress to the London Mathematical Society on 14 November 1929 (Whittaker 1930), according to which Clifford parallelism in a three-dimensional space of constant curvature is a distant parallelism in Einstein's sense. Sen writes down the January 1929 Einstein equations (Einstein 1929b) in the static case, for the metric $ds^2 = V^2 dx_0^2 - \sum_{p,q=1}^{3} g_{pq} dx_p dx_q$, with the constant V representing the velocity of light; he computes the *Vierbein* when the three-metric represents a space of constant curvature and the distant parallelism is Clifford parallelism, proving that this case is exactly that of Minkowski space. As for McVittie, his aim is to compare, for the special case of the gravitational field of a uniform electrostatic field, Einstein's approach (with the January 1929 equations) to the alternative proposed by Levi-Civita in March 1929 (Levi-Civita 1929), using Ricci's tool of orthogonal congruences of lines in an (ordinary) Riemannian space, instead of Einstein's *Vierbein*. McVittie computes a solution in the Levi-Civita framework, showing it to be in agreement with the solution he has already found with the Einstein approach (McVittie 1929) and deduces, for this particular case, a geometrical interpretation of the electromagnetic potential vector in Levi-Civita's theory.

Finally, to grasp what it concretely means to match Einstein's pace, let us follow a group of young MIT mathematicians and physicists: Vallarta, Wiener, Dirk Struik, Nathan Rosen.[92] We have already cited the program of unification between Dirac theory and *Fernparallelismus* presented by Vallarta and Wiener in February 1929 (Wiener and Vallarta 1929b). (In fact, of this program they will publish only the part concerned with solutions of the classical *Fernparallelismus* theory). On 1 March 1929, they send to the *Proceedings of the National Academy of Sciences* a joint paper (Wiener and Vallarta 1929a) on the (non-) existence of a spherically-symmetric static field solution to Einstein's first set of field equations (Einstein 1928a).[93] Supposing the components of the *Vierbein* (in spherical coordinates) to be functions of the radius r alone, and assuming a time symmetry of past and future,[94] they show that both the electromagnetic and the gravitational field vanish. On 26 June 1929, Vallarta, this time alone (Vallarta I), explores the same problem for the second set of Einstein's field equations, those of January 1929; the result for the electrostatic field is the same. Indeed: "[its vanishing is] a consequence of the definition of the electromagnetic potential and is independent of the particular choice of field equations." (Vallarta I, 787). Moreover no Schwarzschild-like solution can be found for the gravitational part.

A blow falls at the New Year; a letter to the editor of *Physical Review* by Meyer Salkover, of the University of Cincinatti, points out an error in Vallarta's paper and exhibits a Schwarzschild solution (Salkover I), a second letter (11 January) completes the study of the solutions (Salkover II). On 3 February, Vallarta acknowledges his mistake (Vallarta II), pointing out, however, that his conclusion for the electrostatic field remains valid. In the meantime, according to Vallarta, Wiener has checked the validity of the usual Schwarzschild solution in the context of Einstein's March 1929 paper (Einstein 1929b) and of Levi-Civita's variant (Wiener's results were apparently never published, perhaps because, as we shall see, he was forestalled by a quicker team).[95]

The game ends on 15 May 1930; a last paper of Vallarta, this time in collaboration with Nathan Rosen, takes over a last set of Einstein field equations, those of August 1929 (Einstein I), both with and without the assumption of time-symmetry. In the second case, they find, up to a change of variables, the Einstein-Mayer solution; they interpret the inherent nonseparability of electric and gravitational fields in the immediate neighborhood of a charged mass in this theory as a possible explanation of nuclear and electronic stability. In the time-symmetric case, they obtain, as in their earlier papers, a pure gravitational solution, though none with charge.

> Thus the existence of an electrostatic field in the unified theory depends on the asymmetry of past and future. We believe that this is the first instance that this asymmetry has been found to have any physical significance in connection with a field theory. The existence of the gravitational field, on the other hand, is apparently not connected with this asymmetry. We may perhaps have found here the fundamental difference, superficial similarity notwithstanding, between the gravitational and the electric field of a charged mass particle.
>
> ... In the absence of a law of motion, not yet discovered, the path of an exploring particle in the unified field cannot be calculated. ... The shift of spectral lines towards the red, on the other hand, does not depend on the law of motion of an exploring particle, but only on the component g_{44} of the Riemann metric. ... The red shift obtained on the basis of the present theory is the same to a first approximation as that predicted on the basis of the 1916 theory (Rosen and Vallarta, 119–120).

The quicker team alluded to above is that composed of the Moscow physicists Igor Tamm and Mikhail Leontovič. Like the MIT group, Tamm also has a program to unify quanta and distant parallelism: first find a generalization of the Dirac equation in spaces with distant parallelism which will serve as an equation of motion (Tamm I and II);[96] then determine solutions of the *Fernparallelismus* field equations to act as a source in the modified Dirac equation (Tamm and Leontovič I and II).

Tamm's point of departure is the usual Dirac wave equation in the absence of an electromagnetic field, $(^s\alpha\, p_s + imc)\psi = 0$ — the Pauli-matrices $^s\alpha$ are, as usual, the components of a constant q-vector and the p_s are the momentum operators. In the presence of an external electromagnetic field, these operators are transformed in the usual Dirac theory by the addition of an interaction term coupling the wave function to the electromagnetic potential. Tamm proposes to treat the problem in spaces with distant parallelism by means of two hypotheses: the components of the q-vector *relative to the Vierbein* will be taken to be constant, and the usual form of the *free* Dirac equation will be assumed to hold in general — in other words, the geometry of the space will automatically take care of the fields. He thus obtains the equation (Tamm II, 653):

$$\left[\alpha^\nu \left(p_\nu + (1-in)iK\Lambda^\lambda_{\nu\lambda}\right) + imc\right]\psi = 0,$$

where n is a real number to be determined, α^ν is $_s h^\nu \cdot {}^s\alpha$, $K = h/2\pi$ is the reduced Planck constant, and the operators p_s are no longer ordinary but rather covariant differential operators.[97] If a proportionality between the trace of the torsion and the electromagnetic potential, $\Lambda^\lambda_{\nu\lambda} = a\Phi_\nu$, is posited, this becomes a Dirac equation with an interaction term; for a specific value of the product na, Tamm recovers the Schrödinger equation up to second order terms.

> The fact that the procedure sketched out really leads to a reasonable wave equation is all the more interesting because the "classical" formulation analogous to the mentioned wave-mechanical hypothesis — that the motion of the electron relative to the *Vierbein* is always uniform — leads to no useful equations of motion. *Thus the wave-mechanical principle appears, in Einstein's theory, to have priority over the principle of the shortest path of geometrical optics* [98] (Tamm I, 290).

That is, Tamm sees general relativity — and its geodesic equation of motion — as the geometrical-optics limit of an intrinsically wave-mechanical *Fernparallelismus* theory.

The next step then is to obtain a particular solution for the *Vierbein*. For this, as we said, Tamm works in collaboration with Leontovič; they present their joint work at the Kharkov conference (I) and extend it in a longer article a month later (II). Here again, they find a static, spherically-symmetric exact Schwarzschild solution of the Einstein field equations of March 1929, which they now interpret as the ground state of the (neutral) hydrogen atom,[99] but none corresponding to a charged particle. Their interpretation of these results is quite optimistic; the non-existence of a charged solution with spherical symmetry, were it to be coupled with the existence of an axially-symmetric solution, would be a reflection of the fact that the electron possesses spin.[100]

> One of us has recently attempted elsewhere [Tamm I] to show how naturally the electron wave equation arises in the new Einstein theory, and has, in addition, put forward the conjecture that, in this theory, the wave-mechanical principle has priority over the principle of the shortest path, so that the equations of motion of a (charged) particle are to be derived from the wave equation by a limiting process. If this conjecture, as well as the conjecture that the solution of the Einstein field equations corresponding to a charged particle accounts for the spin of the elementary charge, should really be confirmed, then the mi-

croscopic interpretation of the Einstein theory would be considerably strengthened[101] (Tamm and Leontowitsch II, 356).

There is no lack of criticism in Kharkov when Tamm and Leontovič present their results, in particular from the Leningrad team of Vladimir Fok and Dimitri Ivanenko, who advocate quite another path towards the unification of Dirac theory and general relativity: to develop a geometry of operators and integrate Dirac matrices as a correcting linear term in the metric.[102]

Several other proposals for the reconciliation between quantum and classical theories are put forward in this year. A few months after Kharkov, in September 1929, Gleb Wataghin underlines that all previous attempts to extend Dirac theory to the framework of general relativity rely on the union of Dirac matrices and Einstein *Vierbein*. But while many authors, as we have seen above, have judged a major advantage of Einstein's new equations precisely the fact that they reduce to the old general relativity equations only to first approximation, Wataghin regards them with disfavor for this very reason, at least so long as no experimental evidence will have come to disconfirm the latter. He himself adopts the theory of distant parallelism in the Levi-Civita form and exhibits a Lagrangian as a sum of three terms, but such that a single variational principle allows him to derive the 4 Dirac, 8 Maxwell and 10 general relativity equations. Thus, *Fernparallelismus* appears here mainly as a convenient technical framework, encapsulating physical equations coming from other theories. But, through the interpretation of the various variables, the computations cast light on an interdependence of the three classes of phenomena; Wataghin concludes, in particular, that the gravitational potentials have an essentially statistical significance.

A last example of the combination of *Fernparallelismus* and quantum physics, Zaikov's work, also witnesses the important effort of assimilation made by the newcomers. Like the previous groups, Zaikov has attempted to follow Einstein's exploration of the theory of distant parallelism, as well as its compatibility with the Dirac wave equation. During the autumn of 1929, however, Zaikov proposes a new path (Zaycoff II, III): extend the theory of distant parallelism with one supplementary dimension and operate directly with the ψ-functions. More precisely, his cylindrical five-dimensional geometry is defined with its fundamental covariant components, $H_{\alpha m} = h_{\alpha m}$, $H_{\alpha 0} = -f_\alpha$, $H_{0m} = 0$, $H_{00} = 1$, where the $h_{\alpha m}$ are as usual defined out of the *Vierbein* and the f_α are proportional to the electromagnetic potentials, such that the $h_{\alpha m}$ and the f_α are independent of the fifth coordinate x^0. Zaikov then introduces the ψ-functions (and their conjugates), also independent of x^0, and proposes a Lagrangian such that the variation of the $h_{\alpha m}$, f_α, ψ and its conjugate, $\bar{\psi}$, produces 28 field equations: 16 of second order in the $h_{\alpha m}$ and first order in the f_α and ψ, describing gravitational and spin phenomena, 4 of the second order in the f_α and of the first in the $h_{\alpha m}$ describing electromagnetism. The 8 complementary equations are of the first order in the $h_{\alpha m}$ and the ψ. In October 1930 (Zaycoff V), however, he switches to the new Einsteinian field equations (Einstein I) and is able to derive from his preceding work an equation of the type $R_{\alpha\beta} - 1/2 g_{\alpha\beta} R + T_{\alpha\beta} = 0$, with the quantities R and T suitably defined (in particular, T depends on the component of the *Vierbein* and on the function ψ), thus mimicking those of general relativity. Unfortunately, the properties of T are very different from those of an energy-momentum tensor, and the conclusion of Zaikov's paper is to call for new concepts to be developed.

The emerging picture of the work centering on Einstein's new theory is thus twofold: on one side, global acute awareness of Einstein's work in progress, but on the other, a constellation of more local debates, joint work and solidarities. There exists no important, very tight, general network of communications among the various protagonists. It is true that the Kharkov proceedings, and the extended versions of the results presented there, are commented and discussed, and the projects of the Soviet groups are faithfully followed by most of the contributors, but no other reciprocal impact is to be seen. We have located a few teams and competitors, in Boston and Princeton, Moscow and Leningrad. Besides the leadership provided by Einstein himself we find more elusive traces of activities inspired by various local leaders, like Whittaker (Sen) or Eddington (McVittie). But we have no evidence from the mutual references in our corpus of, say, direct scientific links between the MIT group and the Princeton geometers, nor is there a specific relationship between the various physicists and mathematicians trying to combine quantum theory and *Fernparallelismus*; neither nationality nor technical orientation are a warrant for effective relationships. A small exception, for the second case, is the constellation around five-dimensional theories, Zaikov referring to the work of Mandel' for instance.[103]

As indicated at the beginning of this section on 1930, the other 18 articles are much more isolated, both scientifically and socially. They compose a digest of most of the programs we have previously met — except that quantum phenomena are, at least as an horizon, a part of more than half these remaining articles. We find among them an attempt to combine various geometries with variable mass (Manev), a general exploration of minimal assumptions for a unified field theory (Whyte), a non-tensorial calculus to integrate electromagnetism, light phenomena and gravitation and reproduce quantum effects (Sevin), a rewriting of quantum theory to fit with Riemannian geometry (Reichenbächer), a criticism of geometrization as anything more than a tool in the construction of unified theories (Band), a multidimensional theory with a strong (Kantian) epistemological component (Rumer) and of course several proposals of one form or another of the affine theories (Novobátzky, Lagunov). It is in this respect quite interesting to remark that the last two articles seem much less connected with the main stream of papers dealing with *Fernparallelismus* than, say, Rumer's papers on multi-dimensional theories.[104] The papers dealing with affine theories refer mainly to the now ancient articles of 1923 by Eddington and Einstein. The status of these "other affine theories" is thus completely different in 1930 from the status of distant parallellism, and even that of five-dimensional theories. It is only fitting that these last two types of theories will be just those on which Einstein will work in the years immediately following.

1930 Corpus

Band, William
I   "A New Relativity Theory of the Unified Physical Field." [Letter to the editor, 26 November 1929]. *Nature* 125 (1930) 130 [P588; J1288] (= *Phys. Rev.* 35 (1930) 115–116 [P816])
II   "A New Unified Field Theory and Wave Mechanics." [Letter to the editor, 28 February 1930]. *Phys. Rev.* 35 (1930) 1015–1016 [P1821]

Einstein, Albert
I   "Auf die Riemann-Metrik und den Fern-Parallelismus gegründete einheitliche Feldtheorie." *Math. Ann.* 102 (1930) 685–697 [J734]
II   "Théorie unitaire du champ physique." [Lectures IHP, November 1929, "rédigées par Al. Proca"]. *Ann. Inst. Poincaré* 1 (1930) 1–24 [P2307]
III "Die Kompatibilität der Feldgleichungen in der einheitlichen Feldtheorie." *SPAW* (1930) 18–23 [P1821; J735]
IV "Professor Einstein's Address at the University of Nottingham." [CR of lecture, 7 June 1930, by I. H. Brose]. *Science* 71 (1930) 608–610 [P1916]
V   "Zur Theorie der Räume mit Riemann-Metrik und Fernparallelismus." *SPAW* (1930) 401–402 [P2671, J738]

Einstein, Albert and Walther Mayer
   "Zwei strenge statische Lösungen der Feldgleichungen der einheitlichen Feldtheorie." *SPAW* (1930) 110–120 [P1821; J736]

Fréedericksz [Frederiks], K. Vsevelod and A. Isakson
   "Einige Bemerkungen über die Feldgeometrie." *Phys. Z.* 30 (1929) 645 [P82]

Grommer, Jakob
   "Eine kleine Bemerkung zur neuen Einsteinschen Feldtheorie." *Phys. Z.* 30 (1929) 645 [P88]

Kunz, Jakob
   "Bewegung von Licht und Materie im Gravitationsfeld." *Phys. Z.* 31 (1930) 83–87 [P1829; J1286]

Lagunoff [Lagunov], B.
   "Über eine Erweiterung der Gleichungen des elektromagnetischen Feldes." *Z. Phys.* 64 (1930) 425–430 [J744]

McVittie, George C.
   "On Levi-Civita's Modification of Einstein's Unified Field Theory." *Phil. Mag.* 8 (1929) 1033–1040 [P413]

Mandel', Heinrich
I   "Über den Zusammenhang zwischen der Einsteinschen Theorie des Fernparallelismus und der fünfdimensionalen Feldtheorie." *Phys. Z.* 30 (1929) 646–648 [P196]
II   "Über den Zusammenhang zwischen der Einsteinschen Theorie des Fernparallelismus und der fünfdimensionalen Feldtheorie." *Z. Phys.* 56 (1929) 838–844 [P196]

Maneff [Manev], Georgi Ivanovič
I   "Le Principe de la moindre action et la gravitation." *CRASP* 190 (1930) 963–965 [P2307; J742]

II   "L'Énergie électromagnétique dans le champ de gravitation." *CRASP* 190 (1930) 1180–1182 [J742]

III   "La Gravitation et l'énergie au zéro." *CRASP* 190 (1930) 1374–1377 [J742]

Northrop, Filmer S. C.
"Two Contradictions in Current Physical Theory and their Resolution." *Proc. Nat. Acad. Sci.* 16 (1930) 55–68 [P820]

Novobátzky, Karl
"Schema einer Feldtheorie." *Z. Phys.* 58 (1929) 556–561 [P412]

Proca, Alexandre
"La Nouvelle Théorie d'Einstein." *Bull. Math. Phys. Bucarest* 1 (1929) 170–176, 2 (1930/31) 15–22 (2 articles) [J738]

Rainich, George Yuri
"Radiation and Relativity. II." *Proc. Nat. Acad. Sci.* 14 (1928) 654–657 [P1115]

Reichenbach, Hans
"Zur Einordnung des neuen Einsteinschen Ansatzes über Gravitation und Elektrizität." *Z. Phys.* 53 (1929) 683–689 [P3]

Reichenbächer, Ernst

I   "Ist Diracs Theorie mit nur zwei Komponenten durchführbar?" [Letter to the editor, 11 September 1929]. *Naturwissen.* 17 (1929) 805 [P727]

II   "Eine wellenmechanische Zweikomponententheorie. I." *Z. Phys.* 58 (1929) 402–424 [P89]

III   "Eine wellenmechanische Zweikomponententheorie. II." *Z. Phys.* 61 (1930) 490–510 [P1395]

IV   "Die Weltfunktion in dem vereinigten Wirkungsintegral der Gravitation, Elektrizität und Materie. I." *Z. Phys.* 65 (1930) 564–570 [J1282]

Rosen, Nathan and Manuel Sandoval Vallarta
"The Spherically Symmetrical Field in the Unified Theory." *Phys. Rev.* (2) 36 (1930) 110–120 [P2671; J740]

Rumer, Yuri Borisovič

I   "Über eine Erweiterung der allgemeinen Relativitätstheorie." *Nachr. Göttingen* (1929) 92–99 [P413 and 1820]

II   "Form und Substanz." *Z. Phys.* 58 (1929) 273–279 [P413]

Salkover, Meyer

I   "The Unified Field-Theory and Schwarzschild's Solution." [Letter to the editor, 31 December 1929]. *Phys. Rev.* (2) 35 (1930) 209 [P727]

II   "The Unified Field Equations and Schwarzschild's Solution. II." [Letter to the editor, 11 January 1930]. *Phys. Rev.* (2) 35 (1930) 214 [P727]

Sen, R. N.
"On the New Field Theory." *Ind. Phys.-Math. Jour.* 1 (1930) 28–31 [J741]

Sevin, Émile
"Introduction d'un vecteur charge électrique. Application à la synthèse des théories de l'électromagnétisme, de la lumière et de la gravitation." *CRASP* 188 (1929) 1603–1604 [P2506]

Swann, William Francis Gray
"Statement in Regard to Professor Einstein's Publications." *Science* 71 (1930) 390–391 [P1821]

Tamm, Igor Evegen'evič
I "Über den Zusammenhang der Einsteinschen einheitlichen Feldtheorie mit der Quantentheorie." *J. Appl. Phys.* 6 (1929) 130–133 [P1823] (= *Proc. Roy. Acad. A'dam* 32 (1929) 288–291)
II "Die Einsteinsche einheitliche Feldtheorie und die Quantentheorie." *Phys. Z.* 30 (1929) 652–654 [P89]

Tamm, Igor Evgen'evič and Mikhail Aleksandrovič Leontowitsch [Leontovič]
I "Über die Lösung einiger Probleme in der neuen Feldtheorie." *Phys. Z.* 30 (1929) 648 [P82]
II "Bermerkungen zur Einsteinschen einheitlichen Feldtheorie." *Z. Phys.* 57 (1929) 354–366 [P412]

Thomas, Tracy Yerkes
I–VI "On the Unified Field Theory." *Proc. Nat. Acad. Sci.* 16 (1930) 761–776, 830–835; 17 (1931) 48–56, 111–119, 199–210, 325–329 (six articles) [J738]

Vallarta, Manuel Sandoval
I "On Einstein's Unified Field Equations and the Schwarzschild Solution." *Proc. Nat. Acad. Sci.* 15 (1929) 784–788 [P413]
II "The Unified Field Theory and Schwarzschild's Solution: A Reply." [Letter to the editor, 3 February 1930]. *Phys. Rev.* (2) 35 (1930) 435 [P1203]

Wataghin, Gleb
I "Sopra un'applicazione della relatività alla meccanica quantica." *Atti Lincei* 10 (1929) 423–429 [P816]
II "Relatività e meccanica ondulatoria." [Abstract, SIF Meeting, Florence, 18–24 September 1929]. *Nuovo Cim.* (NS) 6 (1929) CLVI–CLVII [P1627]

Whyte, Lancelot Law
"Über die Eigenschaften einer einheitlichen physikalischen Theorie. II. Maßstäbe Uhren und eine mögliche Alternative zur Vierkoordinatenbeschreibung." *Z. Phys.* 61 (1930) 274–289 [P1202; J732]

Wiener, Norbert and Manuel Sandoval Vallarta
"On the Spherically Symmetrical Statical Field in Einstein's Unified Theory: A Correction." *Proc. Nat. Acad. Sci.* 15 (1929) 802–804 [P413]

Zaycoff [Zaikov], Gavrilov Raško
I "Zur einheitlichen Feldtheorie." *Z. Phys.* 58 (1929) 280–290 [P88]
II "Fernparallelismus und Wellenmechanik. I." *Z. Phys.* 58 (1929) 833–840 [P292]
III "Fernparallelismus und Wellenmechanik. II." *Z. Phys.* 59 (1929) 110–113 [P292]
IV "Das relativistische Elektron." *Z. Phys.* 61 (1930) 395–410 [P1724]
V "Über die Einsteinsche Theorie des Fernparallelismus." *Z. Phys.* 66 (1930) 572–576 [J1282]

5. CONCLUSION

Before discussing the collective aspects of the production of unified theories, we would like to pause for a moment to revisit two classic questions: the first touches on the role of Einstein, the second on that of quantum theory.

Einstein, with general relativity theory and, to an even greater degree, with his various attempts at unification during our decade and afterwards, is often considered today as the major promoter of the *geometrization* of physics. The articles we have studied here suggest the need to redraw this picture. The attachment of physics to geometry was indeed a controversial topic during this decade, and it was Einstein's name that was frequently put forward as the main target of the project's adversaries and as a rallying banner for its defenders. Moreover, in his more popular works, Einstein often focused on the presentation of a geometrical space-time and its properties — as did the celebrated introductory texts by Weyl and Eddington as well as others in articles addressed to a general audience or devoted to epistemological questions. But if one concentrates on Einstein's technical production, the emphasis is globally different; his interest — and the key point in his interaction with other scientists — was not so much in the geometrical shape of the world per se, as in the choice of field equations.[105] There, in their properties, in the conditions to which they are submitted — in particular their degree of overdeterminacy — is to be found the core of Einstein's *work*. His readiness to abandon, when necessary, the variational principles dear to the Göttingen circle or to leave unsettled problems of the identification between physical quantities and geometrical magnitudes sets him apart from an Eyraud or an Eddington for instance. Even Einstein's increasing eagerness to reach mathematicians seems more indicative of a seeking after complementarity than of a deeply-felt solidarity in *Weltanschauung*.

A second point concerns more directly Einstein's persona. As we have pointed out, Einstein, after 1919 at least, benefited from a universal visibility; his name was known and cited by every other author in our corpus. But the nature of these citations changed with the decade, and in a sense which ran contrary to Einstein's explicit perception of his position. Before 1926, Einstein saw the interest in unified theories as being largely shared by many in the general relativity community:

> The conviction of the essential unity of the gravitational field and the electromagnetic field is firmly established today among the theoretical physicists who work in the field of general relativity theory[106] (1925 Corpus, Einstein I, 414).

But he felt the ground to have largely shifted by the end of the decade:

> As to the way in which the problem [of a unified field theory] may be solved Professor Einstein says that it is a very difficult question to answer, and it has not yet been finished. His colleagues regard his view as a particular craze and do not support it (1930 Corpus, Einstein IV, 610).

Most references during the first half of the decade to Einstein's work in unified field theory, however, occurred as part of a generic name: the "Mie-Hilbert-Weyl-Einstein" approach or, later, the "Weyl-Eddington-Einstein" program. With rare exceptions, articles by Einstein himself in this period were neither developed nor commented on by others. Paradoxically, it was in 1929–1930, when Einstein complained most bitterly of his isolated position, that he, with his unified theory of that time, became the uncontested leader of the domain. As we have shown, almost three-quarters of all the articles of the 1930 corpus relate to Einstein's (and Mayer's) work on *Fernparallelismus*. Both our global data and our more detailed *sondage* thus contradict Einstein's feeling of increasing isolation. That the self-description of a scientist should not be taken at its face-value is a well-admitted historical rule: but at which value then should it be taken? Our study suggest two principal paths to trace this dissonance more accurately and integrate it into a more comprehensive view of Einstein's identity: the first leads to the identification of who counts for him as a 'significant' colleague;[107] the second to a more precise characterization of his program, the emphasis no longer being on a mere geometrical unification of gravitation and electromagnetism, but on the much stronger requirement that matter appear as a consequence of the field theory, with the consequent gradual distancing of Einstein from the main trends in the theoretical physics of the time.

This last suggestion offers a smooth transition to the question of quantum matter. In large part because of Einstein's increasing opposition, quantum theory has been often presented as the alternative to geometrical unified field theories, and its successes a progressive trespass on their territory. What we have seen is different and the demarcation lines are not so clear. It appears that there were never two hermetic programs vying for hegemony, classical and geometrical on one side and quantum on the other. From the beginning of our decade, quantum theory appears, at least as an horizon, for even the most avid promoter of the geometrical approach at the time. The complete Weyl quotation, of which a part begins this article, looks like this:

> A new theory by the author has been added, which … represents an attempt to derive from world-geometry not only gravitational but also electromagnetic phenomena. Even if this theory is still only in its infant stage, I feel convinced that it contains no less truth that Einstein's Theory of Gravitation — whether this amount of truth is unlimited or, what is more probable, is bounded by the Quantum Theory[108] (Weyl 1919, vi).

And very soon quantum elements occur as an effective component in a number of proposals, as we have seen for 1925 in — the very different — Kudar I–III and Vallarta. In fact such attempts occur as early as 1922, in unified theories which are combinations of a geometrical approach and the older quantum theory reworked in various patterns (e.g., Schrödinger 1922; Wilson 1922; Wereide 1923). In this sense, the success of the quantum program, as witnessed in the attempts to integrate it with a theory of gravitation, is at once earlier than usually placed, but less devastating in its impact for classical unification theories.

Moreover, we have seen not one quantum theory, but a variety of quantum approaches: the Bohr-Sommerfeld quantum rules, the Schrödinger-Dirac wave mechanics (though only occasional allusions are made to the matrix mechanics approach); at the end of our decade, it is the budding quantum electrodynamics that is seen as a true alternative to the flagging Dirac-*Fernparallelismus* agenda:

> Until recently there seemed to be little doubt that the connecting link between the unified theory and the quantum theory would be found through some generalization of the Dirac

> equations, as suggested by Wigner, Wiener and Vallarta, Tamm, Fock, Weyl and others. None of these attempts has proved satisfactory and some of them have been shown to be definitely erroneous. An entirely new method of attack, however, has been opened by the quantum electrodynamics of Heisenberg, Pauli, Jordan, and Fermi ...(1930 Corpus, Rosen and Vallarta, 119–120).

Indeed, if quantum theory replaces anything — at least in those articles reviewed in sections devoted to gravitation — it is one form or another of the older theories of matter, and, in particular for our decade, the (classical) theory of the electron. This should be taken not merely in the obvious sense that such quantum theories become the new foundations for matter, but in the sense that quantum approaches take over exactly the various functional roles occupied by the older theories in the unification programs. From 1925 on, they are used, sometimes on an equal footing with gravitation, sometimes as a means of replacing the phenomenological aspects of gravitation or electromagnetic theory by a first-principle theory, sometimes as a source of reduction of one class of phenomena to another.

Let us now return to the questions we raised in the introduction, in particular that of unification theories as collective production. In this respect we have found important modifications during the twenties, modifications in the content and the techniques, of course, but also in the organization of work, for instance in the rhythm and type of publications.

In 1920, the German-language scene was dominant and, on the whole, little disposed to look beyond its borders. A clear epicenter was located around the Hilbert-Weyl program, though there existed a wide variety of alternative proposals. The various directions of research, however, all bore the imprint of the still recent success of general relativity, either in a positive or negative sense.

In 1925, textbooks on general relativity have widely diffused a common set of basic tensorial and Riemannian techniques; there existed a more international, but more scattered, scene — though still largely European — with proposals concerned with the exploration and completion of a relatively limited number of specific theories. In particular, the idea of unification as a geometrical combination, on an equal footing, of Einstein's 1916 theory of gravitation and Maxwell's electromagnetism is well-established, even if neither universally accepted nor necessarily coupled with a reification of geometry. Indeed, especially in Great Britain, there are lively debates on the respective role of geometry and physics, having concrete resonance in scientific work and engaging major figures on both sides. Moreover, quantified matter has entered the picture, as a technical part of several unification programs, and as an alternative to both continuous and classical particulate theories of matter.

In 1930, finally, an overwhelming interest is expressed for a single theory, explored, however, in a variety of directions and at a rapid pace — with a residual interest for a second approach and a collection of isolated projects. The scene is world-wide, the massive arrival of US and Soviet scientists being marked by an emphasis on institutionally centered group work with a strong division of labor. The newcomers pursue however two very different publication policies: the first national, with *Physical Review* and *Proceedings of the National Academy of Science*, the other oriented towards publication in foreign, particularly German and, for short notes, French journals. Quantum theory is widely recognized as an inevitable component of every future unification, even if its nature, its role and its interplay with other phenomena remain variable.

Some of these aspects require a comparative perspective — and thus complementary studies — to be properly appreciated. We can at least underline that taking into account a larger corpus than is usually done allows us to restore the concrete texture of the debates at this early period and to make precise the periodization of the various proposals. Thus, no historiography which selects only the most famous German-language authors (plus Eddington as honorary member) can hope to capture the global dynamics, which requires a knowlege of the standpoints of other groups.

Indeed, a crucial problem in the understanding of historical dynamics lies in the great sensitivity of its models to selection effects. For instance, it would be easy, by picking out appropriate elements, to mimic here a Kuhnian dynamics for the genesis of a new discipline: an initial dispersion of interests, a preliminary coagulation in a range of systematically explored possibilities, and the final emergence of a paradigm, here *Fernparallelismus*. But we know of course that this conclusion does not hold: the flock of sparrows on the Einsteinian *Fern*cake, including Einstein himself, scatters almost immediately. The brief fame of *Fernparallelismus* does not result in a victory in 1930 of a 1925 competition between rival affine theories: there was no such competition and morevoer, Einstein's theory with distant parallelism was not even perceived, at the beginning, as a direct successor of affine ones.

The lack of continuity is apparent at other levels as well. The range of combinations seems molded for most authors far more by the concrete possibilities of available techniques than by a more global conviction concerning the constitution of the world. The goal of unity might be an ideal, its embodied shape is often the consequence of very technical constraints. Links between the various constitutive elements of the proposals then are ephemeral and local; geometry and variational principles, for example, are much less associated in later versions of *Fernparallelismus* than they were in Eyraud's work. To paraphrase Marx as well as Einstein, we lack sufficient overdetermination to suggest a satisfying dynamics at this level.

Moreover, the period of production of many of the scientists engaged in unification programs is short. As a look at the scientific biographies of our authors testifies, most do not remain long in the field.[109] There was no specialist of unification, no "unitarist," as one might have been, in the same period, differential geometer or relativist.

But, while the short active lifetime of a scientist in unification work shows that unification did not constitute a discipline, the very variety of these scientists indicates, strangely enough, a standard, 'normal', research activity.[110] We do not find only a few geniuses and cranks, but all sorts of scientists, at various stages in their careers; some of exceptional rank, a good number more or less well-known, and, in general, regularly productive, most of them in full-time positions as physicists or mathematicians (with the usual exceptions of a few teachers, engineers and unemployed). Nor is unification merely an activity for the elderly; each year, we find contributions originally developed as theses. And the references and other information bear witness to quite a regular flow of exchange and communication.

We have then not the constitution of a discipline,[111] but activity in a respectable area of research. To grasp the nature of this topic during our decade and its evolution, we have to take into account the concrete tensions which structure the configurations of articles we have detected and look at the elements which have been stabilized during this period. Two features, already evoked, would require a larger perspective to be articulated, because of the shift in time of their impact. One, the effect of which is increasingly perceptible in the second half of the decade, concerns the very conception of matter and of its role: quanta evacuate, in *this area of physics*, most other representations or theories. The second feature, appearing only at the very end, is the overall transformation of the research activities and publications, the transformation from the cottage industry of 1920 to the industrial enterprise of the thirties and later.

But the major component has to do with general relativity as the dominant theory of gravitation: although no new (non-cosmological) experimental evidence was found during our decade, our study clearly shows an acceptance of Einstein's 1916 theory — for some, in 1930, even *contra* Einstein. The alternative theories (Whitehead's or Wiechert's for example) we have seen during the first half of the decade have disappeared at its end or have been marginalized. As is confirmed by the place of unified theories in the reviews on gravitation, unification is one important, current way of learning and working in general relativity, during a period the last half of which marks the beginning of the *étiage*, to use Jean Eisenstaedt's term (Eisenstaedt 1986). Paradoxically enough, it is sometimes through the most exotic efforts to go beyond it that a scientific theory consolidates its status.

ACKNOWLEDGEMENTS

We wish to warmly thank the Max-Planck-Institut für Wissenschaftsgeschichte in Berlin which offered its stimulating and efficient hospitality during the preparation of this paper. Urte Brauckmann's assistance in tracing some rare journals has been invaluable. Our very special gratitude goes to Felix A. E. Pirani who helped us obtain several useful documents and carefully read the first draft of this paper.

CNRS - Université de Paris Sud (UMR 8628)
Université de Paris 8

NOTES

* To John Stachel, in affection and comradeship.

1. The expression is Robert D. Carmichael's in a 1926 debate on the theory of relativity (Carmichael 1927, 12).
2. For an interdisciplinary discussion of the role of unification in various domains and, in particular, of the political issues associated with them, see (Galison and Stump 1996).
3. "Dann aber ist eine neue, vom Verfasser herrührende Theorie hinzugefügt worden, welche ... aus der Weltgeometrie nicht nur die Gravitations-, sondern auch die elektromagnetischen Erscheinungen abzuleiten. Steckt diese Theorie auch gegenwärtig noch in den Kinderschuhen, so bin ich doch überzeugt, daß ihr der gleiche Wahrheitswert zukommt wie der Einsteinschen Gravitationstheorie ... " English translation by H. L. Brose in (Weyl 1922, xi).

4. "Einstein's Arbeit ist ein neuer Beitrag zu einem Versuch, den er vor etwa Jahresfrist unternahm — eines neben vielen, vielen anderen, die in den letzten zehn Jahren unternommen wurden. ... Ich glaube, daß durch die Entwicklung der Quantentheorie in den letzten Jahren die Problemlage so verschoben ist, daß man nicht erwarten kann, die gesuchte Einheit zu finden, ohne die materiellen Wellen in das Schema mit einzubeziehen, durch welche die Wellenmechanik die sich bewegenden Materieteilchen ersetzte" (Letter from Hermann Weyl to James Stokley, 3 February 1929).
5. On the history of unified theories, see the pioneering books of Marie-Antoinette Tonnelat (Tonnelat 1965, 1971) and the recent (1985) excellent synthesis by Vladimir Vizgin (quoted here from the English translation: Vizgin 1994), as well as the articles (Goenner 1984; Bergia 1993).
6. See for instance (Garfield 1964; Price 1965; Callon et al. 1986).
7. We prefer this term, as used by Norbert Elias in, for example, *Engagement und Distanzierung*, to the term 'community', which Thomas Kuhn's *Structure of Scientific Revolutions* has made familiar to historians of science, because, as we shall see, even tight relationships do not necessarily imply the emotional commitment or sense of sharing involving human beings as a whole and implicitly conveyed by the latter term.
8. We use the traditional archeological term "sondage," in the usual English sense of trial-trenching as a preliminary to full excavation, see, for example, Mortimer Wheeler's *Archaeology from the Earth*.
9. Quite the contrary: we shall miss out 1921, Vizgin's "pivotal year" for affine theories; 1923, and Cartan's classification of affine spaces; 1926/1927, and the birth of a new and successful quantum program. Nor shall we see most contributions to Kaluza-Klein theory (although other five- and higher-dimensional theories will make their appearance). But as will be seen, we shall be able to see their effects (or lack thereof) in our decade.
10. Within each section, the articles are grouped under specific headings, but these subsections changed continuously, including for instance a heading "Light" or "Quanta" in the section on relativity for certain years.
11. The problems of the *Jahrbuch* during this period have been closely studied by Reinhard Siegmund-Schultze (1993).
12. Quite typically, its publication date is 1932; in fact the change is made first in 1931, with the '1927' volume.
13. The years mentioned correspond to the dates of the reviews and not to the publication dates of the articles. For *Physikalische Berichte*, the volume year corresponds roughly to publication dates in the last half of the previous year and the first half of the nominal year.
14. The traditional view already appeared dubious to Hubert Goenner in his study of German books on relativity (Goenner 1992; see also Eisenstaedt 1986).
15. We are comforted in this choice by the fact that those responsible for classifying the articles have also put into this section, at least at the beginning of our period, attempts to unify various phenomena other than gravitation, and which contest (part of) relativity theory.
16. Although a case might be made for it, we shall not consider general relativistic thermodynamics as a unified theory.
17. It is interesting to note a certain degree of specialization in the reviewers themselves, such as Philipp Frank, who signed the review of almost all these papers at the beginning of the period in the *Jahrbuch*, or Cornelius Lanczos who was in the same privileged situation for the *Berichte* from 1925 on.
18. The year 1930 marks the peak in their production: in 1935, 22% of the relativity papers are devoted to unification and in 1940, 15%, but this proportion should be appreciated against a background of a drastic reduction in absolute numbers; between 1920 and 1940, the number of articles in relativity falls by one-half.
19. As has been explained, by "year," we mean the volume year of *Jahrbuch über die Fortschritte der Mathematik* and *Physikalische Berichte*; their combined coverage corresponds to roughly one and a half years of actual publications.
20. In general, we insist on the distinction between articles and authors. Someone like Einstein, who tried almost every approach to unified (field) theories in turn, and sometimes several in one year, makes the point. In 1920, however, papers by the same author are continuations of each other and, save explicit mention, this distinction will be relaxed here for simplicity.

21. As in each of our years, there are one or two doubtful cases; for instance, we have retained Walter Dällenbach's articles, although they appear to be only a special relativistic extension of Lorentz's 1904 theory, and this on two grounds. Dällenbach announces — too optimistically — the possibility of an "obvious" extension to general relativity and Lorentz's theory itself can be seen as a step towards a global interpretation of natural phenomena, see (Vizgin 1994; Miller 1981). In any case, these borderline cases do not change the general picture.
22. For a start on the rich literature on this subject, see (McCormmach 1970; Miller 1981; Hunt 1991; Darrigol 1996).
23. "Eine Entdeckung ist gemacht worden!"
24. These last represent some of the communications made at the famous meeting in Bad Nauheim, where Einstein had to confront the hostility of Philip Lenard and other tenants of anti-relativistic "German physics," see (Goenner 1993a, 1993b).
25. All the more so because the moment we are looking at offers all too many temptations towards attempting a description in terms of 'national' styles or schools. For a clear vision of the pitfalls in such an approach see the counterexamples in (Warwick 1992–93).
26. See Docs. 230, 232, 236, 240, 413, 438 (De Donder) and 328, 408 (Bateman) in (Schulman et al. 1998). De Donder seemed to see himself as the Lagrange of an Einsteinian Newton, though Einstein was quite critical of De Donder's approach which appeared to him to be an elimination of physics from relativity theory.
27. Chapter IV, note 30 of the third edition (Weyl 1919) mentions Reichenbächer alone, to which chapter IV, note 32 of the fourth edition (Weyl 1921) adds Abraham, Nordström and Wiechert.
28. See (Sánchez-Ron 1999) on Larmor and relativity. Even Eddington's views should not be confused with Einstein's, see (Stachel 1986).
29. See, for different aspects, (Sigurdsson 1991, 1994; Scholz 1995, 1999) and (Vizgin 1994, ch. 3), which also discusses Einstein's and Pauli's reactions.
30. "Dann würden nicht nur die Gravitationskräfte, sondern auch die elektromagnetischen aus der Weltmetrik entspringen; und da uns andere wahrhaft ursprüngliche Kraftwirkungen außer diesen beiden überhaupt nicht bekannt sind, würde durch die so hervorgehende Theorie der Traum des Descartes von einer rein geometrischen Physik in merkwürdiger, von ihm selbst freilich gar nicht vorauszusehender Weise in Erfüllung gehen, indem sich zeigte: die Physik ragt mit ihrem Begriffsgehalt überhaupt nicht über die Geometrie hinaus, *in der Materie und den Naturkräften äußert sich lediglich das metrische Feld*. Gravitation und Elektrizität wären damit aus einer einheitlichen Quelle erklärt."
31. Einstein's commented on this move at Bad Nauheim in these terms: "Since Weyl's theory abandons this empirically grounded category, it deprives the theory of one of its most solid empirical supports and test possibilities." ("Indem die W e y lsche Theorie auf diese empirisch begründete Zuordnung verzichtet, beraubt sie die Theorie einer ihrer solidesten empirischen Stützen und Prüfungsmöglichkeiten.") Einstein *apud* Weyl II: 651.
32. Invariant theory was, of course, Weitzenböck's specialty; in the same year, for example, he published an article doing a similar job for the "Galilei-Newton" group (Weitzenböck 1919/20).
33. His communication was to be criticized the following year by Erwin Freundlich (1920), acting as a defender of Einstein's position, for its neglect of physical contents; relativity, Freundlich would explain, is not a mathematical but a physical theory. In many respects this controversy recalls that between Einstein and the mathematicians Hilbert and Weyl. For the opinions of Einstein, see (Vizgin 1989, 1994, 98–104).
34. "So sind alle physikalischen Gesetze schließlich zurückgeführt auf das einzige Problem der Metrik der ... vier-dimensionalen räumlich-zeitlichen Mannigfaltigkeit. ... Eine der wichtigsten Zukunftsaufgaben, die in dieser Hinsicht ... gestellt ist, ist wohl die Einfügung der *Quantentheorie* in das System der allgemeinen Relativitätstheorie.
Bei der Inangriffnahme dieser Aufgabe müßte die physikalische Axiomatik offenbar an einen Gedanken anknüpfen, den schon ... Riemann ... geäußert hat: daß nämlich das Objekt der Geometrie auch eine *diskontinuierliche Mannigfaltigkeit* sein könnte.... Wäre aber die ... Mannigfaltigkeit selbst diskontinuierlich aufzufassen, dann würde es begreiflich sein, warum die bei bestimmten physikalischen Prozessen auftretende Menge an Wirkung notwendigerweise ein ganzzahliges Vielfaches eines *elementaren Wirkungsquantums* sein müßte."

35. The case of Weitzenböck and Pauli vindicates our reticence to speak of "community": while their articles, as well as Weyl's, were certainly involved in a tight network of exchanges favoring quick responses, there was nonetheless no question of personal commitment of the authors to the theory itself.
36. Vizgin mentions Mie in his list of opponents to Einstein, in particular at Bad Nauheim. We have not found any evidence that points in this direction; on the contrary, at this time, Mie seems quite enthusiastic about Einstein's achievements, see (Illy 1992). His critical comments seem mainly directed against calling Einstein's theory of gravitation a theory of 'general relativity'. Moreover, his name does not appear on the lists provided by the "Hundred Authors against Einstein," see (Goenner 1993b).
37. "wunderbare, vollendet schöne mathematische Struktur"
38. Note that these two form precisely the intersection of our authors with the list of anti-relativists published in (Goenner 1993b). However, their papers show clearly that there was not, in their case at least, any question of negating the importance of Einstein's work.
39. "Hierbei ergibt sich die Massendichte im Gegensatz zu der gewöhnlichen Anschauung nicht als Skalar, sondern als 44-Komponente eines 16gliedrigen Tensors ... . Dies und die Tatsache, daß die $g_{\mu\nu}$ wegen ihrer Abhängigkeit von der Wahl des Koordinatensystems einer freilich eingeschränkten Willkür unterworfen sind, hat mich in der Einsteinschen Theorie nicht befriedigt, und ich habe deshalb in meiner Arbeit: "Grundzüge zu einer Theorie der Elektrizität und der Gravitation" versucht, die Theorie eines skalaren Gravitationspotentials, das ich mit der Lichtgeschwindigkeit identifizierte, aufzustellen, wobei ich an bestimmte Voraussetzungen über die Gravitationserregung durch die Elektronen anknüpfte, die ich — positive und negative — als das einzig Materielle ansah. Die einfachste Fall ... eines einzigen Elektrons hatte ich dabei erledigt und nach Analogie dieses Falles allgemein die Gleichung ... aufgestellt."
40. A scalar theory had, of course, already been proposed by Einstein some years earlier (Einstein 1912a, 1912b) in the context of a theory of the static gravitational field.
41. "Es ist demnach möglich, zu einer ... Lösung des Weltproblems ... zu gelangen, wenn man ... sich damit auf einen realistischen Standpunkt gegenüber dem mehr phänomenalistischen der Relativitätstheoretiker stellt."
42. "Das Fundament der Theorie soll die Annahme sein, daß die molekulare Materie aus elektrischen Teilchen aufgebaut ist. Es wird damit die Elektrisierung als eine Grundeigenschaft aller Bausteine der Materie erklärt. Die Annahme erscheint als die natürliche Folgerung aus den Ergebnissen der molekularphysikalischen Forschung der letzten drei Dezennien. Einst lehrte die Elektrodynamik in der Elektrisierung eine wesentliche Ursache der *Trägheit* kennen, nun soll der Nachweis versucht werden, daß die Elektrisierung auch eine wesentliche Ursache der *Gravitation* ist."
43. The reception of relativity theory at Cambridge has been thoroughly explored by Andrew Warwick (1988; 1992–93), who has also stressed the differences, as well as the links, among these people. Note that, though Bateman was at this time at CalTech (then known as Throop's College) working on hydrodynamics, his earlier career was a typical, though brilliant, Cambridge one, and, in this field at least, he published in a British journal, the *Philosophical Magazine*.
44. For a *mise en contexte* of this publication, see (Sánchez-Ron 1999).
45. It may be revealing to note that, in the journal where they appear, De Donder's and Vanderlinden's articles are classified in the "mathematical physics" rather than the "theoretical physics" section.
46. An instance is Ludwik Silberstein, a Polish physicist whose path crosses a number of countries; at this time he publishes principally in British journals and his work is quite abundantly, though not exclusively, discussed by British and American authors. On Silberstein see (Sánchez-Ron 1992) and, for his later debate with Einstein, (Havas 1993).
47. We will come back soon to these theories. Note that this alignment with particular traditions does not mean that the same author is always restricted to just one. Henri Eyraud, for instance, devotes one note to exploring the framework of Weyl's geometry in 1924 (I) but then turns more systematically to the consequences of Schouten's point of view (II, III).
48. Similar concepts were independently invented by a number of mathematicians. For a history of this topic, see (Reich 1992).
49. Note that Arakatsu defines the covariant derivative of a covariant vector for this second connection (equation (2.7) in his paper) in a way which would imply that the two connections are in fact the same. His definition can be corrected, however, without harm to the conclusions of the article.

50. In fact, one of the most articulate speakers in 1920 against a geometry having no regard for experiment is Einstein himself.
51. It is this dichotomy, for instance, which helps Vizgin to define two distinct research programs in the Lakatosian sense, see (Vizgin 1994, 129).
52. "Entweder man hält die Zeit für noch nicht gekommen und verschiebt seine Lösung, bis die vielleicht gleichen Quellen der noch rätselhafteren, an die Namen von Planck, Einstein und Bohr geknüpften Quantengesetze freigelegt sind.
    Oder man hält die Lösung auf den von Maxwell, Lorentz, Mie, Einstein, Hilbert, Weyl, u. a. geschaffenen Grundlagen für möglich."
53. "Définir la physique quantique au sein de la physique relativiste comme une espèce dans un genre par l'adjonction d'un caractère spécifique." We might mention that Wavre is rather sceptical of the chances of success for this approach that he sees as a last attempt to avoid the discretization called for by quantum theory.
54. Note that the title of the paper, "Einheitliche Feldtheorie … " marks Einstein's first public use of the term 'unified field theory' in connection with this topic.
55. Einstein was still at that time quite concerned with quantum theory. Explaining this new project in a letter to his friend Michele Besso, he writes: "This is then a magnificent opportunity, which should probably correspond to reality. There now arises the question whether this field theory is compatible with the existence of atoms and quanta." ("Dies ist doch eine prachtvolle Möglichkeit, die wohl der Realität entsprechen dürfte. Nun ist die Frage, ob diese Feldtheorie nicht der Existenz der Atome und Quanten vereinbar ist.") (Einstein 1925). On the complex relationship between Einstein and quanta, see (Stachel 1993).
56. This paper seems never to have been published.
57. On Eddington's epistemology, see (Merleau-Ponty 1965; Kilmister 1994a). On the hostile reactions to Eddington and especially Jeans, on these issues, see (Sigurdsson 1996). For Lodge's views at this time, see (Rowlands 1990, 270–290) as well as (Sánchez-Ron 1999).
58. This early example of "numerology" had to be retracted a month later (Rice II) in a letter to the editor of the same review; Rice had misread the length units in which the radius of the universe he used were expressed. It is ironical that the current values of R are just what Rice needed!
59. There are of course several intermediate cases, as illustrated by Arakatsu's article, already discussed.
60. We do not consider here in detail the question of the ontological commitment of these authors, nor the meaning of the word "geometry" (or "physics"), and their relations to the question of matter, for them; we propose to examine these issues elsewhere.
61. This approach, which passed practically unnoticed at the time, except by Einstein himself (Ritter 1993, 142), was rediscovered and prominently featured by John Wheeler and his school in the 1950s and 1960s under the name of "geometrodynamics" (Wheeler 1962); see (Stachel 1974). Note that virtually all of Rainich's publications on the question were in American mathematical journals, which may account for their lack of impact on the scientists we study here.
62. Eyraud, in fact, learned general relativity in Weyl's 1917 course at the ETH in Zurich, where he had been placed by the Red Cross, under Swiss control, as an ex-prisoner of war. We would like to thank M. Gustave Malecot for this information (private communication).
63. "Le potentiel vecteur trouve son expression géométrique dans la torsion."
64. On Temple's style of work and his relation to Whitehead, see (Kilmister 1994b), in particular p. 386. We are grateful to the author and to Felix Pirani for this reference.
65. This manifold is the one associated with measurement, the coordinates being evaluated by means of clocks and rigid rods. Whitehead (1922) had originally considered "true" space to be flat.
66. Such considerations have become, of course, more familiar in a quantum field-theoretic context, under the name of "*CPT* invariance;" see for example (Pais 1986, 525–529).
67. "Wesentlich scheint mir die Erkenntnis zu sein, daß eine Erklärung der Ungleichartigkeit der beiden Elektrizitäten nur möglich ist, wenn man der Zeit eine Ablaufsrichtung zuschreibt und diese bei der Definition der maßgebenden physikalischen Größen heranzieht. Hierin unterscheidet sich die Elektromagnetik grundsätzlich von der Gravitation: deshalb erscheint mir auch das Bestreben, die Elektrodynamik mit dem Gravitationsgesetz zu einer Einheit zu verschmelzen, nicht mehr gerechtfertigt."

68. In addition, a number of papers mixing gravitation and quantum theory are now classified in the section "Quantenlehre" of the reviewing journals and thus do not enter into our corpus.
69. And not during the "summer of 1929," as sometimes suggested in the current literature. As we shall see, the precise dates play a role in the interpretation of the event.
70. The importance of Soviet work around general relativity, quantum physics and unified theories may surprise those who have read of the critical manner in which these theories were supposed to be viewed by orthodox Marxist-Leninist philosophers and politicians in the Soviet Union as early as the late 1920s. While the Russian physicist Yuri Rumer, working in Born's laboratory in Göttingen, felt that conditions were indeed difficult for those working in these fields (Born 1929), the holding of the Kharkov conference and the number of Soviet physicists working in these areas seems to raise some serious doubts. See (Graham 1966).
71. Lively examples of the appeal of this theory for a general public are given in (Pais 1982, 346). However the stir was not limited, as Pais implies, to popular journalists and their readers.
72. The problem of notation, already mentioned, is specially interesting in the case of the *Vierbein*. Einstein himself changed his notations several times, adopting Weizenböck's when this author pointed out to him previous work on similar spaces (Weizenböck 1928), changing again when Cartan's priority was established (Cartan 1930). However, to follow these changes would have been intractable in an article and, on this point, we have chosen to uniformize the notation.
73. The corresponding lines in (Einstein 1928, 219) are incorrect: the reference in the line leading up to eq. 7a should be to eq. 4 and not eq. 5, and the expression for dA^ν includes a mysterious $h^{\nu a}$ instead of h_a^ν. Such typographical errors are not uncommon in Einstein's publications in the *Sitzungsberichte*.
74. Cartan has already been mentioned in the 1925 *sondage* a propos of Eyraud's article, which used an affine space with non-zero torsion and curvature. Indeed, Einstein acknowledged the lack of novelty of his theory in this respect, after this had been pointed out to him by several authors. Cartan himself, in a letter of 8 May 1929, reminded Einstein that he had spoken to him of this very possibility as early as 1922, during Einstein's visit to Paris (Cartan 1929). This letter is the origin of the historical survey article published by Cartan, at Einstein's request, in the 1930 *Mathematischen Annalen* (Cartan 1930), and of an important exchange on the mathematical and physical possibilities of the theory between the two scientists, published in (Debever 1979).
75. " ... eine metrische Kontinuumsstruktur, welche zwischen der Riemannschen und der Euklidischen liegt." Despite the corrections published by several mathematicians, the same point of view is maintained in Einstein's adress at Nottingham, as late as June 1930, transcribed by I. H. Brose in *Science* (Einstein IV).
76. Precisely which laws they yield depends in fact on the version of the theory under consideration; for reasons of space, we shall not enter into this question here.
77. See the reactions of Weyl and Pauli mentioned in (Pais 1982, 347).
78. "Ein Rückgang zu den alten Relativitätstheorien, [vierdimensionalen... wie auch fünfdimensionalen...] scheint jedoch ein für allemal ausgeschlossen zu sein. Man hat mit dem Gedanken an einen Fernparallelismus wirklich einen Erkenntnisschritt gemacht!"
79. For a physicist's survey of the later developments of this topic, see (Kichenassamy 1992).
80. The note is slightly too early to be included in our corpus for 1930, but the continuation of the program is included. See also (Vizgin 1994, 246).
81. For a number of them, in particular Einstein's articles, cf. also the discussions in (Vizgin 1992, 234–255).
82. In particular, it is sometimes useful to distinguish between the date of submission to a journal, or the date of presentation to a conference, and the date of publication.
83. " ... die Auffindung der einfachsten Feldgesetze, welchen eine Riemannsche Mannigfaltigkeit mit Fern-Parallelismus unterworfen werden kann."
84. "Der große Reiz der hier dargelegten Theorie liegt für mich in ihrer Einheitlichkeit und in der hochgradigen (erlaubten) Übereinstimmung der Feldvariablen."

85. "Je höher die Zahl der Gleichungen ist (und folglich auch der zwischen ihnen bestehenden Identitäten), desto bestimmtere, über die Forderung des bloßen Determinismus hinausgehende Aussagen macht die Theorie: desto wertvoller ist also die Theorie, falls sie mit den Erfahrungstatsachen verträglich ist." Not only does Einstein hope to constrain the initial conditions as far as possible, but he equally wants to account for the specific conditions put forward by quantum theory.
86. "Wohl aber erkennt man, daß in der neuen Theorie die Singularitätsfreiheit derjenigen Lösungen verlangt werden muß, die die Elementarpartikeln der Materie darstellen sollen."
87. Among the reasons given in the individual articles to dismiss the previously announced equations, we find objections stemming from Einstein himself as well as criticisms by others, in particular Lanczos and H. Müntz. None of this is mentioned in the IHP lecture, but has to be taken into account in order to understand Einstein's variable mood during this period. Moreover, the identification of the physical quantities with the mathematical elements of the theory is also relaxed.
88. "In der Zwischenzeit habe ich zusammen mit Dr. Mayer viel über den Gegenstand gearbeitet und bin von den damaligen Feldgleichungen abgekommen."
89. "Nach dem Grundgedanken der fünfdimensionalen Theorie ist die Gesamtheit der ∞^4 Weltpunkte nicht als Gesamtheit der ∞^4 Punkte einer Hyperfläche im R_5, sondern als Gesamtheit der ∞^4 Linien (der Kongruenz X_5^i) im zylindrischen R_5 aufzufassen."
90. For a presentation of the Princeton school and their program of a new differential geometry based on paths, see (Eisenhart 1927).
91. Here h_α^i is $h_{i\alpha}$ in Einstein's previous notation.
92. A lively recollection of the group is to be found in (Struik 1989), in particular, p. 172.
93. Their paper in our corpus is a corrected version of this, sent to the same journal on 23 May.
94. These conditions are taken from Eddington's 1923 book on relativity, in his presentation of the classical Schwarzschild solution.
95. In these cases, as well as in the essentially analogous theory in Einstein's January 1929 article, the result is to be expected because the equations lead, in first approximation, to those of general relativity. It should be noted that in December 1929 (Einstein III, 18) Einstein acknowledges an error in his March 1929 paper.
96. See also his note to the French Academy of Sciences of 15 April 1929 (Tamm 1929).
97. This idea was also advocated by, among others, Wiener and Vallarta in their work discussed above.
98. "Die Tatsache, daß das skizzierte Verfahren wirklich zu einer vernünftigen Wellengleichung führt, ist deshalb besonders interessant, weil der zu der erwähnten wellenmechanischen Annahme analoge "klassische" Ansatz: die auf die 4-Beine bezogene Bewegung des Elektrons sei immer gleichförmig, zu keinen brauchbaren Bewegungsgleichungen führt. *Somit erscheint in der Einsteinschen Theorie das wellenmechanische Prinzip dem Prinzipe des kürzesten Weges der geometrischen Optik übergeordnet.*"
99. Lanczos, who reviews the paper in the *Physikalische Berichte*, points out that this result might have been foreseen directly from the field equations, without the need of further computation, see note 95.
100. Besides the interpretation discussed here, another possibility suggested is that these results only point to the necessity of new field equations. We do not have however any evidence that the new field equations derived by Einstein one month later, or indeed any others, renewed their interest in these questions. The search for axial symmetry however was taken up by others, e.g., (McVittie 1930/31), with no satisfactory result.
101. "Einer von uns hat kürzlich zu zeigen versucht, wie ungezwungen die Wellengleichung des Elektrons sich in der neuen Einsteinschen Theorie ergibt, und dabei die Vermutung ausgesprochen, daß in dieser Theorie das wellenmechanische Prinzip dem Prinzip des kürzesten Weges übergeordnet ist, so daß die Bewegungsgleichungen einer (geladenen) Korpuskel durch einen Limesübergang aus der Wellengleichung abzuleiten sind. Wenn diese Vermutung und auch die Vermutung, daß die einer geladenen Partikel entsprechende Lösung der Einsteinschen Feldgleichungen von dem Spin der Elementarladungen Rechenschaft gibt, sich wirklich bestätigen sollte, so wird damit die mikroskopische Deutung der Einsteinschen Theorie weitgehend gestützt sein."

102. These papers, (Fock and Iwanenko 1929a, 1929b) — as well as Fok's own search for a union of Dirac theory and general relativity (Fock 1929a–c) — are missing from our corpus since they were reviewed in the quantum physics section of *Physikalische Berichte*; the same is true of other, similar approaches (Wigner 1929; Weyl 1929a, 1929b). We shall therefore not discuss them here, see (Vizgin 1994) for some of them.
103. In this respect, our study helps to understand the chronology of the reception of Kaluza-Klein theories in the 1920s. What little impact Kaluza's original article (Kaluza 1921) had on the physics community had completely dissipated by 1925, while the echoes of the more influential reworking by Klein (1926) had been only partly drowned out by the tidal wave of distant parallelism after 1929.
104. This difference could be traced to a question of personal relationships. Whereas Zaikov had studied in Göttingen and Berlin, Rumer works at Göttingen at this time in Born's laboratory, Mandel', as we have pointed out, was an active participant at the Kharkov conference and in personal contact with Einstein, Lagunov, on the contrary, is not listed as a participant at Kharkov.
105. Such an emphasis is already noticeable in his development of general relativity, see (Renn and Sauer 1999). Field equations, and their degree of overdetermination, were also Einstein's main interest in his correspondence with Cartan in the thirties (Debever 1979).
106. "Die Überzeugung von der Wesenseinheit des Gravitationsfeldes und des elektromagnetischen Feldes dürfte heute bei den theoretischen Physikern, die auf dem Gebiete der allgemeinen Relativitätstheorie arbeiten, feststehen."
107. In this respect, it is interesting to analyze the responses of his contemporary correspondents, see (Pais 1982, 347). Also telling is Vallarta's commentary, in Norbert Wiener's *Collected Works*, on their joint work on *Fernparallelismus* (Vallarta 1982); according to this account, Einstein's reformulation of his theory was a consequence of his reception of Vallarta's and Wiener's results concerning the first set of field equations, though Einstein never mentions this in his later publications.
108. "Dann aber ist eine neue, vom Verfasser herrührende Theorie hinzugefügt worden, welche ... aus der Weltgeometrie nicht nur die Gravitations-, sondern auch die elektromagnetischen Erscheinungen abzuleiten. Steckt diese Theorie auch gegenwärtig noch in den Kinderschuhen, so bin ich doch überzeugt, daß ihr der gleiche Wahrheitswert zukommt wie der Einsteinschen Gravitationstheorie — mag nun dieser Wahrheitswert ein unbegrenzter sein oder, wie es wohl wahrscheinlicher ist, begrenzt werden müssen durch die Quantentheorie." English translation by H. L. Brose in (Weyl 1922, vii).
109. It would be all the more interesting to look closely at the rare exceptions (besides the much studied Einstein, Reichenbächer, and De Donder), in the perspective of the constitution of individual trajectories and collective scientific production, see (Goldstein 1994) for examples in number theory.
110. We of course lack analogous studies for other contemporary topics in order to appreciate more precisely how 'normal' it was. Certainly, that people generally did not remain in a given area does not seem to us to be a regular feature in physics at the time.
111. It is remarkable in this respect that the varieties of unification are not reduced; in 1930 as in 1920, we have found reductionist projects, attempts to integrate different fundamental phenomena on an equal footing, replacement by one phenomenon of a specific aspect in the theory of another, etc.

REFERENCES

Primary Sources:

Abraham, Max. 1902. "Dynamik des Elektrons." *Nachr. Göttingen*: 20–41.
Bohr, Niels. 1913. "On the Constitution of Atoms and Molecules." *Phil. Mag. (6)* 26:1–25, 470–502, 857–875.
Born, Max. 1929. Letter to Albert Einstein, 12 August 1929. In *Albert Einstein – Max Born: Briefwechsel 1916–1955,* ed. M. Born. Munich: Nymphenburger (1969).
Carmichael, Robert D. 1927. "The Foundation Principles of Relativity." Pp. 1–38 in *A Debate on The Theory of Relativity*. Chicago: Open Court.
Cartan, Élie. 1923–25. "Sur les variétés à connection affine et la théorie de la relativité généralisée." *Ann. Éc. Norm.* 40:325–412; 41:1–25; 42:17–88.

———. 1922. "Sur une généralisation de la notion de coubure de Riemann et les espaces torsion." *CRASP* 174:593–595.

———. 1929. Letter to Albert Einstein, 8 May 1929. In (Debever 1979).

———. 1930. "Notice historique sur la notion de parallélisme absolu." *Math. Ann.* 102:698–706.

De Donder, Théophile. 1921. *La Gravifique einsteinienne*. Paris: Gauthier-Villars.

Dirac, Paul A. M. 1928. "The Quantum Theory of the Electron." *Proc. Roy. Soc. London A*. 117:610–624; 118:351–361.

Eddington, Arthur Stanley. 1921. "A Generalization of Weyl's Theory of the Electro-magnetic and Gravitational Fields." *Proc. Roy. Soc. London A* 99:104–122.

———. 1923. *The Mathematical Theory of Relativity*. Cambridge: Cambridge University Press.

Einstein, Albert. 1912a. "Lichtgeschwindigkeit und Statik des Gravitationsfeldes." *Ann. Phys.* 38:355–369.

———. 1912b. "Zur Theorie des statischen Gravitationsfeldes." *Ann. Phys.* 38:443–458.

———. 1917. "Kosmologische Betrachtungen zur allgemeinen Relativitätstheorie." *SPAW*: 142–152.

———. 1920. "Meine Antwort auf die antirelativitätstheoretische G.m.b.H." *Berliner Tageblatt und Handelszeitung*. 27. August: 1–2.

———. 1923a. "Bietet die Feldtheorie Möglichkeiten für die Lösung des Quantenproblems?" *SPAW*: 359–364.

———. 1923b. "Theory of the Affine Field." *Nature* 112:448–449.

———. 1925. Letter to Michele Besso, 28 July 1925. In *Albert Einstein, Michele Besso. Correspondance 1903–1955*, ed. P. Speziali. Paris: Hermann (1972).

———. 1928a. "Neue Möglichkeit für eine einheitliche Feldtheorie von Gravitation und Elektrizität." *SPAW*: 224–227.

———. 1928b. "Riemann-Geometrie mit Aufrechterhaltung des Begriffs des Fernparallelismus." *SPAW*: 217–221.

———. 1929a. "Einheitliche Feldtheorie und Hamiltonsches Prinzip." *SPAW*: 156–159.

———. 1929b. "Über den gegenwärtigen Stand der Feld-Theorie." Pp. 126–132 in *Festschrift zum 70. Geburtstag von Prof. Dr. A. Stodola*. Zürich and Leipzig: Orell Füssli.

———. 1929c. "Zur einheitlichen Feldtheorie." *SPAW*: 1–7.

———. 1931. Letter to Élie Cartan, 13 June 1931. In (Debever 1979).

Eisenhardt, Luther Pfahler. 1927. *Non-Riemannian Geometry*. (American Mathematical Society Colloquium Publications 8). New York: American Mathematical Society.

Fock [Fok], Vladimir A. 1929a. "Geometrisierung der Diracschen Theorie des Elektrons." *Z. Phys* 57:261–277.

———. 1929b. "L'Équation d'onde de Dirac et la géométrie de Riemann." *J. Phys. Rad.* 10:329–405.

———. 1929c. "Sur les équations de Dirac dans la théorie de relativité générale." *CRASP* 189:25–27.

Fock [Fok], Vladimir A., and Dmitri D. Iwanenko [Ivanenko]. 1929a. "Géométrie quantique linéaire et déplacement parallèle." *CRASP* 188:1479–1472.

———. 1929b. "Zur Quantengeometrie." *Phys. Z.* 30:648–651.

Freundlich, Erwin. 1920. "Zu dem Aufsatz 'Die Physik als geometrische Notwendigkeit' von Arthur Haas." *Naturwiss.* 8:121–127.

Kaluza, Theodor. 1921. "Zum Unitätsproblem der Physik." *SPAW*: 1479–1472.

Klein, Oskar. 1929. "Quantentheorie und fünfdimensionale Relativitätstheorie." *Z. Phys.* 37:895–906.

Larmor, Joseph. 1921. "The Einstein Theory. A Belgian Professor's Investigations." *Times*. 7 janvier: 8.

Levi-Civita, Tullio. 1917. "Nozione di parallelismo in una varietà qualunque e conseguente spezificazione geometrica della curvatura Riemanniana." *Rend. Cir. mat. Palermo* 42:172–205.

———. 1929. "Vereinfachte Herstellung der Einsteinschen einheitlichen Feldgleichungen." *SPAW*: 137–153.

Lorentz, Hendrik Antoon. 1904. "Weiterbildung der Maxwellschen Theorie. Elektronentheorie." Pp. 145–288 in *Encyclopädie der mathematischen Wissenschaften, mit Einschluß ihrer Anwendungen*, vol. 5. *Physik*, part 2., ed. A. Sommerfeld. Leipzig: Teubner.

MacVittie, George C. 1929. "On Einstein's Unified Field Theory." *Proc. Roy. Soc. London* 124:366–374.

———. 1930/31. "Solution with Axial Symmetry of Einstein's Equations of Teleparallelism." *Proc. Edin. Math. Soc.* 2:140–150.

Pauli, Wolfgang, Jr. 1921. *Relativitätstheorie*. Leipzig: B. G. Teubner.

Reichenbächer, Ernst. 1917. "Grundzüge zu einer Theorie der Elektrizität und der Gravitation." *Ann. Phys.* 4:134–173, 174–178.

Schouten, Jan A. 1923. "On a Non-Symmetrical Affine Field Theory." *Proc. Roy. Acad. A'dam* 26:850–857.

———. 1924. *Der Ricci-Kalkül*. (Grundlehren der mathematischen Wissenschaften in Einzeldarstellungen 10). Berlin: Springer.

Schrödinger, Erwin. 1922. "Über eine bemerkenswerte Eigenschaft der Quantenbahnen eines einzelnen Elektrons." *Z. Phys.* 12:13–23.

Silberstein, Ludvik. 1929. "General Relativity Without the Equivalence Hypothesis." *Phil. Mag. (6)* 36:94–128.

Sommerfeld, Arnold. 1919. *Atombau und Spektrallinien*. 1st. Braunschweig: Vieweg.

Tamm, Igor E. 1929. "La Théorie nouvelle de M. Einstein et la théorie des quanta." *CRASP* 188:1598–1600.

Weitzenböck, Roland. 1919/20. "Die Invarianten der Galilei-Newton-Gruppe." *Math. Ann.* 80:75–81.

———. 1928. "Differentialinvarianten in der Einsteinschen Theorie des Fernparallelismus." *SPAW*: 466–474.

Wereide, Thorstein. 1923. "The General Principle of Relativity Applied to the Rutherford-Bohr Atom-Model." *Phys. Rev.* 21:391–396.

Weyl, Hermann. 1919. *Raum-Zeit-Materie*. 3rd. Berlin: Springer.

———. 1921. *Raum-Zeit-Materie*. 4th. Berlin: Springer.

———. 1922. *Space-Time-Matter*. Translated by H. L. Brose. London: Methuen. English translation of (Weyl 1921).

———. 1929a. "Elektron und Gravitation. I." *Z. Phys* 56:323–334.

———. 1929b. "Gravitation and the Electron." *Proc. Nat. Acad. Sci.* 15:323–334.

———. 1952. *Space-Time-Matter*. New York: Dover. 1st American printing of (Weyl 1922).

Wheeler, John Archibald. 1962. *Geometrodynamics*. (Italian Physical Society. Topics of Modern Physics 1). New York: Academic Press.

Whitehead, Alfred North. 1922. *The Principle of Relativity with Applications to Physical Science*. Cambridge: Cambridge University Press.

Whittaker, Edmund T. 1939. "Parallelism and Teleparallelism in the Newer Theories of Space." *J. Lond. Math. Soc.* 5:68–80.

Wiechert, J. Emil. 1916. "Die Perihelbewegung des Merkur und die allgemeine Mechanik." *Nachr. Göttingen*: 124–141.

Wiener, Norbert, and Manuel Sandoval Vallarta. 1929a. "On the Spherically Symmetrical Statical Field in Einstein's Unified Theory of Electricity and Gravity." *Proc. Nat. Acad. Sci.* 15:353–356.

———. 1929b. "Unified Theory of Electricity and Gravitation." *Nature* 123:317. [Letter to the editor, 7 February 1929].

Wigner, Eugen P. 1929. "Eine Bemerkung zu Einsteins neuer Formulierung des allgemeinen Relativitätsprinzips." *Z. Phys.* 53:592–596.

Wilson, William. 1922. "The Quantum Theory and Electromagnetic Phenomena." *Proc. Roy. Soc. London* 102:478–483.

Secondary Sources:

Bergia, Sylvio. 1993. "Attempts at Unified Field Theories (1919–1955). Alleged Failure and Intrinsic Validation/Refutation Criteria." Pp. 274–307 in (Earmann, Janssen, and Norton 1993).

Biezunski, Michel. 1989. "Inside the Coconut: The Einstein-Cartan Discussion on Distant Parallelism." Pp. 315–324 in (Howard and Stachel 1989).

Callon, Michel, J. Law, and A. Rip, eds. 1986. *Mapping the Dynamics of Science and Technology*. London: MacMillan.

Corry, Leo. 1999. "From Mie's Electromagnetic Theory of Matter to Hilbert's Unified Foundations of Physics." Pp. 159–183 in *Studies in the History and Philosophy of Modern Physics*, vol. 30.

Darrigol, Oliver. 1996. "The Electrodynamic Origins of Relativity Theory." Pp. 241–312 in *Historical Studies in the Physical Sciences*, vol. 26.

Debever, Robert, ed. 1979. *Élie Cartan — Albert Einstein. Letters on Absolute Parallelism 1929–1932*. Princeton: Princeton University Press.

Earmann, John, Michel Janssen, and John D. Norton, eds. 1993. *The Attraction of Gravitation: New Studies in the History of General Relativity*. (Einstein Studies 5). Boston: Birkhäuser.

Eisenstaedt, Jean. 1986. "La Relativité générale à l'étiage: 1925–1955." *Archive for History of Exact Sciences* 35:115–185.

———. 1987. "Trajectoires et impasses de la solution de Schwarzschild." *Archive for History of Exact Sciences* 37:275–375.

Eisenstaedt, Jean, and A. J. Kox, eds. 1992. *Studies in the History of General Relativity*. (Einstein Studies 3). Boston: Birkhäuser.

Galison, Peter, and David J. Stump, eds. 1996. *The Disunity of Science. Boundaries, Contexts and Power*. Standford: Stanford University Press.

Garfield, Eugene. 1964. *The Use of Citation Data in Writing the History of Science*. Philadelphia: Institute of Scientific Information.

Goenner, Hubert. 1984. "Unified Field Theories: From Eddington and Einstein up to Now." Pp. 176–196 in *Proceedings of the Sir Arthur Eddington Centenary Symposium*, vol. 1. *Relativistic Astrophysics and Cosmology*, eds. V. de Sabbata and T. M. Karade. Singapore: World Scientific.

———. 1992. "The Reception of the Theory of Relativity in Germany as Reflected in Books Published between 1908 and 1945." Pp. 15–38 in (Eisenstaedt and Kox 1992).

———. 1993a. "The Reaction to Relativity Theory. I. The Anti-Einstein Campaign in Germany in 1920." *Science in Context* 6:107–133.

———. 1993b. "The Reaction to Relativity Theory. III. 'A Hundred Authors Against Einstein'." In . Pp. 107–133 in (Earmann, Janssen, and Norton 1993).

Goenner, Hubert, Jürgen Renn, Jim Ritter, and Tilman Sauer, eds. 1999. *The Expanding Worlds of General Relativity*. (Einstein Studies 7). Boston: Birkhäuser.

Goldstein, Catherine. 1994. "La Théorie des nombres dans les *Notes aux Comptes rendus de l'Académie des sciences* (1870–1914): un premier examen." *Rivista di Storia della Scienza* 2:137–160.

Graham, Loren R. 1966. *Science and Philosophy in the Soviet Union*. New York and Toronto: Random House.

Havas, Peter. 1993. "The General-Relativistic Two-Body Problem and the Einstein-Silberstein Controversy." Pp. 88–125 in (Earmann, Janssen, and Norton 1993).

Howard, Don, and John Stachel, eds. 1989. *Einstein and the History of General Relativity*. (Einstein Studies 1). Boston: Birkhäuser.

Hunt, Bruce J. 1991. *The Maxwellians*. Ithaca [NY]: Cornell University Press.

Illy, Jòzsef. 1992. "The Correspondence of Albert Einstein and Gustav Mie, 1917–1918." Pp. 244–259 in (Eisenstaedt and Kox 1992).

Kilmister, Clive W. 1994a. *Eddington's Search for a Fundamental Theory*. Cambridge: Cambridge University Press.

———. 1994b. "George Frederick James Temple (2 September 1901–30 January 1992. Elected F. R. S. 1943)." *Biographical Memoirs F. R. S*: 385–400.

Kitchenassamy, S. 1992. "Dirac Equations in Curved Space-Time." Pp. 383–392 in (Eisenstaedt and Kox 1992).

McCormmach, Russell. 1970. "H. A. Lorentz and the Electromagnetic View of Nature." *Isis* 61:459–497.

Merleau-Ponty, Jacques. 1965. *Philosophie et théorie physique chez Eddington*. (Annales Littéraires de l'Université de Besançon 75). Paris: Les Belles Lettres.

Miller, Arthur I. 1981. *Albert Einstein's Special Theory of Relativity. Emergence (1905) and Early Interpretation (1905–1911)*. Reading: Addison-Wesley. [Republication: New York: Springer, 1998].

Pais, Abraham. 1982. *'Subtle is the Lord....' The Science and the Life of Albert Einstein*. Oxford: Oxford University Press.

———. 1986. *Inward Bound. Of Matter and Forces in the Physical World*. Oxford: Oxford University Press.

Price, Derek J. 1965. "Networks of Scientific Papers." *Science* 149:510–515.

Pyenson, Lewis. 1979. "Physics in the Shadow of Mathematics: The Göttingen Electron-Theory Seminar of 1905." *Archive for History of Exact Sciences* 21:55–89.

Reich, Karin. 1992. "Levi-Civitasche Parallelverschiebung, affiner Zusammenhang, Übertragungsprinzip: 1916/7–1922/3." *Archive for History of Exact Sciences* 44:77–105.

Renn, Jürgen, and Tilman Sauer. 1999. "Heuristics and Mathematical Representation in Einstein's Search for a Gravitational Field Equation." Pp. 87–125 in (Goenner et al. 1999).

Ritter, Jim. 1993. "Théories unitaires." Pp. 131–191 in *Albert Einstein: Œuvres choisies*, vol. 3. *Relativités II. Relativité générale, cosmologie et théories unitaires*, ed. F. Balibar et al. Paris: Le Seuil.

Rowlands, Peter. 1990. *Oliver Lodge and the Liverpool Physical Society*. (Liverpool Historical Studies 4). Liverpool: Liverpool University Press.

Sánchez-Ron, José M. 1992. "The Reception of General Relativity among British Physicists and Mathematicians." Pp. 57–88 in (Eisenstaedt and Kox 1992).

———. 1999. "Larmor versus General Relativity." Pp. 405–430 in (Goenner et al. 1999).

Scholz, Erhard. 1995. "Hermann Weyl's 'Purely Infinitesimal Geometry'." Pp. 1592–1603 in *Proceedings of the International Congress of Mathematicians, Zürich, Switzerland 1994*. Basel: Birkhäuser.

———. 1999. "Weyl and the Theory of Connections." Pp. 260–284 in *The Symbolic Universe: Geometry and Physics 1890–1930*, ed. J. Gray. Oxford: Oxford University Press.

———, ed. 2001. *Hermann Weyl's Raum-Zeit-Materie and a General Introduction to His Scientific Work*. Basel: Birkhäuser.

Schulman, Robert, Anne J. Kox, Michel Janssen, and Jòzsef Illy, eds. 1998. *The Collected Papers of Albert Einstein, vol. 8. The Berlin Years: Correspondence, 1914–1918*. 2 vols. Princeton: Princeton University Press.

Siegmund-Schulze, Reinhard. 1923. *Mathematische Berichterstattung in Hitlerdeutschland*. Göttingen: Vandenhoeck and Ruprecht.

Sigurdsson, Skúli. 1991. "Hermann Weyl. Mathematics and Physics. 1900–1927." Ph.D. diss., Harvard University, Cambridge [MA].

———. 1994. "Unification, Geometry and Ambivalence: Hilbert, Weyl and the Göttingen Community." Pp. 355–367 in *Trends in the Historiography of Science*, ed. K. Gavroglu et al. Dordrecht: Kluwer.

———. 1996. "Physics, Life and Contingency: Born, Schrödinger, and Weyl in Exile." Pp. 48–70 in *Forced Migration and Scientific Change. Emigré German-Speaking Scientists and Scholars after 1933*, eds. M. G. Ash and A. Söllner. Washington: German Historical Institute and Cambridge: Cambridge University Press.

Stachel, John. 1974. "The Rise and Fall of Geometrodynamics." In *PSA 1972. Proceedings of the 1972 Biennial Meeting, Philosophy of Science Association*, (Boston Studies in the Philosophy of Science 20), eds. K. F. Schaffner and R. S. Cohen. Dordrecht: Reidel.

———. 1986. "Eddington and Einstein." Pp. 225–250 in *The Prism of Science*, ed. E. Ullmann-Margalit. Dordrecht: Reidel.

———. 1993. "The Other Einstein: Einstein Contra Field Theory." *Science in Context* 6:275–290.

Struik, Dirk J. 1989. "MIT Department of Mathematics: Some Recollections." Pp. 163–177 in *A Century of Mathematics in America*, Part III (History of Mathematics 3), ed. P. Duren. Providence: American Mathematical Society.

Tobies, Renate. 1994. "Mathematik als Bestandteil der Kultur. Zur Geschichte des Unternehmens 'Encyclopädie der mathematischen Wissenschaften mit Einschluß ihrer Anwendungen.'" *Mitteilungen der ÖGW* 14:1–90.

Tonnelat, Marie-Antoinette. 1965. *Les Théories unitaires de l'électromagnétisme et de la gravitation*. Paris: Gauthier-Villars.

———. 1971. *Histoire du principe de relativité*. Paris: Flammarion.

Vallarta, Manuel Sandoval. 1982. "Comments on [29f], [29g]." P. 630 in *Norbert Wiener. Collected Works with Commentaries*, vol. III, ed. P. Masani. Cambridge: MIT Press.

Vizgin, Vladimir P. 1985. *Edinye teorii polya v perevoi treti XX veka*. Moscow: Nauka.

———. 1989. "Einstein, Hilbert and Weyl: The Genesis of the Geometrical Unified Field Theory Program." Pp. 300–314 in (Howard and Stachel 1989).

———. 1994. *Unified Field Theories in the First Third of the Twentieth Century*. (Science Networks 13). Basel: Birkhäuser. Translation of (Vizgin 1985).

Warwick, Andrew. 1988. "The Reception of Relativity Theory in Britain." Ph.D. diss., Cambridge University [UK].

———. 1992. "Cambridge Mathematics and Cavendish Physics: Cunningham, Campbell and Einstein's Relativity, 1905–1911." *Studies in the History and Philosophy of Science*. 23: 625–656; 24: 1–25.

SIMON SAUNDERS

INDISCERNIBLES, GENERAL COVARIANCE, AND OTHER SYMMETRIES: THE CASE FOR NON-REDUCTIVE RELATIONALISM *

INTRODUCTION

What is the meaning of general covariance? We learn something about it from the *hole argument*, due originally to Einstein. In his search for a theory of gravity, he noted that if the equations of motion are covariant under arbitrary coordinate transformations, then particle coordinates at a given time can be varied arbitrarily — they are underdetermined — even if their values at all earlier times are held fixed. It is the same for the values of fields. The argument can also be made out in terms of transformations acting on the points of the manifold, rather than on the coordinates assigned to the points. So the equations of motion do not fix the particle positions, or the values of fields at manifold points, or particle coordinates, or fields as functions of the coordinates, even when they are specified at all earlier times. It is surely the business of physics to *predict* these sorts of quantities, given their values at earlier times. The principle of general covariance seemed quite untenable.

It is understandable that Einstein, sometime in 1911, gave up the principle in consequence, but it was an error all the same; four years later, once he had realized his mistake, progress was rapid. Within the year, he was in possession of the full field equations of the general theory of relativity (GTR).[1]

Now, I want to draw attention to a much older argument, due originally to Leibniz. Leibniz argued that given the homogeneity of Newtonian space, the overall positions of particles, over and above their relative positions, can be varied arbitrarily; they too are underdetermined. Likewise, given the Galilean symmetries, overall absolute velocities and orientations are underdetermined. Call these *shift arguments*. In fact, an argument exactly parallel to the hole argument can be formulated in Newtonian gravity (NTG), if one is prepared to make use of the symmetry of the theory under time-dependent boosts (equivalently, under time-dependent uniform gravitational fields). In that case the absolute quantities are all underdetermined even given their values at all earlier times.

Evidently these arguments target quantities of a specific sort - *absolute* quantities. The solution in each case is broadly the same: the physically real properties and relations are the invariant ones, which do not include the absolute quantities. Pretty well all the invariant quantities turn out to be relational ones, of certain specified sorts. And in the case of NTG, insofar as Leibnizians and Newtonians were really opposed on the nature of space, the shift arguments surely come down on the side of Leibniz's relationalism.

Now for my principal claim: it is that the form of relationalism underpinned by these arguments has nothing to do with a *reductionist* doctrine of space or spacetime (the doctrine that space or spacetime has no independent existence independent of matter); that further, this form of relationalism is a systematic and coherent doctrine that can be applied to any exact symmetry in physics, including the more contentious cases of gauge symmetry and the symmetries of constrained Hamiltonian systems.[2] It is, in point of fact, a natural expression of Leibniz's philosophical principles, in particular the principle of identity of indiscernibles (PII).

Relationalism is usually taken to be a reductionist account of space, that space is *nothing more* than the system of spatial relations between actual or possible distributions of matter. Call this *eliminative* relationalism. Leibniz, in his criticism of the Newtonians, did argue for eliminativist relationalism. The links between Leibniz's philosophical principles and contemporary debates in the foundations of spacetiime theories have recently been much discussed, see e.g. (Belot 2001); it is always the eliminativist doctrine that is supposed to follow from his principles. Earman, in his influential study of the absolute-relational debate, uses the term "relationism" to mean exactly this eliminativist doctrine (Earman 1989). The hole argument, in the hands of Earman and Norton (1987), was considered an argument for eliminativist relationalism, on the understanding that space or spacetime in itself is described by the bare manifold, divorced of any fields, metrical or otherwise. They also applied the hole argument to the pregeneral-relativistic theories, to roughly similar ends.

Leibniz's principles in their original form certainly do imply the eliminativist version of relationalism; what has been overlooked is the difference to his principles made by modern logic. It is the PII, understood in the context of modern logic, that I am interested in, and it is non-reductive relationalism that follows from it.

A different sort of objection has been made to Earman's agenda, and specifically to his use of the hole argument. It has been objected, most prominently by Stachel, that the hole argument cannot be taken over to the pregeneral-relativistic case; that, in point of fact, GTR differs radically from any other spacetime theory, in that it alone requires that any method for the identification of points of space be dynamical in origin (Stachel 1993). Pregeneral-relativistic theories, in contrast, permit the use of non-dynamical methods for identifying points of space.

Now, I am sympathetic to Stachel's concern to distinguish GTR from other spacetime theories, but I do not believe there is any such thing as a non-dynamical method for identifying points of space. To be sure there are practical, operational methods, but these are available whatever one's theory; they do not seem to be what Stachel has in mind. This point is important, because the existence of such methods would bear equally on the shift arguments. It would undercut appeal to the PII as well. It is, in many respects, a restatement of Clarke's point of view, as argued against Leibniz in the debate over Newtonian space, in the *Correspondence* (Alexander 1956).

First I will state the hole argument in detail, and then go on to see why Stachel restricts the argument to GTR. After that I will say something about Earman's understanding of relationalism. Only then will I go on to Leibniz's principles, and the PII in its modern guise, and the form of relationalism that follows from it. At the end I will return to Stachel's objections.

1. THE HOLE ARGUMENT ACCORDING TO STACHEL

Stachel denies that the hole argument has any bearing on precursors to GTR; he denies, further, that general covariance is a space-time symmetry, on a par with the familiar symmetries of NTG. So let us use the term "diffeomorphism covariance" instead, to denote covariance under diffeomorphisms (covariance under the smooth mappings on the manifold M given by C^∞ functions $f : M \to M$). It is well-known that the equations of motion of pregeneral-relativistic theories can all be written in diffeomorphism-covariant form. From now on let "general covariance" mean what Stachel means: diffeomorphism covariance along with an additional principle, namely that no *individuating fields* are available, other than those provided by solving the dynamical equations.[3] By "individuating field" Stachel means any set of properties that can be used to uniquely distinguish the points of a manifold. He gives as an example a set of colors, as indexed by hue, brightness and saturation, which provides an individuating field for the points of a 3-dimensional manifold. According to Stachel, in pregeneral-relativistic theories one can always suppose that such a field exists $independent$ of the dynamical system of equations under study. In principle one can always give meaning to the coordinates used in the equations, independent of solving them. Not so in GTR, where there *is* no space-time to be individuated, prior to a particular solution to the equations.

Let me state the hole argument more precisely. I shall use the coordinate-dependent method. Consider a manifold M, equipped with an atlas of charts. With respect to one of these charts, write the metric tensor field g along with the sources ρ of the gravitational field as functions of x (denote generically $D(x)$). Suppose this is done everywhere on M except on an open subset H (passing from one chart to another as necessary), and suppose that locally a system of equations can be given for these fields, supplemented if necessary by appropriate boundary conditions and rules for passing from one chart to another. Let these equations be diffeomorphism-covariant. Let the boundary of H be ∂H. Then, given a solution D to these equations for g and ρ outside of H and on ∂H, the equations do not determine any unique solution in the interior of H.

The proof is by construction. Recall first that for any tensor field ϕ on M and for any diffeomorphism f on M, one can define a new tensor field $f * \phi$ on M, the drag-along[4] of ϕ under f; and that if a system of equations is diffeomorphism covariant, then the drag-along of any solution is also a solution. Let D be as above; then D and $f * D$ are both solutions to the equations of motion, for any f. But $D(x)$ and $(f * D)(x)$ are in general different tensors at one and the same point x (we are interpreting diffeomorphism symmetries as *active* symmetry transformations). If now we choose f so that it is non-trivial inside H, but reduces to the identity on ∂H and outside H, then $D(x) = (f * D)(x)$ for $x \notin H$, but $D(x) \neq (f * D)(x)$ for $x \in H$.

There is of course a way out of the conundrum. Following Stachel, we should pass to the *equivalence class* of solutions under diffeomorphisms, a view which is by now quite standard in the literature, see e.g. (Hawking and Ellis 1973; Wald 1984).[5] Only the equivalence class is physically real. On this understanding, general covariance is invariably an unbroken symmetry, and the physical world is to be described in a diffeomorphic invariant way. The price, however, is that the values of the fields at manifold points, or as functions of a fixed coordinate system, are not physically real. Only the relations among these field values are invariant, so only these relations are real.

This takes some getting used to: the values of fields at points go the way of particle velocities, positions, and directions; only relations among them remain. But can we adopt this point of view if the manifold points can be *independently* individuated? In that case, says Stachel, surely not:

> The fact that the gravitational field equations are [diffeomorphism]-covariant does *not* suffice to allow the physical identification of a class of mathematically distinct drag-along fields. If such an individuating field existed, a relative dragging between metric tensor field and individuating field would be physically significant, and Einstein's hole argument would be ineluctable, (Stachel 1993, 140–141).

If there is an individuating field, which can be used to specify the values of fields at points, but which is not itself subject to the diffeomorphism, then these values will be *changed* by the diffeomorphism; if this individuating field is physically meaningful, then so too are the values of fields at points marked out by it. The hole argument then becomes "ineluctable." Stachel is surely right on this.

But there is an additional and crucial feature of GTR, which none of the classical theories shares; namely, that *in principle* there cannot be any individuating field of this kind. For the metrical field necessarily determines *both* the gravitational field structures, the affine connection and the Riemann tensor, *and* the chronogeometrical structure, the spatiotemporal relations. The coordinates figuring in the field equations *cannot* be antecedently understood as space-time coordinates, for there are as yet *no* space-time relations, not even a topology, prior to solving the equations for the metrical field in GTR.

It is now clear what is wrong with Earman and Norton's (1987) argument. In the pregeneral-relativistic theories one always has an *a priori* chronogeometrical structure. One always knows what the geometry is, independent of obtaining any solution to the equations of motion. So it does make sense, from this perspective, to introduce an *independent* individuating field; there will always be this option in such cases. In fact, Stachel goes on to insist, for this reason the pregeneral-relativistic theories are better stated in terms of their rigid, finite-dimensional groups, as in Klein's *Erlangen* program (what he calls their "affine-space-plus" form). This brings out the true nature of space and space-time according to these theories. But once this is done the hole argument cannot even be applied.

In support of Stachel, one might add that Newton explicitly denied that points of space or instants of time could have different spatiotemporal relations than those which they have, and yet remain the same points; as Maudlin has put it, they have their metrical relations "essentially" (1988). Newton would have denied that transformations that change these essential properties could have any physical meaning (one simply wouldn't be talking about *the same points*). The only transformations that Newton allowed to be actively interpreted (in line with Maudlin's argument) were the isometries, which preserve all the metrical relations among points. So it seems we are back to the rigid symmetries, the translations, rotations, and boosts. And if a rigid symmetry transformation acts as the identity on any open set, it *is* the identity; the hole argument cannot even be posed.

But here Stachel makes an important qualification (1993, 148, and note 10), explicitly limiting his remarks to pregeneral-relativistic theories *other* than NTG — so, essentially, among the fundamental classical dynamical theories, to special relativistic electromagnetism. For as is well known, NTG can be put in a form which reveals a wider class of symmetries. These symmetries were, moreover, recognized by Newton, who saw very well that his theory of gravity could also be applied to the case of uniformly accelerating frames of references — even, in point of fact, that it could be applied when the acceleration varies *arbitrarily* with the time (so long as it is constant in space).[6] Unfortunately this point was not understood by Clarke in the *Correspondence*. There, he supposed this symmetry of NTG only followed given *Leibniz's* principles - and viewed it as a *reductio* of them;[7] nor did Leibniz rectify the error. Nevertheless the fact remains: to a clear thinker, familiar with Newton's *Principia*, circa the time of the *Correspondence*, the following symmetry principle was readily apparent: the equations of motion for the relative distances of a system of bodies, referred to a non-rotating frame of reference, are covariant under the group of transformations:[8]

$$\vec{x} \to R \cdot \vec{x} + \vec{f(t)} \qquad (1)$$

where R is an orthogonal matrix and $\vec{f}$ is a twice differentiable but otherwise arbitrary vector-valued function of the time. If we now choose R as the identity, and $\vec{f}$ so that it is zero prior to some instant t_0 but non-zero thereafter, we have a version of the hole argument: the value of the absolute position of any particle, and its derivatives, after t_0, can be varied arbitrarily, even keeping fixed its values at all previous times, and even keeping fixed the values of all *other* particles at all previous times. Call this the *generalized* shift argument.

The rejoinder, presumably, is that here too a non-dynamical individuating field may be antecedently available, so that the symmetries (1) will have to be abandoned (equivalently, equations for the relative particle configurations will not be judged to tell the whole story).[9] Against it, one wonders what these non-dynamical individuating fields can really amount to. It does not appear that in NTG there ever was a method available for determining differences in absolute positions, and differences in absolute velocities, at different times.

We will return to this question in due course. Here I want only to note the alternative solution: evidently, if no such non-dynamical individuating field is available, then the symmetry (1) can be retained, with no consequent underdetermination of the theory, so long as we acknowledge that the only physically-meaningful quantities are comparisons between positions, velocities and accelerations *at a single time*. In terms of Newton-Cartan spacetime, this follows from the fact that in general no meaning can be given to the decomposition of the connection into an inertial part and a gravitational part.

Evidently both arguments, the hole argument and the generalized boost argument, promote the view that only certain kinds of relational structures are physically real. But this is not relationalism as it is ordinarily considered; let us get clear on the difference.

2. RELATIONALISM AND ELIMINATIVISM

Gravity and geometry are inseparable in GTR. Does this mean that gravity is reduced to geometry? Stachel is as likely to put it the other way round:

> Several philosophers of science have argued that the general theory of relativity actually supports spacetime substantivalism (if not separate spatial and temporal substantivalisms) since it allows solutions consisting of nothing but a differentiable manifold with a metric tensor field and no other fields present (empty spacetimes). This claim, however, ignores the second role of the metric tensor field; if it is there chronogeometrically, it inescapably generates all the gravitational field structures. Perhaps the culprit here is the words "empty spacetime" An empty spacetime could also be called a pure gravitational field, and it seems to me that the gravitational field is just as real a physical field as any other. To ignore its reality in the philosophy of spacetime is just as perilous as to ignore it in everyday life (Stachel 1993, 144).

Stachel hints at a deflationary view of relationalism, similar to mine, without the usual eliminitivism that goes with it. This does not quite mean that he sees no opposing doctrine; indeed, he is prepared to treat "substantivalism" and "absolutism" as synonymous.[10] Absolutism, in the context of NTG, is the doctrine that there is a preferred state of rest, the Gallilean symmetries notwithstanding. By extension we may take it to be the view that there are absolute positions and directions as well: relationalism in my sense and in Stachel's is certainly opposed to absolutism.

The term "substantivalism" is due to Sklar (1974, 161); he used it to mean that space exists independent of matter. The term "substantival" is older, and has a more specific meaning: it was used by Johnson for an account of space in which spatial points themselves have positions, and objects have positions by virtue of occupying points (Johnson 1924, 79). Earman sees the hole argument as an argument against substantivalism in Sklar's sense, taking the bare manifold to represent space in itself, independent of fields altogether. Relationalism, as he understands it, is opposed to substantivalism. Space and time, on this view, must be *eliminated*, as independent entities, in favour of material ones. And this would be a highly non-trivial affair:

> Not a single relational theory of classical motion worthy of the name "theory" and of serious consideration was constructed until the work of Barbour and Bertotti in the 1960s and 1970s. This work came over half a century after classical space-time gave way to relativistic space-time, and in the latter setting a purely relational theory of motion is impossible (Earman 1989, 166).

Moreover, substantivalism in Sklar's general sense is essential to almost every other part of physics:

> ...no detailed antisubstantivalist alternative has ever been offered in place of the field theoretic viewpoint taken in modern physics (*ibid.*).

But is either of these positions well-motivated? Stachel is doubtful that there is any longer a well-defined distinction between matter and space in the first place, for they have both been superseded by the field. Einstein said much the same:

> There is no such thing as an empty space, i.e. a space without field. Space-time does not claim existence on its own, but only as a structural quality of the field (Einstein 1954).

For a systematic discussion of the various ways in which the distinction between matter and space has been weakened by developments in physics over the last two centuries, see (Rynasiewicz 1996a). One has only to consider the ether, the electromagnetic field, the metrical field, and the wave-function of quantum mechanics, to see what he means. And once this distinction is broken down, Rynasiewicz goes on to claim, the old philosophical dispute no longer has any point.

Stachel will hardly be moved by talk of wave-functions, but he will agree with Rynasiewicz in respect of his other examples. Let it be granted that the old distinction between space and matter is no longer clear-cut. Does anything remain to the relationalist position? Is Rynasiewicz right to say that none of the philosophical disputes any longer has bite?

Earman has stated three criteria for relationalism, extracted from the pregeneral-relativistic context. Let us take them one by one.

> R1. All motion is the relative motion of bodies, and consequently, spacetime does not have, and cannot have, structures that support absolute quantities of motion (Earman 1989, 13).

Evidently if material and spatial structures enjoy much the same status, little remains to R1 but the platitude that all motion is relative. We hasten on.

> R2. Spatiotemporal relations among bodies and events are direct; that is, they are not parasitic on relations among a substratum of space points that underlie bodies or space-time points that underlie events (*ibid.*, 13).

Here Earman draws on Johnson's criterion. We should not take this to mean that points of space or spacetime do not exist at all; it is the claim - at least in the pregeneral-relativistic case - that we do not first determine the points at which bodies or events are situated, and deduce their spatiotemporal relations from the spatiotemporal relations among points. This is the reading I shall give it. It is backed up by Earman's third condition:

> R3. No irreducible, monadic, spatiotemporal properties, like "is located at space-time point p" appears in a correct analysis of the spatiotemporal idiom (*ibid.*, 14).

We do not begin with the locations of bodies at points; there are no such irreducible properties. On the other hand, if there are spacetime points at all, we can surely end up with statements of location. I shall take R3 to be laying down a constraint on how such statements are to be made: there can be no appeal to *irreducible monadic properties* to specify the point p.

There is therefore a kernel to the relationalist position, as formulated by Earman in the case of NTG and special relativity, which I shall summarize thus:

> R0. Points of space and spacetime, in NTG and special relativity, are specified by their relations to bodies and events.

Although R0 does not treat space and matter on an even footing, it leaves open the possibility that the difference, such as it is, is due to the symmetries of space and spacetime, for this is a criterion, I say again, abstracted from the pregeneral-relativistic context. As we shall see, granted such symmetries, R0 does indeed follow from Leibniz's amended principles. It is a consequence of non-reductive relationalism.

Earman anticipates one sort of deflationary move, the suggestion that the substantival-relational debate has no physical consequences, but he expects this move to be made broadly on instrumentalist grounds. He grants that there is a corresponding weakened version of relationalism, but he takes this to be entirely trivial:

> ...the relationist can follow either of two broad courses. One, he can decline to provide a constructive alternative field theory and instead take over all of the predictions of field theory for whatever set of quantities he regards as relationally pure. I do not see how this course is any different from instrumentalism. While I believe instrumentalism to be badly flawed, I do not intend to argue that here. Rather, the point is that relationism loses its pungency as a distinctive doctrine about the nature of space and time if it turns out to be nothing but a corollary of a methodological doctrine about the interpretation of scientific theories. Two, the relationist can attempt to provide a constructive alternative to field theory... (Earman 1989).

It will be clear, however, that relationalism as I understand it has nothing to do with instrumentalism; it is on the contrary a form of realism. Neither did Stachel and Rynasiewicz downplay the distinction between space and matter on positivist grounds; they too are realists.

Earman is clearly angling for a connection between relationalism as it presents itself on the basis of the shift and hole arguments, and *Machianism*: his reference to the Barbour-Bertotti theory makes this plain. But all these arguments proceed from *given* symmetries of theories; they are none of them *a priori*. Certainly we do not as yet have any argument for the view that transformations among rotating frames of references should be a symmetry group of a dynamical theory.

That does not mean there are no such arguments. Indeed, Machianism does follow from Leibniz's philosophical principles, in their original form. Let us see how.

3. LEIBNIZ'S PRINCIPLES

In common with the scholastics, Leibniz believed that the description of a thing should describe the thing independent of anything else, not in terms of any actual or possible relations it might have with other things. Call this his *independence thesis*.

True sentences are those which, albeit via a process of analysis, are of subject-predicate form, where the predicate is already contained in the concept of the subject (the "subject" of a sentence is what the sentence is about). They are, as it were, *definitional*, of the form "gold is a yellow metal"; they follow from the description of the thing. For this to be so the subject mentioned, the thing or natural kind, has to be thought of "completely" as already containing within it all the meaningful physical predicates that can properly be assigned to it. Call this his *containment theory of truth*.

Most important of all, *for everything there is a reason*; nothing is to be arbitrary or unexplained. There is, indeed, no choice to be made, unless a meaningful distinction has been drawn. This is Leibniz's celebrated *principle of sufficient reason* (PSR). It is closely related to another of his principles, the *identity of indiscernibles* (PII). According to this, numerically distinct things must differ in some meaningful way, for otherwise there could be no basis for choice among them.

Finally, Leibniz is committed to a weak form of verificationism: physically real differences, if there are any, had better be experimentally detectable, however indirectly. Indeed, by a symmetric object Leibniz understands an object which can be arranged in nominally distinct ways which do not admit any detectable experimental difference.

Given these principles, Leibniz's views follow quite naturally. Eliminativism is among them. Things cannot be located in space because

> Space being uniform, there can be neither any external nor internal reason, by which to distinguish its parts, and to make any choice between them. For, any external reason to discern between them, can only be grounded upon some internal one. Otherwise we should discern what is indiscernible, or choose without discerning (Alexander 1956, 39, Fourth Paper).

External reasons are ruled out by the independence thesis: things have to be thought of independent of anything else (so not in relation to matter). Granted that neither can there be any internal reason to distinguish one part of space over another (because space is uniform) there can be no reason, period, why matter should be located at one part of space rather than another. If the parts of space are real, the PSR will have to be violated. So space is not real.

Here is one of Leibniz's shift arguments, stated in full:

> I say then, that if space was an absolute being, there would something happen for which it would be impossible there should be a sufficient reason. Which is against my axiom. And I prove it thus. Space is something absolutely uniform; and, without the things placed in it, one point of space does not absolutely differ in any respect whatsoever from another point of space. Now from hence it follows, (supposing space to be something in itself, besides the order of bodies among themselves), that 'tis impossible there should be a reason, why God, preserving the same situations of bodies among themselves, should have placed them in space after one certain particular manner, and not otherwise; why everything was not placed the quite contrary way, for instance, by changing East into West.[11] But if space is nothing else, but that order or relation; and is nothing at all without bodies, but the possibility of placing them; then those two states, the one such as it now is, the other supposed to be the quite contrary way, would not at all differ from one another. Their difference therefore is only to be found in our chimerical supposition of the reality of the space in itself. But in truth the one would be exactly the same thing as the other, they being absolutely indiscernible; and consequently there is no room to enquire after a reason of the preference of the one to the other (Alexander 1956, 26, Third Paper).

Evidently the argument applies just as well to the other Galilean symmetries, including boosts, and to the translations.

It is clear that there would be no implication that space is unreal if space did *not* have these symmetries. There would then be internal reasons — inhomogeneities — by which its parts might be discerned. Of course non-homogeneous spaces were not on offer in the early 18th century, but the reasons for that had little or nothing to do with Leibniz's principles. It should also be clear that the argument as stated does need Leibniz's independence thesis. It is this which rules out use of external reasons to distinguish parts of space — relations to material bodies, for instance. If relations to bodies were used Leibniz would say that space would not then qualify as a *bona fide* substance. We would not be considering space as it is in itself, but only in relation to other substances.

Leibniz's independence thesis has a resemblance to essentialism, in contemporary philosophy. Essentialists too insist that there is an important distinction between qualities of objects, namely those used to identify an object, its *essential* qualities, without which it would not be the same object, and those that can change, its *accidental* qualities. We have seen this in play in Maudlin's account of Newton's views on points of space. But the distinction between accidental and essential qualities does not generally line up with Leibniz's distinction, between internal and external relations: the connection is more apparent than real.

Unlike Leibniz's assumptions on the nature of geometry, it is not obvious that the independence thesis can be challenged by any empirical or mathematical discovery. But clearly it *can* be denied; we can allow for a broader notion of "object" whether or not objects will then be things in Leibniz's sense. Neither do we have to insist on a distinction between accidental and essential features or relations. Relationalism, as I shall understand it, is not committed to the independence thesis or to essentialism. It is, further, noncommittal on the existence of any *a priori* symmetries to space or spacetime, and on any *a priori* distinction between space and matter. But it is committed to Leibniz's other metaphysical principles, in particular the PSR and PII.

4. THE IDENTITY OF INDISCERNIBLES

In fact it is hard nowadays to believe in Leibniz's independence thesis. The problem lies not with advances in physics but in logic. Leibniz's logic was based on the subject-predicate form of the proposition. He held that in every meaningful proposition there is a subject of predication, in parallel to the concept of substance as the bearer of properties. Where relations seem to be invoked, in reality one is still attributing a predicate to a single subject of predication: relations, for Leibniz, had to be *reducible* - derivable from the monadic properties of their relata (what Leibniz called *internal*, also sometimes called *intrinsic*, in contrast to *external* or *extrinsic* relations, which may hold or fail to hold independent of any properties of their relata).

Leibniz's views on relations provided a clear basis for Leibniz's independence thesis. But whilst Frege, the founder of modern logic, drew a distinction superficially similar to Leibniz's (but between *object* and *concept*, rather than *subject* and *predicate*), in Frege's philosophy nothing of particular significance attached to 1-place concepts; relations were as fundamental to Frege's logic as was quantification theory;

propositions no longer have to be cast into subject-predicate form to be meaningful;[12] there is nothing wrong with relations *per se*.

Consider again the PII, the principle, roughly speaking, that distinct objects must differ in some qualitative, predicative respect. This is a substantive thesis; a "qualitative" feature of an object can be common to many others; "qualitative identity" is the limiting case of similarity in all respects; qualitatively identical objects might yet be thought to differ numerically - but this is what is ruled out by the PII. The principle insists that there must be qualitative (and physically meaningful) criteria for numerical difference.

It is easy to say what identity is in set theory. Given a set U, it is the binary relation $\{<x,x> : x \in U\}$. But it is evident, since anyway we identify sets purely extensionally (they are defined by their elements), that the set elements themselves are to be given in advance; in Cantor's words, sets are always "composed of definite well-distinguished objects" (1895, 481). Of course the PII may be violated for some language and for some collection of objects; the point is to find a language and a theory in that language which can say what each of them is (which admits no more and no less than what there is).[13] Then the PII will be satisfied.

What then does logic have to say on the matter? If there are only one-place predicates, as Leibniz thought, then the principle is just that objects with the same properties are identical. It has seemed obvious, in consequence, that admitting predicates in two or more variables, one weakens the principle to allow that objects may be counted as distinct if only they differ in their properties *or relations*. This version of the PII is normally called the *weak* principle; the former is called the *strong* principle. But the weak principle as just stated is not what logic dictates.

To see what does follow, from a purely logical point of view, consider the simplest case, a first order language without identity in which all the predicate symbols are explicitly specified.[14] Now introduce the identity sign, and supplement the laws of deduction accordingly, by the following axiom scheme (here 'F' is a letter which can be replaced by any predicate of the language):

$$x = x, \ x = y \rightarrow (Fx \rightarrow Fy). \tag{2}$$

This scheme implies *substitutivity*, the "indiscernibility of identity": terms (variables, logical constants, and functions of such) with the same reference can always be substituted without change of truth value. It is *complete* in the following sense: suppose a complete proof procedure is available for the original language (without identity), meaning that every logically true sentence (true in every model) can be deductively proved; then, supplementing the proof procedure with the scheme (2), every logically true sentence, including those involving the identity sign, can be proved. This is in fact how Gödel proved his celebrated completeness theorem for the predicate calculus with identity (1930).

Any definition of identity which implies (2) will therefore be formally adequate for the purposes of deduction. Here then is how identity can be explained: for any terms 'x', 'y', $x = y$ if and only if for all unary predicates A, binary predicates B, ..., n-ary predicates P, we have:

$$A(x) \longleftrightarrow A(y)$$

$$B(x, u_1) \leftrightarrow B(y, u_1), B(u_1, x) \leftrightarrow B(u_1, y) \tag{3}$$

$$\ldots\ldots, \ldots, \ldots$$

$$P(x, u_1, \ldots, u_{n-1}) \leftrightarrow P(y, u_1, \ldots, u_{n-1}) + \text{permutations}$$

together with all generalizations (universal quantifications) over the free variables $u_1, \ldots, u_{n-1}$ other than x and y. From this (2) obviously follows.

This definition of identity is due to Hilbert and Bernays (1934), and was subsequently defended by Quine (1960); its consequences have not been widely recognized, however. They are straightforward if the language contains only one-place predicates; we then obtain the strong principle of identity as stated above. In the more general case, call two objects *absolutely discernible* if there is a formula with one free variable which applies to the one, but not to the other. There is another way for (3) to be false, so that $x \neq y$: call two objects *relatively discernible* if there is a formula in two free variables which applies to them in only one order. It should be clear that not all relatively discernible objects are absolutely discernible: the relation $x > y$ is true of any distinct real numbers, taken in only one order, although it is impossible to find finite expressions in a finite or countable alphabet which are in 1:1 correspondence with all of the reals.

But this case too falls under the weak principle of identity, since, for any two relatively discernible reals, at least they bear a different relation to each other. Objects can fail to be identity in a third way, however. To see this, suppose $B(x, y)$ is true, and that B is a symmetric predicate (so that $B(x, y)$ iff $B(y, x)$). Evidently B cannot be used to discern objects relatively. But (3) will still fail to hold if only B is *irreflexive* (so $B(x, x)$ is always false), for then there will exist a value of u_1 such that $B(u_1, x)$ is true but $B(u_1, y)$ is false, namely $u_1 = y$. Hence $x \neq y$. Call such objects *weakly discernible*.[15] This is the PII in accordance with modern logic: objects are numerically distinct only if absolutely, relatively, or weakly discernible.

Most of the classical counter-examples to Leibniz's principle, and all of the really convincing ones, turn out to be examples of weak discernibles (for example, Max Black's two iron globes, a certain distance from each other, each exactly the same, in an otherwise empty space) - so are not counter-examples to the PII as just stated. It is true that there is a quantum mechanical counter-example to it, namely elementary bosons all in exactly the same state, but fermions are always at least weakly discernible. Even in the most symmetric case, where the spatial part of the state has exact spherical symmetry, and the spin state is spherically symmetric too (as in the singlet state of two spin $\frac{1}{2}$ particles), fermions are weakly discernible: they satisfy the symmetric but irreflexive relation "... opposite component of spin to ..." As for elementary bosons, with the exception of the Higgs particle, they are all gauge particles: the objects in such cases may well be better considered as the modes of the gauge field, with the number of quanta understood as excitation numbers instead.

Weak discernibles certainly could not be objects that we encounter in any ordinary way. By assumption, there is no physically meaningful predicate that applies to one of them, rather than to any other, so one cannot *refer* to any one of them singly. Call them *referentially indeterminate*.[16] But apart from objects at the microscopic scale, there seems to be little possibility of encountering *macroscopic* weak discernibles, not at least if they are impenetrable; for in that case, in any universe with large-scale asymmetries, spatiotemporal relations with other objects will always differ among them (such objects will invariably be absolutely discernible).

5. NON-REDUCTIVE RELATIONALISM

The new PII differs from the old entirely through the unrestricted use of relations - any that are physically real. Admitting relations in this way, we are clearly abandoning Leibniz's independence thesis. This changes the debate between Leibniz and Clarke. Consider for example Clarke's objection:

> Why this particular system of matter, should be created in one particular place, and that in another particular place; when, (all place being absolutely indifferent to all matter,) it would have been exactly the same thing *vice versa*, supposing the two systems (or the particles) of matter to be alike; there could be no other reason, but the mere will of God (Alexander 1956, 20–21, Clarke, Second Reply).

Clarke denied that the PSR could have the fundamental status that Leibniz ascribed to it. We have seen that in a uniform space the PSR is violated if particles have positions, for there can be no reason why particles should be in one position rather than another. Leibniz took this to be a *reductio* of the view that particles have positions, over and above their relations with each other; but that does not solve the problem just posed by Clarke, given that permutations are symmetries (given that the particles are all exactly alike), for the permutation changes not only the positions of particles, but which particle is related to which. Call it the *permutation argument*. Leibniz's response to it was to deny that there *could* be such a symmetry; it was to reject atomism altogether. But an alternative response *is to allow that bodies can be identified by their relations to one another*; then a particular body is no more than a particular pattern-position. This is the modern, relationalist description of atoms;[17] the price is that we abandon the independence thesis.

In fact, in this application, Clarke was just as committed to the independence thesis as was Leibniz. Neither could allow that the numerical identity of atoms could be settled by appeal to their external relations. And indeed a reduction of sorts has taken place: particles have been replaced by pattern positions, by nodes in a pattern. Taken independent of their relations with one another, one might have thought to still have a collection of objects, whether substances or bundles of properties or whatever; but under the PII that cannot be so.

In the case of diffeomorphism covariance, there is a reduction of a lesser sort: positions in space have been replaced by positions in patterns of values of fields. This time it is not so clear that these positions cannot be thought of as objects in their own right, independent of the patterns of field-values. Manifold points do, after all, still bear relations to one another, even when the metric and other fields are removed, for there remains the differentiable structure of the manifold, as defined by the atlas of charts. The local topology, the open sets and their relations under set membership, as inherited from the usual topology of R^4, is preserved by diffeomorphisms. It is the smoothness of the manifold which is independent of any fields on it; manifold points can be counted as distinct if and only they are contained in disjoint open sets (of course we are assuming the manifold is Hausdorff), and this too is a diffeomorphism-invariant condition, independent of any metrical structure. Were invariance under diffeomorphisms a sufficient condition for a relationship to be real, the manifold points would be counted as objects by the PII independent of their arrangement in patterns of fields.

In the case of the rigid symmetries, there is a reduction of yet another sort: the positions of objects in space have been replaced by the positions of objects relative to other objects — in the first instance, to positions of particles. Absolute positions disappear; under the PII, points in space, considered independent of their relations with other points in space and with material particles, all disappear. But points in space considered independent of matter, but in relation to other points in space, are perfectly discernible (albeit weakly), for they bear non-reflexive metrical relationships with each other. There is no problem for the PSR in consequence; there is no further question as to which spatial point (or manifold point) underlies which pattern-position, for they are only weakly discernible. Only in the case of absolute discernibles — in fact, only in the case of objects absolutely discernible by a subset of all the predicates available in a language — can their be any further question as to which object has which attribute.

In the case of a homogeneous space, spatial points cannot be absolutely discerned by the subset of predicates that apply to them alone. But since we have rejected the independence thesis, there is no reason in principle not to make use of other predicates as well, specifically those which apply to matter and events. Hence R0 follows: points of space and spacetime, in the symmetric case, are to be specified (absolutely discerned) by their relationships with matter and events respectively.[18]

Finally, I come back to the further symmetry principle urged by Mach: transformations to rotating frames of reference. At the end of the *Correspondence*, Clarke did at last challenge Leibniz to give an account of rotations. Rotational motion in NTG is possible even when all the relative distances at each time are exactly the same; this does in fact pose a problem for Leibniz's original principles. We can, indeed, recover a similar perspective to Leibniz's in modern, relationalist, terms, if only we assume that temporal relations are *reducible* — that time is in fact an internal or intrinsic relation (a relation that follow from monadic properties of its relata). For then it would follow that if two spatial configurations of particles or fields are not discernible in themselves (if they are not absolutely discernible), then they cannot be relatively or weakly discernible either. They would then be indiscernible, period, so counted as numerically one. So there could then be *no such thing* as a rotating system of particles or fields, in which the spatial configuration at each time were exactly the same. In other words, relationalism implies Machianism, but only if time is not an external relation.[19]

For Leibniz, of course, no real physical relation can be external, so he was committed to Machianism. Today we may allow that this question — of whether time is an external relation — is open; but it is no longer a logical thesis. Machianism is no longer grounded in logic.

6. INDIVIDUATING FIELDS

A rose by any other name will smell as sweet. Earman has sketched a theory similar to mine; he calls it "resolute substantivalism." In his more recent collaborations with Belot (Belot and Earman 1999, 2001), it is called "sophisticated substantivalism." In their usage, substantivalism means realism. But according to Earman it faces a severe difficulty: if there are "multiple isomorphisms," cases where an object i in one spacetime model (or "world") Σ can be mapped onto distinct objects i_1, i_2, in a second world Σ', by each of two distinct isomorphisms ψ_1, ψ_2, then there is a problem if "identity follows isomorphism" as he puts it. If $i \in \Sigma$ is mapped onto $\psi_1(i) = i_1$, but is also mapped onto $\psi_2(i) = i_2 \neq i_1$, we have a contradiction, for surely identity is an equivalence relation. If so i_1 would be identical to i_2 as well, contrary to hypothesis (Earman 1989, 198–199).

According to Earman the problem is quite intractable; all the most straightforward ways of making sense of this position are, he says, "indefensible." But on our framework there is no such difficulty. Nominally distinct worlds are in 1:1 correspondence with the elements of the symmetry group. If there are two distinct isomorphisms — group elements — then there are two nominally distinct worlds as well. The formal properties of identity parallel the formal properties of the relationship $\sim$, defined as:

$$\Sigma_1 \sim \Sigma_2 \text{ iff } \exists g \in G \text{ such that } g(\Sigma_1) = \Sigma_2. \tag{4}$$

The two distinct isomorphisms therefore map the object i into two nominally distinct worlds: $i_1 = g_1(i) \in \Sigma_1$, $i_2 = g_2(i) \in \Sigma_2$. There is indeed a map $i_1 \to i_2$, since (4) is an equivalence relationship (so long as the symmetry transformations form a group), but it is a map between worlds: $g_3 : \Sigma_1 \to \Sigma_2$ (where $g_3 = g_2 \circ g_1^{-1}$). It is worlds, and objects across worlds, which are identified, not objects within a single world (unless of course they are indiscernible).

Hoeffer (1996) has defended a view similar in some respects to mine. He likewise accepts that spatial points may be referentially indeterminate (that "primitive identity" may fail, as he puts it). Considering space independent of matter, there is no underdetermination as to where matter is to be located (no violation of the PSR), for if there is no primitive identity of spatial points then there is no matter of fact as to which point is to be occupied. With this I agree; but Hoefer denies that spatial points even considered together with matter can be referred to uniquely (by their relation to the matter distribution). He tacitly embraces Leibniz's independence thesis, whilst explicitly denying the PII. He believes his position is opposed to relationalism: his denial of primitive identity appears *ad hoc* in consequence. Certainly he is not in a position to motivate it on logical grounds, as I do, by the PII.

Relationalism, I say again, rejects the independence thesis and is diametrically opposed to absolutism. But it is neutral with respect to the contrast between matter and space. On an even-handed approach to matter and space, if we can use relations to space to individuate material bodies, then surely we can use relations with bodies to individuate parts of space.

Earman and Hoefer will at least agree with me on this: the hole argument is continuous with the arguments used by Newton, Leibniz and Clark. Stachel has argued otherwise, however. I come back to the nature of general covariance. Does the hole argument apply to NTG, written in diffeomorphism covariant form? Do not exactly the same conclusions follow as for GTR?

There are of course certain differences. In GTR, so long as symmetries are lacking (the physically realistic case), we can follow Stachel's suggestion (1993, 156) to specify points in spacetime in purely chronogeometrical terms. We always have available at least four of the fourteen invariants ξ_k of the Riemann tensor (ten of them vanish *in vacuuo*); in the absence of symmetries, these quadruples of real numbers will generally differ at distinct points of M, and we can refer macroscopic objects in space-time to their values. The usual way of speaking of fields at points can be recovered as well. Say "the field ρ has value λ at the point $(\xi_1, \xi_2, \xi_3, \xi_4)$" and write "$\rho(\xi_1, \xi_2, \xi_3, \xi_4) = \lambda$" just when ρ has value λ where the Riemann scalars have values $(\xi_1, \xi_2, \xi_3, \xi_4)$; or, more parsimoniously, that the λ-value for ρ coincides with the ξ-values for the Riemann invariants.[20] Evidently in NTG this construction will not be of much use. The invariants built out of the chronogeometrical tensors will all be constant along the integral curves of the Killing vector fields. In this situation we will have to individuate points of these curves by reference to values of the gravitational field or, equivalently, by reference to material particles - assuming, of course, that these do not in turn have further symmetries. We will have to solve the dynamical equations of motion, and use these solutions to define a dynamical individuating field for space-time points. Throughout we will have to abide by the relational, qualitative approach to predication, tolerating weak discernibles should they arise.

But with all of this Stachel disagrees. He claims that on the contrary (i) The space-time structure of pregeneral-relativistic theories is most simply and economically written in terms of an affine geometry; it is to be defined by the linear transformations of an inhomogeneous affine space, not by a differentiable space-time manifold. (ii) An individuating field is required, granted, but it can be defined completely independent of the dynamical field equations of these theories. (iii) One can still rewrite NTG in diffeomorphism-covariant form, but when one does this the hole argument is completely trivial, and the general covariance employed is a sham (it is "trivial general covariance" corresponding to what I have been calling diffeomorphism covariance). (iv) One cannot apply the hole argument to pregeneral-relativistic theories written in affine-space-plus form (for the reasons we have already considered).

Let me give what ground I can. (i) is perfectly reasonable, but it turns out - with Stachel's qualification when it comes to NTG - that he is speaking here only of electromagnetism, among the fundamental theories (I have made this point already); and even in that case there remain the various shift arguments (but admittedly not the generalized shift argument of NTG): his definition of the individuating field mentioned in (ii) will have to be consistent with these. (ii) is clearly the sticking point; Stachel flatly denies the claim that I am making: that as a matter of course the dynamical equations (for the matter fields) will have to be used to define an individuating field. On (iii) we can guardedly agree, but there remains the generalized shift argument. With (iv) I agree, but with the same proviso as with (i).

(ii) is the decisive point of disagreement; otherwise, given (i), (iii) and (iv), I wish only to see, in Stachel's account, how the shift and permutation arguments are to be dealt with, in accordance with his definition of the individuating field in (ii) (and in the case of NTG, how the generalized shift argument is to be dealt with). How does Stachel define a non-dynamical individuating field? In fact he makes use of the familiar quasi-operationalist approach due to Einstein, in terms of a collection of clocks and rigid bodies (although he disavows any commitment to operationalism thereby). By parallel transport of their direction vectors, as defined by the affine geometry, one sets up a unique tetrad field in spacetime, with its associated coordinate system. He adds that many alternative procedures are possible, for example the use of test particles and light rays "to mention only two other possibilities" (Stachel 1993, 149). Which is used is not the important point, however; rather:

> The important point for present purposes is that the spacetime structures, as well as the individuating field mapped out with the help of these methods, are *independent* of any dynamical fields that are subsequently introduced in the spacetime (*ibid.*).

There are points of contact here with Einstein's view on the matter:

> It is ... clear that the solid body and the clock do not in the conceptual edifice of physics play the part of irreducible elements, but that of composite structures, which must not play any independent part in theoretical physics. But it is my conviction that in the present stage of development of theoretical physics these concepts must still be employed as independent concepts; for we are still far from possessing such certain knowledge of the theoretical principles of atomic structure as to be able to construct solid bodies and clocks theoretically from elementary concepts (Einstein 1921, 236–237).

Einstein is saying that we do not know how to provide a dynamical model of the individuating field, and that in this situation, as a stop-gap, we must treat rulers and clocks as "independent concepts" for determining space-time intervals. Stachel, by contrast, denies that in pregeneral-relativistic theories it is a stop-gap. In such theories there is never any need for such a dynamical model. So long as we confine ourselves to Newtonian or special relativistic theories, we need never inquire as to the dynamical structure of the individuating field; we do not have to solve for the equations of motion, prior to determining what the coordinates mean.

But what exactly are these "independent concepts"? Stachel elsewhere calls them "ideal elements" (1983, 256); he is critical of the "empiricist, operationalist or instrumentalist spirit" according to which " ... the ideal elements initially introduced ... must be immediately ... identifiable with objects used in laboratory tests of the theory." Evidently Stachel is not appealing to any concrete operational procedure, or any given technology. But now one wonders if, like the absolutist, it is a *purely* mathematical coordination that he has in mind. One wonders if they are ideal elements in *Leibniz's* sense. If the procedure has no operational definition, nor is it a matter of using the dynamical theory, then what does it consist in, so as to make any physical sense? One could of course make use of the dynamical individuating field of *some other* theory; this point I freely grant; but it is historically idle, if our concern is with NTG, and it is inconsequential, for surely our interest switches to this other theory.

It is the other theory that will be the foundational one for our understanding of space and spacetime.

Another alternative is to make use of a dynamical individuating field defined by one application of the theory, to give content to the use of coordinates in the context of a *different* application of that same theory. This is, in fact, of methodological importance, because it makes clear why, if available, a non diffeomorphic-covariant formulation of a theory is to be preferred. We may grant that it is more convenient, more simple, to first define a dynamical individuating field, by providing a dynamical model for a given physical system, and then to make use of that particular physical system (a particular system of clocks and physical bodies) in other applications of the theory. (The policy of "divide and rule" is well-named.) What is required, of course, for this to be possible, is that the theory not be *generally* covariant, in Stachel's sense. Such a procedure is *not* applicable in the case of GTR. Is this Stachel's point, in sharply distinguishing other theories from GTR? If so one should not say the individuating field so introduced is non-dynamical; one should say, to coin a term, that the field is *external* to an application. The pregeneral-relativistic theories permit of external individuating fields; they will be dynamical fields all the same.

The treatment of referential indeterminateness is a litmus-test for the meaning of an individuating field more generally. Stachel has no comment to make on indeterminateness in affine-space-plus theories, but he does comment on the symmetric case in GTR:

> If a particular metric tensor field does have some symmetry group, the values of the four invariants [of the curvature tensor] will be the same at all points of an orbit of the symmetry group, so that additional individuating elements have to be introduced to distinguish between such points. (For example, the preferred parameters of the symmetry group, one of which is associated with each of the Killing vector fields that generates the symmetry group, may be used.) However chosen, such additional elements cannot be independent of the metric tensor field since the latter serves to define the orbits in question (*ibid.*, 143).

As it stands these preferred parameters are purely mathematical artifacts. Stachel does not attend to their physical definition. Nowhere, so far as I know, does he acknowledge the possibility of indeterminateness of reference, in the extreme case where likewise the material distribution has the same symmetries (which is likely to follow in GTR, given exact global symmetries of the metric).

7. INDIVIDUATING TIME

Stachel sees no virtue in a *purely* operational non-dynamical individuating field, but others might. It was, in point of fact, a method of great use to astronomy, and to much of 19th century physics. Surely it has a bearing on our topic. I will close with some remarks on this option.

The issues are more straightforward in the case of time, so I will confine myself to that. For the notion of a purely operationally-defined clock, we can do no better than turn to the definition of Heinrich Hertz, that exemplar of 19th century experimental electromagnetism:

> Rule 1: We determine the duration of time by means of a chronometer, from the number of beats of its pendulum. The unit of duration is settled by arbitrary convention. To specify any given instant, we use the time that has elapsed between it and a certain instant determined by a further arbitrary convention.
>
> This rule contains nothing empirical which can prevent us from considering time as an always independent and never dependent quantity which varies continuously from one value to another. The rule is also determinate and unique, except for the uncertainties which we always fail to eliminate from our experience, both past and future (Hertz 1894, 298).

But there is nothing determinate and unique about the time kept by a pendulum clock, except and insofar as it is under theoretical control — exactly, that is, when the uncertainties in its behavior *are* eliminated from our experience. Evidently a good clock is one whose construction is guided by theory — in Hertz's time, mechanical theory. With that, we are on the road towards a dynamical individuating field after all.

The point is clearer if one is talking of the very best clocks, of the sort needed by astronomers. Here one might think that Hertz's parallel definition, of a spatial individuating field, was on the right lines, for he defined position and orientation by reference to the fixed stars. What then of Sidereal Time, time as defined by the diurnal motion of the Earth with respect to the stars? This is a big improvement on any pendulum clock. Of course, we know this will be a good clock because we have good reasons to view the rotation of the earth as uniform, which derive ultimately from NTG; so Sidereal Time has a dynamical underpinning of sorts. But we need hardly solve any equations of motion for it, in order to define this time standard. It is not much of a dynamical individuating field; and it is anyway an external field, in the sense that I have just defined. It is surely grist to Stachel's mill.

But good as Sidereal Time is, it is not good enough for astronomy - so long as we do not model it more precisely, and so long as we insist on treating it as an external individuating field. When we do model it more precisely, we find that the rotation of the earth is *not* perfectly uniform, and that it varies in very complicated ways with respect to a true measure of time. The effect is known as *nutation*. The main contribution to this was discovered by Bradley, in the early nineteenth century, with an amplitude of 9 seconds of arc every 18.6 years. This wobble of the Earth's axis shows up in a periodic shift in all stellar coordinates (declination and right ascension), so as long as the variation is periodic corrections to Sidereal Time can easily be made; the difficulty is that there are many other harmonics (the component studied by Bradley is due to the regression of the nodes of the moon's orbit). To isolate these, there is no alternative to using the full NTG, and thereby a dynamical individuating field. And, because one has to consider the Moon's orbit, and in principle the other celestial bodies as well, it is hardly an external individuating field that we end up with.

By the early 20th century, it was clear that Sidereal Time would not do for astronomy. What then replaced it? The answer was already to be found in Newton's procedure in the *Principia*. Newton's clock was the Solar System, specifically the Earth-Moon system, Jupiter, and the Sun. The *Principia*, as Newton said, was written to distinguish the true from the apparent motions, to tell us how this clock should be read. His procedure made use of no other data but relative distances and the angles of intersection of lines (rays of light). What other data are there? He modeled a part of the system under study to define an individuating field (therefore a dynamical individuating field), to which the coordinates of other quantities could be referred. In this he but followed Galileo. How is one to test the equation $h = \frac{1}{2}gt^2$ for a freely-falling body, when the only clocks available were hour-glasses, sundials and candles? Galileo's answer was simplicity itself. Conclude, from Galileo's mechanics, that this equation is unaffected if the body is given a horizontal component of velocity v; conclude that the horizontal distance d travelled is vt; then h is proportional to d^2. Galileo used a dynamical individuating field.

Newton's procedure was, of course, much more complicated, and it was very poorly understood. It was only with Lagrange's work, a century later, that the theoretical problem was solved completely; and it was another century, following the work of Simon Newcombe in the 1880s, before astronomers were making use of Lagrange's techniques. The result is *Ephemeris Time*; it is this which defined the SI unit of time until very recently. It is time defined as that parameter with respect to which Newton's equations hold good of the observed celestial motions. Other quantities, as functions of time, are then referred to those very motions. It is a dynamical individuating field *par excellence*.

There is a twist to this story. In 1976 Ephemeris Time was replaced by the atomic clock standard.[21] The two are, in fact, fully comparable in accuracy. With that, and for the first time, we are really in a position to do astronomy using a non-dynamical individuating field for the time coordinate. It is non-dynamical, of course, only with respect to *classical* theory; it is a dynamical individuating field from the point of view of quantum electrodynamics.[22]

If Hertz was oblivious to all of this, immersed as he was in a study of the foundations of mechanics,[23] it is hardly a surprise to find that the young Einstein was ignorant of it too. As Stachel says, the idea of a non-dynamical individuating field was one that Einstein had painfully to unlearn. But here he only followed tradition: from ancient times, when Hipparchus and Ptolemy first doubted that the stars were really fixed, the lesson had to be repeatedly learned that there is no perfect individuating field given to us observationally; and from Leibniz's time, that it cannot be given to us as an ideal thing, either. The history of dynamics is in large part the story of how we are to proceed in this situation, and define such a field all the same. A dynamical individuating field.

Universtity of Oxford

NOTES

* In honour of Professor Stachel. He will agree with me on points, if not on substance, but I am indebted to him for his guidance on both topics; and on many others besides.

1. Some consensus on the history has recently been arrived at by a number of historians, among them Renn, Corry, and Norton, as also by Stachel; see (Renn et al. 2002).
2. Hence they have a bearing on the so-called "problem of time" See (Belot and Earman 2001) for a recent review.
3. See (Stachel 1993). Many others have thought that general covariance must mean something more than diffeomorphism-covariance; see e.g. (Wald 1984, 57).

4. For a scalar field, the drag-along is just $(f * \phi)(x) = \phi(f(x))$; we will not need the definition in the case of tensor fields.
5. Although it is not clear that it is shared by all; see e.g. (Weinberg 1989, 4).
6. Newton stated only the weaker result (at Corollary VI, Book 1, *Principia*), although the stronger principle (which includes time-dependent "equal accelerative forces" as he called them) is obvious. He used this symmetry to justify the application of his laws of motion to the Jupiter system, notwithstanding the gravitational influence of the Sun. It is evidently a precursor to Einstein's Principle of Equivalence.
7. (Alexander 1956, Clarke, 3rd Reply); here he talks of moving the "whole material world entire" and "the most sudden stopping of that motion" He returned to this objection in his 4th Reply (*ibid.*, 48), only to confuse it with the very different objection that there are dynamical effects when a *subsystem* of the universe is brought to a sudden halt. Leibniz pointed out the confusion in his 5th Paper (*ibid.*, 74); in his 5th Reply, Clarke restated the objection in its orignal form, complaining that Leibniz "had not attempted to give any answer" (*ibid.*, 105). In fact Libniz had responded as do we, identifying the two motions as indiscernible (*ibid.*, 38).
8. Called the *Newtonian group* by (Ehlers 1973).
9. This was not quite Carke's conclusion; as remarked, he supposed the argument was liscensed only by Leibniz's principles, not Newton's; but it is in line with his position in the face of those symmetry arguments that he did acknowledge applied to NTG.
10. See (Stachel 1993, 154, fn. 2), I shall not follow him in this.
11. We can also interpret Leibniz here to mean spatial inversion, whereupon — supposing it is an exact symmetry — it would follow that the world and its mirror image must be identified. For a defence of this conclusion, and the consequences of parity violation in quantum theory, see my (2002a).
12. It might appear from this that Leibniz's containment theory of truth is in just as much trouble, in modern logic, as is his independence thesis. In general that is correct, but it is possible to restrict the containment theory so that it only applies to *closed* physical systems and to *complete* descriptions of them. Frege's logic is consistent with this.
13. The EPR completeness criterion is one half of this principle: "every element of the physical reality must have a counterpart in the physical theory" (Einstein, Podolsky, and Rosen 1935).
14. There is no difficulty in making the same definition in the second-order case, where one quantifies over predicates or properties and relations, so long as they do not include the identity sign or the relation of identity.
15. Quine missed this category earlier (1960, 230), where he introduced the distinction between absolute and relative discernibles. He subsequently spoke of grades of *discriminability*, rather than discernibilty (Quine 1976), but I will not follow him in this.
16. Not to be confused with Quine's doctrine of *indeterminacy of reference*, which applies to objects whether or not they are weakly discernible. Connections between the symmetries of model theory, as exploited by Quine, and the ones that we have been considering, have been drawn by (Liu 1997) and (Rynasiewicz 1996b); they have been found wanting by (Stachel 2001).
17. It should be evident from this that "particle indistinguishability" figures just as much in classical statistical mechanics as quantum statistical mechanics, see e.g. (Hestines 1970). That raises the question of just why quantum statistics differ at all from the classical case. The answer, of course, lies in the discreteness of available energies.
18. Leibniz might be taken to agree with me on this, when he says: "The parts of time or place, considered in themselves, are ideal things; and therefore they perfectly resemble one another, like two abstract units. But it is not so with two concrete ones, or with two real times, or two spaces filled up, that is, truly actual" (Alexander 1956, 63, Fifth Paper). (But of course, Leibniz did conclude from the fact that the parts of space and time in themselves perfectly resembled one another, that they are ideal, and hence not real; he did not grant that they were still weakly discernible.)
19. Julian Barbour, arch-Machian, has in recent years moved towards just this view of Machianism; see (Barbour 1999).
20. Such methods have been discussed at length by Rovelli (1991), and in the quantum case as well as in classical GTR.

21. Known as *Temps Atomique International*, based on a free-running, data-controlled timescale (Échelle Atomique Libre), formed by combining data from all available high-precision atomic clocks (principally cesium beam standards and hydrogen masers). The accuracy is presently of the order of 1 part in 10^{14}, approximately the same as in Ephemeris Time, but it is likely to be considerably improved on by moving to optical frequencies. For more on astronomical time standards, see (Seidelman 1992).
22. Any discrepancies now, between these two dynamical individuating fields, will require a dynamical model of both kinds of clocks, therefore a theory of quantum gravity.
23. For the defects of Hertz's mechanical principles, see (Saunders 1998, 2002a, 2002b). Because of the equivalence principle, experiments in electromagnetism and, more generally, microscopic physics (including the definition of ETA) are insensitive to the choice of frame, but Hertz could not easily appeal to it; unlike Newton's principles, his did not imply it.

REFERENCES

Alexander, H. 1956. The Leibniz-Clarke Correspondence.
Barbour, J. 1999. *The End of Time*. Oxford: Oxford University Press.
Belot, G. 2001. "The Principle of Sufficient Reason." *The Journal of Philosophy* XCVIII:55–74.
Belot, G., and J. Earman. 1999. "From Metaphysics to Physics." In *From Physics to Philosophy*, eds. J. Butterfield and C. Pagonis. Cambridge: Cambridge University Press.
———. 2001. "Pre-Socratic Quantum Gravity." In *Physics Meets Philosophy at the Planck Length*, eds. C. Callendar and N. Huggett.
Cantor, G. 1895. "Beiträge zur Begründung der transfiniten Mengenlehre." *Mathematische Annalen* 46:481–512.
Earman, J. 1989. *World Enough and Space-Time*. Cambridge: MIT Press.
Earman, J., and J. Norton. 1987. "What Price Substantivalism? The Hole Story." *The British Journal for the Philosophy of Science*, vol. 38.
Ehlers, J. 1973. "Survey of General Relativity Theory." In *Relativity, Astrophysics and Cosmology*, ed. W. Israel. Drodrecht: Reidel.
Einstein, A. 1921. "Geometrie und Erfahrung." In *Erweiterte Fassung des Festvortrages gehalten an der Preussische Akademie*. Berlin: Springer. Translated 1954 by S. Bergmann. Pp. 232–246 in *Ideas and Opinions*. New York: Crown Publishers.
———. 1954. "Relativity and the Problem of Space, Appendix 5." In *Relativity: The Special and General Theory*, 15th ed. Methuen.
Einstein, A., B. Podolsky, and N. Rosen. 1935. "Can Quantum-Mechanical Description of Physical Reality be Considered Complete?" *Physical Review* 47:777–80.
Gödel, K. 1930. "Die Vollständigkeit der Axiome der logischen Funktionenkalküls." *Monatshefte für Mathematik und Physik* 37:349–360. Translated as "The Completeness of the Axioms of the Functional Calculus of Logic". In *From Frege to Gödel*, ed. J. van Heijenoort. Cambridge, Mass.
Hawking, S., and G. Ellis. 1973. *The Large-Scale Structure of Space-Time*. Cambridge: Cambridge University Press.
Hertz, H. 1894. *Die Prinzipien der Mechanik*. London. Translated by Jones, D. and T. Walley in *The Principles of Mechanics*.
Hestines, D. 1970. "Entropy and Indistinguishability." *American Journal of Physics* 38:840–845.
Hilbert, D., and P. Bernays. 1934. *Grundlagen der Mathematik*. vol. 1. Berlin: Springer.
Hoeffer, C. 1996. "The Metaphysics of Space-Time Substantivalism." *The Journal of Philosophy* XCIII:5–27.
Johnson, W. E. 1924. *Logic: Part 3, The Logical Foundations of Science*. Cambridge: Cambridge University Press.
Liu, C. 1997. "Realism and Spacetime: Of Arguments Against Metaphysical Realism and Manifold Realism." *Philosophia Naturalis* 33:243–63.
Maudlin, T. 1988. "The Essence of Space-Time." *Proceedings of the Philosophy of Science Association* 2:82–91.
Quine, W. V. 1960. *Word and Object*. Mass.: Harvard University Press.

———. 1976. "Grades of Discriminability." *Journal of Philosophy* 73:113–116. [Reprinted in *Theories and Things*. Cambridge: Harvard University Press, 1981].

Renn, J., T. Sauer, M. Janssen, J. Norton, and J. Stachel. 2002. "The Genesis of General Relativity: Sources and Interpretation." In *General Relativity in the Making: Einstein's Zurich Notebook*, vol. 1. Dordrecht: Kluwer.

Rovelli, C. 1991. "What is Observable in Classical and Quantum Gravity." *Classical and Quantum Gravity* 8:297–304.

Rynasiewicz, R. 1996a. "Absolute Versus Relational Space-Time: An Outmoded Debate?" *Journal of Philosophy* XCIII:279–306.

———. 1996b. "Is there a Syntactic Solution to the Hole Argument?" *Philosophy of Science* 63:55–62.

Saunders, S. 1998. "Hertz's Principles." In *Heinrich Hertz: Modern Philosopher, Classical Physicist*, ed. R. I. G. Hughes. Kluwer.

———. 2002a. "Leibniz Equivalence and Incongruent Counterparts." forthcoming.

———. 2002b. "Physics and Leibniz's Principles." In *Symmetries in Physics: Philosophical Reflections*, eds. K. Brading and E. Castellani. Cambridge: Cambridge University Press.

Seidelman, P. K., ed. 1992. *Explanatory Supplement to the Astronomical Almanac; a revision to the Explanatory Supplement to the Astronomical Ephemeris and the American Ephemeris and Nautical Almanac*. Mill Valley: University Science.

Sklar, L. 1974. *Space, Time, and Spacetime*. Berkeley: University of California Press.

Stachel, J. 1983. "Special Relativity from Measuring Rods." In *Physics, Philosophy, and Psychoanalysis: Essays in Honor of Adolf Grünbaum*, eds. R. S. Cohen and L. Laudan. Dordrecht: Kluwer.

———. 1993. "The Meaning of General Covariance." In *Philosophical Problems of the Internal and External Worlds: Essays on the Philosophy of Adolf Grünbaum*, eds. J. Earman, A. Janis, G. Massey, and N. Rescher. Pittsburgh: University Press.

———. 2001. "'The Relations Between Things' Versus 'The Things Between Relations': The Deeper Meaning of the Hole Argument." In *Reading Natural Philosophy: Essays in the History and Philosophy of Science and Mathematics to Honour Howard Stein on His 70th Birthday*. Open Court: Chicago and LaSalle.

Wald, R. 1984. *General Relativity*. Chicago: Chicago University Press.

Weinberg, S. 1989. "The Cosmological Constant Problem." *Reviews of Modern Physics* 61:1–23.

DAVID B. MALAMENT

ON RELATIVE ORBITAL ROTATION IN RELATIVITY THEORY*

I want to consider this question within the framework of relativity theory: given two point particles X and Y, if Y is rotating relative to X, does it follow that X is rotating relative to Y? To keep the discussion as simple as possible, I'll allow X and Y to be test particles.

As it stands, the question is ambiguous. Roughly speaking, one wants to say that "Y is rotating relative to (or around) X," at least in the sense I have in mind, if "the direction of Y relative to X" is "changing over time." What must be explained is how to understand the quoted expressions. There is a perfectly straightforward way to do so within Newtonian particle mechanics (section 1), where there is an invariant notion of "time," and "space" is assumed to have Euclidean structure. At all times, there is a well-defined *vector* that points from X to Y, and one can use it to define the angular velocity of Y relative to X.

But the situation is more delicate in relativity theory. Here no such simple interpretation of "relative rotation" is available, and some work is required to make sense of the notion at all. (It seems to me unfortunate that this is often overlooked by parties on both sides when it is debated whether relativity theory supports a "relativist" conception of rotation.) In section 2, I'll consider one way of defining the "angular velocity of Y relative to X" (Rosquist 1980) that does not presuppose the presence of special background spacetime structure (e.g., flatness, asymptotic flatness, stationarity, rotational symmetry), and can be explained in terms of simple (idealized) experimental procedures. I'll also derive an expression for the angular velocity of Y relative to X in the special case where the worldlines of X and Y are (the images of) integral curves of a common background Killing field. Finally, in section 3, I'll turn to the original question.

SECTION 1

For purposes of motivation, let us first consider relative rotation within the framework of Newtonian particle mechanics. Here we can associate with the particles, at every time t, a relative position vector $\vec{r}_{XY}(t)$ that gives the *position of Y relative to X*. (We can think of the vector as having its tail coincident with X and its head coincident with Y.) The inverted vector, $\vec{r}_{YX}(t) = -\vec{r}_{XY}(t)$ gives the position of X

relative to Y at time t. Let us take for granted that the particles never collide (so that $\vec{r}_{XY}(t)$ is non-zero at all times), and consider the normalized vector:

$$\vec{n}_{XY}(t) = \frac{\vec{r}_{XY}(t)}{|\vec{r}_{XY}(t)|}.$$

We can think of it as giving the *direction of Y relative to X* at time t. The *(instantaneous) angular velocity of Y relative to X* at time t is given by the vector cross product:

$$\vec{\Omega}_{XY}(t) = \vec{n}_{XY}(t) \times \frac{d}{dt}(\vec{n}_{XY}(t)).$$

Notice that it is not here presupposed that X is in a state of uniform rectilinear motion. X (and Y too) can wiggle so long as $\vec{n}_{XY}(t)$ has a well-defined derivative. Notice also that if $\vec{n}_{YX}(t)$ and $\vec{\Omega}_{YX}(t)$ are defined in the obvious way, by interchanging the roles of X and Y, then $\vec{n}_{YX}(t) = -\vec{n}_{XY}(t)$ and $\vec{\Omega}_{YX}(t) = \vec{\Omega}_{XY}(t)$. We will be interested in two assertions.

(i) *Y is not rotating relative to X*:

$$\vec{\Omega}_{XY}(t) = 0 \text{ (or, equivalently, } \frac{d}{dt}(\vec{n}_{XY}(t)) = 0) \text{ for all } t.$$

(ii) *Y is rotating relative to X with constant angular velocity* (i.e., in a fixed plane with constant angular speed):

$$\frac{d}{dt}(\vec{\Omega}_{XY}(t)) = 0 \text{ for all } t.$$

What is important for present purposes is that *both assertions are manifestly symmetric in X and Y*.[1] It is the purpose of the present modest note to show that the situation changes, and changes radically, when one passes to the context of general relativity. We show with an example in section 3 that there *it is possible for Y to be non-rotating relative to X, and yet for X to be rotating relative to Y with constant (non-zero) angular velocity*. Moreover, the X and Y in question can be chosen so that the distance between them is constant (according to any reasonable standard of distance). And the distance can be arbitrarily small. (Of course, it remains to explain the interpretation of relative orbital rotation in general relativity on which these claims rest.)[2]

SECTION 2

Let us now turn to the relativity theory. In what follows, let (M, g_{ab}) be a *relativistic spacetime structure*, i.e., a pair consisting of a smooth, connected 4- manifold M, and a smooth semi-Riemannian metric g_{ab} on M of Lorentz signature $(+1, -1, -1, -1)$.[3] Let γ_X and γ_Y be smooth, non-intersecting timelike curves in M representing, respectively, the worldlines of X and Y. (We will not always bother to distinguish between the curves and their images.) We will follow Rosquist (1980), and define at each point on γ_X a vector Ω^a that may be interpreted as the "instantaneous (apparent) angular velocity of Y relative to X."[4]

Imagine that an observer sitting on particle X *observes* particle Y through a tubular telescope. We can take the orientation of his telescope at a given moment to determine the "(apparent) direction of Y relative to X" at that moment; and we can represent the latter as a unit vector, orthogonal to γ_X. In this way, we pass from the curves γ_X and γ_Y to a (normalized, orthogonal) direction field v^a on γ_X. Once we have the field v^a in hand, we are almost done. We can then define Ω^a in terms of v^a in close analogy to the way we previously defined $\vec{\Omega}_{XY}$ in terms of $\vec{n}_{XY}$. We need only replace the "time derivative" of $\vec{n}_{XY}$ with the Fermi derivative of v^a along γ_X. The construction is shown in fig. 1.

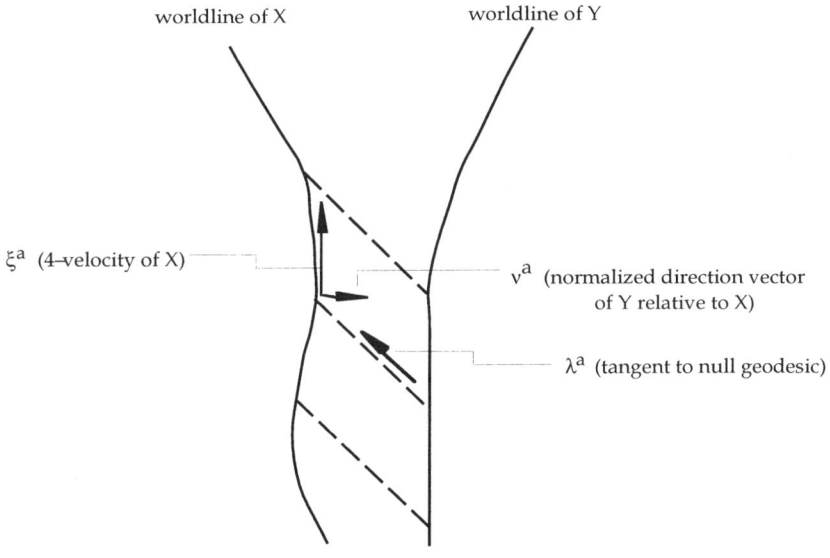

Figure 1.

Let ξ^a be the four-velocity of X, i.e., a future-directed,[5] timelike vector field on γ_X, normalized so that $\xi^a \xi_a = 1$. We assume that given any point p on γ_X, there is (up to reparametrization) a unique future-directed null geodesic that starts at some point (or other) on γ_Y and ends at p. This amounts to assuming that X can always see Y, and never sees multiple images of Y.[6] Let λ^a be the (future directed, null) tangent field to this geodesic (given some choice of parametrization). We arrive at the *direction vector* v^a *(of Y relative to X)* at p by starting with $-\lambda^a$ at that point, then projecting it orthogonal to ξ^a and finally normalizing the resultant vector:

$$v^a = \frac{-\lambda^a + (\lambda^n \xi_n)\xi^a}{(\lambda^m \xi_m)}.$$

(Equivalently, v^a is the unique vector at p in the two-plane spanned by ξ^a and λ^a such that $v^a \xi_a = 0$, $v^a v_a = -1$, and $v^a \lambda_a > 0$.) The Fermi derivative of v^a in the direction ξ^a,

$$(g^a{}_m - \xi^a \xi_m)\xi^n \nabla_n v^m,$$

is just the component of the directional derivative $\xi^n \nabla_n v^a$ orthogonal to ξ^a, i.e., the spatial component of the derivative as determined relative to X. We arrive at the *angular velocity* Ω^a *of Y with respect to X* at each point on γ_X, in effect, by taking the cross product there of v^a with $(g^a{}_m - \xi^a \xi_m)\xi^n \nabla_n v^m$ in the three-plane orthogonal to ξ^a:

$$\Omega^a = -\varepsilon^{abcd} \xi_b v_c ((g_{dm} - \xi_d \xi_m)\xi^n \nabla_n v^m).$$

In analogy to the conditions formulated in section 1, we say

(i′) *Y is not rotating relative to X* if

$\Omega^a = 0$ (or, equivalently,[7] $(g^a{}_m - \xi^a \xi_m)\xi^n \nabla_n v^m = 0$) at all points on γ_X;

(ii′) *Y is rotating relative to X with constant angular velocity* if

$$(g^a{}_m - \xi^a \xi_m)\xi^n \nabla_n \Omega^m = 0 \text{ at all points on } \gamma_X.$$

These conditions have a natural physical interpretation. Consider again our observer sitting on particle X and observing Y through his tubular telescope. Condition (i′) holds iff the orientation of his telescope is constant as determined relative to the "compass of inertia." So, for example, we might position three gyroscopes at X so that their axes are mutually orthogonal.[8] The orientation of the telescope tube at any moment can then be fully specified by the angles formed between each of the three axes and the tube. Condition (i′) captures the requirement that the three angles remain constant. Condition (ii′) captures the requirement that the three gyroscopes can be positioned so that the telescope tube is at all times orthogonal to one of the three, and its angles relative to the other two assume the characteristic, sinusoidal pattern of uniform circular motion (with respect to elapsed proper time).

We now consider the special case where γ_X and γ_Y are integral curves of a background future-directed, timelike Killing field τ^a. In this case, there is a strong sense in which the particles X and Y remain a constant distance apart.[9] To match our notation above, we express τ^a in the form $\tau^a = \tau \xi^a$, with $\xi^a \xi_a = 1$ and $\tau = (\tau^n \tau_n)^{1/2}$. Associated with ξ^a is a *vorticity* (or *twist*) vector field

$$\omega^a = \frac{1}{2} \varepsilon^{abcd} \xi_b \nabla_c \xi_d.$$

We want to derive an expression for Ω^a in terms of ω^a. To do so, we direct attention to the one-parameter group of local isometries $\{\Gamma_s\}$ associated with τ^a, i.e., the "flow maps" of which τ^a is the "infinitesimal generator." Given any one null geodesic segment running from γ_Y to γ_X, it's image under each map Γ_s is another null geodesic segment running from γ_Y to γ_X. (This follows immediately. Since γ_Y and γ_X are integral curves of τ^a, each is mapped onto itself by Γ_s. Since Γ_s is an isometry, it preserves all structures that can be characterized in terms of the metric g_{ab}, and that includes the class of null geodesics.) The collection of maps $\{\Gamma_s\}$ in its entirety, acting on the null geodesic segment, sweeps out a two-dimensional submanifold S, bounded by γ_Y and γ_X, through every point of which there passes a (unique) integral curve of τ^a and a (unique) null geodesic segment running from γ_Y and γ_X (see figure 2).

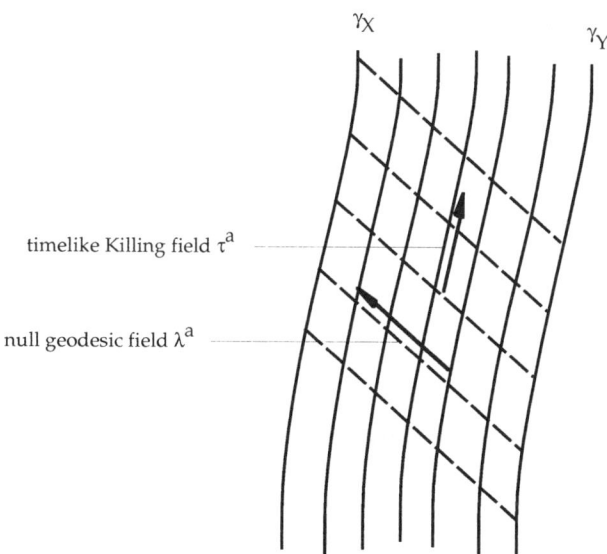

Figure 2.

Thus, we have on S two fields tangent to S: the timelike Killing field τ^a, and a future-directed null geodesic field λ^a ($\lambda^n \nabla_n \lambda^a = 0$ and $\lambda^n \lambda_n = 0$) that is preserved by each map Γ_s, or, equivalently, that is Lie derived by the Killing field τ^a, i.e.,

$$\tau^n \nabla_n \lambda^a - \lambda^n \nabla_n \tau^a = 0.$$

With this equation in hand, it is a matter of routine computation to derive an expression for Ω^a in terms of ω^a.

Proposition:[10] Let S, τ^a, and λ^a be as in the preceding paragraph (and let v^a, λ^a, and ω^a be the corresponding fields on S, as defined earlier in this section). Then, at all points on S,

$$\Omega^a = \omega^a + (v^n \omega_n) v^a.$$

(We have formulated the proposition in terms of the relative velocity of Y with respect to X. But, of course, a corresponding statement holds if the roles of X and Y are reversed. One just has to remember that the reversal brings with it a different two-dimensional submanifold S and a different null field λ^a.)

Proof: Since λ^a, g_{ab} (and τ^a) are Lie derived by the Killing field τ^a, so are all fields definable in terms of them. In particular, v^b is Lie derived by τ^a. Thus,

$$0 = \tau^n \nabla_n v^m - v^n \nabla_n \tau^m = (\tau \xi^n) \nabla_n v^m - v^n \nabla_n (\tau \xi^m)$$
$$= (\tau \xi^n) \nabla_n v^m - \tau v^n \nabla_n \xi^m - \xi^m v^n \nabla_n \tau.$$

So

$$\xi^n \nabla_n v^m = v^n \nabla_n \xi^m + (\tau^{-1}) \xi^m v^n \nabla_n \tau$$

and, hence,

$$(g_{dm} - \xi_d \xi_m) \xi^n \nabla_n v^m = (g_{dm} - \xi_d \xi_m) v^n \nabla_n \xi^m$$

since $(g_{dm} - \xi_d \xi_m) \xi^m = 0$. Therefore,

$$\Omega^a = -\varepsilon^{abcd} \xi_b v_c ((g_{dm} - \xi_d \xi_m) \xi^n \nabla_n v^m) = -\varepsilon^{abcd} \xi_b v_c ((g_{dm} - \xi_d \xi_m) v^n \nabla_n \xi^m)$$
$$= -\varepsilon^{abcd} \xi_b v_c (v^n \nabla_n \xi_d).$$

The final equality follows from the fact that $\varepsilon^{abcd} \xi_b \xi_d = 0$ (since ε^{abcd} is anti-symmetric in the indices 'b' and 'd'). To proceed further, we use the following expression for $\nabla_n \xi_d$ that holds for any unit timelike field ξ^a proportional to a Killing field[11]:

$$\nabla_n \xi_d = \varepsilon_{ndpr} \xi^p \omega^r + \xi_n \xi^m \nabla_m \xi_d.$$

Direct substitution yields:

$$\begin{aligned}\Omega^a &= -\varepsilon^{abcd} \xi_b \nabla_c v^n (\varepsilon_{ndpr} \xi^p \omega^r + \xi_n \xi^m \nabla_m \xi_d) = -\varepsilon^{abcd} \xi_b \nabla_c v^n \varepsilon_{ndpr} \xi^p \omega^r \\ &= -\varepsilon^{dabc} \varepsilon_{dnpr} \xi_b \nabla_c v^n \xi^p \omega^r = 6 \delta^a_{[n} \delta^b_p \delta^c_{r]} \xi_b \nabla_c v^n \xi^p \omega^r \\ &= 6 \xi_b \nabla_c v^{[a} \xi^b \omega^{c]} = \xi_b \nabla_c (v^a \xi^b \omega^c \ldots - v^c \xi^b \omega^a + \ldots) \\ &= (\nabla_c \omega^c) v^a + \omega^a.\end{aligned}$$

(The second equality follows from the fact that $v^n \xi_n = 0$; the fourth from the fact that $\varepsilon^{dabc} \varepsilon_{dnpr} = -6 \delta^a_{[n} \delta^b_p \delta^c_{r]}$. For the latter, see (Wald 1984, 432).) ∎

We claimed above that relativity theory allows for the possibility that there be two point particles X and Y, a constant distance apart, such that Y is non-rotating relative to X, but X is rotating relative to Y with constant (non-zero) angular velocity. Our strategy for producing an example in section 3 is this. We exhibit a spacetime with a future-directed, timelike Killing field $\tau^a = \tau \xi^a$, and two integral curves of the field, γ_X and γ_Y, such that the following conditions hold.

(a) $\omega^a = 0$ on γ_X.

(b) $\omega^a \neq 0$ on γ_Y, but $(g^a_m - \xi^a \xi_m) \xi^n \nabla_n \omega^m = 0$ on γ_Y.

(c) Whether working from γ_Y to γ_X, or from γ_X to γ_Y, the associated future-directed null geodesic field λ^a that is Lie derived by τ^a (as in the construction above) is everywhere orthogonal to ω^a.

This will suffice. Consider the condition in (c). If the connecting null field λ^a is orthogonal to ω^a, then the direction field v^a induced by λ^a is also orthogonal to ω^a:

$$v^a \omega_a = \left(\frac{-\lambda^a + (\lambda^n \xi_n) \xi^a}{(\lambda^m \xi_m)} \right) \omega_a = \xi^a \omega_a = 0.$$

So, by the proposition, $\Omega^a_{(Y \, wrt \, X)} = \omega^a$ on γ_X, and $\Omega^a_{(X \, wrt \, Y)} = \omega^a$ on γ_Y. So by (a) and (b),

$$\Omega^a_{(Y \, wrt \, X)} = 0 \text{ on } \gamma_X, \text{ while}$$

$$\Omega^a_{(X \, wrt \, Y)} \neq 0 \text{ on } \gamma_Y, \text{ but } (g^a_m - \xi^a \xi_m) \xi^n \nabla_n \Omega^m_{(X \, wrt \, Y)} = 0 \text{ on } \gamma_Y,$$

as desired.

SECTION 3

The example we present in this section is a bit artificial. But it does have the virtue of simplicity. It will be relatively easy to identify the necessary elements of structure—the timelike Killing field τ^a, and the integral curves γ_X and γ_Y—and verify that they satisfy conditions (a)-(c). Given how very stringent the conditions are, it is of some interest, perhaps, to have any simple example at all.

In constructing the example, we start with Gödel spacetime (M, g_{ab}) in its entirety and then, at a certain point, shift attention to a restricted model of form $(O, g_{ab|O})$, where O is an open subset of M. The restricted spacetime is, in some respects, much better behaved than the original. In particular, it does not admit closed timelike curves. Indeed, it satisfies the stable causality condition. (But, unlike the original, of course, it is extendible.)

In what follows, we take Gödel spacetime[12] to be the pair (M, g_{ab}), where M is the manifold $\mathbf{R}^4$ and g_{ab} is characterized by the condition that given *any* point p in M, there is a global (adapted) cylindrical coordinate system t, r, φ, y on M such that $t(p) = r(p) = y(p) = 0$ and

$$g_{ab} = 4\mu^2[(dt)_a(dt)_b - (dr)_a(dr)_b - (dy)_a(dy)_b$$
$$+ (sh^4 r - sh^2 r)(d\varphi)_a(d\varphi)_b + \sqrt{2} sh^2 r((dt)_a(d\varphi)_b + (d\varphi)_a(dt)_b)].$$

(We use '$sh\ r$' and '$ch\ r$' to stand for '$\sinh r$' and '$\cosh r$'.) Here $-\infty < t < \infty$, $0 \le r < \infty$, $-\infty < y < \infty$, and $0 \le \varphi < \infty$ with $\varphi = 0$ identified with $\varphi = 2\pi$; μ is an arbitrary positive constant. (We will assume a point p has been chosen, once and for all, and work with the corresponding coordinate system.) The metric g_{ab} is a solution to Einstein's equation

$$R_{ab} - (1/2)g_{ab}R = 8\pi G T_{ab}$$

for a perfect fluid source

$$T_{ab} = \rho \eta_a \eta_b - p(g_{ab} - \eta_a \eta_b),$$

with four-velocity $\eta^a = (2\mu)^{-1}(\partial/\partial t)^a$, mass density $\rho = (16\pi G\mu^2)^{-1}$, and isotropic pressure $p = (16\pi G\mu^2)^{-1}$.

The field $(\partial/\partial t)^a$ is everywhere timelike, and defines a temporal orientation on (M, g_{ab}). The integral curves of $(\partial/\partial t)^a$ will be called "matter lines" (since the four-velocity η^a of the fluid source is everywhere proportional to $(\partial/\partial t)^a$).

In the appendix, we give an explicit expression for a volume element ε^{abcd} on (M, g_{ab}) in terms of coordinates t, r, φ, y. It defines an orientation on (M, g_{ab}).

In Gödel spacetime, $(\partial/\partial t)^a$, $(\partial/\partial y)^a$, and $(\partial/\partial \varphi)^a$ are all Killing fields and so, therefore, are all linear combinations of these fields. We will be interested, specifically, in the field

$$\tau^a = (\partial/\partial t)^a + \sqrt{2}(\partial/\partial\varphi)^a.$$

Since

$$\tau^a\tau_a = 4\mu^2[1 + 2(sh^4r - sh^2r) + 4sh^2r] = 4\mu^2[1 + 2(sh^2r)(ch^2r)],$$

it follows immediately that

(1) τ^a is everywhere timelike.

It is also clear that

(2) the coordinate functions r and y are constant on all integral curves of τ^a.[13]

If the constant value of r is 0, the integral curve is a matter line (since $(\partial/\partial\varphi)^a = 0$ where $r = 0$), characterized by its y value. We call it an "axis curve." If the constant value of r is strictly positive, we can picture it as a helix that wraps around an axis curve (the one with the same y value).[14]

If, as above, we express τ^a in the form $\tau^a = \tau\xi^a$, with $\xi^a\xi_a = 1$, the vorticity field associated with ξ^a comes out to be:

$$(3) \quad \omega^a = \frac{2\sqrt{2}(sh^2r)(ch^2r)}{(4\mu^2)[1 + 2(sh^2r)(ch^2r)]}(\partial/\partial y)^a$$

(The computation requires just a bit of work. We present it in the appendix.) It follows immediately that

(4) $\omega^a = 0 \Leftrightarrow r = 0$.[15]

It also follows that

(5) $(g^a{}_m - \xi^a\xi_m)\xi^n\nabla_n\omega^m = 0$ everywhere.

In fact, the stronger condition $\xi^n\nabla_n\omega^a = 0$ holds everywhere. This follows because $\xi^n\nabla_n r = 0$, by (2) above, and $(\partial/\partial y)^a$ is covariantly constant, i.e., $\nabla_n(\partial/\partial y)^a$.[16]

So

$$\xi^n\nabla_n\omega^a = \frac{2\sqrt{2}(sh^2r)(ch^2r)}{(4\mu^2)[1 + 2(sh^2r)(ch^2r)]}\xi^n\nabla_n(\partial/\partial y)^a = 0.$$

We are now well on our way. If we take γ_X to be any integral curve of τ^a with $r = 0$, and γ_Y to be any one with $r \neq 0$, conditions (a) and (b) listed at the end of section 2 will be automatically satisfied. So it only remains to consider condition (c).

To satisfy the orthogonality constraint in (c), we need to further restrict the choice of γ_X and γ_Y so that the y coordinate function has the same (constant) value on both curves. Let r_o be any positive real number and let y_o be any real whatsoever. Let γ_X be an integral curve of τ^a with constant values $r = 0, y = y_o$, and let γ_Y be one with constant values $r = r_o, y = y_o$. The following conditional claim about null geodesics follows easily.

(6) *If there exists a null geodesic that intersects both γ_Y and γ_X, and if λ^a is the tangent field to the curve, then $\lambda^a \omega_a = 0$ at all points on the curve.*
For assume there is such a curve σ with tangent field λ^a. Since λ^a is a geodesic field, we have $\lambda^n \nabla_n (\lambda^a \kappa_a) = 0$ for all Killing fields κ^a.[17] In particular, taking κ^a to be $(\partial/\partial y)^a$, $\lambda^a (\partial/\partial y)_a$ is constant on σ. But $\lambda^a (\partial/\partial y)_a = -4\mu^2 (\lambda^a \nabla_a y)$. So $\lambda^a \nabla_a y$ is constant on σ. If the constant value of this function were not 0, the value of the coordinate y would have to increase or decrease along σ — contradicting the fact that the initial and final points share the value y_0. So it must be the case that $\lambda^a (\partial/\partial y)_a$ is 0 at all points on σ. But, by (3), ω^a is everywhere proportional to $(\partial/\partial y)^a$. So $\lambda^a \omega_a = 0$ at all points on σ.

Now it only remains for us to consider the existence and uniqueness of null geodesics running between γ_Y and γ_X. But here, for the first time, things get sticky. We want to be able to assert that an observer on one of the particles will see the other at all times, but not see it in more than one position on the celestial sphere. It is a curious fact about null geodesics in Gödel spacetime that this will simply not be the case, in general. It turns out that if $sh\, r_o > 1$ (i.e., if $r_o > \ln(1 + \sqrt{2})$), the observer will not see the other particle at all. And if $sh\, r_o \leq 1$, he will, in general, see multiple images of the other. Roughly speaking, this results from the fact that photons act like boomerangs in Gödel spacetime. Any future or past directed null geodesic that starts at a point on γ_X moves outward (with monotonically increasing r value) until it reaches the critical radius $r_c = \ln(1 + \sqrt{2})$, and then moves inward (with monotonically decreasing r value) until it hits γ_X again; and then the process starts all over.[18] So, it can happen, for example, that two past-directed null geodesics start out in different directions from a point on γ_X, and *both* intersect γ_Y, though at different points. One hits γ_Y on the way out. The other hits it on the return trip in.

To avoid this complication, we now impose the requirement that $r_o < r_c$, restrict attention to the open subset

$$O = \{q \in M: r(q) < r_c\},$$

and consider $(O, g_{ab|O})$ as a spacetime model in its own right (with the temporal orientation and orientation inherited from the original).[19] Then we can make the desired existence and uniqueness claim concerning null geodesics.

(7) *Given any point q_X on γ_X, there is a unique point q_Y on γ_Y such that there exists a future directed null geodesic running from q_Y to q_X; and symmetrically, with the roles of X and Y reversed.*
That this is true follows alone from the qualitative description of past and future directed null geodesics just given (the boomerang effect). We sketch the proof in the appendix.

This puts all the needed pieces of the example together. We now revert to the discussion at the end of section 2.

APPENDIX: NEEDED FACTS ABOUT GÖDEL SPACETIME

(A) Derivation of formula (3) in section 3

Let k be any real number and let τ^a be the Killing field $(\partial/\partial t)^a + k(\partial/\partial\varphi)^a$. If we restrict attention to the (open) region where it is timelike, we can express τ^a in the form $\tau^a = \tau\xi^a$, with $\xi^a\xi_a = 1$. We claim that the vorticity associated with ξ^a (in this region) is given by

$$\omega^a = \frac{\sqrt{2} + k(2sh^2 r - 1) + k^2\sqrt{2}sh^4 r}{(4\mu^2)[1 + k2\sqrt{2}sh^2 r + k^2(sh^4 r - sh^2 r)]}(\partial/\partial y)^a.$$

If $k = \sqrt{2}$, this reduces to (3).

In the derivation, we use the following basic relations:

$$(\partial/\partial t)_a = 4\mu^2((dt)_a + \sqrt{2}sh^2 r(d\varphi)_a)$$

$$(\partial/\partial\varphi)_a = 4\mu^2(\sqrt{2}sh^2 r(dt)_a + (sh^4 r - sh^2 r)(d\varphi)_a)$$

$$\tau_a = 4\mu^2((1 + k\sqrt{2}sh^2 r)(dt)_a + (\sqrt{2}sh^2 r + k(sh^4 r - sh^2 r))(d\varphi)_a)$$

$$\tau^2 = \tau^a\tau_a = 4\mu^2(1 + k2\sqrt{2}sh^2 r + k^2(sh^4 r - sh^2 r)),$$

and we work with the volume element defined[20] by

$$\varepsilon^{abcd} = f(\partial/\partial t)^{[a}(\partial/\partial r)^b(\partial/\partial\varphi)^c(\partial/\partial y)^{d]} \text{ where } f = \frac{-4!}{(16\mu^4)(sh\ r)(ch\ r)}.$$

Since $\varepsilon^{abcd}\tau_b\tau_d = 0$, we have

$$\omega^a = \frac{1}{2}\varepsilon^{abcd}\xi_b\nabla_c\xi_d = \frac{1}{2}\varepsilon^{abcd}\frac{\tau_b}{\tau}\nabla_c\frac{\tau_d}{\tau} = \frac{1}{2\tau^2}\varepsilon^{abcd}\tau_{[b}\nabla_c\tau_{d]}.$$

So we start by deriving an expression for $\tau_{[b}\nabla_c\tau_{d]}$. First, note that since $\nabla_{[c}(dt)_{d]} = 0 = \nabla_{[c}(d\varphi)_{d]}$,

$$\nabla_{[c}\tau_{d]} = 4\mu^2((dt)_{[d}\nabla_{c]}(1 + k\sqrt{2}sh^2 r) + (d\varphi)_{[d}\nabla_{c]}(\sqrt{2}sh^2 r + k(sh^4 r - sh^2 r)))$$

$$= (4\mu^2(k2\sqrt{2}(sh\ r)(ch\ r)(dt)_{[d}(dr)_{c]}) +$$

$$2(sh\ r)(ch\ r)(\sqrt{2} + k(2sh^2 r - 1))d\varphi_{[d}(dr)_{c]}).$$

Hence,

$$\tau_{[b}\nabla_c\tau_{d]} = \tau_{[b}\nabla_{[c}\tau_{d]]} = (16\mu^4)K(dt)_{[b}(dr)_c(d\varphi)_{d]}$$

where

$$K = -k2\sqrt{2}(sh\ r)(ch\ r)(\sqrt{2}sh^2 r + k(sh^4 r - sh^2 r))$$
$$+ (1 + k\sqrt{2}sh^2 r)2(sh\ r)(ch\ r)(\sqrt{2} + k(2sh^2 r - 1))$$
$$= 2(sh\ r)(ch\ r)[\sqrt{2} + k(2sh^2 r - 1) + k^2\sqrt{2}sh^4 r].$$

So,

$$\omega^a = \frac{1}{2\tau^2}\varepsilon^{abcd}\tau_{[b}\nabla_c\tau_{d]} = \frac{1}{2\tau^2}(16\mu^4)K\varepsilon^{abcd}(dt)_{[b}(dr)_c(d\varphi)_{d]}$$

$$= \frac{1}{2\tau^2}(16\mu^4)K\ f(4!)^{-1}(-(\partial/\partial y)^a)$$

$$= \frac{\sqrt{2} + k(2sh^2 r - 1) + k^2\sqrt{2}sh^4 r}{(4\mu^2)[1 + k\ 2\sqrt{2}sh^2 r + k^2(sh^4 r - sh^2 r)]}(\partial/\partial y)^a.$$

(B) Proof sketch of claim (7) in section 3

Let q_X be any point on γ_X and let $\tilde{\lambda}^a$ be any past-directed (non-zero) null vector at q_X such that $\tilde{\lambda}^n \nabla_n y = 0$, Let σ be the (unique) inextendible, past-directed null geodesic starting at q_X whose tangent at that point is $\tilde{\lambda}^a$. Let its tangent field be λ^a. The r coordinate on σ starts at 0 and increases (monotonically) through all values less than r_c. So there is exactly one point q on σ whose r value is r_o.[21] Let the coordinates of q be (t, r_o, φ, y_o). (We know, from the discussion after (5), that $\lambda^n \nabla_n y$ is constant on σ. Since it is 0 at q_X, it must be 0 at all points. So the value of the y coordinate must be y_o at all points on σ.) The point q need not fall on γ_Y.

We have so far considered just one inextendible, past-directed null geodesic starting at q_X along which y has the constant value y_o. But the entire class of these is generated by taking the image of σ under "rotations" of form

$$(t, r, \varphi, y) \rightarrow (t, r, \varphi + \varphi_o, y),$$

i.e., under isometries generated by the Killing field $(\partial/\partial \varphi)^a$. One of these isometric images of σ *does* intersect γ_Y (since there is *some* point q_Y on γ_Y and *some* φ_o such that q_Y has coordinates $(t, r_o, \varphi + \varphi_o, y_o)$). The time reversed, i.e., future-directed, version of this curve qualifies as a null geodesic running from a point q_Y on γ_Y to q_X. So we have established the existence claim in (7). And uniqueness follows easily as well. Suppose σ_1 and σ_2 are both past-directed null geodesics starting at q_X that intersect γ_Y. Then since both arise as images of σ under rotations of the sort just described, and since these maps preserve the value of the coordinate t, the intersection points share a common value of t. But there can *be* only one point on γ_Y having any particular value of t. (*This* follows because γ_Y is a future directed timelike curve, and (see note 19) the coordinate function t is strictly increasing on all such curves.)

The argument for the symmetric claim (with the roles of X and Y interchanged) is very much the same. But now, in addition to considering "rotations" (as above), we also consider "timelike translations" of form

$$(t, r, \varphi, y) \to (t + t_o, r, \varphi, y),$$

i.e., isometries generated by the Killing field $(\partial/\partial t)^a$. Let q_Y be any point on γ_Y. Essentially the same argument as we have just considered shows that given any point on γ_X, there is a unique point on γ_Y such that there exists a *future*-directed null geodesic running from the first point to the second. By moving to the image of this curve under, first, a timelike translation and, then, a rotation, we arrive at a future-directed null geodesic σ that starts at a point q_X on γ_X and ends at q_Y. This gives us existence. For uniqueness, suppose there were a second point q'_X on γ_X and a null geodesic σ' running from q'_X to q_Y. By first sliding σ' up or down so that q'_X is mapped to q_X, and then rotating it, we could generate a future-directed null geodesic that starts at q_X, but ends at a point on γ_Y distinct from q_Y—and this we know is impossible.

University of California

NOTES

[*] It is a pleasure to dedicate this paper to John Stachel, and thank him here for the encouragement and support he has given me over the years. (It is also a pleasure to thank Robert Geroch, Howard Stein, and Robert Wald for helpful comments on an earlier draft.)

1. It should be emphasized that this does not imply that *all* claims about "orbital rotation" are symmetric within the framework of Newtonian physics. For example, let X be a particle sitting at the center of mass of the solar system. The earth and the sun both rotate relative to X (and relative to each other) in our sense; and X rotates relative to both the earth and the sun in that sense. But there is this asymmetry between the motion of X on the one hand, and that of the earth and the sun on the other: X is non-accelerating, while both the earth and the sun have non-zero acceleration vectors that point toward X. This captures *one* sense in which one might say that the earth and sun are rotating around X, but not conversely.

2. The discussion to this point has been cast in terms of textbook Newtonian particle mechanics. It might be asked what, if anything, changes when one passes to the Cartan formulation of Newtonian theory in which gravity is treated as a manifestation of spacetime curvature. (Rather than thinking of point particles as being deflected from their natural straight trajectories by the presence of a gravitational potential, one thinks of them as traversing the geodesics of a non-flat affine connection.) The short answer is that our notion of relative angular velocity carries over in a natural way, and conditions (i) and (ii) remain symmetric. (More problematic is the notion of orbital rotation considered in the preceding note

since it makes reference to the "acceleration" of particles in a gravitational field. But it can be reformulated (in terms of the presence of background spacetime symmetries) and remains asymmetric).)
It would take us too far afield to sort this all out here.
3. Definitions of the technical terms used here and in what follows can be found, for example, in (Wald 1984). (Strictly speaking, a few minor transpositions will be necessary since Wald works with the signature $(-1, +1, +1, +1)$ rather than ours.)
4. We might have written '$\Omega_{XY}{}^a$', but that notation is potentially misleading. The identification indices 'XY' should not be confused with tensor or spinor indices. In what follows, it will usually be clear from context whether we are talking about the angular velocity of Y relative to X, or of X relative to Y. But when there is danger of confusion, we will write '$\Omega^a_{(Y \text{ wrt } X)}$' or '$\Omega^a_{(X \text{ wrt } Y)}$'.
5. In what follows, we assume that (M, g_{ab}) is temporally orientable and a particular temporal orientation has been selected. We also assume that it is orientable and a volume element ε_{abcd} has been selected. (A smooth field ε_{abcd} on M qualifies as a *volume element* if it is completely anti-symmetric ($\varepsilon_{abcd} = \varepsilon_{[abcd]}$) and normalized so that $\varepsilon^{abcd}\varepsilon_{abcd} = -4$!) Neither the assumption of temporal orientability nor orientability is really necessary. We can, alternatively, restrict attention to appropriate local neighborhoods of M. But the assumptions are convenient and, in fact, the spacetime we will use for our example in section 3 (Gödel spacetime) is temporally orientable and orientable.
6. This is a substantive assumption, and will play a role in the presentation of our example in section 3.
7. The equivalence here corresponds perfectly to that in (i) in section 1, and the proof is essentially the same. $\Omega^a = 0$ iff the three vectors ξ^a, v^a, and $((g^a{}_m - \xi^a\xi_m)\xi^n\nabla_n v^m$ are linearly dependent. But since ξ^a and v^a are non-zero, and both v^a and $((g^a{}_m - \xi^a\xi_m)\xi^n\nabla_n v^m$ are orthogonal to ξ^a, this condition holds iff $(g^a{}_m - \xi^a\xi_m)\xi^n\nabla_n v^m$ is proportional to v^a. But $(g^a{}_m - \xi^a\xi_m)\xi^n\nabla_n v^m$ is orthogonal to v^a (since $v_a\xi^a = 0$ and $v_m\xi^n\nabla_n v^m = (1/2)\xi^n\nabla_n(v^m v_m) = (1/2)\xi^n\nabla_n(-1) = 0$). So $\Omega^a = 0$ iff $(g^a{}_m - \xi^a\xi_m)\xi^n\nabla_n v^m = 0$, as claimed.
8. If they are positioned so as to be orthogonal at some initial moment, they will remain so.
9. For example, the distance between them is constant as determined by the time it takes a light signal to complete a round trip passage from one particle to the other and back—as measured by clocks sitting on the respective particles. Indeed, the distance between them is constant according to *any* notion of distance that can be formulated in terms of the spacetime metric g_{ab} and the curves γ_X and γ_Y, since they are all preserved under the flow maps associated with τ^a.
10. The proposition is slightly more general than the one proved in (Rosquist 1980). He worked with a unit timelike vector field ξ^a that is Born rigid (i.e., has vanishing scalar expansion and shear) and geodesic. These two conditions imply that ξ^a is proportional to a Killing field, but not conversely. Rosquist also limited attention to the case where, in our notation, $v^n\omega_n = 0$.
11. Every unit timelike field ξ^a whatsoever satisfies

$$\nabla_n \xi_d = \theta_{nd} + \omega_{nd} + \xi_n \xi^m \nabla_m \xi_d,$$

where

$$\theta_{nd} = h_{(n}{}^r h_{d)}{}^s \nabla_r \xi_s \qquad \omega_{nd} = h_{[n}{}^r h_{d]}{}^s \nabla_r \xi_s \qquad h_{nr} = g_{nr} - \xi_n \xi_r.$$

And every such field satisfies, $\varepsilon_{ndpr}\xi^p\omega^r = \omega_{nd}$ (as one can verify by direct substitution for ω^r in the left side expression). But if $\tau\xi^a$ is a Killing field for some scalar field τ, $\theta_{nd} = 0$. (This follows from Killing's equation, $\nabla_r(\tau\xi_s) + \nabla_s(\tau\xi_r) = 0$.)

12. For an indication of what Gödel spacetime "looks like," see the diagrams in (Hawking and Ellis 1973; Malament 1984).

13. This is equivalent to the claim that $\tau^n\nabla_n r = 0$ and $\tau^n\nabla_n y = 0$. The first equation holds since

$$(\partial/\partial t)^n(\partial/\partial r)_n = 0 = (\partial/\partial\varphi)^n(\partial/\partial r)_n \text{ and, hence,}$$

$$0 = \tau^n(\partial/\partial r)_n = \tau^n(-4\mu^2(dr)_n) = -4\mu^2\tau^n\nabla_n r.$$

The argument for the second equation is similar.

14. This picture, while helpful, is potentially misleading in one respect. As we shall see, a particle whose worldline is one of these helices can qualify as *non-rotating* relative to a particle whose worldline is an axis curve.

15. This fact explains the choice of the coefficient $\sqrt{2}$ in our expression for τ^a. We *want* ω^a to be 0 at points where $r = 0$. As we show in the appendix, the vorticity associated with the general field

$(\partial/\partial t)^a + k(\partial/\partial\varphi)^a$ (where it is timelike) comes out to be

$$\frac{\sqrt{2} + k(2sh^2 r - 1) + k^2\sqrt{2}sh^4 r}{(4\mu^2)[1 + k2\sqrt{2}sh^2 r + k^2(sh^4 r - sh^2 r)]}(\partial/\partial y)^a.$$

This reduces to

$$(4\mu^2)^{-1}(\sqrt{2} - k)(\partial/\partial y)^a$$

at $r = 0$.

16. $\nabla_{(a}(\partial/\partial y)_{b)} = 0$, since $(\partial/\partial y)^a$ is a Killing field; and $\nabla_{[a}(\partial/\partial y)_{b]} = 0$, since

$$\nabla_{[a}(\partial/\partial y)_{b]} = -4\mu^2\nabla_{[a}\nabla_b y = 0.$$

17. We have

$$\lambda^n\nabla_n(\lambda^a\kappa_a) = \lambda^a\lambda^n\nabla_n\kappa_a + \kappa_a\lambda^n\nabla_n\lambda^a = 0.$$

The first term in the sum vanishes because κ_a is a Killing field (and, so, $\nabla_{(n}\kappa_{a)} = 0$). The second does so because λ^a is a geodesic field.

18. See (Lathrop and Teglas 1978) for an analytic characterization of geodesics passing through points where $r = 0$, and see (Hawking and Ellis 1973) for a picture.

19. The coordinate map $t: O \to \mathbf{R}$ qualifies as a global time function on $(O, g_{ab|O})$, i.e., it increases along all future-directed timelike curves. (Hence there cannot be any closed timelike curves in $(O, g_{ab|O})$.) The assertion is equivalent to the claim that the vector field $(\nabla^a t)$ is timelike and future-directed (i.e., $(\nabla^a t)(\nabla_a t) > 0$ and $(\partial/\partial t)^a(\nabla_a t) > 0$) on O. But this is clear since

$$(\nabla^a t)(\nabla_a t) = \frac{1 - sh^2 r}{4\mu^2(1 + sh^2 r)} \text{ on } O,$$

and $(\partial/\partial t)^a(\nabla_a t) = 1$ everywhere. (The expression for $(\nabla^a t)(\nabla_a t)$ follows from the fact that the inverse metric is:

$$g^{ab} = \frac{1}{4\mu^2(sh^2 r + sh^4 r)}[(sh^2 r - sh^4 r)(\partial/\partial t)^a(\partial/\partial t)^b - (sh^2 r + sh^4 r)(\partial/\partial r)^a(\partial/\partial r)^b$$
$$- (sh^2 r + sh^4 r)(\partial/\partial y)^a(\partial/\partial y)^b - (\partial/\partial\varphi)^a(\partial/\partial\varphi)^b + 2\sqrt{2}sh^2 r(\partial/\partial t)^{(a}(\partial/\partial\varphi)^{b)}].)$$

20. Let R be the closed set of points where $r = 0$. Since the fields $(\partial/\partial t)^a$, $(\partial/\partial r)^a$, $(\partial/\partial\varphi)^a$, and $(\partial/\partial y)^a$ are linearly independent on $M - R$, there must exist *some* function f defined on $M - R$ for

which the equation holds. We can determine f, up to sign, with the following calculation:

$$-(4!) = \varepsilon^{abcd}\varepsilon_{abcd} = f^2(\partial/\partial t)^{[a}(\partial/\partial r)^{b}(\partial/\partial\varphi)^{c}(\partial/\partial y)^{d]}$$
$$\bullet\ (\partial/\partial t)_{[a}(\partial/\partial r)_{b}(\partial/\partial\varphi)_{c}(\partial/\partial y)_{d]}$$
$$= f^2(\partial/\partial t)^{[a}(\partial/\partial r)^{b}(\partial/\partial\varphi)^{c}(\partial/\partial y)^{d]}$$
$$\bullet\ (4\mu^2)^4((sh^4r - sh^2r) - 2sh^4r)(dt)_{[a}(dr)_{b}(d\varphi)_{c}(dy)_{d]}$$
$$= -f^2(4\mu^2)^4(sh^4r + sh^2r)\frac{4!}{(4!)^2}.$$

The volume elements on $M - R$ defined by the two solutions for f have well defined limits at points in R. Once those limit values are included, we have a (smooth) volume element on all of M.

21. It is precisely here that the present argument would break down if we had not restricted attention to O.

REFERENCES

Hawking, S. W. and G. F. R. Ellis. 1973. *The Large Scale Structure of Space–Time.* Cambridge: Cambridge University Press.

Lathrop, J. and R. Teglas. 1978. "Dynamics in the Gödel Universe." *Il Nuovo Cimento* 43 B: 162–171.

Malament, D. 1986. "Time Travel in the Gödel Universe." *PSA 1984,* Vol. 2 (Proceedings of the Philosophy of Science Association Meetings, 1984): 91–100.

Rosquist, K. 1980. "Global Rotation." *General Relativity and Gravitation* 12: 649–664.

Wald, R. 1984. *General Relativity.* Chicago: University of Chicago Press.

II

Foundational Issues in Relativity and their Advancement

GEORGE F. R. ELLIS

THE UNIQUE NATURE OF COSMOLOGY

INTRODUCTION

Cosmology has progressed in the last 35 years from a mainly mathematical and philosophical exercise to an important branch of both astronomy and physics, and is now part of mainstream science, with a well—established standard model confirmed by various strands of evidence (Weinberg 1972; Peebles et al. 1991; Coles and Ellis 1997). Nevertheless because of its nature, it is different from any other branch of the natural sciences. The major issue causing the differences is the uniqueness of its object of study—the Universe as a whole—together with its role as the background for all the rest of physics and science, the resulting problems being accentuated by its vast scale and the extreme energies occurring in the early universe. We are unable to manipulate in any way its originating conditions, and there are limitations on our ability both to observe to very distant regions and very early times and also to test the physics relevant at the earliest times. Consequently it is inevitable that (as is the case for the other historical sciences) specific philosophical choices will to some degree shape the nature of cosmological theory, particularly when it moves beyond the purely descriptive to an explanatory role—which move is central to its impressive progress. These philosophical choices will dominate the resulting understanding to the extent that we pursue a theory with more ambitious explanatory aims.

Cosmology is the study of the large-scale structure of the Universe, where 'the Universe' means all that exists in a physical sense (to be distinguished from the *observable universe*, namely that part of the universe containing matter accessible to our astronomical observations). Thus it considers the vast domain of galaxies, clusters of galaxies, quasi-stellar objects, etc., and the nature of their origins. Observational cosmology aims to determine the large-scale geometry of the observable universe and the distribution of matter in it (Hoyle 1960; Kristian and Sachs 1966; Gunn et al. 1978; Sandage et al. 1993), while physical cosmology is the study of interactions during the expansion of the universe from an early hot big bang phase (Peebles 1971; Sciama 1971; Weinberg 1972) and astrophysical cosmology studies the resulting later development of large-scale structures such as galaxies and clusters of galaxies (Peebles 1993; Padmanbhan 1993; Rees 1995). These studies function in a mainly symbiotic way, each informing and supplementing the other to create an overall cosmological theory of the origin and evolution of the physical universe (Bondi 1960; Harrison 1981; Peacock 1999).[1] A unique role of the universe is in creating the environment in which galaxies, stars, and planets develop, thus providing a setting in which local physics and chemistry can function in a way that enables the evolution of life. If this cosmological environment were substantially different, we would not be here—indeed no biological evolution at all would have taken place (Carr and Rees 1979; Davies 1982; Barrow and Tipler 1984; Tegmark 1993). Thus cosmology is of substantial interest to the whole of the scientific endeavor, for it sets the framework for the rest of science.

Cosmology has been transformed in the past decades into a mainstream branch of physics[2] by the linking of nuclear and particle physics theories to observable features of the cosmos (Weinberg 1972; Kolb and Turner 1990; Peacock 1999), and into an important part of astronomy because of the massive flow of new astronomical data becoming available (Harwit 1984; Bothun 1998), particularly at the present time through new ground-based telescopes such as Keck and through satellite observatories such as the Hubble Space telescope (optical and ultra-violet), IRAS (infra-red), ROSAT (x-ray), and COBE (microwave). A series of basic themes of present day cosmology are now well established. Observational support for the idea of expansion from a hot big bang epoch is very strong (Weinberg 1972; Peebles et al.; Coles and Ellis 1997), the linear magnitude-redshift relation for galaxies demonstrating the expansion—the alternative interpretation as gravitational redshifts does not work out because of the linearity of the redshift-distance relation (Ellis et al. 1978), with source number counts and the existence of the black-body Cosmic Background Radiation (CBR) being strong evidence that there was indeed evolution from a hot early stage—particularly important are measurements of the CBR temperature at high redshift, confirming the standard interpretation of this radiation (Meyer 1994). Agreement between measured light element abundances and the theory of nucleosynthesis in the early universe confirms this interpretation (Schramm and Turner 1998), as do detailed observations of phenomena such as gravitational lensing and extragalactic supernovae. This basic theory is robust to critical probing (Borner and Gottlober 1997; Turok 1997).

Much present activity links particle physics interactions during very early stages of the expansion of the universe to the creation of structures by gravitational instability much later (Peacock 1999; Gibbons et al. 1983;[3] Linde 1990; Steinhardt 1995), traces of the early seed fluctuations being accessible to us through present day CBR anisotropy patterns. Following the universe back in time, it is well-established that ordinary physics cannot be extended to arbitrary early times—general relativity theory predicts a singularity, an origin to space and time at the beginning of the universe (Hawking 1973; Tipler et al. 1980), where classical physics breaks down. This can be interpreted as a statement that conditions become so extreme that some (as yet unknown) consistent theory of quantum gravity must be invoked, possibly thereby avoiding an initial physical singularity in one way or another—but still singular from a classical viewpoint. Various ambitious studies aim to describe the origin of the universe itself in scientific terms, and hence to give an explanation for the nature of the initial conditions of the universe, or else to characterize its existence in some earlier exotic state which eventually provides the start of the hot big bang expansion phase (Hawking 1993; Gott 1997). Clearly these extensions are more speculative than the basic theory applicable at lower energies and later times in the universe's evolution.

The study of cosmology is shaped by its unique nature. This uniqueness results from four specific features, namely the uniqueness of the Universe; its role as a background for all the rest of science; its large scale in both space and time; and the extreme high energies occurring in the very early Universe. These features play themselves out in the ongoing interaction between theory and observation in cosmology.

1. THE UNIQUENESS OF THE UNIVERSE

The first and most fundamental issue is that there is only one Universe (Munitz 1962, 1986; McCrea 1960, 1970; Ellis 1991). This essential uniqueness of its object of study sets the cosmology apart from all other sciences. In particular, the unique initial conditions that lead to the particular state of the universe we see were somehow set by the time that physical laws as we know them started governing the evolution of both the universe and its contents, whenever that time may be. Prior to that time physics as we know it is not applicable and our ordinary language fails us because time did not exist, so our natural tendency to contemplate what existed or happened 'before the beginning' (Ellis 1993; Rees 1998) is highly misleading—there was no 'before' then, indeed there was no 'then' then! Talking as if there was is commonplace but quite misleading in trying to understand a scientific concept of 'creation' (Grunbaum 1989).

We cannot alter these unique initial conditions in any way—they are given to us as absolute and unchangeable, even though they are understood as *contingent* rather than necessary (that is, they could have been different while still being consistent with all known physical laws). The implications are that:

1A: *We cannot re-run the universe with the same or altered conditions* to see what would happen if they were different, so we cannot carry out scientific experiments on the universe itself. Furthermore,

1B: *We cannot compare the universe with any similar object,* for none exists, nor can we test our hypotheses about it by observations determining statistical properties of a known class of physically existing universes. Thus consequent on 1A and 1B,

1C: *We cannot scientifically establish 'laws of the universe' that might apply to the class of all such objects,* for we cannot test any such proposed law except in terms of being consistent with one object (the observed universe). This is insufficient: one observational point cannot establish the nature of a causal relation. Indeed the concept of a 'law' becomes doubtful when there is only one given object to which it applies (Munitz 1962, 1986). The basic idea of a physical law is that it applies to a set of objects all of which have the same invariant underlying behavior (as defined by that law), despite the apparent variation in properties in specific instances, this variation resulting from varying initial conditions for the systems on which the law acts. This understanding is tested by physical experiments in which initial conditions for evolution of a set of similar systems are varied, and observations by which the statistical nature of a set of objects of the same broad kind is investigated. Neither is possible in the case of cosmology. All that we can do is observe and analyze the one unique object that exists. Finally,

1D: *Problems arise in applying the idea of probability to cosmology as a whole* — it is not clear that this makes much sense in this context of existence of a single object which cannot be compared with any other existing object. But a concept of probability underlies much of modern argumentation in cosmology. Talk of 'fine tuning' for example is based on use of probability (it is a way of saying something is improbable). This assumes both that things could have been different, and that we can assign probabilities to the set of unrealized possibilities in an invariant way. The issue here is to explain in what sense they could have been different with well-defined probabilities assigned to the different theoretical possibilities, if there is indeed only one universe with one set of initial conditions fixed somehow before physics came into being, or more accurately as physics came into being. As follows from 1C, we cannot scientifically establish laws of creation of the universe that might determine such initial conditions or resulting probabilities. If we use a Bayesian interpretation (Garrett and Coles 1993)[4] the results depend on our prior knowledge which can be varied by changing our initial pre-physics assumptions. There is no reason to believe the usual maximum entropy estimates will mean anything under these circumstances; indeed even the counting of states is not possible in the pre-physics era when states do not exist! Related issues arise concerning the meaning of 'the wave function of the universe,' at the heart of quantum cosmology.

Two comments on the above. First, it is useful to distinguish (Ellis 1993) between the *experimental sciences*—physics, chemistry, microbiology for example—on the one hand, and the *historical and geographical sciences*—astronomy, geology, evolutionary theory for example, on the other. It is the former that are usually in mind in discussions of the scientific method. The understanding in these cases is that we observe and experiment on a class of identical or almost identical objects and establish their common behavior. The problem then resides in just how identical those objects are. Quarks, protons, electrons, and water molecules are presumably indeed all identical to each other, and so have exactly the same behavior. All DNA molecules, frogs, human beings, and ecosystems are somewhat different from each other, but are similar enough nevertheless that the same broad descriptions and laws apply to them; if this were not so, then we would be wrong in claiming they belonged to the same class of objects in the first place.

As regards the geographical and historical sciences, here one explicitly studies objects that are unique (the Rio Grande, the continent of Antarctica, the Solar System, the Andromeda galaxy, etc.) or events that have occurred only once (the origin of the Solar System, the evolution of life on Earth, the explosion of SN1987a, etc.). Because of this uniqueness, comment 1A above applies in these cases also: we can only observe rather than experiment; the initial conditions that led to these unique objects or events cannot be altered or experimented with. However comment 1B does not apply: at least in principle, there is a class of similar objects out there (other rivers, continents, planetary systems, galaxies, etc.) or similar events (the origin of other galaxies, the evolution of other planetary systems, the explosion of other supernovae, etc.) which we can observe and compare with our specific exemplar, also carrying out statistical analyses on many such cases to determine underlying patterns of regularity; and in this respect these topics differ from cosmology. If we truly cannot carry out such analyses—that is, if 1B applies as well in some particular case—then that subject partakes in this respect of the nature of cosmology.

One may claim that the dividing line here is that if we convince ourselves that some large-scale physical phenomenon essentially occurs only once in the entire universe, then it should be regarded as part of cosmology proper; whereas if we are convinced it occurs in many places or times, even if we cannot observationally access them (e.g. we believe that planets evolved around many stars in other galaxies) then study of that class of objects or events can be distinguished from cosmology proper precisely because there is a class of them to study. However cosmology will uniquely set the context within which they occur; astrophysical cosmology is the linking subject studying how these astrophysical processes are related to the uniquely given cosmological setting. Note that the burden of definition shifts to the meaning of the word 'essentially' above, and here is where the importance of classification of more-or-less similar objects and events comes in. I will not pursue this further here, but these series of issues certainly deserve further study; the philosophy of science has not yet tackled them effectively. In any case the upshot is that the subject matter of cosmology is uniquely unique!

The second comment is that some workers have tried to get around this set of problems by essentially denying the uniqueness of the universe. This is done by proposing the physical existence of 'many universes' (Leslie 1989, 1998) to which concepts of probability can be properly applied, envisaged either as widely separated regions of a larger universe with very different properties in each region (Ellis 1979) —as in chaotic inflation for example, (Linde 1990)—or as an ensemble of completely disconnected universes—there is no physical connection whatever between them (Sciama 1993; Tegmark 1993)—in which all possibilities are realized. In my view neither move solves the fundamental issue to be faced. Considering situations that can be described by classical physics, if such 'universes' are directly or indirectly physically connected to us, they are part of our one universe, and the terminology is seriously misleading; the larger whole they all comprise together is indeed unique, and that is the proper subject of cosmology. If it has many varied locations with differing properties, that may indeed help us understand the problems such as the Anthropic issue—some regions will allow life to exist and others will not (Barrow and Tipler 1984; Rees 1998; Leslie 1998)—but the question then is why this unique larger whole has the properties it does. The essential issues remain. Introducing the Many Worlds interpretation of quantum mechanics—in essence: all possible worlds exist as a branching 'multi-verse' (Barrett 1998)—is in principle a way out, but is far from being a widely accepted view and has certainly not been shown correct by experiment, indeed this is probably not possible. Using this concept to explain the properties of the single observed classical universe is certainly problematic until the essential problems at the foundation of quantum mechanics are solved (Isham 1997).

If an ensemble exists with members not connected in any physical way to the observable universe, then we cannot interact with them in any way nor observe them, so we can say anything we like about them without fear of disproof. Thus any statements we make about them can have no solid scientific or explanatory status; they are totally vulnerable to anyone else who claims an ensemble with different properties (for example claiming different kinds of underlying logics are possible in their ensemble, or making claims such as existence of multiple physically effective gods and devils in many universes in their ensemble). The issue of what is to be regarded as an ensemble of 'all possible' universes is unclear; it can be manipulated to produce any result you want by redefining what is meant by this phrase—standard physics and logic have no necessary sway over them; what I envisage as 'possible' in such an ensemble may be denied by you.

Considering the properties of a well-defined hypothetical ensemble of universes is an interesting and indeed valuable exercise (see item 2B below), but we need to treat it as such—namely 'hypothetical.' Claiming they 'exist' in some totally disconnected fashion places an extreme strain on the word 'exist.' Without some ability to check this claim in some way, the use of this word seems vacuous, and certainly lacking in explanatory solidity. The argument that this infinite ensemble actually exists can be claimed to have a certain explanatory economy (Tegmark 1993), although others would claim that Occam's razor has been completely abandoned in favour of a profligate excess of existential multiplicity, extravagantly hypothesized in order to explain the one universe that we do know exists (Davies 1992, 190, 219). Certainly the price is a total lack of testability through either observations or experiment—which is usually taken to be an essential element of any serious scientific theory.

In any case if we take this standpoint, then I suggest the correct move would be to say that the proper subject matter of cosmology is this 'multiverse'—the entire ensemble of universes (precisely because that is then 'all that exists'), which presumably is unique, and we end up where we were before, facing the problems of essential uniqueness. If it is not unique, then there is no explanatory gain; in either case, there is a complete loss of verifiability. For these reasons the move to claim such an ensemble actually exists in a physical sense should be strongly queried; this is very problematic as a proposal for scientific explanation. However consideration of an explicitly hypothetical such ensemble can indeed be useful, as discussed below.

2. THE UNIVERSE AS THE BACKGROUND FOR PHYSICS AND SCIENCE

The underlying program of the standard approach is to use only known local physics, pushed as far as far as possible, to explain the structure of the Universe, giving a solely physical explanation of what we see. The universe provides the environment for all of science, by determining the initial conditions within which all physical laws are constrained to operate. It creates the environment necessary for the evolution of life—heavy elements at the micro-level and habitable planets at the macro level; thus it is the background which enables organic chemistry and biology (and hence sociology, psychology, and postmodernism!) to exist and function. If it were substantially different, for example if it had too short a lifetime or if the background temperature never dropped below 3000K, then life would not be possible (Carr and Rees 1979; Davies 1982; Barrow and Tipler 1984; Tegmark 1993). Thus different initial conditions would lead to a barren universe. To some extent this chain of argumentation can be run the other way, for example perhaps deducing the expansion of the universe from the fact that the night sky is dark—this is 'Olber's paradox' (Bondi 1960; Harrison 1981)—which is in fact a necessary condition for the existence of life as we know it—the biosphere on Earth functions by disposing of waste energy to the heat sink of the dark night sky (Penrose 1989b).

However because of the uniqueness of the universe discussed in the previous section, unlike the rest of physics where the distinction is clear and fundamental,

2A: *We have an essential difficulty in distinguishing between laws of physics and boundary conditions* in the cosmological context of the origin of the universe. Because we cannot vary the initial conditions in any way, as far as we are concerned they are necessary rather than contingent—so the essential distinction between initial conditions and laws is missing (the distinction is clear once the cosmos has come into existence—but we are concerned with 'prior' conditions, as explained above). This is expressed by McCrea as follows:

> When we speak of the other solutions of the equations of stellar structure, besides the one we are interested in at the moment, as representing systems that could exist, we mean that they could exist in the universe as we know it. Clearly no such attitude is possible towards the universe itself (McCrea 1953).

Certainly any proposal for distinguishing between laws of nature and boundary conditions governing solutions to those laws is untestable in this context. Given the feature that the universe is the unique background for all physics, it is then not far-fetched to suggest it is possible the cosmos influences the *nature* of local physical laws rather than just their initial conditions. This has been examined in some depth in three specific cases.

First: **(a)** It might be that there is a time variation in physical constants related to the expansion of the universe, as proposed in the case of the gravitational constant G by Dirac (Dirac 1938) and developed in depth by Brans and Dicke (Brans and Dicke 1961). This kind of proposal is to some degree open to observational test (Cowie and Songaila 1995; Will 1979, 24), and in the cases where it has been investigated it seems it does not occur—the constants are invariant. However testing that invariance is fundamentally important, precisely because cosmology assumes as a ground rule that physics is the same everywhere in the universe. If this were not true, local physics would not guide us adequately as to behavior of matter elsewhere in the universe, and cosmology would become an arbitrary guessing game. In order to be able to proceed in a scientific manner when such variation is proposed, one needs then to hypothesize the manner of such variation, thus the old laws where G was constant are replaced by new laws governing its time variation (Brans and Dicke 1961); the principle of nature being governed by invariant (unchanging) physical laws remains. Thus in the end the proposal is to replace simpler old laws by new more complex ones that express the relation between the expanding universe and local physics. These must then be assumed invariant or we cannot proceed scientifically.

More fundamentally, it is conceivable that **(b)** the local inertial properties of matter are related to the distribution of matter in the distant universe, which provide the origin of inertia—the complex of ideas referred to as Mach's principle (Barbour and Pfister 1995); and **(c)** that the existence of the macroscopic arrow of time in physics—and hence in chemistry, biology, psychology, and for example the social sciences—is related to boundary conditions in the past and future of the universe (the fundamental physical laws by themselves being time symmetric, and so unable to explain this feature) (Davies 1974; Ellis and Sciama 1972; Zeh 1992). This proposal relates existence of the Second Law of Thermodynamics to the nature of the universe itself; a recent argument of this kind is Penrose's claim that the existence of the arrow of time undermines standard inflationary universe models (Penrose 1989a, 1989b).

In each case various proposals have been made as to the possible nature of the deeper underlying laws and conditions that might express such a relation. These proposals are however intrinsically untestable, for the reasons explained above, and so are unlikely to gain consensus, although they serve as a continual fertile source of ideas. In any case, the important conclusion is that it is certainly appropriate for cosmology to consider what would have happened if either the laws of physics or the boundary conditions at the beginning of the universe had been different:

2B: *Cosmology is interested in investigating hypothetical universes where the laws of physics are different from those that obtain in the real universe in which we live*—for this may help us understand why the laws of physics are as they are (a fundamental feature of the real physical universe). This is in stark contrast to the rest of science where we are content to take the existence and nature of the laws describing the fundamental behavior of matter as given and unchangeable, whereas cosmological investigation is interested in the properties of hypothetical universes with different physical behavior (this is of course not the same as assuming an ensemble of such universes actually exists, cf. the discussion in the previous section). Indeed if one wants to investigate issues such as why life exists in the universe, consideration of this larger framework—in essence, a hypothetical ensemble of universes with many varied properties—is essential. One cannot take the existence and nature of the laws of physics (and hence of chemistry) as unquestionable—which seems to be the usual habit in biological discussions on the origin and evolution of life. Consideration of what might have been is the basis not only of science fiction, but also of useful cosmological speculation that may help throw light on what actually is. However we need to be very cautious about any claimed statistics of universes in such a hypothetical ensemble of all possible or all conceivable universes: this runs into the kinds of problems mentioned above, and should be treated with scepticism. We can learn from such considerations the nature of possible alternatives, but not necessarily the probability with which they might occur (if that concept has any real meaning).

3. THE LARGE SCALE OF THE UNIVERSE IN SPACE AND TIME

The problems arising from the uniqueness of the universe are compounded by its vast scale (Hogan 1998), which poses major problems for observational cosmology. The distance to the nearest galaxy is about 10^6 light years, that is about 10^{24}cm (cf. the size of the earth, about 10^9cm). The present size of the visible universe is about 10^{10} light years, that is about 10^{28}cm. This size places major constraints on our ability to observe distant regions (and certainly prevents us experimenting with them). The uniqueness of cosmology in this respect is that it deals with this scale: the largest with which we can have observational contact.

Astronomical observations of sources and background radiation (Bothun 1998) are obtained from telescopes operating at all wavelengths (optical, infrared, ultraviolet, radio, X-ray), giving detailed observations (including visual pictures, spectral information, and polarization measurements) of matter. We can also aspire to use neutrino and gravitational wave telescopes. However distant sources appear both very small and very faint, because of their physical distance, and because their light is highly redshifted (due to the expansion of the universe). Additionally, absorption of intervening matter can interfere with light from distant objects. The further back we look, the worse these problems become; thus our reliable knowledge of the universe decreases rapidly with distance (although the situation has improved greatly owing to the new generation of telescopes and detectors, particularly the Hubble Space Telescope and the COBE satellite). The CBR we detect probes the state of the universe from the time of decoupling of matter and radiation (at a redshift of about 1100) to the present day; this is the most distant matter we can observe by telescopes detecting electromagnetic radiation at any wavelength.

Another source of cosmological information is data of a broadly geological nature (Hoyle 1960); that is, the present day status of rocks, planets, star clusters, galaxies, and so on contains much information on the past history of the matter comprising those objects. Thus we can obtain detailed information on conditions near our spatial position—more accurately, near our past world-line in spacetime (Ellis 1971a)—at very early times if we can interpret this data reliably, for example by relating theories of structure formation to statistical studies of source properties. This involves us in physical cosmology: namely the study of the evolution of structures in the universe, tested by comparison with astronomical observation. Particularly useful are measurements of the abundances of elements which resulted from nucleosynthesis in the Hot Big Bang, and age estimates of the objects we observe.

Now the vast scale of the universe implies we can only view it, considered as a whole, from one spacetime event ('here and now') (Ellis 1971a, 1975). If we were to move away from this spatial position at almost the speed of light for say 10,000 years, we would not succeed in leaving our own galaxy, much less in reaching another one; and if we were to start a long term astronomical experiment that would store data for say 20,000 years and then analyze it, the time at which we observe the universe would be essentially unchanged (because its age is of the order of 10^{10} years: the extra time would make a negligible difference). This is quite unlike other geographic sciences: we can travel everywhere on earth and see what is there. The situation would be quite different if the universe were much smaller. Given its actual scale, where we are now seeing galaxies whose present distance from us is about 10^9 light-years, the effect is as if we were only able to observe the earth from the top of one mountain, and had to deduce its nature from those observations. Furthermore the early universe is opaque to radiation, so it is as if distant mountains were shrouded in cloud. Because we can only observe by means of particles—photons, massless neutrinos, gravitons—travelling to us at the speed of light (and so along light rays lying in our past light cone),

3A: *We can effectively only observe the universe, considered on a cosmological scale, on the one past light cone of one space-time event since the time of decoupling of matter and radiation.*

As a consequence, two interrelated problems occur in interpreting the astronomical observations. The first is that (because we can only view the universe from one point) we only obtain a 2-dimensional projection on the sky of the 3-dimensional distribution of matter in the universe. To reconstruct the real distribution, we need reliable distance measurements to the objects we see. However because of variation in the properties of sources, we lack reliable standard candles or standard size objects to use in calibrating distances, and have to study statistical properties of classes of sources. Second, we necessarily see distant galaxies and other objects at earlier times in their history (where their world lines intersect this past light cone). Thus cosmology is both a geographic and a historical science combined into one: we see distant sources at an earlier epoch, when their properties may have been different. The inevitable lookback time involved in our observations means we need to understand evolution effects which can cause systematic changes in the properties of sources we observe; but we do not have good theories of source evolution. The situation is however improved by availability of the geological type data, which probes events near our past world-line at early times. If we can obtain adequate quality data of this kind at high redshifts, we can use this to probe conditions very early on at some distance from our past worldline; encouraging in this regard are recent determination of element abundances at high redshift.

The further essential point is that the region of the universe we can see from this vantage point is restricted, because a finite time has elapsed since the universe became transparent to radiation, and light can only have travelled a finite distance in that time. As no signal can travel to us faster than light, we cannot receive any information from galaxies more distant than our *visual horizon*—essentially the distance light can have travelled since the decoupling of matter and radiation as the hot early universe cooled down (Ellis and Stoeger 1988; Ellis and Rothman 1993). The key point here is that the universe itself is much bigger than the observable universe. There are many galaxies—perhaps an infinite number—at a greater distance than the horizon, that we cannot observe by any electromagnetic radiation. Furthermore no causal influence can reach us from matter more distant than our *particle horizon*—the distance light can have travelled since the creation of the universe, so this is the furthest matter with which we can have had any causal connection (Rindler 1956; Tipler et al. 1980). We can hope to obtain information on matter lying between the visual horizon and the particle horizon by neutrino or gravitational radiation observatories; but we can obtain no reliable information whatever about what lies beyond the particle horizon. We can in principle feel the gravitational effect of matter beyond the horizon; however we cannot uniquely decode that signal to determine what matter distribution caused it (Ellis and Sciama 1972).

The possible exception to this is if we live in a *small universe*, which has closed spatial sections whose size is smaller than the Hubble scale, so that we have already seen around the universe at least once, and in fact are seeing the same matter over and over again—looking like a much larger universe (like a room with mirrors on all walls, the ceiling, and the floor)—(Ellis 1971b). This is a possibility, and is in principle testable (Ellis and Schreiber 1986; Cornish and Levin 1997; Cornish et al. 1996); however there is at present no solid evidence to show that the real universe is indeed like this (Roukema 1996; Roukema and Edge 1997). But if this is not the case, claims about what conditions are like on very large scales—that is, much bigger than the Hubble scale—are completely unverifiable (Ellis 1975), for we have no evidence at all as to what conditions are like beyond the visual horizon. The situation is like that of an ant surveying the world from the top of a sand dune in the Sahara desert. Her world model will be a world composed only of sand dunes—despite the existence of cities, oceans, forests, tundra, mountains, and so on beyond her horizon. In chaotic inflationary models, it is a definite prediction that the universe will not be like a RW geometry on a very large scale—rather it will consist of many RW-like domains, each with different parameter values, separated from each other by highly inhomogeneous regions outside our visual horizon (Linde 1990). This prediction is just as untestable as the previously prevalent assumption (based on a Cosmological Principle) that the universe is RW-like on such scales (Bondi 1960; Weinberg 1972). Neither can be observationally confirmed or denied. Similarly, it is commonly stated that if we live in a low-density universe and the cosmological constant vanishes, then the universe has infinite spatial sections. However this deduction only applies if the RW-like nature of the universe within the past light cone continues to be true indefinitely far outside it, and the space sections have their 'natural' simply-connected topology—and there is no way we can obtain observational evidence that these are both true. The conclusion is that,

3B: *Unless we live in a 'small universe,' most of the matter in the universe is hidden behind horizons.* This is quite unlike most geographic sciences, where we can see most of what there is, but is similar in some ways to limitations in the historical sciences. The resulting observational problems inside the visual horizon are shared by most of astronomy, particularly extragalactic astronomy—which on large scales blends into observational cosmology. However the verification status of the regions outside the horizons is totally different from the rest of geographic sciences and astronomy—our lack of access to the regions outside the particle horizon being only equalled by our lack of access to the interior of black holes. Our lack of observational access to any totally disconnected universes (by definition, outside the particle horizon!) in a supposed ensemble of such universes is unlike the situation in any other science, as has been emphasized above.

4. THE UNBOUND ENERGIES IN THE EARLY UNIVERSE

The analogous problems for physical cosmology arise because energies occurring in the Hot Big Bang early universe phase are essentially unbounded, so the highest energies we can attain in particle accelerators cannot reach the levels relevant to very early times. The uniqueness of cosmology in this regard is that it is the only science contemplating spacetime regions that have experienced such high energies, and with which we are in intimate causal contact (despite the huge timescales involved—indeed events at those early times determined much of what we see around us today).

The nuclear reactions underlying nucleosynthesis are well understood, and their cross-sections reasonably well-known; the processes of baryosynthesis and quark-gluon recombination are reasonably understood and are on the border of being testable; but physical processes relevant at earlier times are inaccessible to testing by laboratory or accelerator-based experiment. Consequently:

4A: *We cannot experimentally test much of the physics that is important in the very early universe* (this is independent of the issue of creation, considered above: the problem arises after the initial conditions have been set and the universe is running according to invariable physical laws). Hence our understanding of physics at those times has to be based on extrapolation of known physics way beyond the circumstances in which it can be tested. Thus we cannot be confident of the validity of the physics we presuppose then, and this becomes particularly so in the presumed quantum gravity era (and *a fortiori* in considering 'laws' that may have lead to the initiation of the hot big bang and setting of initial conditions for its expansion, cf. the discussion in Section 1 above). Rather than using known physics to predict the evolution of the universe, we end up testing proposals for this physics by exploring their implications in the early universe—which is the only 'laboratory' where we can test some of our ideas regarding fundamental physics at the highest energies (Yoshimura 1988, 293). The problem is we cannot simultaneously do this and also carry out the aim of physical cosmology, namely predicting the evolution of the early universe from known physical theory. Rather,

4B: *We have to extrapolate from known physics to the unknown and then test the implications; to do this, we assume some specific features of known low energy physics are the true key to how things are at higher energies.* The trick is to identify which features of known physics are these key fundamental features: variational principles, broken symmetries and phase changes, duality invariance are candidates, for example. If we confirm our guesses for the relevant physics by their satisfactory implications for the early universe, tested in some suitable way, then this is very impressive progress; but if this is the *only* way we can test the proposed physics then the situation is problematic. If the hypothesis solves *only* the specific issues it was designed to solve and nothing else, then in fact it has little explanatory power, rather it is just an alternative (perhaps theoretically preferable) description of the known situation. This corresponds to the applied mathematics statement that you can fit any data if you allow enough arbitrary constants and functions in your theory. One obtains positive observational support for a particular proposal for the relevant physics only if it that proposal predicts multiple confirmed outcomes (rather than just one), so that a single hypothesis simultaneously solves several different observational issues, or if one can show that no other proposal can give similar cosmological outcomes. The latter is usually not true. Some of the options may be theoretically preferred to others on various grounds; but one must distinguish this from their having observational support. They lack physical power if they have no other testable consequences.

A particular example is the inflationary universe proposal: the supposed inflaton field underlying an inflationary era of rapid expansion in the early universe (Kolb and Turner 1990; Gibbons et al. 1983; Guth 1981; Blau and Guth 1987, 524) has not been identified, much less shown to exist by any laboratory experiment or demonstrated to have the properties required in order that inflation took place as proposed. Because this field ϕ is unknown, one can assign it an arbitrary potential $V(\phi)$, this arbitrariness reflecting our inability to experimentally determine the relevant behavior; but it can be shown that *any* scale evolution $a(t)$ of the universe can be attained by suitable choice of this potential (Ellis and Madsen 1991), and also any desired perturbation spectrum can be obtained by a (possibly different) suitable choice (Lidsey et al. 1997); indeed in each case one can run the mathematics backwards to determine the required potential $V(\phi)$ from the desired outcome. If we could observationally confirm *both* outcomes—the form of $a(t)$ and the fluctuation spectrum—from a single choice of $V(\phi)$, we would have an impressive evidence that this choice was physically correct. The inflationary prediction for $a(t)$ is weakly confirmed by flatness of the universe and solution of the horizon problem in FRW models, but the usually assumed consequence that the density parameter Ω_0 is unity to high accuracy seems not to be true (Coles and Ellis 1997). This can however be fixed by introducing an extra parameter (a cosmological constant or another dynamic field), (Ellis et al. 1991). The fluctuation spectrum is (indirectly) observable through the consequent matter perturbations and associated CBR anisotropies predicted at later times (Peebles 1993; Kolb and Turner 1990; Hu and Sugiyama 1995, 1995a); the observations and theory may agree if one introduces a bias factor b as well as a suitable mixture of cold and hot dark matter, each giving extra freedom in the relation between the initial perturbations and resulting inhomogeneity and CBR spectra. The impressive part is if the predicted CBR anisotropy spectrum agrees with the matter power spectrum (this is currently being tested); but that depends on the physics from tight coupling to the present day, given a suitable initial fluctuation spectrum in the early universe, rather than on the specific hypothesis of an inflationary origin for that spectrum.

The challenge to inflationary theory is to show the data can be fit with less free functions and parameters than data points explained; otherwise the hypothesis that no inflation took place is as viable as far as the data is concerned, although it is not as satisfying from an explanatory viewpoint. The true clincher would be if properties of an inflationary field were predicted from the cosmology side and then confirmed in the laboratory; indeed that would count as one of the great feats of theoretical physics. This may not happen however because of the experimental problems focused on here, resulting because we cannot reproduce all the conditions relevant to very early cosmology on Earth. In all other sciences except astronomy and astrophysics, we are able to access the energies involved because they refer to events on Earth.

5. DETERMINING SPACETIME GEOMETRY: OBSERVATIONAL LIMITS

The unique core business of observational cosmology is determining the large-scale geometry of everything there is, or at least of everything we can observe. One can go about this in a direct manner: trying to determine the geometry of the universe direct from observations (Kristian and Sachs 1966; Ellis et al. 1985). However this is difficult because of the paucity of data (for example we cannot easily determine the transverse velocities of the matter we see), and additionally this approach has little explanatory value, once a spacetime model has been obtained; so the usual option is different: it is to assume a space-time geometry with high symmetry, and then to determine its essential parameters from comparison of theoretical relations with astronomical observations in order to give a good description of the observed universe. A more advanced analysis in addition tries to determine a detailed best-fit between the theoretical spacetime model and the real (inhomogeneous and anisotropic) universe (Ellis and Stoeger 1987; Matravers et al. 1995).

The standard models of cosmology are the Friedmann-Lemaitre (FL) family of universe models expanding from a Hot Big Bang, based on the Robertson-Walker (RW) geometries, that is, space-times that are exactly spatially homogeneous and isotropic everywhere (Robertson 1933; Weinberg 1972; Hawking and Ellis 1973). They are easy to understand, and have tremendous explanatory power; furthermore their major physical predictions (existence of blackbody CBR and specific light element production in the early universe) seem confirmed. Assuming one is talking about a large enough averaging scale, which should be explicitly indicated (Ellis 1984; Stoeger et al. 1987), the issue is, to what degree does observational data uniquely indicate these universe models for the expanding universe geometry?

Considered on a large enough angular scale, **(a)** astronomical observations are very nearly isotropic about us, both as regards source observations and background radiation; indeed the latter is spectacularly isotropic, better than one part in 10^5 after a dipole anisotropy, understood as resulting from our motion relative to the rest frame of the universe, has been removed (Partridge 1995). Because this applies to all observations, this establishes that in the observable region of the universe, to high accuracy both the space-time structure and the matter distribution are isotropic about us (thus there are not major observable matter concentrations in some other universe region). If we could additionally show that **(b)** the source observational relations had the unique FL form (Sandage 1961; Ellis 1971a; Weinberg 1972) as a function of distance, this would additionally establish spatial homogeneity, and hence a RW geometry (Ellis et al. 1985). However because of item 3B above, the observational problems mentioned earlier—specifically, unknown source evolution—prevent us from carrying this through. Astrophysical cosmology could resolve this in principle, but is unable to do so in practice.

Indeed the actual situation is the inverse: taking radio-source number-count data at its face value, without allowing for source evolution, contradicts a RW geometry; it is better fit by a flat spacetime. In the face of this, the usual procedure is to *assume* spatial homogeneity is known some other way, and *deduce* the source evolution required to make the observations compatible with this geometric assumption (Ellis 1975). It is always possible to find a source evolution that will achieve this (Mustapha et al. 1998). Thus attempts to observationally prove spatial homogeneity this way fail; indeed an alternative interpretation would be that this data is evidence of spatial inhomogeneity, i.e. that we live in a spherically symmetric inhomogeneous universe where we are situated somewhere near the centre (as otherwise our observations would not be almost isotropic), with the cosmological redshift being partly gravitational (Ellis et al. 1978), and conceivably with a contribution to the CBR dipole from this inhomogeneity (if we are a bit off-centre).

Most people regard this proposal as very unappealing—but that does not show it is incorrect. One can claim that physical processes such as inflation make existence of almost-RW regions highly likely, indeed much more probable than spherically symmetric inhomogeneous regions. This is a viable argument, but we must be clear what is happening here—we are replacing an observational test by a theoretical argument based on a physical process that may or may not have happened (for there is no definitive observational proof that inflation indeed took place). It will be strongly bolstered if current predictions for the detailed pattern of CBR anisotropy on small scales (Hu and Sugiyama 1995), based on the inflationary universe theory, are confirmed; but that argument will only become rigorous if it is shown that spherically symmetric inhomogeneous models (with or without inflation) cannot produce similar patterns of anisotropy. This is unlikely to be the case, because the acoustic oscillations that lead to the characteristic predicted anisotropy patterns in fact take place after inflation, and can equally happen if suitable initial conditions occur without a previous inflationary phase.

What about alternative routes? Another proposal is (**c**) to use the uniformity in the nature of the objects we see to deduce they must have all undergone essentially the same thermal history, and then to prove from this uniformity of thermal histories that the universe must be spatially homogeneous; for example observations of element abundances at high z are very useful in constraining inhomogeneity (Ellis 1995). However turning this into a proper test of homogeneity has not succeeded so far, indeed it is not clear if this can be done (Bonnor and Ellis 1986). Finally (**d**) if we could show isotropy of all observations about more than two observers, we would prove spatial homogeneity. Now the crucial point has already been made: we cannot observe the universe from any other point, so we cannot establish this observationally. Hence the standard argument is to assume a *Copernican Principle*: that *we are not privileged observers*. This is plausible in that all observable regions of the universe look alike: we see no major changes in conditions anywhere we look. Combined with the isotropy we see about ourselves, this implies that all observers see an isotropic universe, and this establishes a RW geometry (Weinberg 1972; Ellis 1971a; Hawking and Ellis 1973). The result holds if we assume isotropy of all observations; a powerful enhancement was proved by Ehlers, Geren, and Sachs (Ehlers et al. 1995; Hawking and Ellis 1973), who showed that if one assumes simply isotropy of freely-propagating radiation about each observer the result follows from the Einstein and Liouville equations; that is, exact isotropy of the CBR at each point implies an exact RW geometry.

This is currently the most persuasive observationally-based argument we have for spatial homogeneity. A problem is that it is an exact result, assuming exact isotropy of the CBR; is the result stable? Recent work (Stoeger et al. 1995) has shown that indeed it is: almost-isotropy of the CBR everywhere in some region proves the universe geometry is almost-RW in that region. Thus the result applies to the real universe—provided we make the Copernican assumption that all other observers, like us, see almost isotropic CBR. And that is the best we can do. Weak tests of the isotropy of the CBR at other spacetime points come from the Sunyaev-Zeldovich effect (Goodman 1995) and from CMB polarization measures (Kamionkowski and Loeb 1997), giving broad support to this line of argument but not enough to give good limits on spatial inhomogeneity through this line of argument. The observational situation is clear:

5A: *The deduction of spatial homogeneity follows not directly from astronomical data, but because we add to the observations a philosophical principle that is plausible but untestable.* It may or may not be true. The specific features of cosmology characterized above prevent us from attaining certainty through observational testing.

The purpose of the above analysis is not to seriously support the view that the universe is inhomogeneous, but rather to show clearly the nature of the best observationally-based argument by which we can (quite reasonably) justify the assumption of spatial homogeneity. Accepting this argument, the further question is, in which spacetime regions does it establish a RW-like geometry?

The CBR we detect probes the state of the universe from the time of decoupling of matter and radiation (at a redshift of about 1100) to the present day, within the visual horizon. The argument from CBR isotropy can legitimately be applied for that epoch. However it does not necessarily imply isotropy of the universe at much earlier or much later times, because there are spatially homogeneous anisotropic perturbation modes that are unstable in both directions of time; and they will occur in a generic situation. Indeed, if one examines the Bianchi (spatially homogeneous but anisotropic) universes, using the powerful tools of dynamical systems theory, one can show that intermediate isotropisation can occur (Wainwright and Ellis 1996; Wainwright et. al 1998; Goliath and Ellis 1998): despite being highly anisotropic at very early and very late times, such models can mimic a RW geometry arbitrarily closely for an arbitrarily long time, and hence can reproduce within the errors any set of FL-like observations. We can obtain strong limits on the present-day strengths of these anisotropic modes ((Wainwright and Ellis 1996; Wainwright et. al 1998; Goliath and Ellis 1998) from CBR anisotropy measurements and from data on element abundances, the latter being a powerful probe because (being of the 'geological' kind) it can test conditions at the time of element formation, long before decoupling. But however low these observational limits, anisotropic modes can dominate at even earlier times as well as at late times (long after the present). If inflation took place, this conclusion is reinforced: it washes out any information about very early universe anisotropies and inhomogeneities in a very efficient way. As well as this time limitation on when we can regard isotropy as established, there are major spatial limitations. The above argument for homogeneity does not apply far outside the visual horizon, for we have no reason to believe the CBR is highly isotropic there. These limitations have parallels in limitations on what can be tested in other historical sciences: for example the geological record in effect also runs into horizons at early enough times—the earlier record does not survive.

Given that a RW geometry is a good description of the observable universe on a large scale, the further issue is *what are the best-fit parameters that characterize it* (selecting the specific universe we observe from the family of all FL models). Establishing the Hubble constant H_0, deceleration parameter q_0, and density parameter Ω_0 has been the subject of intensive work for the past 30 years (Gun et al. 1978; Sandage et al. 1993). However there is still major uncertainty about their values, essentially because of the observational problems discussed in section B. Because of our lack of adequate theories of historical development for the objects we observe, there are a variety of conflicting estimates for the cosmological parameters, based on different lines of argument (Coles 1997). Many of the methods of estimating Ω_0

depend on studying the growth and nature of inhomogeneities in the universe; this makes them rather model-dependent, and introduces a further set of parameters (describing the nature and statistical properties of the matter distribution) to be determined by observation. To obtain believable answers one has to use informed judgement to decide which methods are more reliable, and give them more weight.

Additionally we encounter instrumental and observational issues generic to all sciences, but with a particular nature in cosmology. In particular, a variety of selection effects interfere with observations, as happens in all historical sciences because some events or objects leave long-lived easily detectable traces and others do not (e.g. as in the fossil record on Earth). In the case of cosmology, some astronomical objects are easy to detect but others are not. Most notably, some kinds of matter emit very little radiation and are not easy to detect by absorption, hence the famous *dark matter problem* (Coles and Ellis 1997; Peacock 1999; Kolb and Turner 1990; Borner and Gottlober 1997; Turok 1997): we do not know the amount of matter in the universe to within an order of magnitude, but we do know that what we can see is between 10% and 1% of all there is. The implication is we do not even know the kind of matter that dominates the dynamics of the universe (the Earth of course is quite atypical of the universe at large; observations of its constitution do not give a good guide to the dominant component in the vast high-vacuum regions between galaxies). Part of the problem is a series of difficult questions regarding how both observations and dynamics depend on the averaging scale assumed in the theoretical description used (Ellis 1984; Stoeger et al. 1987); the answer obtained for Ω_0 may apparently depend on this averaging scale, and this needs careful investigation.

A key issue is estimates of the age of the universe t_0, which is dependent on H_0 and Ω_0, as compared to the age of objects in the universe. For standard models, $t_0 < 1/H_0$. This limit can however be violated in a model with a cosmological constant Λ that dominates the recent expansion of the universe, but this in turn is constrained by deep number counts and gravitational lensing observations.

5B: *The tension between the age of the universe and ages of stars is one area where the standard models are vulnerable to being shown to be inconsistent*, hence the vital need to establish reliable distance scales, basic to estimates of both H_0 and the ages of stars, and good limits on Λ. At present this issue is OK, because of a recent revision of our distance scale estimates (Harris et al. 1998), assisted by evidence that Λ is positive (Perlmutter et al. 1998); but continued vigilance is needed on this front. One can ask similar questions in any of the historical sciences; for example, what specific observation could disprove standard evolutionary theory for life on earth? If there were no such observations, the subject would be of questionable scientific status. Thus it is a plus for cosmology that this issue exists!

6. EXPLAINING HOMOGENEITY AND STRUCTURE: THE ISSUE OF ORIGINS

This is the unique core business of physical cosmology: explaining both why the universe has the very improbable high-symmetry FL geometry on large scales, and how structures come into existence on smaller scales. Clearly only cosmology itself can ask the first question; and it uniquely sets the initial conditions underlying the astrophysical and physical processes that are the key to the second, underlying all studies of origins (Fabian 1989).

Given these astrophysical and physical processes, explanation of the large-scale isotropy and homogeneity of the universe together with the creation of smaller-scale structures means determining the dynamical evolutionary trajectories relating initial to final conditions and then essentially either (**a**) explaining initial conditions or (**b**) showing they are irrelevant. As regards the latter, demonstrating minimal dependence of the large-scale final state on the initial conditions has been the aim of the chaotic cosmology programme (Misner 1968, 1969) and of the inflationary family of theories (Guth 1981; Blau and Guth 1987, 524; Turok 1997; Gibbons et al. 1983; Linde 1990; Steinhardt 1995; Kolb and Turner 1990), which are both partially successful: with or without inflation one can explain a considerable degree of isotropisation and homogenization of the physical universe, but this will not work in all circumstances (Penrose 1989a; Wainwright and Ellis 1996; Wainwright et al. 1998; Goliath and Ellis 1998; Rothman and Ellis 1986). It can only be guaranteed to succeed if initial conditions are somewhat restricted—so this takes us back to the former issue. Inflation then goes on to provide a causal theory of initial structure formation from an essentially homogeneous early state (via amplification of initial quantum fluctuations)—a major success if the details can be sorted out. As explained above, one already runs here into a present inability to verify the initial stages of the proposed underlying physical theory, because of the high energies involved.

The explanation of initial conditions has been the aim of the family of theories one can label collectively as quantum cosmology (Hawking 1993; Gott and Li-Xin Li 1997); however as discussed earlier, here we inevitably reach the limits to what the scientific study of the cosmos can ever say—if we assume that such studies must of necessity involve an ability to observationally or experimentally check our theories. No physical experiment at all can help here because of the uniqueness of the universe, and the feature that no spacetime exists before such a beginning; so brave attempts to define a 'physics of creation' stretch the meaning of 'physics.' Attempts at 'explanation' of a true origin usually seem to depend on assuming a pre-existing set of physical laws that are similar to those that exist once space-time exists, for they rely on an array of properties of quantum field theory and of fields that seem to hold sway independent of the existence of the universe and of space and time (for the universe itself is to arise out of their validity); however there is no clear locus for those laws to exist in or material for them to act on. The manner of their existence or other grounds for their validity in this context are unclear—and we run into the problems noted before: there are problems with the concepts of 'occurred,' 'circumstances' and even 'when'—for we are talking *inter alia* about the existence of spacetime. Our language can hardly deal with this.

Even if a literal creation does not take place, as is the case in various of the present proposals, this does not resolve the underlying issue. Some attempts involve avoiding a true beginning by going back to some form of eternal or cyclic initial state, for example Tolman's series of expansion and collapse cycles (Tolman 1934), proposals for creation of the universe as a bubble formed in a flat space-time (Tryon 1973; Gott 1982), Linde's eternal chaotic inflation (Linde 1990), and Veneziano's re-expansion from a previous collapse phase (Ghosh et al. 1998). It is unclear that these avoid the real problem; it can be claimed they simply postpone facing it, for one now has to ask all the same questions of origins and uniqueness about the supposed prior state to the Hot Big Bang expansion phase. The Hartle-Hawking 'no-boundary' proposal (Hawking 1993) avoids the initial singularity because of a change of space-time signature, and so gets round the issue of a time of creation in an ingenious way, and Gott's causality violation in the early universe (Gott and Li-Xin Li 1997) does the same kind of thing in a different way. However neither can get around the basic problem: *How was it decided that this particular kind of universe would be the one actually instantiated?* Perhaps the most radical proposal is that order including the laws of physics somehow arises out of chaos, in the true sense of that word—namely a total lack of order and structure—but this does not seem fully coherent as a proposal. If the pre-ordered state is truly chaotic and without form, I do not see either how order can arise therefrom when physical action is as yet unable to take place, or even how we can meaningfully contemplate that situation. We cannot assume any statistical properties would hold in that regime, for example; even formulating a description seems well nigh impossible, for that can only be done in terms of concepts that have a meaning only in a situation of some stability and underlying order such as is characterized by physical laws. The same problem arises in every approach:

6A: *A choice between different contingent possibilities has somehow occurred; the fundamental issue is what underlies this choice.* Why does the universe have one specific form rather than another, when other forms seem perfectly possible? The idea of an ensemble of universes is one approach that sidesteps this problem, because by hypothesis all that can occur then has occurred, and Anthropic arguments select the universe in which we live; but the penalty is the complete lack of verifiability discussed above. In my view that means this proposal is a metaphysical rather than scientific one. Item 4B above applies, in extreme form.

Given these problems, any progress is of necessity based on specific philosophical positions, which decide which of the many possible physical and metaphysical approaches is to be preferred. These philosophical positions should be identified as such and made explicit. As explained above, no experimental test can determine the nature of any mechanisms that may be in operation in circumstances where even the concepts of cause and effect are suspect. Initial conditions cannot be determined by the laws of physics alone—for if they were so determined they would no longer be contingent conditions, the essential feature of initial data, but rather would be necessary. A purely scientific approach cannot succeed in explaining this specific nature of the universe.

7. THE EXPLICIT PHILOSOPHICAL BASIS

Consequent on the discussion above, and particularly items 4B and 5A, it follows that

7A: *Unavoidably, whatever approach one may take to issues of cosmological origins, metaphysical issues inevitably arise in both observational and physical cosmology. Philosophical choices are needed in order to shape the theory.*

There is of course always a philosophical basis to any scientific analysis, namely **(i)** adoption of the basic scientific method and a commitment to the attempt to explain what we see as far as possible simply in terms of causal physical laws. This will clearly be true also in cosmology. However we need further explicit philosophical input **(ii)** in order to attain specific geometric models—for example a Copernican principle, as explained above, **(iii)** in order to decide what form physical cosmology should take in the very early universe, for example deciding which physical principle to use as the core of one's extrapolation of known physics to the unknown. Underlying both sets of choices are **(iv)** criteria for satisfactoriness of a cosmological model, which help decide which feature to focus on in formulating a theory. These are discussed in this section. Of particular importance is the issue discussed in the following section, namely **(v)** what is the scope chosen for our cosmological theory? Together with the choice of criteria for a good theory, this is a philosophical decision that will shape the rest of the analysis.

As regards criteria for a good theory, typical would be (Coles and Ellis 1997):

1. *Satisfactory structure:* (a) internal consistency, (b) simplicity (Ockham's razor), and (c) aesthetic appeal ('beauty' or 'elegance');

2. *Intrinsic explanatory power:* (a) logical tightness, (b) scope of the theory—the ability to unify otherwise separate themes, and (c) probability of the theory or model with respect to some well-defined measure;

3. *Extrinsic explanatory power*, or *relatedness*: (a) connectedness to the rest of science, (b) extendability—providing a basis for further development;

4. *Observational and experimental support*, in terms of (a) testability: the ability to make predictions that can be tested; and (b) confirmation, the extent to which the theory is supported by such tests as have been made.

These are all acknowledged as desirable. The point then is that generally in pursuing historical sciences, and in particular in the cosmological context,

7B: *These philosophical criteria will in general come into conflict with each other, and one will have to choose between them to some degree; this choice will shape the resulting theory* (Ellis and Stoeger 1987; Matravers et al. 1995). The thrust of much recent development has been away from observational tests towards strongly theoretically based proposals, indeed sometimes almost discounting observational tests (for example (Coles and Ellis 1997), in the case of the density parameter Ω_0). At present this is being corrected by a healthy move to detailed observational analysis of the consequences of the proposed theories, marking a maturity of the subject. However because of all the limitations in terms of observations and testing [criteria (4)], in the cosmological context we still have to rely heavily on other criteria for choice of our theory, and some criteria that are important in most of science may not really make sense then. This is true of 2(c) in particular, as discussed above; nevertheless many approaches still give the idea of probability great weight. At a minimum, the ways this can make sense need exploration and explication.

Furthermore the meaning of some of the criteria may come into dispute. 1(b) is clearly a case in point ('beauty is in the eye of the beholder'), but 1(c) is also controversial: for example, is the idea of an existent ensemble of universes displaying all possible behaviors simple (because it is a single idea that can be briefly stated), or immensely complex (because that statement hides all the complexities and ambiguities involved in the idea of an infinity of possibilities)? [cf. the discussion above, supporting the latter view]. Criterion 3(a) also can be controversial: most of current cosmology is heavily based on theoretical physics ideas, but other options are possible. An example is Smolin's suggestion, based on ideas of evolutionary biology, of a universe evolving in a Darwinian sense through multiple collapses and re-expansions into new expanding universe regions (Smolin 1992). The result is different in important ways from standard theory precisely because it embodies in one theory three of the major ideas of this century, namely Darwinian evolution through competitive selection, the evolution of the universe in the sense of major changes in its structure associated with its expansion, and quantum theory (through the only partly explicated mechanism supposed for re-expansion out of a collapse into a black hole). Is this a good move or not? Resolving any such disputes necessarily involves philosophical argumentation rather than simply scientific investigation.

Additionally the tenor of scientific understanding may change, altering the balance of what is considered a good explanation and what does not. An example (Ellis 1990) is the way cosmologists strongly resisted the idea of an evolving universe in the 1920's, at a time when biological evolution was in vogue but the idea of continental drift was also being strongly resisted. The change to an appreciation of the explanatory power of an evolving model came later in both cases, but even then in the cosmological case, for either aesthetic or metaphysical reasons, some still sought for a steady state description, resisting the implication of a beginning to the universe. That tendency is still with us today, in the form of models that are eternal in one way or another (e.g. some forms of chaotic inflation). Another example is the change from supposition of underlying order, expressed in the idea of a Cosmological Principle, to a broad supposition of generic disordered conditions, embodied in the ideas of chaotic inflation. The underlying motivation for the change in both cases is metaphysical. It would be helpful to have clear analysis of what criteria of satisfactoriness are promoted by the new theories in each case, and why the result is philosophically preferable to the previously assumed situation.

Such criteria choices also underlie our theories in all the historical and geographic sciences, because of the corresponding limits on testing there also, but they do so to a considerably greater degree in cosmology than the others because of the features discussed in this article. There is nothing wrong with this, provided it is acknowledged and brought out into the open; this then enables us to consider the issue carefully and make the best choices.

8. THE SCOPE OF COSMOLOGY: HOW MUCH SHOULD WE TRY TO EXPLAIN?

To sensibly choose priorities for the criteria just discussed, we need an answer to the question,

8A: *What is the scope we envisage for our cosmological theory?* This is a choice one has to make, as regards both foundations and outcomes. Given a decision on this, one can sensibly debate what is the appropriate philosophical position to use in studying a cosmological theory with that scope.

The study of expansion and structure formation from nucleosynthesis to the present day is essential and well-informed. The philosophical stance adapted is minimal and highly plausible. Physics at earlier times, back to quantum gravity, is important but less well based. The philosophical stance is more significant and more debatable. Developments in the quantum gravity era and before are highly speculative; the philosophical position adopted is dominant because experimental and observational limits on the theory are lacking. One can choose the degree to which one will pursue the study of origins (Fabian 1989) back to earlier and earlier times, and hence the degree to which specific philosophical choices are dominant in one's theory:
The basic underlying cosmological questions are (Ellis 1991):

(1) *Why do the laws of physics have the form they do?* Issues arise such as what makes particular laws work? what guarantees the behavior of a proton, the pull of gravity? What makes one set of physical laws fly rather than another?

(2) *Why do boundary conditions have the form they do?* The key point here (already mentioned above), is how are specific contingent choices made between the various possibilities, whether there was an origin to the universe or not; and the more profound:

(3) *Why do any laws of physics exist?* This relates to unsolved issues concerning the nature of the laws of physics: are they descriptive or prescriptive? Is the nature of matter really mathematically based in some sense, or does it just happen that its behavior can be described in a mathematical way?

(4) *Why does anything exist?* This profound existential question is a mystery whatever approach we take. Finally the adventurous also include in these questions the more profound forms of the contentious Anthropic question (Carr and Rees 1979; Davies 1982; Barrow and Tipler 1984; Tegmark 1993):

(5) *Why does the universe allow the existence of intelligent life?*, which is of somewhat different character than the others and largely rests on them but is important enough to generate considerable debate in its own right.

The status of these questions is philosophical rather than scientific, for they cannot be resolved purely scientifically. How many of them—if any—should we include in our theory? One option is to decide to treat cosmology in a strictly scientific way, excluding all the above questions, because they cannot be solved scientifically. One ends up with a solid technical subject that by definition excludes such philosophical issues This is a consistent and logically viable option; one just accepts the initial data for the universe as given and requiring no further explanation. This logically unassailable position however has little explanatory power; thus most tend to reject it because of criteria 2(b) and 3 above. The second option is to decide that these questions are of such interest and importance that one will tackle some or all of them, even if that leads one outside the strictly scientific arena. This is also a legitimate exercise, provided one follows two basic guidelines.

First, one must avoid the claim that scientific methods alone can resolve these questions: it is essential to respect the limits of what the scientific method can achieve, and acknowledge clearly when arguments and conclusions are based on some philosophical stance rather than purely on scientific argument. If we acknowledge this and make that stance explicit, then the bases for different viewpoints are clear, and alternatives can be argued rationally. One can then use the best current philosophical understanding of the scientific method, for example Imre Lakatos' characterization of the nature of scientific research programmes (Lakatos 1980), as a basis for looking at the alternatives.

Second, in undertaking this task, one must be aware of the limitations of the models of reality we use as our basis for understanding. They are necessarily partial and incomplete reflections of the true nature of reality, helpful in many ways but also inevitably misleading in others. No model (literary, intuitive, or scientific) can give a perfect reflection of reality; so they must not be confused with reality. This understanding does not diminish the utility of these models; rather it helps us use them in the proper way.

Given these guidelines, one can include some or all of these foundational issues within the scope of one's investigation. It is clear then that:

8B: *The cosmological philosophical base becomes more or less dominant in shaping our theory according to the degree that we pursue a theory with more or is less ambitious explanatory aims.* It is here that criteria 2 and 3 above are to some degree in conflict with criterion 4. Thus if we try to explain the origin of the universe itself, these philosophical choices become dominant precisely because the experimental and observational limits on the theory are weak (this can be seen by viewing the variety of such proposals—briefly mentioned above—that are at present on the market). An interesting question is to what degree social and cultural issues are at work implicitly shaping the nature of cosmological theory. It is plausible that this influence is very small if one pursues the more technical approach to the subject (with narrower focus) but becomes more significant if one includes the broader areas in one's consideration.

Finally, if one wants to seriously tackle issues in the relation of cosmology to humanity, an important but controversial issue arises: one must include in one's analysis concepts and data of a broad enough scope to reflect fully the nature of human beings, rather than implicitly or explicitly adopting an over simplistic and excessively greedy reductionism. As well as taking into account that we are complex structures based on the physics and chemistry of organic molecules who have evolved by natural processes in the context of the expanding universe, such attempts must acknowledge our truly human attributes and experience—consciousness and emotion, love and pain, free will and ethical choice. Only if we add to the astronomical cosmological data the much broader range of data relevant to issues of this kind can we hope to obtain a world view of adequate scope to be a worthy theory of humanity and cosmology—that is, of Cosmology in the broad sense that relates fully to philosophy and the humanities as well as to science (Ellis 1993).

If we propose a 'thin' theory that does not reflect human nature and experience adequately, the broader public and our academic colleagues in other disciplines will rightly dismiss it as simplistic and inadequate as a full view of the nature of the universe (usual discussions of the Anthropic Principle do not begin to touch this range of issues—they are in fact discussions of the conditions for existence of complex structures such as an amoeba or virus rather than human beings). The full range of human experience is indeed evidence about the universe both because we exist in the universe, and because we have arisen from it. This kind of study can be undertaken as a perfectly rational project; it is a question of choice as to whether one wants to embark on a study of this much broader scope, or to restrict one's consideration to the physical aspects of cosmology. Confusion will be avoided if one makes quite clear at the outset what is the scope of the theory one wishes to consider.

CONCLUSION

In the end the main claim to uniqueness of cosmology rests in its ability to consider questions regarding origins in the uniquely existing physical universe. These questions can be extended to include ultimate issues such as those mentioned in the last section; these do not have to be included, but they can be if one so wants. Many musings in the rest of science—and particularly popular writings on evolutionary biology—at least implicitly choose to tread on this controversial territory; they should fully take into account the limits and problems considered in this paper, or they will not be taking cosmology seriously and important claims may be flawed.

Essentially the same comment applies to the more philosophical and popular writings of cosmologists: they too must take these limits seriously, and not claim for scientific cosmology more than it can actually achieve or more certainty than is in fact attainable—for claim this will in the long term undermine cosmology's legitimate claim to be a project with solid scientific achievements to its name. That claim can be vigorously defended as regards the 'Standard Model' of cosmology, provided this standard model is characterised in conservative terms (Peebles et al. 1991; Coles and Ellis 1997) so that it is not threatened by relatively detailed shifts in theory or data that do not in fact threaten the core business of cosmology. Further, this defence must take adequate cognizance of the difficult philosophical issues that arise if one pushes the explanatory role of cosmological theory to its limits (Leslie 1998; Ellis 1991). Care in this respect is particularly important because of the unique 'ultimate' nature of cosmology as compared to the rest of physics and science, both in terms of its topic and internal structure, and in terms of its relation to them.

University of Cape Town

NOTES

1. For an excellent current overview, see also (Silk 1997).
2. As shown by the appearance of cosmology in the Particle Physics Summary by the Particle Date Group: see (Barnett et al. 1996) and also (Nilsson et al. 1991).
3. See also (Guth 1977) for a good description of these developments for the non-specialist.
4. See also (Cousins 1995; Edwards 1991).

REFERENCES

Barbour, J. B. and H. Pfister. 1995. *Mach's Principle: From Newton's Bucket to Quantum Gravity.* Basel: Birkhäuser.

Barrett, Jeffrey A. 1998. *Everett's Relative-State Formulation of Quantum Mechanics,* http://plato.stanford.edu/entries/qm-everett/#ManyWorlds.

Barnett, R. M. et al. 1996. *Reviews of Modern Physics* 68:611-732.

Blau, S. K. and A. H Guth. 1987. "Inflationary Cosmology." In *300 Years of Gravitation,* eds. S. W. Hawking and W. Israel. Cambridge: Cambridge University Press.

Barrow, J. and F. Tipler. 1984. *The Anthropic Cosmological Principle.* Oxford: Oxford University Press.

Bondi, H. 1960. *Cosmology.* Cambridge: Cambridge University Press.

Bonnor, W. B. and G. F. R. Ellis. 1986. "Observational Homogeneity of the Universe." *Mon. Not. Roy. Ast. Soc.* 218:605-614.

Borner, G. and S. Gottlober, eds. 1997. *The Evolution of the Universe.* New York: Wiley.

Bothun, Greg. 1998. *Modern Cosmological Observations and Problems.* Taylor and Francis.

Brans, C. and R. H. Dicke. 1961. "Mach's Principle and a relativistic theory of gravitation." *Phys. Rev.* 124:925.

Carr, B. J. and M. J. Rees. 1979. *Nature* 278:605.

Coles, P. and G. F. R. Ellis. 1997. *Is the Universe Open or closed? The Density of Matter in theUniverse.* Cambridge: Cambridge University Press.

Cornish, N. J, and J. J. Levin. 1997. *Phys. Rev. Lett.* 78:998.

Cornish, N. J., D. N. Spergel and G. D. Starkman. 1996. *Phys. Rev. Lett.* 77:215.

Cousins, R. D. 1995. "Why isn't every Physicist a Bayesian?" *Am. Journ. Phys.* 63:398-410

Cowie, L. and A. Songaila. 1995. "Astrophysical limits on the evolution of dimensionless physical constants over cosmological time." *Astrophys. Journ.* 453:596.

Davies, P. C. W. 1974. *The Physics of Time Asymmetry.* London: Surrey University Press.

———. 1982. *The Accidental Universe.* Cambridge: Cambridge University Press.

Davies, Paul. 1992. *The Mind of God.* Simon and Schuster.

Dirac, P. A. M. 1938. "New basis for cosmology." *Proc. Roy. Soc.* A165:199-208.

Edwards, A. W. F. 1991. "Bayesian Reasoning in Science." *Nature* 352:386-38.

Ehlers, J., P. Geren and R. K. Sachs. 1995. "Isotropic solutions of the Einstein-Liouville equations." *Journ. Math. Phys.* 9:1344-1349.

Ellis, G. F. R. 1971a. "Relativistic cosmology." In *General Relativity and Cosmology. Proceedings of the XLVII Enrico Fermi Summer School,* ed. R. K. Sachs. New York: Academic Press.

Ellis, G. F. R. 1971b. "Topology and Cosmology." *Gen. Rel. Grav.* 2:7-21.Ellis, G. F. R. 1975. "Cosmology and Verifiability." *Qu. Journ. Roy. Ast. Soc.* 16:245-264.

———. 1979. "The Homogeneity of the Universe." *Gen. Rel. Grav.* 11:281-289.

———. 1984. "Relativistic Cosmology: its nature, aims and problems." Pp. 215-288 in *General Relativity and Gravitation,* eds. B. Bertotti et al. Reidel.

———. 1990. "Innovation Resistance and Change: the transition to the expanding universe." Pp. 97-114 in *Modern Cosmology in Retrospect,* eds. B. Bertotti, R. Balbinto, S. Bergia and A. Messina. Cambridge: Cambridge University Press.

———. 1991. "Major Themes in the Relation between Philosophy and Cosmology." *Mem. Ital. Ast. Soc.* 62:553-605.

———. 1993. *Before the Beginning: Cosmology Explained.* London: Bowerdean/Marion Boyars.

———. 1995. "Observations and Cosmological Models." Pp. 51-65 in *Galaxies and the Young Universe*, eds. H. Hippelein, K. Meisenheimer and H.-J. Roser. Springer.
Ellis, G. F. R., D. H. Lyth and M. B. Mijic. 1991. "Inflationary Models with Ω not equal to 1." *Phys. Lett.* B271:52.
Ellis, G. F. R., R. Maartens and S. D. Nel. 1978. "The Expansion of the Universe." *Mon. Not. Roy. Ast. Soc.* 184:439-465.
Ellis, G. F. R. and M. Madsen. 1991. "Exact Scalar Field Cosmologies." *Class. Qu. Grav.* 8:667-676.
Ellis, G. F. R., S. D. Nel, W. Stoeger, R. Maartens and A. P. Whitman. 1985. "Ideal Observational Cosmology." *Phys. Reports* 124:315-417.
Ellis, G. F. R. and G. Schreiber. 1986. "Observational and Dynamic Properties of Small Universes." *Phys. Lett.* A115:97-107.
Ellis, G. F. R. and D. W. Sciama. 1972. "Global and Non-global Problems in Cosmology." Pp. 35-59 in *General Relativity*, ed. L. O'Raifeartaigh. Oxford: Oxford University Press.
Ellis, G. F. R. and W. R. Stoeger. 1987. "The Fitting Problem in Cosmology." *Class. Qu. Grav.* 4:1679-1690.
———. 1988. "Horizons in Inflationary Universes." *Class. Qu. Grav.* 5:207.
Ellis, G. F. R. and T. Rothman. 1993. "Lost Horizons." *Am. Journ. Phys.* 61:93.
Fabian, A. C., ed. 1989. *Origins*. Cambridge: Cambridge University Press.
Garrett, A. J. M. and P. Coles. 1993. "Bayesian Inductive Inference and the Anthropic Cosmological Principle." *Comments Astrophys.* 17:23-47.
Ghosh, A., G. Pollifrone, and G. Veneziano. 1998."Quantum Fluctuations in Open Pre-Big Bang Cosmology." *Phys. Lett.* B440:20-27.
Gibbons, G. W., S. W. Hawking and S. T. C. Siklos. 1983. *The Very Early Universe*. Cambridge: Cambridge University Press.
Goliath, M. and G. F. R. Ellis. 1998. "Homogeneous Models with cosmological constant." gr-qc/9811068. Forthcoming in *Phys. Rev. D* (1999).
Goodman, J. 1995. "Geocentrism reexamined." *Phys. Rev.* D52:1821.
Gott, J. R. 1982. "Creation of Open Universes from de Sitter Space." *Nature* 295:304.
Gott, J. R. and Li-Xin Li. 1997. "Can the Universe Create Itself?" *astro-ph./9712344* [this contains references to most of the other work of this kind].
Grunbaum, A. 1989. "The pseudo problem of creation." *Philosophy of Science* 56:373-394 [reprinted in (Leslie 1998)].
Gunn, J. E., M. S. Longair and M. J. Rees. 1978. *Observational cosmology: Saas-Fee Course* 8. Switzerland: Geneva Observatory.
Guth, A. H. 1981. *Phys. Rev.* D 23:347.
———. 1997. *The Inflationary Universe: The Quest for a new Theory of Cosmic Origins*. Addison Wesley.
Harris, W. E., P. R. Durell, M. J. Pierce and J. Seckers. 1998. "Constraints in the Hubble Constant from Observations of the Brightest Red Giant Stars in a Virgo-Cluster Galaxy." *Nature* 395:45.
Harrison, E. R. 1981. *Cosmology: The Science of the Universe*. Cambridge: Cambridge University Press.
Harwit, M. 1984. *Cosmic Discovery*. MIT Press.
Hawking, S. W. 1993. *Hawking on the Big Bang and Black Holes*. World Scientific.
Hawking, S. W. and G. F. R. Ellis. 1973. *The Large Scale Structure of Space-time*. Cambridge: Cambridge University Press.
Hogan, C. J. 1998. *The Little Book of the Big Bang*. Springer/Copernicus.
Hoyle, F. 1960. "Cosmological tests of gravitational theories." In *Rendiconti Scuola Enrico Fermi. XX Corso*. New York: Academic Press.
Hu, W. and N. Sugiyama. 1995. *Phys. Rev.* D 51:2599; *Astrophys. Journ.* 444:489.
Isham, C. J. 1997. *Lectures on Quantum Theory: Mathematical and Structural Foundations*. Imperial College Press.
Kamionkowski, M. and A. Loeb. 1997. "Getting around cosmic variance." *Phys. Rev. D*.
Kolb, E. W. and M. S. Turner. 1990. *The Early Universe*. Addison Wesley.
Kristian, J. and R. K. Sachs. 1966. "Observations in cosmology." *Astrophys. Journ.* 143:379.
Lakatos, Imre. 1980. *The Methodology of Scientific Research Programmes*. Cambridge: Cambridge University Press.

Leslie, J. 1989. "Cosmology: a philosophical survey." In *Universes*. Routledge. First published in *Philosophia* 24 (1994):3-27.

———. ed. 1998. *Modern Cosmology and Philosophy*. Prometheus Books.

Lidsey, J. E., A. R. Liddle, E. W. Kolb, E. J. Copeland, T. Barriero and M. Abney. 1997. "Reconstructing the Inflation potential—an overview." *Rev. Mod. Phys.* 69:373-410.

Linde, A. D. 1990. *Particle Physics and Inflationary Cosmology*. Harwood Academic.

Matravers D. R., G. F. R. Ellis and W. R. Stoeger. 1995. "Complementary approaches to cosmology: Relating theory and observations." *Qu. Journ. Roy. Ast. Soc.* 36:29-45.

McCrea, W. H. 1953. "Cosmology." *Rep. Prog. Phys.* 16:321.

———. 1960. "The Interpretation of Cosmology." *Nature* 186:1035.

———. 1970. "A philosophy for big bang cosmology." *Nature* 228:21.

Meyer, D. M. 1994. "A Distant Space Thermometer." *Nature* 371:13.

Misner, C. W. 1968. "The Isotropy of the Universe." *Astrophys. Journ.* 151:431

———. 1969. "The Mixmaster Universe." *Phys. Rev. Lett.* 22:1071.

Munitz, M. K. 1962. "The Logic of Cosmology." *British Journ. Philos. Sci.* 13:104

———. 1986. *Cosmic Understanding: Philosophy and Science of the Universe*. Princeton: Princeton University Press.

Mustapha, N., C. Hellaby and G. F. R. Ellis. 1998. "Large Scale Inhomogeneity vs Source Evolution: Can we distinguish them?" *Mon. Not. Roy. Ast. Soc.* 292:817-830.

Nilsson, J. S. et al., eds. 1991. *The Birth and Early Evolution of our Universe. Proceedings of Nobel Symposium 19*. World Scientific.

Padmanbhan, T. 1993. *Structure Formation in the Universe*. Cambridge: Cambridge University Press.

Partridge, B. 1995. *3K: The Cosmic Microwave Background Radiation*. Cambridge: Cambridge University Press.

Peacock, J. A. 1999. *Cosmological Physics*. Cambridge: Cambridge University Press.

Peebles, P. J. E. 1971. *Physical Cosmology*. Princeton: Princeton University Press.

———. 1993. *Principles of Physical Cosmology*. Princeton: Princeton University Press.

Peebles, P. J. E., D. N. Schramm, E. L. Turner and R. G. Kron. 1991. "The Case for the Relativistic Hot Big Bang Cosmology." *Nature* 352:769-776.

Penrose, R. 1989a. "Difficulties with Inflationary Cosmology." *Ann. New York Academy of Science*.

———. 1989b. *The Emperor's New Mind*. Oxford: Oxford University Press, ch. 7.

Perlmutter, S. et al. 1998. "Discovery of a Supernova Explosion at Half the Age of the Universe." *Nature* 391:51.

Rees, M. J. 1995. *Perspectives in Astrophysical Cosmology*. Cambridge: Cambridge University Press.

———. *Before the Beginning: Our universe and others*. Simon and Schuster/Touchstone.

Rindler, W. 1956. "Visual horizons in World Models." *Mon. Not. Roy. Ast. Soc.* 116:662 [Despite its name, this paper actually deals with causal horizons].

Robertson, H. P. 1933. Relativistic Cosmology. *Rev. Mod. Phys.* 5:62-90.

Rothman, A. and G. F. R. Ellis. 1986. "Can Inflation Occur in Anisotropic Cosmologies?" *Phys. Lett.* B180:19-24.

Roukema, B. F. 1996. *Mon. Not. Roy. Ast. Soc.* 83:1147

Roukema, B. F. and A. C. Edge. 1997. *Mon. Not. Roy. Ast. Soc.* 292:105.

Sandage, A. 1961. "The Ability of the 200-inch Telescope to Distinguish between Selected World Models." *Astrophys. Journ.* 133:355.

Sandage, A., R. G. Kron and M. S. Longair. 1993. *The Deep Universe: Saas-Fee Advanced Course 23*. Berlin: Springer.

Schramm, D. N. and M. S. Turner. 1998. "Big-Bang Nucleosynthesis Enters the Precision Era." *Rev. Mod. Phys.* 70:303-318.

Sciama, D. W. 1971. "Astrophysical Cosmology." Pp. 183-236 in *General Relativity and Cosmology*, ed. R. K. Sachs. New York: Academic Press.

———. 1993. "Is the Universe Unique?" In *Die Kosmologie der Gegenwart*, eds. G. Borner and J. Ehlers. Serie Piper.

Silk, J. 1997. *A Short History of the Universe*. Scientific American Library.

Smolin, L. 1992."Did the Universe Evolve?" *Class. Qu. Grav.* 9:173-191.

Steinhardt, P. J. 1995. *Int. Journ. Mod. Phys.* A10:1091.

Stoeger, W. R., G. F. R. Ellis and C. Hellaby. 1987. "The Relationship between Continuum Homogeneity and Statistical Homogeneity in Cosmology." *Mon. Not. Roy. Ast. Soc.* 226:373.

Stoeger W., R. Maartens and G. F. R. Ellis. 1995. "Proving Almost-Homogeneity of the Universe: An almost-Ehlers, Geren and Sachs theorem." *Astrophys. Journ.* 443:1-5.

Tegmark, M. 1993. *Ann. Phys.* 270:1-52. [see http://www.hep.upenn.edu/~max/toe.html].

Tipler, F. J., C. J. S. Clarke and G. F. R. Ellis. 1980. "Singularities and Horizons: A Review Article." In *General Relativity and Gravitation*, ed. A. Held. Plenum Press.

Tolman, R. C. 1934. *Relativity, Thermodynamics, Cosmology.* Oxford: Clarendon Press.

Tryon, E. P. 1973. "Is the Universe a Quantum Fluctuation?" *Nature* 246:396.

Turok, N., ed. 1997. *Critical Dialogues in Cosmology.* World Scientific.

Wainright, J., A. A. Coley, G. F. R. Ellis and M. Hancock. 1998. "On the Isotropy of the Universe: Do Bianchi VIIh universes isotropize?" *Class. Qu. Grav.* 15:331.

Wainwright, J. and G. F. R. Ellis, eds. 1996. *The Dynamical Systems Approach to Cosmology.* Cambridge: Cambridge University Press.

Weinberg, S. W. 1972. *Gravitation and Cosmology.* New York: Wiley.

Will, C. M. 1979. "The Confrontation between Gravitational Theory and Experiment." In *General relativity: an Einstein Centenary Survey*, eds. S. W. Hawking and W. Israel. Cambridge: Cambridge University Press.

Yoshimura, M. 1988. "The Universe as a Laboratory for High Energy Physics." In *Cosmology and Particle Physics.* Proceedings of the CCAST (World Laboratory) Symposium ..., eds. Li-Zhi Fang and A. Zee. New York: Gordon and Breach.

Zeh, H. D. 1992. *The Physical Basis of the Direction of Time.* Berlin: Springer Verlag.

LEE SMOLIN

TIME, STRUCTURE AND EVOLUTION IN COSMOLOGY

1. INTRODUCTION

The primary task of theoretical physics is to understand why the world is arranged the way we find it, and not otherwise. As such, we invent theories in which we take some aspects of the world and some principles to be fundamental, and try to understand how everything else can be understood in terms of them. As Einstein emphasized, we have a free choice of which elements of reality and which principles we choose to be fundamental, and which secondary (Einstein 1934). Because of this it can and does happen that at certain steps in the development of science we find it convenient or useful to choose very different starting points, from which very different things can be thought of as fundamental.

Throughout the development of science, from Ptolemy through quantum mechanics, one of the aspects of nature that was always taken to be fundamental is time. The notion of time that underlies every theory in physics and astronomy from the Greeks through special relativity and quantum mechanics is the idea that it is meaningful to think of the universe as being described by a configuration or a state, that describes the way things are at a fixed notion of time. From this point of view, the task of a physical theory is to give the laws that describe how this configuration or state changes in time.

There are two great traditions in physics and philosophy about the nature of time (Barbour 1989b). The one, identified with Newton, is that time is absolute, by which is meant that the properties of time are an intrinsic and fixed aspect of reality that are quite independent of what there is in the world. The second, identified with Leibniz, is that time is only an aspect of the relations of things in the world, and that its properties, if not its very existence, reflect contingent features of the organization of matter in the world.

The Newtonian view shaped the development of physics and cosmology from the 17th Century through the development of quantum mechanics early in the 20th Century, while the Leibnizian view is closely connected with the other great development of 20th Century physics, which is Einstein's theory of general relativity. At the present time, the great problem of theoretical physics is how to combine general relativity and quantum mechanics into one unified theory that could serve as a single framework within which to organize our entire present understanding of nature.

This is a problem that has great implications for our understanding of time, as it involves a confrontation between the Newtonian, absolute, view of time that we find in quantum mechanics and the Leibnizian, relational, view we find in general relativity. My main task in this essay is to explain the roots of this confrontation, and offer some reflections about how it might be resolved.

I will argue here that most of the characteristics of time that have been, since at least the time of Newton, assumed to be fundamental will have to be given up by any theory that unifies general relativity and quantum mechanics. We will see that it is even possible that a case can be made for giving up entirely the idea that time is a fundamental aspect of nature.

The idea that time is an intrinsic and necessary aspect of reality is a very difficult one to give up. One reason for this is that our own conscious experience is rooted so profoundly in a sense of the flow of time and in the sense of the continuity of our own identities over time. Even so, I will argue that there is a strong reason, coming from the problem of joining quantum theory and relativity, why we may have to give up the idea that time is one of the fundamental aspects of the organization of the world.

However, this will not be the end of my argument. For, the main goal of this essay is to propose that there is a notion of time which may play a fundamental role in a theory unifying quantum theory and general relativity. But it is a notion of time which is rather different from what we are used to in physics, as the notions of structure, complexity and evolution are fundamental to it.

I want to make it clear, especially as I am addressing an audience which is primarily made up of philosophers, that I am writing as a theoretical physicist. Even so, this essay is meant to be a contribution to a philosophical tradition, which is that, founded by Leibniz (1973), that seeks to found a picture of the universe completely on a relational concept of space and time. In particular, there are three main comments that I hope to contribute to the discussion of the implications of Leibniz's concepts of space and time.

1) The debate between Newton and Leibniz over whether time (and also space) has an absolute, pre-existing character, or is only an aspect of the relation of things (*ibid.*), is, as I mentioned above, very much with us. I will argue that the central trajectory of twentieth century physics is to replace the Newtonian view of space and time with something very close to Leibniz's original vision. I further will try to argue that many of the main issues that presently confront attempts to make a quantum theory of gravity are tied up with the problem of constructing a Leibnizian view of time. This is true, further, of many of the technical issues, as well as of the conceptual issues.

Beyond its role in defining the problematic, I believe that the Leibnizian vision of time has implications that have not been, so far, strongly appreciated (although they may not be surprising for anyone who has read and thought about the Mondadology.) Among these are:

2) The Leibnizian notion of time only makes sense in a structured universe. This is because without structure, variables which are defined purely by the relations between things become ill defined. This means that any theory of cosmology based on relational notion of space and time must explain why the universe is, in fact, structured

rather than, for example, consisting only of a homogeneous gas in thermal equilibrium.

3) The only detailed and mechanistic theory we have of how structure is generated in nature is Darwin's theory of natural selection. Because it is based on a theory of self-organization and structure formation, time plays a rather different role in the theory of evolution, and in biology in general, than it does in physics. I will close by suggesting that there are reasons to believe that when the problem of quantum gravity is finally solved and we have a consistent, verifiable and believable cosmological theory based on it, the physicists conception of time will look, in some ways, like the notion of time that arises from the theory of natural selection.[1]

This essay is organized in eleven sections. In the next three sections, I discuss the notion of time in, successively, Newtonian mechanics, quantum mechanics and special relativity. In section 5 I give a general discussion of the problem of constructing cosmological theories which is followed by sections 6 and 7 which describe, respectively, the way in which general relativity is based on a relational concept of space and time and the particulars of the way time is described in general relativity. The problem of time in a theory that would combine relativity and quantum theory is treated in section 8; it is here that I give an argument that any notion of time coming out of such a theory must be based on the notion of structure. Section 9 is about various aspects of the problem of structure in the context of cosmology, which serve as preparation for section 10 where a notion of a cosmological time based on the concepts of structure, complexity and evolution is proposed. Concluding remarks are in section 11.

2. TIME IN NEWTONIAN PHYSICS

Much has been written about the concept of time in Newton's *Principia*; here I only want to summarize certain key points in order to set the stage for the later discussions. I would like to discuss four elements of the concept of time in Newtonian physics: its geometrization, its spatialization, its universality and its absolute character.

Perhaps the most interesting of these aspects of Newtonian time is its geometrization. This is the representation of time as being in correspondence to the real number line. This means two things. First, an identification is made between the points of the real number line and the "instants" of time. By means of this correspondence, time is supposed to acquire all of the properties of the real continuum. If we believe this then it seems we must believe that it is meaningful to say that each interval of time contains an uncountable number of instants.

I think it is important to reflect that while the concept of an interval of time has an operational meaning, that corresponds precisely to something that is measurable, it is difficult to know exactly what in nature an instant of time — corresponding to a mathematical point — is supposed to be. For any time fixed by a real clock comes with an uncertainty — and is therefore more like an interval than an instant.

Indeed, one does not have to be a strict operationalist to wonder whether there is anything in nature corresponding to an instant of time. The difficulty, however, with criticizing the correspondence between time and the mathematical continuum is that it seems to be very hard to think of convincing alternatives. Of course, the simplest alternative is that time consists of a set of moments and that the set of such moments is discrete — so that like in a computer, real time consists of a chain of static moments, where causality is expressed by saying that the state of the system at one moment is completely determined, by some law, by the state at the previous moment.

The problem with this view is not only that it is at least as absolute as the Newtonian one. A grave problem is that the simplest attempts to construct theories based on the idea of a discrete time fall apart as soon as one attempts to incorporate the relativity of simultaneity from special relativity. For, in whose reference frame are the discrete moments of simultaneity defined?

It seems to me that what is missing in the concept of a moment of time is an explicitly relational element. In such a picture, a moment of time would have a complexity related to the complexity of the entire universe. I will have more to say about this later.

Closely connected with the geometrization of time is its spatialization. Let us consider a particle moving on a line. When we describe its trajectory by a function $x(t)$ we are implicitly making use of the assumption that both time and space are to be represented by the real number line. This makes it natural to describe the trajectory by a curve in R^2. Sometimes I wonder whether this perhaps isn't the original sin behind some of the difficulties of quantum gravity, for this invented two dimensional space is the first primitive appearance of a *space-time continuum*. This representation of physical reality in terms of a spacetime ascribes to the whole set of events occuring over time many of the properties that we, intuitively, ascribe to space; this is what I mean by the spacialization of time.

Because it seems so natural a step, it is perhaps a shock to recall that, like all mathematical representations in physics, both the continuum R^2 in which we draw the trajectory and the trajectory itself are inventions. They may correspond to something in nature but at the beginning, at least, they are fictions. How astounding fictions they are can be gauged by the length of time it took humanity to invent them.[2] Indeed, perhaps we should reflect deeply before we take this step, because what we are doing is asserting that, at least as far as physics is concerned, both space and time are to be represented by different copies of the same mathematical object. It is a wonderful trick, and it has led to many wonderful things. But, perhaps it really is a trick for, in nature, in our experience, or even in experimental procedure, besides the fact that it is convenient to represent the results of measurements of intervals of both by numbers, is it really the case that space and time have identical, or even very similar, properties?

It is because of the geometrization and the spacialization of time that the step from Aristotle's physics to Newton's physics was much larger than the step from there to Einstein, who, apparently, accepted both of these postulates.

We now come to those aspects of Newtonian time that were modified by relativity theory.

The first of these is the postulate of the universality of the rate of flow of time. This is the assumption that all clocks measure time to flow at the same rate, regardless of where they are and what their states of motion are. This postulate is, as we know now, simply false; it has been shown to be so by experiments that demonstrate that both location with respect to the gravitational field and relative velocity influence the flow of time.

The failure of the postulate of universality was, needless to say, an enormous blow for the classical notion that the world can be described by giving its state at one moment, where the meaning of a moment was a fundamental and universally agreed upon notion. In my opinion, although we have learned in classical physics how to deal with the loss of this idea, we are still struggling to understand the implications it has for quantum physics, which is, conventionally based on the notion of the state of the system at a particular time.

The last basic aspect of Newtonian time is its absolute character. By this is meant the postulate, made explicitly by Newton in the famous Scholium to the *Principia*, that time flows at the same rate, independently of what is in the universe or how it is moving (Newton 1962). According to this notion, it makes sense to talk of the flow of time in an empty universe, or in a universe in which nothing is moving and no change is taking place.

As I mentioned in the introduction, the main theme of this essay is that this point of view is wrong, and cannot be maintained in any theory of cosmology that incorporates relativity and quantum mechanics.

3. TIME IN QUANTUM MECHANICS

Many things about our understanding of nature were altered radically as a result of the transition from classical mechanics to quantum mechanics. Time was not one of them. As I will now try to explain, the notion of time in conventional quantum mechanics is, in all important respects, identical to the Newtonian notion.

The key reason for this is that quantum mechanics was invented to be an extension, or generalization of Newtonian mechanics. More explicitly, the mathematical structure of quantum mechanics mirrors precisely the structure of Hamiltonian mechanics.[3]

It will be useful to mention here several of the points of correspondence between quantum mechanics and Hamiltonian mechanics, paying attention to the role of time in each. In each case the system is described by giving a state, which is supposed to describe anything that can be determined about the system at a particular time. Operationally, what is meant by time in each case is the reading of a clock which is on the wall of the laboratory or on the wrist of the experimenter. Whether the state is a quantum state in Hilbert space or a point in the classical phase space,[4] it depends on the variables of the system *plus* the time variable. As many observers have noted, time thus plays role in these formulations quite different than that played by positions, or other observables. In the quantum description there is no operator for time; this mirrors the fact that in the classical description it is not a coordinate of the phase space.

The reason time is not represented by either a classical observable or a quantum operator is because what is meant by time in these theories is not to be found within the dynamical system described by the notion of a state. *Instead, time in these theories corresponds to something in the world outside of the system being studied.*

In both the classical and quantum cases there is a dynamical equation, which describes how the state of the system evolves with respect to that external time. In the quantum case, this is the Schroedinger equation,

$$i\hbar \frac{d}{dt}|\Psi, t> = \hat{H}|\Psi, t>\tag{1}$$

while in the classical case it is the Hamilton's equations,

$$\frac{dq(t)}{dt} = \frac{dH}{dp},\tag{2}$$

$$\frac{dp(t)}{dt} = -\frac{dH}{dq}.\tag{3}$$

In each case H represents what is called the Hamiltonian. In the quantum case it is an operator and in the classical case it is a function on the phase space. In each case it is the object responsible for generating the evolution of the system in time, where time here is always referring to something outside of the system. In a real and practical sense, because of its role in specifying how a system changes in time, the Hamiltonian is part of the definition of the time of a system. More precisely, the Hamiltonian is the object that tells us how the system changes with respect to time as measured by that clock on the wall of the laboratory, or otherwise outside of the system.

By means of its connection to Hamiltonian dynamics, time in quantum mechanics inherits each of the four characteristics of time in Newtonian physics. Nothing is changed because, as time in Newtonian mechanics referred to something outside of the system being studied, time in quantum mechanics must do the same. The formal structure is the same and so is the experimental situation. Indeed, in each case, the treatment of time reflects the true situation of the experimenter. For, except for the special case of cosmology, which I will come to shortly, it is indeed the case that the clock is not a part of the systems we describe by means of both classical and quantum mechanics.

4. TIME IN SPECIAL RELATIVITY

There is no step in the development of physics that is quite as surprising, or as unprecedented, as the notion of time in Einstein's special theory of relativity. While there was a long tradition of advocacy for dropping the notion of absolute time-which I will be discussing in the next section-to my knowledge there was no preparation in the philosophical literature for what Einstein actually did in 1905. This was to keep the absolute notion of time while dropping its universal aspect. What I mean by this is that in special relativity different observers measure time to flow at different rates, and the differences between them are functions of their states of motion. Thus, in special relativity there is no longer a single notion of time that can be identified with, for example, the time of God. However, it must be stressed that in the sense in which Newton and Leibniz used the term, time is no less absolute in special relativity than it is in Newtonian physics. This is because the time measured by any observer is determined only by the geometry of spacetime and the trajectory of that observer. It is completely independent of whether there is anything else in the universe besides that observer, or, indeed, of any physical property of that observer.

So in special relativity the geometry of spacetime retains its absolute character. It is given a priori by specifying the geometry to be a Minkowski spacetime. This fixes both the local and the topological properties of time. This geometry is independent of what the matter and fields may be doing, and is not arrived at by solving any dynamical equations. The only thing that changes is that there is no single notion of the flow of time in a Minkowski spacetime-different observers measure time to flow at different rates.

Because in special relativity time remains an absolute property of the fixed geometry of spacetime, it turns out to be straightforward to generalize quantum mechanics to special relativistic systems. This generalization is known as quantum field theory, and it is the basis of our understanding of all known fundamental particles and fields with the important exception of the gravitational field. How it is done is the following: In a quantum field theory there are many Hamiltonian operators: one for each inertial observer allowed in the Minkowski spacetime. There are then many Schroedinger equations of the form of (1), each describes how the fields evolve with respect to the time of one of these observers. In terms of the measurement theory and interpretation, everything is essentially as in the special relativistic case: all the possible observers are assumed to be outside the quantum system under study, so that the properties of their clocks are exactly the same as in special relativity.

It is, of course, necessary to specify the relationship between these different Hamiltonian operators. This can be done directly, because there is a fixed transformation group that governs the relationships between observations made by different observers in Minkowski spacetime-the Poincare group. What happens in quantum field theory is that the Poincare group is realized by a group of unitary transformations acting on the state space of the system. Thus the transformations between the quantum measurements made by different intertial observers is treated exactly as in the case of measurements made by observers that differ by simple translations from each other. Because the geometry of spacetime is fixed a priori, there is no essential difficulty in treating time in quantum field theory.

I might mention that there is one problem that has attracted the attention of some people, which is the problem of how the relativity of simultaneity can be made consistent with the notion of collapse of the wavefunction. This collapse is a feature of some, but not all, measurement theories for quantum mechanics. If one believes that collapse of the wavefunction is something that really happens, then there is certainly a problem. One must either specify in what frame this takes place or find a way to understand collapse of the wavefunction in a way that is invariant under Poincare transformations between frames. On the other hand, if one believes that collapse of the wavefunction is only a convenient fiction, but is not something that really happens, there is no need to solve this problem. My personal opinion is that the existence of this problem is probably a strong reason to take a point of view about quantum mechanics that does not require one to believe in the reality of the collapse of the wavefunction.

I would like to close the discussion of time in special relativity by mentioning that many attempts to realize a relational notion of time in the twentieth century find their most difficult challenge to be the problem of incorporating the relativity of simultaneity. There have been a number of such attempts. Most of them have been also attempts to construct a hidden variable theory or a quantum theory of gravity- among these I can mention the work of David Finkelstein (1989, 1992), the spin networks of Roger Penrose (1971, 1979a) and the causal sets of Rafael Sorkin and collaborators (Bombelli et al. 1988). There have also been explicit attempts to construct a classical dynamics consistent with the notion of relational time-the best known of this being the work of Barbour and Bertotti (Barbour 1974, 1975; Barbour and Bertotti 1977).[5]

The one exception to this is general relativity. As I will describe shortly, the great achievement of Einstein concerning time was to find a way to incorporate a relational notion of time in a theory that at the same time incorporated and generalized the relativity of simultaneity of special relativity.

5. THE PROBLEM OF THE CONSTRUCTION OF COSMOLOGICAL THEORIES

Before describing the notion of time in general relativity I would like to make some general observations about the development of twentieth century physics to set the stage for what I will be saying during the rest of this essay. Philosophers and historians of science often assert that general relativity constituted a scientific revolution, as it brought us from one paradigm concerning space and time to another one. There is, however, a problem with this view, which makes me doubt that it is very useful to think that general relativity represents, by itself, a complete scientific revolution. The problem is that in the first third of this century there were two such revolutionary developments-relativity *and* quantum mechanics-each of which separately, and in different ways, overthrew Newtonian physics. The problem is that while relativity and quantum mechanics are each successful in their own domain, we cannot say that the result of either revolution is a new comprehensive theory of physics. The reason is that what is wanted is one theory, not two. It is a far from trivial problem to construct either a conceptual picture of nature, or a consistent mathematical theory that incorporates both relativity and quantum mechanics. To do so is the problem of quantum gravity, which I, and many other theoretical physicists, work on. It is, to put it mildly, a difficult problem.

Many people who have worked on it express frustration with the difficulty of the problem of quantum gravity. However, I believe that this frustration comes from a misunderstanding of what is at stake. The problem of quantum gravity is not a problem of constructing a theory to describe a particular phenomena. It is instead a problem on which rests the basics of our understanding of all of nature. History tells us that scientific revolutions of this magnitude-those that involve the overthrow of a comprehensive theory of nature such as Newton's and its replacement with an equally comprehensive theory-take a very long time. The last time this occurred, which we refer to as the Copernican revolution, it took at least 140 years, counting from the publication of Copernicus's *Revolutionibus* to the publication of Newton's *Principia*. I believe that we are in the midst of a revolution in our conception of nature of this magnitude, and that, rather than being completed revolutions on their own, general relativity and quantum mechanics each represent steps in that revolution. Other steps are occuring now, among these are the tremendous expansion in our understanding of astrophysics and cosmology and the introduction of gauge theories in particle physics. The problem of quantum gravity is, I believe, such a difficult problem because it is exactly the problem of constructing the great synthesis that will complete the revolution.

Arthur Koestler called his book on the Copernican revolution *The Sleepwalkers* (1959) to highlight his observation that participants in such great revolutions often do not know where it is they are going. Indeed, as Koestler describes, the key ideas that afterwards will be said to describe the transition from the old theory to the new theory can be completely outside the thoughts of the main participants. For example, in the case of the Newtonian universe those key ideas include the assertions that the universe is infinite, with no preferred notion of place, that the sun is a star and that everything in the universe is governed by one set of deterministic laws of motion. It is doubtful that either Copernicus, Galileo or Kepler had a conception of the universe that included any of these ideas. It is, indeed, still a bit embarrassing that one of the few sixteenth century figures who did was Bruno.

I believe that we are now in a position very much like that of Galileo, Kepler and their contemporaries. We are far enough away from the beginning of the revolution to know that the old, Newtonian picture is finished. But we are not yet close enough to the end to glimpse, with any certainty, what the key ideas of this revolution will, after it is all over, be seen to be.

One of the things I would like to do in this essay is make some tentative suggestions as to what this current revolution may be about. I put these forward fully knowing that this is a risky business. However, I do so with the belief that it could be useful to try.

In my opinion, the key problem which is behind the current revolution is the problem of trying to construct a theory that could stand as a description of a single, closed universe. All successful existing theories (with the exception of general relativity) are, I would like to argue, necessarily descriptions of only a portion of a universe. That is, each of these theories has mathematical and conceptual elements that necessarily point to the existence of things outside of the system that is described by the theory. In (Smolin 1991) I have argued this point in detail. The argument, put very briefly, is the following: First, in each case, there are mathematical structures that are fixed a priori and are not subject to any dynamical law. These include the fixed spacetime geometries of Newtonian physics, special relativity and both classical and quantum field theory. Also included in this is, I believe, the fixed Hilbert structures and operator algebras of quantum mechanics.

Second, in order that each of these theories have a coherent interpretation, such fixed structures must be taken as referring to the existence of things outside of the system being studied. The paradigmatic arguments for this are Leibniz's and Mach's criticisms of the concept of absolute space and time in Newtonian mechanics (Leibniz 1973; Mach 1866, 1893). The fixed structure of inertial frames, as Mach argues, must be distinguished for some reason. A key role in all such arguments is played by Leibniz's principle of sufficient reason or an equivalent assertion that every choice made by nature must have a reason in terms of perceptible phenomena. This reason can only be something outside of the mechanical system being studied-Mach's guess was the distribution of matter in the universe as a whole.

In quantum mechanics, as usually interpreted, both the operators and the inner product refer implicitly to things that are outside of the system described by the quantum state. There has been an argument going on since Everett's paper of 1957 as to whether quantum mechanics can be consistently reinterpreted so as to make no reference to such outside systems (Everett III 1957, 1973; Wheeler 1957; Geroch 1984). Although the argument is too long to be given here.[6] I believe that the answer is no.

These fixed structures may be called background structures.[7] A theory with background structures may be an important and necessary step in the development of physics. Thus, it is quite possible that Newtonian physics and quantum mechanics were necessary steps in the development of physics. However, in the twentieth century we have discovered that the universe as a whole is a dynamical entity and we have taken on the task of constructing a theory that could stand as a description of a single, closed and dynamical universe. The problem is that a theory that relies on background structures cannot be a theory of the whole universe. If we are to succeed then we must learn how to construct a theory that does not rely on such background structures but still has a coherent and useful interpretation.

I believe that many of the key problems that face us presently in theoretical physics are related directly to this theme. These include most of the problems-both technical and conceptual-faced by attempts to construct a quantum theory of gravity. Beyond this, I believe that a number of other problems that on their face seem unrelated to cosmology will only be understood after we have solved the problem of how to construct a sensible theory of a single, closed universe. These include the problem of the interpretation of quantum mechanics, the problem of how the parameters of the standard model of particle physics are chosen and the problem of galaxy formation and large scale structure formation in the universe.

Let me now return to the subject of this essay, which is the problem of time. Seen against this background, I believe that the key to the problem of time is the construction of a theory that incorporates a relational notion of time, as advocated by Leibniz. Indeed, if we leave aside the problem of quantization, we have several examples of such theories. For general relativity itself is, as Stachel (1989), Barbour (1982, 1987, unpublished ms.), Rovelli (1990, 1991a, 1991c, 1991d) and others have argued, a perfect realization of a relational concept of time. In addition to this, we have a number of model systems, such as that invented by Barbour and Bertotti (Barbour 1974, 1975; Barbour and Bertotti 1977, 1982), that are very useful in understanding the more complicated structure of general relativity.

I thus now turn a discussion of the notion of time in general relativity, as it illustrates both how a relational notion of time can be incorporated in a mathematical theory and the difficulties that are raised by doing so. Indeed, exactly because it is based on a relational notion of time, classical general relativity involves several interpretational difficulties that have only very recently been clarified.

After this I will turn to the main problem currently facing theoretical physics which is connected with the notion of time. This is the problem of the meaning of time in a quantum theory of gravity or cosmology. As I will try then to show, the necessity of giving time a relational or Leibnizian formulation is the key to many of the difficulties facing quantum cosmological theories. Then, in the remaining sections of this essay,

I will argue that the solution to the problem of time in quantum cosmology may lie in a rather different direction than it is usually sought.

6. GENERAL RELATIVITY AS A THEORY BASED ON A RELATIONAL NOTION OF SPACE AND TIME

In the following discussion, I will restrict myself to the case of general relativity applied to cosmology. This means that we assume that the topology of space is closed so that there are no spatial boundaries and no resulting necessity to supplement the equations of the theory with boundary conditions.[8]

For the following it will be useful to explain how, from a mathematical point of view, general relativity in the cosmological context explicitly incorporates a relational notion of space and time. The key point is to understand the role of what is called *diffeomorphism invariance* in the theory. I will now give a nontechnical introduction to this idea.

Let us begin by splitting the mathematical structures involved in the description of space and time into fixed and dynamical structures. The dynamical structures are those elements that are subject to dynamical equations-that is they are determined by solving a set of partial differential equations. Further, there are an infinite number of possible solutions to these equations which represent different states and evolutions of the system. The fixed structures are those mathematical elements which are the same for all spacetimes-the background structures that are necessary for the statement of the dynamical equations of the theory.[9]

In general relativity the dynamical structure includes the metric of spacetime, which is the function that tells us how to measure distance intervals, time intervals and so forth. But it is, in a certain sense, less than that, which is what I want now to explain.

To see the point, I must tell you what the fixed structure is. The specification of the fixed structure begins with the definition of what is called a differential manifold (Wald 1984; Spivak 1970). This is a four dimensional continuum that is defined with the standard topology of four dimensional spaces. That is, we know what the points are and we know which are the open sets that constitute the neighborhoods of points. On this continuum we introduce a further structure which is called the differential structure. The specification of the differential structure is, essentially, the specification of the notion of a derivative. To do this we note that, given the topological structure, we may define functions on the manifold. These are simply maps from the manifold to the real numbers. If we introduce the notion of the partial derivative on the manifold, this allows us to define a class of differentiable functions (those whose derivatives exist everywhere.) The whole structure: the set of points, the topology that makes it a four dimensional continuum, and its family of differentiable functions and set of partial derivatives, is called a differential manifold.

Now, the key point is that this structure is *not* what is taken to represent spacetime. The problem is that as it is constructed it assumes that the individual points of the manifold have a physical meaning and this is, from the relational point of view, absurd. For, if we accept the relational point of view we must admit that we do not know *a*-

priori how to physically distinguish one point of spacetime from another. We could only do so by reference to some observable phenomena-some physical event-taking place there. But those physical phenomena have to be specified by dynamical fields.

What we want to do is to introduce the requirement that the correspondence between the mathematical model-this four dimensional differentiable manifold-and physical reality must done in such a way that points of spacetime are distinguished only by the values of the dynamical fields.

This is accomplished-and it is accomplished completely in general relativity-by the notion of diffeomorphism invariance. Let me define what is meant by a diffeomorphism, in this context. A diffeomorphism is any mapping from the four dimensional continuum to itself that preserves only the topological structure and the definition of differentiable functions. That is, the map can be any one which takes each point of the manifold to another point in a way that preserves these two structures: open sets are taken to open sets and differentiable functions are taken to differentiable functions.

Now, what is taken to be the mathematical object that corresponds to a physical spacetime is not a four dimensional differentiable manifold. It is an *equivalence class of four dimensional manifolds in which two manifolds are considered to be equivalent if there is a diffeomorphism from one to another.* This is, admittedly, a somewhat difficult notion, because for any manifold there are an infinite number of such maps. But it is an essential notion, because it forbids us from attributing any physical meaning to a point of the manifold.

Another way of saying this is that the physical interpretation of any mathematical structure in the theory must be left unchanged by the action of all the diffeomorphisms. Such an interpretation can then be considered to apply to a whole equivalence class of manifolds rather than to any single manifold in the equivalence class. This is what is meant by the notion of the diffeomorphism invariance of a theory.

How then do we give a physical interpretation to the mathematical object which is the equivalence class of manifolds by diffeomorphisms. We do it by adding to the structure additional fields. A key postulate of the theory is that all additional fields that come into the theory must be dynamical. They must be determined by the solution to differential equations. The structure of equivalence classes of differentiable manifolds must be the only fixed or background structure in the theory.

Now, let us suppose we add to the structure a set of fields, which may include the gravitational fields and electromagnetic fields. The key point, which makes the notion useful, is that when we make a diffeomorphism, we require that the values of all the physical fields are taken along by the map. Let me illustrate this with an example. Let us suppose that in one manifold there is a point, which I will call p, where the electromagnetic fields are specified by a particular set of values.[10] I will call this particular collection of values of the components of the field $\mathcal{E}$. Now, let us apply a diffeomorphism. That point, p, is sent by the map to a different point, which I will call q. We then impose the requirement that the fields at q, after the diffeomorphism, take exactly the same values that they took at p, before the diffeomorphism.

Now, let me further suppose that the configuration of the electromagnetic fields are sufficiently complicated that the point p is the unique point at which the fields take exactly the values $\mathcal{E}$. That is, before the diffeomorphism we may distinguish the point

p by saying it is the point at which the fields take those values, $\mathcal{E}$. Now, because of the transformation rule I just specified, after the diffeomorphism q will be the unique point at which the fields take exactly the values $\mathcal{E}$.

Here is, finally, the main point: from the point of view of physics, the only thing we can say is that there exists somewhere in spacetime a point where the electromagnetic field takes the values $\mathcal{E}$. We cannot, without introducing extra unphysical background structure, give any meaning to a statement of whether p or q is that point. Both descriptions are allowed, as are, indeed any of the infinite set of descriptions-each involving a manifold and a set of functions, that are related to these by diffeomorphisms. The only thing that can be said to correspond to the physics is the whole equivalence class of manifolds and fields.

I would like to make four general comments about this situation. First, in his 1905 paper on special relativity Einstein introduced the postulate that the points of the spacetime are related to physical events. This is simply no longer true in general relativity. Events can only be specified by the values of dynamical fields and, as long as we stick to our philosophy of making no use of unphysical and unobservable background fields, these can only be described meaningfully in terms of equivalence classes of manifolds under diffeomorphisms. Let me again repeat for emphasis: spacetime points by themselves have no physical meaning in general relativity.

Second, I have given this entire discussion without ever referring to coordinates. It is true that it is often convenient to use a set of mathematical coordinates to write down solutions to the Einstein equations and that, because all of the fields that occur in the equations are dynamical, those equations are invariant under an arbitrary relabeling of the coordinates. But this is not essential to the discussion. General coordinate invariance is not the same thing as diffeomorphism invariance, and it is the latter, and not the former, that is the key to the physical interpretation of the theory.

The idea that the key is general coordinate invariance gives the mistaken impression that the individual points of the spacetime manifold are meaningful and that it is only the coordinates that are used that are arbitrary. This leads to a lot of confusion because it can be asserted- indeed it is true-that with the introduction of explicit background fields any field theory can be written in a way that is generally coordinate invariant. This is not true of diffeomorphisms invariance, which relies on the fact that in general relativity there are no non-dynamical background fields. Diffeomorphisms, in contrast to general coordinate transformations, are active transformations that take points to other points, so that diffeomorphism invariance is, explicitly, the statement that the points are not meaningful. Both philosophically and mathematically, it is diffeomorphism invariance that distinguishes general relativity from other field theories.

Third, the reader may wonder at the complicated way through which we construct, in general relativity, a mathematical structure corresponding to physical reality. Is it really necessary to go through a two step process in which we first define a rather complicated object-the differentiable manifold, and then assert that physical reality corresponds not to it but only to an infinite equivalence class of such objects? Is it really necessary to invent abstract points, put them all together into a continuum, and then banish them? Surely there must be some easier way to arrive at a relational description of nature. It is clear that in this procedure we have a tension which arises from an attempt to combine in one structure the Newtonian, geometricized notion of space and time with the Leibnizian relational notions. Indeed, the procedure can be described as first listen to Newton, then listen to Leibniz, which is a fair summary of its historical origin. I believe that there must be a simpler, more direct and more physical way to arrive at a relational theory of space and time. The task of finding such a construction is, as will be come clear as I continue, closely related to the problem of quantum gravity.

Finally, note in order for me to assert that a physical point was distinguished by the values of the electromagnetic fields there, it is necessary to assume, as I did, that there is a single spacetime point at which the fields take those particular values. If every point of spacetime is to be distinguished by the values of real physical fields, then each must have a unique value of those fields. This means that the field must have no symmetries-for symmetries are exactly operations that take points to points leaving the values of the fields unchanged. Thus, the interpretation of general relativity I have describe breaks down when applied to solutions that have symmetries.[11]

This is an important and troubling point, because essentially all of the explicit solutions to the Einstein equations that we know have symmetries. This is alright as long as we keep in mind that what we are studying when we study solutions with symmetries are models that may be employed for certain particular purposes. But if we take solutions with symmetry too seriously we can be seriously misled as far as questions of physical interpretation are concerned. As I will discuss in the next section, I believe that a certain amount of confusion about the interpretation of solutions in both classical and quantum cosmology is due to this circumstance.

After this discussion, the reader may wonder, if what is physically meaningful in general relativity is equivalence classes of manifolds and fields under diffeomorphisms, how do we describe a physical observable in the theory? Indeed, it is non-trivial to describe the physical observables in general relativity. Further, because of this it is difficult to give a physically meaningful description of time in general relativity. Indeed, as I will describe in the next section, the problem of how to construct physical observables is the key to the problem of time in both classical and quantum cosmological theories based on general relativity.

7. THE NOTION OF TIME IN CLASSICAL GENERAL RELATIVITY

To discuss the notion of time in general relativity we must first understand how to construct mathematical quantities that could correspond to physically meaningful observables.[12] Let us begin with the concept of an observable in a theory, like general relativity, in which the degrees of freedom are represented by fields. Now, in general, an observable in a field theory is simply some function of the physical fields of the theory. In an ordinary field theory, defined, for example, on Minkowski spacetime, any such function corresponds to a quantity that could, at least in principle, be determined by measurements made on the fields. However, in a diffeomorphism invariant theory such as general relativity, most functions of physical fields do not correspond to physical observables. This is because of the basic principle that the physics is meant to correspond not directly to fields, but to equivalence classes of fields under all diffeomorphisms. This means that *to correspond to something physically meaningful a function of the fields must have the property that it is invariant under the action of diffeomorphisms.*

We will then take this as the basic definition of a physical observable in general relativity: a physical observable is any diffeomorphism invariant function of the fields of the theory.

The problem of giving a physical interpretation to general relativity then rests to a large extent on the problem of constructing such physical observables. Before coming to the implication of this fact for the problem of time, it will be helpful if I make some general remarks on this situation.

It follows directly from what I said in the last section that the value of some field at a point is never a physical observable. Instead, physical observables are usually constructed in one of two ways. The first way is by constructing integrals of some quantity over the whole spacetime manifold. For example, when properly constructed, the average of the curvature over all of spacetime is a physical observable. Observables of this kind are not difficult to construct, but it is often difficult to interpret them in terms of measurements made by observers that live within the spacetime.

The second kind of observable is one that is constructed to express a correspondence or a relation between physical fields. Such observables depend for their construction on the fact that points may be distinguished by the values of physical fields. For example, it does not make sense to ask what the value of the scalar curvature is at the point p, because points have no meaning. But it does make sense to ask what the result is of averaging the value of the scalar field over all points at which the electromagnetic fields take the specific value $\mathcal{E}$. In a typical case, in which there is at most one point at which the values $\mathcal{E}$ are realized, this returns the value of the scalar field at that point. Thus, in this case, this observable can be interpreted as yielding the value of the scalar curvature at some point, which has been distinguished physically.

It is typical of such observables that their interpretation can break down for certain field configurations. One way this can happen is if there is no point at which the value $\mathcal{E}$ is realized. The interpretation will also break down if more than one point has the configuration $\mathcal{E}$. In particular, as I already mentioned, the physical interpretation of such observables will always break down if the configuration has a symmetry.

In spite of this danger, such observables are useful because it is the case that they are well defined for almost every field configurations. Roughly, what is meant by the phrase almost every is that in the space of solutions to the field equations, configurations where the interpretation breaks down, because of the presence of a symmetry or for another reason, are very improbable.[13]

It is this second kind of observable that we use to describe observations that we human beings, who live in a small corner of the universe, may make. If we want to describe in the language of general relativity an observation that we make here on earth, we must put into the theory physical fields that describe and label us. One way to do this is to describe the observer by a particle that is coupled to the gravitational and electromagnetic fields. We may then ask physically meaningful questions about the values of fields on the worldline of that particle.

This brings us exactly to the representation of time in the theory. For one thing that we can do is to build a physical clock into the description of our observer. That is, we put in a dynamical variable that describes the hand of that clock carried by the observer. This dynamical variable is coupled dynamically to the gravitational field and may be interpreted as giving the proper time of the observer carrying the clock. It is then possible to define an observable of the following form: What is the value of the scalar curvature on the worldline of the particle at the moment when the clock carried by the observer reads $t = 1$?

Since we can do this for each value that the observer's clock may take, and we can do this for any physical field that the observer may measure at her location, we can define a family of observers that describe the evolution of what that particular observer sees. These are all physical observables, they are invariant under diffeomorphisms and involve only relations among dynamical variables that are coupled to each other by equations of motion.

This is then one way that time can be represented in general relativity: in terms of the values of physical clocks that are put into the system and coupled dynamically to the gravitational field. It must be admitted that this notion of time is physically meaningful and corresponds directly to observations that we know how to make as observers inside a relational universe. There is then only one additional question that must be asked about time in general relativity: is there any meaning that could be ascribed to time in general relativity besides this one?

Let me quickly rule out one possible type of time that might be defined, which is coordinate time. That is, we might imagine that we can define a time coordinate on the spacetime manifold and use this to define time and evolution. Of course, there is nothing that prevents us from defining a time coordinate; by general coordinate invariance, we are free to define and use any time coordinate we would like to. However, such a time cannot have any physical meaning. The reason follows directly from the requirement that any physical quantity must be invariant under diffeomorphisms. Now, given some choice of a time and space coordinates there will be a diffeomorphism that does the following thing: it takes a point with coordinates $(\vec{x}, t)$ and sends it to the point $(\vec{x}, t + 1)$. That is each point is taken to a point with the same values of the space coordinates and with the time coordinate increased by one second. Now, by diffeomorphism invariance, the value of any physical quantity must be unchanged by this operation. But, this means that no physical observable can be seen to evolve with respect to coordinate time.

This point is both confusing and important. It is true that the values of the local fields, as expressed in terms of the coordinates, will certainly evolve as the value of the time coordinate is increased. This evolution is, as in any dynamical system, governed by the equations of motion of the theory. The problem is that the value of any *physical observable* must be unchanged under any such evolution, because this evolution is described by a diffeomorphism. That is, any physical observable must be a constant of the motion with respect to evolution of the fields in terms of the coordinate time.

This circumstance has caused a lot of confusion in the literature of both classical and quantum cosmology. For it seems that if we take coordinate time seriously as a measure of time we are left with the conclusion that in a diffeomorphism invariant theory physical quantities cannot evolve Some people have stated this as a kind of paradox. But there is no paradox. It is just that coordinates simply cannot supply any physically meaningful notion of time.

We have seen that a meaningful notion of time can be given that depends on the readings of real physical clocks. The values of such observables are independent of any choice of coordinates, indeed, the description I gave of one example: the value of the scalar curvature on the world line of the particle at the point when $t = 17$, does not employ any coordinates. Thus, while it may seem counterintuitive, it follows directly that this quantity must be invariant under any changes of the time coordinate, and hence be a constant of the motion as far as evolution of the fields in coordinate time is concerned.

It is useful to understand how this can happen. Given a particular set of coordinates one may write an expression for this quantity in terms of the physical fields. That expression will be rather complicated because it must take into account the fact that where the particle is when its clock reads $t = 17$ is determined by both the equations of motion that describe its interaction with the gravitational field and where it was at some earlier, initial time. The reason why the whole expression is, in the end, independent of the time coordinate is that both the description of the coupling of the particle to the field and the dependence on the initial conditions will change as the time coordinate is changed. These changes can and do cancel each other, leaving an expression that is independent of the time coordinate.

A rather different approach to the problem of time has been attempted by a number of relativists. This is to uncover in the equations of general relativity some other, more intrinsic notion of time.[14] That is, they have hypothesized that there could be a function of the dynamical variables that describe the gravitational field itself-some function of the metric and curvatures, that could be taken to be a time. The conjecture is that if such a quantity could be isolated the Einstein equations would reduce to a simpler set of equations which could be interpreted as saying that the other dynamical quantities were evolving with respect to this particular quantity, considered as time.

There are two problems with such a proposal. The first is that if one such an intrinsically preferred dynamical variable were discovered there might be no guarantee that there were not others. That is, if the certain components of the gravitational field can be isolated and called, for some reason intrinsic to the equations, time, why should this choice be unique? In order to resolve the uniqueness problem some additional conditions must be imposed. The second problem is that so far no one has succeeded in isolating a time variable from the equations of the theory that corresponds to a reasonable uniqueness condition.

My personal point of view is that there will not be a solution to the problem of time in general relativity along these lines, but whether it is possible or not is a technical problem and a technical result is needed to settle the issue.

Before closing this section, let me return to a problem that I mentioned at the end of the last section: the fact that a physical interpretation of general relativity based on diffeomorphism invariant observables will become ambiguous when applied to spacetimes with symmetries. What are the consequences of this situation for the problem of time?

The problem is that what is called time in most treatments of cosmology based on these symmetric solutions is not directly related to a notion of time that would arise in the full theory. Instead, what is usually called time in many cosmological models is a coordinate which is defined only in the context of a spacetime with a high degree of symmetry. The problem is that to do this is to take a particular solution to the Einstein equations as a background structure for the rest of physics; it is to replace the absolute spacetime of Newton or Minkowski with, for example, the one of the Friedman-Robertson-Walker solution.

This is the wrong thing to do because general relativity, time can refer only to something that is well defined in terms of physical observables of the theory and those must be functions on the space of the possible configurations of the gravitational field.

This does not mean that a particular solution, such as the Friedman- Robertson-Walker solution cannot have a limited use in observational cosmology.[15] But it does mean that it cannot be the basis for an understanding of any questions of principle in either classical or quantum cosmology.

So, in conclusion, at least at the present time, the only useful concept of time that exists in general relativity, in the cosmological context, is the one I have described: the time as measured by a physical clock that is part of the universe and is dynamically coupled to the rest of it. Time in general relativity is a completely relational quantity, it is precisely no more and no less than an aspect of the relationships holding between observable phenomena.

8. THE PROBLEM OF TIME IN QUANTUM COSMOLOGY

I would like now to turn to the problem of time in quantum cosmology. As opposed to the problem of time in classical cosmology, which I believe is completely solved by the point of view I described in the previous section, the notion of time in quantum cosmology is still an open problem. Moreover, the question of time is not only open, it is, for reasons that I will try to explain here, absolutely central. At present, many of the key difficulties confronting attempts to construct a quantum theory of cosmology can be directly traced to the problem of time.

My first purpose in this section will then be to try to explain what the key problems are facing attempts to construct a quantum theory of cosmology and how they involve the problem of time. After this, I will discuss a few, but by no means all, of the ideas that are presently being pursued concerning the notion of time in quantum cosmology.

Let me begin by saying that what we require by a quantum theory of cosmology is, potentially, quite broad. Such a theory must be capable of being applied and interpreted consistently when applied to a whole, single universe. It must agree with what is already known, which means that the theory should have, at least for some states, a classical limit in which the physics is well approximated by classical general relativity. It must be a further property of that limit that the conventional formulation of quantum mechanics in a background spacetime is recovered as the description of the behavior of matter fields in small regions of the universe.

In particular, it is not necessary that a quantum theory of cosmology correspond to a quantization, in the usual sense, of general relativity or any other classical gravitational theory. It is possible that this is the right way to construct the theory, but it is possible also that quantum cosmology requires a radical starting point, which is in no sense closely related either to general relativity or conventional quantum mechanics. A further possibility is that the correct quantum theory of cosmology shares some features of conventional quantum mechanics, but eschews others, replacing them by structures that make sense only in the cosmological case. All three possibilities are actively being pursued at the present time.

In this section I will concentrate mainly on the most conservative possibility because it is the one that, by definition, we can say the most about. I will describe a point of view advocated by the Italian physicist Carlo Rovelli, which is based on the application to the quantum theory of the view of time in classical relativity that I described in the previous section (Rovelli 1990, 1991a, 1991c, 1991d). I will then contrast this with another proposal, due to the British physicist and philosopher Julian Barbour, in which quantum cosmology has some of the mathematical structure of quantum mechanics, but has a different set of interpretational rules that make sense only when applied to the universe as a whole (Barbour 1992a, 1992b).

If quantum cosmology is to be constructed as the quantization of some classical theory, whether it is general relativity or any other theory, we may expect that, as in conventional quantum mechanics, the notion of time in the quantum theory should follow closely the notion of time in the classical theory. From the discussion of the last sections, we know that the key to the problem of time in classical general relativity is the fact that time must be described in terms of physical observables which are invariant under diffeomorphisms of the spacetime. Thus, the key to the problem of time in any theory of quantum cosmology that results directly from a quantization of general relativity must be to understand what are the quantum operators that represent these classical physical observables.

To describe the ideas that are presently being considered to solve this problem, and put them in the context of the historical development of the subject, I must first say a few words about the history of the problem of quantization of general relativity.

The failure of perturbative or semiclassical approaches to quantum gravity

There are three approaches to the problem of the quantization of general relativity that have been well studied. The first is what is called the perturbative approach. In this approach one assumes that the gravitational field is in a state which can be described as small quantum fluctuations around a single classical solution. This classical solution is usually taken to be Minkowski spacetime, but it need not be. One decomposes the field into a background part that represents the fixed classical solution and a part which represents gravitational radiation moving on this background. One then applies one of the standard methods of quantization to the field that represents the gravitational waves.

It is well understood by this time that this procedure fails to lead to a sensible quantum theory when applied to general relativity or any other field theory of gravitation. The reason is essentially that the assumption that the effects of the radiation are small cannot be consistently maintained, because the uncertainty principle turns out to imply that, as one probes on smaller and smaller scales the effects of fluctuations on the gravitational field are more and more important. This is because the gravitational field couples to energy and the smaller scales that are probed the larger are the uncertainties in energy.

The perturbative point of view has been considered by many people to be, in any case, unsatisfactory, because it evades the central problem, which is how to construct a quantum theory in the absence of a fixed background geometry for spacetime. That is, in the full theory of quantum gravity, the metric field should become an operator and the Einstein equations become operator equations. Since the key point of general relativity is that there are no background fields and the geometry is entirely described by dynamical fields, part of the problem of quantum gravity is the problem of constructing a quantum field theory in the absence of a fixed background structure. The failure of perturbative approaches to yield a sensible theory means that we cannot evade facing this problem.

I should spend a word on one perturbative approach which has received a lot of attention during the last eight years, which is string theory. String theory is, up till now, understood entirely in the context of a perturbation theory around a fixed classical geometry. It does seem likely that it yields a sensible perturbation theory (although to my knowledge this is still unproven past the second order of the perturbation theory). This is the reason that so much interest, justifiably, has been devoted to it. The problem is that there turn out to be a great many-more than a million- string theories that are equally consistent at the perturbative level. Each of these makes different predictions concerning the dimensionality, history and particle content of the universe, as well as the strengths of the various interactions. Without an additional principle which will make it possible to select one of these out from the others, string theory is, apparently, completely non-predictive. Furthermore, it is clear that a principle that would allow one to be selected over the others could only arise at a level beyond the perturbation theory.

Thus, the work on string theory has led to the same conclusion as the work on general relativity. One needs a formulation that goes beyond perturbation theory in order to construct a sensible and useful quantum theory of cosmology.

At this point, there is no definite formulation of string theory beyond perturbation theory. But, during the last six years a great deal has been learned about the quantization of general relativity, from a nonperturbative point of view.

There are two main approaches to the quantization of general relativity from an exact, or nonpertubative point of view. These correspond to the two main methods of quantization, which are the canonical method and the path integral method. I will describe the progress of each of these, in turn, and their prospects for resolving the problem of time.

A first nonperturbative approach: path integrals

I begin with the path integral method. The idea here is to represent quantum amplitudes as infinite dimensional integrals, corresponding to the path integral formulation originally proposed by Feynman. Until recently this approach seemed extremely difficult on technical grounds, but in the last year some spectacular results have been announced by Agishtein and Migdal (1992a, 1992b),[16] which show that such calculations may be carried out numerically using supercomputers. The idea in these calculations is that the infinite dimensional integrals may be approximated by representing the geometry of spacetime by a discrete structure of four dimensional simplices.[17] Each simplex measures one Planck length[18] on each side.

The main limitation of this approach is that with present day supercomputers only very small universes may be considered, that are about ten Planck lengths in each direction. However, within this limitation it seems that very accurate results may be achieved. Most importantly, there is good evidence that the calculations make sense, so that the failure of the perturbative constructions seems not be an indication that the exact theory does not exist.

However, it is not enough to compute numbers with computers. One must give the results of the calculations a sensible physical interpretation in terms of some definite formulation of quantum gravity. In conventional quantum theories, path integrals can be shown to be equivalent to expressions in the canonical theory. In the case of quantum cosmology one must then decide if the interpretation is to be made through the canonical theory or in some different way. If the interpretation is to be made through the canonical theory, then the results of the path integrals must be shown to represent either wavefunctions that live in some Hilbert space or expectation values of particular states and operators in that space.

At present, the problem of the interpretation of the results of the path integral calculations is unresolved. It is, however, clear that the path integral methods have, at least in principle, one great advantage when it comes to the difficult problem of representing physically meaningful observables in the quantum theory. This is that as the path integral method works always with four dimensional spacetimes, it is simple to express diffeomorphism invariant observables in this framework.

Even if we do calculations with the path integral formalism, it is likely that we will need to understand the canonical quantum theory to interpret the results. I thus now turn to a discussion of this approach.

Quantum gravity and cosmology from the point of view of canonical quantization

While the progress on the path integral quantization has come only in the last year, there has been steady progress on the canonical quantization of the theory over the last six years. The program of constructing a canonical quantization of general relativity is a very old one, it was, in fact, initiated by Dirac in the late 1940s (1959, 1964). However, the recent progress began with a reformulation of the Einstein equations by Abhay Ashtekar (1986, 1987), that greatly simplified the equations of the theory. As a result of this there has been steady progress on some aspects of the problem. There are also real conceptual difficulties, that are centrally related to the problem of time that are at present impeding further progress.

To explain what progress has been made and what are the roots of the difficulties that remain, I need to explain some things about the process of canonical quantization and how it is applied to general relativity.

The starting point for canonical quantization is, as I mentioned above, the Hamiltonian formulation of the classical theory. This Hamiltonian formulation is closely tied to the notion of the state of the system and its evolution in time. In the Hamiltonian formulation, the state of the system at any moment of time is described by giving what are called the canonical coordinates and the canonical momenta of the system. The canonical coordinates correspond to the position of a particle in particle mechanics, or to the instantaneous configuration of a field in field theory. The canonical momenta correspond to the time derivatives of the coordinates, they generalize the notion of momentum which, in ordinary particle mechanics is proportional to the velocity.[19]

Now, it in not obvious at first that the Einstein equations, which are written in terms of the geometry of four dimensional spacetime, can be expressed in the formalism of Hamiltonian mechanics. The problem is that the Hamiltonian theory requires the notion of the state of the system at a given time, and we know that, because of diffeomorphism invariance, no physical meaning can be given to an instant of time in general relativity. It was the great achievement of a number of scientists working in the 1950s and 1960s to invent a way to describe general relativity as a theory of something evolving in time which also is consistent with the diffeomorphism invariance of the theory (Bergmann 1956a, 1956b; Dirac 1959, 1964; DeWitt 1967; Arnowitt, Deser, and Misner 1960).

How this was done was, more or less, the following: On each spacetime one picks an arbitrary set of time and space coordinates. Using these coordinates one splits the spacetime manifolds into sequences of spatial manifolds. Each spatial manifold is a three dimensional space that consists of all the points that share the same value of the time coordinate in some spacetime.

The idea of the Hamiltonian form of general relativity is to think that what the theory is describing is how these arbitrarily defined spatial slices evolve in time, where by time is meant only the arbitrary time coordinate. That is, we have to generate, from a given spatial slice with the appropriate fields on it, a one parameter family of such slices which we will call its evolution and then put all the slices back together to make a spacetime. The canonical coordinates of the theory turn out to be certain fields that describe the geometry of these three dimensional spatial slices. The canonical momenta turn out to be certain geometrical quantities that essentially measure how fast these fields change as we evolve in the arbitrary time coordinate. There is thus a Hamiltonian, that tells us, by equations that look a lot like (2) and (3) how these fields change in terms of the coordinate time.

Now, the reader is undoubtedly asking, is this not a strange thing to do since at the beginning we admit that how we choose the time coordinate, and hence how we split up the spacetime in terms of an evolving set of three dimensional spaces is arbitrary. Conversely, many different sequences of spatial slices will actually represent the same spacetime. The neat part of the trick, however, is that it turns out that the arbitrariness of the slicing has quite definite consequences for the equations of the Hamiltonian formalism. It turns out that certain of the field equations of the theory can be interpreted as consistency conditions that guarantee that the different ways of slicing up the spacetime into evolutions of spatial slices really do go back together to represent the same spacetime. These equations are called the constraint equations of the theory.

They are certain conditions that the canonical coordinates and momenta must satisfy at each point of the spatial manifold. They are the central equations of the Hamiltonian formulation of general relativity, at both the classical and the quantum level.

There is a further consequence of the fact that the time coordinate of the theory is arbitrary. This is that when the constraints are satisfied the Hamiltonian, which tells us by equations analogous to (2) and (3) how the geometry of the three spaces evolve in time, actually is equal to zero.[20] This vanishing of the Hamiltonian means that the theory has no notion of total energy. This may be surprising, but it is a direct consequence of the diffeomorphism invariance of the theory, as I will now explain.

The total energy of a system is closely related to how fast it evolves, as seen from a clock outside the system. The vanishing of the total energy turns out to be, when one traces it through the formalism, a direct consequence of the fact that there is no clock standing outside the universe with respect to which the speed of the evolution of the universe can be measured. That is, it is impossible to have two distinct universes which are evolving through the same set of three geometries, but just at different speeds. By diffeomorphism invariance any two such spacetimes must be considered to be equivalent. Since there can be no notion of the speed of the evolution of the whole universe, there can be no concept of the energy of the whole universe.

So, to summarize, in the hamiltonian formulation of classical general relativity one works with a three dimensional manifold. The dynamical variables of the theory are a set of configuration variables that describe the geometry of this three dimensional space and a set of momentum variables that tell us how this three dimensional geometry is evolving in time. One then has Hamilton's equations of motions, analogous to (2) and (3), that tell us how to evolve these dynamical variables, given that we specify initially the three dimensional geometry and its momenta. This gives us a family of three geometries, one for each value of a coordinate time, that taken together can be considered to constitute a four dimensional spacetime. Finally, the initial three geometry and momentum variables must satisfy a set of equations, called the constraint equations, that express the invariance of the theory under diffeomorphisms of spacetime.

Once a theory has been expressed in hamiltonian form, there is a relatively well known procedure by means of which the corresponding quantum theory can be constructed. Just like in conventional quantum mechanics, states in the theory are to be represented by elements of a linear vector space. These can be represented as functions of the configuration variable, which in this case is the three dimensional geometry of space. That is, a quantum state of the system gives a probability amplitude for space to have any given three dimensional geometry.

Now, because the energy vanishes there is no Schroedinger's equation. This is in any case good, as there is no external notion of time so the d/dt on the left hand side of (1) can have no meaning. Indeed, the only thing that t could refer to in such an equation is the time measured by some time coordinate, and by diffeomorphism invariance the theory, and hence the quantum states, should be independent of the time coordinate. Thus, instead of the equation (1) one finds in the quantum theory the equation

$$\frac{d}{dt}|\Psi\rangle = 0 = H|\Psi\rangle \tag{4}$$

This equation is the expression in the quantum theory of one of the constraints of the theory. It expresses the fact that the physics must be independent of the time coordinates. It is sometimes called the Wheeler-DeWitt equation.

In addition to this equation, we must also impose the condition that the states be invariant under the action of the diffeomorphisms of the three dimensional space. Together, these two conditions express in the quantum theory the four dimensional diffeomorphism invariance of the original classical theory.

The space of states that satisfy these conditions are called the physical states of the theory. They make up a linear space, which is called the physical state space. Once the physical state space has been found, the theory must be completed by the construction of two additional structures. The first is the observables-these must be represented by linear operators acting on the physical state space. The second is the inner product, which must be specified on the physical state space in order to compute expectation values and give the theory a probability interpretation.[21]

This, then, is the setup of any quantum theory of cosmology that would come about from the canonical quantization of classical general relativity. We may now raise the question of how time is to be treated within the framework of such a theory.

The problem of the missing time in the canonical approach to quantum cosmology

The reader may note that there is no variable corresponding to time anywhere in the framework I've just described. Furthermore, not only is there no time in the formalism, *there is no four dimensional spacetime manifold anywhere in the basic structure of the theory.* There is a three dimensional spatial manifold, whose geometry is the dynamical variable of the theory. But nowhere in the basic structure I've just described does there appear a four dimensional manifold. This may be puzzling as this theory is the result of the quantization of general relativity, which itself is supposed to be a theory of spacetime. The question we want to ask is: where has the spacetime manifold gone?

The answer is that the spacetime manifold has gone to the same place in the quantum theory of gravity that the trajectory of the electron has gone in the quantum mechanics of the atom. Given most quantum states of the electron it is impossible to deduce any quantity that could be called the trajectory of the electron. All that the state can furnish is a probability that the electron can be found at different locations in space. Similarly, from most quantum states of the gravitational field, no notion of spacetime can be inferred. All that the state can give us is probabilities that the three geometry of the manifold (or, rather the fields that describe it) take various values.

Now, as the reader may know, there are special quantum states of the electron that allow one to construct an approximate classical description in terms of a trajectory. These are called semiclassical states. They arise, for example, when the action of the system is very large compared to Planck's constant.[22] Similarly, there are quantum states of the gravitational field that allow an approximate classical description, which will be in terms of a spacetime manifold.

Thus, as in ordinary quantum mechanics there is a correspondence principle and a classical limit, which allows us to draw a spacetime description out of certain, special, semiclassical states of the gravitational field. However, apart from its role in the classical limit, the quantum theory of gravity gives us, and indeed, has no place for, a notion of a four dimensional spacetime geometry.

Now, if the notion of spacetime emerges from the quantum theory of cosmology only for certain states, and only in the classical limit, we want to ask whether the whole concept of time is subject to the same limitations. That is, is time something that emerges from the theory only in the classical limit? This is a possibility that must be considered seriously. If this is the case then for most quantum states that the universe could be in it would be simply meaningless to speak of time. On the other hand, it might instead be that the quantum theory of gravity has within it some other, more intrinsic notion of time, which makes sense for all states of the theory and not just for those close to the classical limit.

This question is unresolved at present and both possibilities have been advocated by different people in the field, and sometimes by the same people at different times! We will shortly go on to discuss two proposals, one on each side of this question. When we do it will be important to keep one central fact in mind: if the theory has some intrinsic notion of time that makes sense beyond the classical limit it cannot be found in the quantum state alone, as the quantum state just gives amplitudes for the three geometry to take different values. Instead, we must look for an intrinsic notion of time in the two structures on which the conventional interpretation of quantum theory is based: the operators and the inner product.

However, before introducing these two proposals about time, it will be helpful if say a little about what is the present status is of the attempt to construct the quantum theory of gravity along these lines.

Geometry comes from topological relations in
a diffeomorphism invariant quantum theory

One of the key discoveries of the last five years is that it is actually possible to exactly solve the quantum constraint equations that express the diffeomorphism invariance of the classical theory and so construct explicitly a large number of members of the physical state space (Jacobson and Smolin 1988; Rovelli and Smolin 1988, 1990; Ashtekar 1991; Rovelli 1991b; Smolin 1992b). These states turn out to have a very beautiful description. They may be described very simply in terms of certain familiar topological structures, which are knots. Consider a number of pieces of string, with the ends tied up so they form loops. These may be knotted and linked together in many inequivalent ways. Here, we regard two knots as equivalent if they may be deformed into each other by smooth deformations in which the strings are not cut or pulled through each other. It turns out that there are actually an infinite number of such inequivalent ways to knot and link any finite number of strings together, these are called the knot classes.[23]

It is then very simple to state one of the basic results of canonical quantum gravity: For each knot class there is a quantum state of the gravitational field (Rovelli and

Smolin 1988, 1990; Ashtekar 1991; Rovelli 1991b; Smolin 1992b). Thus, what has been achieved is a very simple characterization of the state space of the theory. Note, moreover, that as knots are entirely characterized by their topology, the result is described in a way that is diffeomorphism invariant: the specification of the topology of a knot involves no coordinates or background fields.

This result is, in every sense, satisfactory. We have an infinite dimensional space of possible quantum states of the universe. Each quantum state is a superposition of basis states and each basis state can be labeled by a particular knot. Moreover, as there are very simple knots, we know that some of these states have very simple structures. And, as knots can be arbitrarily complicated, we have states that contain an arbitrary amount of complexity. Intuitively, we expect that the simple knots correspond to simple quantum universes, that are of the order of a few Planck units in size, while the very large and complicated knots correspond to large and complicated universes. Indeed, there is strong, although indirect evidence for this interpretation, which leads us to a picture in which large universes such as are own are represented by three dimensional woven structures of great extent and complexity. We have learned, in particular, that only states based on very complicated woven structures can play a role in the classical limit (Ashtekar, Rovelli, and Smolin 1992).[24]

These results tell us one way in which geometry can be represented in a diffeomorphism invariant quantum theory. We see that the classical geometry emerges in the classical limit from a quantum state space which is based purely on topological structures. The requirement of diffeomorphism invariance and the the discrete nature of the quantum states work together to give a very simple picture of a purely quantum mechanical description of spacetime geometry that is also purely relational.

Unfortunately, although we have a construction of the physical state space of the theory, more remains to be done before we have a construction of a quantum theory of gravity. What remains to be done is to find the operators that represent the physical observables and to construct the inner product on this space. These problems are closely related to each other. They are very difficult, and only a little progress has been made concerning them. Moreover, as I will now explain, both of these problems are intimately related to the problem of time.

The problem with the construction of physically meaningful observables in quantum gravity

As I mentioned above, if the theory of quantum gravity contains some notion of time which is more fundamental that the one that emerges in the classical limit, this notion must have something to do with the operators and the inner product. Furthermore, as we discussed in the previous section, the notion of time in the classical theory is completely bound up with the problem of the observables. Thus, one way to resolve the problem of time in the quantum theory would be to construct operators that correspond to the classical observables that we described in the last section that encode the observations made by an observer carrying a dynamical clock inside universe. Expectation values of such operators, taken with a particular state, could be interpreted as giving the expected outcome of measurements made by those observers, at the times labeled by the clocks they are carrying, in a universe described by that quantum state.

This proposal has been most completely and forcefully advocated by Carlo Rovelli and it has been widely discussed during the last several years (Rovelli 1990, 1991a, 1991c, 1991d). To explain its present status I must first explain why it is difficult in general to construct the physically meaningful operators and the inner product.

I will begin with the inner product. In the conventional formulations of quantum mechanics and quantum field theory one uses the fixed, background structure of the spacetime geometry to pick out the inner product. In quantum mechanics the inner product is determined by requiring that it lead to conservation of probability, under time evolution as measured by the observers clock outside of the system. In quantum field theory the inner product is picked out by requiring it to be invariant under the group of Poincare transformations. Thus, in each case, the inner product is closely related to the notion of time in the theory, and that notion is one of time measured by clocks external to the system.

In the case of quantum cosmology, there is no fixed background structure and there are no clocks which are external to the system. Thus, the usual criteria for selecting the inner product cannot be applied in this context. New criteria are needed.

Ashtekar and Rovelli have proposed one such criterion, which is potentially applicable in the case of quantum cosmology (Ashtekar 1991; Rovelli 1991b; Smolin 1992b). This is that the inner product should be picked out in order to realize the condition that physical measurements yield real, rather than complex, numbers. That is, the classical observables of the theory are real, in the sense that they must express the fact that the results of measurements of distance and time intervals must yield real numbers. It must then be the case that the quantum operators that correspond to these classical observables are hermitian operators, so that their eigenvalues and expectation values are real. This turns out to be a condition on the choice of the inner product: it must be chosen so that the operators that correspond to physical observables that are real are hermitian.

There are good reasons to believe that this is the correct condition to pick out the inner product in cases, such as quantum cosmology, where there is no background structure or external notion of time to determine it. However, note that to implement it, we need to know which quantum operators correspond to the physical observables of the classical theory. Because of this, the problem of picking the inner product in quantum cosmology depends on our having solved the problem of the physical observables in the theory. Let us then turn to a discussion of that problem.

In the canonical theory it is very difficult to construct the operators that represent physical observables. The problem is that the operator form of the theory is closely connected to the canonical, or Hamiltonian, form of the classical theory and it is very difficult to construct the physical observables in that context. To understand why this is the case, I must remind you of what I said above: the canonical formulation is very closely tied to the idea of evolution, in some background coordinate time, of the geometry of a three dimensional manifold.

Now, let us recall from the discussion of the previous section the argument that led to the conclusion that *any physical observable must be a constant of motion of the theory.* This followed directly from the basic requirement that physical observables invariant under spacetime diffeomorphisms because the motion from one spatial surface to another, each defined by one value of coordinate time, can be accomplished by a diffeomorphism. In order to be invariant under spacetime diffeomorphisms, a function of the canonical coordinates and momenta must be independent of which spatial surface it is evaluated on.

Now, the problem becomes clear! The Einstein equations are complicated, nonlinear partial differential equations, and it is very non-trivial to find their constants of motion. Indeed, only a literal handful are known, and these were, for the cosmological case, all found in the last year (Goldberg, Lewendowski, and Stornaiolo 1992; Jacobson and Romano 1992). But, as we discussed in the previous section, to characterize the physics of the theory we need an infinite number of physical observables.

It is clear from the previous discussion of the example of measurements made by an observer riding on a particle that the difficulty is not with the notion of physically meaningful observables or with the notion of time, in either the classical or the quantum theory. It is only with the problem of representing them in terms of the canonical coordinates and momenta on a necessarily arbitrary spatial surface. The question that naturally emerges is then: is there a way to represent these physical observables in the quantum theory in a way that does not require us to first express them classically in terms of canonical coordinates and momenta?

A positive answer to this question would be worth a great deal. At present several different ideas are being tried to approach this problem. Until one of them works, or until another way to construct the operators that represent physical observables is found, this problem will remain the chief difficulty blocking progress in constructing a quantum theory of cosmology through a quantization of general relativity.

Because of the difficulty of this problem, it is natural to wonder whether there is a way in which the theory could be completed, and given a sensible physical interpretation, without a solution to the problem of the physical observables. I would like to mention here one such idea, due to Julian Barbour, as it involves taking a radically new point of view about the nature of time.

Time as an emergent and contingent property of certain universes: the proposal of Barbour

The proposal of Julian Barbour (1992a, 1992b) is the clearest expression of a point of view that has been explored by a number of people who worked on quantum cosmology, beginning with the early work of Charles Misner (1970, 1972). To explain it, it is convenient, although not completely necessary, to include matter in the description of the universe. For simplicity I will assume that the configuration of the matter is described by some fields which live on the three manifold which is taken to represent space, which I will denote ϕ. A quantum state of the universe can then be represented by a wavefunction, which depends on both ϕ and the three geometry. Denoting the latter by g, the state can then be represented by functions $\psi[g, \phi]$. As before, to be physically meaningful a state must be invariant under four dimensional diffeomorphisms. As before, this means that it is invariant under diffeomorphisms of the spatial manifold and is also a solution to the Wheeler-DeWitt equation (4).

Barbour proposes that it is possible to make an interpretation of such states without having to construct operators to represent the physical observables. Instead, he wants to base the interpretation directly on a generalization of the idea that the absolute value squared of the state function gives a probability measure for the system to be found somewhere in its configuration space. To implement this he needs to introduce some probability measure on the configuration space in question. This is an infinite dimensional space which consists of all possible three geometries together with all possible configurations of the fields ϕ. Actually it is not exactly that, but equivalence classes of those configurations under three dimensional diffeomorphisms. I will call this space $\mathcal{C}$, for the configuration space.

Let me also introduce the notation $d\mu(g, \phi)$ to denote the probability measure. Using it, the basic statement of Barbour's interpretation is the following:

The Universe consists of an ensemble of configurations (g, ϕ), of the three dimensional geometry and matter fields. This ensemble is taken to be actual, that is, physical reality is asserted to consist of a collection of such configurations. Consider, now, some particular region of this configuration space, which I will denote $\mathcal{R}$. Given the quantum state, $\Psi[g, \phi]$, the quantum theory of gravity then allows us to compute the probability that if I pick one member of this ensemble out at random it will fall into any such region $\mathcal{R}$. This is given by,

$$P(\mathcal{R}) = \frac{\int_{\mathcal{R}} d\mu[g, \phi] |\Psi[g, \phi]|^2}{\int_{\mathcal{C}} d\mu[g, \phi] |\Psi[g, \phi]|^2} \tag{5}$$

This the complete statement of the theory, according to the interpretation proposed by Barbour. Note that there are no operators in the statement of the interpretation. Note also that time is nowhere mentioned. The Universe is asserted to consist simply of an ensemble of configurations. Each configuration corresponds to the description of a moment of time in conventional physics. Thus, in this theory, there is no time, but there are moments. There is, in fact, nothing but moments. What is missing is any structure of ordering or causality to tie the different moments together. Barbour asserts that these structures are not intrinsic aspects of nature. Nor, he asserts, are they necessary to give the theory a consistent interpretation. They will emerge, but only in the classical limit of the theory.

What Barbour is proposing challenges not only the notions of time we have become used to in physics. It challenges our everyday assumptions about the nature of time and the nature of our conscious experiences. For what Barbour is proposing is that time is not a fundamental aspect of reality. There is, according to this view, no flow of time, there are no trajectories, there are no evolutions. Our subjective experience of the flow of time, he wants to say, is an illusion. What we experience is only moments, (or, perhaps, a moment). The existence of other moments, with properties related to the present one by notions like future, past, causality, etc. is, he is asserting, only conjectural. This moment, as you are reading this, is one of the collections of configurations which make up reality. The you that I am addressing is part of that moment, part of that configuration. Perhaps there are in the ensemble other moments, corresponding to yesterday, or tomorrow, or five minutes ago, in each of which a being very much like yourself is also reading this essay. Given certain features of this moment and the laws of physics, it may be possible to deduce the probabilities that those other moments are in the ensemble. But, there is in general no ordering and no structure relating the moments except their occurrence together in the ensemble. And, it is only in a probabilistic sense that the almost-you's in the other moments can be asserted to be related to the you in this moment.

Before we can accept this interpretation there are two questions that must be answered. First, how is the usual notion of time in physics to be recovered from the theory? Second, how are our subjective impressions of the nature of time, as we experience it, to be explained? Barbour has answers to these questions. While they may not be easy to accept, I believe they are logically coherent.

Barbour asserts, first, that the notion of time can only be recovered in the semiclassical limit. For most possible quantum states $\Psi[g, \phi]$ no notion of time can be recovered. If any of the configurations in such an ensemble happened to contain beings like you and me, they would have some experience of their world, as it were, instant by instant, but they would have no sense of the flow of time, or of memory. Only in those special states for which a classical limit can be defined can a notion of time be recovered.

This is possible because in the classical limit it is possible to isolate degrees of freedom that correspond to both the configuration and momenta of the system as a function of some coordinate time. That is, in this limit, the notion of a trajectory can be recovered, just as it can in the classical limit of ordinary quantum mechanics. This trajectory can be shown to satisfy, not exactly, but up to a certain degree of approximation, the classical equations of motion of the system. In our particular case this means that one can introduce, for these special states, a sequence of configurations labeled by some parameter time, t, which may be written $[g(t), \phi(t)]$ and that such sequences will make up a four dimensional manifold, on which some matter fields are evolving in such a way that the classical Einstein equations are satisfied to a certain degree of approximation.

Furthermore, these special states that we call semiclassical have the property that they give appreciable probability only to those configurations that are on (or more properly, near) these classical trajectories. That is, the ensemble of moments that correspond to such a semiclassical state, is dominated (in terms of statistical weight), by those that can be ordered according to a classical time parameter.

Exactly how this is done need not concern us here, it is discussed in many places in the literature on quantum gravity and quantum cosmology (Kuchar 1992). The problem of how to take the classical limit of quantum cosmology has received a lot of attention in the last few years,[25] and is still not entirely settled. But there is good reason to believe that there will exist quantum states which are solutions to the Wheeler-DeWitt equation which have a good classical limit (Ashtekar, Rovelli, and Smolin 1992).[26]

Now we come to what is, for me, perhaps the most original and the most important part of Barbour's proposal. This is that while, in any such a scheme,[27] the existence of the classical limit is necessary for the recovery of our notion of time, it is not sufficient.

This is because quantum cosmology is meant to be a theory of the whole universe, while Newtonian mechanics and quantum mechanics are theories of only a portion of the universe, which does not include the clock carried by the observer. Thus, if time is to be recovered in the classical limit from a theory of quantum cosmology, that theory must tell us not only that there is a time parameter with respect to which certain quantities evolve approximately according to the classical equations of motion. The theory must also tell us why there are clocks in the universe that measure this classical time and measuring devices that can record the results of their interactions with things external to them and store these records reliably for some amount of evolution by that classical time.

What Barbour proposes is that the configurations that dominate the ensembles associated with semiclassical states are characterized not only by the fact that they admit an approximate ordering in terms of a classical notion of time. He hypothesizes that they have an additional property, which is that they are sufficiently structured so as to allow the existence of clocks and observers. In particular, he hypothesizes that the configurations that dominate the ensembles are of a type that he calls *time capsules*. A time capsule is defined to be a classical configuration of the universe that contains structures which allow us to deduce information about other configurations that, in such a semiclassical state, would be said to be in the past of it. In short, a time capsule is a configuration that can be read as a record of the past.

Barbour argues that it is the presence of time capsules in the configurations we see around us that are responsible for both our ability to do classical physics and for our subjective impressions about time. In particular, he claims that both the subjective impression we have about the flow of time and the impression we have about the continuity of our own identity, as well as the identity of objects in the world, is due to the existence of time capsules. That is, it is because what we are experiencing at this moment is a configuration that contains structures which code memories that we have, at this moment, the impression that we have an existence that is continuous in time.

Barbour's theory thus requires certain assumptions about the structure of consciousness. What is required is actually very minimal, it is only that associated with structures in each configuration is something we would like to call the conscious experience of beings described by that configuration. Given this, he asserts that the impression of the flow of time and of the identity of ourselves and our conscious experience over time is not a necessary property of that consciousness. This is because in his interpretation all that exists in physical reality is the ensemble of moments, represented mathematically by the configurations (g, ϕ). All else is a contingent property which is related to the particular ensemble, and hence to the quantum state which tells us the statistical distribution of the elements of a particular ensemble. In particular, he is asserting that all of the properties of time that we usually consider intrinsic to the structure of the world, including both those that we use in classical physics and those that we usually assume are the fundamental properties of subjective time as we experience it, are contingent. What this means is that he is asserting that a quantum theory of gravity will be able to explain these aspects of time as a consequence of the assertion that the quantum state that describes the universe is a solution to the Wheeler DeWitt equation (4).

Thus, the answer to both of the questions that we raised above, how the notion of time in classical physics is to be recovered, and how our subjective impressions about the flow and the continuity of time are to be recovered, rely on the assertion that at least some of the quantum states that solve the Wheeler DeWitt equation have the property that they are dominated by configurations that have time capsules. As a result, for Barbour, that the universe is structured sufficiently that its configurations can indeed be considered to be time capsules becomes a fact of key significance for the quantum theory of gravity. From his point of view, to explicate why the universe is so structured becomes a key problem for the theory of quantum gravity and quantum cosmology. I believe that there is something very right about this assertion, and I will devote the remaining three sections of this essay to examining it. I would like to then end this section by discussing two points.

Concluding remarks about the problem of time in quantum cosmology

First, and most importantly, both Rovelli's and Barbour's proposals are based on the assertion that the only cases in which time can be given meaning in quantum cosmology are those in which the universe is so structured as to contain subsystems that can be considered clocks. Thus, under both proposals the notion of time is a contingent property of a universe, and the particular properties of time are dependent on the form of the quantum state.

Second, it is interesting to note that whether the notion of time in quantum cosmology will in the end look like Rovelli's proposal, or Barbour's proposal, or some mixture of the two, it will have none of the four characteristics that I ascribed to the Newtonian notion of time in section 2, above. Not only will time in quantum gravity not be universal or absolute, it will not be geometricized or spatialized. In either case, the extent to which time can be put into a one to one correspondence with the real number line is a purely contingent property of the quantum state of the universe and the observable that is used to define the time. For some states and usually for some finite intervals it may be possible to make such a correspondence. But, in fact, because the clock must be, when considered approximately as an isolated system, a quantum system with finite energy, there is a limit to the accuracy with which any clock can resolve intervals of time. This means that the idea that the moments of time are in any meaningful sense in one to one correspondence with the points of the real continuum is a purely classical idealization.

Let me clarify a possible confusion about this. It is true that in Barbour's proposal the elements of physical reality are asserted to be a collection of configurations representing "moments." However, there is no particular need for the number of such moments to be uncountable. Furthermore, the claim that for certain special "semiclassical states" the moments that dominate the ensemble can be put in an ordering corresponding to time is only meant in an approximate sense. Because wavefunctions always have, in the semiclassical approximation, some small amplitudes outside the classical region there will always be configurations in the ensemble that cannot be so ordered. Further, the size of the time intervals for which the ordering can be well defined is clearly a function of the complexity of the configurations that dominate the ensemble-the more information each configuration contains that can be used, in the sense of a time capsule, to describe the probable past, the more accurately the ordering can be made. This means that the accuracy with which time can be resolved into intervals depends on the complexity of the universe. From the Leibnizian point of view, this is certainly a completely satisfactory resolution.

It seems that in the twentieth century physics has taken apart, property by property, the notion of time from Newtonian physics. For, if the development of the notion of time from Newton to Einstein is sometimes seen as the discovery that what fundamentally exists is a spacetime, in quantum cosmology we see all of a sudden that this notion is not really fundamental; spacetime is an approximate and contingent concept that emerges only in the classical limit. Thus, quantum cosmology seems to reverse several centuries of development during which time seemed to be more and more like space; we come finally in quantum cosmology to the conclusion that time is, after all, not really very much like space. The only property of space that survive in quantum cosmology is the topology of the initial three dimensional manifold on which the canonical quantization is based. All other spatial relations are deduced from topological relations of the quantum state, i.e. from the connectedness of the knots. Time enters in a completely different way, as a contingent property that is dependent on the existence of structures that behave as physical clocks.

This deconstruction (if I may use the word) of all the aspects of Newtonian time comes about by taking seriously the point of view of Leibniz: that space and time are to be considered purely relational quantities. As far as I know, there is no more ironic development in the whole history of science.

9. THE PROBLEM OF STRUCTURE IN COSMOLOGY

In these last sections, I want to explore the implications of the idea that the notion of time in cosmology must be dependent on the fact that the universe has structure. I will argue that if we take this idea seriously we are led to a notion of time that is rather different than any notion of time in physics since at least the time of Galileo. In particular, I will argue that if we are to have a notion of time which is to make sense in a quantum theory of cosmology, that notion must involve fundamentally notions of structure and complexity.

The point of view that I will sketch here is quite speculative, and I must emphasize at the beginning that a great deal of science has yet to be done if we are to demonstrate that it is either consistent or useful. It is a point of view that has been growing in my mind for the last several years, and for me, its motivation comes from several different directions. I will try to sketch here the view and those motivations for it that are connected with the problem of time.[28]

As I can give here only a sketch of the arguments, I will organize them under a series of subheadings.

The Universe does in fact have a great deal of structure

At the time that the first relativistic cosmological models were invented, it was perhaps reasonable to assume the truth of the so-called cosmological principle. This holds that there is a scale, which is much shorter than the radius of the universe, above which it is a good approximation to regard the distribution of matter as homogeneous and isotropic.

The period during which the implications of this principle for cosmology were first explored coincides, more or less, with the period in which it was definitively established that the universe contains a large number of galaxies outside of our own. Since that time a great deal has been learned about the distribution of these galaxies. Indeed, presently the study of the large scale distribution of the galaxies is one of the most active and exciting areas of cosmology. However, in spite of the fact that the subject is very much under-development the trend is clear. This is that the evidence for the cosmological principle is weak, and is growing weaker as more data comes in. It has been the case for quite some time, and continues to be the case, that as larger and larger scales are probed, structure and inhomogeneities are observed at the largest scale consistent with the resolution of the data (Coleman and Pietronero 1992).[29]

Instead of a homogeneous distribution what is seen is a structure of sheets and voids, with large inhomogeneous flows. There are even suggestions of periodic structures in the surveys of the distributions that probe the largest scales (Broadhurst et al. 1990). It is even possible that it may be appropriate to describe the distribution over a large range of scales as fractal, which is equivalent to the statement that there is structure on every scale (Coleman and Pietronero 1992).

Furthermore, even on much smaller scales than that of the large scale distribution of galaxies, the trend seems to be that astronomers are finding much more structure than was originally seen. For example, individual spiral galaxies turn out to be highly structured systems. Much more than just a collection of stars held together by their mutual gravitational attraction, a spiral galaxy seems more and more to be a system, in which gas of various kinds, dust and stars together comprise a system that maintains an organized steady state analogous to the organization of an ecological system.[30] The interstellar medium seems to consist of a number of components each with different densities, temperatures and chemical compositions, which, separated by rather sharp boundaries, exist together in a steady state. This state is maintained by the activity of great cycles in which matter and energy flow between the stars and the gas and dust of the inter-stellar medium. It seems, further, that the organization of the cycles and the distributions of matter and energy in the different components of the galaxy are maintained by feedback loops that control the rates at which stars of different masses are formed.

It is true that the 2.7° black body radiation is, to a very excellent approximation, in thermal equilibrium. However, it seems that the matter is organized in a way which is not close to thermal equilibrium. Nor is there any reason to believe that the distribution of matter on scales of galaxies and above is becoming more uniform, or is tending towards thermal equilibrium. Indeed, the basic problem of cosmology at the present time may be expressed by saying that we lack an explanation of why the radiation in the universe is in equilibrium while the matter is in an apparently structured, far from equilibrium, state.

It is surprising, from the point of view of physical law, that the universe has so much structure in it.

The original founders of the field of statistical mechanics, Boltzmann and Gibbs, wrestled with the problem of why the universe has not yet come to thermal equilibrium. Working in the context of an infinite and immortal Newtonian universe, they could only conclude that the present situation is a fluctuation, and that the universe is destined to return to equilibrium. This gave rise to the common view of a dead universe, hostile to life, whose unavoidable destiny is to end by the return to equilibrium-what was called the "heat death of the universe."

In spite of the transition to relativistic cosmology, it remains a surprising fact that the matter in the universe has not yet come to thermal equilibrium. I would like to examine briefly the reasons why this is surprising.

The expectation that the universe should come to thermal equilibrium arises, I believe, from two sources. First, from the second law of thermodyamics, which states that closed systems come to equilibrium. Furthermore, most closed systems come rather quickly to equilibrium; studies of the dynamics of systems close to equilibrium suggest that the time scale to come to equilibrium is governed by the longest time scale of the fundamental processes of the system. In the case of the universe, this longest fundamental time scale might be most generously taken to be the time it takes light to cross a galaxy, or go between galaxies.[31] On cosmological scales even these times, on the order of a few hundred thousand years, are very short.[32]

The second source of this expectation is that this is what is predicted by simple cosmological models in which both the matter and geometry are represented by homogeneous distributions, in accord with the cosmological principle. In many commonly studied models, the matter is usually represented by a combination of simple fluids and quantum fields.

I believe it may be the case that such models are too simple, in certain essential ways, to explain the observed structure in the universe. The explanation for structure may come from elements that are left out in such models. These elements include the degrees of freedom of the gravitational field itself and the nonequilibrium statistical mechanics of gravitationally bound systems. In the limited space I have here I would like to mention briefly how putting these elements could play a role in understanding the fact that the universe is structured.

Structure is a common feature of systems that are far from thermodynamic equilibrium

If the universe is a highly structured system, and there is no evidence of a rapid approach to equilibrium, then the proper context for its description may be far from equilibrium-rather than equilibrium-thermodynamics. A picture of how systems maintain themselves far from equilibrium for long periods of time has been developed by a number of people, including Prigogine (1967, 1980) and Morowitz (1968). The starting point of this picture is that such systems cannot be closed. They must have sources and sinks of energy, which are themselves stable over long time scales. Such a system maintains a fixed amount of energy, but it is necessary that energy flows through it from the source to the sink. Because such systems are open, the second law of thermodynamics cannot be applied to them. Instead, what happens, at least in some cases, is that such systems develop structure because the transport of energy from the sources to the sinks requires the setting up of chemical cycles in which the energy is stored in chemical bonds.

This description certainly applies to the biosphere, in which case the source of energy is the sun and the sink is outerspace (Morowitz 1968; Lovelock 1988). I believe it may also applies to spiral galaxies, in which case the source of energy is primarily the nuclear energy produced in stars and the sink is, again, energy radiated into space.

However, the application of the picture of Prigogine and Morowitz to a particular system requires more than just a source and sink of energy. That system must be capable of forming metastable subsystems which are stable on long time scales. In the biosphere, these are the organic molecules out of which life is made. In the case of galaxies, these structures are clearly stars.

Indeed, stars play a key role in keeping the universe out of equilibrium. Not only do they provide sources of energy that are stable for cosmological time scales, it is in stars that carbon, oxygen and the other organic elements are produced that are necessary for the formation of complex and stable structures from atoms.[33] Thus, a key question that must be asked if we want to understand why the universe has so much structure is why stars that are stable on such long time scales exist. I will return to this question in the next section.

Given this framework, it is tempting to ask whether the entire universe might be thought of as a far from equilibrium thermodynamic system. Normally, a closed system could not be a stable far from equilibrium system, because it contains, by definition, no sources or sinks of energy. However, the universe is a system whose evolution and large scale structure are governed by gravitation, and there is a large body of evidence that systems in which gravitation is important do not behave like conventional thermodynamic systems. I now go on to discuss some of this evidence.

Systems whose physics is dominated by gravitation do not come to thermal equilibrium

There are several independent pieces of evidence that statistical systems whose physics is dominated by gravitation do not evolve towards unique and structureless equilibrium configurations. I only have space to list them here.

1) Gravitationally bound systems have negative specific heat. This means that when one takes energy out of them their temperature increases, rather than the reverse (Lyndon Bell and Lyndon Bell 1977; Penrose 1979b). One consequence of this is that one can always increase the entropy of a system by splitting it into subsystems, each of which is more tightly bound, but which are widely separated from each other. Because of this a gravitationally bound system can become more inhomogeneous as it increases its entropy.

2) In numerical experiments which simulate a gravitationally bound system of particles moving under their mutual Newtonian gravitational attraction, there is generally observed a tendency to develop structures such as clumps and voids, rather than an approach to a homogeneous distribution.[34]

3) Gravitational radiation cannot be made to come to thermal equilibrium in a finite time, as long as the matter it interacts with satisfies the condition that its energy density is positive (Smolin 1984b; Smolin 1985). This means that gravitational waves carry information about their sources throughout the history of the universe; that information is not lost due to thermalization as it is for electromagnetic radiation.

Thus, it is at least plausible to conjecture that if we had a good theory of the statistical mechanics of self-gravitational systems, it would show that they have a tendency to generate structure as they evolve, rather than destroy structure, as the second law requires of all nongravitational systems. Unfortunately, the statistical mechanics of gravitational systems is not well enough developed to be able to formulate or demonstrate a theorem to this effect (Rovelli 1993). But, it is intriguing to wonder whether the fact that structure is necessary for the notion of time in gravitational theories is connected to the fact that gravitational systems seem to be systems that organize and structure themselves.

There is, indeed, a rather general argument that suggests that general relativity requires that its configurations be structured in order to have an unambiguous physical interpretation at both the classical and the quantum level. I would like to next sketch this argument, as I believe that it points up a general characteristic of theories based on a relational concept of space and time. This is that such theories require that the configurations they describe are structured if they are to have consistent physical interpretations.

Diffeomorphism invariant systems require structure to be well defined

In this subsection I would like to motivate the following conjecture: *A physical interpretation of general relativity, possibly coupled to matter, which is based on the use of diffeomorphism invariant observables will, when applied to the case of a closed universe, be inconsistent or ambiguous unless the configuration of matter and gravitational fields in the universe is sufficiently structured so as to allow the points of spacetime to be distinguished by the values of the fields.* Although I do not have a mathematical proof for this conjecture, I believe that its plausibility can be established, essentially by a kind of counting argument. I will sketch this argument here.

We know from analyzing the hamiltonian formulation of the theory that the gravitational field has four independent degrees of freedom per spacetime point. This implies that in the classical theory, a complete specification of the state of the field requires four numbers per point of some initial spacelike surface.[35] If we couple general relativity to other fields than there are additional degrees of freedom. For simplicity, I will continue to use the example of general relativity coupled to the electromagnetic field. In this case we know that the electromagnetic field adds four degrees of freedom so that, all together, a complete specification of the state of the system requires that we specify eight numbers per point.

Now, if the theory has a good physical interpretation then there should be a sufficient number of physical observables to completely specify the state of the system. Thus, we require that the theory give us eight well defined observables per point of space. Further, if those observables are to be useful, we would like them to be local. That is, we would like to be able to interpret them as statements about observable correlations of the values of fields at spacetime points. If this is the case then it must be possible to use the values of the components of the field to distinguish all the points of the spacetime from each other. This means that any two observers, each sitting at two arbitrary points in spacetime, must see different electromagnetic and gravitational fields.

Let me give this property a name. I will call a universe, described by a solution to the Einstein equations coupled with some matter fields, *Leibnizian* if it has the property that any two points in spacetime may be distinguished from each other by the values of the fields at those points.[36]

My claim is then that a solution to the coupled Einstein-matter equations must be Leibnizian if a sufficient number of physical observables are to be defined so that the classical state of the system can be completely specified.[37] This means that in a non-Leibnizian solution some of the physical degrees of freedom must become degenerate or singular. If this is the case then if the theory can be translated into a description in terms of purely physical, diffeomorphism invariant variables, the new invariant formulation of the theory may not allow the description of non-Leibnizian universes.

Structure arises naturally from model systems based on relational notions of space

I would like to offer a last piece of evidence for the claim that the requirement that the universe be described by a theory based on a relational view of space and time necessarily leads to a description of a structured universe. This is based on work that Julian Barbour and I have been carrying out over the last several years (Smolin 1991; Barbour 1989a; Barbour and Smolin 1992). In this work we have been interested in trying to construct models of dynamical systems that are purely relational. The systems we are studying are characterized by the property that the state of the system must be characterized entirely in terms of relationships between the dynamical degrees of freedom. In the context of these studies the concept of a Leibnizian configuration arises naturally; for example in a system of particles a Leibnizian configuration is one in which every particle can be uniquely identified in terms of its relations with other particles.

In these studies we have found that it is very useful to introduce a quantity which measures how easily an element of it can be characterized in terms of its relations with the other elements. We call such a quantity the *variety* of the system. The variety can be defined to be the negative of the amount of information that is necessary to characterize each particle in the system uniquely in terms of its relations with the others.

We have studied the configurations of systems with high and low variety in several different models involving distributions of particles in one and two dimensions. In each case we have carried out numerical simulations to construct systems of high variety and compare them with randomly generated systems. In the Figures, I show an example of the result of such a simulation for a two dimensional case, which was carried out by Nick Benton. The variety in this case was defined the following way. In any such two dimensional distribution of N points, each particle may be said to have its own view of the configuration. This can be given by a configuration of $N - 1$ points on the circle, which represent the angles in which one must look from see each of the others. Each view is then coarse grained by dividing the circle up into a number (in this case 8) of equal sectors. A view is then a partition of the $N - 1$ points into these 8 sectors. To define the variety one compares the N views, by pairs. For each pair one defines the difference between the views to be the minimum of the sum of the differences between the numbers of points in each sector, as the two circles are rotated with respect to each other. The variety is then computed by summing these differences over all the pairs.

It is obvious from the definition that if a configuration has any symmetries, the variety will be relativity low. We found by numerical simulation that random configurations tend to have a similar, moderate value of the variety. Systems with high variety, on the other hand, were distinguished from both random configurations and configurations with symmetries by the presence of a high degree of structure.

Concluding remarks about the relationship between structure and diffeomorphism invariance

In this section I have given two very different reasons to expect that configurations of the gravitational field may in general be expected to be highly structured. The first is the tendency of the long ranged attractive nature of the gravitational field to produce gravitationally bound subsystems, and hence structure. The second is that structure follows from the requirement that the diffeomorphism invariant degrees of freedom distinguish each point from each other by the configuration of the fields. It is natural to ask whether there is a relationship between these two very different arguments. As we know that, in four dimensions, the the long ranged and attractive nature of the gravitational force is closely connected with the diffeomorphism invariance of the theory, it is not impossible that a deep connection may emerge between these two different reasons to expect that general relativity is a theory of a structured universes. An investigation of this question must be based on a treatment of the statistical mechanics of the gravitational field that takes into account its diffeomorphism invariance. This is not a subject that is very well developed, to my knowledge there exists so far only one paper about it (Rovelli 1993).

In this connection, it is interesting to point out that if the notion of time depends on the existence of subsystems with the properties of clocks, then the notion of time also requires a universe which is not in equilibrium, as there can be no good clocks in an equilibrium distribution. Thus, the relational point of view leads us to the conclusion that the notion of time that one uses in classical or quantum cosmology must be dependent on the notion of time from thermodynamics, at least in the sense that if there is no thermodynamic arrow of time (which arises only when a system is out of equilibrium) there can be no physical basis for any other notion of time. Perhaps all of these considerations taken together suggest that the long sought for connection between general relativity, quantum physics and thermodynamics, suggested by black hole thermodynamics and other results, could be found by studying the non-equilibrium, rather than the equilibrium, thermodynamics of the gravitational field.

10. STRUCTURE, EVOLUTION, AND TIME IN COSMOLOGY

In this section I would like to suggest that there is a notion of time which is available when we have a structured system, in the case that that structure is the result of the action of natural law. This is that in nature, a structured system must have a *history*, which is a description of its evolution from a less structured initial configuration through a succession of progressively more structured configurations to its present configuration. The reason for this is that structured configurations are very improbable. (If they were probable we would see structure in equilibrium configurations.) An explanation for how an improbable configuration is established must then tell us how it developed from an *a priori* more probable configuration. Any such explanation involves a sequence of configurations ordered according to some measure of structure or complexity. Such a sequence then provides a notion of time.

I will begin the argument of this section by asking a question: is gravitation sufficient to explain the structure we see around us in the universe? That is, if I postulate a universe described by general relativity coupled to some matter fields, will I expect in most cases to see a universe with the large amount of structure that we see around us?

Of course, it is hard to answer such general questions cleanly. But I believe that the answer is no. To explain why I need to formulate the question in a more precise way. At present, the best description that we can give of the laws of physics is to say that we live in a universe whose spacetime geometry is described by general relativity and whose particle content and forces are described by a quantum field theory. The quantum field theory which is presently believed to best describe nature[38] has a number of free parameters that characterize which particles and which interactions exist as well as the masses of the particles and the strengths of their coupling to those interactions. There are about twenty of these parameters. At present we don't know why they have the values they do; it is the goal of much current work in elementary particle theory to explain how they are fixed.

Let me then imagine an ensemble of possible universes that differ by the values taken by these parameters. To describe them I can consider that the possible values of the parameters make a space, which I will call the parameter space, denoted $\mathcal{P}$. A universe is described by a point of this space and an ensemble of universes by a probability distribution on it. We can define various functions on the space which represent how various properties of a universe can depend on the parameters. The interpretation of such a function will be statistical; it will tell us the value of that property averaged over all universes in the ensemble which share the same parameters. Examples of such functions are the expected value of the maximal spatial volume the universe attains, the expected value of its lifetime, and the expected value of the number of black holes created during the lifetime of the universe. Let me consider a function on the parameter space which measures the complexity of a universe. This can be done, for example, by measuring the variety, which seems to be a good measure of complexity. It is interesting to ask how the expected value of the complexity of the universe depends on the parameters. Of course, it is difficult to answer that question generally, given what we know presently. But there are arguments that suggest that the parameters of our universe are at, or close to, a local maximum of the complexity. These are based on our ability to construct arguments about what would happen to physics if we make changes in the parameters from their present values. These arguments suggest that, at least for a region of the parameter space around the parameters of our universe, a universe would not attain the complexity of our universe.

The evidence for this assertion is based on a series of arguments that suggest that the structures we see around us depend, for their existence and stability, on the parameters having, within rather narrow limits, the values we observe them to have.

To illustrate this, I will concentrate on a question I raised in the previous section: why do stars exist and why are they stable on cosmological time scales? It is quite remarkable that the time scale given by the lifetime of a typical star, which is a number that can be expressed only in terms of fundamental constants, is of the same order of magnitude as the Hubble time, which sets[39] the time scale for the expansion of the universe as a whole. This is possible only because a very large dimensionless number can be constructed from the fundamental constants of physics-this is the ratio between the proton mass and the Planck mass.[40] It is this very large ratio that allows stars to be fueled by nuclear processes with a time scale of the light crossing time of a nuclei, but to burn stably for ten billion years. Another way to say this is that the extreme smallness of the gravitational constant makes it possible to construct stable bound systems out of a great many fundamental particles without their collapsing into a black hole.

Thus, the existence of stars depend on the ratio of the proton mass to the Planck mass being a very small number. Were it, say, as large as one part in a million rather than one part in 10^{19}, stars would not exist.

This is not the only small ratio on which the existence of stars depend. Another number that must lie in a narrow range for stars to exist is the difference in the masses between the proton and the neutron. This is a small number, although not as small as the one we just considered because when measured in units of the proton mass itself it is about one and a half parts in a thousand. I will not give the argument here (it is given in (Smolin 1992a)), but it is the case that if the mass of the neutron were raised by a few more parts in a thousand of the proton mass, there would be no stable atomic nuclei besides the simple proton. A world in which the proton was the only stable nuclei would be drastically different from our own. As there would be no nuclear fusion there would be no stars and, obviously, there would be no chemistry. There would be hot self-gravitating clouds of gas, that would gradually cool, radiating their gravitational self-energy, but there would likely be very many less of them and they would be hot for very short times compared to the lifetimes of stars.

There are still other examples of this kind. Fred Hoyle discovered a long time ago that carbon would not be copiously produced by stars were it not for the existence of a coincidence in the nuclear energy levels of carbon and berylium; a change in several different constants by a small amount removes this coincidence (Hoyle).

I believe that these arguments establish that for most values of the parameters stars would not exist. As stars play an important role in the structuring of our present universe, by providing long lived and stable sources of energy to keep the interstellar and perhaps also the intergalactic mediums out of equilibrium, I believe it is at least plausible that complexity is a strongly varying function on the parameter space and that over most of the parameter space the expected complexity is less than that of our own universe.

If this is the case then we must ask why is it that the parameters of our universe fall in a region that is so favorable for structure. To the extent that clocks can only exist in a universe that is out of thermal equilibrium and sufficiently structured, an answer to this question must be part of an understanding of the problem of time.

One can imagine several different kinds of explanation that might be offered to answer a question like why the parameters of the laws of physics take the values they do. I would like to discuss two different types of explanations that have been the subject of much investigation. These are an explanation in terms of a hypothesized fundamental unified theory and an explanation using the anthropic principle.

During the last fifteen years, much of the research done in elementary particle theory has been motivated by the search for a unified theory of all of the interactions. Because such a theory must include a quantum theory of gravity we know that it would contain one dimensional scale, which is the Planck length. As a result, this scale, which I remind the reader is 10^{-33}cm, becomes the natural scale at which phenomena described by the theory are simple. The working assumption that has guided the search for a unified theory is that there is some very simple and beautiful mathematical theory that describes physics at this scale from which physics at all larger scales is to be deduced. It is also believed by many people working on this problem that there is only one, or at most a very few theories that are equally consistent and beautiful, so that once one has been constructed there will be no need for further explanation of why one theory has been picked over the others.

This one unique theory, if it exists, must then fix the values of all of the parameters mentioned above. We may then ask what kind of explanation such a theory would be able to give of why it is the case that the parameters fall into the region of parameter space that leads to a very large and structured universe. The problem is that given that the Planck scale is so far removed from the scale of nuclear and atomic physics it is difficult to see how this could come out as much more than a coincidence. It is very difficult to see how satisfying some mathematical condition of consistency at the Planck scale could have anything to do with the existence or nonexistence of a large number of stable nuclei, whose scale is twenty orders of magnitude larger. If such is the case, I think it could be only be considered to be an argument for a universe created by an intelligent god.

I have come more and more to the conclusion that it is very unlikely that such a theory exists. In my personal case, the failure of string theory to, so far, yield such a theory made a tremendous impression. However, once one focuses on the problem of why the universe is so structured, the idea of a fundamental Planck scale theory that fixes the parameters of physics into the narrow ranges necessary for structure seems, to me, as unlikely as the possibility that the biologists will discover a mathematical theory of DNA that tells us that the only stable structures are those that correspond to the organisms we see around us.

A second type of explanation for why the parameters take the values they do that has received a lot of attention is the anthropic principle.[41] This principle states that this must be the way we find the Universe, because we are in it, and were it not this way we would not exist. My personal view is that this is not any kind of explanation, it is only a restatement of the problem. At least as much as the postulate of a fundamental Planck scale theory, it leaves the question of the why the parameters are such as to allow the universe to be as structured as ours is outside the domain of rational explanation. What we would like is a theory that is based on a conventional dynamical or causal mechanism that explains to us how the parameters came to have the values that they do and which further explains why the parameters fall into the narrow ranges required for a structured universe.

If we reject these two alternatives, the question we must ask is: what kind of theory could do this for us?

We can see immediately that such a theory must have certain characteristics. First, it is necessary that the numbers that we understand as parameters in conventional quantum field theory be subject to dynamics. It must be either possible that they differ in different universes described by different solutions to the theory or that they can change as a function of time in a given solution. Second, the theory must provide a dynamical mechanism that fixes the parameters. Third, there should be a mechanism by means of which the expression of the parameters at large scales, such as the existence or non existence of stars, can have a causal influence on the mechanism by which the values are selected.

It is interesting to ask whether there is anywhere in natural science a theory of this kind. I know of only one such example, it is the theory of natural selection in biology.

Not only is natural selection the only well developed example of such a theory in science, an argument can be given that it is the only possible example of such a theory. This argument is given by Richard Dawkins (1986), I will summarize it here.[42] In both biology and the present case the situation can be described in the following way. We have a very large parameter space; in the case of biology it is the space of all possible DNA sequences, in fundamental physics it is the parameter space, $\mathcal{P}$. A given system is represented by a point, and an ensemble of systems is represented by a distribution function on this space. We may consider, as we did for the case of cosmology, a function on this space, which measures the average complexity of a representative system constructed with these parameters (genes or coupling constants). This function varies a great deal over the space, in particular for most of the parameter space the complexity will be low while the regions of parameters that lead to high complexity are of rather small measure. This is because for most genes sequences and most choices of fundamental parameters the system does not live very long measured in the appropriate fundamental units of time. (Here live can be taken in both cases to mean maintain a state far from thermal equilibrium.)

The fact we want to explain is that we observe a system, or systems, whose parameters are in a region corresponding to high complexity. A key point is that because the volume of such regions is very small, if the parameters are chosen randomly it is extremely improbable that they will be in such a region. At the same time, we do not want to invoke the fact of complexity directly in the causal dynamics of the system, as we want to avoid explanation by final causes. So we rule out Lamarkian explanations or any other such explanation in which the parameters evolve subject to a causal dynamics that seeks to extremize the complexity.

The invention of Darwin and Wallace is that one can have an explanation for why it is probable to find such a system in the region of high complexity that is completely causal. There are two key ideas in their theory. The first is random variation, followed by selection. However, this is not sufficient by itself to make it probable for the system to move from a random starting point of low complexity to a final point of high complexity. It is very improbable to jump, in one step, from a bacteria to a whale. The second key idea is then that, to make it probable, this evolution must be accomplished in a large number of small steps, in each one of which the complexity is increased by a small amount.

Dawkins's point, which I agree with, is that this is the only known way in which structured complexity can be generated, over time, from a system that begins with much less structure, if we allow ourselves only the conventional kinds of explanation through some causal mechanism.

In this light, it is then interesting to attempt to construct an analogy in which the space of the possible values of the genes of living organisms is analogous to the parameter space of quantum field theory and the expression of the genes in the actual organisms is analogous to the observed phenomena of physics and astronomy. This leads us to ask whether or not a mechanism analogous to natural selection could operate in the universe which would serve to select the parameters of physics and cosmology in the narrow range that allows stars to exist. Such a mechanism, were it to be found, would serve as a causal explanation for why the parameters of the laws of physics take unnatural values in a narrow range that allows a universe as complex as ours to exist.

Such mechanisms can be proposed, as I will now describe. I will describe two briefly, they are treated in more detail elsewhere. They are, admittedly, outlandish; my purpose in mentioning them here is not to convince you of their plausibility, but only to demonstrate, by example, that such explanations are possible. After describing them I will return to the problem of time.

The first condition that is required for such an explanation, that the parameters are mutable, is indeed a feature of several currently popular theories in elementary particle physics. What is then needed is only a selective mechanism to determine which values of the parameters are actually expressed in our universe. One such mechanism requires only accepting another hypothesis long popular among theoretical physicists- that each singularity inside of a black hole is turned by quantum effects into a tunnelling event that results in the birth of a new universe. These two ideas can be combined to yield a mechanism for natural selection of the parameters of quantum field theory if one adjoins the hypothesis (quite natural in this context) that during the tunneling event leading to the birth of a new universe the parameters undergo a small, random, variation.

To see that a mechanism of natural selection then follows, consider the ensemble of universe that have come into existence after many generations of universal reproduction. If we denote the parameters collectively by p, then the ensemble can be represented by a distribution $\rho(p)$ on the parameter space $\mathcal{P}$. Let us then have a function $N(p)$ which measures the expected number of black holes in a universe with the parameters p. It is then easy to demonstrate that after many generations ρ is concentrated around the local maxima of $N(p)$. This is a mechanism for natural selection that selects universes that have many black holes.

If we then assume that we live in a typical universe (such an assumption is always necessary when employing such an essentially statistical explanation), it then follows that the parameters of our universe are such as to extremize (at least locally in $\mathcal{P}$) the number of black holes. This could provide an explanation for why the laws of physics are such that stars exist, because at least one way to make many black holes is to have many stars.

We may also note that this explanation could be tested empirically, because it has as a corollary that most changes in the parameters will lead to a decrease in the expected number of black holes in the universe. The possibility of carrying out such a test is discussed in (Smolin 1992a).

A second kind of mechanism might operate during the era of nucleosynthesis, in a cold big bang model (Rees 1972, 1978; Layzer and Hively 1973; Carr 1977, 1981; Carr, Bond, and Arnett 1984; Teresawa and Sato 1985). In such a model (but not in the more commonly accepted hot big bang model), the energetics of nucleosynthesis can have an effect on the large scale structure of the universe. One can then invent a mechanism in which universes, or regions of the universe, blow up and thus live a long time compared to nuclear time scales, if the parameters are such that the helium and deuterium nuclei are stable (Smolin 1992a).

These examples demonstrate at least that a mechanism for the evolution of the parameters through natural selection can be invented. Moreover, on the scale of ideas commonly discussed these days in cosmology these ideas are not really too far out. If the reader will accept it as plausible that something like this might have occurred I would like to return to the problem of time, for I believe that the implications of such a theory for that problem are profound.

The reason is that any explanation of evolution by natural selection is based essentially on an analysis of probabilities. Once it is admitted that what is to be explained is why the system in question has such an improbable amount of structured complexity, the question is to invent a mechanism by which this could have been shaped by natural law. The background for any such explanation must be based on the idea that the system in question is embedded in an open far from equilibrium thermodynamic system; this is certainly the case for biology and I argued above it is plausible that the a gravitationally bound system, such as the universe as a whole can be seen this way. But, to explain the high level of structure that we see around us it is not sufficient to invoke non-equilibrium thermodynamics. We must explain why it is probable that the open system evolved such a high level of complexity.

The key point is that, as I mentioned above, it is never probable for a system to jump in one step from a homogeneous to a highly structured configuration. The reason is that the exploration of the parameter space by the system must be random if the explanation is to not invoke final causes, and the region of the parameter space in which one finds stable structure is small. However, it is probable that the system reach a highly structured state by evolving through a large number of small steps, in each of which the system randomly explores a small region of the parameter space.

This is the reason that any structured system whose structure is the result of self-organization from a less structured system under natural law has a history. It is this history that, I would like to claim, must be the ultimate origin of the notion of time in a relational universe.

11. CONCLUSIONS

In this paper I have described several different aspects of the problem of time in a universe described by general relativity and quantum mechanics. I would like here to summarize the entire argument and by so doing highlight the way that the different parts of it come together to suggest the possibility of a physical notion of time based on the evolution of structure.

In the earlier sections of this paper I showed that time in quantum cosmology must be a relational concept, whose description is based on correspondences or correlations between physical degrees of freedom of the universe. This is necessary because the conventional notions of time in physics correspond to the readings of a clock outside of the system being described, and there is, by definition, no such clock outside the system in the case of cosmology. As I then explained, this means that, at least as far as is known, the definition of time in quantum cosmology must be contingent, in the sense that a particular universe must have certain characteristics if there is a concept of time that can usefully be applied to its description. The main such characteristic is that a universe must be complex enough to have metastable subsystems which behave (necessarily to some degree of approximation) like ideal clocks. Furthermore, the universe should be structured enough that its configurations contain what Barbour calls "time capsules."

I thus concluded that the question of time in quantum cosmology must rest on the problem of structure. If we are to understand why time is such a fundamental aspect of the universe and our experience of it, we must understand why our universe is sufficiently structured as to allow clocks and time capsules to exist. I then raised the question of why the universe is in fact as structured as it is. Closely connected to this is the question of why the matter in the universe is not, an enormously long time (measured in fundamental units) after the universe's apparent origin, in thermal equilibrium. I argued that this depended on two facts about the universe. First, the existence of gravitation, and second, the parameters of physical theories being in very narrow ranges that allow structures of great complexity to be stable over times comparable to the age of the universe.

Finally, I argued in the last section that the only kind of explanation we know of for the existence of systems with structured complexity, which is based on the conventional assumptions of causality, is incremental evolution by random variation followed by natural selection. Any other kind of explanation must either assume what is to be shown, which is that the universe is complex enough for clocks and time capsules to exist, or rest on a coincidence such as the assumption that there is one possible consistent theory of Planck scale physics with certain characteristics *and* that this one theory happens to predict that the coupling constants of the interactions we observe at enormously larger scales are in the narrow ranges that allows stars, clocks and time capsules to exist.

Let me grant that this conclusion must be considered tentative, at the very least for its distance from conventional cosmological ideas. Let us, however, ask, what implications follow about the nature of time if we accept it. I believe there are several, I close this article by listing them.

1) The notion of evolution through a series of steps, each differing incrementally from the previous one, provides a notion of time which is purely relational as it is based on a concrete property of the configuration of the universe: its complexity.

2) Furthermore, if we make structure a measure of time we have a way of ordering, at least partially, the ensemble of different configurations that, according to Barbour's proposal, are given appreciable probability by the quantum state of the universe. Thus, we do not need any *a priori* ordering of configurations as long as the laws of nature (expressed in quantum cosmology by the Wheeler-DeWitt equation) provide an explanation for why the probable configurations are sufficiently structured.

Thus, if we tie time to structure and complexity, we may find out that the laws of nature predict that time is a universal aspect of phenomena because the occurrence of a succession of structured configurations turns out to be a necessary consequence of those laws. Moreover, a deterministic notion of causality emerges exactly where we need it to, in the classical limit. In addition, in this picture the cosmological notion of time must be in close correspondence with the thermodynamic notion.

Furthermore, this picture can work both within the evolution of a single universe and within the evolution of the whole ensemble needed to make sense of the proposal involving black holes I described in the last section. Indeed, from the point of view of Barbour's proposal, the distinction between a single universe, which may be described approximately by a single classical spacetime, and the whole ensemble is not fundamental-it emerges only in the classical limit. That is, in the context of Barbour's proposal, we can give definite meaning to the ensemble of many classical universes employed there.

At least in the context of the evolution of a single universe, the idea of using complexity as a measure of time fits well also with Rovelli's notion of time, as there is nothing in his proposal to specify *how* time is measured by physical clocks.

3) In the course of this essay I mentioned two types of explanation for the occurrence of structured complexity in a system described by general relativity. These were the formation of structures in open, far from equilibrium, statistical mechanical systems and natural selection through random variation. There is a strong suspicion that these two stories are closely related to each other. That is, natural selection through random variation may be a general description of how open systems far from equilibrium find, by random exploration of their phase spaces, the metastable structures that characterize what have been variously called dissipative structures or self-organized systems. This suspicion underlies the search for an understanding of natural selection, and biology in general, as an aspect or consequence of far from equilibrium thermodynamics.

In this essay I have suggested that both ideas: natural selection and the self-organization of far from equilibrium systems may have roles to play in our understanding of cosmology. What I am proposing is that such a general theory that ties together natural selection and self-organized systems, will also become the basis for our understanding of structure and time in cosmology.

However, what I am proposing is, in a sense, more even than that. Because, such a theory, if it is to be a theory of cosmology, must fit into the framework of our understanding of space and time in general relativity. The key idea that underlies the description of space and time in general relativity is diffeomorphism invariance, which is, in turn, an expression of the old Leibnizian ideal of a purely relational theory of space and time. Furthermore, as I've argued here, it is a necessary consequence of a diffeomorphism invariant description-or more generally any purely relational description-of a closed universe, that the universe be structured, because the physical observables that characterize it are defined unambiguously only in the case that it is possible to distinguish between the views of the universe as seen from any two points in spacetime.

To summarize, a theory of a relational universe must be a theory of a structured universe. Furthermore, if that theory is to be complete, it must explain how it is that the universe is, in fact, sufficiently structured to allow such a description to make sense. That is a relational theory of the universe must not only use structure and complexity in the construction of its concept of time, it must be a theory of how that structure and complexity comes to exist.

John Wheeler sometimes quotes Einstein as praising Newton for having the judgement and the courage to forego the vision of Descartes and Leibniz of a universe based solely on relational notions of space and time (Wheeler). Given the conceptual and mathematical equipment that was available at that time, Leibniz could not possibly have succeeded in founding a consistent and useful physics or cosmology on purely relational principles. A whole set of ideas was missing; among these was the fantastic realization of Einstein that a relational theory of space and time could be constructed by identifying the gravitational field with the geometry of spacetime.

In this essay I have tried to explain how the struggle to construct a relational theory of space and time underlies the major issues we wrestle with in our attempts to construct a quantum theory of gravitation and cosmology. I do not know how close we are to the realization of such a theory. It is easy to imagine that there are still missing elements about which we are as ignorant as Leibniz was of the metric tensor and the equivalence principle. The chain of thought that I have tried to sketch in this essay leads me to believe that one of those missing elements must be a theory of how the universe has come to be as structured as it apparently is. If this is the case then I believe we must try to take seriously what the biologists have learned over the last one hundred years. This is that there is only one kind of theory that is capable of explaining from first principles, and without going outside of the usual ideas of causality, how a system with a high amount of structured complexity can come to exist. This is a theory of evolution through random variation and incremental natural selection.

Granted, the idea of applying ideas from biology to cosmology may be a dangerous one. But, if we begin with the idea that what we want is a theory of the cosmos that is based on the idea that space and time are relational quantities, is there any other framework, besides the evolution of structure through random variation and incremental selection, that is capable of explaining, from first principles, how a universe can come to exist that is complex enough to account for all the richness that the phenomena of time has in our universe?

Pennsylvania State University

NOTES

1. This original version of this essay was presented at a conference Tempo nella scienzia e nella filosofia organised in Napoli by Prof. E. Agazzi and I understand is eventually to appear in Italian translation in the procedings of that conference. Some of the ideas first developed here were later incorporated in my book, The Life of the Cosmos (OUP and Weidenfeld and Nicolson, 1997). I am grateful for the opportunity to publish this essay in the original English, and I can think of no better tribute to the example and influence of John Stachel's life and work than to dedicate this essay to him. For me and others of my generation, John Stachel has provided a rare example of how one may successfully combine science and philosophy in a single quest for truth and for a better world.
   This essay was my first exploration of some ideas that later become the core of the argument in my book, *Life of the Cosmos* (Oxford Univerity Press, New York and Weidenfeld and Nicolson, 1998).
2. I think it is likely that one reason that physics is so difficult to teach is that we forget how genuinely new and difficult these concepts that underlie the representation of time in Newtonian physics really are for people who are seeing them for the first time.
3. For those unfamiliar with it, this is a particular formulation of mechanics, due to the 19th century Scottish mathematician, Hamilton.
4. The phase space is a mathematical space whose coordinates are the positions and the momenta of each of the particles or degrees of freedom of the system. For example, for a system of N particles moving in 3 dimensional space it has $6N$ dimensions.
5. Barbour and Bertotti find a very interesting way to deal with the problem of the relativity of simultaneity (1982).
6. It is given in (Smolin 1984a, 1991).
7. In (Smolin 1991) I called them ideal elements.
8. In many cases it is useful to model an isolated system in a larger universe in general relativity. This is done by choosing boundary conditions such that the gravitational field, and hence the curvature of spacetime, fall off as we recede from the system. While this is a useful approximation in many circumstances, to choose such boundary conditions essentially restores an external, Newtonian, notion of time to the theory. To consider what general relativity has to say about time in the context of cosmological theories, we can ignore such cases and restrict the discussion to closed systems, as I will do here.
9. It must be admitted that the ambition to eliminate background structures from our description of nature can be fulfilled in any given theory only imperfectly. What is important for our discussion is how much of the notion of time is in the background structure and how much is in the dynamical structure.
10. Exactly how the fields are specified is a technical problem that I do not need to go into here.
11. This is, of course, exactly what Leibniz called the identity of the indiscernible; it is a direct consequence of having based the interpretation of the theory on relational ideas.
12. I do not know who is the originator of the point of view that I describe here. I learned it from conversations with a number of people including Abhay Ashtekar, Julian Barbour, Karel Kuchar and Carlo Rovelli. It is certainly implicit in many of the early works on quantum cosmology such as those by Peter Bergmann (1956a, 1956b), Bryce DeWitt (1967) and Charles Misner (1970, 1972).
13. This can be given a precise formulation in terms of measure theory, I do not do this here.
14. This point of view is well described in a review by Karel Kuchar (1992).

15. Although, as I describe in section 9, it seems that the assumption of homogeneity that otherwise justifies its use is less and less supported by the observational evidence (Coleman and Pietronero 1992).
16. For other works in this direction, see also (DeWitt et al. 1991).
17. These are the four dimensional versions of triangles and tetrahedrons.
18. The Planck length is the basic unit that sets the scale of phenomena in a quantum theory of gravity. It is given by $\sqrt{\hbar G/c^3}$ and is approximately equal to 10^{-33} cm.
19. The key idea behind the specification of the coordinates and the momenta is that if one gives them at one moment of time one one has sufficient information to predict their values for all time using the equations of motion of the theory. One must specify two quantities, the coordinates and momenta, because Newton's law and its generalizations determine the second time derivatives of the coordinates. In this case there are two quantities: the coordinates and their first time derivatives that must be specified initially, and which are henceforth determined by the equations of motion. The idea of the Hamiltonian formalism is to express the same physics in terms of only first time derivatives. That is why the Hamilton's equations have the form in (2) and (3).
20. Let me say a word about this as there is a possible confusion here. To say the Hamiltonian is equal to zero is not the same thing as to say there is no Hamiltonian. The Hamiltonian is some definite function, whose derivatives may be taken as in (2) and (3) to find the evolution equations. This function has the property that when the constraints are satisfied the equation that says that the function vanishes is also satisfied.
21. It is interesting to note that these constructions can be done, and the theory completely written down and solved exactly, in a model system which is general relativity with space taken to have 2 rather than 3 dimensions (Witten 1988; Ashtekar et al. 1989; Ashtekar 1991; Smolin 1989). In this theory there are no propagating gravitational waves so that there are actually only a finite number of degrees of freedom. Further, the problem of time can be completely solved in this theory, and it can be done in any of the different ways that are described here (Carlip 1990, 1992).
22. Note that this is a necessary, but not a sufficient, condition. The superposition of two semiclassical states, each based on a different trajectory, is not itself a semiclassical state.
23. The theory of how to classify knots is an intriguing and active branch of mathematics; for a good introduction see (Kauffman 1991).
24. For details see (Smolin 1992).
25. See, for example, the sections on the semi-classical interpretation in the review (Kuchar 1992).
26. For details see (Smolin 1992).
27. Which is based on using the quantum state directly to construct a probability density.
28. A more complete exposition of this view will be contained in a book (Smolin 1997).
29. A nice introduction to the field can be found in (Rubin and Coyne 1988).
30. See, for example, (Franco and Cox 1983; Franco and Shore 1984; Ikeuchi and Tanaka 1984; Wyse and Silk 1985; Dopita 1985; Parravano 1988; Parravano, Rosenzweig, and Teran 1990; Hensler and Burkert 1990; Elmegreen 1992; Seiden and Schulman 1990, 1986).
31. Of course, this is generous; during the early evolution of the universe, the relevant time scales were much shorter. Furthermore, as there seem at present to be difficulties with the models that were proposed to account for the presently seen large scale distributions as the result of fluctuations of a earlier thermalized distribution, it may be necessary to explain why equilibrium was not established during each of the earlier epochs of the universe.
32. For example, one problem in the theory of galaxies which is still not completely resolved is the question of why the spiral arms of galaxies persist on cosmological time. It may be easier to account for this in the context of a model that treats a spiral galaxy as a far from equilibrium system that is in a stable state governed by feedback mechanisms (see note 30).
33. It is possible that carbon is necessary also for the self-regulation processes that govern the star formation rate and thus produces the spiral structures of galaxies. This is because carbon dust grains and organic chemical reactions are hypothesized to play a role in the cooling and feedback processes in the molecular clouds within which star formation takes place (see note 30).
34. For an introduction, see (Peebles 1971). For a review, see the contribution by M. Davis and G. Efstathiou in (Davis and Efstathiou 1988).

35. In the quantum theory, exactly half as many are required, because of the uncertainty principle. The argument as I am about to make it applies equally well to either the classical or the quantum theory.
36. This definition is due to Julian Barbour, in a different context (Barbour 1989a; Barbour and Smolin 1992). Note that an analogous definition may be given in quantum cosmology. We may call a quantum state of a quantum cosmology *Leibnizian* if all the expectation values of operators that measure relational variables of the sort described here are distinct.
37. As I do not have a proof for this claim, I will offer some very weak evidence for it. This comes from the recent progress in classical relativity which seems to suggest that as we get closer to specifying the theory in terms of gauge invariant variables, we find that what we have is a form of the equations that is equivalent to general relativity for most, but not all configurations. The correspondence between the gauge invariant and the conventional description always holds for generic fields, but in some cases it fails for configurations with symmetries, such as Minkowski spacetime. This is the case for the Capovilla-Dell-Jacobson variables (Capovilla, Dell, and Jacoboson 1989, 1991, 1991), which result from a solution of four out of seven of the constraint equations of quantum gravity. The formulation of Rovelli and Newman, which eliminates six of the seven constraints also has this property (Newman and Rovelli 1992). The conjecture I am making here is that when a gauge invariant form of the dynamics of general relativity is constructed, it will be equivalent to the conventional form of general relativity for Leibnizian solutions, but the equivalence will fail for solutions with symmetries.
38. It is sometimes called the standard model, it is based on a Yang-Mills theory with gauge group $SU(3) \times SU(2) \times U(1)$.
39. Or, more properly, bounds from below.
40. The mass that is made from Planck's constant, Newton's constant and the speed of light.
41. A good review of the subject is given in (Barrow and Tipler 1989).
42. In Dawkin's words: "The theory of evolution by cumulative natural selection is the only theory we know of that is in principle *capable* of explaining the existence of organized complexity. Even if the evidence did not favour it, it would *still* be the best theory available" (Dawkins 1986, 317).

REFERENCES

Agishtein, and A. A. Migdal. 1992a. *Phys. Lett. B* 278:42–50.
———. 1992b. "Simulations of 4-Dimensional Simplical Quantum-Gravity as Dynamic Triangulation." *Modern Physics Letters A* 7:1039–1061.
Arnowitt, R., S. Deser, and C. W. Misner. 1960. "The Dynamics of General Relativity." *Phys. Rev.* 117:1595. [Reprinted 1962 in *Gravitation, An Introduction to Current Research*, ed. L. Witten. New York: Wiley].
Ashtekar, A. V. 1986. "New Variables for Classical and Quantum Gravity." *Physical Review Letters* 57(18):224 2247.
———. 1987. *Physical Review D* 36:1587.
———. 1991. *Non-Perturbative Canonical Gravity. Lecture notes prepared in collaboration with Ranjeet S. Tate*. Singapore: World Scientific Books.
Ashtekar, A. V., V. Husain, C. Rovelli, J. Samuel, and L. Smolin. 1989. "2+1 quantum gravity as a toy model for the 3+1 theory." *Class. and Quantum Grav.*, pp. L185–L193.
Ashtekar, A. V., C. Rovelli, and L. Smolin. 1992. "Weaving a Classical Metric with Quantum Threads." *Physical Review Letters* 69(2):237–240.
Barbour, J. B. "Historical background to the problem of inertia." Unpublished manuscript.
———. 1974. "Relative-distance Machina Theories." *Nature* 249:328. [Erratum: *Nature* 250(5467):606].
———. 1975. *Nuovo Centimo* 26B:16.
———. 1982. *Brit. J. Phil. Sci.* 33:251.
———. 1987. "Leibnizian Time, Machian Dynamics and Quantum Gravity." In *Quantum Concepts of Space and Time*, eds. C. J. Isham and R. Penrose. Oxford: Oxford University Press.
———. 1989a. *Found. Physics* 19:1051–73.
———. 1989b. *Absolute or Relative Motion, Vol. 1: The Discovery of Dynamics*. Cambridge: Cambridge University Press.

———. 1992a. In *Proceedings of the NATO Meeting on the Physical Origins of Time Asymmetry*, eds. J. J. Halliwell, J. Perez-Mercader, and W. H. Zurek. Cambridge: Cambridge University Press.

———. 1992b. "Time and the interpretation of quantum gravity." Syracuse University Preprint.

Barbour, J. B., and B. Bertotti. 1977. *Nuovo Centimo* 38B:1.

———. 1982. *Proc. Roy. Soc. Lond. A* 382:295.

———. 1989. *Proc. Roy. Soc. Lond.* A 382:295.

Barbour, J. B., and L. Smolin. 1992. "Extremal variety as the foundation of a cosmological quantum theory." Syracuse University Preprint.

Barrow, J., and F. Tipler. 1989. *The Anthropic Principle*. Oxford: Oxford University Press.

Benton, Nick. personal communication.

Bergmann, P. G. 1956a. *Helv. Acta. Suppl.* 4:79.

———. 1956b. *Nuovo Cimento* 3:1177–1185.

Bombelli, L., J. Lee, D. Meyer, and R. D. Sorkin. 1988. *Physics Letters* 60:655.

Broadhurst, R., R. S. Ellis, D. C. Koo, and A. S. Szalay. 1990. "Large-Scale distributions of Galaxies at the Galactic Poles." *Nature* 343(6260):726–728.

Capovilla, R., J. Dell, and T. Jacoboson. 1989. *Phys. Rev. Lett.* 63:2325.

———. 1991. *Class. and Quant. Grav.* 8:59.

Capovilla, R., J. Dell, T. Jacoboson, and L. Mason. 1991. *Class. and Quant. Grav.* 8:41.

Carlip, S. 1990. *Phys. Rev. D* 42:2647.

———. 1992. *Phys. Rev. D* 45:3584. [UC Davis preprint UCD-92-23].

Carr, B. J. 1977. *M. N. R. A. S.* 181:293.

———. 1981. *M. N. R. A. S.* 195:669.

Carr, B. J., J. R. Bond, and W. D. Arnett. 1984. *Ap. J.* 227:445.

Coleman, P., and L. Pietronero. 1992. "The fractal structure of the universe." *Physics Reports* 213:311–389.

Davis, M., and G. Efstathiou. 1988. In *Large-Scale Motions in the Universe: A Vatican Study Week*, eds. V. C. Rubin and G. V. Coyne. Princeton: Princeton University Press.

Dawkins, R. 1986. *The Blind Watchmaker*. New York: W. W. Norton.

DeWitt, B. S. 1967. *Phys. Rev.* 160:1113.

DeWitt, B. S., E. Myers, R. Harrington, and A. Kapulkin. 1991. *Nucl. Phys. B* (Proc Supp.) 20:744.

Dirac, P. A. M. 1959. *Phys. Review* 114:924–930.

———. 1964. *Lectures on Quantum Mechanics*. Belfer Graduate School of Science, Yeshiva University, New York.

Dopita, M. A. 1985. "A law of star formation in disk galaxies: Evidence for self-regulating feedback." *Astrophys. J.* 296:L1–L5.

Einstein, A. 1934. "On the method of theoretical physics." In *A. Einstein: Essays in Science*. New York: Philosophical Library.

Elmegreen, B. G. 1992. "Large Scale Dynamics of the Interstellar Medium." In *Interstellar Medium, processes in the galactic diffuse matter*, eds. D. Pfenniger and P. Bartholdi. Springer Verlag.

Everett III, H. 1957. *Rev. Mod. Phys.* 29:454.

———. 1973. In *The Many Worlds Interpretation of Quantum Mechanics*, eds. B. S. DeWitt and N. Graham. Princeton: Princeton University Press.

Finkelstein, D. 1989. "Quantum Net Dynamics." *International Journal of Theoretical Physics* 28:441–467.

———. 1992. "Higher-Order Quantum Logics." *International Journal of Theoretical Physics* 31:1627–1630.

Franco, J., and D. P. Cox. 1983. "Self-regulated star formation in the galaxy." *Astrophys. J.* 273:243–248.

Franco, J., and S. N. Shore. 1984. "The galaxy as a self-regulated star forming system: The case of the OB associations." *Astrophys. J.* 285:813–817.

Geroch, R. 1984. "The Everett Interpretation + Quantum Mechanics." *Nous* 18:617.

Goldberg, J., J. Lewendowski, and C. Stornaiolo. 1992. "Degeneracy in the loop variables." *Commun. Math. Physics* 148:337.

Hasse, M., M. Kriele, and V. Perlick. 1996. "Caustics of Wavefronts in General Relativity." *Class. Quatum Grav.* 13:1161.

Hensler, G., and A. Burkert. 1990. "Self-regulated star formation and evolution of the interstellar medium." *Astrophys. and Space Sciences* 171:149–156.

Hoyle, F. Unpublished.
Ikeuchi, A. Habe, and Y. D. Tanaka. 1984. "The interstellar medium regulated by supernova remnants and bursts of star formation." *MNRAS* 207:909–927.
Jacobson, T., and J. Romano. 1992. Maryland preprint.
Jacobson, T., and L. Smolin. 1988. *Nucl. Phys. B*, vol. 299.
Kauffman, L. 1991. *Knots and Physics*. Singapore: World Scientific.
Koestler, A. 1959. *The Sleepwalkers*. London: Penguin.
Kuchar, K. V. 1992. "Time and interpretations of quantum gravity." In *Proceedings of the 4th Canadian Conference on General Relativity and Relativistic Astrophysics*, eds. G. Kunstatter, D. Vincent, and J. Williams. Singapore: World Scientific.
Layzer, D., and R. M. Hively. 1973. *Ap. J.* 179:361.
Leibniz, G. W. 1973. ""The Monadology" and "On the principle of indicernibles," from the Clark-Leibniz Correspondence." In *Leibniz, Philosophical Writings*, ed. G. H. R. Parkinson. Translated by M. Morris and G. H. R. Parkinson. London: Dent.
Lovelock, J. 1988. *Gaia: A New Look at Life on Earth, The Ages of Gaia*. New York: W. W. Norton and Co.
Lyndon Bell, D., and R. M. Lyndon Bell. 1977. *Mon. Not. R. Astron. Soc.* 181:405.
Mach, E. 1866. *Fichtes Zeitschrift für Philosophie* 49:227.
———. 1893. *The Science of Mechanics*. London: Open Court.
Misner, C. W. 1970. "Classical and quantum dynamics of a closed universe." In *Relativity*, eds. M. Carmelli, S. I. Fickler, and L. Witten. New York: Plenum.
———. 1972. "Minisuperspace." In *Magic withoiut Magic: John Archibald Wheeler*, ed. J. Klauder. San Fransisco: Freeman.
Morowitz, H. 1968. *Energy Flow in Biology*. Academic Press.
Newman, E. T., and C. Rovelli. 1992. "Generalized Lines of Force as the Gauge-Invariant Degrees of Freedom for General Relativity and Yang-Mills Theory." *Physical Review Letters* 69(9):1300–1303.
Newton, I. 1962. *Mathematical Principles of Natural Philosophy and the System of the World*. University of California Press.
Parravano, A. 1988. "Self-regulating star formation in isolated galaxies: thermal instabilities in the interstellar medium." *Astron. Astrophys.* 205:71–76.
Parravano, A., P. Rosenzweig, and M. Teran. 1990. "Galactic evolution with self-regulated star formation: stability of a simple one-zone model." *Astrophys. J.* 356:100–109.
Peebles, P. J. E. 1971. *Physical Cosmology*. Princeton: Princeton University Press.
Penrose, R. 1971. In *Quantum Theory and Beyond*, ed. T. Bastin. Cambridge: Cambridge University Press.
———. 1979a. In *Advances in Twistor Theory*, eds. L. P. Hughston and R. S. Ward. San Francisco: Pitman.
———. 1979b. "Singularities and Time Asymmetry." In *General Relativity, An Einstein Centanary Survey*, eds. S. W. Hawking and W. Israel. Cambridge: Cambridge University Press.
Prigogine, I. 1967. *Introduction to the Thermodynamics of Irreversible Processes*. New York: Interscience.
———. 1980. *From Being to Becoming: time and complexity in the physical sciences*. San Fransisco: Freeman.
Rees, M. J. 1972. "Origin of Cosmic Microwave Background Radiation in a Chaotic Universe." *Phys. Rev. Lett.* 28:1669.
———. 1978. *Nature* 275:35.
Rovelli, C. 1990. "Quantum Mechanics without Time — A Model." *Phys. Rev. D* 42:2638.
———. 1991a. In *Conceptual Problems of Quantum Gravity*, eds. A. Ashtekar and J. Stachel. Boston: Birkhauser.
———. 1991b. "Ashtekar Formulation of General Relativity and Loop-Space Nonperturbative Quantum-Gravity — A Report." *Classical and Quantum Gravity* 8:1613–1676.
———. 1991c. "Quantum Reference Systems." *Class. Quantum Grav.* 8:317–331.
———. 1991d. "Time in Quantum Gravity." *Phys. Rev. D* 43:442.
———. 1993. "Statistical mechanics of the gravitational field and the thermodynamic origin or time." *Class. Quantum Grav.* 10(8):1549–1566.
Rovelli, C., and L. Smolin. 1988. "Knots and quantum gravity." *Phys. Rev. Lett.* 61:1155.
———. 1990. "Loop representation for quantum General Relativity." *Nucl. Phys. B* 133:80.

Rubin, V. C., and G. V. Coyne, eds. 1988. *Large-Scale Motions in the Universe: A Vatican Study Week*. Princeton: Princeton University Press.

Seiden, P. E., and L. S. Schulman. 1986. "Percolation and galaxies." *Science* 233:425–431.

———. 1990. "Percolation model of galactic structure." *Advances in Physics* 39:1–54.

Smolin, L. 1984a. "On quantum gravity and the many worlds interpretation of quantum mechanics." In *Quantum theory of gravity (the DeWitt Feschrift)*, ed. Steven Christensen. Bristol: Adam Hilger.

———. 1984b. "The thermodynamics of gravitational radiation." *Gen. Rel. and Grav.* 16:205.

———. 1985. "On the intrinsic entropy of the gravitational field." *Gen. Rel. and Grav.* 17:417.

———. 1989. "Loop representation for quantum gravity in 2+1 dimensions." In *Proceedings of the John's Hopkins Conference on Knots, Tolopoly and Quantum Field Theory*, ed. L. Lusanna. Singapore: World Scientific.

———. 1991. "Space and time in the quantum universe." In *Proceedings of the Osgood Hill conference on Conceptual Problems in Quantum Gravity*, eds. A. Ashtekar and J. Stachel. Boston: Birkhauser.

———. 1992a. "Did the Universe evolve?" *Classical and Quantum Gravity* 9:173–191.

———. 1992b. "Recent developments in nonperturbative quantum gravity." In *The Proceedings of the 1991 GIFT International Seminar on Theoretical Physics: "Quantum Gravity and Cosmology"*. held in Saint Feliu de Guixols, Catalonia, Spain: Singapore: World Scientific.

———. 1997. *The Life of the Cosmos*. Oxford University Press and Weidenfeld and Nicolson.

Spivak, M. 1970. *A Comprehensive Introduction to Differential Geometry, Vol. 1*. Publish or Perish Press.

Stachel, J. 1989. "Einstein's search for general covariance 1912–1915." In *Einstein and the History of General Relavity*, vol. 1 of *Einstein Studies*, eds. D. Howard and J. Stachel. Boston: Birkhauser.

Teresawa, N., and K. Sato. 1985. *Ap. J.* 294:9.

Wald, R. S. 1984. *General Relativity*. University of Chicago Press.

Wheeler, J. A. personal communication.

———. 1957. "Assessment of Everett's Relative State Formulation of Quantum Theory." *Rev. Mod. Phys.* 29:463.

Witten, E. 1988. *Nucl. Phys. B* 311:46.

Wyse, R. F. G., and J. Silk. 1985. "Evidence for supernova regulation of metal inrichment in disk galaxies." *Astrophys. J.* 296:11–15.

JAMES L. ANDERSON

TIMEKEEPING IN AN EXPANDING UNIVERSE

Abstract. The time-keeping property of a model 'electric' clock in an Einstein-deSitter universe is examined. The dynamics of such a clock is derived from the Einstein-Maxwell field equations using a modified version of the Einstein-Infeld-Hoffmann (EIH) surface integral method for obtaining equations of motion for compact sources of the gravitational and electromagnetic fields. The results show that such a model is only approximately an 'ideal' clock and differs from one by terms of the order of (clock period/Hubble time)^2.

1. INTRODUCTION

The original formulation of the general theory of relativity included an assumption regarding the measurement of time intervals, namely that ideal clocks measure the proper time along their trajectory as computed using the gravitational field tensor $g_{\mu\nu}$ as a Riemannian metric in the line element

$$ds^2 = g_{\mu\nu}dx^\mu dx^\nu. \tag{1}$$

Unfortunately, this assumption has a kind of circularity associated with it: how do you know if a candidate clock is an ideal clock. Answer, it measures proper time. How do you know what the proper time is? Answer, it is the time measured by an ideal clock. There is, in fact, no way to tell *ab initio* if a candidate clock is in fact an ideal clock. Since all real clocks are physical systems their behavior will in general be affected by the presence of gravitational and other fields. If, for example, tidal forces acting on the clock are comparable to its internal forces then the timekeeping properties of the clock will be modified. Likewise, the frequency of an 'electric' clock, e.g. a hydrogen atom, will be modified in the presence of an electric or magnetic field. In fact, one does not have to introduce the proper time-ideal clock hypothesis any more that one has to introduce the geodesic hypothesis into general relativity and therefore one can dispense with the metric interpretation of $g_{\mu\nu}$.

In order to determine the properties of clocks and measuring rods in a gravitational field without additional assumptions, one makes use of the ideas first developed by (Einstein, Infeld, and Hoffmann (EIH) 1939; Einstein and Infeld 1940, 1949). Later work by myself extended the EIH procedure to include radiation (Anderson 1987) and electromagnetic interactions (Anderson 1997) using the methods of multiple time scales (an extension of the original EIH slow-motion approximation) and the method of matched asymptotic expansions (Nayfeh 1973). Later these methods were extended to deal with motion in an expanding universe (Anderson 1995). In the earlier works the fields produced by their sources were treated as perturbations on a 'flat' gravitational field and yielded, in the lowest orders of approximation, equations of motion containing Newtonian and Coulomb interactions. Higher orders of approximation produced radiation reaction forces in these equations. In the last work cited here the source fields were treated as perturbations on an Einstein-deSitter gravitational field. The resulting equations of motion were used to derive the dynamics of simple clock models consisting of two gravitationally or electromagnetically bound compact sources. Both of these model clocks were shown to measure cosmic time in the lowest order of approximation used and hence, to this order, behaved as ideal clocks. In this work I have extended these calculations to obtain the next order effects of the cosmic expansion which cause these clocks to deviate from ideal clock behavior.

2. EIH PROCEDURE

General relativity is unique among classical field theories in yielding equations of motion for the sources of the various fields in the theory. In classical electrodynamics one needs to postulate the form of the Lorentz force as well as the form of the inertial force in the equations of motion. Neither such postulate is necessary in general relativity. Furthermore, the point singularity problems one encounters in other field theories are absent in this theory. Given these virtues it is all the more strange that Einstein's last great contribution to general relativity, the work he did together with L. Infeld and B. Hoffmann, is so little known or referred to in the literature. For example, a description of the EIH method appears in only one early text on general relativity by Peter Bergmann who was an assistant of Einstein's. Bergmann (1942) and only a handful of papers in the literature make use of it. It is therefore perhaps not out of place to speculate for a moment on why this is so. A number of factors may have contributed. For one thing the initial paper was published at the beginning of the second World War when the scientific community began turning its attention to the military needs of the combatants. Then too, the original paper was published in the Annals of Mathematics and the two subsequent papers by Einstein and Infeld appeared ten years later in the Canadian Journal of Physics, neither journal being widely read by physicists. Some readers were probably scared off by the complexity of the calculations. The complete details were never published, the reader being referred to an archive at the Institute for Advanced Study. In addition, some puzzling procedures were used, e.g., the introduction of fictitious dipoles, that later proved to be unnecessary. Perhaps too, general relativists in those days were somewhat disdainful of the approximate nature of the EIH results, being more concerned with exact results and deriving theorems. Furthermore. after 1949, Einstein did no further work on the problems of motion and Infeld, after returning to Poland, abandoned the EIH approach in favor of his "good" delta functions. Also, at the time these papers appeared the attention of the majority of physicists had turned from general relativity, with its lack of experimental predictions, to the new and fruitful field of quantum mechanics. As a consequence of one or more of these and possibly other factors, the EIH method is largely unused by modern day workers in the field of general relativity.

EIH derived their equations of motion by integrating the field equations over a two dimensional surface surrounding each compact source. Since the integrals never extended over the sources the question of their exact nature never arose. In particular it was unnecessary to assume that they were point sources. These integrals are most easily obtained by using a form of the field equations given by Landau and Lifshitz (1985, 282) which are

$$U^{\mu\nu\rho}{}_{,\rho} = \Theta^{\mu\nu}, \tag{2}$$

where $U^{\mu\nu\rho}$ is a so-called superpotential, antisymmetric in ν and ρ, and a function of $g_{\mu\nu}$ and its first derivatives and

$$\Theta^{\mu\nu} = (-g)(T^{\mu\nu} + t^{\mu\nu}_{LL}). \tag{3}$$

In this latter equation $g = \det(g_{\mu\nu})$, $t^{\mu\nu}_{LL}$ is the Landau-Lifshitz energy-stress pseudo-tensor and $T^{\mu\nu}$ is the energy-stress tensor of any other fields present. Because the integrals used in EIH are surface integrals surrounding the sources of these fields and the gravitational field one does not have to include the contributions of these sources to $T^{\mu\nu}$. It is for this reason that one avoids the usual difficulties associated with the introduction of point sources such as those encountered in Dirac's derivation of the radiation reaction force in electrodynamics (Dirac 1938). One only needs to assume that the sources are compact so that the results are valid, for example, both for neutron stars and black holes.

Because of the antisymmetry of $U^{\mu\nu\rho}$ in its last two indices, $U^{\mu rs}{}_{,s}$ is a curl whose integral over any closed spatial 2-surface vanishes identically. As a consequence, integration of eq. 3 over such a surface surrounding a source gives

$$\oint (U^{\mu r 0}{}_{,0} - \Theta^{\mu r}) n_r \, dS = 0, \tag{4}$$

where n_r is a unit surface normal. It is this last equation that yields the equations of motion of the source contained within the integral.

To actually obtain equations of motion from the above integral one must use solutions of the field equations to evaluate the integrands appearing in them. These solutions however can only be obtained by an approximation scheme. In their original work, EIH introduced what they called the slow-motion approximation to obtain these solutions. As I discussed in (Anderson 1987), a systematic application of this approach requires the use of the method of multiple time scale analysis. To apply it one identifies the different time scales associated with the motion under consideration and expands the fields in powers (or more complicated functions) of the ratios of these time scales. If one is interested in motion in a 'flat' background field $g_{\mu\nu} = \mathrm{diag}(1, -1, -1, -1)$ as in the original EIH papers and in (Anderson 1987; Anderson 1997), then two time scales enter. One is T_L, the light travel time across the system and the other is T_S, a characteristic time such as a period associated with the system. The ratio ε_S is then used as the dimensionless parameter in the expansions one constructs - hence the appellation 'slow-motion.' In addition one also introduces, in addition to t, a slow time t_S and assumes that the fields are functions of these two times. (Actually, one needs to introduce even slower time scales such as $\varepsilon^2 t$ in higher orders of the approximation.)

If however one is interested in motion in an expanding universe then one must use one or another of the gravitational fields associated with such universes. In this case, another time scale enters into the problem, namely, the Hubble time t_H and one can construct another small parameter $\varepsilon_H = t_L/t_H$ to use in one's approximation scheme so that one has a double expansion. However, since for most systems $\varepsilon_H \ll \varepsilon_S$ the effects of the expansion can be separated out from the slow-motion effects without too much trouble.

3. EQUATIONS OF MOTION

In this paper I will not attempt to give the details of the derivation of the equations of motion in an expanding universe using the EIH procedure and its extensions since they are somewhat complicated and instead refer to reader to the references below. To simplify the results I will use the Einstein-deSitter gravitational field given by

$$g_{\mu\nu} = \text{diag}\left[1, -R^2(t), -R^2(t), -R^2(t)\right] \quad (5)$$

where $R(t) = (t/t_0)^{2/3}$, t is the cosmic time and t_0 is the current age of the universe. In this case $t_H = R/\dot{R}$ where a dot over a quantity refers to differentiation with respect to t. Furthermore, in order to simplify the results I have assumed as a clock model a classical hydrogen atom and neglected the gravitational interaction between its constituents. Had I used a gravitational clock the results would have differed only by numerical factors of the order unity. The resulting equation for the 'electron' of the clock, with the proton fixed at the origin takes the form

$$m\mathbf{x}_{\tau\tau} + 2\varepsilon m\mathbf{x}_{t\tau} + \varepsilon^2 m\mathbf{x}_{tt} + 2\varepsilon m \frac{R_t}{R}\mathbf{x}_\tau = -\frac{1}{R^3}\frac{e^2}{x^3}\mathbf{x}. \quad (6)$$

where $\varepsilon = \varepsilon_H/\varepsilon_S$ and the subscripts on the electron coordinate $\mathbf{x}$ denotes differentiation with respect to t and the fast time $\tau = t/\varepsilon$. These equations are correct to $\mathcal{O}(\varepsilon_S^2)$, that is, to Newtonian order and to all orders in ε_H. The inclusion of post-Newtonian terms do not change qualitatively the results obtained below.

In order to construct a uniform approximate solution to these equations of motion it is necessary to employ two approximation schemes in conjunction. One of these involves a multiple time expansion in t and τ. The other approximation involves the method of stretched coordinates. In this approximation one makes a change of variable from τ to s given by

$$s = \tau(1 + \varepsilon\alpha_1(t) + \varepsilon^2\alpha_2(t) + \cdots). \quad (7)$$

One also expands the position coordinate $\mathbf{x}(s, t, \varepsilon)$ according to

$$\mathbf{x}(s, t, \varepsilon) = \mathbf{x}_0(s, t) + \varepsilon\mathbf{x}_1(s, t) + \varepsilon^2\mathbf{x}(s, t) + \cdots \quad (8)$$

and substitutes into eq. 6. The sum of the lowest order (ε^0) terms in eq. 6 will vanish if

$$m\mathbf{x}_{ss} = -\frac{1}{R^3}\frac{e^2}{x^3}\mathbf{x}. \quad (9)$$

For simplicity I will assume a circular orbit so that

$$x_0 = z(t)\cos\{\omega(t)s\}, \quad y_0 = z(t)\sin\{\omega(t)s\} \quad \text{and} \quad z_0 = 0 \tag{10}$$

where ω and z satisfy

$$R^3 z^3 \omega^2 = \frac{e^2}{m}. \tag{11}$$

and their dependence on t in the next order of approximation.

The next order (ε^1) equation has the form

$$2x^i_{0ts} + 2\alpha_1 x^i_{0ss} + 2\frac{R_t}{R}x^i_{0s} + x^i_{1ss} = -\frac{e^2}{mR^3}\left(\frac{x^i_1}{x_0^3} - 3\frac{x_0^i x_0^j x_1^j}{x_0^5}\right). \tag{12}$$

In order to obtain a uniform expansion in ε it is necessary to require that x^i_1 not contain secular terms, that is, terms that grow in time without bound. The first three terms in this equation will lead to such growth so we must require that their sum be zero. This will be the case provided that

$$\omega_t = 0, \quad \alpha_1 = 0 \quad \text{and} \quad (Rz\omega)_t = 0. \tag{13}$$

Combining this result with eq. 10 we find to this order of accuracy that

$$\omega = \omega_0 \quad \text{and} \quad z = z_0/R. \tag{14}$$

where ω_0 and z_0 are constants. As far as $\mathbf{x}_1$ is concerned, we will take it to be equal to zero to avoid unnecessary complications. Consequently, we can conclude that to this order of accuracy our electric clock measures the cosmic time t. Also, if we take Rz to be a measure of our clock size, we see that it does not partake of the cosmic expansion. Thus, to this order, our clock behaves like an ideal clock.

Finally we come to the order (ε^2) equation. Again one must set equal to zero those terms that would lead to secular growth in $\mathbf{x}_2$. This condition gives

$$2\alpha_2 \mathbf{x}_{0ss} + 2\frac{R_t}{R}\mathbf{x}_{0t} = 0. \tag{15}$$

As a consequence we find that

$$\alpha_2 = \frac{1}{2\omega_0^2}\frac{R_{tt}}{R} \tag{16}$$

When this result is substituted back into the expression (7) for s and that in turn is substituted into the zero order solutions (10) one obtains finally the result

$$x_0 = \frac{z_0}{R}\cos[\omega_0\{1 - \frac{\varepsilon^2}{2\omega_0^2}\frac{R_{tt}}{R}\}\tau] \tag{17}$$

and similarly for y_0 with the cos replaced by the sin. Thus we see that the effect of the expansion on the clock frequency is to alter it by a term of order ε^2 whose coefficient varies slowly with the cosmic time t and to this extent our clock will no longer measure proper time. For a real hydrogen atom this term however is so small as to be completely negligible.

4. CONCLUSION

There are two positions one can take concerning the role of measuring devices in a physical theory. One is that such devices lie outside the province of the theory and their properties must be postulated. Such a view is common in quantum mechanics, in which measuring devices are held to be classical devices, and leads to a number of problems associated with the so-called measurement problem. The other position holds that measuring devices are physical systems whose behavior must be describable within the framework of the theory of the systems they are designed to measure. When the measurement process is itself taken to be describable by quantum mechanics many of the measurement problems go away. In the case of general relativity I have tried to show how one can calculate the properties of physical systems that can serve as clocks directly from the field equations of this theory. Such an approach has, I believe, two important consequences. For one, one is able to dispense with the notion of ideal clocks and rods as primative objects in the theory.. The other is that it allows one to calculate to what extent and to what accuracy a putative clock measures time. In the example given here we see that if its period is small compared with the Hubble time then the clock behaves as an ideal clock and does not feel the effects of the expansion of the universe.

Stevens Institute of Technology

REFERENCES

Anderson, J. L. 1987. "Gravitational radiation damping in systems with compact components." *Phys. Rev. D* 36:2301.
———. 1995. "Multiparticle dynamics in an expanding universe." *Phys. Rev. Lett.* 75:3602.
———. 1997. "Asymptotic conditions of motion for radiating charged particles." *Phys. Rev. D* 56:4675.
Bergmann, P. G. 1942. *Introduction to the theory of relativity*. New York: Prentice Hall.
Dirac, P. A. M. 1938. "Classical theory of radiating electrons." *Proc. Roy. Soc. (London)* A167:148.
Einstein, A., and L. Infeld. 1940. "Gravitational equations and the problem of motion. II." *Ann. Math., Ser. 2* 40:455.
———. 1949. "On the motion of particles in general relativity theory." *Can. J. Phys.* 3:209.
Einstein, A., L. Infeld, and B. Hoffmann. 1939. "Gravitational equations and the problem of motion." *Ann. Math., Ser. 2* 39:65.
Landau, L. D., and E. M. Lifshitz. 1985. *The Classical Theory of Fields*. 4th ed. Oxford: Pergamon Press.
Nayfeh, Ali. 1973. *Perturbation Methods*. New York: John Wiley & Sons.

JÜRGEN EHLERS, SIMONETTA FRITTELLI, EZRA T. NEWMAN

GRAVITATIONAL LENSING FROM A SPACE-TIME PERSPECTIVE*

Abstract. We want to discuss gravitational lensing as far as presently possible from the point of view of spacetime geometry without use of perturbation theory or a background metric. The intent is to construct a conceptual framework so that lensing theory fits covariantly into general relativity and then to see how the usual perturbation approach is related to it.

1. INTRODUCTION

The study of gravitational lensing has a long and illustrious history (Schneider, Ehlers, and Falco 1992). The earliest references go back as far as Newton (1704), who raised the question of whether bodies would have an effect on light. Cavendish and Soldner, among others, in the late 1700's took up the question, calculating the bending angle from Newtonian gravitational theory using the corpuscular theory of light. The modern theory of lensing begins with and rests on the General Theory of Relativity. Using his equivalence principle and assuming a flat physical 3-space, Einstein in 1911 computed the same bending angle (Einstein 1911) as his predecessors and derived in the following year the main formulae for gravitational lensing by a spherical body (Renn, Sauer, and Stachel 1997) without publishing his results. In 1915, employing his gravitational field equation, Einstein found that space curvature increases the bending angle to a value twice as large as the "Newtonian" one (Einstein 1915, 831), a famous result which was roughly verified in 1919 by the solar eclipse expeditions led by Andrew Crommelin and Arthur Eddington and which has by now been tested with an accuracy of 10^{-3} (Lebach et al. 1995). The issue of the feasibility of extending the solar type observation to other lensing objects, stars or galaxies, remained dormant until 1936 when Einstein, approached by the Czech engineer Mandl, returned to the issue and published his old results, incorporating of course the factor 2. Thinking only of lensing by stars and underestimating future advances in observational techniques Einstein concluded that (though in principle it was there) lensing from other astronomical objects would never be observed. Lenses afford "perfect tests of general relativity that are unavailable," as H. N. Russell ironically put it (Russell 1937). However within one year, Zwicky realized that galaxies would make excellent lenses and predicted that lensing should, in fact, be observed — and that if it was not observed, that would constitute a disproof of general relativity. In the mid 1960s Refsdal, realizing that quasars, discovered in 1963, would make ideal distant "point" sources to be lensed and that the technology was at hand, gave a strong impetus for the observers to begin their search. It took another 15 years until in 1979 D. Walsh, R. F. Carswell and R. J. Weymann announced the detection of the first lensing candidate (1979), the "double quasar," whose lensing nature was observationally confirmed subsequently. Since then the field has burgeoned with close to 1.000 papers published a year — both observational and theoretical — on lensing.

Lensing observations are no longer used simply as another test of general relativity — instead they have become a major research tool for cosmological/astronomical discoveries and investigations (Schneider, Ehlers, and Falco 1992; Wambsganss 1998). Examples of their use abound. Assumed models of the mass distribution within lenses (stars, galaxies, clusters of galaxies) can be tested by the observation of positions, intensities, shapes and numbers of the multiple images of more distant sources. Dark matter can be "weighed," the Hubble constant can be estimated from arrival time differences of rays from a source along different paths. Because of the "magnifying effects" of focusing lensed galaxies have been observed (with spectra) at far larger distances than unlensed ones. Statistical lensing provides estimates of additional cosmological parameters.

Since the lensing aspects of general relativity have become such an "applied" subject, most theoretical work on lensing has been associated with the detailed connection of observations with the structure of sources and lenses. For this purpose, it appears to be a safe assumption that the combined use of linearized perturbations off either Minkowski spacetime or Friedmann models, geometrical optics, small-angle and thin lens approximations, is quite adequate for all contemporary applications. (Speculations on future use of non-linear terms in the applications seem unnecessary now.) From a practical point of view there can not be much of an argument against this — it appears to fill all the needs of astrophysicists — researchers will find and use the tools needed to solve their problems and the standard toolkit of approximation methods works well. However we believe that there is a value (both for the clarification of conceptual issues and for the future comparison with observations) to look at gravitational lenses from a more fundamental theoretical perspective — i.e., to consider what questions can be asked and answered (in principle) when a background space-time is not employed and perturbation theory is not used.

We would thus like to consider a completely *general* situation for the discussion of lensing issues and see how observational quantities can be possibly introduced without background concepts.

Specifically we consider an arbitrary spacetime, i.e. a four dimensional manifold $\mathfrak{M}$ with a Lorentzian metric g_{ab} and time-like world line, $\mathfrak{L}$, representing the history of an observer. On $\mathfrak{M}$ we take some arbitrary distribution of sources (of light); they could be represented as either luminous point sources moving on time-like world lines or as (spatially) extended sources described in space-time as time-like worldtubes. The intrinsic luminosity of these sources could also be time varying. From this point of view gravitational lenses manifest themselves in the dependence of the metric on the energy-momentum distribution of the lenses via Einstein's field equation;

in other words, the lens parameters, masses, multipole moments, lens positions, time variablility, etc., are all hidden in the properties of the metric g_{ab}.

The question then is what can the observer "see," — and how that fits into the theoretical constructs associated with the underlying ideas we have of space-time geometry from general relativity. To try to fully answer this question is far too ambitious for us to contemplate — though perhaps even raising it is of some value. In any case we will try to give some partial answer and in the process give a broader view of what is gravitational lensing. The approach that we will take is highly idealized: if a quantity is in principle measurable we will treat it as measurable, and if a calculation is in principle doable then we will treat it as doable.

In section 2 we briefly described what our observer can "see." In section 3 we give a definition of an (exact) gravitational lens mapping and the time of arrival function. This discussion is based on the assumption that the space-time metric is known at least in a domain containing the past light cones of an observer during some interval of proper time on his world line, together with the null geodesics generating those cones. Where and how magnification (flux increase) and multiple images arise is described. Section 4 is devoted to the applications of these ideas to extented sources where a curious (superficially paradoxical) result becomes readily apparent. The connection between the exact ideas described here and the usual linear perturbation ideas are described in section 5.

There is no intent here to give an overview of the contemporary conventional view of lensing theory — we do not even discuss any observational results.[1] Our intent here is to try to view lensing theory as fully as possible from the point of view of the four dimensional geometry of general relativity.

2. WHAT CAN BE SEEN?

Most information about the universe reaches us in the form of electromagnetic radiation. This holds in particular for gravitational lens phenomena. The typical scales involved in lensing are such that the radiation may well be treated in the short wave, geometrical optics approximation. Accordingly we are interested here exclusively in light rays, idealized as null geodesics, that reach the observers world line from the past: The observer can only see his/her local past light-cone at any one time. All information must come from that channel. The observer can identify directions on his celestical sphere and measure angles between them. He can register the arrival of rays (from different sources, points or extended), he can measure frequencies and intensities and if watched over a period of time, their variations.

In order to make theoretical sense of these observations, i.e., to associate them to the distant sources and the (lensing) properties of the intervening space-time, we must follow the observed rays backwards in time. In other words we will be interested (essentially) only in the one parameter family of the past light-cones of the observer, and thus our fundamental tool for the study of lensing will be the analysis of these past cones.

As we said earlier we will assume that the metric tensor g_{ab} of $\mathfrak{M}$ is known. We also assume that the maximally extended, past-directed null geodesis ending at the world line $\mathfrak{L}$ of the observer are given. Let $\mathfrak{L}$ be given parametrically by

$$x^a = X_0^a(\tau) \tag{1}$$

where τ is the observers proper time. The one parameter family of the observers past light-cones, $\mathfrak{C}(\tau)$, is given (parametrically) by

$$x^a = X^a(\tau, \eta, s) \tag{2}$$

where the complex stereographic coordinate η labels the points on the observers celestial sphere $\mathfrak{S}(\tau)$, i.e. it corresponds to null directions at $x^a = X_0^a(\tau)$ while s is an affine parameter along each of the geodesics labeled by (τ, η). We fix s uniquely by demanding that $s = 0$ at the observer, s increases towards the past and equals ordinary distance from the observer close to the latter. Then the (null) tangent vector to the geodesics is

$$L^a(\tau, \eta, s) = \partial_s X^a \neq 0, \quad \text{and} \quad L_a \partial_\tau X^a = 1.$$

Derivatives of X^a with respect to the other parameters are Jacobi fields. In particular, if η is varied along a vector ξ tangent to the sphere $\mathfrak{S}(\tau)$, one obtains a Jacobi field

$$M^a(\tau, \eta, \xi, s) = \xi \frac{\partial X^a}{\partial \eta} + \bar{\xi} \frac{\partial X^a}{\partial \bar{\eta}}.$$

Note that equation (2), which will be our fundamental relationship, can be viewed in two ways: (i) it can be thought of as the local description of the past null cone in the local coordinates x^a where, whenever a new coordinate patch is reached by the geodesics, the new coordinates x'^a are introduced by the relevant coordinate transformation or (ii) it can be thought of as representing, in Penrose's abstract index notation, the null cones $\mathfrak{C}(\tau)$ intrinsically, regardless of coordinates. (In fact, the mapping $(\tau, \eta, s) \mapsto x^a$ given by equation (2) for fixed τ is the restriction of the exponential map at $X_0^a(\tau)$ to the tangent past null cone at that point.) We will adopt the latter view.

3. THE LENS MAPPING AND THE ARRIVAL TIME FUNCTION

Before describing what is meant by the lens mapping and the arrival time function, we want to consider some geometrical properties of equation (2) and describe a few physical facts related to it.

In contrast to the arguments τ, η of X^a, the affine parameter s is not observable. It describes how far back one has to follow a null geodesic before reaching a particular event x^a on $\mathfrak{C}(\tau)$, e.g., a source. This raises the question whether s is related to some, at least indirectly observable distance.

Geometrically, one kind of distance of an event x^a on $\mathfrak{C}(\tau)$ from the apex is the area distance r_A, defined by constructing a narrow pencil of generators of $\mathfrak{C}(\tau)$ filling a solid angle at the apex, taking the ratio of its cross sectional area A_s at x^a and the size ω_0 of the solid angle measured by the observer at the apex, and taking the limit

$$r_A \equiv \lim \left(\frac{A_s}{\omega_0}\right)^{\frac{1}{2}} \quad (3)$$

for vanishingly small ω_0. This r_A can be computed on all light rays by means of the Jacobi vectors M^a (Frittelli, Newman, and Silva-Ortigoza 1999), resulting in

$$r_A = R_A(\tau, \eta, s). \quad (4)$$

Starting at $s = 0$, r_A will first increase, but because of ray focusing it may reach a maximum, then possibly decrease and become zero at s_1, corresponding to the first point conjugate to the apex on the geodesic considered. Quite generally, r_A vanishes exactly at conjugate points. Exactly at these points the mapping given by equation (2) for fixed τ is not an immersion, i.e. there and only there $\mathfrak{C}(\tau)$ is not a smooth hypersurface. In the context of general relativity and optics these points are said to form the caustic of $\mathfrak{C}(\tau)$, but see the remark 2 below on terminology. At the caustic, the light "cone" has 2-dimensional edges. Moreover, in general the mapping given by equation (2) for fixed τ is not injective; $\mathfrak{C}(\tau)$ may intersect itself, see figures 1 and 2.

For fixed τ, η, the function (4) is continuous for all s, but fails to be C^1 at conjugate points where it has strict minima (Seitz, Schneider, and Ehlers 1994). Between its successive extrema, it can be inverted "patchwise,"

$$s_{(\alpha)} = S_{(\alpha)}(\tau, \eta, r_A), \quad (5)$$

with (α) the patch label.

Another distance of an event x^a on $\mathfrak{C}(\tau)$ from $X_0^a(\tau)$ is defined by interchanging the roles of the two events, using the future light cone of x^a, a pencil proceeding from there to $X_0^a(\tau)$, etc., obtaining a distance r'_L which, of course, can also be calculated by means of Jacobi vectors, this time those which vanish at x^a. According to Etherington (Etherington 1933), r_A and r'_L are related by[2]

$$r'_L = (1+z)r_A, \quad (6)$$

where $1 + z = \frac{d\tau_o}{d\tau_s}$ is the (strictly positive) ratio between proper time intervals at observer (at $X_0^a(\tau)$) and source (at $X^a(\tau, \eta, s)$), related by light signals and measurable as red shift $z = \frac{\lambda_0 - \lambda_s}{\lambda_s}$.

Let us turn attention to energy-related observables. Photon conservation and ray kinematics imply that the (bolometric) flux F observed at $X_0(\tau)$ with red shift z, due to a source with luminosity L at $X^a(\tau, \eta, s)$ is given by

$$F = \frac{L}{4\pi(1+z)^2 r'^2_L} = \frac{L}{4\pi r_L^2} \quad (7)$$

where $r_L = (1+z)r'_L$ is called the luminosity distance of the source from the observer. Combining equation (7) with the definition (3) of r_A and using equation (6), one obtains the important result that the surface brightness (or integrated intensity, defined as energy/(area x time x solid angle)) I_0 at the observer is related to that at the source I_s, along a ray by

$$I_0 = \frac{F}{\omega_0} = \frac{\frac{L}{4\pi A_s}}{(1+z)^4} = \frac{I_s}{(1+z)^4}, \qquad (8)$$

provided the photons of the ray do not interact with intervening matter.

The foregoing results show:

(i) If the flux F and the red shift z of a source are measured and if the luminosity L of the source can be inferred (e.g. from spectral features or a light curve), then r_A is determined via equations (6) and (7), i.e., indirectly measurable.

(ii) If the true area A_S of an extended source can be determined and the corresponding solid angle ω_O is measured, r_A is again indirectly measurable, by equation (3). A_S can in principle be determined by measuring curves of constant surface brightness in the image, i.e. on $\mathfrak{S}(\tau)$, and relating them via equation (8) to such curves on the source's surface, where they enclose — for a particular type of source — a definite intrinsic area A_S.

In view of these (known) considerations we henceforth take r_A to be measurable and, provided one could detect the patch a source is in (by the observation of multiple images or via intergalactic absorption; if a source has more than one image, one image will not belong to the first two patches), s can be replaced by r_A via equation (5).

Remark 1. Though it will not be of immediate relevance to the remainder of this work, we nevertheless want to remark that the ideas discussed here fit into a much larger framework, namely the theory of Lagrangian and Legendrian manifolds and maps, developed by V. I. Arnold (1980, 1986) and his colleagues. The basic idea is that, instead of working only in the space-time itself, all the structures should be lifted to the phase-space, i.e., to the cotangent bundle, $T^*\mathfrak{M}$ over the space-time. In general multivalued functions and singular structures (as, e.g., occur on $\mathfrak{C}(\tau)$ at conjugate points) become single-valued and smooth in the bundle. As our basic objects are the null geodesics that generate the past cones, they are objects to be lifted. The edges and self intersections of the light cone mentioned in the previous discussion disappear when lifted. Thus in addition to the parametric version of the family of past cones, i.e., $x^a = X^a(\tau, \eta, s)$, we also consider the associated momentum four-vector field $p_a = g_{ab}\frac{dX^a}{ds} = g_{ab}L^b(\tau, \eta, s)$. This set of eight equations defines a smooth four dimensional submanifold, P, globally parametrized by (τ, η, s), on the (eight-dimensional) phase-space, $T^*\mathfrak{M}$, of the (x^a, p_a). It is not difficult to show that the canonical symplectic form $dp_a \wedge dx^a$ of $T^*\mathfrak{M}$ vanishes when restricted to P, thus P is a "Lagrangian" submanifold. The projection of P to the space-time, given by equation (2) and referred to as a Lagrange map, is simply our construction of the family of past cones from the observer. The singularities of this map are the wavefront singularities. Though this point of view is extremely valuable in general, we will make no further use of it here.

We now want to first give a very general definition of a lens mapping and time of arrival equation and then later to specialize it.

First we temporarily consider our basic equation (2) in terms of local coordinates x^a in some region intersecting our null cones $\mathfrak{C}(\tau)$. We assume that these coordinates x^a are such that three of them, x^i, are space-like and one, t, is time-like, i.e., $\frac{\partial}{\partial x^i}$ are space-like vectors and $\frac{\partial}{\partial t}$ is time-like. We then rewrite equations (2) as

$$t = X^0(\tau, \eta, s), \tag{9}$$

$$x^i = X^i(\tau, \eta, s). \tag{10}$$

In addition we assume that the x^i have been chosen such that they, or some of them, label the world lines of source points and that t measures proper time along them. (An alternate treatment would be not to specialize the x^i as constant on the sourcelines but to describe the sources parametrically, i.e. $x^i = X^i_s(y^I, \tau_s)$ with constant y^I.) With this interpretation we consider equations (9), (10) again intrinsically as giving the time of emission, t, and the source world lines (x^i), respectively, in terms of τ, η, s, and we call (9) the time of arrival function and (10) the generalized lens mapping. Used with equations (4), i.e. with $s_{(\alpha)} = S_{(\alpha)}(\tau, \eta, r_A)$, the first equation describes the time of emission of radiation from a source in terms of observable quantities, its arrival time and direction (τ, η) and "distance" r_A, while the second describes the world-line of that source which the observer sees at time τ in the direction η at distance r_A,

$$x^i = X^i_{(\alpha)}(\tau, \eta, r_A). \tag{11}$$

Of fundamental importance to lensing theory is the question of invertibility of the lens mapping; in other words can one uniquely write

$$\eta = N(x^i, \tau), \tag{12}$$

$$r_A = R(x^i, \tau)? \tag{13}$$

Or again stated in a different way, at an observers time τ, is there a unique image direction η and "distance" r_A for a specific source at the position x^i? For small values of s or r_A the answer is yes — but as s increases the gravitational field refocuses (in the backwards direction) the rays so that eventually they cross and generically develop caustics (sharp edges in each of the past cones) which results in multiple images of the same source. In figure 1, we see the world line of a source intersecting one past cone in three places, yielding three images. In the same figure we can see the self-intersections (cross-overs) of the past rays and the formation of the caustic. The study of multiple images is one of the main occupations of lensing experts. Here we have assumed that the "lenses" are known — i.e. hidden in the assumed model of the space-time geometry; in practice it is the properties of the multiple images that are used to determine the space-time geometry.

As we have explained, a global inverse of the mapping defined by equation (10) (τ fixed) cannot be expected to exist in general. However, local invertibility at (τ, η, s) is equivalent to that mapping being a local diffeomorphism of $\mathfrak{C}(\tau)$ onto the 3-manifold (x^i) of source world lines, and that holds if and only if $r_A(\tau, \eta, s) > 0$, as was stated above. If this is the case, the equations (12) and (13) provide the direction η and distance r_A where the observer will find the source with world line x^i at time τ.

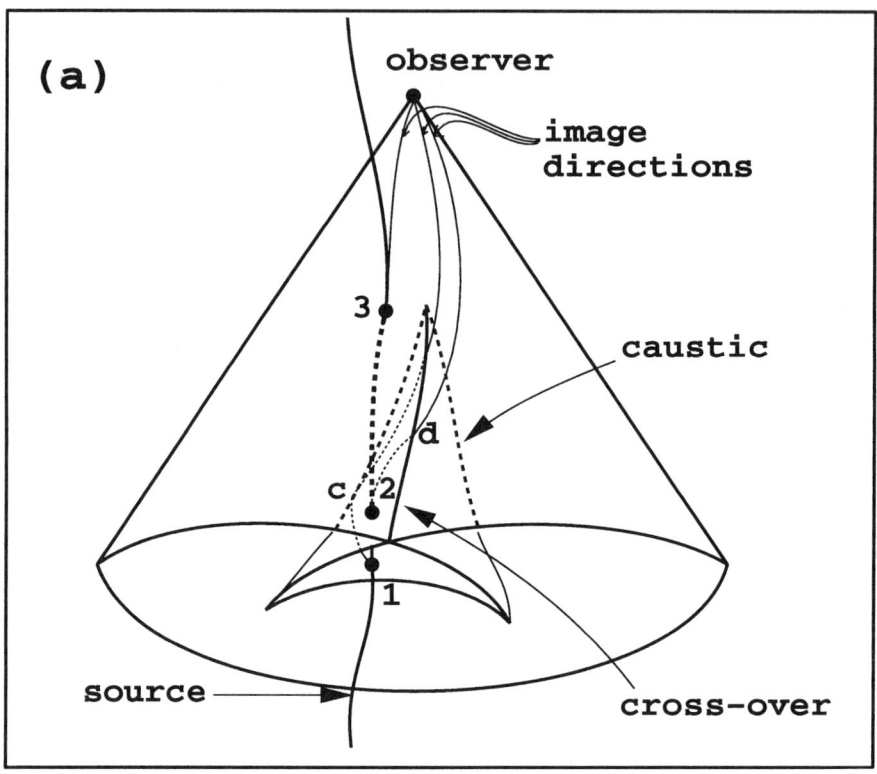

*Figure 1a. The past lightcone of an observer intersecting the worldline of a source at three points, and the three resulting image directions observed. The worldline of the source intersects the inner sheet of the lightcone at **1**, the right outer sheet at **2**, the left outer sheet at **3** and then continues outside of the cone. The lightray emitted at **1** first moves on the inner sheet, touches the caustic at **c** and continues on the right outer sheet to finally reach the observer. The ray emitted at **2** moves on the right sheet to the observer. The ray emitted at **3** moves on the left sheet to the observer. The ray from **2** passes the cross-over line at **d**.*

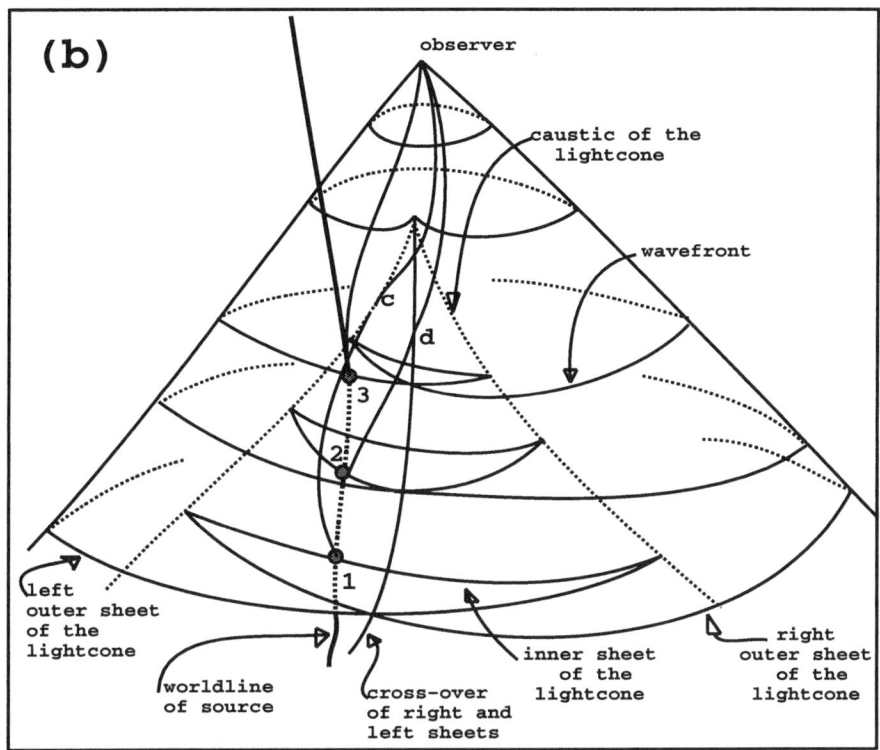

Figure 1b. A wavefront in the observer's past lightcone, moving backwards in time, intersects the source first at 3, then at 2 and finally at 1, in the order in which its three constituent sheets reach the source's location in space.

As we have just seen, it is the condition

$$r_A(\tau, \eta, s) = 0 \qquad (14)$$

which characterizes the critical points where the mapping (10) can not be uniquely inverted. The image of the set of critical points for fixed τ is called (at least in GR and optics) the caustic of $\mathfrak{C}(\tau)$, and the union of these caustics for variable τ is called the caustic of the family $\{\mathfrak{C}(\tau)\}$ of light cones or wave fronts (compare remarks 1 and 2.)

The value of s at which the geodesic (τ, η) has its βth conjugate point depends on τ and η; we may thus write the solutions of equation (14) as

$$s = S_{(2\beta)}(\tau, \eta). \qquad (15)$$

Parametric representations of these caustic 3-surfaces, which meet at 2-dimensional edges and may intersect each other, are

$$x^a = X^a(\tau, \eta, S_{(2\beta)}(\tau, \eta)). \qquad (16)$$

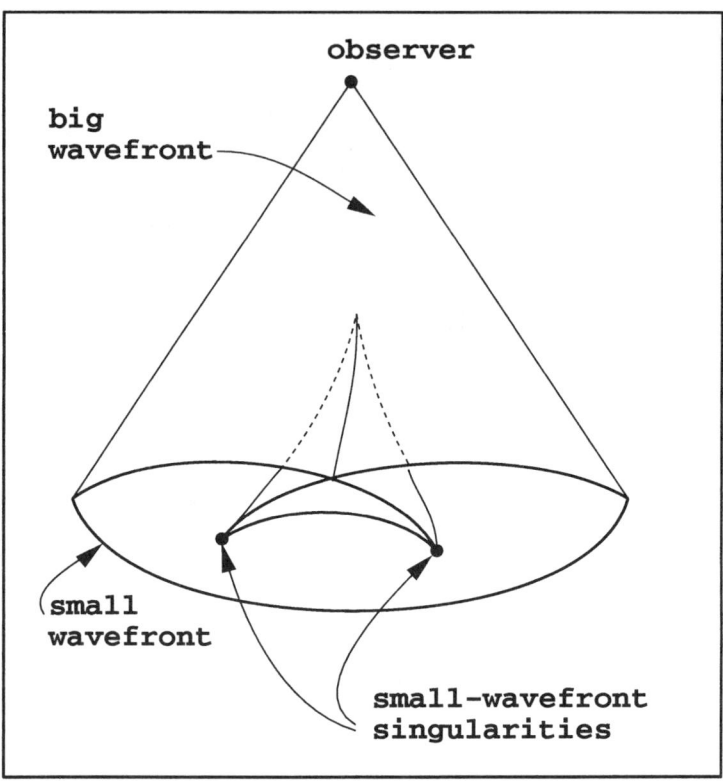

Figure 2a. A small wavefront and its singularities in the lower-dimensional representation.

Remark 2. We would like to comment on some questions of nomenclature — a subject for which there is not universal agreement. Characteristic (or null) "surfaces" (i.e. the level surfaces defined by a function, $S(x^a)$ that satisfies $g^{ab}\partial_a S \partial_b S = 0$ or its global generalisation) are strictly speaking not surfaces since they can self intersect and split into several branches; Arnold (Arnold 1986) refers to them as "big wavefronts." Thus our one parameter family of past light-cones from $\mathfrak{C}(\tau)$ is a one parameter family of "big wavefronts." A three-surface that slices through a big wavefront. e.g. $X^0 = const.$, defines a "small (two-dimensional) wavefront." equation (16) defines the caustic "three-surface" in the space-time of the family of big wavefronts. If in equation (16), the parameter τ is held constant. (i.e., if a particular past-cone or big wavefront is chosen) the resulting region (two-dimensional) is referred to as a "big wavefront singularity"; it is the intersection of the caustic with the past cone labeled by τ. Finally, the intersection of a "big wavefront singularity" with a generic three surface defines a "small wavefront singularity," i.e., the singularities of the small wavefront. These singularities are curves. See figure 2. There is a complete and simple

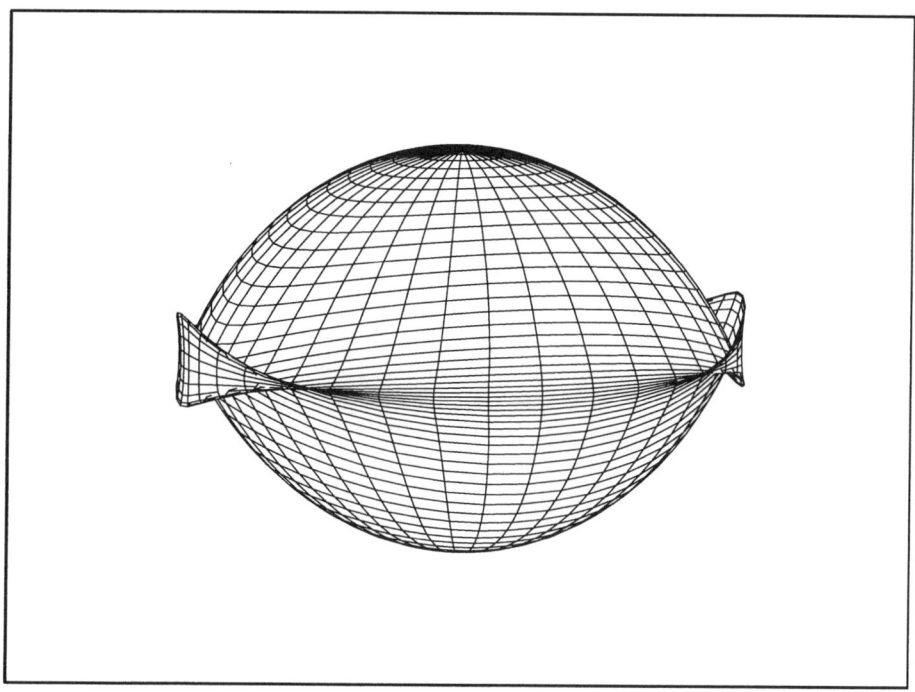

Figure 2b. A small wavefront in the appropriate three-dimensional representation, showing cusp ridges and swallowtails.

classification (Arnold 1980) of both the stable caustics and (small and big) wavefront singularities. We will not explore this here. We stick to the nomenclature of Arnold's, with the exception that we call the "big wavefront singularities" also "caustics," in order to be in agreement with usage in GR and optics.

We now want to illustrate the uses of the physical relations (7), (8) and of the lensing laws (9), (10) or their analogs in terms of r_A.

Suppose a source is (conjectured to be) seen in two images with fluxes F_1, F_2 and the same red shift. Then, from (7),

$$\frac{F_1}{F_2} = \left(\frac{r_A^{(2)}}{r_A^{(1)}}\right)^2. \tag{17}$$

If the source is extended and visible in two resolved images and if corresponding areas can be identified in the images (in terms of isophotes, e.g. using (8)) and are observed to subtend solid angles ω_1, ω_2, then, from (3),

$$\frac{\omega_1}{\omega_2} = \left(\frac{r_A^{(2)}}{r_A^{(1)}}\right)^2. \tag{18}$$

Thus, the ratio of the angular sizes of the images equals the flux ratio. Both the size of an image and its flux depend on the amount of (in general astigmatic) focusing of light bundles by a "lens" and intervening matter. In the first case it concerns a beam centred on the observer, in the second case, one centred on the source, the equality of the two kinds of magnification being due to the "reciprocity law" (6). These effects enable astronomers to observe remote galaxies which would not be visible without magnification. This is the magnification effect of lensing, stripped of the (convenient, but in principle unnecessary) comparison with fictitious "unlensed" sources in a homogeneous background universe. If, in particular, the world line of a source intersects the caustic of $\mathfrak{C}(\tau)$, the observer will see, at time τ, a bright, short "signal" since r_A passes through zero.

Assume next that a source event at x^a — e.g., a supernova outburst — is seen in two images. Then the equations

$$X^a_{(\alpha_1)}(\tau_1, \eta_1, r_A^{(1)}) = X^a_{(\alpha_2)}(\tau_2, \eta_2, r_A^{(2)}) \tag{19}$$

must hold. They impose four conditions on the observables $\tau_i, \eta_i, r_A^{(i)}$, given g_{ab} and the $\mathfrak{C}(\tau)$. Again, in practice equations (19) or rather their approximate analogs, are used to determine lens parameters on which the metric depends. In practice, the directions η_i, fluxes and/or angular sizes and sometimes time delays $\tau_1 - \tau_2$, are observable.

4. THE SOURCE SURFACE

The discussion just given of lensing is more general than the usual treatment. To get closer to the standard approach we introduce the idea of a "source surface"; in the usual treatment this is treated as a "source plane." The source surface is taken as a time-like hyper-surface $\mathfrak{T}$ of space–time. It will represent either the surface of an extended light source evolving in time or an arbitrary two-dimensional surface that is evolving in time, which has been chosen with some special physical or mathematical attribute — such as being at some fixed "distance" (e.g. fixed z, r_A or s) from the observer, on which arbitrary sources reside. Analytically it will be described by

$$x^a = X_T^a(y^1, y^2, t) = X_T^a(y^A, t) \tag{20}$$

where $y_A^a = \partial X_T^a/\partial y^A$ are two space-like vectors and $T^a = \partial X_T^a/\partial t$ is time-like. It is convenient to think of $\mathfrak{T}$ as the product $\mathfrak{N} \times \mathfrak{R}$, with $\mathfrak{N} = \{y^A\}$, a 2-dimensional space of possible source positions and with $\mathfrak{R}$ corresponding to proper time on the world lines with tangent T^a. The plan is to study the intersections of the past cones, $\mathfrak{C}(\tau)$, from the observer $\mathfrak{L}(\tau)$, with $\mathfrak{T}$; i.e.,

$$X_T^a(y^A, t) = X^a(\tau, \eta, s). \tag{21}$$

For those rays from $\mathfrak{L}(\tau)$ that intersect $\mathfrak{T}$, these four equations have solutions of the form

$$y^A = Y^A(\tau, \eta), \qquad (22)$$
$$t = V(\tau, \eta), \qquad (23)$$
$$s = S(\tau, \eta). \qquad (24)$$

This can be seen from the fact that each ray, (τ, η), which intersects $\mathfrak{T}$, does so at some specific point (y^A, t) and at some specific affine distance s. If a ray intersects $\mathfrak{T}$ more than once we only consider its intersection at the smallest value of s.

In this narrower context we define equation (22) as the "restricted" lens mapping, giving the "spatial position" on $\mathfrak{N}$ of a source as seen by the observer at time τ in the η direction, while equation (23) is the "restricted" time of arrival equation; the emission time t, for arrival at the observer at τ in the η direction. s can be related to r_A via equation (5). Essentially all work on lensing is related in some way to equations (22) and (23). Again an important issue is the inversion of the lens mapping; given a set y^A what are the different possible observation directions at time τ? For sufficiently small s it is unique while for larger s, in general, there will be multiple images. Analytically the pathologies of the imaging arise from the vanishing of the Jacobian of equation (22), i.e. from

$$\frac{\partial(y^1, y^2)}{\partial(\eta^1, \eta^2)} = 0. \quad (\eta = \eta^1 + i\eta^2). \qquad (25)$$

For each τ this equation defines the critical curves on the η sphere, {i.e., $\eta = \eta(\tau, \omega)$, with ω, a curve parameter} which when mapped, via equation (22), to $\mathfrak{N}$ yield the caustic curves on $\mathfrak{N}$ given by

$$y^A = Y^A(\tau, \eta(\tau, \omega)). \qquad (26)$$

From the work of H. Whitney (1955), it is known that there are only two types of stable singularities of maps $\mathbb{R}^2 \to \mathbb{R}^2$, namely folds and cusps. The critical set consists of non-intersecting, smooth curves, while the caustic set consists of piecewise smooth curves which may have cusps, and which possibly intersect each other. See figure 3.

A valuable method for analyzing the effects of the caustic on image formation is to consider on $\mathfrak{N}$, a hypothetical curve — a "source curve" — (e.g. an isophotal curve on an extended source or the trajectory of a point source) and study the inverse mapping of the source curve to the celestial sphere of the observer. The morphology of the mapping, as the source curve is deformed to cross the caustic at both the folds and the cusps, is fairly complicated but well studied (Berry 1987). For an example see figure 4. If, during such a deformation, a point on the source curve leaves the three-image region passing through the cusp, *three* images become very bright and merge, at the corresponding critical curve, into a single image which remains very bright immediately after crossing. If, on the other hand, a point passes through a fold, *two* images brighten up, merge at the critical curve and disappear, while one image, which remains separated from the critical curve, persists without becoming particularly bright. A similar story holds if, instead of considering the deformation of

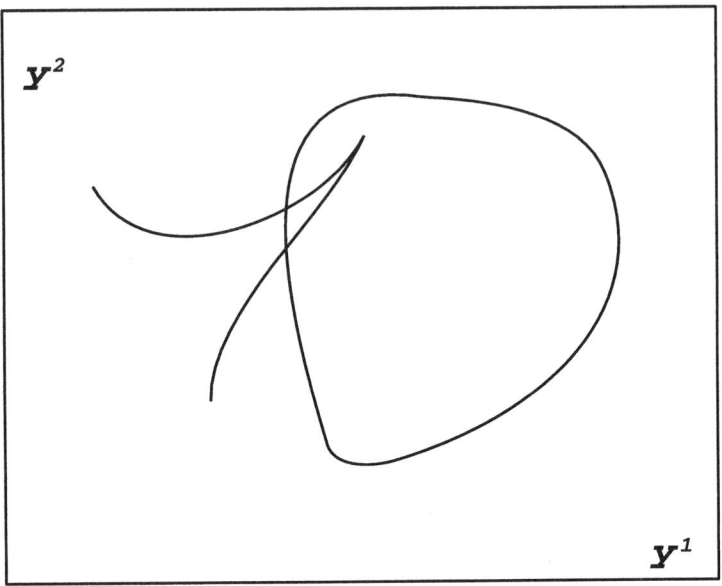

Figure 3. Typical caustics.

source curves seen by the observer at *one* time, one traces the images of a single point source seen by the observer during some intervall of time, and if during that intervall some sheets of the caustic of $\mathfrak{C}(\tau)$ intersect the world line of the source.

Several important points should be emphasized. There are (theoretically) always an odd number of images of a point source (except if an emission event is situated on the caustic) — though observationally some might either not reach the observer, having been blocked by other matter, or be to weak to be seen.

We have just been discussing the lens equation, equation (22), the mapping from the observers celestial sphere $\mathfrak{S}(\tau)$ to $\mathfrak{N}$. Of perhaps greater geometric significance is the mapping from the celestial sphere to the source surface, the world-tube $\mathfrak{T}$. Geometrically the image is the intersection of a past cone from $\mathcal{L}$ with $\mathfrak{T}$ and is given analytically by equations (22) and equation (23), the lens and time of arrival equations. It defines — piece-wise — a two surface in $\mathfrak{T}$, *a small wave-front*. The mapping from the two-space $\mathfrak{S}(\tau)$ into the three space $\mathfrak{T}$ is an example of what Arnold calls a Legendrian mapping. The stable singularities of this map, defined by the drop below two of the rank of the map $\eta \mapsto (y^A, t)$, and referred to as wave-front singularities, are curves (images of the critical curves of the lens map (in $\mathfrak{S}(\tau)$)) but now in $\mathfrak{T}$ instead of in $\mathfrak{N}$. They are classified into two types, cusp ridges and swallow-tails and project to $\mathfrak{N}$ as folds and cusps. See figure 5. One can see from this perspective, why sources close to caustics have bright images: there is an increase in density of source points that radiate into the associated image point.

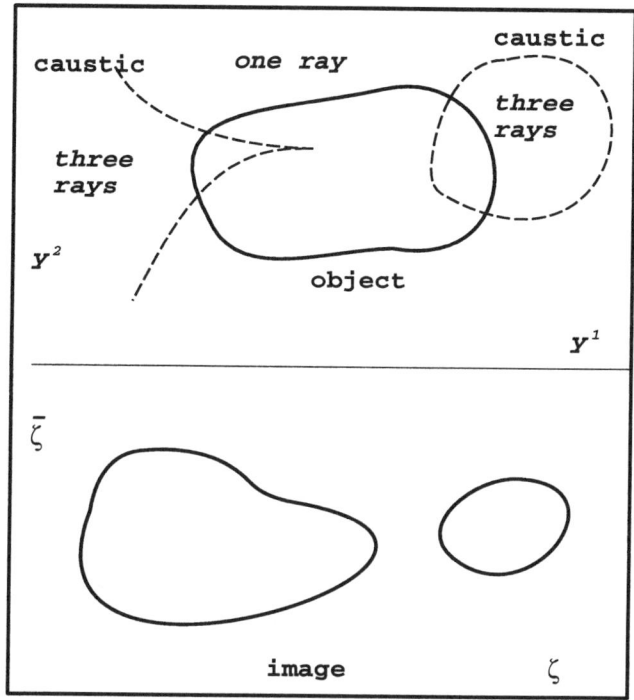

Figure 4. The appearance of the image of an extended object lying across caustic lines (adapted from M. Berry (1987)).

Remark 3. There is a curious and surprising observation that can be made concerning these small wavefronts. An extended source emits light, in the form of a cone of lightrays, continuously from each one of the points in its surface. Generally, at least one ray from each point will reach the observer, although the rays that do reach the observer may arrive at different times. The observation is that given any source surface $\mathfrak{T}$ and any observer on the curve $\mathfrak{L}$, there is a (preferred) two-dimensional spacelike section of the source surface such that the rays emitted perpendicularly to it, all arrive at the observer at the same time τ. Though at first this seems difficult to visualize — how do the rays "know" to leave perpendicularly in order to reach the observer simultaneously? —, in actuality it is rather simple to see that this is so. Any displacement confined to a null surface is either a null displacement (along a null geodesic) or is spacelike and orthogonal to the null rays. Thus any cross section (a small wavefront) of any of the past cones (as long as it is not tangent to a null direction) must be spacelike and orthogonal to the rays and hence, in particular, the intersection of any past cone with the source surface is spacelike and orthogonal to the rays. The preferred section of the source surface is thus obtained as the intersection of the past cone and the source surface.

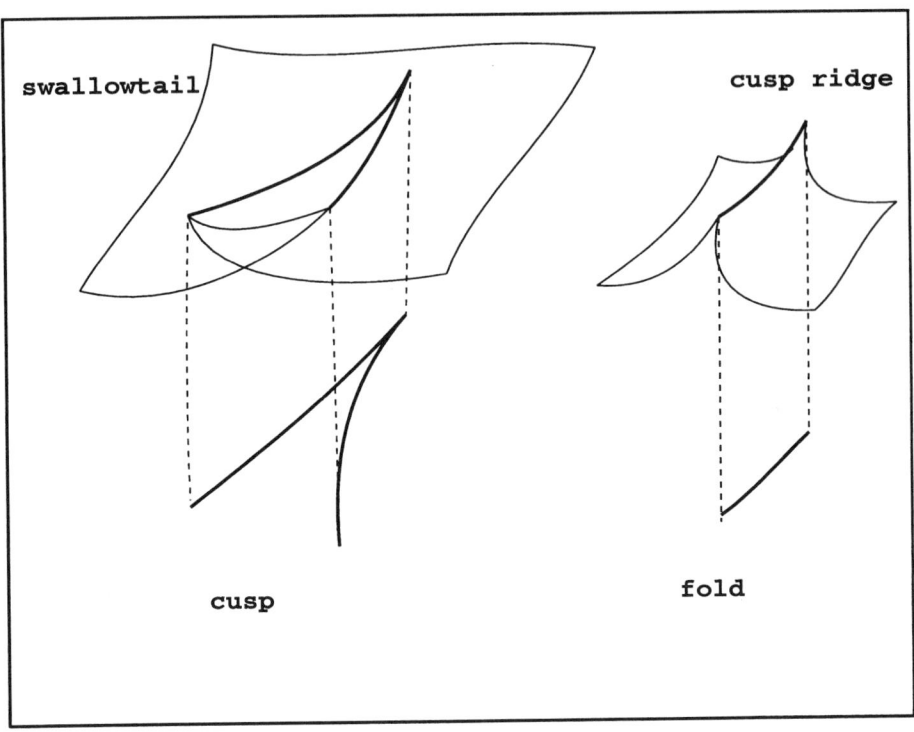

Figure 5. Wavefront singularities and their projections.

5. RELATIONSHIP TO THE STANDARD APPROXIMATE LENS THEORY

While the considerations in the preceding sections provide a clear qualitative space-time picture of lensing, they do not give concrete quantitative relations between the lens observables and the properties of sources and lenses which give rise to those relations. To obtain such relations one needs (i) solutions to Einstein's field equation representing inhomogeneous mass distributions modelling stars, galaxies and clusters of galaxies and resembling, on average, an expanding Friedmann model, and (ii) the past null geodesics ending on the world line of an observer ("us"), quite apart from details of astrophysical modelling of sources, lenses and the intergalactic medium. Obviously analytical solutions to these problems which are both exact and sufficiently realistic to be astrophysically useful are not available; even numerical work on lensing, as far as we are aware, is based on the approximate theory, not on the full, underlying GR.

The purpose of this section is first to outline the main assumptions underlying the approximate lens theory and then to relate the equations of that theory to the exact ones given in the foregoing sections. The following account is based mainly on references

(Schneider, Ehlers, and Falco 1992; Seitz, Schneider, and Ehlers 1994; Sasaki 1993; Holz and Wald 1998).

The first and most fundamental assumption on which standard lens theory is (implicitly or explicitly) based is that a metric of the form

$$ds^2 = -(1+2\phi)dt^2 + (1-2\phi)a^2(t)d\sigma_k^2, \tag{27}$$

in which $d\sigma_k^2$ is the metric of a Riemannian 3-space $\mathfrak{P}$ of constant curvature $k \in \{1, -1, 0\}$, is sufficient to describe all lens phenomena.

If it is assumed that ϕ is small compared to one, that it varies in time much more slowly than in space and that the scale on which it varies is much smaller than the Hubble scale $a/\dot{a}$, then Einstein's field equation for pressureless matter with density ρ and a four velocity not deviating much from ∂_t in equation (27), can be satisfied approximately (in the sense of perturbation theory) by requiring that $a(t)$ satisfies Friedmann's equation

$$3(\frac{\dot{a}}{a})^2 = 8\pi\bar{\rho} - 3\frac{k}{a^2} + \Lambda, \tag{28}$$

$$a^3\bar{\rho} = const. \tag{29}$$

for the spatially averaged density $\bar{\rho}$ and the Poisson equation

$$a^{-2}\Delta\phi = 4\pi(\rho - \bar{\rho}) \tag{30}$$

holds, with $\bar{\phi} = 0$.[3] The symbol Δ denotes the Laplace operator associated with the metric $d\sigma_k^2$. (Additional laws have to hold for the motion of matter (Takada and Futamase 1999), but they will not be needed here.) It is important that the density constrast $\rho - \bar{\rho}/\bar{\rho}$ need not be small for this approximation to be valid; in realistic applications that contrast is, in fact, very large compared to unity at some places.

The points (x^i) of the 3-space $\mathfrak{P}$ introduced in connection with equation (27) label the world lines of particles or "fundamental observers" who participate in the mean motion of matter, the Hubble flow.

To proceed, one needs to make an assumption about the mass distribution in the model universe. The simplest one — the only one to be considered here — is that matter consists of small "clumps" separated by vacuum. In other words, the support of the density ρ is assumed to consist, at cosmic times relevant to lensing, of compact regions in $\mathfrak{P}$, small compared to their distances and slowly moving relative to the mean cosmic motion. (Other assumptions can and have been made; for simplicity we do not consider them here.)

Since the source of ϕ is not ρ, but $\rho - \bar{\rho}$, whose support at time t generically is the whole, possibly non-compact slice $t = const.$ of $\mathfrak{M}$, ϕ is not "localized" at and near the clumps. However, if one transforms the global metric of equation (27) to coordinates T, X^a which are Fermi coordinates of the underlying Friedmann metric with respect to a fundamental observer close to a particular clump, the metric takes the form (Holz and Wald 1998)

$$ds^2 = -(1+2\Psi - \frac{1}{3}\Lambda\vec{X}^2)dT^2 + (1-2\Psi - \frac{1}{6}\Lambda\vec{X}^2)d\vec{X}^2, \tag{31}$$

if terms of higher than second order in the X^a are neglected, where now

$$\Delta \Psi = 4\pi\rho, \quad \Psi = \phi + \frac{2\pi}{3}\rho\vec{X}^2, \tag{32}$$

Δ denoting the ordinary Laplacian in the X^a–coordinates. Under "realistic" conditions equation (31) is a good approximation in any domain whose radius is much smaller than the Hubble radius, and even the Λ-terms are negligible in such a domain. Thus, one can cover $\mathfrak{M}$ by overlapping "nearly Newtonian" domains, "held together" by the global metric (27). The advantages of (31), (32) compared to (27), (30) is that, in each domain containing and surrounding a clump, the total density ρ has compact support, and Ψ can be taken to be the standard Newtonian potential which decreases far from the clump. (A mathematical problem may be the fitting together of the local ϕ's, resulting from the Ψ's, into one smooth and globally small ϕ with vanishing global average, at least for "open" models.)

After this preparation we turn to lensing. In the approximate treatment one considers (apart from statistical lensing) not the whole light cone $\mathfrak{C}(\tau)$ of the observation event $X_0^a(\tau)$, but only a narrow angular sector of it corresponding to a small neighbourhood of a point η_0 on the celestial sphere $\mathfrak{S}(\tau)$. One parametrizes the relevant directions in terms of small vectors $\vec{\Theta}$ tangent to $\mathfrak{S}(\tau)$ at η_0, as indicated in figure 6, and one approximates the corresponding part of the space $\mathfrak{C}(\tau)$ of source positions (introduced below (20)) as a plane.

In order to determine the lens mapping (22), one strategy, developed independently in (Seitz, Schneider, and Ehlers 1994; Sasaki 1993), is to first integrate (backwards) the geodesic deviation equation

$$\ddot{M}^a = R^a{}_{bcd}L^bL^cM^d$$

along a light ray from the observation event $X_0^a(\tau)$ to a source position y^A on the source surface $\mathfrak{N}$. The result relates the infinitesimal separation dy^A (of two neighbouring light rays, and thus) of two points on $\mathfrak{N}$, to the corresponding angle $d\Theta^A$ at the observer,

$$dy^A = Z^A{}_B(\Theta^c)d\Theta^B.$$

In a second step, integration of this total differential equation should then give the lens mapping, equation (22). Within the framework described above in connection with equations (27)–(32), each of the two steps is carried out under a simplifying assumption.

The first assumption is that in the empty regions between observer and lens and between lens and source, the contribution to the geodesic deviation due to the conformal curvature — the astigmatic Weyl focusing (Penrose 1966) — is negligible compared to that due to Ricci curvature (anastigmatic focusing). In other words, it is assumed that a narrow light beam picks up shear only when passing the (thin) lens which is represented by an effective surface mass distribution on the lens plane. Under this (questionable) assumption, one obtains (Seitz, Schneider, and Ehlers 1994; Sasaki 1993)

$$dy^A = D_S d\Theta^A - \frac{2mD_{LS}}{D_L}\frac{\partial^2 \tilde{\psi}}{\partial \Theta^A \partial \Theta^B}d\Theta^B. \tag{33}$$

In this equation, D_S, D_L, D_{LS} denote the area distances of the source S and the lens L from the observer O and of S from L, respectively, computed for shearfree light beams in the corresponding empty regions between "clumps." The deflection potential

$$\tilde{\psi}(\vec{\Theta}) = \int \tilde{\Sigma}(\vec{\Theta}') ln|\vec{\Theta} - \vec{\Theta}'| d^2\Theta' \tag{34}$$

arises from the "localized" Weyl tensor near the lens, computed with equations (31), (32). It depends on the mass distribution of the lens, projected into the lens plane; $\tilde{\Sigma} d^2\Theta'$ denotes the fraction of the lens mass contained in the solid angle $d^2\Theta'$, seen from the observer. The vector

$$\vec{\alpha} = \frac{2m}{D_L} \frac{\partial \tilde{\psi}}{\partial \vec{\Theta}}$$

is the deflection angle of a ray, see figure 7. Equation (33) contains the differential deflection $\partial \vec{\alpha}/\partial \vec{\Theta}$, due to the tidal field of the lens. m is the Schwarzschild radius of the lens.

The second assumption, presumably less critical, is that in the narrow angular sector of $\mathfrak{C}(\tau)$ used to model a lens phenomenon, the D's may be taken to be constant. Then, integration of equation (33) leads to

$$y^A = D_S \Theta^A - \frac{2mD_{LS}}{D_L} \frac{\partial \tilde{\psi}}{\partial \Theta^A},$$

where the integration constant has been absorbed into the choice of origin in the source plane. Using the angle $\vec{\beta} = D_S^{-1}\vec{y}$ — see figure 6 — we finally rewrite the restricted lens mapping $\vec{\Theta} \mapsto \vec{\beta}$ compactly as

$$\vec{\beta} = \vec{\Theta} - \Theta_E^2 \frac{\partial \tilde{\psi}}{\partial \vec{\Theta}}, \tag{35}$$

in which

$$\Theta_E \equiv \sqrt{\frac{2mD_{LS}}{D_L D_S}} \tag{36}$$

is called the generalized Einstein angle.

The equations (34)-(36) are valid also for lensing in a flat background spacetime. To see this, one may begin with eqs. (31), (32), globally on $\mathbb{R}^4$, with $\Lambda = 0$ and $\bar{\rho} = 0$, and obtain eqs. (34)–(36) with obvious modifications of the preceeding arguments. This "local" version of lensing applies, e.g. to microlensing of stars in the Magellanic clouds by stars in our Milky Way galaxy or its halo.

In the cosmological case, the unobservable "empty cone distances" in eqs. (35), (36) can be related to the observable red shifts z_S, z_L of source and lens and to the parameters H_0, Ω_0, Λ of the Friedmann model underlying the metric (27), if one additionaly assumes that in empty regions between clumps the relation between affine parameter s and red shift z is (nearly) the same as in the background model. Then the generalized Einstein angle can be expressed as

$$\Theta_E^2 = 2mH_0 F(z_L, z_S, \Omega_0, \Lambda); \tag{37}$$

the complicated function F has been evaluated in (Schneider, Ehlers, and Falco 1992).

The time delay between two images of a source, say image 1 and image 2, can be computed similarly from the difference of the path lengths and the Shapiro delays due to the gravitational potential well of the lens. It turns out that image 2 arrives later than image 1 by the amount

$$\Delta t_{12} = m(1+z_L)\left\{(\vec{\Theta}_1 - \vec{\Theta}_2)\cdot\left(\frac{\partial\tilde{\psi}}{\partial\vec{\Theta}}(\vec{\Theta}_1) + \frac{\partial\tilde{\psi}}{\partial\vec{\Theta}}(\vec{\Theta}_2)\right) - 2(\tilde{\psi}(\vec{\Theta}_1) - \tilde{\psi}(\vec{\Theta}_2))\right\} \tag{38}$$

The formulae (34)–(38) are the *main tools of approximate lens theory*. Given a "background" cosmological model through $(H_0, \Omega_0, \Lambda_0)$ and a projected mass density $\tilde{\Sigma}(\vec{\Theta})$ of a lens, these equations provide the relations between the observables z_L, z_S, the image directions $\vec{\Theta}$ corresponding to each source position $\vec{y}$ (see figure 6), the time delays of "signals" arriving in the images, the flux ratios of the images (via equations (17) and (35), see below), and for each resolvable extended source, the intensities of the images in terms of the intrinsic intensity of the source (via equation (8)).

The lens mapping and the time of arrival function have been generalized to the case when light deflection occurs at several, well-separated "lenses."[4] For the purpose of weak statistical lensing one can work directly with equations (27)–(32) and the Jacobi equation for thin light bundles without using the lens equation (35), as shown in (Holz and Wald 1998). We will not deal with these extensions of the theory here.

The preceding brief review of standard lens theory illustrates the dependence of the lensing observables on the mass distributions of the lenses which has been assumed, but not exhibited in the previous sections.

The D's of the approximate theory which occur in eqs (35), (36) refer to shearfree light beams in empty regions between clumps, while r_A in secs. 3 and 4 refers to the full metric. Within the approximation, the "real" or "perturbed" area distance of a source point from the observer is not D_S but follows from (35) to be

$$r_A = D_S|\det(\delta_{AB} - \Theta_E^2\partial_{AB}\tilde{\psi})|^{\frac{1}{2}}. \tag{39}$$

This formula corresponds to equation (4); the former η corresponds to the present $\vec{\Theta}$, τ is now fixed and therefore not displayed, and the dependence on the affine parameter s is contained in the dependence of D_S and D_{LS} on s. Under the various assumptions made in this section, the "Dyer-Roeder distances" D_S and D_{LS} can be computed as functions of H_0, Ω_0, Λ and s.[5] Because of equation (39) and the meaning of the previous equations (3), (6) and (8), the quantity

$$\mu \equiv \{\det(\delta_{AB} - \theta_E^2\partial_{AB}\tilde{\psi})\}^{-1} \tag{40}$$

is called the "signed magnification." Its absolute magnitude gives the angular and flux magnification of an image relative to the (fictitious) "unlensed" source, and its sign tells whether the image is equally ($\mu > 0$) or oppositely ($\mu < 0$) orientated as its original. Clearly, μ has a meaning only in perturbation theory and is not observable. The ratios of the μ-values for different images of a source are, however, observable as

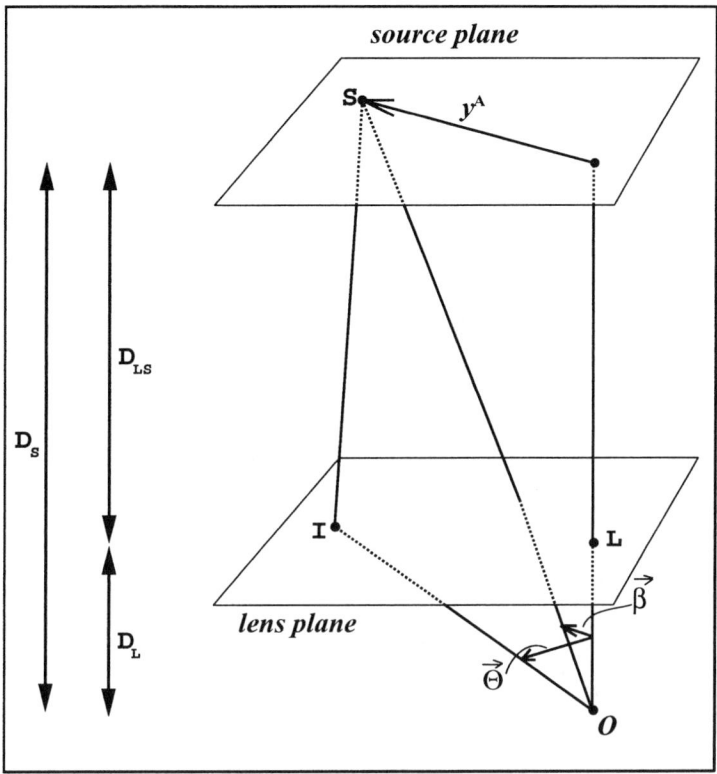

Figure 6. Illustration of the lens mapping. Shown are light paths ≡ null geodesics projected into 3-space. O: observer, L: lens, S: source, I: deflection point, OIS: deflected, "perturbed" light path, OS: unperturbed light path.

relative magnifications and relative parities of these images; recall equations (17) and (18).

The caustic of the mapping (35) is given by the vanishing of the determinant in equation (39), since under the relevant conditions, $D_S > 0$. In the approximation considered that caustic is that of a "small," spatial wave front. If the observer position and the lens are fixed, but the source plane is varied, one obtains the caustic of the light cone $\mathfrak{C}(\tau)$ as the union of the various small caustics. In figure 7, the null geodesic generators of $\mathfrak{C}(\tau)$ are represented as broken straight lines parametrized by $\vec{\Theta}$, and the whole "two dimensional" caustic of $\mathfrak{C}(\tau)$ can be visualized — as far as its differential topology is concerned — as a many-sheeted, self intersecting "surface" in $\mathbb{R}^3$.

The relationship between two images, considered in connection with equation (19), is now obtained by eliminating $\vec{\beta}$ from the equations (35) for $\vec{\Theta}_{(1)}$ nd $\vec{\Theta}_{(2)}$ The time delay can then be computed from equation (38). To see the connection with equation (19), one has to take into account that, during an observation, the source positions

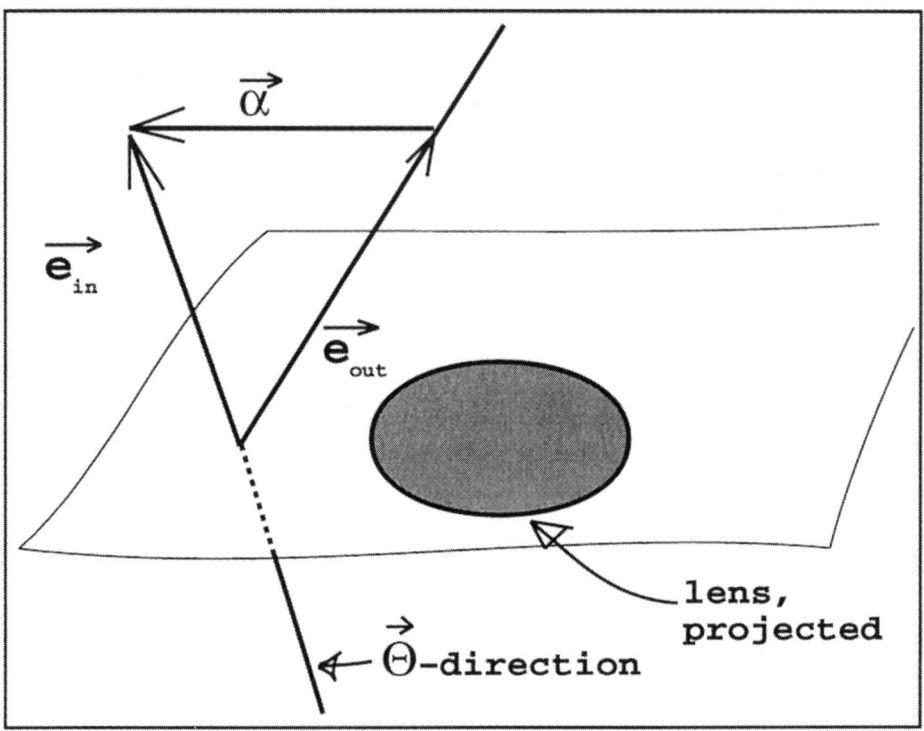

Figure 7. The deflection angle vector field $\vec{\alpha}(\vec{\Theta})$ in the lens plane, $\vec{\alpha} = \vec{e}_{in} - \vec{e}_{out}$.

do not change appreciably, so in equations (19) for $a = i$, the τ's can be forgotten. The resulting three equations contain the six real variables contained in $\eta_1, \eta_2, r_A^{(1)}, r_A^{(2)}$. The approximate theory provides the solution by giving $\vec{\Theta}_2$ as a function of $\vec{\Theta}_1$ and the r_A's by equation (39). equation (19) for $a = 0$ is taken care of by (38). — The relations discussed here become considerably more complicated in the multiple-lensing case; they have been treated exhaustively and elegantly in (Petters 1996).

Can one "derive" the approximate theory from the underlying exact one? In a mathematical sense, i.e., with controlled error estimates, no — that would require a much better quantitative understanding of generic solution classes of Einstein's field equation than the one which is available now. One may, however, be able to test the equations of the approximate theory numerically, using the exact equations in discretized form. Unfortunately, apart from the cases of the (neutral or charged) Kerr spacetimes and that of a static, spherically symmetric constant density star surrounded by vacuum, whose lensing properties have been determined exactly (Kerscher 1992; Kling and Newman 1999), no other tractable exact examples seem to be at hand. So, in lensing as in other cases where a complicated theory is used to make testable predictions, one is confronted with the fact: The qualitative picture can be understood using the exact theory; to obtain quantitative statements one has to resort to "plausible" approximations.

ACKNOWLEDGEMENT

E. T. Newman thanks the NSF for support under grant #92-05109.

NOTES

* Submitted for publication in January 2000.

1. For a systematic presentation of lens theory which contains a brief history of it and covers the major astrophysical applications up to 1991, see (Schneider, Ehlers, and Falco 1992); for later work consult (Wambsganss 1998) and the refs. given there.
2. For a proof of (6) see, e.g., (Schneider, Ehlers, and Falco 1992, 114).
3. See, e.g., (Holz and Wald 1998; Takada and Futamase 1999).
4. See, e.g., (Schneider, Ehlers, and Falco 1992; Seitz, Schneider, and Ehlers 1994).
5. See, e.g., (Schneider, Ehlers, and Falco 1992).

REFERENCES

Arnold, V. J. 1980. *Mathematical Methods of Classical Mechanics*. Berlin: Springer Verlag.
———. 1986. *Catastrophy Theory*. Berlin: Springer Verlag.
Berry, M. V. 1987. "Disruption of Images: the Caustic Touching Theorem." *J. Opt. Soc. Amer.* A4:561.
Einstein, A. 1911. "Über den Einfluß der Schwerkraft auf die Ausbreitung des Lichtes." *Annalen der Physik* 35:898.
———. 1915. "Erklärung der Perihelbewegung des Merkur aus der allgemeinen Relativitätstheorie." *Sitzungsber. Preuß. Akad. Wiss.*
Etherington, J. M. H. 1933. "On the definition of distance in general relativity." *Phil. Mag.* 15:761.
Frittelli, S., E. T. Newman, and G. Silva-Ortigoza. 1999. "The Eikonal Equation in Asymptotically Flat Spacetimes." *J. Math. Phys.* 40(2):1041–1056.
Holz, D. E., and R. M. Wald. 1998. "A new method for determining cumulative gravitational lensing effects in inhomogeneous universes." *Phys. Rev.* D 58:063501.
Kerscher, Th. F. 1992. "Lichtkegelstrukturen statischer, sphärisch-symmetrischer Raumzeiten mit homogenen Zentralkörpern." Diploma thesis, Univ. München.
Kling, T., and E. T. Newman. 1999. "Null Cones in Schwarzschild Geometry." *Phys. Rev.* D 59:124002.
Lebach, D. E., et al. 1995. "Measurement of the Solar Gravitational Deflection of Radio Waves using VLBI." *Phys. Rev. Lett.* 75:1439.
Penrose, R. 1966. "General-relativistic energy flux and elementary optics." Pg. 259 in *Essays in Honor of Václav Hlavatý*, ed. B. Hoffmann. Bloomington: Indiana Univ. Press.
Petters, A. O. 1996. "Mathematical Aspects of Gravitational Lensing." Pg. 1117 in *Proc. 7th M. Großmann Meeting on GR*, vol. B, eds. R. T. Jantzen and G. M. Kaiser. Singapore: World Scientific.
Renn, J., T. Sauer, and J. Stachel. 1997. "The Origin of Gravitational Lensing." *Science* 275:184.
Russell, H. N. 1937. *Scientific American*, February, 76.
Sasaki, M. 1993. "Cosmological Gravitational Lens Equation: Its Validity and Limitation." *Progr. Theor. Phys.* 90:753.
Schneider, P., J. Ehlers, and E. E. Falco. 1992. *Gravitational Lenses*. New York, Berlin, Heidelberg: Springer Verlag.

Seitz, S., P. Schneider, and J. Ehlers. 1994. "Light Popagation in arbitrary spacetimes and the gravitational lens approximation." *Class. Quantum Grav.* 11:2345.

Takada, M., and T. Futamase. 1999. "Post-Newtonian Lagrangian perturbation approach to the large scale structure formation." *Phys. Rev. D*.

Walsh, G., R. F. Carswell, and R. J. Weymann. 1979. *Nature* 279:381.

Wambsganss, J. 1998. "Gravitational Lensing in Astronomy." *Living Reviews in Astrophysics*, March.

Whitney, H. 1955. "Mappings of the plane into the plane." *Ann. Math.* 62:374.

C. V. VISHVESHWARA

RIGIDLY ROTATING DISK REVISITED*

1. INTRODUCTION

In a very interesting article entitled 'Einstein and the Rigidly Rotating Disk', John Stachel (1980) has traced the genesis of the general theory of relativity based on the concept of a curved spacetime. As he remarks, the rigidly rotating disk 'seems to provide a "missing link" in the chain of reasoning that led him (Einstein) to the crucial idea that a nonflat metric was needed for a relativistic treatment of the gravitational field'. The chain of reasoning begins essentially with space and time being combined into spacetime as a consequence of the special theory of relativity. After that, equivalence principle connecting gravitation to accelerated frames was the first step towards the formulation of the general theory of relativity. Now comes the rigidly rotating disk as the "missing link". The inertial force, namely the centrifugal force, experienced by observers fixed on the rotating disk is identified as equivalent to a gravitational field. Einstein showed that these observers find the ratio of the circumference to the radius of a circle around the axis of rotation to be greater than 2π. This immediately leads to the conclusion that the geometry associated with the rotating disk is non-Euclidean. In other words, the presence of gravitation leads, in analogy with the Gaussian theory of surfaces, to a metrical theory of four dimensional curved spacetime. The general theory of relativity is born and the inertial forces lie dormant under the cover of curved spacetime. Although all gravitational phenomena are considered using general spacetime metrics that are solutions to the Einstein field equations, the basic concepts arising in the case of the rigidly rotating disk are still important. Some of these concepts can be formulated precisely and fruitfully in the case of stationary, axisymmetric spacetimes. Such a spacetime can be considered to be a broad generalization of that of the rigidly rotating disk. For instance, in contrast to the observers fixed on the rotating disk, there are the rest observers outside the disk. The idea and the existence of global rest observers in an arbitrary stationary axisymmetric spacetime become fundamentally important for the study of physical phenomena. It is often claimed that one of the major consequences of general relativity is the abolition of the notion of forces. On the other hand, analogues of inertial forces in general relativity have been recently formulated. They can be used in analysing the effects of rotation inherent to the spacetime as well as in the case of orbiting particles. There are two other phenomena that are of considerable interest in studying rotation, namely gyroscopic precession and

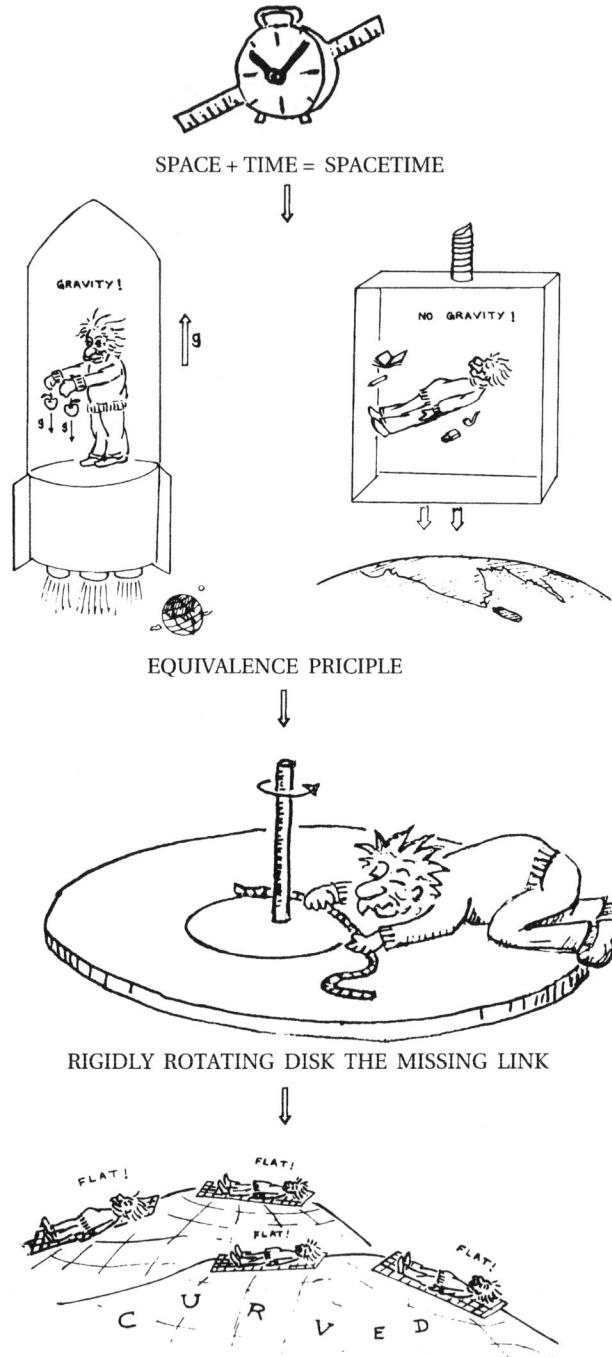

SPACE + TIME = SPACETIME

EQUIVALENCE PRICIPLE

RIGIDLY ROTATING DISK THE MISSING LINK

CURVED SPACETIME WITH LOCAL FLATNESS.

gravito-electromagnetism. All these three can be elegantly inter-connected utilizing spacetime symmetries.

Whereas the considerations above may be viewed as arising from the generalization of the rotating disk, the rigidly rotating frame associated with the disk has recently emerged in the most practical and unexpected form. This is in connection with the Global Positioning System. One wonders what Einstein might have thought of a machinery, making use of his good old rotating disk, that is not merely utilitarian but also carries with it the ominous potential for destructive usage.

2. RIGIDLY ROTATING DISK IN FLAT SPACETIME

Within the framework of general relativity, one can study the rigidly rotating disk. For this purpose, let us start with the flat spacetime. The line element in cylindrical coordinates is given by

$$ds^2 = dt^2 - dr^2 - r^2 d\phi^2 - dz^2 \tag{1}$$

with $c = G = 1$. Static observers in the inertial frames of reference at each space point follow the Killing vector field $\xi^a = \delta_0^a$ This field is orthogonal to the hypersurfaces $t = $ constant and they can synchronise their clocks with respect to time t since $\xi_a = (1, 0, 0, 0) = t_{,a}$. In order to go over to the disk rotating at the uniform angular speeed ω, we make the coordinate transformation,

$$t = t', \quad r = r', \quad \phi = \phi' + \omega t', \quad z = z'. \tag{2}$$

Then the line element for the spacetime geometry of the rotating disk is given by,

$$ds^2 = \left(1 - \omega^2 r'^2\right) dt'^2 - 2\omega r'^2 d\phi' dt' + d\sigma'^2, \tag{3}$$

where $d\sigma'^2 = dr'^2 + r'^2 d\phi'^2 + dz'^2$.

We note two new features in the line element (3) in comparison with the original one (1). We have $g_{00} = (1 + 2\Phi_c)$, where the centrifugal potential $\Phi_c = -\frac{1}{2}\omega^2 r^2$. There is also the term $g_{03} = -\omega r'^2$ which shows the rotation inherent to the spacetime of the disk. Stationary observers fixed at spatial points on the disk follow the Killing vector field adapted to them, $\overline{\xi}^a = \delta_{t'}^a$. Then $\overline{\xi}_a = (g_{00}, 0, 0, g_{03})$ and therefore is not hypersurface orthogonal. If it were, it could have been expressible in the form $\overline{\xi}_a = \alpha \beta_{,a}$, where α and β are scalar functions. On the other hand, consider the vector field $\overline{\chi}^a = \overline{\xi}^a - \frac{g_{03}}{g_{33}}\overline{\eta}^a$, where $\overline{\eta}^a = \delta_{\phi'}^a$ is the Killing vector field giving rise to rotational symmetry about the $z' - axis$. It is easy to verify that $\overline{\chi}^a = (1, 0, 0, -\omega)$ and $\overline{\chi}_a = (1, 0, 0, 0) = t_{,a}$. Hence $\overline{\chi}^a$ is orthogonal to $t = constant$ hypersurfaces. In fact $\overline{\chi}^a$ is nothing but the original Killing vector field ξ^a of the inertial observers expressed in the rotating coordinates.

To sum up, stationary observers fixed on the rigidly rotating disk follow worldlines along t' which is no longer global synchronous time for these observers. However they can define a hypersurface orthogonal timelike vector field which turns out to be the $t - lines$ of the inertial observers in the flat spacetime. Also, the metric component g_{00} contains the Newtonian centrifugal potential Φ_c from which one can construct the centrifugal force acting on the observers fixed to the disk as measured by the inertial observers.

In the case of a rigidly rotating frame, one can trivially recover the original nonrotating flat spacetime by the global transformation inverse to the one given in equation (2). In a genuine stationary spacetime with rotation, this is impossible although locally one can obtain a flat Minkowskian metric. Nevertheless, the ideas pertaining to the rigidly rotating disk can be generalized to an arbitrary stationary spacetime with axial symmetry as we shall see in the next sections.

3. THE GLOBAL REST FRAME

As has been pointed out, an axisymmetric stationary spacetime is the broad generalization of the rotating disk. Two classic examples of such a spacetime is the Schwarzschild and the Kerr. The Schwarzschild spacetime is static, or equivalently, possesses no inherent rotation since the source, eg. the black hole, is non-rotating. On the other hand, the Kerr spacetime is stationary and has rotation. In both cases it is important to identify the global rest frame akin to the frames of observers at rest outside the rotating disk.

The Schwarzschild metric is written as

$$ds^2 = \left(1 - \frac{2m}{r}\right) dt^2 - \left(1 - \frac{2m}{r}\right)^{-1} dr^2 - r^2(d\theta^2 + \sin^2\theta d\phi^2) \quad (4)$$

where $m = \frac{MG}{c^2}$; M = Mass; $c = G = 1$.

It admits the Killing vector fields $\xi^a = \delta_0^a$ and $\eta^a = \delta_3^a$ correponding respectively to time symmetry and axial or rotational symmetry with respect to $z - axis$. They commute with each other and their scalar product $\xi^a \eta_a = g_{03} = 0$ since it is a static spacetime. As the metric is spherically symmetric there exist two more rotational Killing vectors which would be absent in a spacetime which is only axially symmetric. The timelike Killing vector field, in its covariant form, can be written as:

$$\xi_a = g_{00} t_{,a}. \quad (5)$$

This shows that ξ^a forms a hypersurface orthogonal congruence ie., orthogonal to surfaces $t = constant$. It is therefore irrotational. In otherwords, observers following the timelike Killing field trajectories with $(r, \theta, \phi) = constant$ have no mutual rotation and, further more, share a common synchronous time t. At any given moment, $t = constant$ is the entire three dimensional space as seen by these observers. These are the rest observers and their spatial frame of reference is the rest frame.

Let us now consider the more complicated case of the stationary spacetime with axial symmetry as exemplified by the Kerr line element :

$$ds^2 = \left(1 - \frac{2m}{\Sigma}\right) dt^2 + 2\frac{2mar}{\Sigma} \sin^2\theta dt d\phi$$
$$- (a^2 + r^2 \frac{2ma^2 r}{\Sigma} \sin^2\theta) \sin^2\theta d\phi^2 - \frac{\Sigma}{\Delta} dr^2 - \Sigma d\theta^2 \quad (6)$$

Where m = Mass, $J = ma$ = Angular Momentum, $\Sigma = r^2 + a^2 \cos^2\theta$, $\Delta = m^2 - 2mr + a^2$. It admits the two Killing fields $\xi^a = \delta^a_0$ and $\eta^a = \delta^a_3$ giving rise as before to the time and rotational or axial symmetries respectively. They commute with each other, but their scalar product $\xi^a \eta_a = g_{03} \neq 0$ since the spacetime has inherent rotation. This shows up in the fact that the vorticity of the ξ^a congruence is non-zero:

$$\omega^a_\xi = \frac{1}{\sqrt{-g}} \epsilon^{abcd} \xi_b \xi_{c;d} \neq 0 \tag{7}$$

where ϵ^{abcd} is the completely antisymmetric Levi-Civita symbol. Consequently ξ^a field is not globally hypersurface orthogonal and the stationary observers are not rest observers. On the other hand, consider the vector field

$$\chi^a \equiv \xi^a - \frac{(\xi^b \eta_b)}{(\eta^c \eta_c)} \eta^a \tag{8}$$

which is the projection of ξ^a orthogonal to η^a. The vorticity of the χ^a congruence is zero. The frames adapted to χ^a were called locally non-rotating reference frames (LNRF) by Bardeen (1970). Actually χ^a forms a hypersurface orthogonal congruence.

$$\chi_a = \left[\xi^b \xi_b - \frac{(\xi^b \eta_b)^2}{(\eta^c \eta_c)} \right] t_{,a} \tag{9}$$

Observers following χ^a are in fact the global rest observers and frames adapted to them form the global rest frames. Properties of χ^a congruence and the global rest frames were studied in detail by Greene, Schucking and Vishveshwara (1975). The rest observers revolve in circular orbits with the angular speed $\omega_0 = -\frac{\xi^b \eta_b}{\eta^c \eta_c} = -\frac{g_{03}}{g_{33}}$, whereas ξ^a— observers are stationary fixed at spatial points! The former share a common synchronous time t. If two circular light rays start at the same moment from a point in opposite directions, they will return to that point at different moments of time with respect to the stationary observer. On the other hand, the rest observer will receive them at the same moment, since in a given interval of time one ray travels longer than the other and the rest observers will have moved exactly by this distance by then. This motion is a manifestation of the so-called inertial frame dragging. When the spacetime has orthogonal transitivity, ie., there are no cross terms between (t, ϕ) and (r, θ), $g_{tr} = g_{t\theta} = g_{\theta\phi} = g_{r\phi} = 0$, one can show that χ^a becomes null on a surface which is itself null. This surface is identified as the stationary event horizon. This happens for the Killing field ξ^a in the case of static spacetimes like the Schwarzschild. One can also show that the surfaces given by $t = constant$ are maximal.

Physical phenomena are studied within the global rest frames especially since extended objects can be defined on $t = constant$ hypersurfaces. As we shall see, the general relativistic analogues of inertial forces can be covariantly defined with respect to the global rest frames.

4. ROTATIONAL EFFECTS

There are three important aspects of rotational effects that appear in a stationary axisymmetric spacetime.

4.1. Gravito-Electromagnetism

There is a striking similarity between electromagnetism and gravitation. This shows up through the metric component g_{00} which behaves like the electrostatic potential, while $g_{0\alpha}(\alpha = 1 - 3)$ acts as the magnetic potential. Or, the timelike Killing vector field ξ_a plays the role of the electromagnetic four potential. This has been studied extensively in the weak field approximation.[1]

4.2. Gyroscopic Precession

As is well known, gyroscopic precession is an important phenomenon which brings out rotational effects. Not only does it find applications in various contexts, but also has been put forward as a candidate for testing the general theory of relativity itself. In flat spacetime Thomas precession is improtant in atomic physics. Spacetime curvature, as in the case of the Schwarzschild spacetime, is incorporated into the geodesic precession, e.g. Fokker-De Sitter precession. The source angular momentum gives rise to additional effects in the case of stationary axisymmetric spacetimes. In the weak field limit this is related to the Lense-Thirring effect. A consequence of this is that a gyroscope carried by a static observer in a static spacetime does not precess whereas a gyroscope carried by a stationary obsever in a stationary spacetime does undergo precession. Gyroscopic precession has also been studied by several authors in the past.[2]

4.3. Inertial Forces

Recently there has been a great deal of interest in the general relativistic analogues of inertial forces. We shall be considering them in some detail in the next section.

All the three aspects of rotation could be studied on a rigidly rotating disk. As has been already mentioned, we study them in the more general framework of a stationary axisymmetric spacetime. Utilizing the two inherent symmetries, we shall indicate how these three aspects can be inter-related in an elegant and covariant manner.

5. INERTIAL FORCES IN GENERAL RELATIVITY

First of all, one may ask why inertial forces be considered in the context of the general theory of relativity at all. After all, one of the major steps taken by the theory was to banish the notion of forces. Nevertheless, there are advantages in constructing the analogues of inertial forces in general relativity. To begin with, the question 'why?' may very well be countered by the response 'why not?'. After all we are more familiar with forces, both conceptually and physically than, for instance, spacetime curvature and geodesics. But more pertinently, these forces serve as an aid in analysing rotational effects, especially in identifying those engendered by the stationary spacetime with rotation in contrast to the static case. Furthermore, as has been mentioned earlier, inertial forces can be inter-related to gyroscopic precession and gravito electro-magnetism bringing out the essential unity among these three different approaches to the study of rotation in general relativity.

The original motivation for defining the inertial forces may perhaps be traced to the observation of Abromowicz and Lasota (1986), see also Abromowicz and Prasanna (1990) regarding the centrifugal force reversal in the Schwarzschild spacetime. In this spacetime there exists a circular null geodesic, or a photon orbit, at $r = 3m$. It was pointed out in (Abramowicz and Lasota 1986), that rockets moving at different speeds in circular timelike orbits at $r = 3m$ have no relative acceleration. They need the same engine thrust to move. This was attributed to the fact that they need to overcome only the gravitational pull, the centrifugal force being zero for all of them. This was considered to be a natural consequence of a light ray being a straight line, in the present case, the circular path at $r = 3m$ being a 'straight line' in a curved spacetime. This centrifugal force reversal at $r = 3m$ led to the definition and study of the general relativistic analogues of inertial forces.

Next, Abromowicz, Carter and Lasota (1988) formulated what they termed as the optical reference frame. Essentially, g_{00} is factored out in the line element of a stationary axisymmetric spacetime and particle dynamics studied in the conformal spacetime which is identified with the optical reference frame. They showed that the spatial trajectories of the null geodesics of the original four dimensional spacetime happen to be geodesics in the optical reference frame. Further, the four acceleration of particles following geodesics in the original spacetime, when projected onto the conformal spacetime, splits into terms that may be identified as inertial forces. This formalism was followed by the covariant definition of inertial forces by Abromowicz, Nurowski and Wex (1993). This may be summarized as follows in the case of stationary spacetimes with axial symmetry.

The globally hypersurface orthogonal field defined in equation (5) is written as

$$\chi^a = e^\phi n^a; \quad n^a n_a = 1. \tag{10}$$

That is, n^a is the four velocity of the rest observers and $e^\phi = (\chi^b \chi_b)^{\frac{1}{2}}$ is the normalization factor. Any arbitrary four velocity u^a may be split into

$$u^a = \gamma(n^a + v\tau^a); \quad u^a u_a = 1 \tag{11}$$

where τ^a is a unit vector orthogonal to n^a and v is the spatial velocity of the particle along this direction. Denote by a tilde over any tensor its projection orthogonal to n^i. Then the four acceleration a^i projected thus is made up of the four analogues of inertial forces as follows.

$$m\tilde{a}^i = G^i + Z^i + C^i + E^i \tag{12}$$

where $m =$ particle mass.
We have then

$$\text{Gravitational force } G_k = \phi_{,k}$$
$$\text{Centrifugal force } Z_k = -(\gamma v)^2 \tilde{\tau}^i \tilde{\nabla}_i \tilde{\tau}_k \quad (13)$$
$$\text{Euler force } E_k = -\dot{V}\tilde{\tau}_k$$
$$\text{Coriolis-Lense-Thirring force } C_k = \gamma^2 v X_k$$

where

$$\dot{V} = (ve^\phi \gamma)_{,i} u^i$$
$$X_k = n^i (\tau_{k;i} - \tau_{i;k}) \quad (14)$$
$$\phi_{,k} = -n^i n_{k,i}$$
$$\tilde{\tau}_k = e^\phi \tau_k$$

In the case of particles in circular orbits with uniform motion the Euler forces $E^i = 0$. In the case of static spacetimes the Coriolis-Lense-Thirring force $C^i = 0$.

Let us return to the phenomenon that motivated the above formulation of the inertial forces, namely the centrifugal reversal at the circular photon orbit in the Schwarzschild spacetime. Since this orbit is viewed as a straight line, one would expect that a gyroscope carried along this orbit would not precess. This happens to be true. This raises the question whether inertial forces and gyroscopic precession are related to each other in stationary axisymmetric spacetimes. As we shall see, this is in fact the case. First we shall briefly outline the covariant description of gyroscopic precession for this purpose.

6. GYROSCOPIC PRECESSION

A comprehensive treatment of gyroscopic precession in a stationary, axisymmetric spacetime has been given by Iyer and Vishveshwara (1993). This work makes use of the Frenet-Serret formalism which provides a geometric, invariant description of a one dimensional curve. In a four dimensional spacetime, the Frenet-Serret tetrad $e_{(a)}$ is transported along a timelike curve which is characterized by the three scalars κ, τ_1 and τ_2 which are termed the curvature, the first and the second torsions respectively. The curvature κ is the magnitude of the four acceleration of the particle following the timelike trajectory. The torsions τ_1 and τ_2 are directly related to gyroscopic precession. Consider a tetrad $f_{(a)}$ Fermi-Walker transported along the particle trajectory. Identify $e_{(0)}$ and $f_{(0)}$ with the four velocity u^a of the particle. The triad $f_{(\alpha)}$ is physically realized by three mutually orthogonal gyroscopes. In general $f_{(\alpha)}$ or equivalently the gyroscopes precess with respect to $e_{(\alpha)}$. The rate of gyrosocopic precession is given by

$$\omega_g^a = -(\tau_2 e_{(1)}^a + \tau_1 e_{(3)}^a). \quad (15)$$

Determination of the tetrad $e_{(a)}$ and the parameters τ_1 and τ_2 along a particle trajectory provides a complete description of gyroscopic precession with respect to the comoving Frenet-Serret frame.

7. APPLICATION TO AXIALLY SYMMETRIC STATIONARY SPACETIMES AND COVARIANT CONNECTIONS

Let us consider a particle following the Killing vector field,

$$\zeta = \xi + \Sigma_i \omega^i \eta_i \qquad (16)$$

where ξ is the timelike Killing vectors, η_i the spatial Killing vector that the spacetime may admit and ω^i are constants. Although our formalism is applicable to such a general case, we shall specialize to a circular orbit with $\eta_i \equiv \eta$ the axial Killing vector, so that the particle describes a circular orbit with angular speed ω. The four velocity of the particle is then

$$u^a \equiv e^a_{(0)} = e^\phi (\xi^a + \omega \eta^a) \qquad (17)$$

where e^ϕ is the normalization factor. The Frenet-Serret triad $e^a_{(1)}, e^a_{(2)}, e^a_{(3)}$ will be aligned respectively along the radial direction, tangential to the orbit and normal to the orbital plane. The covariant derivative of the tetrad components with respect to the proper time (denoted by an overhead dot) satisfies a Lorentz-like equation,

$$\dot{e}^a_{(i)} = F^a_{\ b} e^b_{(i)}; \quad F_{ab} \equiv e^\psi (\xi_{a;b} + \omega \, \eta_{a;b}). \qquad (18)$$

Killing vector fields ξ^a and η^a obey the equations.

$$V_{a;b} + V_{b;a} = 0 \text{ and } V_{a;b;c} = R_{abcd} V^d. \qquad (19)$$

Consequently we have

$$F_{ab} = -F_{ba}; \quad \dot{F}_{ab} = 0; \quad \dot{\kappa} = \dot{\tau}_1 = \dot{\tau}_2 = 0. \qquad (20)$$

Therefore F_{ab} is like a homogeneous electromagnetic field tensor. In fact the motion along Killing trajectories is remarkably similar to the motion of charged particles in a homogeneous electromagnetic field as was demonstrated by Honig, Schucking and Vishveshwara (1974). Acceleration

$$a^i = \kappa e^i_{(1)} = F^i_{\ j} e^j_{(0)} \qquad (21)$$

is analogous to the electric field. Similarly one can show that

$$\Omega^i = {}^*F^i_{\ j} e^j_{(0)}, \qquad (22)$$

where * the star denotes the dual. This is like the magnetic field which produces spin precession in an electro-magnetic field. Hence there is a striking resemblance to electro-magnetic phenomena which is the essence of gravito electro-magnetism.

It is straight forward to compute κ, τ_1, τ_2 and $e_{(\alpha)}$ in terms of powers of F_{ab} and $e_{(0)}$. Therefore, gyroscopic precession can be described in a covariant manner for Killing trajectories and can be related to gravito-electromagnetism. This has been done in detail in reference (Iyer and Vishveshwara 1993).

The formalism developed for inertial forces can also be applied to the Killing orbits, once again, in a straight forward manner. This has been done in detail by Rajesh Nayak and Vishveshwara (1998). Furthermore, in this paper covariant connections have been established among the inertial forces, gyroscopic precession and gravito-electromagnetism. As we have seen, gyroscopic precession is represented by τ_1 and τ_2, while the inertial forces are given by the gravitational force G_i, the centrifugal force Z_i and the Coriolis force C_i. In the case of Killing orbits it can be shown (*ibid.*) that

$$\tau_1 \sim a \cdot (Z + \alpha C)$$
$$\tau_2 \sim a \times (Z + \alpha C), \qquad (23)$$

where a is the acceleration and α is a scalar function involving the Killing vector fields. Thus inertial forces and gyroscopic precession are inter-related.

Now, acceleration $a^i = F^i_j u^j$ and $Z^i + \alpha C^i \sim^* F^i_j u^j$, so that a and the combination $(Z + \alpha C)$ are equivalent to gravito-electric and gravito-magnetic fields E and B respectively. Therefore, we have the relations

$$\tau_1 \sim E \cdot B$$
$$\tau_2 \sim E \times B$$

This is exactly similar to the case of charged particles moving in a homogeneous electromagnetic field (Honig, Schücking, and Vishveshwara 1974). Thus the interelations among the three formalisms can be established in a covariant and elegant manner. The formals are considerably simplified in the case of static spacetimes for which the Coriolis force $C^i = 0$.

The results of reference (Rajesh Nayak and Vishveshwara 1998) shed further light on some of the observations that had been made earlier. Equation (23) shows that, in static spacetimes, ($C^i = 0$), gyroscopic precession and centrifugal force undergo simultaneous reversal when they become zero. It can also be shown (*ibid.*) that this happens at a photon orbit. This phenomenon had been studied earlier in the case of the Schwarzschild spacetime (Abramowicz and Lasota 1986; Abramowicz and Prasanna 1990) and also by Rajesh Nayak and Vishveshwara (1997) in the case of Ernst spacetime. There is no such correlation in the case of stationary spacetimes when $C^i \neq 0$. This had been noticed by Sai Iyer and Prasanna (1993) for the Kerr spacetime and by Rajesh Nayak and Vishveshwara (1996) in the case fo Kerr-Newman spacetime.

8. BACK TO THE ROTATING DISK

In recent years, the rigidly rotating disk has made a dramatic comeback. This is in the context of the Global Positioning System. For a lucid account of GPS and the relativistic effects involved we refer the reader to the article by Neil Ashby (1998). The users of GPS are fixed on the rotating earth. One of the relevant processes to be considered is the Einstein synchronization of clocks carried by these observers to set up a network of synchronized clocks. This is achieved by means of light propogation as seen by these earth bound observers. The simplest model used for carrying out calculations in this regard makes use of the rigidly rotating frame and the resulting line element

(Langevin metric) given in the equation (3) with $\omega = \omega_E = 7.292115 \times 10^{-5} rad/s$, the rotation rate of the earth. A light pulse travelling between two rigidly rotating observers takes an extra amount of time to catch up with the moving receiver. This path dependent contribution, known as the Sagnac Effect, depends on the metric component g_{03}. For a light pulse traversing the equatorial circumference this additional time delay amounts to about 207 nanoseconds. This is a significant amount considering the accuracy of the GPS system. Probably, no one had dreamt of a situation in which general relativity became significant on a terrestrial scale that too in such a utilitarian manner!

Within the full machinery of the general theory of relativity, the rigidly rotating disk has invaded the area of exact solutions as well. Neugebauer and Meinel (Neugebauer 1996) have derived an exact solution for a rigidly rotating disk of dust. This is of significance on two counts. Firstly, no exact global solution of Einstein's field equations has been found for a rotating distribution such as a rotating star. Secondly, the solution may serve as a model for astrophysical objects such as galaxies and accretion disks. In any event, the spacetime of the disk exhibits novel features that make it interesting in its own right. For instance, as the angular speed of rotation is increased, the exterior of the disk undergoes a phase transition and becomes the extreme Kerr solution. Another unusual feature is the formation of ergospheres as the physical parameters are varied. These and other aspects of the spacetime associated with the disk makes it worthy of further studies.

In this article, we have tried to trace briefly the evolution of the rigidly rotating disk. As we have seen, it started out as the 'missing link' in the chain of reasoning that led Einstein from the special theory of relativity to the general theory building up a spacetime picture of gravitation. Aspects inherent to the spacetime of the disk find their natural generalization in the phenomena characteristic of axisymmetric, stationary spacetimes. The disk metric has recently found amazing application in the GPS system. And, finally, one has now an exact solution for the disk made up of dust exhibiting interesting features. For all we know, there may be surprises still in store for us. A ride on the rigidly rotating disk cannot but turn you on!

ACKNOWLEDGEMENT

I thank Ms. G. K. Rajeshwari and Mr. Rajesh Nayak for their valuable help in the preparation of the manuscript.

Indian Institute of Astrophysics

NOTES

* I have known John for some three decades now as a very close friend and an esteemed colleague. It is a great pleasure to dedicate this article to him with affection and my warmest best wishes.

1. See (Bini, Carini, and Jantzen 1997a, 1997b; Embacher 1984; Thorne, Price, and Macdonald 1986; Jantzen, Carini, and Bini 1992; Ciufolini and Wheeler 1995).

2. See (Abramowicz, Nurowski, and Wex 1995; Barrabes, Boisseau, and Israel 1995; de Felice 1991, 1994; Rindler and Perlick 1990; Semerak 1995, 1996, 1997).

REFERENCES

Abramowicz, M. A., B. Carter, and J. P. Lasota. 1988. *Gen. Rel. Grav.* 20:1173.
Abramowicz, M. A., and J. P. Lasota. 1986. *Am. J. Phys.* 54:936.
Abramowicz, M. A., P. Nurowski, and N. Wex. 1993. *Class. Quantum. Grav.* 10:L183.
———. 1995. *Class. Quantum Grav.* 12:1467.
Abramowicz, M. A., and A. R. Prasanna. 1990. *Mon. Not. R. Astr. Soc.* 245:720.
Ashby, Neil. 1998. "Relativistic Effects in the Global Positioning System." In *Gravitation and Relativity at the Turn of the Millenium*, eds. N. Dadhich and J. V. Narlikar. Pune, India.
Bardeen, J. M. 1970. *Astrophys. J.* 162:71.
Barrabes, C., B. Boisseau, and W. Israel. 1995. *Mon. Not. R. Astr. Soc.* 276:432.
Bini, D., P. Carini, and R. T. Jantzen. 1997a. *Int. J. Mod. Phys.* D 6(1):1–38.
———. 1997b. *Int. J. Mod. Phys.* D 6(2):143–198.
Ciufolini, I., and J. A. Wheeler. 1995. *Gravitation and Inertia*. Princeton: Princeton Univ. Press.
de Felice, F. 1991. *MNRAS* 252:197.
———. 1994. *Class. Quantum Grav.* 11:1283.
Embacher, F. 1984. *Found. Physics* 14:721.
Greene, R. D., E. L. Schücking, and C. V. Vishveshwara. 1975. *J. Math. Phys.* 16:153.
Honig, E., E. L. Schücking, and C. V. Vishveshwara. 1974. *J. Math. Phys.* 15:774.
Iyer, B. R., and C. V. Vishveshwara. 1993. *Phys. Rev.* D48:5706.
Iyer, S., and A. R. Prasanna. 1993. *Class. Quantum. Grav.* 10:L13.
Jantzen, R. T., P. Carini, and D. Bini. 1992. *Ann. Phys.* 215(1):1–50.
Neugebauer, Gernot. 1996. "Gravitostatics and Rotating Bodies." In *General Relativity*, eds. G. S. Hall and J. R. Putnam. Institute of Physics Publishing.
Rajesh Nayak, K., and C. V. Vishveshwara. 1996. *Class. Quantum. Grav.* 13:1173.
———. 1997. *Gen. Rel. Grav.* 29:291.
———. 1998. *Gen. Rel. Grav.* 30:593.
Rindler, W., and V. Perlick. 1990. *Gen. Rel. Grav.* 22:1067.
Semerak, O. 1995. *N. Cimento* B 110:973.
———. 1996. *Class. Quantum Grav.* 13:2987.
———. 1997. *Gen. Rel. Grav.* 29:153.
Stachel, John. 1980. "Einstein and the Rigidly Rotating Disk." In *Gen. Rel. Grav.*, ed. A. Held. Plenum Press.
Thorne, K. S., R. H. Price, and D. A. Macdonald, eds. 1986. *Black Holes: The Membrane Paradigm*. New Haven: Yale Univ. Press.

RAY A. D'INVERNO

DSS 2+2

1. MOTIVATION

The ADM 3+1 formalism is well known and central to much of general relativity. It takes its name first from the initials of the original authors Arnowitt, Deser and Misner (see, for example, the review article (Arnowitt et al. 1962)), and the 3+1 refers to the decomposition of space-time into families of 3-dimensional spacelike hypersurfaces which are threaded by a 1-dimensional timelike fibration. The formalism has been key to our understanding of the important Cauchy problem or initial value problem in general relativity. In addition, most of numerical relativity has, to date, been based on it and it has also been the main starting point for a canonical quantization programme.

Despite its considerable track record of achievement the 3+1 formalism has two main drawbacks. First of all the formalism breaks down if the foliation goes null. Yet null foliations and the associated characteristic initial value problem are extremely important in their own right. They seem the natural vehicle for studying gravitational radiation problems, since gravitational radiation propagates along the null geodesics which rule the null hypersurfaces. They arise when investigating the asymptotics of radiating isolated systems, since $\mathcal{J}^+$ and $\mathcal{J}^-$ are both null hypersurfaces. Again, they are central to cosmology, since we gain information about the universe from observations along our past null cone.

Another major limitation of the 3+1 approach is that the initial data is not freely specifiable, but must satisfy the constraints. (It was John Stachel who originally taught me how to do the following counting.) If we restrict attention to the vacuum case there are 10 unknowns, namely the components of the 4-dimensional metric, and 10 vacuum field equations. However, 4 of the unknowns may be prescribed arbitrarily because of the 4-fold coordinate freedom, leaving 6 components of the 4-metric freely specifiable. Moreover, the field equations are not independent but satisfy 4 differential constraints through the contracted Bianchi identities. The Lichnerowicz lemma reveals that if the constraints are satisfied initially and the evolution equations hold generally, then the constraints are satisfied for all time by virtue of the contracted Bianchi identities. The initial data consist of specifying 6 components of the 3-metric together with the corresponding 6 components of the extrinsic curvature subject to 4 constraints on the initial slice, which leaves 8 variables freely specifiable. However, there exists a 3-fold coordinate freedom within the initial slice. This can be used, for example, to specify 3 of the components of the 3-metric. This leaves 5 variables free. Finally there is a condition which describes the embedding of the initial slice into the 4-geometry. This is a little harder to see, but examples of conditions like maximal slicing and constant mean curvature slicing are essentially constraints between the 3-metric and the extrinsic curvature which encode the nature of the embedding. Such a constraint finally reduces the number of freely specifiable data to 4. These can be thought of, from a Lagrangian point of view, as being $2qs$ and $2\dot{q}s$, that is 2 pieces of information encoded in the metric and 2 pieces in its time derivative (or equivalently 2 pieces of information in the extrinsic curvature). It is in this sense that we say the gravitational field has two dynamical degrees of freedom. However, the precise nature of the freely specifiable data in the 3+1 case is rather intricate, since it involves extracting conformal factors and decomposing tensors into their transverse traceless parts. Moreover, there appear to be 6 evolution equations in this formulation rather than the 2 you would expect for a system possessing 2 degrees of freedom.

The DSS 2+2 formalism directly addresses these last two points. The initials are taken from the work of the original authors d'Inverno, Stachel and Smallwood (d'Inverno and Stachel 1978; d'Inverno and Smallwood 1980) and the 2+2 refers to the decomposition of spacetime into two families of spacelike 2-surfaces. The formalism was originally designed to cope with both characteristic and mixed initial value problems, including 3+1 as a special case. It provides integration schemes in which the initial data are unconstrained. Moreover, these initial data—the so-called conformal 2-structure—identifies the gravitational degrees of freedom explicitly in simple geometrical terms and their evolution is governed by precisely 2 dynamical equations. The formalism is manifestly covariant and there exist versions which cover both holonomic and non-holonomic decompositions of space-time.

In this article we shall review the DSS 2+2 formalism and look at two more recent applications of it to Ashtekar variables and the use of principal null directions in numerical relativity. The formalism is, not surprisingly, more complicated than the corresponding 3+1 formalism because it covers 6 separate cases. The formalism has not been exploited as much as one might have expected given the extensive use of the 3+1 formalism, and perhaps it is this added appearance of complexity which has produced something of a stumbling block for potential users. In fact the 3+1 and 2+2 formalisms have much in common. To illustrate this, we shall compare in detail the decomposition of the metric in the two formalisms and show that they are very similar, although there is, in some sense, a doubling of the metric variables in the latter case. For simplicity we shall restrict attention to vacuum field equations, although the introduction of matter does not significantly alter the story.

2. ELEMENTS OF ADM 3+1

We adopt the index conventions that Greek indices run from 0 to 3, early Latin indices $(a, b, \ldots)$ run from 0 to 1, middle Latin indices $(i, j, \ldots)$ run from 2 to 3, upper-case Latin indices $(A, B, \ldots, I, J, \ldots)$ run from 1 to 3 (see the notation used in (d'Inverno and Vickers 1995)). Let M be a four-dimensional orientable manifold with metric g of signature $(+1, -1, -1, -1)$. We start with a foliation of spacetime into a family of spacelike hypersurfaces Σ given by

$$\phi(x^\alpha) = \text{constant}. \tag{1}$$

Then $w_\alpha = \phi_{,\alpha}$ is the covariant normal to the family and $w^\alpha = g^{\alpha\beta} w_\beta$ is the contravariant normal formed from it with the contravariant metric. We define u^α to be the unit normal to the foliation, which is necessarily proportional to w^α, and, being a unit normal, satisfies

$$u^\alpha u_\alpha = 1. \tag{2}$$

We use the unit normal to define the projection operator

$$B_\alpha{}^\beta = \delta_\alpha{}^\beta - u_\alpha u^\beta, \tag{3}$$

which can be used to project tensors into the foliation. Clearly $B_\alpha{}^\beta u^\alpha = 0$ by (2) and (3). Moreover, we can project the 4-dimensional metric into the foliation Σ to obtain the induced 3-dimensional metric ${}^3g_{ab}$, namely

$${}^3g_{\alpha\beta} = B_\alpha{}^\gamma B_\beta{}^\delta g_{\gamma\delta}. \tag{4}$$

Using (2) and (3) we find that

$${}^3g_{\alpha\beta} = g_{\alpha\beta} - u_\alpha u_\beta, \tag{5}$$

or, equivalently,

$${}^3g^{\alpha\beta} = g^{\alpha\beta} - u^\alpha u^\beta. \tag{6}$$

Next, we consider any rigging vector field t^α which transvects the foliation (i.e. lies nowhere in the foliation). Then we can decompose it into a component parallel and a component orthogonal to u^α, specifically

$$t^\alpha = n^\alpha + b^\alpha = Nu^\alpha + b^\alpha, \tag{7}$$

where N is a proportionality factor called the lapse and b^α is called the shift vector and satisfies

$$b^\alpha u_\alpha = 0. \tag{8}$$

Thus, from (6) and (7), we find

$$g^{\alpha\beta} = \frac{1}{N}(t^\alpha - b^\alpha)\frac{1}{N}(t^\beta - b^\beta) + {}^3g^{\alpha\beta}. \tag{9}$$

Finally, we introduce coordinates

$$(x^\alpha) = (x^0, x^I) = (t, x^1, x^2, x^3), \tag{10}$$

adapted to the foliation and the rigging, in which the hypersurfaces $\Sigma(t)$ are the constant time slices

$$t = \text{constant}, \tag{11}$$

and the rigging vector field is $\partial/\partial t$. In these coordinates, since the induced 3-metric and the shift vector are purely spatial, we can write

$$^3g^{\alpha\beta} = \delta_I{}^\alpha \delta_J{}^\beta \, ^3g^{IJ}, \tag{12}$$

say, and

$$b^\alpha = \delta_I{}^\alpha b^I. \tag{13}$$

We therefore find from (9), (12) and (13) that the contravariant metric can be written in the form

$$g^{\alpha\beta} = \begin{pmatrix} \dfrac{1}{N^2} & -\dfrac{b^I}{N^2} \\ \dfrac{-b^I}{N^2} & {}^3g^{IJ} + \dfrac{b^I b^J}{N^2} \end{pmatrix}. \tag{14}$$

Defining the induced covariant metric ${}^3g_{IJ}$ by

$$^3g_{IJ}\,{}^3g^{JK} = \delta_I{}^K, \tag{15}$$

then we can raise and lower Greek indices with ${}^3g^{IJ}$ and ${}^3g_{IJ}$, respectively, and so finally the covariant metric is

$$g_{\alpha\beta} = \begin{pmatrix} N^2 + {}^3g_{IJ} b^I b^J & {}^3g_{IJ} b^J \\ {}^3g_{IJ} b^J & {}^3g_{IJ} \end{pmatrix}, \tag{16}$$

since (14) and (16) satisfy

$$g_{\alpha\beta} g^{\beta\gamma} = \delta_\alpha{}^\gamma. \tag{17}$$

The extrinsic curvature is given by

$$K_{\alpha\beta} = -2 B_\alpha{}^\gamma B_\beta{}^\delta \nabla_{(\gamma} n_{\delta)}. \tag{18}$$

We define the intrinsic covariant derivative D by

$$D_\alpha = B_\alpha{}^\beta \nabla_\beta, \tag{19}$$

so that

$$D_\alpha {}^3g_{\beta\gamma} = 0, \tag{20}$$

from which it follows that D is the 3-covariant derivative operator defined in terms of the induced 3-metric ${}^3g_{IJ}$. The 3+1 formalism then consists of breaking up tensors into their respective 3+1 components. However, we need one more piece of machinery, which was first suggested by John Stachel in his thesis (Stachel 1962), to couch the equations in their most geometrically transparent form and that it is to encode the time derivatives into Lie derivatives with respect to a timelike vector field. Following York (York 1979, 83) we choose the vector field $n^\alpha = Nu^\alpha$. We introduce an abbreviated notation

$$\perp R_{\delta\gamma\beta u} = B^\lambda_\delta B^\mu_\gamma B^\nu_\beta u^\tau R_{\lambda\mu\nu\tau}, \tag{21}$$

which extends in an obvious manner. The various projections of the Riemann tensor then give rise to the Gauss, Codazzi and Mainardi equations, which are respectively

$$\perp R_{\alpha\beta\gamma\delta} = {}^3R_{\alpha\beta\gamma\delta} + K_{\alpha\gamma}K_{\beta\delta} - K_{\alpha\delta}K_{\beta\gamma}, \tag{22}$$

$$\perp R_{\alpha\beta\gamma u} = D_\beta K_{\alpha\gamma} - D_\alpha K_{\beta\gamma}, \tag{23}$$

$$\perp R_{\alpha u\beta u} = N^{-1}\mathcal{L}_n K_{\alpha\beta} + K_{\alpha\gamma}K^\gamma{}_\beta + N^{-1}D_\alpha D_\beta N. \tag{24}$$

Restricting attention to the vacuum field equations, we obtain the Hamiltonian and momentum constraints and the dynamical equations, which are respectively

$$2G_{uu} = {}^3R - K_{\alpha\beta}K^{\alpha\beta} + (K_\alpha{}^\alpha)^2, \tag{25}$$

$$\perp G^{\alpha u} = D_\beta(K^{\alpha\beta} - {}^3g^{\alpha\beta}K_\gamma{}^\gamma), \tag{26}$$

$$\perp R_{\alpha\beta} = \mathcal{L}_n K_{\alpha\beta} + D_\alpha D_\beta N - N({}^3R_{\alpha\beta} - 2K_{\alpha\gamma}K^\gamma{}_\beta + K_{\alpha\beta}K^\gamma{}_\gamma). \tag{27}$$

We shall look at the corresponding breakup in the DSS 2+2 formalism in section V.

3. THE CHARACTER OF THE 2+2 DECOMPOSITION

In his thesis (Stachel 1962), John Stachel considered the initial value problem from two points of view. In the analytic viewpoint we start from a given spacetime in a particular coordinate system and we then use the geometry to analyze the metric and the various tensors formed from it. In the constructive viewpoint, we start from a bare manifold, introduce a foliation or foliations, impose some initial data and then use the field equations to generate a solution (locally) which in turn imparts a geometrical character to the foliation or foliations. The 2+2 formalism decomposes space-time into two families of spacelike 2-surfaces. We can view this as a constructive procedure in which an initial 2-dimensional submanifold S_0 is chosen in a bare manifold, together with two vector fields v_1 and v_2 which transvect the submanifold everywhere (fig. 1).

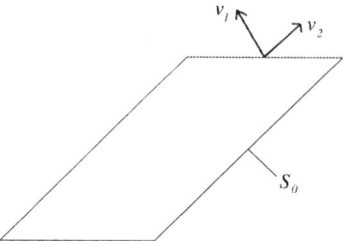

Fig. 1. 2-dimensional submanifold and two transvecting vector fields.

The two vector fields can then be used to drag the initial 2-surface out into two foliations of 3-surfaces. The character of these 3-surfaces will depend in turn on the character of the two vector fields. Since the two vector fields may each separately be null, timelike or spacelike this gives rise to six different types of decomposition. The three most important cases are the double null foliation (fig. 2), the null-timelike foliation (fig. 3) and the timelike-spacelike foliation (fig. 4).

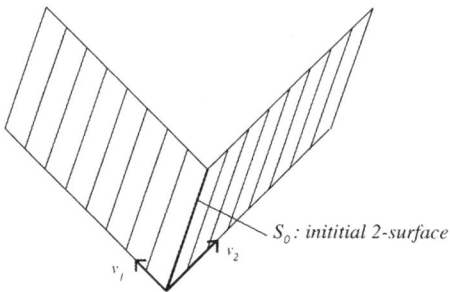

Fig. 2. Double null foliation.

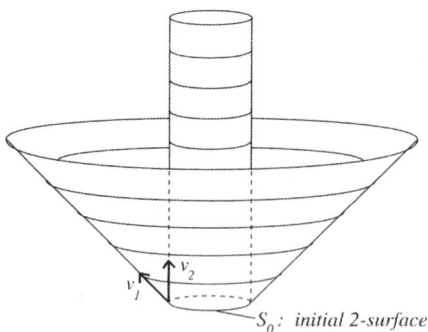

Fig. 3. Null-timelike foliation.

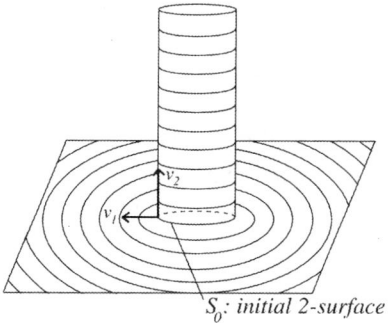

S_0: initial 2-surface

Fig. 4. *Timelike-spacelike foliation.*

There are a number of ways of then constructing a 2+2 formalism, but the most elegant way of proceeding is to introduce a formalism which is manifestly covariant and which uses projection operators and Lie derivatives, or generalization of Lie derivatives, associated with the two vector fields. The resulting formalism is called the DSS 2+2 formalism (Smallwood 1983; d'Inverno and Smallwood 1980; d'Inverno 1984, 221). When the vector fields are of a particular geometric character, then this can be refined further into a 2+(1+1) formalism. Finally, in analogy to the conformal approach of the 3+1 formalism, one extracts a conformal factor from the spacelike 2-geometries which then enables one to explicitly isolate the gravitational degrees of freedom. Let us start by looking at the breakup of the metric and compare it with the 3+1 breakup.

4. THE 2+2 METRIC DECOMPOSITION

A foliation of codimension two can be described by two closed 1-forms n^0 and n^1. Thus locally

$$dn^a = 0 \Leftrightarrow n^a = d\phi^a. \tag{28}$$

The two 1-forms generate hypersurfaces defined by

$$\{\Sigma_0\} : \phi^0(x^\alpha) = \text{constant},$$

$$\{\Sigma_1\} : \phi^1(x^\alpha) = \text{constant},$$

respectively. These hypersurfaces define a family of 2-surfaces $\{S\}$ by

$$\{S\} = \{\Sigma_0\} \cap \{\Sigma_1\}.$$

We restrict attention to the case when $\{S\}$ is spacelike and denote the family of two dimensional timelike spaces orthogonal to $\{S\}$ at each point by $\{T\}$ (fig. 5). Let n_a be the dyad basis of vectors dual to n^a in $\{T\}$, so that

$$n_a{}^\alpha n^b{}_\alpha = \delta_a^b. \tag{29}$$

Since n^a is a 1-form basis for $\{T\}$, the vectors n_a form a basis of vectors for the span of $\{T\}$. Note that, in general, $[n_0, n_1] \neq 0$ so that $\{T\}$ does not form an integrable distribution. If, however, the Lie bracket vanishes then $\{T\}$ forms a 2-dimensional subspace of M and is said to be holonomic. We use the n_a to define a 2×2 matrix of scalars N_{ab} by

$$N_{ab} = g_{\alpha\beta} n_a{}^\alpha n_b{}^\beta, \tag{30}$$

with inverse N^{ab}. We may use N_{ab} to relate n^a and n_a since

$$n_a{}^\alpha = g^{\alpha\beta} N_{ab} n^b{}_\beta, \tag{31}$$

and

$$n^a{}_\alpha = g_{\alpha\beta} N^{ab} n_b{}^\beta. \tag{32}$$

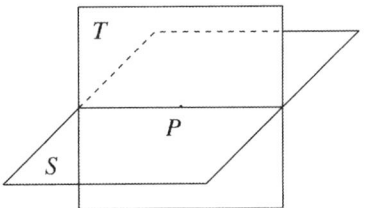

Fig. 5. The timelike 2-space $\{T\}$ orthogonal to $\{S\}$ at P

We define projection operators into $\{S\}$ and $\{T\}$ by

$$B^\alpha_\beta = \delta^\alpha_\beta - n_a{}^\alpha n^a{}_\beta, \tag{33}$$

$$C^\alpha_\beta = n_a{}^\alpha n^a{}_\beta. \tag{34}$$

The 2-metric induced on $\{S\}$ is given by the projection

$$\begin{aligned}{}^2 g_{\alpha\beta} &= B^\gamma_\alpha B^\delta_\beta g_{\gamma\delta} \\ &= B_{\alpha\delta} B^\delta_\beta \\ &= B_{\alpha\beta}.\end{aligned}$$

Similarly, the 2-metric induced on $\{T\}$ is given by the projection

$$h_{\alpha\beta} = C^{\gamma}_{\alpha} C^{\delta}_{\beta} g_{\gamma\delta}$$
$$= C_{\alpha\gamma} C^{\gamma}_{\beta}$$
$$= C_{\alpha\beta}.$$

Note that the dyad components of $h_{\alpha\beta}$ are just N_{ab} since

$$h_{ab} = h_{\alpha\beta} n_a{}^{\alpha} n_b{}^{\beta} = g_{\alpha\beta} n_a{}^{\alpha} n_b{}^{\beta} = N_{ab}.$$

In particular, the elements N_{00} and N_{11} define the lapses of $\{S\}$ in $\{\Sigma_0\}$ and $\{\Sigma_1\}$, respectively. We next choose a pair of vectors E_a which connect neighboring 2-surfaces in $\{S\}$ satisfying

$$n^a{}_{\alpha} E_b{}^{\alpha} = \delta^a_b, \tag{35}$$

which defines E_a up to an arbitrary shift vector b_a, i.e.

$$E_a = n_a + b_a, \tag{36}$$

with

$$n^a{}_{\alpha} b_c{}^{\alpha} = 0. \tag{37}$$

Although, in general, the n_a do not commute, it is always possible to choose b_a so that $[E_0, E_1] = 0$. Thus, each E_a is tangent to a congruence of curves in $\{\Sigma_a\}$ parametrized by $\phi^a(x^{\alpha})$.

We may, therefore, choose adapted coordinates in which $\phi^0(x^{\alpha}) = x^0$, $\phi^1(x^{\alpha}) = x^1$ with x^2 and x^3 being constant along the congruence of curves. Then in these coordinates

$$n^0 = dx^0, \quad n^1 = dx^1,$$

and

$$E_0 = \frac{\partial}{\partial x^0}, \quad E_1 = \frac{\partial}{\partial x^1},$$

so that

$$n_0 = E_0 - b_0 = (1, 0, b^i{}_0),$$
$$n_1 = E_1 - b_1 = (0, 1, b^i{}_1).$$

The details of the decomposition are exactly analogous to the procedure for obtaining the 3+1 decomposition of the metric. The resulting 2+2 decomposition of the contravariant metric is

$$g^{\alpha\beta} = \begin{pmatrix} N^{ab} & -N^{ab}b^i \\ -N^{ab}b^i{}_b & {}^2g^{ij} + N^{ab}b^i{}_a b^j{}_b \end{pmatrix}, \tag{38}$$

where $N^{ab} = N^{ab}$ by the choice of n^a. The inverse is given by

$$g_{\alpha\beta} = \begin{pmatrix} N_{ab} + {}^2g_{ij}b^i{}_a b^j{}_b & {}^2g_{ij}b^j{}_a \\ {}^2g_{ij}b^j{}_a & {}^2g_{ij} \end{pmatrix}, \tag{39}$$

where

$$N_{ab}N^{bc} = \delta_a^c. \tag{40}$$

We compare and contrast this form of the covariant metric with the corresponding form in the 3+1 case, namely equation (16). Then in the 2+2 case the lapse function becomes a 2×2 lapse matrix and there are now two shift vectors. The situation is illustrated schematically in figs. 6 and 7.

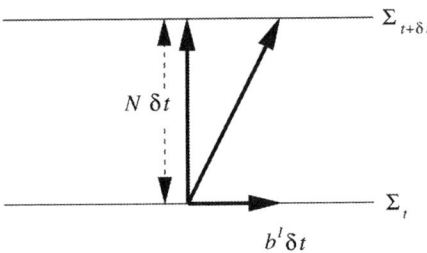

Fig. 6. Two neighboring spacelike hypersurfaces in a 3+1 decomposition.

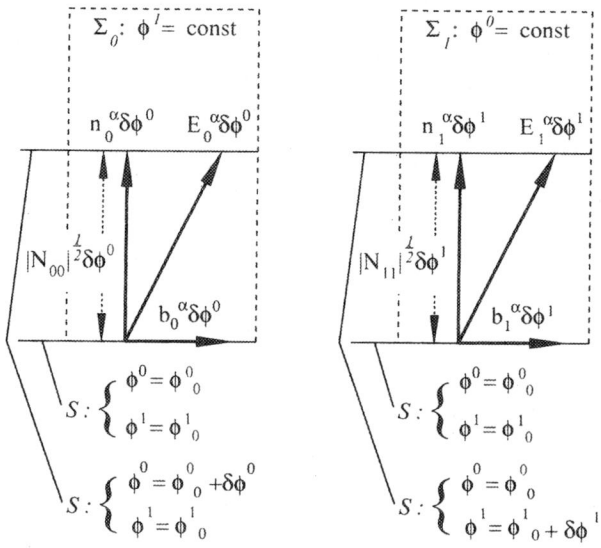

Fig. 7. Two neighbouring spacelike 2-surfaces belonging to $\{S\}$ in a 2+2 decomposition foliating the hypersurfaces Σ_0 and Σ_1.

5. DSS 2+2

We shall largely follow the conventions of d'Inverno and Smallwood (1980). In particular, we shall insert the 4 prefix explicitly for 4-dimensional quantities, but drop the 2 prefix for 2-dimensional quantities. It turns out that there is a duality between the 2-spaces $\{S\}$ and $\{T\}$, and to underline this we use a $*$ prefix to denote quantities in $\{T\}$ which are natural analogues of quantities in $\{S\}$. We use a manifestly 4-dimensional covariant notation, although we can easily introduce adapted coordinates to regain the formulae of the last section. Using this notation, the 2-metric induced on $\{S\}$ is given by

$$g_{\alpha\beta} = B_\alpha^\gamma B_\beta^{\delta\,4}g_{\gamma\delta}, \tag{41}$$

and the 2-metric induced on $\{T\}$ is given by

$$*h_{\alpha\beta} = C_\alpha^\gamma C_\beta^{\delta\,4}g_{\gamma\delta}, \tag{42}$$

using (33) and (34). Then for any vector $v^\alpha = B_\beta^\alpha v^\beta$ in $\{S\}$ covariant differentiation in $\{S\}$ is denoted by

$$\nabla_\gamma v^\alpha = B_\gamma^\mu B_\tau^\alpha({}^4\nabla_\mu v^\tau). \tag{43}$$

Similarly, for any vector $w^\alpha = C_\beta^\alpha w^\beta$ in $\{T\}$ covariant differentiation in $\{T\}$ is denoted by

$$^*\nabla_\gamma w^\alpha = C_\gamma^\mu C_\tau^\alpha (^4\nabla_\mu w^\tau). \tag{44}$$

It is then straightforward to show that the connections arising from ∇ and $^*\nabla$ are the metric connections of the metrics g and $^*h$, respectively.

We define the Riemann tensor in $\{S\}$ in the usual manner

$$R_{\delta\gamma\beta}{}^\alpha v^\beta = 2\nabla_{[\delta}\nabla_{\gamma]} v^\alpha, \tag{45}$$

which satisfies

$$R_{\delta\gamma\beta}{}^\alpha n_a{}^\beta = 0. \tag{46}$$

However, the Riemann tensor in $\{T\}$ has a more complicate definition due to the anholonomicity of $\{T\}$, namely

$$^*R_{\delta\gamma\beta}{}^\alpha w^\beta = 2\,^*\nabla_{[\delta}{}^*\nabla_{\gamma]} w^\alpha + 2n_a{}^\alpha \Omega_{\delta\gamma}{}^\varepsilon \nabla_\varepsilon w^a, \tag{47}$$

where

$$w^a = n^a{}_\alpha w^\alpha; \tag{48}$$

that is they are the dyad components of w^α. The object of anholonomicity can be defined in terms of the Lie derivative with respect to $n_a{}^\alpha$ by

$$\Omega_{\delta\gamma}{}^\varepsilon = B_\alpha^\varepsilon n^a{}_\delta n^c{}_\gamma \left(-\frac{1}{2} \mathcal{L}_{n_a} n_c{}^\alpha\right), \tag{49}$$

which vanishes if and only if $\{T\}$ is holonomic. The analogue of (46) now becomes

$$B_\mu^{\beta *} R_{\delta\gamma\beta}{}^\alpha = 0. \tag{50}$$

We next introduce curvature tensors of valence 3 defined in $\{S\}$ by

$$H_{\delta\gamma}{}^\alpha = B_\delta^\lambda B_\gamma^\mu{}^4\nabla_\lambda B_\mu^\alpha, \tag{51}$$

and in $\{T\}$ by

$$L_{\delta\gamma}{}^\alpha = C_\delta^\lambda C_\gamma^\mu{}^4\nabla_\lambda C_\mu^\alpha. \tag{52}$$

The 2-surfaces $\{S\}$ have 2 extrinsic curvatures (since they are embedded in 4-dimensional space-time), or second fundamental forms, defined by

$$h^{\beta\gamma}{}_a = \frac{1}{2} \mathcal{L}_{n_a} g^{\beta\gamma}. \tag{53}$$

Then the dyad components of H are given by

$$H_{(\delta\gamma)}{}^a = h_{\delta\gamma}{}^a, \tag{54}$$

and

$$H_{[\delta\gamma]}{}^a = 0. \tag{55}$$

If we define the quantities

$$l_{dc}{}^\alpha = -\frac{1}{2}\nabla^\alpha N_{dc}, \tag{56}$$

and

$$l^\alpha = l_e{}^{e\alpha}, \tag{57}$$

then the dyad components of L are given by

$$L_{(dc)}{}^\alpha = l_{dc}{}^\alpha, \tag{58}$$

and

$$L_{[dc]}{}^\alpha = -\Omega_{dc}{}^\alpha. \tag{59}$$

We shall also make use of the alternating quantity $\varepsilon_{ab} = \varepsilon_{[ab]}$ where

$$\varepsilon_{01} = \varepsilon^{01} = 1, \text{ rest zero}, \tag{60}$$

from which we find

$${}^*\nabla_a \varepsilon_{bc} = 0, \tag{61}$$

and

$$\Omega_{dc}{}^\alpha = \varepsilon_{dc}\Omega^\alpha, \tag{62}$$

since $\Omega_{dc}{}^\alpha$ has only one non-vanishing independent component, namely $\Omega_{01}{}^\alpha = \Omega^\alpha$. Finally, we define the conformal 2-structure of $\{S\}$ by

$$\tilde{g}_{\alpha\beta} = \gamma^{-1} g_{\alpha\beta}, \tag{63}$$

where

$$\gamma = |g_{\alpha\beta}|, \tag{64}$$

and

$$\tilde{g}^{\alpha\beta} = \gamma g^{\alpha\beta}. \tag{65}$$

We also define the corresponding conformal extrinsic curvatures by

$$\tilde{h}^{\beta\gamma}{}_a = \frac{1}{2}\mathcal{L}_{n_a}\tilde{g}^{\beta\gamma}. \tag{66}$$

We then find that

$$\tilde{h}_\alpha{}^\alpha{}_a = 0, \tag{67}$$

and

$$h_{\beta\gamma a} = \gamma \tilde{h}_{\beta\gamma a} + \frac{1}{2} g_{\beta\gamma} h_a, \qquad (68)$$

where

$$h_a = h_{\alpha\ a}^{\ \alpha}. \qquad (69)$$

We have now defined all the necessary quantities to effect a 2+2 decomposition of all the equations. In analogy to (21), we introduce an abbreviated notation

$$\perp {}^4 R_{\delta\gamma\beta a} = B_\delta^\lambda B_\gamma^\mu B_\beta^\nu n_a^\tau \ {}^4 R_{\lambda\mu\nu\tau}, \qquad (70)$$

which extends in an obvious manner. For example, we can decompose the Riemann tensor to obtain the equations of Gauss and Codazzi for $\{S\}$ and $\{T\}$, which are

$$\perp {}^4 R_{\delta\gamma\beta a} = R_{\delta\gamma\beta a} + 2 h_{[\delta|\beta|}^{\ e} h_{\gamma]ae}, \qquad (71)$$

$$\perp {}^4 R_{\delta\gamma\beta a} = 2\nabla_{[\delta} h_{\gamma]\beta a} + 2 L_a^{\ e} {}_{[\delta} h_{\gamma]\beta e}, \qquad (72)$$

$$ {}^4 R_{dcba} = {}^* R_{dcba} + 2 L_{[d|b|}^{\ \varepsilon} L_{c]a\varepsilon} - 2\Omega_{dc}^{\ \varepsilon} L_{ba\varepsilon}, \qquad (73)$$

$$\perp {}^4 R_{dcba} = 2 {}^* \nabla_{[d} L_{c]ba} + 2 h_a^{\ \varepsilon} {}_{[d} L_{c]b\varepsilon} - 2\Omega_{dc}^{\ \varepsilon} h_{\varepsilon ab}. \qquad (74)$$

We can then go on to decompose the Ricci tensor

$$ {}^4 R^{\gamma\beta} = {}^4 R^{\varepsilon\gamma\beta}_{\ \ \varepsilon}, \qquad (75)$$

and the Einstein tensor, which we write as

$$G^{\gamma\beta} = {}^4 R^{\gamma\beta} - \frac{1}{2} {}^4 g^{\gamma\beta} \ {}^4 R. \qquad (76)$$

We also split the projections into their trace and trace-free parts

$$\perp G = \perp G_\varepsilon^{\ \varepsilon}, \qquad (77)$$

$$\gamma^{-1} \perp \tilde{G}^{\gamma\beta} = \perp G^{\gamma\beta} - \frac{1}{2} {}^4 g^{\gamma\beta} \perp G. \qquad (78)$$

It is then easy to show that the projections of the Ricci and Einstein tensors are related by

$$\perp \tilde{G}^{\gamma\beta} = {}^4 \tilde{R}^{\gamma\beta}, \qquad (79)$$

$$\perp G = - {}^4 R_e^{\ e}, \qquad (80)$$

$$\perp G^{c\beta} = \perp {}^4 R^{c\beta}, \qquad (81)$$

$$G^{cb} = {}^4R^{cb} - \frac{1}{2}N^{cb}(\perp {}^4R + {}^4R_e{}^e), \qquad (82)$$

$$G_e{}^e = -\frac{1}{2}{}^4R. \qquad (83)$$

We eventually obtain for the 2+2 decomposition of the Einstein tensor

$$\perp \tilde{G}^{\gamma\beta} = {}^*\nabla_e \tilde{h}^{\gamma\beta e} - 2\tilde{h}^{\gamma\varepsilon e}\tilde{h}^\beta{}_{\varepsilon e} - \tilde{h}^{\gamma\beta e}h_e + \gamma \nabla^{(\gamma}l^{\beta)} - \gamma l^{ef(\gamma}l_{ef}{}^{\beta)}$$

$$- 2\gamma N^{-2}\Omega^{(\gamma}\Omega^{\beta)} - \frac{1}{2}\tilde{g}^{\gamma\beta}\nabla_\varepsilon l^\varepsilon + \frac{1}{2}\tilde{g}^{\gamma\beta}l^{af}{}_\varepsilon l_{ef}{}^\varepsilon + \gamma N^{-2}\tilde{g}^{\gamma\beta}\Omega_\varepsilon \Omega^\varepsilon, \quad (84)$$

$$\perp G = -{}^*R - {}^*\nabla_e h^e + \frac{1}{2}h_e h^e + \tilde{h}^{\varepsilon\beta e}h_{\varepsilon\beta e} - \nabla_\varepsilon l^\varepsilon - 4N^{-2}\Omega_\varepsilon \Omega^\varepsilon, \qquad (85)$$

$$\perp G^{c\beta} = N^{-2}\varepsilon^{ce}[{}^*\nabla_e \Omega^\beta - 2(\tilde{h}^\beta{}_{\varepsilon e} + \delta^\beta_e h_e)\Omega^\varepsilon] - \nabla^\varepsilon \tilde{h}^\beta{}_{\varepsilon e} + \frac{1}{2}\nabla^\beta h^c$$

$$+ 2\tilde{h}^\beta{}_{\varepsilon e}l^{c\varepsilon\varepsilon} - \tilde{h}^\beta{}_\varepsilon{}^c l^\varepsilon - \frac{1}{2}h^c h^\beta - {}^*\nabla_e l^{c\varepsilon\beta} + {}^*\nabla^c l^\beta, \qquad (86)$$

$$G^{cb} = N^{-2}\varepsilon^{ce}\varepsilon^{bf}\nabla_{(e}h_{f)} + 2N^{-2}\varepsilon^{e(c}l^{b)}{}_e{}^\varepsilon \Omega_\varepsilon + \frac{1}{2}h^c h^b + \tilde{h}^{\varepsilon\theta c}\tilde{h}_{\varepsilon\theta}{}^b$$

$$+ \nabla_\varepsilon l^{cb\varepsilon} - 2l^{ce\varepsilon}l^b{}_{e\varepsilon} - l^{cb\varepsilon}l_\varepsilon + N^{cb}\Big(\frac{1}{2}\tilde{h}^{\varepsilon\theta e}\tilde{h}_{\varepsilon\theta e} + \frac{3}{4}h^e h_e$$

$$- \nabla_\varepsilon l^\varepsilon + \frac{1}{2}l^{ef\varepsilon}l_{ef\varepsilon} + \frac{1}{2}l^\varepsilon l_\varepsilon + N^{-2}\Omega_\varepsilon \Omega^\varepsilon + -\frac{1}{2}R\Big), \qquad (87)$$

$$G_e{}^e = -{}^*\nabla_e h^e + h_e h^e - \nabla_\varepsilon l^\varepsilon + l^{ef\varepsilon}l_{ef\varepsilon} + 2N^{-2}\Omega_\varepsilon \Omega^\varepsilon - R, \qquad (88)$$

where

$$N = [-\det N_{ab}]^{1/2}. \qquad (89)$$

Note that the equations (84)-(88) contain 2, 1, 3, 3 and 1 independent components, respectively. Finally, the contracted Bianchi identities break up into the equations

$$B_\beta^{a4}\nabla_\varepsilon{}^4G^{\beta\varepsilon} = {}^*\nabla_e\perp{}^4G^{ea} - (2h^a{}_{\varepsilon e} + \delta^a_\varepsilon h_e)\perp{}^4G^{e\varepsilon} + \nabla_\varepsilon\perp G^{\alpha\varepsilon}$$

$$- l_\varepsilon\perp G^{\alpha\varepsilon} + l_{ef}{}^\alpha G^{ef} = 0, \qquad (90)$$

$$n^a_\beta{}^4\nabla_\varepsilon{}^4G^{\beta\varepsilon} = {}^*\nabla_e{}^4G^{ea} - h_e G^{ea} + \nabla_\varepsilon\perp G^{ae} - (2L_e{}^a{}_\varepsilon + \delta^a_\varepsilon l_\varepsilon)\perp{}^4G^{e\varepsilon}$$

$$+ h_{\varepsilon\theta}{}^a\perp G^{\varepsilon\theta} = 0. \qquad (91)$$

6. 2+2 IVPS

We are now in a position to address the uniqueness problem: Given a solution to the vacuum field equations in M, together with particular foliations (which determine $\{S\}$) and fibrations, then what data, on an initial 2-surface S_0 and either or both of the hypersurfaces intersecting in S_0 (which we denote by $\{\Sigma_a\}_0$), are necessary and sufficient to determine that solution in some neighborhood of S_0? The 2+2 decomposition of the 4-dimensional metric is given by

$$^4g^{\alpha\beta} = \gamma^{-1}\tilde{g}^{\alpha\beta} + N^{ab}(E_a{}^\alpha - b_a{}^\alpha)(E_b{}^\beta - b_b{}^\beta), \qquad (92)$$

so that the metric is a functional of the following quantities

$$^4g^{\alpha\beta} = {}^4g^{\alpha\beta}(\gamma, \tilde{g}^{\alpha\beta}, N^{ab}, b_a{}^\alpha). \qquad (93)$$

This has the requisite 10 degrees of freedom consisting of 1 degree in γ, 2 in $\tilde{g}^{\alpha\beta}$, 3 in N^{ab} and 4 in $b_a{}^\alpha$. So given a foliation $\{S\}$ and the fibrations ($E_a{}^\alpha$) then a knowledge of the 10 functions in (93) will enable us to reconstruct the metric through (92). We may choose four of these functions arbitrarily (subject to maintaining the correct signature of $^4g^{\alpha\beta}$), since the field equations are invariant under a 4-dimensional gauge group. This gauge group gives us the freedom to specify some (though not all) of the metrical properties of, and relations between, the foliation $\{S\}$ and the fibrations. The details of the subsequent analysis depends on the nature of the IVP under consideration.

In outline, the procedure is:
(i) Choose 4 of the functions in (93) arbitrarily.
(ii) Use the contracted Bianchi identities to analyze the status of the field equations.
(iii) Choose the remaining gauge quantities.
(iv) Determine a formal integration scheme.

The scheme then indicates how the field equations propagate the six remaining field variables into some region of M in the neighborhood of the initial S_0. The analysis leads to the identification of the freely specifiable initial data, and hence the dynamical variables of the theory. Of course, in a gauge theory like general relativity, this information (at least at the level of the metric) may be chosen to lie in different parts of the metric depending on how the gauge freedom is exploited. The 2+2 formalism reveals, however, that it may always be chosen to reside in the conformal 2-structure $\tilde{g}^{\alpha\beta}$.

From the point of view of numerical relativity we can use the various formal integration schemes to construct a solution from some given initial data. In doing so, however, we are effectively making the (very restrictive) assumption that the solution is analytic and so may be developed in a formal Taylor series. If ϕ^a are the local coordinates of the foliation $\{\Sigma_a\}$ then, given any geometric object F_Λ, its Taylor series may be written in the form

$$F_\Lambda = \exp\{\phi^a \mathcal{L}_{E_a}\} F_\Lambda |_{S_0}. \tag{94}$$

The actual details of the analysis are very different for the various types of IVPs (d'Inverno and Smallwood 1980). But the key point to note is that the dynamical equations consist of the two equations which propagate the quantities $\tilde{h}^{\alpha\beta}{}_0$, or equivalently $\tilde{g}^{\alpha\beta}$, which are precisely the conformal 2-structure. In general, choosing a particular form for the lapse matrix N^{ab} is really equivalent to carrying out a 2+(1+1) decomposition, which is a refinement of a 2+2 decomposition. However, references (Smallwood 1983; d'Inverno and Smallwood 1980) involve a more general formalism in which $\{T\}$ is kept much more on the same footing as $\{S\}$. Moreover, the Lie derivatives with respect to the two directions within $\{T\}$ are encoded in a special 2-dimensional derivative called D_a. The general formalism can accommodate the 4 separate cases when $\{S\}$ and $\{T\}$ are each either holonomic or anholonomic. Since there are 6 different types of IVP, this means that the formalism can handle all the 24 different cases of holonomic and anholonomic IVPs. In every case the 3-dimensional initial data can be chosen to be the conformal 2-structure. In (d'Inverno and Stachel 1978; Smallwood 1983) a Lagrangian formulation is presented which, in particular, gives insight into the special character of the break up of the field equations. For example, it is shown that the conformal 2-structure generates the dynamical equations in the sense that

$$\frac{\delta \mathcal{L}_G}{\delta \tilde{g}^{\alpha\beta}} = \text{Dynamical equations}, \tag{95}$$

where $\mathcal{L}_G$ is the Einstein Lagrangian $\sqrt{-g}R$. We have only essentially discussed the question of uniqueness in the somewhat restrictive case of analytic solutions. However, there are some theorems known on the existence and stability for the double null CIVP (Mueller zum Hagen and Seifert 1977).

The 2+2 approach served as a starting point for work on formulating general relativity as a first order hyperbolic system (Stewart 1983) and related work in numerically simulating colliding plane gravitational waves (Corkhill and Stewart 1983). As far as applications of the 2+2 approach to numerical relativity are concerned, there is the work of the Southampton group on mixed IVPs based on the null-timelike foliation of space-time (see section 9), that of Gnedin and Gnedin on the Cauchy horizon in the Reissner-Nordstrom black hole (Gnedin and Gnedin 1993), and the work of Brady et al. (Brady et al. 1996) on a version of a covariant 2+2 formalism, which is aimed at investigating the Cauchy horizon of a Kerr black hole. There is work of Hubner on a 2+2 formalism for investigating conformal infinity (Hubner 1996). We have mentioned a 2+2 Lagrangian formulation of general relativity. There are two Hamiltonian formulations in existence by Torre (Torre 1986) and Hayward (Hayward 1993). There is also the related but separate work based on the null cone formalism of Winicour and collaborators (Isaacson et al. 1983; Gómez and Winicour 1992). This is a degenerate case of a null-timelike IVP, in which the innermost timelike cylinder collapses to a worldline. This approach has led to extensive applications in numerical relativity. In the next section we discuss work on a 2+2 decomposition of Ashtekar's new variables (d'Inverno and Vickers 1995), which is preliminary work also aimed at a Hamiltonian formulation, and in the final section we look at an application to computing principal null directions.

7. 2+2 DECOMPOSITION OF ASHTEKAR VARIABLES

In a paper of Goldberg et al. (Goldberg et al. 1992; also D. C. Robinson's contribution to this volume), Ashtekar's 3+1 canonical formulation of Einstein's theory of gravitation is reformulated in terms of a time parameter whose level sets are null hypersurfaces. We refer to this briefly as the null 3+1 formulation. The motivation for their work is two-fold: null hypersurfaces provide a natural vehicle for studying gravitational radiation in asymptotically flat space-times, and in a null canonical formalism the true degrees of freedom and the observables of the theory may be more easily isolated. In the original Ashtekar formulation (Ashtekar 1987) the configuration variables are taken to be the components of a complex SO(3) connection on a 3-manifold S and the conjugate momenta are the components of a triad of vector densities of weight one on S. The resulting field theory may be interpreted as a complex version of general relativity in which Einstein's equations are satisfied by a complex metric on a real 4-dimensional manifold. Unlike the standard 3+1 formulation, the resulting Ashtekar Hamiltonian and field equations turn out to be polynomial in the field variables. Moreover, the resulting canonical formulation has a number of features in common with other gauge theories. However, as shown by Ashtekar (Ashtekar 1991), the same results as those obtained from the standard 3+1 formalism may subsequently be obtained by imposing conditions on the complex phase space which ensure the reality of the metric.

In (d'Inverno and Vickers 1995) we consider a 2+2 formulation of the Ashtekar variables, focusing on the important case of a double null foliation of space-time. Let $(e_{\boldsymbol{\alpha}})$ be a NP (Newman-Penrose) type tetrad for g (where tetrad indices will be written in bold) with dual basis $(\theta^{\boldsymbol{\alpha}})$ so that

$$ds^2 = \theta^{\mathbf{0}} \otimes \theta^{\mathbf{1}} + \theta^{\mathbf{1}} \otimes \theta^{\mathbf{0}} - \theta^{\mathbf{2}} \otimes \theta^{\mathbf{3}} - \theta^{\mathbf{3}} \otimes \theta^{\mathbf{2}}$$

$$= \eta_{\boldsymbol{\alpha}\boldsymbol{\beta}} \theta^{\boldsymbol{\alpha}} \otimes \theta^{\boldsymbol{\beta}}, \tag{96}$$

where

$$\eta_{\alpha\beta} = \begin{pmatrix} 0 & 1 & 0 & 0 \\ 1 & 0 & 0 & 0 \\ 0 & 0 & 0 & -1 \\ 0 & 0 & -1 & 0 \end{pmatrix}. \tag{97}$$

For real Lorentzian metrics θ^0 and θ^1 are real and θ^3 is the complex conjugate of θ^2. We choose an orientation so that the volume form is given by

$$V = -i\theta^0 \wedge \theta^1 \wedge \theta^2 \wedge \theta^3, \tag{98}$$

and consequently the components of the Levi-Civita tensor with respect to this basis, $\varepsilon_{\alpha\beta\delta\gamma}$ and $\varepsilon^{\alpha\beta\delta\gamma}$, satisfy

$$\varepsilon_{0123} = \varepsilon^{0123} = -i. \tag{99}$$

A general basis of 1-forms for the metric may be written in the form

$$\theta^a = \mu^a{}_b dx^b + \alpha^a{}_i(dx^i + s^i{}_b dx^b), \tag{100}$$

$$\theta^i = v^i{}_j(dx^j + s^j{}_a dx^a). \tag{101}$$

The four 2×2 matrices $\mu^a{}_b$, $v^i{}_j$, $\alpha^a{}_i$, $s^i{}_a$ constitute the 16 degrees of freedom corresponding to the 10 metric degrees of freedom and the 6 Lorentz degrees of freedom of $L_\uparrow^+$. The dual basis is then given by

$$e_a = u_a{}^b \left(\frac{\delta}{\delta x^b} - s^i{}_b \frac{\delta}{\delta x^i} \right), \tag{102}$$

$$e_i = v_i{}^j \left(\frac{\delta}{\delta x^j} + \alpha^a{}_i u_a{}^b s^j{}_b \frac{\delta}{\delta x^j} - \alpha^a{}_i u_a{}^b \frac{\delta}{\delta x^b} \right), \tag{103}$$

where the 2×2 matrices $u_a{}^b$ and $v_i{}^j$ are defined to be inverses of $\mu^a{}_b$ and $v^i{}_j$, respectively, so that

$$u_a{}^b \mu^a{}_c = \delta^b_c, \qquad u_a{}^c \mu^b{}_c = \delta^b_a, \tag{104}$$

$$v_i{}^j v^i{}_k = \delta^j_k, \qquad v_i{}^k v^j{}_k = \delta^j_i, \tag{105}$$

and, for convenience, we define the 2×2 matrices $\alpha^a{}_i$ and $s^i{}_a$ by

$$\alpha^a{}_i = v_i{}^j \alpha^a{}_j, \tag{106}$$

$$s^i{}_a = v^i{}_j s^j{}_a. \tag{107}$$

The determinants of the matrices $\mu^a{}_b, v^i{}_j, u_a{}^b, v_i{}^j, \alpha^a{}_i, \bar{\alpha}^a{}_i, s^i{}_a$ and $\bar{s}^i{}_a$ are denoted by $\mu, v, u, v, \alpha, \bar{\alpha}, s$, and $\bar{s}$, respectively. Then the volume form is given by

$$V = -i\mu v dx^0 \wedge dx^1 \wedge dx^2 \wedge dx^3. \tag{108}$$

Note that since $\theta^0, \theta^1, \theta^2, \theta^3$ are linearly independent then $V \neq 0$, and it follows that $\mu \neq 0$ and $v \neq 0$ and hence $\mu^a{}_b$ and $v^i{}_j$ are each invertible. In the case of a real Lorentzian metric with $\theta^3 = \bar{\theta}^2$, μ is real and v is purely imaginary so that V is real. We choose the orientation so that $-i\mu v$ is positive. Then a single null slicing condition with null coordinate $x^0 = p = $ constant, say, requires

$$^4g^{00} = {}^4g^{\mu\nu}p_{,\mu}p_{,\nu} = 2(u_1{}^0 u_0{}^0 - \alpha^a{}_2 u_a{}^0 \alpha^b{}_3 u_b{}^0) = 0, \tag{109}$$

and a double null slicing condition with null coordinates $x^0 = p = $ constant and $x^1 = q = $ constant, say, requires additionally

$$^4g^{11} = {}^4g^{\mu\nu}q_{,\mu}q_{,\nu} = 2(u_1{}^1 u_0{}^1 - \alpha^a{}_2 u_a{}^1 \alpha^b{}_3 u_b{}^1). \tag{110}$$

The basis of self-dual 2-forms S^A is defined to be

$$\{S^A\} = \{S^1, S^2, S^3\} = \left\{\frac{1}{2}(\theta^1 \wedge \theta^0 + \theta^3 \wedge \theta^2), \theta^1 \wedge \theta^2, \theta^3 \wedge \theta^0\right\}. \tag{111}$$

As is normal in the Ashtekar formalism, we define a densitized version of S^A by introducing the quantities

$$\Sigma_A{}^{\alpha\beta} = \frac{1}{2}\varepsilon^{\alpha\beta\gamma\delta}S^B{}_{\gamma\delta}g_{AB}. \tag{112}$$

The SO(3)-invariant metric

$$g_{AB} = \begin{pmatrix} 2 & 0 & 0 \\ 0 & 0 & -1 \\ 0 & -1 & 0 \end{pmatrix}, \tag{113}$$

and its inverse

$$g^{AB} = \begin{pmatrix} \frac{1}{2} & 0 & 0 \\ 0 & 0 & -1 \\ 0 & -1 & 0 \end{pmatrix}, \tag{114}$$

are used to lower and raise uppercase Latin indices. We next decompose the SO(3) connection 1-forms Γ^A

$$\Gamma^A = \Gamma^A{}_\mu dx^\mu = A^A{}_i dx^i + B^a{}_a dx^a, \tag{115}$$

and the curvature 2-forms R^A

$$R^A = d\Gamma^A + \eta^A{}_{BC}\Gamma^B \wedge \Gamma^C, \tag{116}$$

where $\eta_{ABC} = \eta_{[ABC]}$ and $\eta_{123} = 1$. It is useful to introduce the differential operator D_i which annihilates g_{AB} and is the restriction to S of the four-dimensional self-dual SO(3) covariant derivative. Its action on a SO(3)-valued function f^A is given by the equation

$$D_i f^A = f^A{}_{,i} + 2\eta^A{}_{BC} A^B{}_i f^C, \tag{117}$$

and it acts covariantly on SO(3)-valued vectors and vector densities.

In place of the Einstein-Hilbert action, we work with the complex first-order action (Samuel 1987; Jacobsen and Smolin 1987)

$$S = \int \mathcal{L} dx^0 \wedge dx^1 \wedge dx^2 \wedge dx^3 = \int R^A \wedge S^B g_{AB}, \tag{118}$$

where $\mathcal{L}$ is the Lagrangian density. Variation with respect to θ^α gives the source-free Einstein equations

$$G^\alpha{}_\beta = 0, \tag{119}$$

and variation with respect to Γ^A gives the first structure equation for the SO(3) connection

$$dS^A + 2\eta^A{}_{BC}\Gamma^B \wedge S^C = 0. \tag{120}$$

A direct calculation reveals that the Lagrangian density can be written in the form

$$\mathcal{L} = -[R^A{}_{01}\Sigma_A^{01} + R^A{}_{23}\Sigma_A^{23} + (D_i \Sigma_A^{ai})B^A{}_a + \Sigma_A^{ai} A^A{}_{i,a} - (\Sigma_A^{ai} B^A{}_a)_{,i}]. \tag{121}$$

Naively, one might expect the configuration space for this Lagrangian density to consist of the 30 variables

$$\Sigma_A^{\alpha\beta}, A^A{}_i, B^A{}_a.$$

However, we can consider the $\Sigma_A^{\alpha\beta}$ as being expressible in terms of tetrad quantities and so, from a tetrad viewpoint, the configuration space consists of the 28 variables

$$\mu^a{}_b,\ \alpha^a{}_i,\ s^i{}_a,\ v^i{}_j,\ A^A{}_i,\ B^A{}_a.$$

There is clearly some redundancy in the first formulation, and so we need to find the constraints that must exist between the $\Sigma_A^{\alpha\beta}$. This is most easily achieved by employing a frame adapted to the foliation in which e_2 and e_3 are tangent to $\{S\}$ and hence the α's vanish, namely

$$\alpha^a{}_i = 0. \tag{122}$$

Thus, by (106), the double bold α's must vanish as well. This is not a restriction on the metric but purely on the frame. Instead of the full 6-parameter group of proper orthochronous Lorentz transformations, the condition (122) reduces the group to the 2-parameter subgroup of spin and boost transformations. The condition will simplify ensuing formulae without constraining the formalism, and so we shall adopt it for the remainder of the section. An advantage of the 2+2 formulation is that one can work with an adapted frame without also choosing a null foliation. This is because (122) simply requires e_0 and e_1 to be orthogonal to $\{S\}$, and hence they must lie in $\{T\}$, but it does not require e_a to be normal to x^a = constant, ($a = 1$ or 2). Thus (122) does not imply $^4g^{00} = {}^4g^{11} = 0$. This is different from the 3+1 formulation, where choosing an adapted frame also leads to $^4g^{00} = 0$, and naively one only obtains nine of the ten Einstein equations. This difficulty was circumvented by Goldberg et al. (Goldberg et al. 1992) who imposed the adapted frame condition (which produces null slicing) using a Lagrange multiplier.

In an adapted frame, the explicit expressions for the $\Sigma_A^{\alpha\beta}$ are

$$(\Sigma_1^{01}, \Sigma_2^{01}, \Sigma_3^{01}) = (-\nu, 0, 0), \tag{123}$$

$$(\Sigma_1^{23}, \Sigma_2^{23}, \Sigma_3^{23}) = (-\mu - \tilde{s},\ \mu^0{}_0 s^3{}_1 - \mu^0{}_1 s^3{}_0,\ \mu^1{}_1 s^2{}_0 - \mu^1{}_0 s^2{}_1), \tag{124}$$

$$(\Sigma_1^{ai}, \Sigma_2^{ai}, \Sigma_3^{ai}) = -\varepsilon^{ab}\varepsilon^{ij}(s^3{}_b v^2{}_j - s^2{}_b v^3{}_j,\ \mu^0{}_b v^3{}_j,\ -\mu^1{}_b v^2{}_j), \tag{125}$$

where $\varepsilon^{ab} = \varepsilon^{[ab]}$, $\varepsilon^{01} = 1$ and similarly for ε^{ij}. We can consider (125) as a system of equations for determining the four $v^i{}_j$ in terms of the twelve Σ_A^{ai}. Since this is an overdetermined system, there must exist eight constraints between the Σ_A^{ai}. These turn out to be

$$C^i \equiv \mu^0{}_a \Sigma_2^{ai} = 0, \tag{126}$$

$$\tilde{C}^i \equiv \mu^1{}_a \Sigma_3^{ai} = 0, \tag{127}$$

$$C^i{}_a \equiv \mu \Sigma_1^{bi}\varepsilon_{ba} - s^2{}_a \mu^1{}_c \Sigma_2^{ci} + s^3{}_a \mu^0{}_c \Sigma_3^{ci}. \tag{128}$$

The constraints are then solved by using the explicit expressions for the $\Sigma_A^{\alpha\beta}$. We find that the solution for $v^i{}_j$ is given by

$$v^2{}_j = \mu^{-1}\mu^0{}_a \Sigma_3^{ai}\varepsilon_{ij}, \tag{129}$$

$$v^3{}_j = \mu^{-1}\mu^1{}_a \Sigma_2^{ai}\varepsilon_{ij}. \tag{130}$$

One also has

$$\mu\nu = \Sigma_2^{ai}\Sigma_3^{bj}\varepsilon_{ab}\varepsilon_{ij}. \tag{131}$$

As is usual in the 2+2 formalism we introduce the conformal factor v as an additional degree of freedom which gives the additional constraint

$$\hat{C} \equiv \Sigma_2{}^{ai}\Sigma_3{}^{bj}\varepsilon_{ab}\varepsilon_{ij} - \mu v = 0. \tag{132}$$

In an adapted frame the double null slicing conditions (109) and (110) reduce to the two conditions

$$u_1{}^0 u_0{}^0 = 0, \tag{133}$$

$$u_1{}^1 u_0{}^1 = 0. \tag{134}$$

Since we are assuming that μ and v are invertible, we must have

$$u = u_0{}^0 u_1{}^1 - u_1{}^0 u_0{}^1 \neq 0,$$

and so the conditions (133), (134) require either $u_1{}^0 = 0$ and $u_0{}^1 = 0$ or $u_0{}^0 = 0$ and $u_1{}^1 = 0$. Interchanging the coordinates x^0 and x^1 is simply equivalent to interchanging the conditions. So, without loss of generality, we shall take the double null slicing conditions to be

$$u_1{}^0 = u_0{}^1 = 0, \tag{135}$$

or, equivalently, in terms of $\mu^a{}_b$

$$\mu^0{}_1 = \mu^1{}_0 = 0. \tag{136}$$

We are now in a position to write the Lagrangian density in its final 2+2 form. In a similar way as happens with a single null slicing condition in (Goldberg et al. 1992), if we impose the double null slicing conditions immediately then we will lose some of the field equations. So, instead, we impose the double null slicing conditions by employing Lagrange multipliers ρ and $\tilde{\rho}$. Similarly, if we choose to regard all of the mixed $\Sigma_A{}^{ai}$ as configuration variables (as we shall), then we need to employ additional Lagrange multipliers to impose the constraints (126)-(128). Then the final form for the Lagrangian density is

$$\mathcal{L} = -[vR^1{}_{01} + (\mu+\tilde{s})R^1{}_{23} + (\mu^0{}_1 s^3{}_0 - \mu^0{}_0 s^3{}_1)R^2{}_{23} + (\mu^1{}_0 s^2{}_1 - \mu^1{}_1 s^2{}_0)R^3{}_{23}$$

$$+ (D_i\Sigma_A{}^{ai})B^A{}_a + \Sigma_A{}^{ai}A^A{}_{i,a} + \lambda_i C^i + \tilde{\lambda}_i \tilde{C}^i + \hat{\lambda}\hat{C} + \lambda^a{}_i C^i{}_a + \rho\mu^0{}_1 + \tilde{\rho}\mu^1{}_0]$$

$$+ (\Sigma_A{}^{ai}B^A{}_a)_{,i}. \tag{137}$$

In this equation, $R^A{}_{01}$ and $R^A{}_{23}$ are to be regarded as given in terms of $A^A{}_i$ and $B^A{}_a$ through (115) and (116). In this formulation, the configuration space contains the 33 variables

$$\nu, \mu^a{}_b, s^i{}_a, \Sigma^{ai}_A, A^A{}_i, B^A{}_a,$$

and the 11 Lagrange multipliers

$$\hat{\lambda}, \lambda_i, \tilde{\lambda}_i, \lambda^a{}_i, \rho, \tilde{\rho}.$$

Since all our considerations are purely local, we can drop the divergence term in the Lagrangian density (137). Variations with respect to the 12 variables $A^A{}_i$ and $B^A{}_a$ produce the first structure equations. We then consider the field equations which result from variation with respect to the remaining configuration space variables and the Lagrange multipliers. These latter set of equations are then used to eliminate the Lagrange multipliers in the usual way. The remaining equations then fall into three sets, namely, those arising from variation with respect to $\mu^a{}_b$, $s^i{}_a$, and the 4 independent components of Σ^{ai}_A, which we take to be $\Sigma_2{}^{1i}$ and $\Sigma_3{}^{0i}$. In fact, an independent derivation of the resulting field equations can be obtained by using a variational calculation based on the Einstein-Hilbert action

$$I = \int_\Omega L d\Omega = \int_\Omega \sqrt{-g} g^{\mu\nu} R_{\mu\nu} d\Omega = \int_\Omega \sqrt{-g} G^{\mu\nu} g_{\mu\nu} d\Omega, \qquad (138)$$

where

$$g_{\mu\nu} = \eta_{\alpha\beta} \theta^\alpha{}_\mu \theta^\beta{}_\nu. \qquad (139)$$

Before embarking on any attempt to quantize a field theory, it is very important to have an understanding of the underlying dynamical structure. Unfortunately, this does not have a clear unambiguous meaning because it depends, at least to some extent, on the choice of field variables used to describe the Lagrangian. Furthermore, there are added complications if one chooses a 3+1 formulation with a null time parameter. One of the features of the usual 2+2 formulation is that it isolates in a natural way the dynamical degrees of freedom of the gravitational field as residing in the conformal 2-metric. In the Ashtekar version of the 2+2 formalism described here things are somewhat more complicated because of the gauge freedom of the tetrad.

The Lagrange multipliers $\hat{\lambda}, \lambda_i, \tilde{\lambda}_i$ enforce the constraints $\hat{C}, C^i, \tilde{C}^i$ which enables one to eliminate ν and four of the Σ^{ai}_A from the field equations. The Lagrange multipliers $\lambda^a{}_i$ enforce the constraints $C^i{}_a$ which allows us to eliminate $\Sigma_1{}^{ai}$ form the field equations. Thus the twelve degrees of freedom in the adapted frame may be described by $\Sigma_2{}^{1i}; \Sigma_3{}^{0i}; \mu^a{}_b$, and $s^i{}_a$. Ten of these degrees of freedom correspond to the usual metrical degrees of freedom, while the remaining two correspond to the spin and boost gauge freedom (see (145)-(148) below) in the choice of a tetrad adapted to the foliation. One can now use the results of the variational calculation of the field equations to employ the analysis of (d'Inverno and Smallwood 1980) to identify the rôle of the various field equations in terms of a CIVP. Such an analysis reveals the following grouping:

$$\left.\begin{array}{ll} \dfrac{\delta L}{\delta \Sigma_2{}^{1i}}, & \dfrac{\delta L}{\delta \Sigma_3{}^{0i}} \quad \text{dynamical} \\[1em] \dfrac{\delta L}{\delta s^i{}_0}, & \dfrac{\delta L}{\delta \mu^a{}_0} \quad \text{hypersurface} \end{array}\right\} \text{main,}$$

$$\begin{array}{ll} \dfrac{\delta L}{\delta s^i{}_1}, & \dfrac{\delta L}{\delta \mu^a{}_1} \quad \text{subsidiary.} \end{array}$$

Now the relationship between the standard 2+2 and Ashtekar 2+2 variables turns out to be

$$N_{ab} = 2\mu^0{}_{(a}\mu^1{}_{b)} = \begin{pmatrix} 2\mu^0{}_0\mu^1{}_0 & \mu^0{}_0\mu^1{}_1 + \mu^0{}_1\mu^1{}_0 \\ \mu^0{}_0\mu^1{}_1 + \mu^0{}_1\mu^1{}_0 & 2\mu^0{}_1\mu^1{}_1 \end{pmatrix}$$

$$= \begin{pmatrix} \mu^1{}_0 & \mu^0{}_0 \\ \mu^1{}_1 & \mu^0{}_1 \end{pmatrix} \begin{pmatrix} \mu^0{}_0 & \mu^0{}_1 \\ \mu^1{}_0 & \mu^1{}_1 \end{pmatrix}, \tag{140}$$

$$b^i{}_a = s^i{}_a, \tag{141}$$

$$g_{ij} = -2v^2{}_{(i}v^3{}_{j)} = -\begin{pmatrix} 2v^2{}_2v^3{}_2 & v^2{}_2v^3{}_3 + v^2{}_3v^3{}_2 \\ v^2{}_2v^3{}_3 + v^2{}_3v^3{}_2 & 2v^2{}_3v^3{}_3 \end{pmatrix}$$

$$= -\begin{pmatrix} v^3{}_2 & v^2{}_2 \\ v^3{}_3 & v^2{}_3 \end{pmatrix} \begin{pmatrix} v^2{}_2 & v^2{}_3 \\ v^3{}_2 & v^3{}_3 \end{pmatrix}. \tag{142}$$

(It is intriguing to note that the 2-metric g_{ij} factorizes in some sense, which may well have some deeper significance in a future quantization programme.) It is not therefore surprising that the dynamical degrees of freedom reside in the $\Sigma_2{}^{1i}$ and $\Sigma_3{}^{0i}$ when one considers their relationship to the conformal 2-structure. Using (142) the conformal 2-structure is given by

$$\tilde{g}_{ij} = -\frac{2i}{v}v^2{}_{(i}v^3{}_{j)}, \tag{143}$$

(where the i arises because det $g_{ij} = -v^2$). If we use (129)-(130) to write the v's in terms of the Σ's and impose the null slicing condition (135), (136) then we find that

$$V\tilde{g}_{ij} = 2\varepsilon_{i(k}\varepsilon_{l)}\Sigma_3{}^{0k}\Sigma_2{}^{1l}. \tag{144}$$

Thus $\Sigma_2{}^{1i}$ and $\Sigma_3{}^{0i}$ contain essentially the same information as $v^i{}_j$ and encode the densitized conformal 2-structure. On the other hand, we still have the spin and boost gauge freedom in a tetrad adapted to the foliation, namely

$$e_0 \mapsto \lambda\tilde{\lambda} e_0, \tag{145}$$

$$e_1 \mapsto \frac{1}{\lambda\tilde{\lambda}} e_1, \tag{146}$$

$$e_2 \mapsto \frac{\lambda}{\tilde{\lambda}} e_2, \tag{147}$$

$$e_3 \mapsto \frac{\lambda}{\tilde{\lambda}} e_3, \tag{148}$$

(where $\tilde{\lambda} = \bar{\lambda}$ in the real case). This freedom is equivalent to rescaling $\Sigma_j{}^{ai}$ while leaving $\tilde{g}_{ij}$ invariant. Thus, in this formulation there are four dynamical degrees of freedom, but two of these simply represent the gauge freedom associated with the adapted frame. In current work we are investigating how to proceed to a Hamiltonian formulation of the theory.

8. DSS 2+2 PNDS

Most work to date in numerical relativity has been based on numerical codes using the ADM 3+1 formalism. Moreover these codes are based on the use of a numerical grid covering a finite region of space, complemented by ad-hoc boundary conditions which invariably generate a spurious reflection at the boundary. The far zone behavior of central systems has also been studied using a different technique based on null hypersurfaces, which allows the grid to extend to infinity; but this method cannot be applied to the strong-field interior. The Southampton numerical relativity group is using a combination of the two methods which is based on using two different regions with space and time coordinates in each suitably chosen so as to combine their advantages. The resulting method has become known as CCM (Cauchy-characteristic matching). This technique provides a globally generated spacetime and replaces the ad-hoc boundary conditions at the edge of a finite numerical grid by a compactified characteristic region which provides the correct behavior for waves at the boundary and so eliminates spurious reflections. Information is passed between the inner and outer grids via an interface region which takes appropriate account of the different coordinate systems used in the interior and exterior (see, for example, the review article of Winicour (1998)). One of the key differences about the CCM technique is that the formalism employed in the characteristic region is based on a null-timelike DSS 2+2 decomposition of space-time. CCM may turn out to be important for another reason because some recent work (Bishop et al. 1997) has indicated that if one requires greater and greater accuracy then a CCM code provides the optimal route.

The Southampton CCM is a long-term project which consists of four stages: (i) demonstrating the efficacy of the CCM method for the wave equation; (ii) cylindrically symmetric and spherically symmetric systems with one spatial degree of freedom; (iii) axially symmetric systems with two spatial degree of freedom; (iv) general systems with three spatial degrees of freedom. In each case there is the need to consider separately the vacuum and non-vacuum cases. We are now at the stage where we have completed constructing an axially symmetric code.

The project began in a preliminary paper in which we showed how the method could be consistently applied to a scalar wave in Minkowski space-time in non-flat coordinates (Clarke and d'Inverno 1994). In the second stage, we began by constructing an asymptotically flat vacuum cylindrical code (Clarke et al. 1965; Dubal et al. 1995). Unfortunately, one of the major problems confronting numerical relativity is that there are very few exact solutions known which can be used to calibrate numerical codes. It was, however, possible to show that the code produced excellent agreement with an exact vacuum solution due to Weber and Wheeler. In addition, following Piran et al. (Piran et al. 1985), it was applied to ingoing and outgoing Gaussian wave packets possessing either polarization mode. The cylindrical code was subsequently extended and tested against an exact two-parameter family of cylindrically symmetric vacuum solutions possessing both gravitational degrees of freedom due to Piran, Safier and Katz (Piran et al. 1986). The extension was necessary because the family diverges at future null infinity. This provides a more rigorous test of the code since it involves passing derivative information across the interface (d'Inverno 2000).

Visualization is of central importance in numerical relativity. A constant complaint made by relativists who do not work directly in the field of numerical relativity is that visualization often involves displaying coordinate dependent information such as metric components. It is not the metric tensor which directly tracks the gravitational field, but rather the Riemann tensor. In vacuum the Riemann tensor reduces to the Weyl tensor and in matter regions it encodes the Ricci tensor, which directly relates to the energy-momentum of the sources present. Again, the components of the Riemann tensor are coordinate-dependent, but the physical information may be extracted from it by constructing the frame components in some canonically defined frame. In some pioneering work (Gunnarsen et al. 1995), Gunnarsen et al. propose using the principal null directions (PNDs) as an interpretive tool in numerical relativity. In vacuum, these provide nearly as much information as the entire Riemann tensor and this information is gauge-invariant. Moreover, the peeling theorem of Sachs means that PNDs are natural constructs to investigate asymptotically flat radiative spacetimes (Sachs 1962). They present a method which assumes as input the intrinsic metric and extrinsic curvature of a spacelike hypersurface and use the d'Inverno-Russell-Clark algorithm (d'Inverno and Russell-Clark 1971) to compute the PNDs. The method is therefore clearly adapted to standard 3+1 numerical relativity. We outline the method below.

PNDs are complex solutions of the quartic

$$\Psi_4 z^4 + 4\Psi_3 z^3 + 6\Psi_2 z^2 + 4\Psi_1 z + \Psi_0 = 0, \tag{149}$$

where the Ψ's are defined in a Newman-Penrose null frame $[l^\alpha, n^\alpha, m^\alpha, \bar{m}^\alpha]$ by

$$\Psi_0 = -C_{\alpha\beta\gamma\delta} l^\alpha m^\beta l^\gamma m^\delta, \tag{150}$$

$$\Psi_1 = -C_{\alpha\beta\gamma\delta} l^\alpha n^\beta l^\gamma m^\delta, \tag{151}$$

$$\Psi_2 = -C_{\alpha\beta\gamma\delta} l^\alpha m^\beta \bar{m}^\gamma n^\delta, \tag{152}$$

$$\Psi_3 = -C_{\alpha\beta\gamma\delta} l^\alpha n^\beta \bar{m}^\gamma n^\delta, \tag{153}$$

$$\Psi_4 = -C_{\alpha\beta\gamma\delta} n^\alpha \bar{m}^\beta n^\gamma \bar{m}^\delta. \tag{154}$$

Restricting attention to the vacuum case where the Riemann tensor is equal to the Weyl tensor, then we first decompose the Weyl tensor into its electric part $E_{\alpha\beta}$ and magnetic part $B_{\alpha\beta}$. In the 3+1 case these parts are defined by the Mainardi and Codazzi projections of the Riemann tensor given in equations (24) and (23), namely

$$E_{\alpha\beta} = \perp C_{\alpha u \beta u} = N^{-1} \mathcal{L}_n K_{\alpha\beta} + K_{\alpha\beta} K^\gamma{}_\beta + N^{-1} D_\alpha D_\beta N, \tag{155}$$

$$B_{\alpha\beta} = -\frac{1}{2}\varepsilon_{\alpha u}{}^{\lambda\mu} \perp C_{\lambda\mu\beta u} = -\frac{1}{2}\varepsilon_{\alpha u}{}^{\lambda\mu}(D_\mu K_{\lambda\beta} - D_\lambda K_{\mu\beta}) = \varepsilon_\alpha{}^{\lambda\mu} D_\lambda K_{\mu\beta}, \tag{156}$$

where the tensor $\varepsilon_{\alpha\beta\gamma} = \varepsilon_{[\alpha\beta\gamma]}$ satisfies $\varepsilon_{\alpha\beta\gamma} \varepsilon^{\alpha\beta\gamma} = 3!$ Then using the dynamical vacuum equations (27) we can eliminate the Lie derivative term in (155) and rewrite the electric part as

$$E_{\alpha\beta} = N^{-1}\{-D_\alpha D_\beta N + N(^3R_{\alpha\beta} - 2K_{\alpha\gamma}K^\gamma{}_\beta + K_{\alpha\beta}K^\gamma{}_\gamma)\}$$

$$+ K_{\alpha\beta}K^\gamma{}_\beta + N^{-1} D_\alpha D_\beta N \tag{157}$$

$$= {}^3R_{\alpha\beta} - K_{\alpha\gamma}K^\gamma{}_\beta + K_{\alpha\beta}K^\gamma{}_\gamma. \tag{158}$$

It follows from the Hamiltonian and momentum constraint equations (25) and (26) that $E_{\alpha\beta}$ and $B_{\alpha\beta}$ are both trace-free and symmetric.

The next step is to choose a unit vector $\hat{z}^\alpha$ on a slice Σ and to decompose $E_{\alpha\beta}$ and $B_{\alpha\beta}$ into components along and perpendicular to $\hat{z}^\alpha$. We set

$$e = E_{\alpha\beta} \hat{z}^\alpha \hat{z}^\beta, \tag{159}$$

$$e_\alpha = E_{\beta\gamma} \hat{z}^\beta (\delta^\gamma_\alpha - \hat{z}_\alpha \hat{z}^\gamma), \tag{160}$$

$$e_{\alpha\beta} = E_{\gamma\delta}(\delta^\gamma_\alpha - \hat{z}_\alpha \hat{z}^\gamma)(\delta^\delta_\beta - \hat{z}_\beta \hat{z}^\delta) + \frac{1}{2} e s_{\alpha\beta}, \tag{161}$$

$$b = B_{\alpha\beta} \hat{z}^\alpha \hat{z}^\beta, \tag{162}$$

$$b_\alpha = B_{\beta\gamma} \hat{z}^\beta (\delta^\gamma_\alpha - \hat{z}_\alpha \hat{z}^\gamma), \tag{163}$$

$$b_{\alpha\beta} = B_{\gamma\delta}(\delta_\alpha^\gamma - \hat{z}_\alpha \hat{z}^\gamma)(\delta_\beta^\delta - \hat{z}_\beta \hat{z}^\delta) + \frac{1}{2}b s_{\alpha\beta}, \qquad (164)$$

where

$$s_{\alpha\beta} = {}^3g_{\alpha\beta} - \hat{z}_\alpha \hat{z}_\beta. \qquad (165)$$

Finally we set

$$\Psi_0 = (-e_{\alpha\beta} + J_\alpha{}^\gamma b_{\beta\gamma}) m^\alpha m^\beta, \qquad (166)$$

$$\Psi_1 = \frac{1}{\sqrt{2}}(e_\alpha - J_\alpha{}^\gamma b_\gamma) m^\alpha, \qquad (167)$$

$$\Psi_2 = \frac{1}{2}(-e + ib), \qquad (168)$$

$$\Psi_3 = \frac{1}{\sqrt{2}}(e_\alpha - J_\alpha{}^\gamma b_\gamma) \bar{m}^\alpha, \qquad (169)$$

$$\Psi_4 = (-e_{\alpha\beta} + J_\alpha{}^\gamma b_{\beta\gamma}) \bar{m}^\alpha \bar{m}^\beta, \qquad (170)$$

where $J_\alpha{}^\beta = \varepsilon_\alpha{}^\beta{}_\gamma \hat{z}^\gamma$ is a rotation by 90 degrees in the plane orthogonal to $\hat{z}^\alpha$, and $m^\alpha = 1/\sqrt{2}(\hat{x}^\alpha - i\hat{y}^\alpha)$ for some pair of orthogonal unit vectors which span that plane. Since everything is defined in terms of the induced metric on Σ and the extrinsic curvature of Σ (which are known at each stage of a 3+1 numerical evolution), we have presented an algorithm for computing the projections of PNDs within Σ.

In some recent work, d'Inverno and Vickers have extended this method to both a null 3+1 and a 2+2 formulation. We will not discuss the details here since they are rather extensive. This work has involved some quite deep theoretical investigation of canonical forms for Riemann tensors. As an example of its application, we have utilized this extended method to first investigate the 2-parameter family of vacuum spacetimes of Weber and Wheeler. Here the computed PNDs are stable and can be used to track the ingoing and outgoing cylindrical gravitational waves. We believe that this method may well prove to play an important role in the future for visualization in numerical relativity quite generally.

9. FUTURE WORK

James Vickers and I plan to revisit the 2+2 formalism and see if there is a way of making it yet more geometrically transparent in the hope that it might thereby become more accessible to potential users. The work on null 3+1 and 2+2 versions of the Gunnarsen et al. algorithm for computing PNDs is close to completion and we need to find out how well it performs as a new visualization tool in numerical relativity. We also plan to construct a Hamiltonian formulation of Ashtekar variables following the null 3+1 programme of Goldberg and Robinson (see their contributions to this volume). Much of the work to date, as John Stachel has pointed out, is really

2+(1+1) in character. He has discovered a way of generalizing the concept of Lie derivative which he plans to use to develop what he calls a "true 2+2" formalism (Stachel 1987). Perhaps this will ultimately become the "DSS 2+2" formalism.

ACKNOWLEDGEMENTS

Although John Stachel and I only ever wrote one paper together, it was the key paper on 2+2 and we both went on separately to develop our ideas on the formalism and its applications. Working with John was both the most exciting and most inspiring research collaboration I have ever been involved with. Part of that collaboration was a short period together in Paris, which was a whirlwind of meetings with distinguished academics and friends of John, train journeys, restaurants, book shops, French films and John's jokes. We even made time to take part in a demonstration. Anyone else who has worked with John will know of the sheer exhilaration of being in his company. I must also thank him for the patience he showed me, since my thought processes could never keep up with his, yet he never tired of explaining his ideas to me. It was a privilege being in the company of someone with such a deep-thinking and wide-ranging mind.

University of Southampton

REFERENCES

Arnowitt, R., S. Deser and C. W. Misner. 1962. In *Gravitation*, ed. L. Witten. New York: Wiley.
Ashtekar, A. 1987. *Phys. Rev. D*, 36:1587.
———. 1991. *Lectures on Non-perturbative Canonical Gravity* (Notes prepared in collaboration with R. Tate). Singapore: World Scientific.
Bishop, N.T., R. Gómez, P. R. Holvorcem, R. A. Matzner, P. Papadopoulos, and J. Winicour. 1997. *J. Comp. Phys.* 136:236.
Brady, P. R., S. Droz, W. Israel, and S. M. Morsink. 1996. *Class. Quantum Grav.* 13:2211.
Clarke, C. J. S. and R. A. d'Inverno. 1994. *Class. Quantum. Grav.* 11:1463.
Clarke, C. J. S., R. A. d'Inverno, and J. A. Vickers. 1995. *Phys. Rev. D* 52:6863.
Corkhill, R. N. and J. M. Stewart. 1983. *Proc. Roy. Soc.* A386:373.
Dubal, M. R., R. A. d'Inverno, and C. J. S. Clarke. 1995. *Phys. Rev. D* 52:6868.
Friedrich, H. and J. M. Stewart. 1983. *Proc. Roy. Soc.* A385:345.
Gnedin, M. L. and N. Y. Gnedin. 1993. *Class. Quantum Grav.* 10:1083.
Goldberg, J. N., D. C. Robinson, and C. Soteriou. 1992. *Class. Quantum Grav.* 9:1309.
Gómez, R. and J. Winicour. 1992. In *Approaches to Numerical Relativity*, ed. R. A. d'Inverno. Cambridge: Cambridge University Press.
Gunnarsen, L., H-A. Shinkai, and K-I. Maeda. 1995. *Class. Quantum Grav.* 12:133.
Hayward, S. A. 1993. *Class. Quantum Grav.* 10:779.
Hubner, P. 1996. *Phys. Rev. D* 53:701.
d'Inverno, R. A. 1984. In *Problems of collapse and numerical relativity*, eds. D. Bancel and M. Signore. Dordrecht: Reidel.
d'Inverno, R. A., M. R. Dubal, and E. A. Sarkies. 2000. *Class. Quantum Grav.* 17:3157.
d'Inverno, R. A. and R. A. Russell-Clark. 1971. *J. Math. Phys.* 12:1258.
d'Inverno, R. A. and J. Smallwood. 1980. *Phys. Rev. D* 22:1233.
d'Inverno, R. A. and J. Stachel. 1978. *J. Math. Phys.* 19:2447.
d'Inverno, R. A. and J. A. Vickers. 1995. *Class. Quantum Grav.* 12:753.

Isaacson, R. A., J. S. Welling, and J. Winicour. 1983. *J. Math. Phys.* 24:1824.
Jacobsen, T. and L. Smolin. 1987. *Phys. Lett.* 196B:39.
Müller zum Hagen, H. and J. Seifert. 1977. *Gen. Rel. Grav.* 8:259.
Piran, T., P. N. Safier, and J. Katz. 1986. *Phys. Rev. D* 34:4919.
Piran, T., P. N. Safier, and R. F. Stark. 1985. *Phys. Rev. D* 32:3101.
Sachs, R. K.1962. *Proc. Roy. Soc. London* A270:103.
Samuel, J. 1987. *Pramana J. Phys.* 28:L429.
Smallwood, J. 1983. *J. Math. Phys.* 24:599.
Stachel, J. 1962. *Lie derivatives and the Cauchy problem in the General Theory of Relativity.* Ph.D. Thesis. Stevens Institute of Technology, New Jersey.
———. 1987. "Congruences of Subspaces." In *Gravitation and Geometry*, eds. W. Rindler and A. Trautmann. Naples: Bibliopolis.
Torre, C. G. 1986. *Class. Quantum Grav.* 3:773.
Winicour, J. 1998. "Characteristic evolution and matching." *Living Reviews in Relativity* 1998-5, http://www.livingreviews.org/.
York Jr., J. W. 1979. In *Sources of Gravitational Radiation*, ed. L. Smarr. Cambridge: Cambridge University Press.

DAVID C. ROBINSON

GEOMETRY, NULL HYPERSURFACES AND NEW VARIABLES*

Abstract. Hamiltonian formulations of general relativity employing null hypersurfaces as constant time hypersurfaces are discussed. New variables approaches to the canonical formalism for general relativity are reviewed.

1. INTRODUCTION

Over the last fifty years the Hamiltonian formulation of general relativity has been the subject of many investigations. These have been intimately related to the study of initial value problems, cosmological models, asymptotically flat systems, gravitational waves and more recently black holes. However, the principal preoccupation of much of the research in this area has been the attempt to formulate a quantum theory of gravity. Attempts to construct a comprehensive quantum theory which would naturally incorporate gravity, such as M-theory (Gibbons 1998) and its precursor supersymmetric string theory, are currently being vigorously and profitably pursued. However, in the last decade, there have also been significant advances in a new approach to the canonical quantization of general relativity. Research initiated by Ashtekar in the 1980's, and his introduction of new canonical variables, focussed attention on connection rather than metric dynamics (Ashtekar 1986, 1987, 1988, 1991). Ashtekar's reformulation of Hamiltonian general relativity encouraged the application to gravity of techniques which had become well established in gauge theory. Development of these ideas has led to the establishment of loop quantum gravity and advances in a rigorous approach to a non-perturbative theory of quantum gravity. For recent reviews see (Rovelli 1997, 1997a). This programme has generated a significant amount of new and interesting activity in the canonical quantization approach to quantum general relativity. Currently it provides the weightiest complement to the ambitious M-theory programme.

The new variables canonical formalism, introduced by Ashtekar, can be obtained by a complex canonical transformation from a geometrodynamical approach to Hamiltonian general relativity. Alternatively it can be constructed, via a Legendre transformation, by starting with a complex, chiral, first order Lagrangian (Ashtekar 1986, 1987, 1988, 1991; Peldan 1994). The new canonical variables are the components of a triple of weight one vector densities and an SO(3)-valued connection on a three manifold. The Hamiltonian is a polynomial of low degree but complex-valued.

It is completely constrained. The seven first class constraint terms appearing in it generate canonical transformations corresponding to space-time diffeomorphisms and internal rotations. Their Poisson bracket algebra is not a Lie algebra. In order to recover the real phase space, and real general relativity, reality conditions have to be imposed. However both the constraint equations and the reality conditions are low degree polynomial equations. When the reality conditions are satisfied and the real vector densities are linearly independent, the three manifold can be embedded as a space-like hypersurface in a real Lorentzian four manifold. The connection can then be interpreted as the self-dual part of the Levi-Civita (spin) connection of the metric. The polynomial nature of the formalism, together with the emphasis on connection rather than metric dynamics, have suggested new developments in both classical and quantum theory. In recent years, real Lorentzian and Euclidean connection based versions of the formalism have been employed in the quantum gravity programme. However only the (initially) complex formalism, which is geometrically natural in the Lorentzian context, will be discussed in this paper.

A central aim of this paper is to review a version, based on null rather than space-like hypersurfaces, of Ashtekar's approach to classical canonical general relativity (Goldberg, Robinson, and Soteriou 1991a, 1992; Goldberg and Soteriou 1995; Goldberg and Robinson 1998). Salient reasons for constructing the null hypersurface new variables formalism include the successful use of null hypersurfaces in canonical formulations of field theories on Minkowski space-time (Sundermeyer 1982; Henneaux and Teitelboim 1992). Of course, in general relativity space-time geometry is dynamical, not merely flat and kinematical as it is Minkowski space-time. Hence it does not play merely a background role. This gives rise to technical difficulties as well as conceptual problems, such as the identification of observables and the problem of time (Ashtekar and Stachel 1991). Nevertheless the use of null hypersurfaces has also proven extremely useful in general relativity. Noteworthy and relevant here are studies of gravitational radiation and asymptotically flat systems. Useful comments on work using null hypersurfaces can be found in the recent paper by Bartnik (Bartnik 1998). A key reason for the successful use of null hypersurfaces in the study of field theories is that constraints arising in the relevant characteristic initial value problems can be integrated more easily than those arising in corresponding Cauchy problems. The true degrees of freedom can be exposed more explicitly when null hypersurfaces, rather than space-like hypersurfaces, are used.

Amongst the geometrical features which distinguish null hypersurface based formulations from space-like hypersurface based approaches to Hamiltonian gravity, are the following. First, the geometry of null hypersurfaces, unlike that of space-like hypersurfaces, is not metric. The singular metric induced on a null hypersurface from the space-time has maximal rank two, and the null geodesic generators of the null hypersurface are characteristic vector fields of the singular metric. Connections are also induced onto null hypersurfaces from connections on space-time, but they are not Levi-Civita connections of a three metric. In the initial value or Hamiltonian formalism the relevant structure group becomes the (4 real parameter) null rotation subgroup of the Lorentz group, rather than the three parameter special orthogonal group. Partially as a consequence, the formalism can appear less manifestly covariant than it is in the space-like case. The focussing effect of gravity causes the null geodesic generators of null hypersurfaces to cross over and to form caustics. At such points null hypersurfaces fail to be (immersed) smooth submanifolds of space-time (Friedrich and Stewart 1983; Hasse, Kriele, and Perlick 1996). Generically, topologically simple null hypersurfaces, such as the null planes and null cones of Minkowski space-time, are not available. Problems related to the differential and metric structure arise also in the space-like hypersurface formalisms, but they tend to be more overt in the null hypersurface case.

This paper will deal mainly with local aspects of canonical formalisms. Thus hypersurfaces will be treated as sub-manifolds with a fixed differential structure. Although the emphasis will be on local theory, it should be noted that it is usually assumed (at least implicitly) in null hypersurface Hamiltonian general relativity, that the global context is one in which the null hypersurface leaves of the space-time foliation are non-compact. It is also implicitly assumed here that the condition of asymptotic flatness (or asymptotically cosmological boundary conditions) can be imposed, and that future null infinity can be defined (Chrusciel, MacCallum, and Singleton 1995). Normally here, fields are considered to be defined on outgoing null hypersurfaces.

In the next section a brief review of research related to null hypersurface canonical formulations of general relativity will be presented. This aims only to provide a context within which the results of later sections can be placed. It does not pretend to be comprehensive. The third section will contain an outline of Ashtekar's new variables Hamiltonian formulation of general relativity. This will be presented in such a way that the similarities to, and differences from, the null hypersurface based new variables formalism are highlighted. It will suffice in this paper to consider details of the two formalisms when there are no matter fields coupled to gravity. The null hypersurface new variables canonical formalism will be discussed in the final section. In particular a Hamiltonian system which is completely equivalent to Einstein's vacuum equations will be presented. The structure of the constraints and reality conditions will also be reviewed. Finally the space-time interpretation will be outlined. The body of results on the null hypersurface based new variables canonical formalism is neither as large, nor as developed, as that obtained within the conventional framework where the level sets of the time function are space-like. However the null hypersurface new variables Hamiltonian formulation is more complete than other null hypersurface based canonical formulations of general relativity.

The notation and conventions of (Goldberg, Robinson, and Soteriou 1992) will be followed without major exception. Lower case Latin indices i, j, k sum and range over 1, 2, 3 and indicate components with respect to coordinate bases. Upper case Latin indices A, B sum and run over 1, 2, 3 and label components of SO(3) Lie-algebra valued geometrical objects. Lorentzian manifolds will have four-metrics with signature $(1, -1, -1, -1)$.

2. NULL HYPERSURFACE CANONICAL FORMALISMS

Dirac pointed out in (Dirac 1949), that it is physically natural to use a time parameter with level sets which are null hypersurfaces. A particularly relevant example of the use of this method of describing time evolution in Minkowski space-time is provided by one approach to the canonical formalism for source free Maxwell and Yang-Mills theory. This uses a first order complex Lagrangian, with the potential (connection) and self-dual part of the field (self-dual curvature) as field variables. A Hamiltonian is constructed via a 3+1 decomposition of Minkowski space-time using outgoing null cones or null hyperplanes. It incorporates many of the features which are encountered in the general relativistic null hypersurface new variables formalism, some of which will be discussed in the fourth section. It is also sufficiently simple that it can be carried out completely which, to date, is not the case for general relativity. Consequently it is instructive to outline the results in the case of the Yang-Mills field, with internal symmetry group a real n-parameter semi-simple Lie group (Soteriou 1992). A 3+1 decomposition of the first order Lagrangian, using a null hyperplane foliation, leads by inspection to a complex Hamiltonian defined on a null hyperplane. There are 2n first class constraints and 4n second class constraints. The Yang-Mills field equations are equivalent to the constraint and evolution (Hamilton's) equations. The canonical transformations which correspond to the infinitesimal internal gauge transformations are generated by the first class constraints. The first class constraint algebra represents the internal symmetry Lie algebra. Second class constraints arise because null hypersurfaces are being used. The second class constraint equations relate physically redundant variables. Solution of these leads to the elimination of an equal number of canonically conjugate pairs of phase space variables. On the resulting partially reduced phase space there remain only the first class constraints, reflecting the gauge invariance of the theory. An alternative and equivalent way of handling second class constraints is to introduce Dirac brackets, either directly or via the starred variables procedure of Bergmann and Komar, see e.g. (Sundermeyer 1982). This can be simpler than solving the second class constraint equations directly and is the procedure followed in (Soteriou 1992). There the completely reduced phase space is realized as a Poisson space by gauge fixing so that all constraints become second class. The Poisson structure on the completely reduced 2n dimensional phase space is defined by the Dirac bracket and the constraints can be set equal to zero in the Hamiltonian. The dynamics are determined by the Dirac brackets of the 2n independent degrees of freedom with the reduced Hamiltonian. The imposition of reality conditions enables the real theory to be recovered and a reduced phase space quantization to be effected. For details in the similar but simpler case of the Maxwell field see (Goldberg, Robinson, and Soteriou 1991; Goldberg 1991, 1992).

Early investigations of null hypersurface based canonical formulations of general relativity were carried out mainly within geometrodynamical frameworks and dealt with the vacuum equations; for example see (Aragone and Chela-Flores 1975; Gambini and Restuccia 1978; Goldberg 1984, 1985, 1986; Torre 1986; Aragone and Koudeir 1990). Since a null hypersurface time parametrization was used there were second class constraints as well as the first class constraints associated with gauge freedom.

They were non-polynomial in nature. These non-polynomial canonical formalisms, their constraint structures and their dynamics, were not as straightforward to analyze as had been hoped. The complete formulation of the dynamics on the completely reduced phase space was not satisfactorily achieved. Because this work used as a coordinate t, a time function with level sets which were assumed to be null hypersurfaces, consequently the inverse metric components g^{tt} were assumed to be zero. Hence the Einstein equations, $G_{tt} = 0$ could not be naturally derived, either in the 3+1 decomposed Lagrangian or in the Legendre transformation related Hamiltonian formalism. This research did however provide some useful technical and conceptual insights. It suffices to mention here two significant differences which were revealed between the constraint structure of spacelike hypersurface based and null hypersurface based canonical formalisms. Torre, using a 2+2 formalism in an investigation of a null hypersurface-based geometrodynamical formulation, gave a particularly clear and comprehensive discussion of these (Torre 1986). In the space-like case the Hamiltonian is a linear combination of first class constraints. In the null-hypersurface based case it was found that, while the Hamiltonian is again a linear combination of constraints, not all of them are first class. Furthermore in the space-like case the scalar constraint which generates time translations out of the hypersurface is first class. However there are no compact deformations which map a null hypersurface to a neighboring null hypersurface. The choice of a null time parameter is effectively a choice of gauge. As a consequence it was found that the scalar constraint was second class.

The comparative tractability of constraints on null hypersurfaces, the low degree polynomial nature of the new variables formalism first introduced by Ashtekar, and the focus on connection rather than metric dynamics, suggested that a new variables formalism based on null hypersurfaces might be more successful than previous null hypersurface Hamiltonian constructions. Unlike the Hamiltonian systems mentioned earlier, all the Einstein equations arise naturally in the new formulation (Goldberg, Robinson, and Soteriou 1991, 1992; Goldberg and Soteriou 1995; Goldberg and Robinson 1998). This is achieved by adding Lagrange multiplier terms to a Lagrangian density derived from the self-dual first order Lagrangian mentioned above (Ashtekar 1986, 1987, 1988; Ashtekar and Tate 1991; Peldan 1994). The corresponding Euler Lagrange equations then include all the Einstein (and matter) field equations plus equations which identify level sets of constant time as null hypersurfaces. The Hamiltonian, obtained by a Legendre transformation, consequently also contains all this information. It is completely constrained but not all the constraints are first class. There are a large number of second class constraints as there always are in null hypersurface based formalisms. Analysis of the constraint structure reveals it to be rather different from that of the space-like hypersurface based Hamiltonian formalism. The first class constraints form a Lie algebra and the scalar constraint is second class. This is a distinctive feature of this formalism; the Lie algebra can be interpreted as a representation of the semi-direct product of the null rotation sub-group of the Lorentz group and the diffeomorphisms within the null hypersurface. Both constraints and reality conditions are polynomial of low degree. In addition to the work on the vacuum case, coupled Einstein-matter systems have been analyzed by Soteriou (1992). He extended the vacuum formalism to gravitational systems which included scalar fields, Yang-Mills fields and Grassmann valued Dirac fields. In the latter a two-component spinor formalism was used. Such extensions are essentially unproblematic, as had previously been shown in the space-like case (Ashtekar and Tate 1991; Ashtekar, Romano, and Tate 1989). Boundary conditions have been examined (Goldberg and Soteriou 1995), and some progress has been made towards the elimination of the second class constraints (Goldberg and Robinson 1998). The construction of the completely reduced phase space (ibid.), has yet to be completed as successfully as it has been in, for example, the Yang-Mills canonical formalism discussed above. Consequently, the linearization of the theory about a Minkowski space-time was investigated, using both Minkowski null hyperplanes and outgoing Minkowski null cones (Soteriou 1992). For a geometrodynamical formulation see (Evans, Kunstatter, and Torre 1987). In this case, as might be expected, the programme could be completed along lines similar to those outlined above for Yang-Mills fields in Minkowski space-time. Despite the fact that not all problems have been resolved in practice, this canonical formalism has been developed more fully than any other null hypersurface based Hamiltonian formulation of general relativity; in part this is because of its polynomial nature. In the first approaches to the null hypersurface, new variables canonical formalism (Goldberg, Robinson, and Soteriou 1991a, 1992; Soteriou 1992), the full Dirac-Bergmann algorithm (Sundermeyer 1982; Henneaux and Teitelboim 1992), was applied. A large phase space resulted. The framework used was broad enough to ensure that the complete structure of the new formulation could be properly analyzed. Once this was clearly understood a streamlined version, which employs only connection and vector density triad components as canonically conjugate variables, could be used (Goldberg and Soteriou 1995; Goldberg and Robinson 1998). It is that formalism which will be discussed in more detail in section 4.

In recent years there have been a number of other approaches to new variables formulations of general relativity on a null hypersurface. These include the application of multisymplectic techniques, as used in jet bundle formulations of classical field theories, to the new variables null hypersurface formalism. In this work the properties of multi-momentum maps and their relation to the constraint analysis are discussed (Esposito and Stornaiolo 1997; Stornaiolo and Esposito 1997). The Hamiltonian formulation of general relativity on a null hypersurface has also been considered within the context of teleparallel geometry (Maluf and da Rocha-Neto 1999). Finally it should be noted that a 2+2 (Lagrangian) formulation using new variables has been considered by d'Inverno and Vickers (1995). This work focuses on the important case of a double null foliation of space-time and their discussion mirrors, to an extent, the null hypersurface new variables formalism. It includes a discussion of the relationship between different formalisms.

3. NEW VARIABLES AND SPACELIKE HYPERSURFACES

Let M be a real three dimensional manifold with local coordinates x^i. Let $\Sigma_A^i \partial_i$ be three weight one vector densities tangent to M, and let $A_i^A dx^i$ be a $so(3)$-valued connection 1-form on M with curvature 2-form R^A. The new variables Hamiltonian for vacuum general relativity can be written as

$$H = -2\int d^3x (1/2\mathbf{N}\mathscr{H} + N^i \mathscr{H}_i + B^A G_A). \tag{1}$$

Here $\mathbf{N}$, N^i and B^A are Lagrange multipliers, with $\mathbf{N}$ a weight minus one density. The canonically conjugate dynamical variables are taken to be the pair (A_i^A, Σ_A^i) and the only non-zero Poisson brackets between these variables are

$$\{A_i^A(x), \Sigma_B^j(y)\} = \delta_B^A \delta_i^j \delta(x,y). \tag{2}$$

The Hamiltonian density, Lagrange multipliers and dynamical variables can be regarded as being complex valued so that the phase space is complex. The scalar, vector and Gauss constraint equations are

$$\begin{aligned}\mathscr{H} &\equiv i\varepsilon_A{}^{BC} R_{ij}{}^A \Sigma_B^i \Sigma_C^j = 0, \\ \mathscr{H}_i &\equiv -R_{ij}{}^A \Sigma_A^j = 0, \qquad G_A \equiv -i D_i \Sigma_A^i = 0.\end{aligned} \tag{3}$$

Here

$$R^A = -1/2 R_{ij}^A dx^i \wedge dx^j = (\partial_i A_j^A + \varepsilon_{BC}{}^A A_i^B A_j^C) dx^i \wedge dx^j, \tag{4}$$

D_i is the covariant derivative of the connection A_i^A and $\varepsilon_{ABC} = \varepsilon_{[ABC]}$ with $\varepsilon_{123} = 1$. In the basis used the SO(3)-invariant metric is given by

$$\eta_{AB} = \text{diag}(-1,-1,-1). \tag{5}$$

All these constraints are first class although the constraint algebra is not a Lie algebra. Appropriately combined they generate canonical transformations corresponding to space-time diffeomorphisms and (small internal) gauge rotations. They are transparently low degree polynomials in the canonical variables. Dynamical evolution is determined by Hamilton's equations

$$\begin{aligned}\partial_t \Sigma_A^i &= \{\Sigma_A^i, H\} \\ \partial_t A_i^A &= \{A_i^A, H\}.\end{aligned} \tag{6}$$

When the vector densities Σ_A^i are linearly independent they determine a 3-metric on M, with inverse metric

$$g^{ij} = -v^{-2} \eta^{AB} \Sigma_A^i \Sigma_B^j, \tag{7}$$

where

$$v^2 = -i\varepsilon_{ijk}\Sigma_1^i\Sigma_2^j\Sigma_3^k. \tag{8}$$

In this case, on the four manifold $M \times \mathbb{R}$ with local coordinates (t, x^i), the tetrad of 4-vector fields

$$e_0 = v^{-1}\mathbf{N}^{-1}(\partial_t - N^i\partial_i) \quad \text{and} \quad e_A = -iv^{-1}\Sigma_A^i\partial_i \tag{9}$$

forms an orthonormal basis for a 4-metric with inverse $e_0 \otimes e_0 + \eta^{AB}e_A \otimes e_B$, so

$$\partial^2/\partial s^2 = v^{-2}\mathbf{N}^{-2}\partial_t \otimes \partial_t - v^{-2}\mathbf{N}^{-2}N^i(\partial_t \otimes \partial_i + \partial_i \otimes \partial_t)$$
$$+ 1/2(v^{-2}\mathbf{N}^{-2}N^iN^j - v^{-2}\eta^{AB}\Sigma_A^i\Sigma_B^j)(\partial_i \otimes \partial_j + \partial_j \otimes \partial_i). \tag{10}$$

The volume 4-form is $v^2\mathbf{N}dt \wedge dx^1 \wedge dx^2 \wedge dx^3$. The 1-form,

$$\Gamma^A = A_i{}^A dx^i + B^A dt, \tag{11}$$

corresponds to the self-dual part of the Levi-Civita (spin) connection of this metric. The constraint and evolution equations given above then correspond to Einstein's vacuum equations for the metric. The pull back of the curvature of this connection, to a leaf of the foliation by level sets of t, can then be identified with the 2-form R^A. Real Lorentzian general relativity can be recovered when polynomial reality conditions, given by requiring that $\Sigma_A^i\Sigma_B^j\eta^{AB}$ and $\{\Sigma_A^i\Sigma_B^j\eta^{AB}, H\}$ be real, are imposed on the phase space variables. Agreement with the standard *ADM* geometrodynamical version of Hamiltonian relativity follows when the constraint, evolution and reality equations are satisfied. The standard geometrodynamical field variables can then be identified with the three metric g_{ij} (with Levi-Civita $so(3)$-valued connection $\gamma_i^A dx^i$), and the symmetric extrinsic curvature tensor $K_{ij} = -v^{-1}g_{ik}\Sigma_A^k(A_j{}^A + \gamma_j{}^A)$, on a space-like hypersurface given by a level set of t (Ashtekar 1986, 1987, 1988; Ashtekar and Tate 1991).

4. NEW VARIABLES CANONICAL FORMALISM FOR GENERAL RELATIVITY ON A NULL HYPERSURFACE

In the new variables, null hypersurface Hamiltonian formalism, presented in (Goldberg and Soteriou 1995), the canonically conjugate phase space variables are again defined by the components of a triple of weight one vector densities Σ_A and the components of an $so(3)$ valued connection one form A^A on a three manifold M. The Poisson bracket relations are given as in eq. (2). The complex Hamiltonian is, however, now given by

$$H = \int d^3x(\mathbf{N}\mathcal{H} + N^i\mathcal{H}_i - B^A G_A - \mu_i \mathcal{C}^i - \rho\alpha^2), \tag{12}$$

The functions $\mathbf{N}$ (a scalar density of weight minus one), N^i, B^A, v^i, μ_i and ρ are treated as Lagrange multipliers and

$$\mathcal{H} \equiv v^i(R_{ij}^{\ 1}\Sigma_3^{\ j} + R_{ij}^{\ 2}\Sigma_1^{\ j}) \equiv v^i\phi_i,$$

$$\mathcal{H}_i \equiv -R_{ij}^{\ A}\Sigma_A^{\ j}, \quad G_A \equiv D_i(\Sigma_A^{\ i}), \qquad (13)$$

$$\mathcal{C}^i \equiv \Sigma_2^{\ i} + \alpha v^i.$$

The evolution equations are given by Hamilton's equations, as in eq. (6).

The Hamiltonian density is a linear sum of the product of Lagrange multipliers with terms which all vanish when constraint equations and multiplier equations are satisfied. The constraints, $\mathcal{C}^i = 0$, $\alpha^2 = 0$, and the multiplier equation $\mu_i v^i = 0$, arise from Lagrange multiplier terms originating in the Legendre transformation related Lagrangian formalism. Added to a chiral Lagrangian for Einstein's vacuum equations they ensure that this is a correctly set null hypersurface based formalism, and that the canonical formalism contains all the Einstein gravitational (and, when present matter) field equations. Here α is not treated as an independent dynamical variable; to do so would merely result in further trivial second class constraints which would lead to its ultimate elimination from a reduced phase space. It is set equal to zero after Poisson brackets have been computed. Its vanishing mirrors the fact that this is a null hypersurface based formalism and implies, together with the constraints, $\mathcal{C}^i = 0$, the vector density degeneracy condition, $\Sigma_2^{\ i} = 0$. The latter condition mirrors the degeneracy of a basis of self-dual 2-forms pulled back to a null hypersurface in space-time. The second multiplier equation above, $\mu_i v^i = 0$, is required in order to obtain all the Einstein equations from the formalism. It arises naturally from the Lagrangian formalism. The vanishing of ϕ_i, the coefficients of the multipliers Nv^i, results in three constraint equations. The scalar, vector and Gauss constraint equations are given by $\mathcal{H} = 0$, $\mathcal{H}_i = 0$ and $G_A = 0$, as in section 3. It can be seen from the above that $\mathcal{H} = v^i\phi_i$ and $\Sigma_1^{\ i}\phi_i = \Sigma_3^{\ i}\mathcal{H}_i$. Hence there are eleven independent constraints above. Propagation of the constraints leads to secondary conditions on Lagrange multipliers. An important example of the latter, exemplifying differences from the results of section 3, follows from the propagation of $\mathcal{C}^i = 0$, that is evolution of the vector density degeneracy condition. This leads to three secondary conditions on the multipliers v^i and B^3, given by

$$\chi^i \equiv 2\delta_2^{\ B}D_j\{Nv^{[i}\Sigma_A^{\ j]}(\delta_B^{\ 1}\delta_3^{\ A} + \delta_B^{\ 2}\delta_1^{\ A})\} - 2A_j^{\ 3}N^{[i}\Sigma_1^{\ j]} - B^3\Sigma_1^{\ i} = 0. \quad (14)$$

The eleven constraints comprise six second class constraints and five first class constraints. The independent first class constraints can be identified with

$$G_1 = 0; \quad G_2 = 0; \quad \mathcal{H}'_i \equiv \mathcal{H}_i - A_1^{\ B}D_j(\Sigma_B^{\ j}) = 0. \qquad (15)$$

There are 18 phase space variables, 6 second class constraints and 5 first class constraints. Consequently, at each point of M, there are 2 independent degrees of phase space freedom. The canonical transformations generated by the constraints functions G_1 and G_2 correspond to self dual null rotations, reflecting the fact that the formalism is chiral (self-dual), complex and null hypersurface based. The constraints $\mathcal{H}'_i$ generate diffeomorphisms in M. The scalar constraint is second class. A null hypersurface time gauge is effectively set within the formalism and the problem of time does not appear in the way it does in the space-like formalism. The algebra of first class constraints is a genuine Lie algebra, in contrast to the spacelike based formalism in section 3. The formalism is again polynomial.

As was the case in the previous section, when a non-degeneracy condition is satisfied, the constraint, evolution and multiplier equations are equivalent to the Einstein vacuum equations. More explicitly, when the weight one vector density fields tangent to M, $v^i \partial_i$, $\Sigma_1^{\ i} \partial_i$, $\Sigma_3^{\ i} \partial_i$, are linearly independent so that

$$v^2 \equiv \Sigma_1^{\ i} v^j \Sigma_3^{\ k} \varepsilon_{ijk} \tag{16}$$

is non-zero, a (non-degenerate) four metric can be constructed on $M \times \mathbb{R}$. Its inverse is given by

$$\partial^2/\partial s^2 = e_0 \otimes e_1 + e_1 \otimes e_0 - e_2 \otimes e_3 - e_3 \otimes e_2 \tag{17}$$

and

$$\begin{aligned} v^2 \partial^2/\partial s^2 = &-2\alpha \mathbf{N}^{-2} \partial_t \otimes \partial_t + (2\alpha \mathbf{N}^{-2} N^i - \mathbf{N} \Sigma_1^{\ i})(\partial_t \otimes \partial_i + \partial_i \otimes \partial_t) \\ &+ (\alpha \mathbf{N}^{-2} N^i N^j - v^i \Sigma_3^{\ j} - \mathbf{N}^{-1} \Sigma_1^{\ i} N^j)(\partial_i \otimes \partial_j + \partial_j \otimes \partial_i). \end{aligned} \tag{18}$$

Here the natural basis of vector fields is a null tetrad given by

$$e_0 = v^{-1} \mathbf{N}^{-1}(\partial_t - N^i \partial_i), \quad e_1 = -\alpha v^{-1} \mathbf{N}^{-1} \partial_t + (\alpha v^{-1} \mathbf{N}^{-1} N^i - v^{-1} \Sigma_1^{\ i}) \partial_i, \tag{19}$$

$$e_2 = v^{-1} v^i \partial_i, \qquad e_3 = -v^{-1} \Sigma_3^{\ i} \partial_i,$$

and the volume 4-form is $-i \mathbf{N} v^2 dt \wedge dx^1 \wedge dx^2 \wedge dx^3$. When $\alpha = 0$, the level sets of t are null hypersurfaces and the null tetrad (and its null co-frame) is adapted to the null hypersurface foliation with e_1 tangent to the null geodesic generators of the null hypersurfaces. The connection 1-form, given as in eq. (11), is the self-dual part of the Levi-Civita (spin) connection of the metric. When the remaining multiplier equations, the constraint and the evolution equations are satisfied, the metric satisfies Einstein's vacuum equations. Real relativity can be extracted from the complex phase space description by imposing appropriate polynomial reality conditions (Goldberg, Robinson, and Soteriou 1992; Soteriou 1992). It suffices to record here that v^i can then be chosen to be the complex conjugate of $\Sigma_3^{\ i}$; $\Sigma_1^{\ i}$, v and $\mathbf{N}$ can be chosen to be pure imaginary; finally N^i can be chosen to be real. The null tetrad then satisfies the standard reality conditions: e_0 and e_1 real, e_2 and e_3 complex conjugates.

Starting from the Hamiltonian formalism the properties of the constructed spacetimes emerge from those of the phase-space geometry. The form of the Hamiltonian determines whether M should be embedded in a space-time as a space-like or null hypersurface. Space-time properties such as asymptotic flatness, and hypersurface related properties such as crossovers and caustics, arise from solutions of the differential equations of the canonical formalism subject to boundary conditions. The latter have not been discussed in this paper but extension of this work to include fields on future null infinity and hence a complete global formulation is, in principle, straightforward, (c.f. calculations by Goldberg in a geometrodynamical framework in (Goldberg 1986) and discussions of boundary conditions in (Goldberg and Soteriou 1995)). The formalisms discussed in sections 3 and 4 encompass theories which may be real or complex and which allow the possibility of degenerate inverse metric densities. The latter arise when $v^2 = 0$ and can be identified with the expressions for $v^2 \partial^2 / \partial s^2$ in eq. (10) and eq. (18). This possibility has been discussed in the spacelike case, e.g. in (Bengtsson and Jacobson 1997), and merits investigation in the null hypersurface formalism.

In order to obtain a complete and satisfactory canonical formalism the constraints must be solved and Dirac brackets calculated in one of the ways indicated in section 2. The progress that has been made in this area (Goldberg and Soteriou 1995; Goldberg and Robinson 1998) indicates that the canonically conjugate reduced complex phase space variables are A_2^B and Σ_D^2, with $B = 3$ and $D = 3$. In the linearized theory the imposition of the reality conditions, and the complete fixing of coordinate and gauge freedom, leads to the breaking of the chiral nature of the formalism. The completely reduced phase space and final Hamiltonian are real and the Poisson structure is provided by the Dirac bracket. These calculations, by Soteriou (1992), support the conclusion that, after removal of the gauge and coordinate freedom, the two real degrees of freedom on the completely reduced phase space can be identified with the real and imaginary parts of the connection component A_2^3. The latter has a spacetime interpretation as the shear of the null hypersurface geodesic ray generators. Work in progress, on the geometry of null hypersurfaces and connections with values in the Lie algebra of the group of null rotations, may help clarify these matters.

The approach to the new variables, null hypersurface canonical formalism presented above incorporates the components of a self-dual connection as dynamical variables. The Lagrange multiplier techniques used in (Goldberg, Robinson, and Soteriou 1992; Goldberg and Soteriou 1995) could also be employed to develop a complete, real, null hypersurface based, geometrodynamical canonical formalism. It would expected that this would be related to the above by a complex canonical transformation.

King's College London

ACKNOWLEDGEMENTS

I would like to thank Lee McCulloch, Chrys Soteriou and Josh Goldberg for useful comments.

NOTE

*   John Stachel has contributed significantly to the study of many technical, conceptual and historical problems in general relativity. His hospitality in Boston enabled me to start investing this approach to Hamiltonian general relativity. It is a pleasure to dedicate this paper to him.

REFERENCES

Aragone, C. and J. Chela-Flores. 1975. *Nuovo Cimento* 25B:225.
Aragone, C. and A. Koudeir. 1990. *Class. Quantum Grav.* 7:1291.
Ashtekar, A. 1986. *Phys. Rev. Lett.* 57: 2244.
———. 1987. *Phys. Rev.* D36: 1587.
———. 1988. *New Perspectives in Canonical Gravity.* Naples: Bibliopolis.
Ashtekar, A., J. D. Romano, and R. S. Tate. 1989. *Phys. Rev.* D40:2572.
Ashtekar, A. and J. Stachel, eds. 1991. *Conceptual problems of quantum gravity.* Boston: Birkhäuser.
Ashtekar, A. and R. Tate. 1991. *Lectures on Non-Perturbative Canonical Gravity.* Singapore: World Scientific.
Bartnik, R. 1997. *Class. Quantum Grav.* 14:2185.
Bengtsson, I. and T. Jacobson. 1997. *Class. Quantum Grav.* 14:3109.
Chrusciel, P. T., M. A. H. MacCallum, and D. B. Singleton. 1995. *Phil. Trans. Roy. Soc. Lond.* A 350:113.
Dirac, P. A. M. 1949. *Rev. Mod. Phys.* 21:392.
Esposito, G. and C. Stornaiolo. 1997. *Nuovo Cimento* B112:1395.
Stornaiolo, C. and G. Esposito. 1997. *Nucl. Phys. Proc. Suppl.* 57:241.
Evans, D., G. Kunstatter, and C. Torre. 1987. *Class. Quantum Grav.* 4:1503.
Friedrich, H. and J. M. Stewart. 1983. *Proc. R. Soc.* A 385:345.
Gambini, R. and A. Restuccia. 1978. *Phys. Rev.* D17:3150.
Gibbons, G. W. 1998. *Quantum Gravity/String/M-theory as we approach the 3rd Millennium,* gr-qc/9803065. [Proceedings of General Relativity and Gravitation, GR15, Puna, India 1997.]
Goldberg, J. N. 1984. *Found. Phys.* 14:1211.
———. 1985. *Found. Phys.* 15:439.
———. 1986. In *Gravitational Collapse and Relativity (Proceedings of the XIV Yamada Conference).* eds. H. Sato and T. Nakamura. Singapore: World Scientific.
———. 1991. *Gen. Rel. and Grav.* 23:1403.
———. 1992. *J. Geom.and Phys.* 8:163.
Goldberg, J. N., D. C. Robinson, and C. Soteriou. 1991. In *9th Italian Conference on General Relativity and Gravitational Physics, Capri (Napoli), September 25 - 28, 1990 (Peter G. Bergmann Festschrift),* ed. R. Cianci et al. Singapore: World Scientific.
———. 1991a, in *Gravitation and Modern Cosmology;* vol. in honor Peter Gabriel Bergmann's 75. birthday, eds. A. Zichichi, V. de Sabbata, and N. Sanchez. New York: Plenum.
———. 1992. *Class. Quantum Grav.* 9:1309.
Goldberg, J. N. and C. Soteriou. 1995. *Class. Quantum Grav.* 12:2779.
Goldberg, J. N. and D. C. Robinson. 1998. *Acta Phys. Polonica B.* 29:849.
Hasse, W., M. Kriele, and V. Perlick. 1996. *Class. and Quantum Grav.* 13:1161.
Henneaux, M. and C. Teitelboim. 1992. *Quantisation of Gauge Systems.* Princeton, N.J.: Princeton U. Press.
d'Inverno, R.A. and J. A. Vickers. 1995. *Class. Quantum Grav.* 12:753.
Maluf, J. W. and J. F. da Rocha-Neto. 1999. *Gen. Rel. and Grav. 31:173.*
Peldan, P. 1994. *Class. Quantum Grav.* 11:1087.
Rovelli, C. 1997. *Strings, loops and others: a critical survey of the present approaches to quantum gravity,* gr-qc/9803024. [Proceedings of General Relativity and Gravitation, GR15, Puna, India 1997.]
———. 1997a. "Loop Quantum Gravity," gr-qc/9710008,
http://www.livingreviews.org/Articles/Volume1/1998-1rovelli/.
Soteriou, C. 1992. *Canonical Formalisms, New Variables and Null Hypersurfaces.* Ph.D. Thesis. University of London.
Sundermeyer, K. 1982. "Constrained Dynamics: with Applications to Yang-Mills Theory, General Relativity, Classical Spin, Dual String Model." *Lecture Notes in Physics* 169. Berlin/New York: Springer-Verlag.
Torre, C. 1986. *Class. Quantum Grav.* 3:773.

JERZY F. PLEBAŃSKI AND MACIEJ PRZANOWSKI

ON VACUUM TWISTING TYPE-N AGAIN*

Abstract. Cartan's structure equations for the vacuum twisting type-N are investigated and some gauges for the null tetrad and coordinates are given.

1. INTRODUCTION

Searching for the vacuum twisting type-N solutions is one of the most intriguing and important problems in general relativity. All vacuum non-twisting type-N solutions are known (Kramer et al. 1980; Plebański 1979; García Díaz and Plebański 1981). However, these solutions are singular and, consequently, cannot describe gravitational radiation far away from a bounded source. The only known twisting solution, due to Hauser (1974, 1978; Ernst and Hauser 1978; Kramer et al. 1980), is also singular. Although many authors have tried to find some new twisting solution[1] every effort ends with the Hauser solution only. Moreover, the results of the works (Stephani 1993; Finely, Plebański, and Przanowski 1997; Bičak and Pravda 1998) show that the approximation methods are here very questionable. Therefore, perhaps one should review the twisting type-N problem again, starting from the very beginning in order to find the optimal gauge for the tetrad and for the coordinates. This is the aim of our paper.

First, in the introduction, we state the problem and write the main equations. Then, in section 2, the general analysis of those equations is given and some general forms of the null tetrad and of the connection are presented. Further analysis and the search for the optimal gauge is the subject of section 3.

To start with, assume that there exists the gauge for the null tetrad in which the connection 1-form is given by

$$\Gamma_{42} = -dY \quad \Gamma_{12} + \Gamma_{34} = 0 \quad \text{and} \quad \Gamma_{31} = -\mu \, dY \tag{1}$$

where μ and Y are some functions on the space-time and

$$dY \wedge d\overline{Y} \neq 0, \tag{2}$$

where overbar stands for the complex conjugation.

In this paper we use the results and notation of the works, (Plebański 1979; Debney, Kerr, and Schild 1969; Plebański 1974; Plebański and Przanowski 1988). So in terms of the null tetrad e^a, $a = 1,\ldots,4$, $\overline{e^1} = e^2$, $\overline{e^3} = e^3$, $\overline{e^4} = e^4$ the space-time metric ds^2 takes the following form

$$ds^2 = g_{ab}\, e^a \otimes e^b$$

$$(g_{ab}) = \begin{pmatrix} 0 & 1 & 0 & 0 \\ 1 & 0 & 0 & 0 \\ 0 & 0 & 0 & 1 \\ 0 & 0 & 1 & 0 \end{pmatrix} \qquad (3)$$

The second Cartan structure equations for the connection given by (1.1) yield

$$C^{(5)} = C^{(4)} = C^{(3)} = C^{(2)} = 0, \qquad R = 0 \text{ and } C_{ab} = \nu\, \delta_a^3\, \delta_b^3 \qquad (4)$$

and

$$-2d\mu \wedge dY = C^{(1)} e^3 \wedge e^1 + \nu e^2 \wedge e^3 \qquad (5)$$

where ν is some function.

As we are going to deal with the type N we must put

$$C^{(1)} \neq 0. \qquad (6)$$

From the generalized Goldberg-Sachs theorem (Plebański 1974, 1979) one gets that if the conditions (4) and (6) hold then

$$\nu \neq 0 \implies \Gamma_{424} = 0,$$

$$\nu = 0 \implies \Gamma_{424} = \Gamma_{422} = 0. \qquad (7)$$

Consequently, the twist

$$\rho = \operatorname{Im}(-\Gamma_{421}) \qquad (8)$$

is non-zero if

$$e^3 \wedge de^3 \neq 0. \qquad (9)$$

Our intent is to find the space-time metric under the assumptions (1), (2), (6) and (9). The first Cartan structure equations read

$$de^1 = \overline{\mu} e^3 \wedge d\overline{Y} + e^4 \wedge dY$$

$$de^3 = -e^1 \wedge d\overline{Y} - e^2 \wedge dY$$

$$de^4 = -\mu e^1 \wedge dY - \overline{\mu} e^2 \wedge d\overline{Y} \qquad (10)$$

Then eq. (5) wedged by sides with it complex conjugate implies

$$d\mu \wedge d\overline{\mu} \wedge dY \wedge d\overline{Y} = 0 \qquad (11)$$

and therefore the four functions $\{\mu, \overline{\mu}, Y, \overline{Y}\}$ are dependent. Now an important remark should be made.

From the work (Plebański 1979) it follows that for any vacuum (or with pure radiation matter field) type-N solution with $\Gamma_{421} \neq 0$ there exits the basic gauge for the null tetrad in which (1) and (2) hold. Therefore our considerations are valid in all generality. In the present paper we deal with the vacuum case only.

2. THE GENERAL ANALYSIS OF THE CARTAN STRUCTURE EQUATIONS.

We assume
$$\nu = 0 \tag{12}$$

Hence, using also (5) and (6) one gets
$$-2d\mu \wedge dY = C^{(1)} e^3 \wedge e^1 \neq 0 \tag{13}$$

i.e.,
$$d\mu = \mu_{,1} e^1 + \mu_{,3} e^3$$
$$dY = Y_{,1} e^1 + Y_{,3} e^3 \tag{14}$$

where the symbol ",a" stands for the derivative in the e_a direction. The condition (13) permits us to solve (14) for e^1 and e^3 obtaining

$$e^1 = \frac{2}{C^{(1)}} (Y_{,3} d\mu - \mu_{,3} dY)$$

$$e^3 = \frac{-2}{C^{(1)}} (Y_{,1} d\mu - \mu_{,1} dY) \tag{15}$$

From (14) with (1) and (9) one infers that
$$Y_{,1} \neq 0. \tag{16}$$

Then inserting e^3 given by (15) into the condition (9) we find that
$$\mu_{,1} \neq 0. \tag{17}$$

Wedging the de^1 equation from (10) by sides one has
$$de^1 \wedge de^1 = 2\bar{\mu} dY \wedge d\bar{Y} \wedge e^3 \wedge e^4 = 2\bar{\mu} |Y_{,1}|^2 e^1 \wedge e^2 \wedge e^3 \wedge e^4 \neq 0. \tag{18}$$

But we can also find $de^1 \wedge de^1$ using the formula (15) for e^1. Then we quickly arrive at the conclusion that for the condition (18) to be satisfied it is necessary that
$$Y_{,3} \neq 0 \text{ and } \mu_{,3} \neq 0. \tag{19}$$

Consider de^4 as given by (10). Employing also (14) one gets
$$e^4 \wedge de^4 = -\mu Y_{,3} e^1 \wedge e^3 \wedge e^4 - \bar{\mu}\bar{Y}_{,3} e^2 \wedge e^3 \wedge e^4 \tag{20}$$

Consequently, by (19)
$$e^4 \wedge de^4 \neq 0. \tag{21}$$

On the other hand (10) and (14) imply
$$de^4 \wedge de^4 \neq 0. \tag{22}$$

From (21) and (22) and from the Darboux theorem it follows that there exist real functions r, s and t such that

$$e^4 = dr + sdt \text{ and } e^4 \wedge de^4 = dr \wedge ds \wedge dt \neq 0. \tag{23}$$

We try to use these three variables (r, s, t) as the three out of four independent coordinates.
As $e^3 \wedge de^4 = 0$ (see (10) with (14)) we get by (23) that

$$e^3 \wedge ds \wedge dt = 0. \tag{24}$$

Hence, e^3 must be of the following form

$$e^3 = A\, ds + B\, dt \tag{25}$$

where A and B are some real functions. From (10) and (14) one has

$$de^3 \wedge de^3 = 0 \tag{26}$$

Therefore, remembering the basic assumption (9) and employing the Darboux theorem we arrive at the result that there exist real functions u, v and w such that

$$e^3 = du + vdw \text{ and } du \wedge dv \wedge dw \neq 0. \tag{27}$$

It is an easy matter to show that without any loss of generality one can assume that

$$ds \wedge dt \wedge dw \neq 0. \tag{28}$$

With this assumption, comparing (25) and (27) we conclude that

$$u = u(s, t, w), \quad v = -u_w, \quad A = u_s \text{ and } \quad B = u_t \tag{29}$$

where $u_w \equiv \frac{\partial u(s,t,w)}{\partial w}$ ... etc.
Consequently

$$e^3 = u_s ds + u_t dt = du - u_w dw. \tag{30}$$

Now the twist condition (9) yields

$$e^3 \wedge de^3 = -du \wedge du_w \wedge dw = -\frac{\partial(u, u_w)}{\partial(s, t)} ds \wedge dt \wedge dw \neq 0. \tag{31}$$

We will show that Y and μ are the funtions of the indepedent coordinates (s, t, w) only i.e., $Y = Y(s, t, w)$ and $\mu = \mu(s, t, w)$. Indeed, from (10) with (14) one quickly finds that $e^3 \wedge de^3 \wedge dY = 0$
But this because of (31) implies

$$ds \wedge dt \wedge dw \wedge dY = 0. \tag{32}$$

Hence, $Y = Y(s, t, w)$. Analogously,

$$e^3 \wedge de^3 \wedge d\mu = 0 \Longrightarrow ds \wedge dt \wedge dw \wedge d\mu = 0, \tag{33}$$

and it means that $\mu = \mu(s,t,w)$. Therefore, our statement that $Y = Y(s,t,w)$ and $\mu = \mu(s,t,w)$ holds true.

Then from (14), (16) and (30) we get

$$e^3 \wedge dY \wedge d\overline{Y} = i\Omega ds \wedge dt \wedge dw = |Y_{,1}|^2 \, e^1 \wedge e^2 \wedge e^3 \neq 0 \tag{34}$$

where

$$\Omega := i\left\{ u_t \frac{\partial(Y,\overline{Y})}{\partial(s,w)} - u_s \frac{\partial(Y,\overline{Y})}{\partial(t,w)} \right\} = \overline{\Omega}. \tag{35}$$

Wedging (34) with e^4 given by (23) we arrive at the conclusion that

$$dr \wedge ds \wedge dt \wedge dw \neq 0. \tag{36}$$

It means that (r,s,t,w) *can serve as the space-time coordinates*.

In the next point we consider the tetrad member e^1 in terms of these coordinates. From (14) with (16) we are able to represent e^1 in the following form

$$e^1 = (r+z)dY + Ce^3 \tag{37}$$

where Z and C are some functions.
It is an easy matter to show that

$$Z = Z(s,t,w) \quad \text{and} \quad C = C(s,t,w). \tag{38}$$

Indeeed, substituting (23) and (37) into the first of (10) equations one gets

$$dZ \wedge dY + Cde^3 + dC \wedge e^3 = \overline{\mu} e^3 \wedge d\overline{Y} + sdt \wedge dY. \tag{39}$$

This last equation by (30) leads to the condition

$$Y_w \, dZ \wedge ds \wedge dt \wedge dw = 0. \tag{40}$$

But as Ω defined by (35) satisfies the condition $\Omega \neq 0$ (see (34)) the function $Y_w \neq 0$. Consequently, from (40) it follows that $Z = Z(s,t,w)$ i.e, $Z_r = 0$.
Wedging eq. (39) with $dY \wedge d\overline{Y}$ we get by (34)

$$i\,\Omega dC \wedge ds \wedge dt \wedge dw = 0 \Longrightarrow C = C(s,t,w). \tag{41}$$

Hence (38) holds true.

Summing up: we have proved that in terms of the coordinates (r,s,t,w) the null tetrad reads

$$e^1 = (r+Z)dY + Ce^3 = \overline{e^2}$$
$$e^3 = u_s ds + u_t dt$$
$$e^4 = dr + sdt \tag{42}$$

where the complex functions Z, Y, C and the real function u depend on the coordinates (s,t,w) only.

Consider the de^3 equation from (10) using our tetrad (42). Straightforward calculations show that all the information one can extract from this equation is contained in the following relations

$$Z - \overline{Z} = \frac{1}{i\Omega} \frac{\partial (u, u_w)}{\partial (s, t)} \neq 0 \tag{43}$$

and

$$C = \frac{1}{i\Omega} \frac{\partial (u_w, Y)}{\partial (s, t)} \neq 0. \tag{44}$$

Then the de^4 equation is equivalent to the following condition

$$\mu = \frac{Y_w}{\frac{\partial (Y, u_w)}{\partial (s, t)}} \neq 0. \tag{45}$$

Now we deal with de^1 equation. Simple manipulations lead to the following conditions

$$\frac{\partial (Z, Y)}{\partial (s, t)} + \frac{\partial (C, u)}{\partial (s, t)} + \overline{\mu} \frac{\partial (\overline{Y}, u)}{\partial (s, t)} + sY_s = 0 \tag{46}$$

$$\frac{\partial (Z, Y)}{\partial (w, s)} + (Cu_s)_w + \overline{\mu}\overline{Y}_w u_s = 0 \tag{47}$$

$$\frac{\partial (Z, Y)}{\partial (w, t)} + (Cu_t)_w + \overline{\mu}\overline{Y}_w u_t + sY_w = 0. \tag{48}$$

Finally, we are going to extract all the information contained in the second Cartan structure equations (13). Comparing terms on both sides of (13) standing at $ds \wedge dt$, $ds \wedge dw$ and $dt \wedge dw$ one arrives at the relations

$$\frac{\partial(\mu, Y)}{\partial (s, t)} + \frac{1}{2} (r + Z) \, C^{(1)} \frac{\partial (u, Y)}{\partial (s, t)} = 0 \tag{49}$$

$$\frac{\partial (\mu, Y)}{\partial (s, w)} + \frac{1}{2} (r + Z) \, C^{(1)} \, Y_w \, u_s = 0 \tag{50}$$

$$\frac{\partial (\mu, Y)}{\partial (t, w)} + \frac{1}{2} (r + Z) \, C^{(1)} \, Y_w \, u_t = 0. \tag{51}$$

Remembering that $C^{(1)} \neq 0$ and (see (43) and (45))

$$Y_w u_s \neq 0 \neq Y_w u_t \tag{52}$$

one can rewrite (50) and (51) in the form of one equation

$$u_t \frac{\partial (\mu, Y)}{\partial (s, w)} - u_s \frac{\partial (\mu, Y)}{\partial (t, w)} = 0 \tag{53}$$

and the formula for $C^{(1)}$

$$C^{(1)} = -\frac{2}{(r + Z) Y_w u_s} \frac{\partial (\mu, Y)}{\partial (s, w)} \neq 0. \tag{54}$$

Then the straightforward calculations show that (53) and (54) with (52) imply (49). This closes our general analysis of the first and the second Cartan structure equations for the twisting vacuum type-N.

Gathering: we have the null tetrad defined by (42) and the coordinates (r, s, t, w). Three complex functions Z, Y, C and the real function u depend on (s, t, w) only. Finally, the relations (43) to (48), (53) and (54) should be satisfied. In the next section we are going to look for the tetrad gauge and the coordinate system which seem to be optimal for further analysis.

3. LOOKING FOR THE OPTIMAL GAUGE FOR THE NULL TETRAD AND COORDINATES.

First we perform the γ-gauge of the tetrad (42) and then the σ-gauge (see Plebański 1979, 1974) to obtain the following null tetrad e'^a

$$e^1 \xrightarrow{\gamma} e^1 + \bar{\gamma} e^3 \xrightarrow{\sigma} e^{i\varphi}\left(e^1 + \bar{\gamma} e^3\right) = e'^1 = \overline{e'^2}$$

$$e^3 \xrightarrow{\gamma} e^3 \xrightarrow{\sigma} e^\omega e^3 = e'^3$$

$$e^4 \xrightarrow{\gamma} e^4 - \gamma e^1 - \bar{\gamma} e^2 - \gamma\bar{\gamma} e^3 \xrightarrow{\sigma} e^{-\omega}\left(e^4 - \gamma e^1 - \bar{\gamma} e^2 - \gamma\bar{\gamma} e^3\right) = e'^4 \quad (55)$$

where γ and $\sigma = \omega + i\varphi$ are arbitrary complex functions. The connection 1-form $\Gamma'^a{}_b$ with respect to the new tetrad e'^a is given by

$$\begin{aligned}
\Gamma'_{42} &= -e^\sigma dY, \\
\Gamma'_{12} + \Gamma'_{34} &= -2\gamma dY + d\sigma, \\
\Gamma'_{31} &= e^{-\sigma}\left\{-\left(\mu + \gamma^2\right) dY + d\gamma\right\}.
\end{aligned} \quad (56)$$

Now leaving the coordinates s, t and w unchanged we change the coordinate r as follows

$$p := e^{-\omega} r - K \quad (57)$$

where K is a real function. Substituting (42) and (57) into (55), employing also (56) one finds the null tetrad e'^a to read

$$\begin{aligned}
e'^1 &= -(p + X)\,\Gamma'_{42} + De'^3 = \overline{e'^2} \\
e'^3 &= e^\omega(u_s ds + u_t dt) \\
e'^4 &= dp + p\Gamma'_{34} + \tilde{e}'^4
\end{aligned} \quad (58)$$

where

$$X := K + e^{-\omega} Z, \quad D := (C + \bar{\gamma})\,e^{-\bar{\sigma}}$$

$$\begin{aligned}
\tilde{e}'^4 &:= dK + K\Gamma'_{34} + e^{-\omega} sdt - e^{-\omega}\left(\gamma Z dY + \bar{\gamma}\bar{Z}d\bar{Y}\right) \\
&\quad - \left(\gamma C + \bar{\gamma}\bar{C} + \gamma\bar{\gamma}\right) e^{-2\omega} e'^3.
\end{aligned} \quad (59)$$

Then by (43) we have

$$X - \overline{X} = e^{-\omega}\left(Z - \overline{Z}\right) \neq 0 \tag{60}$$

and it means that the existence of the twist implies that the imaginary part of X must be different from zero.

To proceed further we assume *that the functions γ, σ and K depend on the co-ordinates (s, t, w) only.* Simple considerations show that choosing appropriately the functions $K = K(s,t,w)$ and $\omega = \omega(s,t,w)$ (leaving $\gamma = \gamma(s,t,w)$ and $\varphi = \varphi(s,t,w) = \overline{\varphi}$ still arbitrary) one can bring the 1-form $\tilde{e}'^4$ to the following form

$$\tilde{e}'^4 = H\, e'^3 \tag{61}$$

where $H = H(s,t,w)$ is some real function.[2]

Another possibility is also available. Namely, it can be shown that one can choose K and γ so that

$$\tilde{e}'^4 = 0. \tag{62}$$

In this case σ is still arbitrary complex function.

Consider now the Cartan structure equations in terms of the null tetrad e'^a given by (58). Straightforward but tedious calculations lead to the following result:
The second Cartan structure equations read

$$d\Gamma'_{42} + \Gamma'_{42} \wedge \left(\Gamma'_{12} + \Gamma'_{34}\right) = 0,$$

$$d\left(\Gamma'_{12} + \Gamma'_{34}\right) + 2\Gamma'_{42} \wedge \Gamma'_{31} = 0,$$

$$d\Gamma'_{31} + \left(\Gamma'_{12} + \Gamma'_{34}\right) \wedge \Gamma'_{31} = \frac{1}{2}\tilde{C}'^{(1)} e'^3 \wedge \Gamma'_{42} \tag{63}$$

where (see (57), (59) and (54))

$$\tilde{C}'^{(1)} := -(p + X)\, C'^{(1)} = -(p+X)\, e^{-2\sigma}\, C^{(1)}$$

$$= \frac{2e^{-\omega - 2\sigma}}{Y_w u_s} \frac{\partial(\mu, Y)}{\partial(s, w)} \neq 0. \tag{64}$$

Observe that $\tilde{C}'^{(1)} = \tilde{C}'^{(1)}(s, t, w)$.
The first Cartan structure equations give

$$\left\{-dX - X\Gamma'_{34} + (X - \overline{X})\, D\Gamma'_{41} + D\overline{D}e'^3 + \tilde{e}'^4\right\} \wedge \Gamma'_{42}$$

$$- e'^3 \wedge \left\{dD - D^2\Gamma'_{41} + D\left(-\Gamma'_{12} + \Gamma'_{34}\right) - \Gamma'_{32}\right\} = 0,$$

$$de'^3 + (X - \overline{X})\, \Gamma'_{42} \wedge \Gamma'_{41} - e'^3 \wedge \left(D\Gamma'_{41} + \overline{D}\Gamma'_{42} - \Gamma'_{34}\right) = 0,$$

$$d\tilde{e}'^4 - \left(-X\Gamma'_{42} + De'^3\right) \wedge \Gamma'_{31} - \left(-\overline{X}\Gamma'_{41} + \overline{D}e'^3\right) \wedge \Gamma'_{32}$$

$$-\tilde{e}'^4 \wedge \Gamma'_{34} = 0. \tag{65}$$

The essential feature of the Cartan structure equations (63) and (65) consists in its 3-dimensional nature. Namely they contain only objects defined on a 3-dimensional real manifold M_3 with the coordinates (s, t, w) (Of course our considerations are purely *local*.) Therefore the natural question arises if eqs. (63), (64) and (65) define some interesting 3-dimensional geometry. To be more precise, let the functions K and ω be such that the relation (61) holds.

Taking H in the form of

$$H = \frac{1}{2} \varepsilon h^2, \quad h = \bar{h} \neq 0, \quad \varepsilon = \pm 1 \tag{66}$$

and substituting the null tetrad (58) into (3) one gets the space-time metric ds^2 to be

$$ds^2 = 2 \left\{ -(p+X) \; \Gamma'_{42} + De'^3 \right\} \otimes_s \left\{ -(p+\overline{X}) \, \Gamma'_{41} + \overline{D} e'^3 \right\}$$

$$+ 2e'^3 \otimes_s \left(dp + p\Gamma'_{34} + \frac{1}{2} \varepsilon h^2 e'^3 \right). \tag{67}$$

Assuming $p = 0$ we obtain the following metric ds_o^2 on the 3-dimensional manifold M_3

$$ds_o^2 = 2E^1 \otimes_s E^2 + \varepsilon E^3 \otimes E^3$$

$$E^1 := -X\Gamma'_{42} + De'^3 = \overline{E^2}, \quad E^3 := he^3$$

$$E^1 \wedge E^2 \wedge E^3 = h \, |X|^2 \; e'^3 \wedge \Gamma'_{42} \wedge \Gamma'_{41} \neq 0. \tag{68}$$

Now one can ask if eqs. (63), (64) and (65) are equivalent to structure equations of some 3-dimensional geometry on (M_3, ds_o^2) (perhaps Riemannian, Riemann-Cartan or Weyl geometry). This very interesting problem will be considered elsewhere.

After this comment we come back to our considerations on the optimal gauge for the null tetrad and coordinates. We know that choosing appropriately the real functions K and ω we can obtain (61). But, as γ and $\varphi = \bar{\varphi}$ are still arbitrary one can do much better. Namely, it is an easy matter to show that we are able to find the functions $K = K$, γ, $\omega = \bar{\omega}$ and $\varphi = \bar{\varphi}$ and a new real coordinate q defined by

$$w = L(q, s, t), \quad \dot{L} := L_q \neq 0 \tag{69}$$

where L is a real function, in such a manner that the relation (61) holds and also

$$\Gamma'_{42q} \, dq = -dq \quad \text{and} \quad \left(\Gamma'_{12q} + \Gamma'_{34q} \right) dq = 0. \tag{70}$$

Thus we have the real coordinates (p, q, s, t) and the null tetrad (we omit the primes!)

$$e^1 = -(p+X)\Gamma_{42} + De^3 = \overline{e^2}$$

$$e^3 = \phi ds + \psi dt$$

$$e^4 = dp + p\Gamma_{34} + He^3 \tag{71}$$

with X, D, $\phi = \bar{\phi}$, $\psi = \bar{\psi}$ and $H = \bar{H}$ being the functions of (q, s, t) only, while simultaneously we know that

$$\Gamma_{42} = -dq + Fds + Gdt$$

$$\Gamma_{12} + \Gamma_{34} = Pds + Qdt \tag{72}$$

where F, G, P and Q are complex functions of (q, s, t).

Inserting (72) into the second Cartan structure equations (63) one arrives at the following results (remember that the primes are omitted!)

$$P = \dot{F} \text{ and } Q = \dot{G} \tag{73}$$

$$G_s - F_t + F\dot{G} - \dot{F}G = 0 \tag{74}$$

$$\Gamma_{31} = \frac{1}{2}\ddot{F}ds + \frac{1}{2}\ddot{G}dt - M\Gamma_{42} \tag{75}$$

$$\frac{1}{2}\widetilde{C}^{(1)}e^{(3)} \wedge \Gamma_{42} = \left\{-dM - 2M\left(\dot{F}ds + \dot{G}dt\right) + \frac{1}{2}\left(\dddot{F}ds + \dddot{G}dt\right)\right\} \wedge \Gamma_{42}, \tag{76}$$

where $M = M(q, s, t)$ is a complex function and the overdot stands for the derivative with respect to q i.e., $\dot{F} \equiv F_q$, $\dddot{F} \equiv F_{qqq}$ etc.

Gathering: we have found the null tetrad (71) and the connection 1-form defined by

$$\Gamma_{42} = -dq + Fds + Gdt$$

$$\Gamma_{12} + \Gamma_{34} = \dot{F}ds + \dot{G}dt$$

$$\Gamma_{31} = \frac{1}{2}\ddot{F}ds + \frac{1}{2}\ddot{G}dt - M\Gamma_{42}. \tag{77}$$

The complex functions F, G and M should satisfy eqs. (74) and (76) which contain all the information extracted from the second Cartan structure equations (63). Then, of course, the first Cartan structure equations (65) must be also satisfied. We hope that our gauge, if not optimal, is very close to the optimal one.

To justify this hope we show how easy matter is to find the general non-twisting but diverging type-N solution (i.e, the Robinson-Trautman type-N solution) when our gauge is used.

To this end put

$$F = -i. \tag{78}$$

Consequently, eq. (74) reads

$$G_{\bar{\xi}} = 0, \quad \xi := q + is. \tag{79}$$

Therefore, the function G takes the form of

$$G = f(\xi, t) \tag{80}$$

Now assume that

$$M = 0, \quad \phi = 0, \quad X = 0 \quad \text{and} \quad \psi \neq 0. \tag{81}$$

Then, by (77) with (78), (80) and (81) one gets

$$\Gamma_{42} = -d\xi + f dt$$

$$\Gamma_{12} + \Gamma_{34} = f_\xi dt$$

$$\Gamma_{31} = \frac{1}{2} f_{\xi\xi} dt \tag{82}$$

and this is the canonical form of the connection for the non-twisting and diverging case.[3]

Now (76) implies

$$\tilde{C}^{(1)} = \psi^{-1} f_{\xi\xi\xi} \quad \Longrightarrow \quad C^{(1)} = -(p\psi)^{-1} f_{\xi\xi\xi}. \tag{83}$$

Straightforward calculations show that the first Cartan structure equations (65) give

$$H = \psi^{-1} g(t), \quad g = \bar{g} \tag{84}$$

$$D = \psi^{-1} \psi_{\bar{\xi}} \tag{85}$$

$$D_{\bar{\xi}} + D^2 = 0 \tag{86}$$

$$D_\xi + D\bar{D} + H = 0. \tag{87}$$

Inserting (85) into (86) we get

$$\psi_{\bar{\xi}\bar{\xi}} = 0 \quad \Longrightarrow \quad \psi_{\xi\xi} = 0. \tag{88}$$

Then substituting (84) and (85) into (87) one obtains

$$g(t) = -\psi_{\xi\bar{\xi}}. \tag{89}$$

Finally, (88) with (89) lead to the following solution

$$\psi = -g(t)\,\xi\bar{\xi} + b(t)\xi + \overline{b(t)}\,\bar{\xi} + a(t) \neq 0 \tag{90}$$

where $g = g(t)$ and $a = a(t)$ are arbitrary real functions and $b = b(t)$ is any complex function. Thus we arrive at the general Robinson-Trautman type-N solution with the complex $f = f(\xi, t)$ being an arbitrary structural function ($f_{\xi\xi\xi} \neq 0$) and with $\psi = \psi(\xi, \bar{\xi}, t) = \bar{\psi} \neq 0$ given by (90).[4]

At the end of our paper we elaborate eq. (76). One quickly finds that this equation implies

$$\frac{1}{2}\tilde{C}^{(1)} e^3 + dM + 2M \left(\dot{F} ds + \dot{G} dt \right) - \frac{1}{2} \left(\ddot{F} ds + \ddot{G} dt \right)$$

$$= \lambda\,\Gamma_{42} = \lambda\,(-dq + F ds + G dt) \tag{91}$$

where λ is some complex function. Then as $e_q^3 dq = 0$ we infer from (91) that

$$\lambda = -\dot{M} \tag{92}$$

Therefore (91) with (92) give (see also (71))

$$\frac{1}{2}\widetilde{C}^{(1)}\phi + M_s + 2M\dot{F} + \dot{M}F - \frac{1}{2}\dddot{F} = 0,$$

$$\frac{1}{2}\widetilde{C}^{(1)}\psi + M_t + 2M\dot{G} + \dot{M}G - \frac{1}{2}\dddot{G} = 0. \tag{93}$$

Finally, we get

$$\psi\left(M_s + 2M\dot{F} + \dot{M}F - \frac{1}{2}\dddot{F}\right) = \phi\left(M_t + 2M\dot{G} + \dot{M}G - \frac{1}{2}\dddot{G}\right),$$

$$\widetilde{C}^{(1)} = -2\frac{1}{\phi^2 + \psi^2} \times \left(\phi\left(M_s + 2M\dot{F} + \dot{M}F - \frac{1}{2}\dddot{F}\right)\right.$$
$$\left. + \psi\left(M_t + 2M\dot{G} + \dot{M}G - \frac{1}{2}\dddot{G}\right)\right) \neq 0 \tag{94}$$

Concluding: Eqs. (74) and (94) contain all the information extracted from the second Cartan structure equations (63). The analysis of the first Cartan structure equations (65) and looking for the solutions will be considered elsewhere.

ACKNOWLEDGEMENTS

One of us (M. P.) is grateful to the staff of Departamento de Física, CINVESTAV México, D. F., for their warm hospitality. This work was partially supported by CONACyT and CINVESTAV (México) and KBN (Poland).

Centro de Investigación y de Estudios Avanzados del IPN
Technical University of Lódz

NOTES

* It is a pleasure for Jerzy Plebański to dedicate this paper to his friend John Stachel.

1. See (Plebański 1979; McIntosh 1985; Stephani and Herlt 1985; Chinea 1988, 1998; Plebański and Przanowski 1991; Ludwig and Yu 1992; Finely, Plebański, and Przanowski 1994; Plebański, Przanowski, and Formański 1998).
2. Note that the null tetrad (58) with $\widetilde{e}'^4$ given by (61) resembles very much the null tetrad defined in the works (Plebański 1979; García Díaz and Plebański 1981) for the case of non-twisting $(e^3 \wedge de^3 = 0)$ but diverging $(e^3 \wedge \Gamma_{42} \wedge \Gamma_{41} \neq 0)$ vacuum type-N (the Robinson-Trautman solution). As it has been shown this latter tetrad leads inmediately to the Robinson-Trautman solution. Therefore we hope that, analogously, for the twisting case the null tetrad (58) with (61) will play an important role.
3. Compare with (Plebański 1979; García Díaz and Plebański 1981).
4. For details see (Plebański 1979; García Díaz and Plebański 1981).

REFERENCES

Bičak, J., and V. Pravda. 1998. *Class. Quantum Grav.* 15:1539.
Chinea, F. J. 1988. *Phys. Rev.* D 37:3080.
———. 1998. *Class. Quantum Grav.* 15:367.
Debney, G. C., R. P. Kerr, and A. Schild. 1969. *J. Math. Phys.* 10:1842.
Ernst, F. J., and I. Hauser. 1978. *J. Math. Phys.* 19:1816.
Finely, J. D. III, J. F. Plebański, and M. Przanowski. 1994. *Class. Quantum Grav.* 11:157.
———. 1997. *Class. Quantum Grav.* 14:489.
García Díaz, A., and J. F. Plebański. 1981. *J. Math. Phys.* 22:2655.
Hauser, I. 1974. *Phys. Rev. Lett* 33:1112. [Erratum: 1525].
———. 1978. *J. Math. Phys.* 19:661.
Kramer, D., H. Stephani, M. MacCallum, and E. Herlt. 1980. *Exact Solutions of Einstein's Field Equations.* Berlin: VEB Deutscher Verlag der Wissenschaften; Cambridge: Cambridge University Press.
Ludwig, G., and Y. B. Yu. 1992. *Gen. Relativ. Grav* 24:93.
McIntosh, C. B. G. 1985. *Class. Quantum Grav* 2:87.
Plebański, J. F. 1974. "Spinors, Tetrads and Forms." Unpublished monograph, Mexico: CINVESTAV.
———. 1979. "Type-N Solutions of $G_{\mu\nu} = -\rho k_\mu k_\nu$ with null k_μ." Mexico: Unpublished notes.
Plebański, J. F., and M. Przanowski. 1988. *Acta Phys. Polonica* B19:805.
———. 1991. *Phys. Lett.* A 152:257.
Plebański, J. F., M. Przanowski, and S. Formański. 1998. *Phys. Lett.* A 246:25.
Stephani, H. 1993. *Class. Quatum Grav.* 10:2187.
Stephani, H., and E. Herlt. 1985. *Class. Quantum Grav.* 2:L63.

JOSHUA N. GOLDBERG

QUASI-LOCAL ENERGY

Abstract. An expression for quasi-local energy is suggested. The calculation looks locally at an Alexandrov neighborhood whose boundary is defined by the intersection of the future and past cones from nearby time-like separated points. The calculation uses the canonical formalism on a null cone and defines the quasi-local energy as a two-surface integral over the convergence of the past cone minus a similar integral over a surface in Minkowski space. This definition is suggested by the results of an analysis at null infinity.

1. INTRODUCTION

A local energy density cannot be defined in general relativity. However, there have been a number of attempts to define a quasi-local energy. That is, one does not ask for an energy density, but rather one asks for the energy content within a volume of space defined by a closed spatial two-surface. This notion was introduced by Roger Penrose in connection with his twistor program (1982). In Minkowski space, the energy and momentum are contained in the 10 component angular momentum twistor (Penrose and Rindler 1986). Penrose's idea is to construct a corresponding quantity for the gravitational field using solutions of the twistor equation $\nabla^{(A} \omega^{B)} = 0$. For most cases, this does lead to a satisfactory result for energy-momentum, but not for angular momentum. However, the twistor equation does not have an easily visualizable meaning and therefore there have been a number of other proposals (Brown and York 1993; Dougan and Mason 1991; Hayward 1993). Perhaps the formulation closest to that of Penrose is that of Dougan and Mason (1991). But there the spinors which are defined to be holomorphic are equally elusive in their interpretation.

Here I wish to propose a definition which clearly depends on the light bending properties of mass. It is a definition which is derived directly from the Einstein equations and mirrors the calculations at null infinity which leads to the Trautman-Bondi-Sachs (TBS) energy. The particular path taken follows the reexamination of the gravitational radiation field at null infinity (Goldberg and Soteriou 1995). In the reexamination, the analysis was done with the null surface canonical formalism using an anti-self-dual connection and densitized triad as phase space variables (Goldberg, Robinson, and Soteriou 1992) as in the Ashtekar formalism on a space-like surface (Ashtekar 1991). In this approach one follows the Bondi-Sachs (Bondi, Van Der Berg, and Metzner 1962; Sachs 1962) and Newman-Penrose (1962) calculations. Thus, we impose data on an outgoing null surface, integrate the hypersurface equations, and then examine the time development off of future null infinity. One of the time development equations describes the rate of change of mass (energy) and leads to a definition of the global energy. It turns out that the mass is defined by the $1/r^2$ part of the convergence μ of the generators of $\mathcal{I}^+$.

The idea, then, is to redo this calculation in a small finite domain bounded by outgoing and incoming null cones - an Alexandrov neighborhood. Again one imagines giving data on the outgoing null cone, carrying out the integrations of the hypersurface equations, and examining the time development of the convergence of the null rays on the incoming surface. A problem arises here which does not appear at null infinity. At null infinity, the Minkowski value of μ which is $\sim 1/r$ drops out of the calculation and therefore out of the definition of the mass. It is not clear that the same occurs on a finite null boundary.

In the following section, the general formalism is set up with a discussion of the results at null infinity. The Alexandrov neighborhood will be set up and the conservation equations will be discussed in section 3. The problem with an appropriate zero for the quasi-local energy will also be discussed in section 3. Finally, in section 4 there is a brief discussion of the results.

2. FORMALISM

In the analysis at null infinity we introduced a null tetrad based on a foliation of null surfaces $t = $ constant:

$$\begin{aligned}
\theta^0 &= (N + \alpha \nu^1{}_i N^i) dt + \alpha \nu^1{}_i dx^i \\
\theta^{\mathbf{a}} &= \nu^{\mathbf{a}}{}_i (N^i dt + dx^i) \\
e_0 &= N^{-1}(\partial_t - N^i \partial_i) \\
e_{\mathbf{a}} &= (v_{\mathbf{a}}{}^i + \alpha_{\mathbf{a}} N^{-1} N^i)\partial_i - \alpha_{\mathbf{a}} N^{-1}\partial_t,
\end{aligned} \tag{1}$$

where $\alpha_{\mathbf{a}} = \alpha \delta^1{}_{\mathbf{a}}$ and $\nu^{\mathbf{a}}{}_i v_{\mathbf{b}}{}^i = \delta^{\mathbf{a}}{}_{\mathbf{b}}$. (All indices have the range 1–3 and repeated indices sum. Bold face indices refer to the one forms and tetrads. The signature is -2 and $-iN\nu$ is positive, where ν is the determinent of $\nu^{\alpha}{}_i$.) The connection coefficients are defined by

$$d\theta^\alpha = \theta^\beta \wedge \omega^\alpha{}_\beta \tag{2}$$

The self-dual components of the connection are

$$+\omega^{\alpha\beta} = \frac{1}{2}(\omega^{\alpha\beta} - \frac{i}{2}\eta^{\alpha\beta}{}_{\mu\nu}\omega^{\mu\nu})$$

and are represented by Γ^A, $(A, B, \cdots = 1-3)$:

$$\Gamma^1 := \frac{1}{2}(\omega^{01} + \omega^{23}), \quad \Gamma^2 := \omega^{21}, \quad \Gamma^3 := \omega^{03}. \tag{3}$$

From these we obtain the self-dual components of the Riemann tensor as

$$\frac{1}{2} R^A = d\Gamma^A + \eta^A{}_{BC} \Gamma^B \wedge \Gamma^C. \tag{4}$$

In a $3+1$ decomposition, we have

$$\Gamma^A = A^A{}_i dx^i + B^A dt, \tag{5}$$

and

$$\begin{aligned} R^A{}_{ij} &= 2A^A{}_{[i,j]} + 2\eta^A{}_{BC} A^B{}_{[j} A^C{}_{i]}, \\ R^A{}_{0i} &= D_i B^A - A^A{}_{i,0}. \end{aligned} \tag{6}$$

The derivative operator D acts on the index A as

$$D_i f^A := \partial_i f^A + 2\eta^A{}_{BC} A^B{}_i f^C.$$

The $SO(3)$ invariant metric

$$g_{AB} = \begin{pmatrix} 2 & 0 & 0 \\ 0 & 0 & -1 \\ 0 & -1 & 0 \end{pmatrix}$$

and its inverse,

$$g^{AB} = \begin{pmatrix} \tfrac{1}{2} & 0 & 0 \\ 0 & 0 & -1 \\ 0 & -1 & 0 \end{pmatrix}$$

are used to raise and lower the uppercase Latin self-dual, triad indices.

When written in terms of the self-dual Riemann tensor, the Lagrangian leads to $(A^A{}_i, \Sigma_A{}^i)$ as canonical variables where

$$\Sigma_1{}^i = -\nu v_1{}^i, \quad \Sigma_2{}^i = -\alpha v^i, \quad \Sigma_3{}^i = -\nu v_3{}^i, \quad v^i = \nu v_2{}^i. \tag{7}$$

In addition, there are a number of functions which act as Lagrange multilpiers: B^A, v^i, α and μ_i. From the Lagrangian, we get the following constraints:

$$\begin{aligned} \mathcal{H}_0 &:= v^i \left(R^1{}_{ij} \Sigma_3{}^j + R^2{}_{ij} \Sigma_1{}^j \right) = 0, \\ \mathcal{H}_i &:= -R^A{}_{ij} \Sigma_A{}^j = 0, \\ \mathcal{G}_A &:= D_i \Sigma_A{}^i = 0, \\ \phi_i &:= R^1{}_{ij} \Sigma_3{}^j + R^2{}_{ij} \Sigma_1{}^j = 0 \\ \mathcal{C}^i &:= \Sigma_2{}^i + \alpha v^i = 0. \end{aligned} \tag{8a}$$

(note that $\phi_i v^i = \mathcal{H}_0$ and $\phi_i \Sigma_3{}^i = \mathcal{H}_i \Sigma_1{}^i$) and conditions on the Lagrange multipliers:

$$\alpha = 0, \quad v^i \mu_i = 0, \quad \mu_i \Sigma_1{}^i = R^1{}_{i0} \Sigma_3{}^i - R^1{}_{ij} \Sigma_3{}^i N^j, \tag{8b}$$

$$\chi^i := 2\delta^B{}_2 D_j \left(\underset{\sim}{N} v^{[i} \Sigma_A{}^{j]} Q^A{}_B \right) - 2A^3{}_j N^{[i} \Sigma_1{}^{j]} - B^3 \Sigma_1{}^i = 0, \tag{8c}$$

$$Q^A{}_B := \delta^A_3 \delta^1_B + \delta^A_1 \delta^2_B.$$

The last equation, χ^i, results from the propagation of the constraint $\mathcal{C}^i = 0$ which defines the propagation of $\Sigma_2{}^i = 0$.

In addition, the Hamiltonian equations of motion for the dynamical variables are

$$A^1{}_{i,0} = D_i B^1 + N^j R^1{}_{ij} - \underset{\sim}{N} v^j R^2{}_{ij}, \tag{9a}$$

$$A^2{}_{i,0} = D_i B^2 + N^j R^2{}_{ij} - \mu_i, \tag{9b}$$

$$A^3{}_{i,0} = D_i B^3 + N^j R^3{}_{ij} - \underset{\sim}{N} v^j R^1{}_{ij}, \tag{9c}$$

$$\Sigma_1{}^i{}_{,0} = 2D_j\left(\underset{\sim}{N} v^{[i} \Sigma_A{}^{j]} Q^A{}_1\right) - 2D_j(N^{[i}\Sigma_1{}^{j]}) - 2B^3 \Sigma_3{}^i, \tag{9d}$$

$$\Sigma_3{}^i{}_{,0} = 2D_j\left(\underset{\sim}{N} v^{[i} \Sigma_A{}^{j]} Q^A{}_3\right) - 2D_j(N^{[i}\Sigma_3{}^{j]})$$
$$+ 2B^1 \Sigma_3{}^i + B^2 \Sigma_1{}^i. \tag{9e}$$

In (Goldberg and Soteriou 1995) we assumed the space-time to be asymptotically Minkowskian. But, to solve the equations, we must impose coordinate and gauge conditions. The surfaces t = constant form a foliation of null surfaces near null infinity and give us one coordinate. Space-like cuts of these surfaces are assumed to have the topology S^2. Hence, the generators can be labeled by the angular coordinates (θ, ϕ). We follow the convention of Bondi and Sachs (Bondi, Van Der Berg, and Metzner 1962; Sachs 1962) in choosing the coordinate r along the null generators to be the luminosity distance, so that

$$-ir^2 \sin\theta \, \eta_{1jk} \, v^j \Sigma_3{}^k = \nu^2, \tag{10}$$

and we can set

$$\Sigma_1{}^i = -ir^2 \sin\theta \delta^i{}_1. \tag{11}$$

Then the null rotations generated by $\mathcal{G}_1$ and $\mathcal{G}_2$ allow us to fix $\Sigma_3{}^i$ tangent to the surfaces r = constant, that is, $\Sigma_3{}^1 = 0$, and then to set $A^1{}_1 = 0$. The latter is equivalent to setting $\epsilon = 0$ in the Newman-Penrose formalism (1962). Reality conditions then tell us that v^i must be set equal to $\bar{\Sigma}_3{}^i$ at the end of the calculation.

For the initial data we give $(A^3{}_2, \Sigma_3{}^2)$ and then the integration proceeds in the same order as in (Goldberg, Robinson, and Soteriou 1992). In the course of integrating the constraint equations on $u = 0$, a number of r-independent functions are introduced. After applying the reality conditions and consistency with the Minkowski space limit, only three complex functions remain: $C^1{}_a$ associated with $A^1{}_a$ and $\mathcal{M}$ associated with $A^2{}_a$. The time development of these functions is then determined by the propagation equations for $A^1{}_a$ and $v^a A^2{}_a$, the conservation equations. And because of the Bianchi identities, that is the only information coming from those equations (Bondi, Van Der Berg, and Metzner 1962; Sachs 1962). The $C^1{}_a$ are associated with angular momentum and $\mathcal{M}$ is the mass aspect. Thus, the conservation equations give us information about the rate of change of angular momentum and mass. Note that $\Sigma_3{}^i A^1{}_i = \nu\alpha$, $v^i A^1{}_i = -\nu\beta$, and $A^2{}_a v^a = \nu\mu$. Here α, β, and μ are the NP coefficients defining the connection on the sphere and the convergence of the null rays. However, at $\mathcal{I}^+$ it is only the $1/r^3$ part of α, β, and the $1/r^2$ part of μ which are relevant. Here we will be concerned only with the mass which is defined as the integral over the two-surface $u = constant$ on $\mathcal{I}^+$. Its time derivative is non-positive (ibid.).

3. QUASI-LOCAL ENERGY

We wish to mirror the calculation at null infinity in an Alexandrov neighborhood. Thus we consider two close timelike separated points, p and q with q to the future of p. Thus the neighborhood, $\mathcal{A}$, is defined as the intersection of the causal future of p and the causal past of q. We further assume that p and q are sufficiently close that there is a unique timelike geodesic connecting them. Let t be the proper time along the geodesic with p the point at $t = 0$ and q the point at $t = t_0$. Label the future null cones with origin on the geodesic by t. The past cone $\mathcal{N}^+$ from q cuts these cones and together with $t = 0, \mathcal{N}^-$, defines $\mathcal{A}$. Now consider the past null cones from the geodesic. They intersect the future cones in a folliation of two-surfaces. Label these surfaces by the coordinate r, which we take to be the luminosity distance: the area of the two-surface equals $4\pi r^2$. Label the null rays on $\mathcal{N}^+$ by the angles (θ, ϕ). Thus, we have (t, r, θ, ϕ) as coordinates in $\mathcal{A}$. With these coordinate conditions, the tetrad in (1) becomes $(\mathbf{a}, \mathbf{b}, a, b, \cdots = 2, 3)$

$$\begin{aligned}
\theta^0 &= dt, \\
\theta^1 &= \nu^1{}_1 (N^1 dt + dr), \\
\theta^{\mathbf{a}} &= \nu^{\mathbf{a}}{}_a (dx^a + N^a dt) \\
e_0 &= \partial_t - N^i \partial_i, \\
e_1 &= v_1{}^1 \partial_r, \\
e_{\mathbf{a}} &= v_{\mathbf{a}}{}^a \partial_a.
\end{aligned} \tag{12}$$

In fixing the above tetrad, we have used the r-dependent null rotations to align $v_{\mathbf{a}}{}^a$ tangent to the two-surfaces. With these coordinate and tetrad conditions, e_0 and e_1 are tangent to the generators of $\mathcal{N}^+$ and $\mathcal{N}^-$ respectively. The only remaining freedom is an r-independent phase of the $e_{\mathbf{a}}$.

From

$$-ir^2 \sin\theta \, \eta_{1jk} v^j \Sigma_3{}^k = \nu^2. \tag{13a}$$

and and our fixing $v_1{}^i$ tangent to the outgoing null rays, we find

$$\Sigma_1{}^i = -ir^2 \sin\theta \, \delta^i{}_1. \tag{13b}$$

Our tetrad choice is such that $A^1{}_1 = 0$ and from the guage constraint $\mathcal{G}_2$ we get $A^3{}_1 = 0$. We further assume that the strong energy condition holds so that for an arbitrary null vector we have

$$T^{\mu\nu} k_\mu k_\nu > 0, \qquad k_\mu k^\mu = 0.$$

If we are to integrate the equations, we give as initial data $A^3{}_2$ and $\Sigma_3{}^2$. Then we can proceed to solve the equations as in (Goldberg and Soteriou 1995). However, for our present interests it is unnecessary to do so. We are interested in the information contained in the conservation equation (9b). Therefore, we may assume that we have a known space-time. Thus, we know all the tetrad and connection coefficients in some coordinate system. Then, given p and q, we form the needed transformation $x^\mu \rightarrow$

(t, r, θ, ϕ) in the neighborhood $\mathcal{A}$. The Bianchi identities tell us that the equations for $\dot{A}^1{}_a$ and $v^a \dot{A}^2{}_a$ need be satisfied only at one value of r. Therefore, in the following, these equations are assumed to be evaluated on the surface $S = \mathcal{N}^- \cap \mathcal{N}^+$.

$$\dot{A}^1{}_a = B^1{}_{,a} + A^2{}_a B^3 - A^3{}_a B^2$$
$$+ N^j [A^1{}_{a,j} - A^1{}_{j,a} + A^2{}_j A^3{}_a - A^3{}_j A^2{}_a] \tag{14a}$$
$$+ 4\pi\kappa T^1{}_a - v^b [A^2{}_{a,b} - A^2{}_{b,a} + A^1{}_b A^2{}_a - A^2{}_b A^1{}_a]$$

$$v^a \dot{A}^2{}_a = v^a \{ B^2{}_{,a} - 2A^1{}_a B^2 + 2A^2{}_a B^1$$
$$+ N^j [A^2{}_{a,j} - A^2{}_{j,a} - 2A^1{}_j A^2{}_a + 2A^2{}_j A^1{}_a] \tag{14b}$$
$$- 4\pi\kappa\nu T^1{}_0 \}$$

In the above we have made use of the condition $v^i \mu_i = 0$. Also, since the calculation is local, there may be matter present and that is indicated by the matter tensor on the right hand sides.

At $\mathcal{I}^+$, Bondi and Sachs identify the equations for $\dot{A}^1{}_a$ with conservation of angular momentum with the help of the conformal Killing vectors of the sphere. It is still open how one can extend this definition locally. Furthermore, this definition has difficulties even at $\mathcal{I}^+$. As noted in section 2, the $A^1{}_a$ are related to the N-P spin coefficients α and β.

Here we will restrict our considerations to the energy equation, (9b), which we write as ($v^a = \nu v_2{}^a$)

$$\frac{\partial (v_2{}^a A^2{}_a)}{\partial t} = \dot{v}_2{}^a A^2{}_a + v^i \{ B^2{}_{,i} - 2A^1{}_i B^2 + 2A^2{}_i B^1$$
$$+ N^j A^2{}_{i,j} - A^2{}_{j,i} - 2A^1{}_j A^2{}_i + 2A^2{}_j A^1{}_i] \} \tag{15}$$
$$- 4\pi\kappa\nu T^1{}_0.$$

Now, $v_2{}^a A^2{}_a = \mu$, where μ is the spin coefficient for the convergence of the null rays on $\mathcal{N}^+$. At $\mathcal{I}^+$, it is the $1/r^2$ part of μ which is identified with the mass aspect. The Minkowski space contribution is of order $1/r$ and drops out of the calculation in (Goldberg, Robinson, and Soteriou 1992). Clearly, the difference in the convergence from its Minkowski space value is a measure of the mass enclosed. At $\mathcal{I}^+$, the energy is defined by the integral of the $1/r^2$ part of μ over a two-surface cut. For our choice of coordinates and tetrad vectors, the Bianchi identities tell us that when all the other equations have been satisfied, (14) need be satisfied only at one value of r. Therefore, we would like to define the quasi-local mass by

$$4\pi\kappa M = \mathrm{Re} \left\{ \oint_S [v_2{}^a A^2{}_a] r^2 \sin\theta \, d\theta \, d\phi \right\}, \tag{16}$$

where S is the intersection of $\mathcal{N}^- \cap \mathcal{N}^+$. Note that any matter present which flows into the surface contributes to the rate of change of the convergence. That is contained in (15).

While the above expression is finite, presumably, it still contains the convergence which would be present for such a sphere and incoming null surface in Minkowski space. To determine an appropriate subtraction, one can embed the surface S isometrically in Minkowski space (d'Inverno and Vickers 1995). Since p and q are connected by a geodesic, the embedding surface may be chosen to be a space-like plane with time-like normal equal to the tangent vector to the geodesic at p. So doing takes into account the distortion of the rays due to a Lorentz transformation. Then one calculates the convergence of the incoming rays normal to the embedded surface. Since the surface will not in general be a sphere, the rays will not form a cone. But, in the neighborhood of S, the convergence will be well defined. Therefore, the above expression for the quasi-local mass should be modified by subtracting the corresponding integral in Minkowski space.

4. CONCLUSION

What we have presented above is a suggestion of an approach to quasi-local mass which is explicitly derived from the field equations. We have used the formalism developed in (Goldberg and Soteriou 1995; Goldberg, Robinson, and Soteriou 1992), but the resulting idea is not tied to the formalism. The important point is to recognize the importance of the convergence of incoming rays in defining the energy or mass confined in a finite domain. This is precisely what is done by astronomers for estimating the mass of intervening matter by measuring - or estimating - the bending of light by distant galaxies. There are other suggestions in the literature, but their derivations are either ad hoc (Hayward 1993) or contain some choice which does not have a direct physical meaning (Dougan and Mason 1991). Recent work (Brown, Lau, and York) mentioned earlier, comes closest to the point of view expressed here.

It is a pleasure to dedicate this paper to John Stachel. We have had many opportunities to discuss energy and conservation laws. At one point I considered using his 2+2 formalism for the discussion in this paper. It is certainly similar, particularly as developed by Ray d'Inverno and James Vickers (1995).

ACKNOWLEDGEMENTS

I want to thank David Robinson and John Madore for discussions and to thank Lionel Mason for his critical remarks.

Syracuse University

REFERENCES

Ashtekar, A. 1991. *Lectures on Non-perturbative Canonical Gravity (notes prepared in collaboration with R. Tate)*. Singapore: World Scientific.
Bondi, H., M. Van Der Berg, and A. Metzner. 1962. *Proc. Roy. Soc.* A 269:21.
Brown, D., and J. York. 1993. *Phys. Rev.* D 47:1407.
Brown, J. D., S. R. Lau, and J. York. preprint gr-qc/9810003.
d'Inverno, R., and J. Vickers. 1995. *Class. Quantum Grav,* 12:753.
Dougan, A. J., and L. Mason. 1991. *Phys. Rev. Lett.* 67:2119.

Goldberg, J. N., D. C. Robinson, and C. Soteriou. 1992. *Class. Quantum Grav.* 9:1309.
Goldberg, J. N., and C. Soteriou. 1995. *Class. Quantum Grav.* 12:2779.
Hayward, G. 1993. *Phys. Rev.* D 47:3275.
Newman, E. T., and R. Penrose. 1962. *J. Math. Phys.* 3:566.
Penrose, R. 1982. *Proc. Roy. Soc.* A 381:53.
Penrose, R., and W. Rindler. 1986. *Spinors and Space-Time*. Cambridge: Cambridge University Press.
Sachs, R. 1962. *Proc. Roy. Soc.* A 270:103.

REINALDO J. GLEISER AND PATRICIO S. LETELIER

SPACE-TIME DEFECTS:
OPEN AND CLOSED SHELLS REVISITED

Abstract. Space-times whose Riemann-Christoffel curvature tensors are null, except on a timelike hypersurface are considered. These geometries with distributional curvature tensor can be interpreted as space-time defects with zero associated Newtonian mass. The method to generate axially symmetric defects is studied paying special attention to the global aspects of the spacetime that contains the shell. We find shells that connect the *interiors* of two spacetimes making a compact space locally isometric to Minkowski space. Shells connecting two exteriors are also analized. These can be interpreted as examples of wormholes in Minkowski space. Furthermore we study some new cases of shells with the shapes of cones and hyperboloids. We show that these shells can be built with crossed cosmic strings.

1. INTRODUCTION

The formation of structures predicted by theories of the early universe based in the spontaneous symmetry breaking of some unifying group has been the focus of some attention. In particular, cosmic strings are produced in the breaking of an $U(1)$ symmetry. They are good candidates to seed the formation of galaxies and also to model the observed anisotropy in the microwave background. Cosmic walls are associated to the breaking of a discrete symmetry, and may decay later forming cosmic strings (Vilenkin and Shellard 1994).

Geometrically, in the zero width limit, straight cosmic strings and plane symmetric domain walls are characterized by spacetimes with metrics that have null Riemann-Christoffel curvature tensor everywhere except on the lines that represent the strings and on the planes of the walls (*ibid.*). In other words, they have a generalized conic structure. In general, topological defects are not always characterized in this way, e.g. the spacetime outside a global $U(1)$ loop is not flat.

Motivated by these consideration in the first paper of this series (Letelier and Wang 1995), the generation of space-times whose curvature is nonzero only on a surface was studied in a rather systematic way. Solutions to the Einstein equations that represents shells of matter with spherical, plane, cylindrical, and disk like symmetry were considered in particular. The methodology used was that of the Lichnerowicz-Taub theory of distribution (Lichnerowicz 1971; Taub 1980), that is mathematically sound for distributions with support in hypersurfaces. In general, the theory of distributions in curved spacetimes presents some drawbacks (Geroch and Traschen 1987). The most commonly used description of cosmic walls is based in Israel's theory of thin shells of matter (Israel 1966, 1967). This theory takes as a departure point the study of the extrinsic geometry of the surfaces that describe the shells of matter via Gauss-Codazzi equations. In principle, both approaches give the same results and are complementary (Taub 1980). In fact, the equivalence of Darmois-Israel and distributional methods for thin shells in general relativity has been explicitly demonstrated (Mansouri and Khorrami 1996).

In a second article (Letelier 1995a), the same matter was dwelt in, presenting a collection of spacetime defects with different shapes: parabolic, oblate (prolate) spheroidal, and toroidal. In these two works the geometric interpretation of the solutions as well as some other global aspects were almost not touched.

In the present paper we study in some detail the method of generation of axially symmetric defects paying special attention to the global aspects of the spacetime that contains the shell. In particular, we find shells that connect the *interiors* of two spacetimes making a compact space locally isometric to Minkowski space. The restricted case of two-centered spheres has been previously analyzed, in the context of cosmological models (Lynden-Bell, Katz, and Redmount 1989), and also, in relation with Israels's junction conditions (Goldwirth and Katz 1995). Shells connecting two exteriors are also studied. These can be interpreted as examples of wormholes (Morris and Thorne 1988; Frolov and Novikov 1990; Morris, Thorne, and Yurtsver 1988; Hawking 1992) in Minkowski space. Furthermore we study some new cases of shells with the shapes of cones and hyperboloids, that can be considered as formed by crossed cosmic strings.

2. GENERAL METHOD

Since shells correspond to stress - energy - momentum tensors with support on a timelike hypersurface, we may naturally use methods from the theory of distributions, as applied in the description of distribution valued curvature tensors, to give a general prescription for the construction of axially symmetric shells in Minkowski spacetime. By axially symmetric we mean, as usual, that the spacetime contains a Killing vector $\partial/\partial\phi$, whose orbits are closed. Axially symmetric spacetimes are naturally adapted to the presence of "cosmic strings" i.e., spacetime defects along the symmetry axis. These "cosmic strings" in turn, are characterized by the existence of a "conic singularity" associated to a "deficit angle" structure in the metric. On this account, we shall assume that the shells (defects) separate two regions of spacetime, which we indicate respectively by $\mathcal{M}_1$ and $\mathcal{M}_2$, each of which is a portion of Minkowski spacetime (with, possibly, a deficit angle), that can be covered with (or, is contained in) a chart where the metric takes the form,

$$ds_i^2 = -dt_i^2 + dr_i^2 + A_i^2 r_i^2 d\phi_i^2 + dz_i^2 \ , \quad i = 1, 2 \tag{1}$$

where the index i refers respectively to $\mathcal{M}_1$ and $\mathcal{M}_2$. $A_i \neq 1$ corresponds to a non vanishing deficit angle.

To obtain a spacetime where the distribution method is applicable, we define the location of the shell by introducing a hypersurface Σ that is a common boundary for

$\mathcal{M}_1$ and $\mathcal{M}_2$. We also require continuity of the metric, in the sense that the metric induced on Σ from $\mathcal{M}_1$ should be the same as that induced on Σ from $\mathcal{M}_2$.

We consider first the definition of Σ as seen from $\mathcal{M}_1$. We shall restrict to static, axially symmetric shells. In this case Σ is described by a function $f(z_1)$, such that Σ is the set of points with

$$r_1 = f(z_1) \qquad (2)$$

for all ϕ_1 and t_1. This introduces a natural parametrization on Σ in terms of z_1, ϕ_1 and t_1.

Similarly, from the point of view of $\mathcal{M}_2$, the hypersurface Σ is described by a function $h(z_2)$, such that on Σ

$$r_2 = h(z_2). \qquad (3)$$

Again, from the point of view of $\mathcal{M}_2$, a natural parametrization on Σ is given in terms of z_2, ϕ_2 and t_2.

The corresponding induced metrics on Σ are

$$d\sigma_1 = -dt_1^2 + \left[1 + (f')^2\right] dz_1^2 + A_1^2 f^2 d\phi_1^2 \qquad (4)$$

and

$$d\sigma_2 = -dt_2^2 + \left[1 + (h')^2\right] dz_2^2 + A_2^2 h^2 d\phi_2^2 \qquad (5)$$

where $f' = df/dz_1$ and $h' = dh/dz_2$. The continuity condition for the induced metric implies now $t_1 = t_2 = t$, and we impose $\phi_1 = \phi_2 = \phi$. We remark now that if the construction is at all possible, there should exist a map $z_2 = \zeta(z_1)$, defined on Σ. Then, the equality of the induced metrics requires

$$1 + [f'(z_1)]^2 = [1 + [h'(\zeta)]^2] \left(\frac{d\zeta}{dz_1}\right)^2 \qquad (6)$$

$$[f(z_1)]^2 = A^2 [h(\zeta)]^2 \qquad (7)$$

where ζ stands for $\zeta = \zeta(z_1)$, and $A = A_2/A_1$. In what follows we shall assume that $A_1 = 1$. The case $A_1 \neq 1$ may always be recovered by a redefinition of ϕ.

If we assume that $f(z_1)$ and A are given, the functions $h(\zeta)$ and $\zeta(z_1)$ are obtained as the solutions of the system (6), (7). If they exist and are real the construction of the metric is complete.

There are several possibilities and types of solution, depending on the choice of A and the ranges of r_i, as we shall see in the following subsections.

2.1. The interior-interior and exterior-exterior cases

The system (6), (7), considered as a whole, admits a very simple set of solutions that lead to non trivial shells. If we set $A = 1$, we obtain a solution by setting $\zeta = z_1$ and $f(z_1) = h(\zeta(z_1))$, which implies also that $r_2 = r_1$ on Σ. This solution is *non-trivial* if we impose the condition that the range of r_2 on $\mathcal{M}_2$ is the same as that of r_1 on $\mathcal{M}_1$, as can be seen by carrying out the procedures for obtaining $T_{\mu\nu}$ that we describe in the next section. The spacetime consists of two replicas of a certain region of

Minkowski spacetime, joined through the "topological defect" Clearly, this works for arbitrary f, although certain restrictions hold if we also impose the condition that the surface energy density should be positive. The *interior–interior* case corresponds to those regions describing the interior of a shell, while in the *exterior–exterior* case, the shell provides the inner boundary of two identical open regions, a situation somewhat resemblant of that of a wormhole. Particular examples can be found in (*ibid.*).

We also remark that, in principle, this case does not require axial symmetry, and the matching of two identical replicas of Minkowski spacetime can be carried out through somewhat arbitrary surfaces, which can be restricted to some extent by the condition of positivity of the surface energy density.

2.2. The interior-exterior case

This is the case where $A \neq 1$. Assuming that f and $A \neq 1$ are given, we may use (7) to eliminate h in (6). We have,

$$\frac{d\zeta}{dz_1} = \sqrt{1 - \frac{1-A^2}{A^2}[f'(z_1)]^2}. \tag{8}$$

Assuming that $d\zeta/dz_1 \neq 0$, solving (8) we may also find z_1 as a function of ζ. Then, h is given by

$$h(\zeta) = \frac{1}{A} f(z_1(\zeta)). \tag{9}$$

It can be seen that for this construction to work, it is necessary that the argument of the square root in (8) be non negative. This implies the restriction on f

$$[f'(z_1)]^2 \leq \frac{A^2}{1-A^2}. \tag{10}$$

Therefore, for any given $A < 1$, we have an upper bound on $|f'|$. In particular, an immediate consequence of (10) is that the inside of a sphere cannot be matched to a flat exterior metric, for in this case, for a sphere of radius R, we have $f(z_1) = \sqrt{R^2 - z_1^2}$, and f' is not bounded for $|z_1|$ near R.

In the next sections we give some examples where the construction is possible, but we first discuss the construction of the energy-momentum tensor $T_{\mu\nu}$.

2.3. The distributional stress-energy-momentum tensor

The results of the previous section give the relation between the functions $f(z_1)$ and $h(\zeta)$ that characterize the surface where the spacetime defect is located. The function $\zeta(z_1)$ (and its inverse), provide the necessary connection between points on this surface as seen from each of its sides. Although this proves the existence of a continuous metric, in the application of distributional methods we still need an explicit form for the overall metric, in a chart where its components are continuous functions of the coordinates, in the neighbourhood of all points, even those on Σ. We may use the previous results on continuity to construct this new chart. We consider first $\mathcal{M}_1$ and define new coordinates n, w by the relations

$$r_1 = B_1(w)n + f(w) \quad , \quad z_1 = w + B_2(w)n \tag{11}$$

where $n \leq 0$, and the range of w is, in principle, unrestricted. The hypersurface Σ corresponds to $n = 0$, but it is easily seen that $\partial/\partial n$ is not, in general, orthogonal to Σ. A simple choice that makes $\partial/\partial n$ orthogonal to Σ is $B_1 = \epsilon_1$, with $\epsilon_1 = \pm 1$, and $B_2 = -\epsilon_1 f'$. $\epsilon = 1$ corresponds to the case where r_1 decreases as we move away from Σ, while if $\epsilon = -1$, r_1 increases as we recede from this hypersurface. The metric in $\mathcal{M}_1$ is then given by

$$\begin{aligned}ds_1^2 &= -dt^2 + (1 + f'^2)dn^2 + 2nf'f''dndw \\ &+ (f + \epsilon_1)^2 d\phi^2 + [f'^2 + (1 - \epsilon_1 nf'')^2]dw^2\end{aligned} \tag{12}$$

where f and its first and second derivatives are given as functions of w.

Similarly, we consider in $\mathcal{M}_2$ a coordinate transformation of the form

$$r_2 = B_3(w)n + h(\zeta(w)) \quad , \quad z_2 = \zeta(w) + B_4(w)n \tag{13}$$

where, $n \geq 0$, and, in accordance with Equations (6) and (7), the functions $h(\zeta(w))$, and $\zeta(w)$ satisfy the equations

$$1 + [f'(w)]^2 = [1 + [h'(\zeta(w))]^2]\left(\frac{d\zeta}{dw}\right)^2 \tag{14}$$

$$[f(w)]^2 = A^2[h(\zeta(w))]^2. \tag{15}$$

We may again check that the conditions of continuity of the metric, and orthogonality of $\partial/\partial n$ to Σ are satisfied if we choose

$$B_3 = \epsilon_2 \zeta' \quad , \quad B_4 = -\epsilon_2 \zeta' h' \tag{16}$$

where $\epsilon_2 = \pm 1$. Again, we remark that r_2 increases with increasing n for $\epsilon = 1$, while we have the opposite situation for $\epsilon = -1$.

With these choices, and taking into account (14) and (15), we finally obtain

$$\begin{aligned}ds_2^2 &= -dt^2 + (1 + f'^2)dn^2 + 2nf'f''dndw + (f + \epsilon_2 nA\zeta')^2 d\phi^2 \\ &+ \left[1 + f'^2 - \frac{2\epsilon_2 nf''}{A\zeta'} + \frac{n^2}{A^2}(A^2\zeta''^2 + f''^2)\right]dw^2\end{aligned} \tag{17}$$

where

$$\zeta' = \sqrt{1 - \frac{1 - A^2}{A^2}[f'(w)]^2} \tag{18}$$

and $\zeta'' = d\zeta'/dw$. It is clear now that if we consider a chart where n takes values in a neighbourhood of $n = 0$, and the metric is given by (12) for $n \leq 0$ and by (17) for $n \geq 0$, the overall metric is continuous, with discontinuous first derivatives. We may then apply the distribution theory for the curvature tensor and obtain the Einstein tensor. We find that the non zero components of the Einstein tensor are given by

$$G_{\phi\phi} = 2f^2 f'' \frac{\epsilon_2 - \epsilon_1 \sqrt{A^2 - (1-A^2)f'^2}}{\sqrt{A^2 - (1-A^2)f'^2}(1+f'^2)^2} \delta(n) \quad (19)$$

$$G_{ww} = 2\frac{\epsilon_1 - \epsilon_2 \sqrt{A^2 - (1-A^2)f'^2}}{f} \delta(n) \quad (20)$$

$$G_{tt} = -\frac{1}{f^2} G_{\phi\phi} - \frac{1}{1+f'^2} G_{ww}. \quad (21)$$

The physical interpretation of the corresponding components of the energy - momentum tensor is made simpler by replacing n by proper distance along $\partial/\partial n$, which we indicate by s, and by referring the components of $T_{\mu\nu}$ to an orthonormal basis. This amounts to replacing $\delta(n)$ by $\sqrt{1+f'^2}\delta(s)$, and taking out a factor f^2 in $G_{\phi\phi}$ and a factor $1+f'^2$ in G_{ww}. We also take $8\pi\kappa = 1$. The final result is

$$T_{\hat\phi\hat\phi} = 2f'' \frac{\epsilon_2 - \epsilon_1 \sqrt{A^2 - (1-A^2)f'^2}}{\sqrt{A^2 - (1-A^2)f'^2}(1+f'^2)^{3/2}} \delta(s) \quad (22)$$

$$T_{\hat w \hat w} = 2\frac{\epsilon_1 - \epsilon_2 \sqrt{A^2 - (1-A^2)f'^2}}{f\sqrt{1+f'^2}} \delta(s) \quad (23)$$

$$T_{\hat t \hat t} = -T_{\hat\phi\hat\phi} - T_{\hat w \hat w} \quad (24)$$

where the "hats" over ϕ and w indicate components with respect to an orthonormal basis on Σ. The last equation shows, as expected, that the "Newtonian mass" of the shell vanishes (Vilenkin and Shellard 1994).

We remark that in this construction, both the geometry of the shell, and the surface stress-energy-momentum are completely specified once the function f, and the constants A, and ϵ_i are given. We need to solve equations (6) and (7) only if we want to know what the hypersurface Σ looks like as seen from $\mathcal{M}_2$.

In the next section we give some examples where the construction is carried out.

3. SOME EXAMPLES

3.1. *The general interior-interior case*

As indicated this is the case where $\epsilon_1 = 1$ and $\epsilon_2 = -1$. If we further require that $\mathcal{M}_2$ is a regular interior, we must choose $A = 1$. We then have

$$T_{\hat\phi\hat\phi} = -\frac{4f''}{(1+f'^2)^{3/2}} \delta(s) \quad (25)$$

$$T_{\hat w \hat w} = \frac{4}{f\sqrt{1+f'^2}} \delta(s) \quad (26)$$

$$T_{\hat t \hat t} = -T_{\hat\phi\hat\phi} - T_{\hat w \hat w}. \quad (27)$$

Therefore, we have only surface tension and a positive energy density if $f'' \leq 0$. In particular this holds for a large class of smooth regular closed shells, including spheres (this case is considered below) and ellipsoids with axial symmetry.

We may also have pure tension and positive energy density if f'' is essentially equal to zero, except on a bounded region. An example would the case of two oppositely oriented cones connected through a smooth "throat."

There are, clearly many other possibilities, which we shall not discuss here.

3.2. Spheres

Spherical shells can only be constructed in the interior - interior or exterior - exterior cases. Here we have

$$r_1 = f(z_1) = \sqrt{r_0^2 - z_1^2} \tag{28}$$

where r_0 is the radius of the sphere. A regular interior implies $A = 1$ and it is easy to check that

$$T_{\hat{\phi}\hat{\phi}} = \frac{4}{r_0}\delta(s) \tag{29}$$

$$T_{\hat{w}\hat{w}} = \frac{4}{r_0}\delta(s) \tag{30}$$

$$T_{\hat{t}\hat{t}} = -\frac{8}{r_0}\delta(s). \tag{31}$$

Namely, we obtain, as espected, a constant positive energy density, and a constant isotropic surface tension, see also (Letelier and Wang 1995).

3.3. Cylinders

This is the general case where $r_1 = r_0$, with r_0 a constant. In this case $f = r_0$ and $f' = f'' = 0$. This implies that we may take $\zeta = z_1$, and $h(\zeta) = r_0/A$. We also have $T_{\hat{\phi}\hat{\phi}} = 0$.

This still leaves open the choice of ϵ_1 and ϵ_2. We consider first $\epsilon_1 = 1$. In this case $\mathcal{M}_1$ is the region inside a cylinder of radius r_0. We have

$$T_{\hat{w}\hat{w}} = 2\frac{1 - \epsilon_2 A}{f}\delta(s) \tag{32}$$

$$T_{\hat{t}\hat{t}} = -T_{\hat{w}\hat{w}}. \tag{33}$$

We have now two choices for ϵ_2. If $\epsilon_2 = -1$, the region $\mathcal{M}_2$ corresponds to (part of) the interior of a cylinder of radius r_0/A. The surface energy density is positive for all values of A, but if $A > 1$, we end up with a negative "deficit angle" on the symmetry axis. For $A < 1$, the region $\mathcal{M}_2$ corresponds to a "cosmic string" spacetime that extends from the symmetry axis up to radius r_0/A. Finally, if $A = 1$, what we obtain is the matching of the interiors of two identical cylinders of radius r_0, both with regular symmetry axes. The surface energy density is equal to $2(1 + A)/r_0$, and is also equal to the tension along the generatrixes.

If we choose $\epsilon_2 = 1$, the region $\mathcal{M}_2$ is that external to a cylinder of radius r_0. Positivity of the energy density requires $A < 1$. As result we have a deficit angle $\delta = (1-A)2\pi$ in the external region. Therefore, what we have obtained is a matching of the region inside of a cylinder of radius r_0 with part of the region exterior to a cylinder of radius r_0/A. (We recall that $\delta \neq 0$ implies that a "wedge" has been cut out from the external region). This result is, of course, the same as that obtained in (Letelier and Wang 1995; Tsoubelis 1989).

In the case $\epsilon_1 = -1$, the region $\mathcal{M}_1$ is that external to a cylinder of radius r_0. The only new case is that where $\epsilon_2 = 1$. This is the *external - external* case. It is easily seen that in this case the energy density is negative. We have a wormhole type of spacetime in this case, we shall be back to this point later.

We shall not consider further cases with $\epsilon_1 = -1$, so that in the remaining examples $\mathcal{M}_1$ will correspond to a regular region, interior to the shell.

3.4. Cones

We define a cone by

$$f(z_1) = a|z_1| \tag{34}$$

where a is a real constant. We may then choose $\epsilon_2 = 1$, and

$$\zeta = \frac{\sqrt{A^2 - (1-A^2)a^2}}{A} z_1 \tag{35}$$

provided that $A^2 \geq (1-A^2)a^2$, and, therefore,

$$h(\zeta) = \frac{a}{\sqrt{A^2 - (1-A^2)a^2}} |\zeta|. \tag{36}$$

If we choose the range of r_1 as $0 \leq r_1 \leq |az_1|$, and $r_2 \geq a|z_2|/\sqrt{A^2 - (1-A^2)a^2}$, we have the matching of the region inside of a cone of aperture angle $\alpha = \tan^{-1}(a)$, with part of the region external to a cone with aperture angle $\alpha' = \tan^{-1}(a/\sqrt{A^2 - (1-A^2)a^2})$. We notice that α' approaches $\pi/2$ as the limiting condition $A^2 = (1-A^2)a^2$ is approached.

From the general expressions for $T_{\mu\nu}$ we obtain $T_{\phi\phi} = 0$, and, therefore, $T_{\hat{t}\hat{t}} = -T_{\hat{w}\hat{w}}$, which can be interpreted as indicating that the conical shell is generated by a continuous distribution of cosmic strings, distributed along the generatrices of the cone. The surface energy density is given by

$$\sigma = 2\frac{1 - \sqrt{A^2 - (1-A^2)a^2}}{a|z_1|\sqrt{1+a^2}} \tag{37}$$

and is positive if $A < 1$. We notice that σ is singular for $z_1 = 0$, corresponding to the apex of the cone.

For $A = 1$ we get a trivial result if $\epsilon_2 = 1$, but for $\epsilon_2 = -1$, where $\mathcal{M}_2$ is a copy of $\mathcal{M}_1$, we find

$$\sigma = \frac{4}{a|z_1|\sqrt{1+a^2}} \tag{38}$$

corresponding to the *interior-interior* case of two identical cones.

3.5. Hyperboloids

Hyperboloids are generated by the rotation of a curve of the form

$$r = a\sqrt{z^2 + b^2}. \tag{39}$$

This defines the function f for the hyperboloids. We then have

$$\frac{d\zeta}{dz} = \sqrt{\frac{c^2 z^2 + b^2}{z^2 + b^2}}. \tag{40}$$

where

$$c^2 = 1 - \frac{(1 - A^2)a^2}{A^2} \tag{41}$$

and we require $c^2 \geq 0$ in order to have $d\zeta/dz$ well defined for all z. The stresses (tensions), and the (positive) energy density are then given by

$$T_{\hat{\phi}\hat{\phi}} = \frac{2ab^2}{A\sqrt{c^2 z^2 + b^2}} \frac{(\sqrt{z^2 + b^2} - A\sqrt{c^2 z^2 + b^2})}{[(1 + a^2)z^2 + b^2]^{3/2}} \delta(s) \tag{42}$$

$$T_{\hat{w}\hat{w}} = \frac{2(\sqrt{z^2 + b^2} - A\sqrt{c^2 z^2 + b^2})}{a\sqrt{z^2 + b^2}\sqrt{(1 + a^2)z^2 + b^2}} \delta(s) \tag{43}$$

$$T_{\hat{t}\hat{t}} = -T_{\hat{\phi}\hat{\phi}} - T_{\hat{w}\hat{w}}. \tag{44}$$

Although, in general, we cannot write $\zeta(z)$, (or its inverse) in closed form, it is straightforward to obtain a power series expansion for $z(\zeta)$, around $\zeta = 0$ assuming that $z(0) = 0$. The first non vanishing terms are

$$z = \zeta + \frac{1 - c^2}{6b^2}\zeta^3 + \frac{(c^2 - 1)(13c^2 - 1)}{120b^4}\zeta^5 + \dots \tag{45}$$

Replacing this expansion in f, we obtain an expansion for h,

$$h(\zeta) = \frac{ab}{A} + \frac{a}{2Ab}\zeta^2 - \frac{(4c^2 - 1)a}{24Ab^3}\zeta^4 + \frac{(88c^4 - 44c^2 + 1)a}{760Ab^5}\zeta^6 + \dots \tag{46}$$

which should be compared with

$$f(z) = a\sqrt{z^2 + b^2} = ab + \frac{a}{2b}z^2 - \frac{a}{8b^3}z^4 + \frac{a}{16b^5}z^6 + \dots \tag{47}$$

The external shape of the shell is not, therefore, that of (a part of) a hyperboloid. It reduces to a hyperboloid only in the trivial limit $A = 1$ (where we also have $c^2 = 1$).

A particular case where we obtain closed expressions, is the limiting case $c^2 = 0$. Here we have

$$\zeta = b \ln(z/b + \sqrt{z^2/b^2 + 1}) \tag{48}$$

and

$$z = b \sinh(\zeta/b). \tag{49}$$

Replacing in h we find

$$h(\zeta) = \frac{b}{\sqrt{1-A^2}} \cosh\left(\frac{\zeta}{b}\right) \qquad (50)$$

which clearly shows the different shapes of the surfaces $r_1 = f(z_1)$ and $r_2 = h(z_2)$.

These results have an interesting interpretation. The hyperboloids belong in the family of ruled surfaces, i.e., surfaces that can be generated by the motion of straight lines. In particular, for hyperboloids, there are *two* families of straight lines, tangent to the hyperboloid in all their points, that cross each other in "wire wastepaper basket" fashion. We may therefore think of the hyperboloids in these examples as being generated by a continuous distribution of straight cosmic strings, that cross each other in such a way that the resulting tension has components in both the $\partial/\partial\phi$ and $\partial/\partial w$ directions. It is well known, however, that when we have two cosmic strings that cross each other at an angle, if we adapt our coordinate system to one of them, so that it looks straight, the other appears to be "bent" on account of the "cone" singularity nature of the spacetime associated to a cosmic string. In our case, this "bending" is continuous, and gives as a result that if from the inside of the shell the strings look straight, they appear to be "bent" from the outside, which corresponds to the fact the curvature of the external surface is larger than that of the internal one.

4. OPEN SHELLS

So far we have restricted our treatment to the cases where the condition

$$[f'(z_1)]^2 \leq \frac{A^2}{1-A^2} \qquad (51)$$

was satisfied. In this Section we analyze what happens if this condition is violated. Consider first the case when $[f'(z_1)]^2 \to A^2/(1-A^2)$ as $z_1 \to \infty$. In this case we may write

$$[f'(z_1)]^2 = \frac{A^2}{1-A^2}(1 - g(z_1)) \qquad (52)$$

where $g(z_1) > 0$, and $g(z_1) \to 0$ for large z_1. We then have

$$\zeta(z_1) \simeq \int^{z_1} \sqrt{g(z_1')} dz_1'. \qquad (53)$$

Here we have two possibilities, depending on the behaviour of $g(z_1)$ for large z_1. If $g(z_1)$ is such that the integral in (53) diverges, the external side of the shell extends to all values of z_2, and, since for large z_1 we have

$$f(z_1) \simeq \sqrt{\frac{A^2}{1-A^2}} z_1, \qquad (54)$$

from the relation

$$h(\zeta) = \frac{1}{A} f(z_1(\zeta)) \qquad (55)$$

we find that, for large z_1, r_2 is also divergent, and the shell approaches a conical shape on both sides.

On the other hand, if $g(z_1)$ goes to zero sufficiently fast, the integral in (53) is convergent, and ζ, (and therefore z_2), approaches a finite value $\zeta_\infty = z_2 - \infty$, as $z_1 \to \infty$. We notice, however, that, since (54) holds also in this case, r_2 is unbounded as z_2 approaches $z_2 - \infty$. Thus, the shell approaches a conical shape from inside, but an infinite disk-like shape from outside.

Similar considerations hold in the limit $z_1 \to -\infty$. Thus, this cases are essentially similar to that where $[f'(z_1)]^2 < A^2/(1-A^2)$ for all z_1, in the sense that the resulting spacetime is covered by the union of two charts, $\mathcal{M}_1$ and $\mathcal{M}_2$, and the hypersurface Σ.

A different situation results if we have $[f'(z_1)]^2 = (A^2/(1-A^2)$ for some finite value of z_1. Suppose this equality holds for $z_1 = z_1^0$. Then, if we assume that $f(z_1)$ is differentiable for $z_1 = z_1^0$, we have

$$f(z_1) = f(z_1^0) + \sqrt{\frac{A^2}{1-A^2}}(z_1 - z_1^0) + B^2(z_1 - z_1^0)^2 + \mathcal{O}[(z_1 - z_1^0)^3] \qquad (56)$$

where B is some constant. Then, a simple computation shows that

$$\zeta(z_1) \simeq C_1^2(z_1^0 - z_1)^{3/2} + C_2 \qquad (57)$$

where C_1 and C_2 are some constants, for $z_1 < z_1^0$, near $z_1 = z_1^0$.

This result implies that the hypersurface Σ is bounded both in $\mathcal{M}_1$ and in $\mathcal{M}_2$. Therefore, we may naturally extend, say $\mathcal{M}_1$, so that we go around the border of Σ, and see the "other side of the shell" but staying all the while in $\mathcal{M}_1$, and not going *through* Σ. If, instead, we do move through the shell, but starting on its inner side on $\mathcal{M}_1$, we end up in $\mathcal{M}_2$. Similarly, we may start in $\mathcal{M}_2$, and either go through the shell, or around it, and end up in different subspaces. We may reconcile all these observations by adding a third chart $\mathcal{M}_3$, that is reached when we cross the shell from the "outside" in $\mathcal{M}_1$. The shape of the shell in $\mathcal{M}_3$ is the same as that in $\mathcal{M}_2$, so that the matching conditions are satisfied, but it is easy to check that surface energy density is now negative. Crossing the shell from the "outside" in $\mathcal{M}_1$, we reach the "inside" in $\mathcal{M}_3$. Now, if we cross the shell from the "outside" in $\mathcal{M}_3$, we reach a copy of $\mathcal{M}_1$, which we indicate with $\mathcal{M}_4$, in such a way that the "outside" of the shell in $\mathcal{M}_3$ corresponds to the "inside" of the shell in $\mathcal{M}_4$. This corresponds to a shell with positive energy density. We may finally complete the picture by imposing that when we cross the shell from the "outside" in $\mathcal{M}_4$, we reach the "inside" of the shell in $\mathcal{M}_2$. This time the shell appears with negative energy density.

In this process, we have actually added not only the two charts $\mathcal{M}_3$ and $\mathcal{M}_4$, but also three more matching hypersurfaces. Namely, besides Σ, common to $\mathcal{M}_1$ and $\mathcal{M}_2$, we have Σ_{13}, common to $\mathcal{M}_1$ and $\mathcal{M}_3$, Σ_{34}, common to $\mathcal{M}_3$ and $\mathcal{M}_4$, and Σ_{42}, common to $\mathcal{M}_4$ and $\mathcal{M}_2$. The resulting spacetime has, therefore, a rather non trivial topology, besides containing shells with negative surface energy density.

5. FINAL COMMENTS

In the exterior-exterior case limited by a cylindrical surface, we have that this surface acts as the throat of a wormhole, i.e., its connects to asymptotically flat spacetimes. The fact that the energy density is negative in this case is a well known fact of static wormholes (Morris and Thorne 1988). We have that nonstatic wormholes do not always need to violate the energy conditions, see for instance (Wang and Letelier 1995). One can also do, in a similar way, constructions with toroidal and spheroidal shells like in (Letelier 1995a).

The conical shells as well as the hyperbolic shell (ruled surface) that can be considered as formed by usual straight cosmics strings, i.e., in other words as continuous limits of the metric that represents several crossed cosmic strings (Letelier and Gal'tsov 1993). Of course the cylindrical shell considered here can be also thought as the continuous limit of the metric that represents several parallel strings that is a special case of the one above mentioned.

In this work we have constructed some new static shells and mainly made a geometric interpretation of a variety of closed and open shells. The case of nonstatic shells will be treated in a separate paper.

In this series of papers the defects we dealt with are curvature defects, i.e., spacetimes whose curvature tensor is a Dirac type of distribution with support on a hypersurface. One can likewise consider other type of defects in which the torsion is not null as in a usual Riemannian manifold, but it is also a distribution. This is the analog of the screw dislocation defect in solids, see for instance (Letelier 1995b). Also defects in pure torsion geometries, Weizemböck spaces, has been considered (Letelier 1995c). Global properties of defect space-times can also be studied using loop variables or holonomy transformations (Bezerra and Letelier 1996).

ACKNOWLEDGEMENTS

This work was supported in part by research grants from CONICET, CONICOR, SECYT - UNC, FAPESP and CNPq. R. J. G. is grateful for the hospitality extended to him while visiting the Departamento de Matemática Aplicada - IMECC, where part of this work was done. R. J. G. is a member of CONICET.

Universidad Nacional de Córdoba
Universidade Estadual de Campaninas

REFERENCES

Bezerra, V. M., and P. S. Letelier. 1996. *J. Math. Phys.* 12:6271.
Frolov, V. P., and I. D. Novikov. 1990. *Phys. Rev. Lett.* 42:1057.
Geroch, R., and J. Traschen. 1987. *Phys. Rev.* 35:1017.
Goldwirth, D. S., and J. Katz. 1995. *Class. Quantum Grav.* 12:769.
Hawking, S. 1992. *Phys. Rev. D* 46:603.
Israel, W. 1966. *Nuovo Cimento* 44B:1.
———. 1967. *Nuovo Cimento* 48B:463.

Letelier, P. S. 1995a. *J. Math. Phys.* 36:3043.

———. 1995b. *Class. Quantum Grav.* 12:471.

———. 1995c. *Class. Quantum Grav.* 12:2221.

Letelier, P. S., and D. V. Gal'tsov. 1993. *Class. Quantum Grav.* 10:L101.

Letelier, P. S., and A. Wang. 1995. *J. Math. Phys.* 36:3023.

Lichnerowicz, A. 1971. *C. R. Acad. Sci.* 273:528. [1973. Pg. 93 in *Symposia Matematica*, vol. XII. Istituto Nazionale di Alta Matematica, Bologna.].

Lynden-Bell, D., J. Katz, and I. H. Redmount. 1989. *Mon. Not. R. Astron. Soc.* 239:201.

Mansouri, R., and M. Khorrami. 1996. *J. Math. Phys.* 37:5672.

Morris, M. S., and K. S. Thorne. 1988. *Am. J. Phys.* 56:395.

Morris, M. S., K. S. Thorne, and U. Yurtsver. 1988. *Phys. Rev. Lett.* 61:1446.

Taub, A. H. 1980. *J. Math. Phys.* 21:1423.

Tsoubelis, D. 1989. *Class. Quantum Grav.* 6:101.

Vilenkin, A., and E. P. S. Shellard. 1994. *Cosmic strings and other topological defects*. Cambridge.

Wang, A., and P. S. Letelier. 1995. *Prog. Theor. Phys.* 94:L137.

S. DESER

DIMENSIONALLY CHALLENGED GRAVITIES*

Abstract. I review some ways through which spacetime dimensionality enters explicitly in gravitation. In particular, I recall the unusual geometrical gravity models that are constructible in dimensions different from four, especially in D=3 where even ordinary Einstein theory is "different," e.g., fully Machian.

Once unleashed by general relativity, dynamical geometry has become a fertile playground for generalization in many directions beyond Einstein's D=4 Ricci-flat choice. This trend has intensified with string theory, where D=10 is normal as are (higher curvature power) corrections to the Einstein action. There are many other reasons to study different dimensions; here is one: As I became aware, thanks to John, Einstein already foresaw (1914, 1079; 1957) the potential danger of letting geometry be at the mercy of field equations, in particular worrying about spaces with closed timelike curves, but also optimistically hoping that they would be forbidden in "physically acceptable" matter contexts (this is *not* a tautology since acceptable means having decent stress tensor). Although the best-known examples, such as Gödel's universes (Gödel 1949; van Stockum 1937), fall in this class, it is in the simpler setting of D=3 ("planar") gravity that they have recently been studied on an industrial scale,[1] and have yielded Einstein's hoped-for taboo in a clear way. More generally, one can learn about D=4 Einstein's virtues from studying different D's, and the different sorts of models they support. What is more, we are very likely to be embedded in a world, which, if it has any classical geometry at all, is likely to have as many as eleven dimensions! In this short excursion, I can only point out some recent examples of theories that I have been involved with directly; equations and further references will have to be found in the citations.

Let us begin with some remarks about ordinary Einstein theory in the smaller worlds of D<4. One does not normally think of the curvature components as being dimension-dependent, but we all know that in D=3, Einstein and Riemann tensors have the same number of components and indeed are equivalent, since $G^\mu_\nu = -\frac{1}{4}\epsilon^{\mu\alpha\beta}\epsilon_{\nu\lambda\sigma}R_{\alpha\beta}{}^{\lambda\sigma}$. Strangely, it was a long time before the import of this, that outside sources, spacetime is flat was appreciated! Philosophically, D=3 Einstein theory presents the Machian dream in its purest form: there are no gravitational excitations, so geometry is entirely – and locally – determined by matter. There is a field-current identity: Riemann (being Einstein) equals stress tensor. So the picture that emerges is that this planar world consists of patches of Minkowski space glued together at the sources (most simply discrete point particles, representing parallel strings in a D=4 Einstein world). The 1-particle conical space solution is amusing enough (Staruszkiewicz 1963) but things really get to be fun for two or more stationary or, better still, moving ones (Deser, Jackiw, and 't Hooft 1984). If a cosmological constant is present, it's even more fun as the patches are constant curvature spacetimes (Deser and Jackiw 1984). In that case (for negative cosmological constant) it is also possible to have black holes by suitable identifications of points (Banados et al. 1993). Time-helical structures, requiring identification of times in a periodic way (as well as the space gluings) arise for stationary, rotating, solutions and lead to the whole gamut of possible closed timelike curves and, as mentioned, a clear arena to examine whether they can be physically generated. But D=3 can be more amusing still, for it permits (as does any odd-dimensional space) the construction of different invariants, the Chern–Simons (CS) terms. These are the gravitational analogs of the simple electrodynamics (or Yang–Mills) $\int A \wedge F$ structures that in turn arise from the next higher dimensional topological invariants, such as $F_{\mu\nu} {}^\star F^{\mu\nu} \equiv \partial_\mu(\epsilon^{\mu\nu\alpha\beta} A_\nu F_{\alpha\beta})$ in the abelian context. Here we have the Pontryagin invariant $R^\star R$ instead. Varying these gravitational CS terms with respect to the metric leads to a tensor, because the integral (if not the CS integrand) is gauge invariant. In D=3, this is the famous Cotton tensor, $C^{\mu\nu} \equiv \epsilon^{\mu\alpha\beta} D_\alpha(R^\nu_\beta - \frac{1}{4}\delta^\nu_\beta R)$, discovered long before general relativity (Cotton 1898); $C^{\mu\nu}$ is the conformal tensor in D=3, replacing the (identically vanishing) Weyl curvature. It is a symmetric, traceless, identically conserved quantity, although it superficially seems to be none of these. Its interest lies not so much for generating a theory of gravity in its own right (it could at best only couple to traceless sources) but as an added term to the Einstein one. Being of third derivative order, it has a coefficient with relative inverse length or mass dimension (in Planck units) to that of the Einstein action. This mass is in fact that of small excitations (of helicity ±2) of the metric about flat space: adding CS has restored a degree of freedom absent in either R or CS alone. This combined theory (Deser, Jackiw, and Templeton 1982a, 1982b), called topologically massive gravity (TMG) for obvious reasons, has many other wondrous properties and unsolved aspects. First, despite being a higher derivative theory, it has no unitarity or ghost problems; it may even be finite as a quantum theory, although that is still an open mathematical problem (Deser and Yang 1990). If so, it might really have some lessons for us, for it would be unique in this respect amongst truly dynamical gravity models without ghosts (unlike four-derivative theories) but with a dimensional coupling constant; pure Einstein D=3 theory is renormalizable (Witten 1988; Deser, McCarthy, and Yang 1989) but that doesn't count — it is non-dynamical. Second, TMG, at least in its linearized guise (Deser 1990) acts to turn its sources into anyons; that is, a particle can acquire any desired spin simply by coupling to TMG. But, thirdly, no-one has succeeded as yet in finding the simplest possible, "Schwarzschild" solution to the nonlinear model, i.e., a circularly symmetric time-independent (we don't even know if there's a Birkhoff theorem) exterior geometry that obeys the $G_{\mu\nu} + m^{-1} C_{\mu\nu} = 0$ equations. Although CS-like terms can be constructed for higher odd-D spaces, they have not been studied much because they have no linearized, kinematical, effects beyond D=3 because they are of higher powers in an expansion about flat space. There are both strong similarities and differences between TMG and its spin 1 counterpart, topologically massive Yang–Mills theory. The most striking difference is that in the quantum theory, the coefficient of the CS term in "TM–YM" must necessarily be quantized (Deser, Jackiw, and Templeton 1982a; Deser, Jackiw, and Templeton 1982b), but not that of TMG (Percacci 1987).

But the twists in D=3 gravity do not stop there: there is yet another "CS-ness" present. Once it is noted that, in Einstein gravity, spacetime is flat outside sources, one realizes that this is just the same as what happens to abelian or nonabelian vector fields in their pure CS models: the field equations are just $^\star F^\mu \equiv \frac{1}{2} \epsilon^{\mu\nu\alpha} F_{\nu\alpha} = 0$, so that the field "curvature" also vanishes here [in both cases, the full "curvatures" are proportional to currents, $^\star F^\mu = J^\mu$]. Indeed, there is an equivalence (except for some interesting find print) between the two models and one can formally recast the Einstein action and equations into non-abelian vector field CS form in terms of the dreibein and spin connections. So this is yet another vast subject straddling two ostensibly different types of theory.[2]

We will not descend much to D=2, another well-studied subject (Jackiw and Teitelboim 1984), because there is no Einstein gravity there at all: only the Ricci scalar is non-vanishing, being the "double-dual" of Riemann, while the Einstein tensor vanishes identically. As usual, D=2 is different from all other dimensions in this respect (it is also here that Maxwell theory ceases to have excitations); some sort of additional scalar field is required to assure the Hilbert action from just being a dull Euler topological invariant, and this departs from the realm of pure geometry.

What about dimensions beyond D=4? This becomes a generic area where the differences from D=4 are more quantitative than qualitative. Still, there are some amusing points to be noted. For example, consider the Gauss–Bonnet invariant $R^\star R^\star$, defined in D=4. There, it is a total divergence and hence irrelevant to field equations. However, in higher dimensions, it can still be defined by writing it out in terms of metrics; for example we all know it is proportional to the combination $(R^2_{\mu\nu\alpha\beta} - 4R^2_{\mu\nu} + R^2)$. As a Lagrangian, it is seemingly dangerous to unitarity of excitations because of its fourth derivative order. In fact, there is no danger, because (say about flat space) this combination is a total divergence in its leading quadratic order in $h_{\mu\nu} \equiv (g_{\mu\nu} - \eta_{\mu\nu})$ in *any* D. Thus, this is a "safe" class of alternative actions, say when added to Einstein's. Some of their solutions, e.g., Schwarzschild-like ones, have been studied to see whether they are better or still unique. For example, there can be cosmological solutions without an explicit cosmological constant (Boulware and Deser 1985; Deser and Yang 1989), which is not necessarily a good thing, physically.

In quantum field theory, a powerful tool has been the "large N limit" of Yang–Mills theory in which the number of flavors (internal degrees of freedom: A^a_μ, $a = 1\ldots N$) is sent to infinity. The equivalent in gravity, rather naturally, is the dimensionality D of spacetime, over which the "internal" index a of the vielbein e^a_μ ranges. As far as I know, the literature here consists of but one brave paper (Strominger 1981), which however did not get far. This seems to me a worthy subject of study also by classical relativists, who have instead mostly considered what is in some ways the opposite, ultra-local, limit (Isham 1976; Teitelboim 1980). There must be some simplifying aspects as the number of degrees of freedom $\frac{1}{2}D(D-3)$, rises quadratically, and the Newtonian potential that enters in the Schwarzschild metric behaves as $r^{-(D-3)}$.

There are also auxiliary quantities that are interestingly dimension-dependent; we encountered some of them in current work on D=11 supergravity (Deser and Seminara 1999, 2000). I will not transgress further into the superworld here, except to say that it is absolutely amazing that a) Einstein gravity always has a "Dirac square root" for all D≤11, i.e., can always be consistently supersymmetrized without the need for higher spins or more than one graviton, and b) that this possibility stops (Nahm 1978) at D=11 and c) that cosmological terms are allowed for all D<11, but forbidden (Bautier et al. 1997) at D=11. I certainly do not know of any "Riemannian" differences between D=11 and D=12 for example! In our case, we needed to find a basis of monomial local invariants made up from Riemann or Weyl tensors at a given quartic order. Their number had a natural "generic" bound (Fulling et al. 1992) at D=8, whereas D=4 is always the most degenerate dimension (I drop D<4 here). This sort of property is also of importance when studying gravitational trace anomalies (Deser and Schwimmer 1993; Deser 1996, 2000) in which identities arising from antisymmetrizing an expression over more indices than there are dimensions essentially generate all the strange looking identities between tensors, such as $C^{\mu\lambda\alpha\beta}C^{\nu}_{\lambda\alpha\beta} = \frac{1}{4}g^{\mu\nu}C^2$ in D=4, from antisymmetrizing expressions such as $C^{\mu\nu}_{[\mu\nu}C^{\alpha\beta}_{\alpha\beta}X^{\lambda}_{\lambda]}$, where X is arbitrary and brackets indicate antisymmetizations over 5 indices. In the anomaly context we are actually interested, using so-called dimensional regularization, in spaces of dimension differing from an integer by an infinitesimal parameter, still another unlikely departure from D=4, but one that has its own unlikely set of geometrical rules.

In conclusion, I have tried to indicate that the list of interesting, useful and even important consequences to be drawn from excursions away from our favorite, Einstein, action in its D=4 world is substantial and by no means complete.

Brandeis University

NOTES

* It is a pleasure to dedicate this little travelogue/catalogue of exotic gravity models to John Stachel, whose loyalty to the D=4 Einstein cause is too steadfast to be subverted by reading it.
  This work was supported by NSF grants PHY93-15811 and PHY99-73935. I am grateful to my collaborators in our explorations of the areas discussed here.

1. (Gott 1991; Deser, Jackiw, and 't Hooft 1992; Carroll, Farhi, and Guth 1992; 't Hooft 1992; Deser and Jackiw 1992).
2. For a review see e.g. (Achucarro and Townsend 1986).

REFERENCES

Achucarro, A., and P. Townsend. 1986. *Phys. Lett.* B180:89.
Banados, M., M. Henneaux, C. Teitelboim, and J. Zanelli. 1993. *Phys. Rev.* D48:1506.
Bautier, K., S. Deser, M. Henneaux, and D. Seminara. 1997. *Phys. Lett.* B406:49.
Boulware, D., and S. Deser. 1985. *Phys. Rev. Lett.* 55:2656.
Carroll, S., E. Farhi, and A. Guth. 1992. *Phys. Rev. Lett.* 68:263, (E) 3368.
Cotton, E. 1898. *C.R. Acad. Sci. Paris* 127:349.

Deser, S. 1990. *Phys. Rev. Lett.* 64:611.
———. 1996. *Helv. Phys. Acta* 69:570.
———. 2000. *Phys. Lett.* B479:315–320.
Deser, S., and R. Jackiw. 1984. *Ann. Phys.* 153:405.
———. 1992. *Comments Nucl. Part. Phys.* 20:337.
Deser, S., R. Jackiw, and G. 't Hooft. 1984. *Ann. Phys.* 152:220.
———. 1992. *Phys. Rev. Lett.* 68:267.
Deser, S., R. Jackiw, and Templeton. 1982a. *Phys. Rev. Lett.* 48:975.
———. 1982b. *Ann. Phys.* 140:372.
Deser, S., J. McCarthy, and Z. Yang. 1989. *Phys. Lett.* B222:61.
Deser, S., and A. Schwimmer. 1993. *Phys. Lett.* B309:279.
Deser, S., and D. Seminara. 1999. *Phys. Rev. Lett.* 82:2435.
———. 2000. *Phys. Rev.* D62:084010.
Deser, S., and Z. Yang. 1989. *Class. Quant. Grav.* 6:L83.
———. 1990. *Class. Quant. Grav.* 7:1603.
Einstein, A. 1914. *Berliner Berichte*, pp. 1030–1085.
———. 1957. "Einstein's Reply to Criticism." Pp. 663–688 in *Albert Einstein: Philosopher–Scientist*, ed. P. Schilpp. New York: Tudor.
Fulling, S.A., R.C. King, B.G. Wybourne, and C.J. Cummins. 1992. *Class. Quant. Grav.* 9:1151.
Gödel, K. 1949. *Rev. Mod. Phys.* 21:447.
Gott, J.R. 1991. *Phys. Rev. Lett.* 66:1126.
Isham, C. J. 1976. *Proc. Roy. Soc.* A351:209.
Jackiw, R., and C. Teitelboim. 1984. In *Quantum Theory of Gravity*, ed. S. Christensen. Bristol: Adam Hilger.
Nahm, W. 1978. *Nucl. Phys.* B135:145.
Percacci, R. 1987. *Ann. Phys.* 177:27.
Staruszkiewicz, A. 1963. *Acta Phys. Pol.* 24:735.
Strominger, A. 1981. *Phys. Rev.* D24:3082.
't Hooft, G. 1992. *Class. Quant. Grav.* 9:1335.
Teitelboim, C. 1980. In *General Relativity and Gravitation*, ed. A. Held. New York: Plenum.
van Stockum, W. J. 1937. *Proc. R. Soc. Edin.* 57:13.
Witten, E. 1988. *Nucl. Phys.* B311:46.

WLODZIMIERZ M. TULCZYJEW

A NOTE ON HOLONOMIC CONSTRAINTS*

1. INTRODUCTION

This note is a preliminary account of research undertaken jointly with G. Marmo of Napoli and P. Urbanski of Warsaw.

We propose a new description of dynamics of autonomous mechanical systems which includes the momentum-velocity relation. This description is formulated as a variational principle of virtual action more complete than the Hamilton Principle. The inclusion of constraints in this description is the main topic of the present note. We give examples and models of constraints in variational formulations of statics and dynamics of autonomous systems.

A complete description of the dynamics of a mechanical system involves both external forces and momenta. In a fixed time interval the dynamics is a relation between the motion of the system in configuration space, external forces applied to the system during the time interval, and the initial and final momenta. This relation is derived from a variational principle which involves variations of the end points. Constrained systems are idealized representations of unconstrained systems. Such idealizations are appropriate when forces at our disposal are unable to move the configuration of the system away from a subset of the configuration space by a perceptible distance. This description fits at least the case of holonomic constraints. We believe that constraints should be imposed on virtual displacements. Holonomic constraints are usually interpreted as restrictions on configurations of a mechanical system. Nonholonomic constraints are additional restrictions imposed on velocities. This traditional terminology is not adapted to our concept of constraints as imposed on virtual displacements and only indirectly affecting configurations and velocities. Our concept of nonholonomic constraints makes perfect sense for static systems even if velocities do not appear in the description of such systems. We will use the terms *configuration constraints* and *velocity constraints* instead of *holonomic* and *nonholonomic* constraints.

2. GEOMETRIC STRUCTURES

Let Q be the Euclidean affine space of Newtonian mechanics. The model space for Q is a vector space V of dimension 3. The Euclidean structure is represented by a metric tensor $g: V \to V^*$. The space V^* is the dual of the model space. The canonical pairing is a bilinear mapping

$$\langle\,,\,\rangle\colon V^* \times V \to \mathbb{R}. \tag{1}$$

We denote by $q_1 - q_0$ the vector associated with the points q_0 and q_1. We write $q_1 = q_0 + v$ if $v = q_1 - q_0$. The norm $\|v\|$ of a vector $v \in V$ is defined by

$$\|v\| = \sqrt{\langle g(v), v\rangle}. \tag{2}$$

The derivative of a function $F\colon Q \to \mathbb{R}$ is the mapping

$$\mathrm{D}F\colon Q \times V \to \mathbb{R}\colon (q, v) \mapsto \frac{d}{ds}F(q+sv)|_{s=0}. \tag{3}$$

The first and second derivatives of a differentiable curve $\gamma\colon \mathbb{R} \to Q$ are mappings $\dot\gamma\colon \mathbb{R} \to V$ and $\ddot\gamma\colon \mathbb{R} \to V$.

The tangent bundle $\mathsf{T}Q$ is identified with $Q \times V$, the cotangent bundle T^*Q is identified with $Q \times V^*$, the second tangent bundle $\mathsf{T}^2 Q$ is identified with $Q \times V \times V$, the iterated tangent bundle $\mathsf{TT}Q$ is identified with $Q \times V \times V \times V$, and the tangent of the cotangent bundle TT^*Q is identified with $Q \times V^* \times V \times V^*$. We have the projections

$$\tau_Q\colon \mathsf{T}Q \to Q\colon (q, \dot q) \mapsto q, \tag{4}$$

$$\tau_{\mathsf{T}Q}\colon \mathsf{TT}Q \to \mathsf{T}Q\colon (q, \dot q, \delta q, \delta\dot q) \mapsto (q, \dot q), \tag{5}$$

$$\mathsf{T}\tau_Q\colon \mathsf{TT}Q \to \mathsf{T}Q\colon (q, \dot q, \delta q, \delta\dot q) \mapsto (q, \delta q), \tag{6}$$

and the canonical involution

$$\kappa_Q\colon \mathsf{TT}Q \to \mathsf{TT}Q\colon (q, \dot q, \delta q, \delta\dot q) \mapsto (q, \delta q, \dot q, \delta\dot q). \tag{7}$$

For each subset C of Q we have the *tangent set*

$$\begin{aligned}\mathsf{T}C = \{(q, \delta q) &\in \mathsf{T}Q;\ \text{there is a curve }\gamma\colon \mathbb{R} \to Q\\ &\text{such that }\gamma(0) = q, \mathrm{D}\gamma(0) = \delta q,\\ &\text{and }\gamma(s) \in C\ \text{if}\ s \geq 0\}\end{aligned} \tag{8}$$

The space

$$\overset{\circ}{\mathsf{T}}Q = Q \times \overset{\circ}{V} = \{(q, v) \in \mathsf{T}Q;\ v \neq 0\} \tag{9}$$

is the tangent bundle with the zero section removed.

3. STATICS OF A MATERIAL POINT

We consider the statics of a material point in the Euclidean affine space Q of Newtonian physics.

An element $(q, \delta q)$ of $\mathsf{T}Q$ is a *virtual displacement* and an element (q, f) of T^*Q represents an *external force*. The evaluation

$$\langle (q, f), (q, \delta q)\rangle = \langle f, \delta q\rangle \tag{10}$$

of an external force $(q, f) \in \mathsf{T}^*Q$ on a virtual displacement $(q, \delta q) \in \mathsf{T}Q$ is the *virtual work* performed by an external device controlling the configuration of the system.

Admissible displacements form a subset $C^1 \subset \mathsf{T}Q$. If $(q, \delta q)$ is an admissible displacement, then $(q, k\delta q)$ is again an admissible displacement for each number $k \geq 0$. The set C^1 represents constraints imposed on virtual displacements. Implicitly it restricts *admissible configurations* to the set

$$C^0 = \{q \in Q;\ (q, \delta q) \in C^1 \text{ for some } \delta q \in V\}. \tag{11}$$

The inclusion $C^1 \subset \mathsf{T}C^0$ is usually satisfied. We say that constraints are *configuration constraints* if $C^1 = \mathsf{T}C^0$. The set $C = C^0$ itself is called a *configuration constraint*. A *simple two-sided configuration constraint* is an embedded submanifold $C \subset Q$

There is a function

$$\sigma \colon C^1 \to \mathbb{R} \tag{12}$$

assigning to each admissible virtual displacement the virtual work that an external device has to perform in order to effect this displacement. This *virtual work function* is differentiable on $C^1 \cap \mathring{\mathsf{T}}Q$ and positive homogeneous in the sense that

$$\sigma(q, k\delta q) = k\sigma(q, \delta q) \tag{13}$$

if $k \geq 0$. A typical example of a virtual work function is the mapping

$$\sigma \colon C^1 \to \mathbb{R} \colon (q, \delta q) \mapsto DU(q, \delta q) \tag{14}$$

derived from an *internal energy function* U defined in a domain large enough to make the derivative $DU(q, \delta q)$ meaningful. In the case of a configuration constraint $C \subset Q$ it is enough to have the internal energy defined on C. The function

$$\sigma \colon C^1 \to \mathbb{R} \colon (q, \delta q) \mapsto \rho(q)\|\delta q\| \tag{15}$$

represents virtual work due to friction.

The response of the system to external control is represented by a set $S \subset \mathsf{T}^*Q$ of external forces satisfying the *principle of virtual work*

$$\langle f, \delta q \rangle \leq \sigma(q, \delta q) \text{ for each virtual displacement } (q, \delta q) \in C^1. \tag{16}$$

The set S is the *constitutive set* of the system. It can be viewed as the list of possible configurations of the system together with external forces compatible with these configurations. If $\sigma(q, -\delta q) = -\sigma(q, \delta q)$ and the constraints are two-sided, then the principle of virtual work assumes the simpler form

$$\langle f, \delta q \rangle = \sigma(q, \delta q) \text{ for each virtual displacement } (q, \delta q) \in C^1. \tag{17}$$

Example 1: Let a material point be constrained to a circular hoop with the center at $q_0 \in Q$ and radius a in the plane orthogonal to a unit vector $n \in V$. We have two-sided configuration constraints

$$\begin{aligned}
C^0 &= \{q \in Q;\ \langle g(q - q_0), n \rangle = 0, \|q - q_0\| = a\}, \tag{18}\\
C^1 &= \mathsf{T}C^0 = \{(q, \delta q) \in \mathsf{T}Q;\ \langle g(q - q_0), n \rangle = 0, \|q - q_0\| = a, \tag{19}\\
&\qquad \langle g(\delta q), n \rangle = 0, \langle g(\delta q), u(q) \rangle = 0\},
\end{aligned}$$

where $u(q)$ is the unit vector $(q - q_0)\|q - q_0\|^{-1}$. The constitutive set

$$S = \{(q, f) \in \mathsf{T}^*Q;\ q \in C^0, \langle f, \delta q \rangle = 0 \text{ for each } (q, \delta q) \in C^1\} \quad (20)$$
$$= \{(q, f) \in \mathsf{T}^*Q;\ \langle g(q - q_0), n \rangle = 0, \|q - q_0\| = a,$$
$$\langle f, u(q) \rangle = 0, \langle f, n \rangle = 0\}$$

represents the statics of the system without friction and the constitutive set

$$S = \{(q, f) \in \mathsf{T}^*Q;\ q \in C^0, \langle f, \delta q \rangle \leq \rho \|\delta q\| \text{ for each } (q, \delta q) \in C^1\} \quad (21)$$
$$= \{(q, f) \in \mathsf{T}^*Q;\ \langle g(q - q_0), n \rangle = 0, \|q - q_0\| = a,$$
$$\langle f, u(q) \rangle^2 + \langle f, n \rangle^2 \leq \rho^2\}$$

takes constant friction into account.

Example 2: Let a material point be constrained to the exterior of a solid ball with the centre at $q_0 \in Q$ and radius a. In its displacements on the surface of the ball the point encounters friction proportional to the component of the external force pressing the point against the surface. Correct representation of the statics of the point is obtained with one-sided constraints

$$C^0 = \{q \in Q;\ \|q - q_0\| \geq a\}, \quad (22)$$
$$C^1 = \{(q, \delta q) \in \mathsf{T}Q;\ \|q - q_0\| \geq a, \quad (23)$$
$$\langle g(\delta q), u(q) \rangle \geq \nu \sqrt{\|\delta q\|^2 - \langle g(\delta q), u(q) \rangle^2} \text{ if } \|q - q_0\| = a\}$$

and the constitutive set

$$S = \{(q, f) \in \mathsf{T}^*Q;\ q \in C^0, \langle f, \delta q \rangle \leq 0 \text{ for each } (q, \delta q) \in C^1\} \quad (24)$$
$$= \{(q, f) \in \mathsf{T}^*Q;\ \|q - q_0\| \geq a, f = 0 \text{ if } \|q - q_0\| > a,$$
$$\nu \langle f, u(q) \rangle + \sqrt{\|f\|^2 - \langle f, u(q) \rangle^2} \leq 0 \text{ if } \|q - q_0\| = a\},$$

where $u(q) = (q - q_0)\|q - q_0\|^{-1}$. The constraints in this example are not configuration constraints.

Example 3: Let i, j, and k be mutually orthogonal unit vectors and let q_0 be a point. Let one-sided configuration constraints be specified by

$$C^0 = \{q \in Q;\ \langle g(q - q_0), i \rangle \geq 0, \langle g(q - q_0), j \rangle \geq 0\} \quad (25)$$

and

$$C^1 = \mathsf{T}C^0 \quad (26)$$
$$= \{(q, \delta q) \in \mathsf{T}Q;\ \langle g(q - q_0), i \rangle \geq 0, \langle g(q - q_0), j \rangle \geq 0,$$
$$\langle g(\delta q), i \rangle \geq 0 \text{ if } \langle g(q - q_0), i \rangle = 0,$$
$$\langle g(\delta q), j \rangle \geq 0 \text{ if } \langle g(q - q_0), j \rangle = 0\ \}.$$

The statics of a material point not subject to internal forces is represented by the constitutive set

$$
\begin{aligned}
S &= \{(q,f) \in T^*Q;\ q \in C^0, \langle f, \delta q \rangle \leq \text{ for each } (q, \delta q) \in C^1\} \quad (27)\\
&= \{(q,f) \in T^*Q;\ \langle g(q-q_0), i \rangle \geq 0, \langle g(q-q_0), j \rangle \geq 0, \langle f, k \rangle = 0,\\
&\quad \langle f, i \rangle = 0 \text{ and } \langle f, j \rangle \leq 0\\
&\quad \text{if } \langle g(q-q_0), j \rangle = 0 \text{ and } \langle g(q-q_0), i \rangle \neq 0,\\
&\quad \langle f, j \rangle = 0 \text{ and } \langle f, i \rangle \leq 0\\
&\quad \text{if } \langle g(q-q_0), i \rangle = 0 \text{ and } \langle g(q-q_0), j \rangle \neq 0,\\
&\quad \langle f, i \rangle \leq 0 \text{ and } \langle f, j \rangle \leq 0\\
&\quad \text{if } \langle g(q-q_0), i \rangle = 0 \text{ and } \langle g(q-q_0), j \rangle = 0\}
\end{aligned}
$$

Example 4: In terms of the vectors i, j, and k and the point q_0 of the preceding example we define one-sided configuration constraints by

$$C^0 = \{q \in Q;\ \langle g(q-q_0), i \rangle \leq 0 \text{ or } \langle g(q-q_0), j \rangle \leq 0\} \quad (28)$$

and

$$
\begin{aligned}
C^1 &= TC^0 \quad (29)\\
&= \{(q, \delta q) \in TQ;\ \langle g(q-q_0), i \rangle \leq 0 \text{ or } \langle g(q-q_0), j \rangle \leq 0,\\
&\quad \langle g(\delta q), i \rangle \leq 0 \text{ if } \langle g(q-q_0), i \rangle = 0 \text{ and } \langle g(q-q_0), j \rangle \neq 0,\\
&\quad \langle g(\delta q), j \rangle \leq 0 \text{ if } \langle g(q-q_0), j \rangle = 0 \text{ and } \langle g(q-q_0), i \rangle \neq 0,\\
&\quad \langle g(\delta q), i \rangle \leq 0 \text{ or } \langle g(\delta q), j \rangle \leq 0\\
&\quad \text{if } \langle g(q-q_0), i \rangle = 0 \text{ and } \langle g(q-q_0), j \rangle = 0\ \}.
\end{aligned}
$$

The statics of a material point not subject to internal forces is represented by the constitutive set

$$
\begin{aligned}
S &= \{(q,f) \in T^*Q;\ q \in C^0, \langle f, \delta q \rangle \leq \text{ for each } (q, \delta q) \in C^1\} \quad (30)\\
&= \{(q,f) \in T^*Q;\ \langle g(q-q_0), i \rangle \leq 0 \text{ or } \langle g(q-q_0), j \rangle \leq 0, \langle f, k \rangle = 0\\
&\quad \langle f, i \rangle = 0 \text{ and } \langle f, j \rangle \geq 0\\
&\quad \text{if } \langle g(q-q_0), j \rangle = 0 \text{ and } \langle g(q-q_0), i \rangle \neq 0,\\
&\quad \langle f, j \rangle = 0 \text{ and } \langle f, i \rangle \geq 0\\
&\quad \text{if } \langle g(q-q_0), i \rangle = 0 \text{ and } \langle g(q-q_0), j \rangle \neq 0,\\
&\quad \langle f, i \rangle \geq 0 \text{ and } \langle f, j \rangle \geq 0\\
&\quad \text{if } \langle g(q-q_0), i \rangle = 0 \text{ and } \langle g(q-q_0), j \rangle = 0\}.
\end{aligned}
$$

4. MODELING CONFIGURATION CONSTRAINTS IN STATICS

We believe that constraint static systems are idealized representations of unconstrained systems. The magnitude of the force that an external device can apply to a static system is limited and instruments used to observe displacements have a limited resolution. Idealizations take these limitations into account. We restrict the analysis to configuration constraints. A definition of configuration constraints will be based on the assumption that the norms of external forces at our disposal have an upper bound F and that displacements of distances less than d can not be detected.

Let C be a subset of Q. We denote by $d(q, C)$ the distance of a point $q \in Q$ from C. If $C \subset Q$ is an embedded submanifold or a submanifold with smooth boundary, then for each configuration q sufficiently close to C there is a unique point $q_C \in C$ nearest to q. The distance $d(q, C) = \|q - q_C\|$ of q from C is a well defined function in a neighbourhood of C. If q is not in C, then $q_C \neq q$ and the unit vector $e(q) = (q - q_C)\|q - q_C\|^{-1}$ is orthogonal to C or the boundary of C at q_C.

Let $C \subset Q$ be an embedded submanifold or a submanifold with smooth boundary and let $S \subset \mathsf{T}^*Q$ be the constitutive set of a static system derived from the principle of virtual work

$$\langle f, \delta q \rangle \leq \sigma(q, \delta q) \tag{31}$$

for each virtual displacement $(q, \delta q) \in \mathsf{T}C$.

A model of this static system is constructed by choosing a function $\overline{\sigma}$ on $\mathsf{T}Q$ such that σ is the restriction of $\overline{\sigma}$ to $\mathsf{T}C$ and replacing the original principle of virtual work by the principle

$$\langle f, \delta q \rangle \leq \overline{\sigma}(q, \delta q) + kd(q, C)\langle g(e(q)), \delta q \rangle \tag{32}$$

for each virtual displacement $(q, \delta q)$.

The term $kd(q, C)\langle g(e(q)), \delta q \rangle$ is the directional derivative $DK(q, \delta q)$ of the elastic internal energy function

$$K(q) = \frac{k}{2}(d(q, C))^2 \tag{33}$$

defined in the neighbourhood of C in which the distance function $d(q, C)$ is well defined. The inequalities

$$kd(q, C) \leq \langle f, e(q) \rangle + \sigma(q, -e(q)) \tag{34}$$

and

$$kd(q, C) \leq |\langle f, e(q) \rangle| + |\sigma(q, -e(q))| \tag{35}$$

are derived from the principle of virtual work by setting $\delta q = -e(q)$. We will assume that $F \ll kd$ and expect that the inequality $|\sigma(q, -e(q))| \ll kd$ is satisfied. These inequalities together with $|\langle f, e(q) \rangle| \leq \|f\| \leq F$ result in $d(q, C) \ll d$. It follows that using external forces at our disposal we can not induce the material point to assume configurations at noticeable distances away from C. It also follows that within the limits imposed by $\|f\| \leq F$ the component $\langle f, e(q)\rangle$ is arbitrary. Examples will be used to clarify details and present variations of this construction.

Example 5: Let C be the set C^0 of example 1. We obtain the equation

$$(d(q, C))^2 = \|q - q_0\|^2 - 2a\sqrt{\|q - q_0\|^2 - \langle g(q - q_0), n\rangle^2} + a^2 \qquad (36)$$

for the distance $d(q, C)$ of q from C if this distance is less than a. If $d(q, C) \neq 0$, then

$$e(q) = \frac{q - q_0 - \langle g(q - q_0), n\rangle n}{\sqrt{\|q - q_0\|^2 - \langle g(q - q_0), n\rangle^2}} \qquad (37)$$

is the unit vector orthogonal to C at the point $q' \in C$ closest to q pointing from q' to q. Let a function $\bar{\sigma}: \mathsf{T}Q \to \mathbb{R}$ be defined by $\bar{\sigma}(q, n) = 0$, $\bar{\sigma}(q, q - q_0) = 0$, and $\bar{\sigma}(q, \delta q) = \rho\|\delta q\|$ if $\langle g(\delta q), n\rangle = 0$ and $\langle g(\delta q), q - q_0\rangle = 0$. The unconstrained system represented by the principle of virtual work

$$\langle f, \delta q\rangle \leq \bar{\sigma}(q, \delta q) + kd(q, C)\langle g(e(q)), \delta q\rangle \qquad (38)$$
for each virtual displacement $(q, \delta q)$

is a model of the constrained system of example 1. It follows from the principle of virtual work that $f = kd(q, C)g(e(q)) + f'$, where the component f' satisfies relations $\langle f', n\rangle = 0$, $\langle f', q - q_0\rangle = 0$, and $\|f'\| \leq \rho$. If $k \to \infty$, then $d(q, C) \to 0$. Any value can be obtained for the component $kd(q, C)g(e(q))$ as $k \to \infty$ and $d(q, C) \to 0$. This is in agreement with the principle of virtual work of example 1.

Example 6: Let C be the set C^0 of example 2 and let $u(q) = (q - q_0)\|q - q_0\|^{-1}$. The distance $d(q, C)$ is equal to $\|q - q_0\| - a$. A function $\sigma: \mathsf{T}Q \to \mathbb{R}$ is defined by $\sigma(q, \delta q) = 0$ if $\|q - q_0\| \geq a$ and

$$\sigma(q, \delta q) = -kd(q, C)\langle g(u(q)), \delta q\rangle + kd(q, C)\nu\sqrt{\|\delta q\|^2 - \langle g(u(q)), \delta q\rangle^2} \qquad (39)$$

if $\|q - q_0\| < a$. The principle of virtual work

$$\langle f, \delta q\rangle \leq \sigma(q, \delta q) \text{ for each virtual displacement } (q, \delta q) \in \mathsf{T}Q \qquad (40)$$

implies the following relations for the external force f. If $\|q - q_0\| \geq a$, then $f = 0$. If $\|q - q_0\| < a$, then $f = -kd(q, C)g(u(q)) + f'$ with $\langle f', u(q)\rangle = 0$ and $\langle f', \delta q\rangle \leq kd(q, C)\nu\|\delta q\|$ if $\langle g(u(q)), \delta q\rangle = 0$. If $k \to \infty$, then $d(q, C) \to 0$. The component $\langle f, u(q)\rangle = -kd(q, C)$ can have any negative limit and $\langle f', \delta q\rangle \leq -\langle f, u(q)\rangle\nu\|\delta q\|$. This is in agreement with the principle of virtual work of example 2.

Example 7: A model for the system in example 3 can be easily constructed even if the boundary of the set $C = C^0$ is not smooth. The distance $d(q, C)$ is defined by

$$d(q, C) = -\langle g(i), q - q_0\rangle \qquad (41)$$

if $\langle g(j), q - q_0\rangle \geq 0$ and $\langle g(i), q - q_0\rangle < 0$,

$$d(q, C) = -\langle g(j), q - q_0\rangle \qquad (42)$$

if $\langle g(i), q - q_0 \rangle \geq 0$ and $\langle g(j), q - q_0 \rangle < 0$, and

$$d(q, C) = \sqrt{\langle g(i), q - q_0 \rangle^2 + \langle g(j), q - q_0 \rangle^2} \tag{43}$$

if $\langle g(i), q - q_0 \rangle < 0$ and $\langle g(j), q - q_0 \rangle < 0$. A vector field $e(q)$ is defined is defined outside of C by $e(q) = -i$ if $\langle g(j), q - q_0 \rangle \geq 0$ and $\langle g(i), q - q_0 \rangle < 0$,

$$e(q) = -j \tag{44}$$

if $\langle g(i), q - q_0 \rangle \geq 0$ and $\langle g(j), q - q_0 \rangle < 0$ and

$$e(q) = (\langle g(i), q - q_0 \rangle i + \langle g(j), q - q_0 \rangle j)(d(q, C))^{-1} \tag{45}$$

if $\langle g(i), q - q_0 \rangle < 0$ and $\langle g(j), q - q_0 \rangle < 0$. A function σ on TQ is defined by $\sigma(q, \delta q) = 0$ if $q \in C$ and $\sigma(q, \delta q) = k \langle g(e(q)), \delta q \rangle$ if $q \notin C$. The constitutive set of example 3 is obtained from the principle of virtual work

$$\langle f, \delta q \rangle = \sigma(q, \delta q) \text{ for each virtual displacement } (q, \delta q) \in TQ \tag{46}$$

with $k \to \infty$ and $d(q, C) \to 0$.

Example 8: The construction of the model in the preceding example followed exactly the prescription given at the beginning of the present section. This construction can not be directly applied to the set $C = C^0$ of example 4. It can be applied to the modified set

$$C_\varepsilon = C_0 \setminus \{q \in Q; \langle g(i), q - q_0 \rangle < r, \langle g(j), q - q_0 \rangle < r,$$
$$(\langle g(i), q - q_0 \rangle - r)^2 + (\langle g(j), q - q_0 \rangle - r)^2 > r\}. \tag{47}$$

The original set C_0 is obtained as the limit as $r \to 0$.

5. KINEMATICS OF AUTONOMOUS SYSTEMS AND SCLERONOMIC CONSTRAINTS

Motions of a material point in the Euclidean affine space Q are curves $\xi \colon I \to Q$ parameterized by time t in an open interval $I \subset \mathbb{R}$. We have the *tangent prolongation* $(\xi, \dot{\xi}) \colon I \to TQ \colon t \mapsto (\xi(t), \dot{\xi}(t))$ and the *second tangent prolongation* $(\xi, \dot{\xi}, \ddot{\xi}) \colon I \to T^2Q \colon t \mapsto (\xi(t), \dot{\xi}(t), \ddot{\xi}(t))$ of a motion ξ.

Variational formulations of analytical mechanics require the concept of a *virtual displacement* of a motion. A virtual displacement of a motion ξ is a mapping $(\xi, \delta\xi) \colon I \to TQ \colon t \mapsto (\xi(t), \delta\xi(t))$. This mapping is obtained from a homotopy

$$\chi \colon \mathbb{R} \times I \to Q. \tag{48}$$

The base curve $\chi(0, \cdot)$ is the motion ξ. The virtual displacement is the mapping

$$(\xi, \delta\xi) \colon I \to TQ \colon t \mapsto t\chi(\cdot, t)(0). \tag{49}$$

A mapping $(\xi, \dot{\xi}, \delta\xi, \delta\dot{\xi}) \colon I \to TTQ \colon t \mapsto (\xi(t), \dot{\xi}(t), \delta\xi(t), \delta\dot{\xi}(t))$ is obtained from a virtual displacement $(\xi, \delta\xi)$ as the composition $\kappa_q \circ (\xi, \delta\xi, \dot{\xi}, \delta\dot{\xi})$ of the tangent prolongation $(\xi, \delta\xi, \dot{\xi}, \delta\dot{\xi})$ with the involution κ_Q. Virtual displacements are subject to constraints. All considered versions of constraints can eventually be reduced to differential equations formulated in terms of a subset $C^{(1,1)} \subset \mathsf{TT}Q$ such that if $(q, \dot{q}, \delta q, \delta\dot{q}) \in C^{(1,1)}$, then $(q, \dot{q}, k\delta q, k\delta\dot{q}) \in C^{(1,1)}$ for each number $k \geq 0$. An *admissible virtual displacement* $(\xi, \delta\xi)$ is required to satisfy the condition

$$(\xi(t), \dot{\xi}(t), \delta\xi(t), \delta\dot{\xi}(t)) \in C^{(1,1)} \tag{50}$$

for each $t \in I$. This condition implies conditions

$$(\xi(t), \dot{\xi}(t)) \in C^{(0,1)}, \tag{51}$$

$$(\xi(t), \delta\xi(t)) \in C^{(1,0)}, \tag{52}$$

and

$$\xi(t) \in C^{(0,0)} \tag{53}$$

for each $t \in I$. Sets $C^{(0,1)}$, $C^{(1,0)}$, and $C^{(0,0)}$ are defined by

$$C^{(0,1)} = \left\{ (q, \dot{q}) \in \mathsf{T}Q; \ (q, \dot{q}, \delta q, \delta\dot{q}) \in C^{(1,1)} \text{ for some } (\delta q, \delta\dot{q}) \in V \times V \right\}, \tag{54}$$

$$C^{(1,0)} = \left\{ (q, \delta q) \in \mathsf{T}Q; \ (q, \dot{q}, \delta q, \delta\dot{q}) \in C^{(1,1)} \text{ for some } (\dot{q}, \delta\dot{q}) \in V \times V \right\}, \tag{55}$$

and

$$C^{(0,0)} = \left\{ q \in Q; \ (q, \dot{q}) \in C^{(0,1)} \text{ for some } \dot{q} \in V \right\}. \tag{56}$$

Condition (51) is a differential equation for the motion $\xi \colon I \to Q$. Constraints are usually discussed it terms of this equation. The inclusion

$$C^{(0,1)} \subset \mathsf{T}C^{(0,0)} \tag{57}$$

must be satisfied since it is a necessary integrability condition for the equation (51). Constraints will be called *configuration constraints* if

$$C^{(0,1)} = \mathsf{T}C^{(0,0)}. \tag{58}$$

Velocity constraints are said to be *linear* if the set $C^{(0,0)}$ is a submanifold of Q and $C^{(0,1)}$ is a distribution on this submanifold. Linear constraints are said to be *holonomic* if $C^{(0,1)}$ is integrable in the sense of Frobenius. Configuration constraints are a special case of holonomic constraints. Sets $C^{(1,0)}$ and $C^{(1,1)}$ are not usually discussed directly even if information contained in the *velocity constraints* $C^{(0,1)}$ is not sufficient for the application of variational methods. The condition (50) is equivalent to

$$(\xi(t), \delta\xi(t), \dot{\xi}(t), \delta\dot{\xi}(t)) = \mathsf{t}(\xi, \delta\xi)(t) \in \kappa_Q(C^{(1,1)}) \subset \mathsf{TT}Q. \tag{59}$$

It is a differential equation for the virtual displacement $(\xi, \delta\xi) \colon I \to \mathsf{T}Q$. The inclusion

$$C^{(1,1)} \subset \kappa_Q(\mathsf{T}C^{(1,0)}) \tag{60}$$

is a necessary integrability condition for this equation.

Two different methods of constructing the set $C^{(1,1)}$ from the velocity constraints $C^{(0,1)}$ are found in an article of Arnold, Kozlov, and Neishtadt (1993).

1. In *vaconomic mechanics* the natural construction

$$C^{(1,1)} = \mathsf{T}C^{(0,1)} \qquad (61)$$

is used. This construction is the result of the differential equation

$$\mathsf{t}\chi(s,\cdot)(t) \in C^{(0,1)} \qquad (62)$$

imposed on curves $\chi(s,\cdot)$ also with $s \neq 0$. See [Arn] for modifications of this construction necessary when virtual displacements vanishing at the ends of a time interval are required. With these modifications the formula (61) is still valid. The set $C^{(1,0)}$ is the tangent set $\mathsf{T}C^{(0,0)}$ of $C^{(0,0)}$.

2. The *d'Alembert-Lagrange* principle is based on the inclusion

$$C^{(1,1)} \subset \overline{C^{(1,1)}} = \left\{ (q, \dot{q}, \delta q, \delta \dot{q}) \in \mathsf{TT}Q;\ (q, \dot{q}, 0, \delta q) \in \mathsf{T}C^{(0,1)} \right\}. \qquad (63)$$

This construction derives from the condition

$$\mathsf{t}\chi(\cdot\, t)(s) \in C^{(1,0)} \qquad (64)$$

for each $t \in I$ and each s and the condition

$$\mathsf{t}\chi(0,\cdot)(t) \in C^{(0,1)} \qquad (65)$$

imposed on the base curve $\xi = \chi(0,\cdot)$ but not the curves $\chi(s,\cdot)$ for $s \neq 0$. The set $\overline{C^{(1,1)}}$ may not represent an integrable differential equation (59) for $(\xi, \delta\xi)$. The set $C^{(1,1)}$ is the integrable part of $\overline{C^{(1,1)}}$. If $C^{(0,1)}$ is a vector subbundle of $\mathsf{T}Q$, then $C^{(1,0)} = C^{(0,1)}$. If $C^{(0,1)}$ is an affine subbundle, then $C^{(1,0)}$ is the model bundle. In both cases

$$\overline{C^{(1,1)}} = \left\{ (q, \dot{q}, \delta q, \delta \dot{q}) \in \mathsf{TT}Q;\ (q, \dot{q}) \in C^{(0,1)}, (q, \delta q) \in C^{(1,0)} \right\} \qquad (66)$$

and

$$C^{(1,1)} = \overline{C^{(1,1)}} \cap \mathsf{T}(\tau_{\mathsf{T}Q}(\overline{C^{(1,1)}})). \qquad (67)$$

For configuration constraints both construction give the same result

$$C^{(1,1)} = \mathsf{TT}C^{(0,0)}. \qquad (68)$$

6. DYNAMICS OF UNCONSTRAINED AUTONOMOUS SYSTEMS

We see four possible formulations of dynamics:

A. Dynamics of a material point can be specified as a collection of *boundary value relations with external forces* associated with time intervals. A boundary value relation for a time interval $[a, b] \subset \mathbb{R}$ is a relation between an arc $\xi \colon [a, b] \to Q$, a mapping $(\xi, \varphi) \colon [a, b] \to \mathsf{T}^*Q$, and two covectors $(\xi(a), \pi(a))$ and $(\xi(b), \pi(b))$. The mapping (ξ, φ) represents the *external force* applied to the material point along the arc ξ the covectors $(\xi(a), \pi(a))$, $(\xi(b), \pi(b))$ are the *initial momentum* and *final momentum*. It is convenient to consider the arc $\xi \colon [a, b] \to Q$ and the mapping $(\xi, \varphi) \colon [a, b] \to \mathsf{T}^*Q$ the restrictions to the interval $[a, b]$ of mappings $\xi \colon I \to Q$ and $(\xi, \varphi) \colon I \to \mathsf{T}^*Q$ defined on an open interval $I \subset \mathbb{R}$ containing $[a, b]$. The covectors $(\xi(a), \pi(a))$ and $(\xi(b), \pi(b))$ will be considered values of a mapping $(\xi, \pi) \colon I \to \mathsf{T}^*Q$ at the ends of the interval. The two mappings $(\xi, \varphi) \colon I \to \mathsf{T}^*Q$ and $(\xi, \pi) \colon I \to \mathsf{T}^*Q$ can be combined in a single mapping

$$(\xi, \varphi, \pi) \colon I \to Q \times V^* \times V^*. \tag{69}$$

An element

$$((\xi, \varphi) \colon [a, b] \to \mathsf{T}^*Q, (\xi(a), \pi(a)), (\xi(b), \pi(b))) \tag{70}$$

of the boundary value relation $D_{[a,b]}$ for an interval $[a, b]$ satisfies the *virtual action principle*

$$\int_a^b \langle \varphi(t), \delta\xi(t) \rangle dt - \langle \pi(b), \delta\xi(b) \rangle + \langle \pi(a), \delta\xi(a) \rangle$$
$$= \int_a^b \left(\lambda(\xi(t), \dot\xi(t), \delta\xi(t), \delta\dot\xi(t)) - m\langle g(\dot\xi(t)), \delta\dot\xi(t) \rangle \right) dt. \tag{71}$$

for each virtual displacement $(\xi, \delta\xi) \colon [a, b] \to \mathsf{T}Q$ obtained as a restriction to $[a, b]$ of a virtual displacement $(\xi, \delta\xi) \colon I \to \mathsf{T}Q$. The term

$$m\langle g(\dot\xi(t)), \delta\dot\xi(t) \rangle \tag{72}$$

is the derivative $DT(\xi(t), \dot\xi(t), \delta\xi(t), \delta\dot\xi(t))$ of the *kinetic energy* function

$$T \colon \mathsf{T}Q \to \mathbb{R} \colon (q, \dot q) \mapsto \frac{m}{2} \|\dot q\|^2 \tag{73}$$

of a material point with mass m. The function $\lambda \colon \mathsf{TT}Q \to \mathbb{R}$ represents the virtual action of internal forces. For the sake of simplicity we assume that λ is a linear form (a linear function of $(\delta q, \delta\dot q)$). The proposed principle of virtual action is more general than the Hamilton Principle. Note that the virtual displacements $(\xi(a), \delta\xi(a))$ and $(\xi(b), \delta\xi(b))$ of the end points of the arc do not vanish. Variational principles with variations of end points but without external forces were considered by Schwinger. The momentum-velocity relation is a law of physics and is a part of dynamics of a material point. This relation is included in the Schwinger version of the principle of virtual action but not in the Hamilton Principle.

Examples of the virtual action function λ include the function

$$\lambda\colon \mathsf{TT}Q \to \mathbb{R}\colon (q,\dot q,\delta q,\delta\dot q)\mapsto e\langle A(q),\delta\dot q\rangle + e\langle DA(q,\delta q),\dot q\rangle \tag{74}$$

for a charged particle in a magnetic field derived from the vector potential $A\colon Q\to V^*$ and the function

$$\lambda\colon \mathsf{TT}Q \to \mathbb{R}\colon (q,\dot q,\delta q,\delta\dot q)\mapsto \gamma\langle g(\dot q),\delta q\rangle \tag{75}$$

for a material point immersed in a viscous medium. In the first of these examples the function $\lambda(q,\dot q,\delta q,\delta\dot q)$ is the derivative $D\alpha(q,\dot q,\delta q,\delta\dot q)$ of the function $\alpha\colon \mathsf{T}Q\to\mathbb{R}\colon (q,\dot q)\mapsto e\langle A(q),\dot q\rangle$. We will continue the analysis assuming that the function λ is of the simpler type

$$\lambda\colon \mathsf{TT}Q\to\mathbb{R}\colon (q,\dot q,\delta q,\delta\dot q)\mapsto \langle\mu(q),\delta q\rangle, \tag{76}$$

where μ is a mapping from $\mathsf{T}Q$ to V^*. The principle of virtual action assumes the simpler form

$$\int_a^b \langle\varphi(t),\delta\xi(t)\rangle dt - \langle\pi(b),\delta\xi(b)\rangle + \langle\pi(a),\delta\xi(a)\rangle \tag{77}$$
$$= \int_a^b \Big(\langle\mu(\xi(t)),\delta\xi(t)\rangle - m\langle g(\dot\xi(t)),\delta\dot\xi(t)\rangle\Big)\,dt.$$

Equivalent versions of this variational principle

$$\int_a^b \langle\varphi(t),\delta\xi(t)\rangle dt - \langle\pi(b),\delta\xi(b)\rangle + \langle\pi(a),\delta\xi(a)\rangle \tag{78}$$
$$= \int_a^b \Big(m\langle g(\ddot\xi(t)),\delta\xi(t)\rangle + \langle\mu(\xi(t)),\delta\xi(t)\rangle\Big)\,dt$$
$$- m\langle g(\dot\xi(b)),\delta\xi(b)\rangle + m\langle g(\dot\xi(a)),\delta\xi(a)\rangle$$

and

$$\int_a^b \Big(\langle\varphi(t)-\dot\pi(t),\delta\xi(t)\rangle - \langle\pi(t),\delta\dot\xi(t)\rangle\Big)\,dt \tag{79}$$
$$= \int_a^b \Big(\langle\mu(\xi(t)),\delta\xi(t)\rangle - m\langle g(\dot\xi(t)),\delta\dot\xi(t)\rangle\Big)\,dt$$

are easily derived by using the identities

$$\int_a^b \Big(\langle\dot\pi(t),\delta\xi(t)\rangle + \langle\pi(t),\delta\dot\xi(t)\rangle\Big)\,dt = \int_a^b \frac{d}{dt}\langle\pi(t),\delta\xi(t)\rangle dt \tag{80}$$
$$= \langle\pi(b),\delta\xi(b)\rangle - \langle\pi(a),\delta\xi(a)\rangle$$

and

$$\int_a^b \Big(m\langle g(\ddot\xi(t)),\delta\xi(t)\rangle + m\langle g(\dot\xi(t)),\delta\dot\xi(t)\rangle\Big)\,dt \tag{81}$$
$$= \int_a^b m\frac{d}{dt}\langle g(\dot\xi(t)),\delta\xi(t)\rangle dt$$
$$= m\langle g(\dot\xi(b)),\delta\xi(b)\rangle - m\langle g(\dot\xi(a)),\delta\xi(a)\rangle.$$

B. Dynamics can be specified as the collection $\mathcal{D}$ of curves

$$(\xi, \varphi, \pi) \colon I \to Q \times V^* \times V^* \tag{82}$$

defined on open intervals $I \subset \mathbb{R}$ with the property that for each time interval $[a, b] \subset I$ the arc $(\xi, \varphi)|[a, b]$ and the covectors $(\xi(a), \pi(a))$ and $(\xi(b), \pi(b))$ are in the boundary relation $D_{[a,b]}$.

C. Dynamics can be specified as differential equations

$$\varphi(t) - \dot{\pi}(t) = \mu(\xi(t)) \tag{83}$$

and

$$\pi(t) = g(\dot{\xi}(t)) \tag{84}$$

for mappings $(\xi, \varphi, \pi) \colon I \to Q \times V^* \times V^*$. These equations will be denoted by $\dot{D}$.

D. Dynamics can be specified as differential equations

$$\varphi(t) = mg(\ddot{\xi}(t)) + \mu(\xi(t)) \tag{85}$$

and

$$\pi(t) = g(\dot{\xi}(t)) \tag{86}$$

for mappings $(\xi, \varphi, \pi) \colon I \to Q \times V^* \times V^*$. These equations will be denoted by E.

Of the four formulations of dynamics version **A** is fundamental. The family of curves $\mathcal{D}$ and the differential equations $\dot{D}$ and E introduced in **B**, **C**, and **D** are auxiliary objects.

Differential equations $\dot{D}$ and E are obviously equivalent.

We show that the family $\mathcal{D}$ is the set of solutions of the equations $\dot{D}$. The proof is based on version (79) of the principle of virtual action. If a curve $(\xi, \varphi, \pi) \colon I \to Q \times V^* \times V^*$ is in $\mathcal{D}$ and $(\xi, \delta\xi) \colon I \to TQ$ is an arbitrary virtual displacement, then the equality (79) holds for all intervals $[a, b] \subset I$. It follows that the equality

$$\langle \varphi(t) - \dot{\pi}(t), \delta\xi(t) \rangle - \langle \pi(t), \delta\dot{\xi}(t) \rangle = \langle \mu(\xi(t)), \delta\xi(t) \rangle - m\langle g(\dot{\xi}(t)), \delta\dot{\xi}(t) \rangle \tag{87}$$

holds at each $t \in I$. This implies that equations $\dot{D}$ are satisfied due to arbitrariness of the vectors $\delta\xi(t)$ and $\delta\dot{\xi}(t)$. Conversely if $(\xi, \varphi, \pi) \colon I \to Q \times V^* \times V^*$ is a solution of $\dot{D}$, then the equality (87) holds in I with an arbitrary displacement $(\xi, \delta\xi) \colon I \to TQ$. The validity of the principle of virtual action for each time interval $[a, b] \subset I$ is established by integration. Hence, (ξ, φ, π) is in $\mathcal{D}$.

It follows from the definition of $\mathcal{D}$ that this family is constructed from the boundary value relations. We show that elements of boundary value relations can be constructed from elements of $\mathcal{D}$. Let $[a, b]$ be a time interval included in an open interval $I \subset \mathbb{R}$ and let $(\xi, \varphi, \pi) \colon I \to Q \times V^* \times V^*$ be a mapping such that

$$(\xi, \varphi)|[a, b], (\xi(a), \pi(a)), (\xi(b), \pi(b)))$$

is in $D_{[a,b]}$. It follows from version (78) of the principle of virtual action that the mapping (ξ, φ) satisfies the equation

$$\varphi(t) = mg(\ddot{\xi}(t)) + \mu(\xi(t)) \tag{88}$$

in $[a, b]$ and that
$$\pi(a) = g(\dot{\xi}(a)) \text{ and } \pi(b) = g(\dot{\xi}(b)). \tag{89}$$
Let mappings $\varphi': I \to V^*$ and $\pi': I \to V^*$ be defined by
$$\varphi'(t) = mg(\ddot{\xi}(t)) + \mu(\xi(t)) \tag{90}$$
and
$$\pi'(t) = mg(\dot{\xi}(t)). \tag{91}$$
The mapping $(\xi, \varphi', \pi'): I \to Q \times V^* \times V^*$ is in $\mathcal{D}$ since it is a solution of E. The boundary value data extracted from this mapping are in the boundary value relation since $(\xi, \varphi')|[a, b] = (\xi, \varphi)|[a, b]$, $\pi'(a) = \pi(a)$, and $\pi'(b) = \pi(b)$.

7. DYNAMICS WITH CONFIGURATION CONSTRAINTS

We list four possible formulations of dynamics with constraints analogous to the four formulations in the preceding section. As defined in Section 5 an admissible virtual displacement is a mapping $(\xi, \delta\xi): I \to \mathsf{T}Q$ satisfying the condition
$$(\xi(t), \dot{\xi}(t), \delta\xi(t), \delta\dot{\xi}(t)) \in C^{(1,1)} = \mathsf{TT}C^{(0,0)} \tag{92}$$
for each $t \in I$.

A. Dynamics of a material point can be considered a collection of *boundary value relations* associated with time intervals. An element
$$((\xi, \varphi): [a, b] \to \mathsf{T}^*Q, (\xi(a), \pi(a)), (\xi(b), \pi(b))) \tag{93}$$
of the boundary value relation $D_{[a,b]}$ for an interval $[a, b]$ satisfies the *virtual action principle*
$$\int_a^b \langle \varphi(t), \delta\xi(t) \rangle dt - \langle \pi(b), \delta\xi(b) \rangle + \langle \pi(a), \delta\xi(a) \rangle$$
$$= \int_a^b \left(\langle \mu(\xi(t)), \delta\xi(t) \rangle - m \langle g(\dot{\xi}(t)), \delta\dot{\xi}(t) \rangle \right) dt. \tag{94}$$
for each admissible virtual displacement $(\xi, \delta\xi): [a, b] \to \mathsf{T}Q$ obtained as a restriction to $[a, b]$ of an admissible virtual displacement $(\xi, \delta\xi): I \to \mathsf{T}Q$. The mapping μ is defined on $C^{(1,0)} = \mathsf{T}C^{(0,0)}$. The condition $\xi(t) \in C^{(0,0)}$ for each $t \in I$ is implied.

There are again the equivalent versions of this variational principle
$$\int_a^b \langle \varphi(t), \delta\xi(t) \rangle dt - \langle \pi(b), \delta\xi(b) \rangle + \langle \pi(a), \delta\xi(a) \rangle$$
$$= \int_a^b \left(m \langle g(\ddot{\xi}(t)), \delta\xi(t) \rangle + \langle \mu(\xi(t)), \delta\xi(t) \rangle \right) dt$$
$$- m \langle g(\dot{\xi}(b)), \delta\xi(b) \rangle + m \langle g(\dot{\xi}(a)), \delta\xi(a) \rangle \tag{95}$$

and

$$\int_a^b \Big(\langle\varphi(t)-\dot\pi(t),\delta\xi(t)\rangle - \langle\pi(t),\delta\dot\xi(t)\rangle\Big)\,dt$$
$$= \int_a^b \Big(\langle\mu(\xi(t)),\delta\xi(t)\rangle - m\langle g(\dot\xi(t)),\delta\dot\xi(t)\rangle\Big)\,dt. \quad (96)$$

B. Dynamics can be specified as the collection $\mathcal{D}$ of curves

$$(\xi,\varphi,\pi): I \to Q \times V^* \times V^* \quad (97)$$

defined on open intervals $I \subset \mathbb{R}$ such that for each time interval $[a,b] \subset I$ the arc $(\xi,\varphi)|[a,b]$ and the covectors $(\xi(a),\pi(a))$ and $(\xi(b),\pi(b))$ are in the boundary relation $D_{[a,b]}$.

C. Dynamics can be specified as the differential equation

$$\langle\varphi(t)-\dot\pi(t),\delta\xi(t)\rangle - \langle\pi(t),\delta\dot\xi(t)\rangle = \langle\mu(\xi(t)),\delta\xi(t)\rangle - \langle g(\dot\xi(t)),\delta\dot\xi(t)\rangle \quad (98)$$

to be satisfied by a curve $(\xi,\varphi,\pi): I \to Q \times V^* \times V^*$ at each $t \in I$ and for each $(q,\dot\xi(t),\delta\xi(t),\delta\dot\xi(t)) \in C^{(1,1)}$. This is equivalent to equations

$$\langle\varphi(t)-\dot\pi(t),\delta\xi(t)\rangle = \langle\mu(\xi(t)),\delta\xi(t)\rangle \quad (99)$$

and

$$\langle\pi(t),\delta\xi(t)\rangle = \langle g(\dot\xi(t)),\delta\xi(t)\rangle \quad (100)$$

satisfied at each $t \in I$ for each $(\xi(t),\delta\xi(t)) \in C^{(1,0)}$.

D. Dynamics can be specified as the differential equations

$$\langle\varphi(t),\delta\xi(t)\rangle = \langle mg(\ddot\xi(t)) + \mu(\xi(t)),\delta\xi(t)\rangle \quad (101)$$

and

$$\langle\pi(t),\delta\xi(t)\rangle = \langle g(\dot\xi(t)),\delta\xi(t)\rangle \quad (102)$$

satisfied by a mapping $(\xi,\varphi,\pi): I \to Q \times V^* \times V^*$ at each $t \in I$ and each $(\xi(t),\delta\xi(t)) \in C^{(1,0)}$.

The four formulations are valid for configuration constraints and are equivalent as in the case of unconstrained systems. The situation is much more complex in the case of more general constraints. The method of models could be a tool for testing the validity of different formulations. We will apply this tool to the momentum-velocity relation. Note that the usual momentum-velocity relation

$$\pi(t) = g(\dot\xi(t)) \quad (103)$$

is replaced by the equation (102). We will attempt a justification of this modification of the momentum-velocity relation based on models of configuration constraints.

8. MODELS OF AUTONOMOUS SYSTEMS WITH CONFIGURATION CONSTRAINTS

Let a material point of mass m be constrained to a plane $C^{(0,0)} \subset Q$ passing through a point q_0 and orthogonal to a unit vector n. We will assume that there are no internal forces and no external forces are applied. The constraint will be modeled by a strong internal elastic force $k\langle g(q-q_0), n\rangle g(n)$. Let an initial momentum $(\xi(a), \pi(a))$ such that $\xi(a) \in C^{(0,0)}$ and $\langle \pi(a), n\rangle = 0$ be applied to the point. The solution of the dynamical equations will be the mapping $(\xi, \varphi, \pi) \colon R \to Q \times V^* \times V^*$ with $\xi(t) = \xi(a) + m^{-1}g^{-1}(\pi(a))(t-a)$, $\varphi(t) = 0$, and $\pi(t) = \pi(a)$. If the constraint is replaced by the elastic force the solution mapping will be the same. Let now the initial momentum have a non zero component $\langle \pi(a), n\rangle$. For the unconstrained model the solution is the mapping (ξ, φ, π) with

$$\xi(t) = \xi(a) + m^{-1}\left(g^{-1}(\pi(a)) - \langle \pi(a), n\rangle n\right)(t-a) \\ + \omega^{-1}\langle \pi(a), n\rangle n \sin\omega(t-a), \tag{104}$$

$\varphi(t) = 0$,

$$\pi(t) = \pi(a) + \langle \pi(a), n\rangle g(n)(\cos\omega(t-a) - 1) \tag{105}$$

and $\omega = \sqrt{k/m}$. The oscillation may be invisible since the amplitude $\omega^{-1}\langle \pi(a), n\rangle$ may be small due to the high value of ω. The component $\langle \pi(a), n\rangle g(n)\cos\omega(t-a)$ of the momentum transverse to the plane $C^{(0,0)}$ depends on the initial value $\langle \pi(a), n\rangle g(n)$ changes rapidly and is arbitrary within certain limits. This component can be detected by making the material point collide with an unconstrained mass. Time dependent external forces and curvature of constraint set $C^{(0,0)}$ may even cause the transverse component of momentum influence the visible part of the motion along the constraint. The element of the boundary value relation for the idealized constrained system is composed of the mapping $(\xi, \varphi) \colon [a, b] \to Q \times V^*$ with

$$\xi(t) = \xi(a) + m^{-1}\left(g^{-1}(\pi(a)) - \langle \pi(a), n\rangle n\right)(t-a),$$

$\varphi(t) = 0$, and covectors $(\xi(a), \pi(a))$ and $(\xi(b), \pi(b))$ satisfying the equality $\pi(b) - \langle \pi(b), n\rangle g(n) = \pi(a) - \langle \pi(a), n\rangle g(n)$. The transverse component $\langle \pi(a), n\rangle g(n)$ of the initial momentum is arbitrary. Due to the rapidity of oscillations the final value of the transverse component $\langle \pi(b), n\rangle g(n)$ of final momentum should be considered arbitrary and independent of the initial value.

This analysis based on a purely elastic model suggests that the virtual action principle

$$\int_a^b \langle \varphi(t), \delta\xi(t)\rangle dt - \langle \pi(b), \delta\xi(b)\rangle + \langle \pi(a), \delta\xi(a)\rangle$$
$$= \int_a^b \left(m\langle g(\ddot{\xi}(t)), \delta\xi(t)\rangle + \langle \mu(\xi(t)), \delta\xi(t)\rangle\right) dt$$
$$- m\langle g(\dot{\xi}(b)), \delta\xi(b)\rangle + m\langle g(\dot{\xi}(a)), \delta\xi(a)\rangle \tag{106}$$

for each admissible virtual displacement $(\xi, \delta\xi): [a, b] \to TQ$ is appropriate for material points subject to configuration constraints. If this formulation of dynamics with configuration constraints is adopted, then the momentum-velocity relation

$$momentum = mass \times velocity$$

is no longer valid. Velocity is the rate of change of configuration. The component of velocity transverse to the constraint set is zero. This is not true of the transverse component of momentum. Along directions tangent to the constraint the usual momentum-velocity relation holds. Other models may be found appropriate in certain situations. When a rigid bar is struck with a hammer it starts an invisible vibration detectable through the sound it emits. The sound is due to momentum (and energy) transfer to air molecules colliding with the surface of the bar. The vibration will eventually die away although it may continue for a long time if the bar is placed in vacuum. The inevitable damping can be taken into account by supplementing the elastic force $k\langle g(q - q_0), n\rangle g(n)$ with a viscous damping force $\gamma \langle g(\dot\xi(t)), n\rangle g(n)$. It may be appropriate to consider the transverse component of the initial momentum arbitrary and the transverse component of the final momentum effectively reduced to zero. In cases of relatively short time intervals the effect of the damping can be ignored. In cases of strong damping and relatively long time intervals. It may be correct to assume that the transverse component of the initial momentum is completely absorbed in an essentially inelastic collision with the suspension and not transferred to the material point. In such cases the boundary value relation will be a solution of the Hamilton principle with external forces:

$$\int_a^b \langle \varphi(t), \delta\xi(t)\rangle dt = \int_a^b \left(m\langle g(\ddot\xi(t)), \delta\xi(t)\rangle + \langle \mu(\xi(t)), \delta\xi(t)\rangle \right) dt \qquad (107)$$

for each admissible virtual displacement $(\xi, \delta\xi): [a, b] \to TQ$ with $\delta\xi(a) = 0$ and $\delta\xi(b) = 0$. The equalities $\pi(a) = g(\dot\xi(a))$ and $\pi(b) = g(\dot\xi(b))$ supplement the variational principle.

Istituto Nazionale di Fisica Nucleare

NOTE

* In honour of my friend John Stachel.

REFERENCE

Arnold, V. I., V. V. Kozlov, and A. I. Neishtadt. 1993. "Aspects of Classical and Celestial Mechanics." In *Dynamical Systems III*, ed. V. I. Arnold. New York: Springer-Verlag.

DANIEL H. WESLEY AND JOHN A. WHEELER

TOWARDS AN ACTION-AT-A-DISTANCE CONCEPT OF SPACETIME*

John Stachel and action at a distance? Direct connection or not as there may be between these two themes, physics, with its ever-growing reach, provides one. Vivid in the memory of one of us (JAW) is the seminar in which Richard Feynman reported his work with JAW, defending the 1922 argument of Tetrode that "The sun would not radiate if it were alone in space an no other bodies could absorb its radiation..." Einstein, Pauli, and Wigner are among those who attended the seminar, but no one expected any of them to try a similar replacement of Einstein's field theory of gravitation by a Newtonian direct action account. Did Einstein himself, who had given us the standard field theory of gravity in place of Newtonian direct action, ever attempt the reverse: to sweep out the space-time continuum and replace it by pure direct coupling between particle and particle? That is a question of the history of science, of Einstein thoughts and Einstein records. For an answer to that question, we will look to John Stachel, immersed as he is in the Einstein Papers, and working with them day after day. Who does not rejoice in that life work of Stachel's, witnessed not least on the fly leaf of *The Collected Papers of Albert Einstein*.[1]

So, trust Stachel to answer our question. What, if anything, did Einstein ever do or say about such a theme as "sweeping out space and time from between the particles," and replacing the space-time of general relativity by particle-to-particle interaction? *Space and time are not things, but orders of things*. If these words of old call to a new and deeper conception of space and time than Einstein's publications give us, then in what direction are we to look? For a reply, many today would point to string theory or other new and exotic developments. But if we are not yet ready to accept them, with all of their new elements, then where can we look?

Here, we consider how we may economize today's account of spacetime by dropping fields, instead of adding them. The concept of the field is a very old one in physics. One of the earliest problems faced by physical theory was the explanation of the phenomenon of "action at a distance." Issac Newton's theory of universal gravitation is very successful at predicting the motions of planets and satellites, provided one makes the assumption that every mass can exert a force on distant masses through some invisible mechanism, the *field*. The idea may is expressed mathematically by defining a quantity at every point of space that allows us to calculate the forces on test particles.

As physics progressed, the number of fields, and their role in physical theory, increased greatly. In the general theory of gravitation, classical field theory finds its most powerful expression; the very properties of space and time are given by the metric field $g_{\mu\nu}$. In this work, we consider whether the field is a necessary part of the expression of physical law, and how, in the case of gravity theory, rethinking our notion of the field involves reconsidering our usual notions of space and time.

It is not new to formulate a physical law without introducing the field as a separate physical entity. Electrostatics, for example, accustoms us to account for the forces on a family of electric charges either by consideration of the electric field generated by those charges, or equally well in terms of direct action between charge and charge, without need of any field at all. However, that field-free formulation appears to fail when any one of the charges undergoes acceleration, because it experiences the well-attested force of radiation reaction, which cannot be caused by the other charges present. This circumstance appears to demand that we attribute to the field an existence of its own. Or, it appeared to make this demand until, in 1922, H. Tetrode argued that "The sun would not radiate if it were alone in space and no other bodies could absorb its radiation," suggesting that it is indeed possible to account for the force of radiation reaction by considering only distant charges.

In 1945 JAW and Richard Feynman gave a formal treatment of Tetrode's idea. This analysis provided a quantitative "absorber theory of radiation," extending a 1938 result of Dirac, which accounted for the force of radiation reaction in terms of both "advanced" and "retarded" electromagnetic fields. Is it possible to do the same for gravity, and eliminate all direct reference to spacetime or any comparable field concept, as all direct reference to an "electromagnetic field" disappears in the Fokker-Tetrode-Schwarzschild action principle for electromagnetism (17).

J. V. Narlikar, writing in 1968 on the general correspondence between field theories and theories of direct interparticle action, showed that any field-plus-particle theory described by an action principle of the type (13) admits translation to a direct particle action theory of the form (15). But, the Hilbert variational principle for gravitation is not quadratic in the fields. Thus, it does not admit an action principle of the form (13), and does not admit of any simple translation into a direct-action principle in manner similar to that of electromagnetism.

Lacking any formal means of translating the Einstein-Hilbert action principle from "field language" to "particle language," we advocate a combined strategy: (1) Use particle-borne wristwatches plus light flashes or radar pulses between particle and nearby particle, (2) Use the resulting numbers to derive particle-to-particle separation and local Riemann curvature, (3) Put together these local curvatures to make a complete spacetime, as one fits together the pieces of a smashed porcelain vase to reconstruct the original object. As for this third, most difficult (and most interesting!) part of the enterprise, we have no prescription to offer. In this paper, we discuss the first two parts of this strategy, and make some comments on how the third part may be realized.

THE FIELD CONCEPT

The most elementary application of the field concept is that of a bookkeeping device, a function that assigns a force to a given charge or mass as a function of its position, velocity, or other characteristics. The introduction of the field enables the physicist to conceptually separate the action of particle on particle into their action on, and response to, the field.

As the theory of fields developed, it became clear that they provide a natural mechanism for enforcing "local" physics. By this, we mean the idea that the dynamics of a particle may be described in terms of qualities of its local environment. As the field was used more extensively to enforce these local laws, it became a dynamical quantity in its own right.

LOCAL CAUSALITY

In the classical theories of gravity and electrodynamics, the field is more than just a bookkeeping device. These fields help enforce *local causality*, which circumvents the conceptual problems of action-at-a-distance. Consider the following two equivalent presentations of Newton's gravitational theory, in which the force per unit mass $\vec{F}$ is given by,

$$\vec{F} = -G \int \rho(\vec{r}) \frac{\hat{r}}{r^2} d^3\vec{r} \tag{1}$$

$$\vec{\nabla} \cdot \vec{F} = -4\pi G \rho. \tag{2}$$

Both equations imply the same inverse-square law of force.[2] The first formulation reflects the action-at-a-distance of one body on another. The second formulation predicts the same physical result, but also lends itself to an explanation of action-at-a-distance in terms of laws that act only in *local* regions of space.

This may be seen even more clearly by an analogy. If we define the potential via the relation $-\vec{\nabla}\phi = \vec{F}$, we obtain the Poisson equation,

$$\nabla^2 \phi = 4\pi G \rho. \tag{3}$$

The same equation describes the static deformation ϕ of an elastic membrane with elasticity proportional to G^{-1} and mass density ρ. In the nonstatic case, the deformations of the membrane obey a wave equation, similar in some respects to the D'Alembert equation obeyed by the electromagnetic 4-vector potential A_μ.

In the "local interaction" picture, a mass influences the field in its immediate vicinity, in a manner quantified by (2), much as a mass deforms a local region of our elastic membrane. The influenced region further influences regions infinitesimally more distant from the mass, and so on. The influence of the mass is propagated, from the location of the source mass to the location of the test mass, through fully *local* interactions of the field with itself.

From the point of view of equation (2), *properties of the mass distributions distant from the point under consideration need not enter into the picture.* It is true that the total force on a particle depends on field originating at possibly large distances. The local, differential properties of the field, however, depend only on the source density at the point of investigation. Furthermore, in the case of gravitational fields, the equivalence principle forbids us to measure "absolute" gravitational acceleration of particles. In this case, everything we may measure about particle trajectories *is* determined by the local mass distribution.

AUTOMATIC CONSERVATION OF THE SOURCE

The introduction of the field as a dynamical entity in its own right has consequences for the expression of physical laws. We may generalize equations such as (2) and (3) to obtain a schematic form of classical field equations:

$$\text{(force on particle)} = \tilde{k} \cdot \text{(field)} \tag{4}$$

$$\text{(field derivatives)} = k \cdot \text{(source density)}. \tag{5}$$

Equations of this type are found in the field theories of classical gravity and electromagnetism, as well as in general relativity. When the field equations are expressed in the powerful language of differential geometry, physical conservation laws, such as the conservation of charge, may be "enforced" through geometrical concepts. As an elementary example, Maxwell's equations, when expressed in differential form, automatically imply the conservation of charge. This may be seen when the Maxwell equations are written[3] in terms of the Faraday-Maxwell 2-form $\mathbf{F} \equiv \frac{1}{2} F_{\mu\nu} \mathrm{d}x^\mu \wedge \mathrm{d}x^\nu$ and its dual $^*\mathbf{F}$,

$$\mathbf{dF} = 0 \qquad \mathbf{d}^*\mathbf{F} = 4\pi^*\mathbf{J}. \tag{6}$$

Here **d** is the usual exterior differential operator. If we calculate the divergence of $^*\mathbf{J}$ by applying **d** to the second equation, we find that $\mathbf{d}^*\mathbf{J} = \mathbf{d}^{2*}\mathbf{F} \equiv 0$. The vanishing of this divergence is automatically enforced as a consequence of the *geometric* identity $\mathbf{d}^2 \equiv 0$.

This identification has important consequences for conservation laws. For one, it implies that the naive expression of charge conservation ("total charge of the Universe is constant"),

$$\int Q \, d^3x = \text{const}, \tag{7}$$

is slightly misleading It must be replaced by a different law that enforces local conservation. Change in the local charge density cannot be compensated for by corresponding changes in the distant charge density. Instead, changes in local charge density must be accounted for by the divergence of local currents,

$$J^\mu{}_{;\mu} = 0. \tag{8}$$

While (8) implies (7), they are not equivalent. Again the field enforces a *local* description of physics.

The conservation of *charge* is built-in, as it were, to Maxwell's equations, as a geometrically conserved source. These same equations can be made to imply the local conservation of energy as well, provided we assign a stress-energy field to the electromagnetic fields. When we search for a tensor that is (1) bilinear in the dynamic quantities (fields),[4] and (2) divergent to exactly the extent required to compensate for the change in the current stress-energy due to interaction with the field,

$$T^{\mu\nu}{}_{;\nu} = -F^{\mu\alpha}J_\alpha, \qquad (9)$$

we find the proper assignment is,

$$T_{\mu\nu} = \frac{1}{4\pi}\left(F^{\mu\alpha}F^\nu{}_\alpha - \frac{1}{4}g^{\mu\nu}F_{\alpha\beta}F^{\alpha\beta}\right) \qquad (10)$$

The loss in energy by charge-carrying bodies ("work against the electromagnetic field") is compensated for by the *local* loss of field energy, as defined in (10). This assignment leads to experimentally verifiable predictions. The famous force of radiation reaction may be calculated by imposing local energy conservation, and calculating the energy carried away by the radiation fields of a charge.

We find a similar "automatic conservation of the source" in the theory of general relativity. In this case, the principle of automatic conservation of the source is applied to stress-energy itself. Indeed, the Einstein equations are uniquely defined when one postulates that

(1) Gravitational effects are caused by curvature in spacetime,

(2) The field equation for gravity are expressed in a form similar to (5), that is, terms containing field derivatives (in this case the connection coefficients of the spacetime geometry) are proportional to the local stress-energy density.

Thus, one begins with the general form,

$$G_{\mu\nu} = \kappa T_{\mu\nu}, \qquad (11)$$

and searches for a tensor $G_{\mu\nu}$ that has the proper relations to "natural" measures of curvature, eg, $G_{\mu\nu}$

(1) Vanishes in flat spacetime,

(2) Is built from the Riemann curvature tensor and the metric,

(3) Is linear in the Riemann tensor.

One then imposes the "automatic conservation of the source." In order to make (11) mathematically consistent, our tensor $G_{\mu\nu}$ must share some of the properties of the stress-energy tensor, namely $G_{\mu\nu}$ is

(1) Symmetric,

(2) Has identically vanishing divergence.

These considerations lead directly to the usual form

$$R_{\mu\nu} - \frac{1}{2}g_{\mu\nu}R = \kappa T_{\mu\nu} \qquad (12)$$

for the field equations for gravitation.[5] The ideas of local interaction and local physics are closely intertwined in equations such as (5) and (12). If interactions were not taken to be local in character, we should not be able to write down expressions for the forces in terms of the fields at all. In addition if conservation laws were not local, the field equations would not be mathematically consistent.

QUANTIZATION

Quantum theory describes particles in a field-like manner. Conversely, fields may themselves be described in a particulate, or "quantized," manner. The local, field picture of interaction allows one to start from the Lagrangian density associated with a field equation, and treat the fields as quantum systems. Currently, we believe that the correct description of natural forces is found within the framework of quantized field theories. The fact that no satisfactory quantum description of the gravitational field has yet been found is widely believed to indicate a failure in methodology, and not a failure of quantum theory. Such a prescription reinforces the idea of local physical law through describing the action of one particle on another by means of a medium that spans the space between them.

IS PHYSICS ENTIRELY LOCAL?

A major advantages of the field concept is the natural way in which the fields enforce local physical laws. This begs the question: is physics local? One is reminded of an argument against quantum theory advanced by Einstein, Podolsky, and Rosen in a well-known paper (1935). The authors assert that the quantum-theoretical process of the reduction of the wave packet (collapse of the wavefunction upon measurement) is not physically realistic, since it allows distant measurements to affect local physics. The implicit nonlocality of this process, they argue, is at odds with the idea that physics should be fundamentally local.

In a situation such as this, one is also reminded of the words of Ernst Mach,

> The most important result of our reflection is, *that precisely the apparently simplest mechanical principles are of a very complicated character, that these principles are founded on uncompleted experiences, nay on experiences that never can be fully completed, that practically, indeed, they are sufficiently secured, in view of the tolerable stability of our environment, to serve as the foundation of mathematical deduction, but that they can by no means themselves be regarded as mathematically established truths but only as principles that not only admit of constant control by experience, but actually require it* (Mach 1902, 2:VI, 9).

As has been evidenced by many experimental tests, the view of nature espoused by Einstein et. al. is not quite correct. Various experiments have shown that distant measurements can affect local phenomena. That is, *nature is not described by physical laws that are entirely local*. Effects from distant objects *can* influence local physics. While we do not consider quantum gravitational theories here, this example from quantum theory serves to illustrate that it may be useful to expand our notions regarding what types of physical law are "allowed."

DIRECT ACTION-AT-A-DISTANCE

Here, we consider the question of formulating physical laws without introducing the field as a separate physical entity. The idea that local physics, and electromagnetic radiative phenomena in particular, involved interactions with distant bodies was given a formal treatment by Feynman and Wheeler (1945). In their paper, this idea was developed quantitatively into the "absorber theory of radiation." We review the concepts involved and consider how they may be applied to the description of gravitational interactions.

WHEELER-FEYNMAN ELECTRODYNAMICS

Wheeler and Feynman extended an interesting result obtained by Dirac (1938). In his paper, Dirac observed that the wave equations of electromagnetism admits solutions corresponding to waves propagating not only forward in time, but backward as well.

Dirac obtained the following prescription for calculating the force of radiation reaction on a moving charge. First, let the motion of the charge be given. From this, calculate the advanced and retarded electromagnetic fields, with the boundary conditions that the advanced field contains only incoming waves, and the retarded field only outgoing waves, both of which vanish at infinity. Then, calculate a new field, equal to one-half the difference between the advanced and retarded fields. This field is everywhere finite, and when evaluated in the neighborhood of the charge, gives precisely the force of radiation reaction.

Dirac's prescription is a clearly defined, and yields the correct answer. Despite this, the mystery of the origin of the "half-advanced," "half-retarded" field is left unresolved. In their 1945 paper, Wheeler and Feynman considered the action of the advanced field of the source on a distant absorbing medium. They found that the retarded field of the absorber, when added to the retarded field of the source itself, is equal to precisely the field postulated by Dirac.

In effect, they swept the electromagnetic field from between the charged particles and replaced it with "half-retarded, half-advanced direct interaction" between particle and particle. It was the high point of this work to show that the standard and well-tested force of reaction of radiation on an accelerated charge is accounted for as the sum of the direct actions on that charge by all the charges of any distant complete absorber. Such a formulation enforces global physical laws, and results in a quantitatively correct description of radiative phenomena, *without* assigning stress-energy to the electromagnetic field.

CORRESPONDENCE BETWEEN FIELD AND DIRECT INTERACTION THEORIES

Dirac's prescription is interesting for several reasons. Conceptually, while this theory requires us to rethink some of our notions of causality, it does not violate them. More importantly, the theory starts with the minimum necessary framework upon which electrodynamics can be built. There are no entities other than the particles; the field itself is removed from consideration, either as a "automatic conserver of the source," or as the conveyor of stress-energy from the accelerating charge.

In addition to the conceptual simplicity of the theory, it is also more convenient mathematically. One need not calculate the dynamics of the field, a complex dynamic quantity with an infinite number of degrees of freedom; only the particles, with their finite number of degrees of freedom.[6] Can such a procedure be extended to other types of fields?

Hoyle and Narlikar (1974) have pointed out that there is a general, mathematical correspondence between field theories and direct-action theories. They quote a general result obtained by Narlikar (1968). Let a field be described by a tensor field ϕ of rank N, and let the dynamics of the field in interaction with particles be given by an action principle of the type

$$\int L[\phi]\sqrt{-g}\, d^4x + \sum_i \int I[\phi,(i)]\, dX^{(i)}, \qquad (13)$$

where L is a bilinear invariant built from ϕ and its first derivatives, and $dX^{(i)}$ is an appropriate measure of the configuration space of the ith particle. $I[\phi,(i)]$ is an invariant describing the interaction of the particle with the field, of the form

$$I[\phi,(i)] = g \cdot \langle \phi, \xi^{(i)} \rangle. \qquad (14)$$

Here, $\xi^{(i)}$ is a rank-N tensor depending only on the worldline of the ith particle, and g is a coupling constant. To the field theory given by (13) and (14) there exists a direct-action theory of the form

$$\frac{1}{2}\sum_{i,j} g^2 \int\int \langle \xi^{(i)}, \bar{G}(X^{(i)}, X^{(j)}) \cdot \xi^{(j)} \rangle\, dX^{(i)} dX^{(j)}. \qquad (15)$$

Here, $\bar{G}(,)$ is the symmetric[7] Green's function of the field equation derived from (13).[8] Such a direct-action theory gives the direct field due to particle j at the point X as

$$\phi(X) = g\int \bar{G}(X, X^{(j)}) \cdot \xi_{(j)}\, dX^{(j)}. \qquad (16)$$

We recognize that electrodynamics is a field theory of the form (13). Radiative effects may be calculated using the effects of fields on distant particles, and radiation reaction may be calcuated using the procedure of Wheeler and Feynman.[9] Electrodynamics has its field-free expression in the Fokker-Tetrode-Schwarzschild action principle,

$$I_{em} = \sum_i m_i \dot{z}_i^\mu \dot{z}_{i\mu} \qquad (17)$$

$$+ \sum_{i,j \neq i} e_i e_j \int \dot{z}_i^\mu(\tau) \dot{z}_{j\mu}(\bar{\tau}) \bar{D}(z_i, z_j) \, d\tau d\bar{\tau}.$$

Such an action principle yields the half-advanced, half-retarded direct interactions employed by Wheeler and Feynman in their analysis of electrodynamics.

ANALOGY WITH GRAVITATIONAL THEORY

The description of gravity in general relativity does not admit an expression in the above manner. The first failure occurs in (13), where the Hilbert variational principle for gravitation is not quadratic in the fields. Just the same, the theory of general relativity is itself a field theory. There is a formal similarity between the equations of motion in the conventional electromagnetic field theory,

$$\dot{p}^\mu = e F^\mu{}_\nu u^\nu, \qquad (18)$$

and the equations of motion of masses under the influence of gravity,

$$\dot{u}^\mu = -\Gamma^\mu{}_{\sigma\tau} u^\sigma u^\tau. \qquad (19)$$

The two theories share some of the same conceptual features as well. Might it be possible to introduce a field theory similar to that of Wheeler-Feynman electromagnetism for gravity? Before we get too involved, we must note that the unique nature of gravitation among field theories requires us to discuss several considerations that were not relevant in the case of electrodynamics.

THE EQUIVALENCE PRINCIPLE

Formally, the metric plays a role in gravitation theory similar to that of the vector potential in electrodynamics. The "field" of gravitational theory is, however, fundamentally distinct from the electromagnetic field in several key ways.

In the case of electromagnetism, the strength of a particle's interaction with the field is proportional to the particle's charge, while the acceleration of the particle in response to the field is determined by the mass of the particle. The charge q and the mass m of a given body are independent, and so the electrodynamics of a system of particles is determined when these two numbers are specified for each particle.

We may translate these ideas into gravitational theory, and specify that each mass point has a "gravitational mass" that determines its contribution to the field, and an "inertial mass" that determines the dynamics of the particle under the influence of a gravitational field. Unlike electrodynamics, the gravitational mass and inertial mass are found to be proportional to an exceedingly high degree of precision. Gravitational theories must therefore follow one of two routes. The first method is to assume that the proportionality is a mere coincidence, and to introduce the aforementioned gravitational and inertial masses. The other possibility is to accept the striking nature of such a result as reflecting an important physical principle. Einstein chose the latter path, and incorporated this *equivalence principle* directly into the structure of the general theory of relativity. By his casting the gravitational field as a fundamental *geometric* property of spacetime, all masses in a given region of spacetime "feel" the same geometry and automatically experience the same acceleration. A theory of gravitation that does not involve the metric field should incorporate this equivalence principle in a similarly elegant way, or risk taking a step backwards and ruining the simplicity of this approach.

This equivalence principle, and the association of gravitation with geometry, has other consequences as well. Electromagnetic theory presupposes that we are able to measure times, distances, and other physical quantities in a manner unaffected by the system we are studying. In general relativity, this is not possible. Gravitational theory is intricately connected with our basic ability to measure space and time. The field "works" through distortions of the network of "rods and clocks" that we use to parameterize events in spacetime. Gravitational effects, because of their universal nature, are intertwined with our means of measuring them. The interrelationship of the field we wish to study and the means of measurement significantly complicates the problem of formulating a field-free theory of gravitation.

TOWARDS A FIELD-FREE GRAVITATION

Once we have removed spacetime, what are we to replace it with? Theodore Roosevelt used to advise, "Do what you can with what you have where you are." We report here the line of reasoning to which this injunction has led us. First, having nothing but particles and their histories to work with, we parameterize the history of each particle by a scalar. Thus a monotonically increasing parameter α distinguishes successive points (events) on the worldline of particle a; β events on particle b, γ events on c, etc. We have as yet no tool to measure the passage of time on particle a or any other particle, still less any tool to measure the distance between particle and particle.

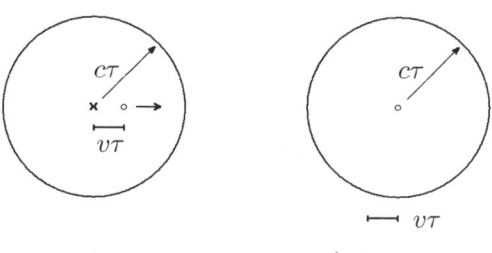

Figure 1. The role of observation in special relativity. Both scientist in laboratory frame (left) and observer moving with the source (right) observe an emitted pulse of light to take the form of a sphere of radius $c\tau$ a time τ after emission.

A field-free gravitational theory would have to be expressed in terms of whatever tool is chosen to map out events on the observer's worldline, as well as other nearby worldlines. Of course, this introduces the question of the proper tool to use. This situation is similar to that found in special relativity. The theory holds that, while there is no canonical reference frame, fig. 1, every reference frame is related to every other one by a set of coordinate transformations (the Lorentz group). Similarly, a field-free gravitational theory must make reference to a system of measurements between events, and relate different groups of such measurements.

RODS AND CLOCKS FOR MEASURING GRAVITY

The usual means of measuring space and time are already too complex for our use. Mechanical rods and clocks are affected by a host of factors. The analysis of these effects (with an aim to "correcting" for them) can take us far afield into quantum theory or solid-state physics. There is, however, a means of measuring spacetime whose analysis requires only the relativistically invariant theory of electromagnetism. Furthermore, this method enables one to measure directly intervals in spacetime. Such a "clock and measuring rod" is provided by the "ideal clock" discussed by Marzke and Wheeler (1964). An observer may use such a device to measure times, distances, or indeed any type of spacetime interval.

The ideal clock may be constructed as follows. Let an observer free-fall along a geodesic γ_1. The observer may construct a parallel path γ_2 (not necessarily a geodesic) using the Schild's ladder construction, as discussed in (Misner, Thorne, and Wheeler 1973). This construction enables one to trace out a trajectory parallel to a given trajectory, in this case the observer's geodesic. The advantage of the Schild's ladder construction is that it is entirely geometric, and requires only the tracing of lightlike trajectories in space and time.

In order to perform measurements, a beam of light is made to bounce back and forth between the two trajectories. Each bounce of the light at a point on γ_1 is considered one "tick" of the clock. Measurements of space and time are made in terms of multiples of these clock ticks (see fig. 2).

The gravitational field is measured through so-called *tidal effects*. The equivalence principle forbids one to measure the absolute gravitational acceleration of a body. Instead, one may set up two test particles at positions x^μ and $x^\mu + \xi^\mu$, and allow them to fall freely. Gravitational acceleration (in geometric language, spacetime curvature) manifests itself as a change in the relative 4-velocities of the particles, $\dot{\xi}^\mu$. In Riemannian geometry, this change may be given as a function of the 4-velocity of the observer and the Riemann curvature tensor $R^\mu{}_{\nu\sigma\tau}$,

$$\ddot{\xi}^\mu + R^\mu{}_{\nu\sigma\tau} u^\nu \xi^\sigma u^\tau = 0, \tag{20}$$

In the rest frame of one of the particles, this equation takes the form,

$$\ddot{\xi}^\mu + R^\mu{}_{0\sigma 0} \xi^\sigma = 0. \tag{21}$$

An observer who wishes to measure all of the components of the Riemann tensor need only send pairs of "ideal clocks" out in various directions, each pair equipped to

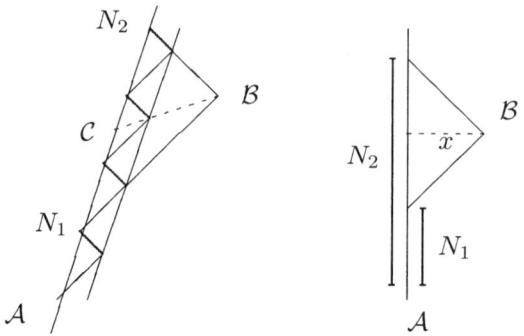

Figure 2. *The ideal clock of Marzke and Wheeler. The interval between the points A and B is the product $N_1 N_2$: for a proof consider the right-hand diagram, and the relation*
$$(t - x)(t + x) = t^2 - x^2.$$

measure its separation using the interval measurement system described above. One member of each pair emits a light signal at a time α^-, which reflects from the other member of the pair at a time β, and is received by the first member of the pair at a time α^+. The tabulation of these times forms the fundamental raw data upon which observations are built, and against which theory must be tested.

Although a single "clock" may infer information about the position of a distant particle, perhaps by a combination of distance and directional measurements, better information may be gleaned by using a set of such devices in combination. Each device may be used to measure the interval between itself and the distant particle at various points along its trajectory. Conceptually, such a measurement places the particle under consideration on one of many hypersurfaces of constant interval which foliate the local regions of space time.[10] Four such devices, operating together and with different 4-velocities, will localize the test particle on the intersection of four such hypersurfaces. The fact that all of the devices will agree on a consistent localization of the particle is a feature of Riemannian geometry that has not been shown to be violated thus far.[11]

In such a measurement scheme, certain observed quantities depend on the observer's history during the measurement process. Let our observer introduce a coordinate system in which the coordinate lines are formed from the trajectories of freely falling masses and light rays (geodesics). All points connected to a given point C on the observer worldline, by geodesics whose tangents at C are spacelike, are causally disconnected from C. In this scheme, however, the measured position of the distant particle is a function of the observer's trajectory between the points of emission and reflection of the observer's "radar signal."

Consider the following situation: The observer on the first particle, traveling along trajectory AB, finds the points C and D to be separated by a purely spacelike interval. Now, let us make an identical copy of the spacetime geometry and particle trajectories. In the second version, some kind of localized disturbance (perhaps a gravitational wave packet traveling through the area) affects our observer between the points C and B. The observer is kicked on to another geodesic $C\tilde{B}$, fig. 3.

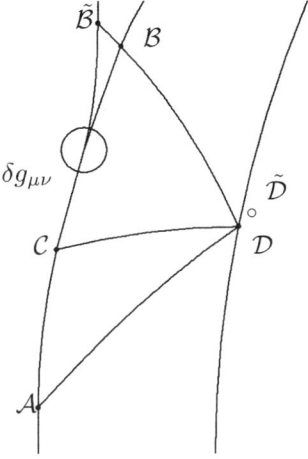

Figure 3. Indeterminacy in measurements in spacetime.

Now, when the observer receives the return signal from the distant particle, he concludes that the particle occupied a different position at a different time than in the previous case. The properties of the spacetime metric near the observed point have not changed, and yet its observed trajectory is altered. Causally, the perturbation $\delta g_{\mu\nu}$ cannot causally affect points C and D, and yet the measurement of spacetime properties near D must ultimately refer to the observer trajectory throughout the measurement process, a fundamentally nonlocal effect.

INTERACTIONS

In a spirit similar to our method of analyzing spacetime via particle-to-particle signals, we may discuss interactions of particles in terms of "signals." The analogue of a light signal between an observer and a test particle is the propagated fields, whose influence travels at a finite speed. The point masses "recognize" each other in this way before there can be any talk of interaction. In keeping with the time-symmetric character of elementary action-at-a-distance electrodynamics, we adopt a time-symmetric interaction protocol. This is also suggested by the results of Narlikar on the correspondence between field and direct particle-particle theories, which might lead us to expect that the final mathematical formulation of a direct particle-particle theory of gravity would involve both advanced and retarded interactions. Also, our considerations of the act of measurement suggests that we should consider the state of the observer at both the retarded and advanced points of action.

Thus, to every point or event α on the worldline of particle a, we associate two events on the worldline of particle b, $\beta^+(\alpha)$ and $\beta^-(\alpha)$, respectively the retarded and advanced points of action of a on b. These *associators* evidently satisfy trivial identities of the form

$$\alpha^-(\beta^+(\alpha)) = \alpha^+(\beta^-(\alpha)) = \alpha \qquad (22)$$

These ideas may be mathematically expressed through a one-half advanced, one-half retarded gravitational theory.

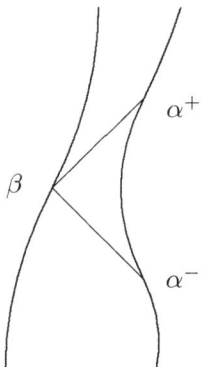

Figure 4. Associators

COMMENTS ON A FIELD-FREE FORMULATION OF LINEARIZED GR

Following this line of reasoning, one may construct a "toy theory" of gravity that incorporates many of the above elements A simple way to do this is to square the velocity terms in (17) and neglect the initial kinetic term. When a variation is carried out, one ends up with a field-free theory of gravitation that is formally identical to linearized general relativity.

Interestingly enough, the equation which results results in an equation of motion *without additional constants*. The usual gravitational constant appears as rank-2 tensor which depends on the mass and velocity distribution in the distant universe. This type of theory is similar to ideas discussed by Sciama (1957) and others. The identification of these terms is found by Wheeler to be reasonable by order-of-magnitude arguments (Wheeler 1964). Additional discussions on similar theories that depend on the mass distribution "at infinity" may be found elsewhere (Ciufolini and Wheeler 1995, chap. 7; Hoyle and Narlikar 1974).

The dependence of the theory on the mass distribution at infinity is somewhat of a difficulty. This is especially problematic when, in the case of this theory, the interaction lagrangian depends on mass-energy *currents* instead of scalar densities. While the isotropy of the distant universe (the "cosmological principle") may be invoked to estimate the contribution from the total mass scalar density, the presence of current terms of the form $\dot{z}_m \dot{z}_n$, requires detailed assumptions about the relationships between the components of the mass-energy currents at infinity. Unlike the Feynman-Wheeler electrodynamics, which requires only that "infinity" contains a perfect absorber, this "toy" theory depends on a detailed way on the dynamics at infinity.

CONCLUSION

If we aim to "sweep out" spacetime in the conventional sense, it is only so that spacetime may be resurrected, not as another field to be studied, but as a network of observations and events. A true field-free expression of gravity will involve more than just an analogue of expressions for field-free lagrangians for other fields. Instead, what will be needed is a revolution in our fundamental conception of what spacetime is. John Stachel, with his colleagues, has given us the opportunity to understand the man who first gave us our modern conception of a gravitational field theory. It is hoped that, by freeing the theory from the constraints of its field formulation, a deeper understanding of gravitation may be achieved. Such an understanding, however, will require a profound rethinking of the meaning of space and time.

Princeton University
University of Texas

NOTES

* In honour of Professor Stachel. He will agree with me on points, if not on substance, but I am indebted to him for his guidance on both topics; and on many others besides.

1. One there was, to be sure, who contested Princeton University Press's choice of Stachel as editor. That was Otto Nathan, executor of the Einstein estate, who, according to the book of Highfield and Carter, evidently feared what Stachel would turn up. Nathan brought an unsuccessful legal action against the Press, in which one of us (JAW) well remembers testifying on behalf of Stachel and the Press against the barrage of disparagement from Nathan's lawyer.
2. Actually, one must specify appropriate boundary conditions (the vanishing of the field at infinity) in the case of (2), and so the second equation in some sense contains "less" information than the first.
3. An equivalent statement may be made in the the usual multivariable calculus. Combining the Maxwell equations $\vec{\nabla} \cdot \vec{E} = \rho$ and $\vec{\nabla} \times \vec{B} = \dot{\vec{E}} + \vec{J}$, with the identity $\vec{\nabla} \cdot \vec{\nabla} \times \equiv 0$ yields the law of charge conservation, $\frac{\partial \rho}{\partial t} + \vec{\nabla} \cdot \vec{J} = 0$.
4. This quadratic dependence is suggested in the quadratic dependence of kinetic energies for other dynamical quantities, such as particles in classical mechanics.
5. See (Misner, Thorne, and Wheeler 1973), chap. 17 for additional details.
6. Of course, in the case of continuous source distributions, there is an infinite number of degrees of freedom.
7. $\bar{G}(,)$ obeys the relation $\bar{G}(X^{(i)}, X^{(j)}) = \bar{G}(X^{(j)}, X^{(i)})$. Thus, $\bar{G}(,)$ must also be invariant ("symmetric") under time reflection.
8. Typically, we may exclude the $i = j$ terms from the summation. A brief discussion of methods to handle this "self-energy" is given by Barut (1980).

9. It should be noted that, while stress-energy is not assigned to the field per se, when one considers a combined action $I_{em} + I_{gravity}$, variations in the $g_{\mu\nu}$ yield a divergence-free object identical in structure to the usual electromagnetic stress-energy tensor. This occurs because the geometry along the line of propagation depends on the metric; when the metric is varied, the $\bar{G}(,)$ is varied as well.
10. That is, a region small enough so that the paths of geodesics that begin with different velocities do not cross (a necessary condition for the uniqueness of assigning coordinates).
11. See (Marzke and Wheeler 1964) for additional discussions on this point, and on the Marzke-Wheeler clock in general.

REFERENCES

Barut, A. O. 1980. *Electrodynamics and Classical Theory of Fields and Particles*. New York: Dover Publications.

Ciufolini, Ignazio, and John Archibald Wheeler. 1995. *Gravitation and Inertia*. Princeton: Princeton University Press.

Dirac, P. A. M. 1938. "Classical theory of radiating electrons." *Proc. Roy. Soc. London*, pp. 148–168.

Einstein, Albert, Boris Podolsky, and Nathan Rosen. 1935. "Can the Quantum-Mechanical Description of Reality be Considered Complete?" *Phys. Rev.* 47:777–80.

Hoyle, F., and J. V. Narlikar. 1974. *Action at a Distance in Physics and Cosmology*. San Francisco: W. H. Freeman.

Mach, Ernst. 1902. *The Science of Mechanics*. 2nd ed. London: Open Court.

Marzke, Robert F., and John A. Wheeler. 1964. "Gravitation as Geometry–I." Pp. 40–64 in *Gravitation and Relativity*, eds. Hong-Yee Chiu and William F. Hoffman. New York: W. A. Benjamin.

Misner, Charles W., Kip S. Thorne, and John Archibald Wheeler. 1973. *Gravitation*. New York: W. H. Freeman.

Narlikar, J. V. 1968. "On the general correspondence between field theories and the theories of direct interparticle action." *Proc. Camb. Phil. Soc.* 64:1071.

Sciama, Dennis. 1957. "Inertia." *Sci. Am.* 196 (2): 99–109.

Wheeler, John A. 1964. "Mach's Principle as Boundary Condition for Einstein's Equations." Pp. 303–350 in *Gravitation and Relativity*, eds. Hong-Yee Chiu and William F. Hoffman. New York: W. A. Benjamin.

Wheeler, John Archibald, and Richard Phillips Feynman. 1945. "Interaction with the Absorber as the Mechanism of Radiation." *Rev. Mod. Phys.* 17 (2): 157–181.

III

Foundational Issues in Quantum Physics and their Advancement

MARA BELLER

INEVITABILITY, INSEPARABILITY AND *GEDANKEN* MEASUREMENT

In Dostoyevsky's novel "Notes from the Underground" the hero expresses the suffocating deterministic vision of Enlightenment Science:

> "You cannot revolt: this is two times two equals four! Nature does not ask you, she cares nothing about your wishes and whether you like or dislike her laws. You must accept the way Nature is ..."

How strangely reminiscent these lines are of those advanced by the founders of the Copenhagen Interpretation. To hope for a causal quantum physics is identical to the hope that $2 \times 2 = 5$ (Heisenberg 1958, 132)—Bohr and Heisenberg ridiculed David Bohm's "hidden variable" program. "*We must accept the way nature is and not prescribe to God not to play dice*" Bohr replied, disposing Einstein's objections. While heralding the breakdown of determinism, the founders of the Copenhagen Interpretation relentlessly argued that the indeterminism of quantum mechanics is "inevitably" determined. Their writings are permeated with such oxymoronic expressions as "necessity of acausality" or "inevitability of indeterminism." Bohr and Heisenberg advanced their arguments by an analysis of numerous Gedanken experiments, illustrating the wholeness, acausality, and impossibility of an objective realistic description in the quantum domain.

The notion of quantum wholeness is common to Bohr's and to Bohm's interpretation. Yet Bohm's interpretation is causal, while Bohr denies the possibility of a causal interpretation of quantum formalism. Moreover, in Bohr's numerous illustrations of thought experiments, the notions of wholeness and acausality are inseparably linked to each other. How is it possible? What kind of holism does the Copenhagen Interpretation advance, and what can we learn about the argumentative strategies of Copenhagen orthodoxy? In this paper I will only offer a partial survey of these issues. A full discussion is given in (Beller 1999).

Heisenberg's uncertainty paper (Heisenberg 1927) was aimed to provide a satisfying interpretation of the new quantum theory. Today there are many interpretations of quantum theory—the Copenhagen, the many world interpretation, the realistic statistical one, the (non-local) hidden variables one, the modal interpretation, the quantum-logical interpretation. Yet there is no agreement on the basic question: what does it mean to interpret a mathematical-physical theory?

The minimal instrumentalist interpretation of quantum theory contains quantization algorithm for any given observable (it determines what values are possible in measurements) and statistical algorithm (it gives the probability of a possible measurement result). Yet such a minimal instrumentalist interpretation serves more as a refuge when one is pressed by the paradoxes and contradictions (especially in the case of the Copenhagen interpretation) than a consistent, committed interpretive stand. Most physicists, Bohr and Heisenberg included, wanted more: some feeling of understanding, of insight, of explanation about the kind of the world that the quantum formalism describes. The need for such a metaphysical grasp is not merely psychological, but a social one as well—the power of successful explanation, and the power of an effective legitimization and dissemination of a theory are interconnected.

One need not be a naive realist to appreciate the need for a satisfying interpretation beyond the minimal instrumentalistic one. Van Fraassen formulated the issue of interpretation in the following way: *"What would it be like for this theory to be true, and how could the world possibly be the way this theory says it is?"* (van Fraassen 1980, 242). Heisenberg had his own succinct definition of interpretation: to understand a theory means "daraus klug zu werden"—to get a sense out of a theory, to get a feeling of enlightenment and insight. In his letters preceding the formulation of uncertainty principle, Heisenberg expressed his desire "klug zu werden" more than once. Yet the sense of this expression changes—originally Heisenberg's major voice was that of denial of the possibility of quantum ontology (the only enlightenment one could hope for is to understand the transition from macro to microdomain); after Schrödinger's challenge Heisenberg, more ambitiously, began the search for a deeper understanding. In a broad sense, the Göttingen-Copenhagen physicists sought to understand the "naturalness" of agreement between theory and experiment. For Born, a Leibnizian preestablished harmony existed between the world of mathematical symbols and the observed laboratory facts. This Leibnizian notion was retained in Bohr's Como lecture, which was woven around the "harmony" between definition and observation.

Heisenberg's letter to Pauli (Feb. 27, 1927; Hermann, et al. 1979), as well as the uncertainty paper, is built around "making sense out" of the uncertainty formula in numerous thought-experiments.

Heisenberg had no definite notion of interpretation, nor did he need one. All he aimed at, wrote Heisenberg in the abstract to the uncertainty paper, is "*to show how microscopic processes can be understood*" through quantum mechanics (Heisenberg 1927, 62). Heisenberg's goal was to "illustrate" quantum theory by discussion of thought experiment (ibid.). Yet in order to effectively disseminate the new theory, and to make it palatable, Heisenberg had to offer not simply a satisfying "illustration," but an intuitable (*anschauliche*) one. How can one claim intuitiveness to such an abstract, unintuitive formalism? At the very end of his letter to Pauli, Heisenberg makes a hesitant suggestion: perhaps one can say that one understands physical laws in an intuitive way, when one can immediately say in each experimental case what should occur –in this way the sense of understanding is advanced (Feb. 27, 1927; Hermann, et al. 1979). After Pauli's support, this hesitant suggestion transformed into the new confident redefinition of intuitability by which Heisenberg opened the uncertainty paper:

"We believe we understand the physical content of a theory when we can see its qualitative experimental consequences in all simple cases and when at the same time we have checked that the application of the theory never contains inner contradictions" (Heisenberg 1927, 62).

Heisenberg's uncertainty paper conquered a permanent place in the annals of science and in history of philosophy, not because of his arguments of *Anschaulichkeit*, but because of its radical message of the "final" overthrow of the law of causality.

What is the ultimate reason for Heisenberg's "final" overthrow of causality? The mere fact of employment of statistical formalism for the derivation of the uncertainty principle does not, of course, signify any predilection to "essential" acausality—this is the reason that Born, Jordan and Pauli all considered the problem of causality open before Heisenberg's uncertainty paper. Is quantum mechanics "characteristically" a statistical theory, as Born and Jordan claim? Or is statistics only "brought in by experiments," as Dirac maintained? Heisenberg's resolution was to bring to a full attention the impossibility to know the present, and therefore to predict the future, thus disposing of any hopes of determinism. How can one argue that the present cannot be known exactly? One can point out, of course, that it follows from the uncertainty formula which is a mathematical consequence of quantum mechanics. The persuasiveness of the overthrow of acausality depends thus on the status of the "finality" of quantum theory. In 1927, when this highly abstract and unconventional theory had yet to prove its mettle, such argument would convince only the believers.

Heisenberg therefore needed some arguments showing *"how the world is the way quantum mechanics says it is"* without direct appeal to the formalism. This is the reason Heisenberg preferred Dirac's stand, seeking to understand how *"statistics is introduced by experiments."* Heisenberg's search resulted in the introduction of the (misleading) disturbance imagery. The disturbance imagery underlied Heisenberg's analysis of position measurement. The validity of the uncertainty formula seemed to follow simply and inevitably from the analysis of conditions of experiment, without any recourse to quantum mechanical formalism. Yet, in the description of the γ-ray *Gedanken* experiment, Heisenberg committed a trivial error, which both Bohr and Dirac (who was at the time in Copenhagen) alerted Heisenberg to (May 16, 1927). Heisenberg treated both photons and electrons as regular point particles, and argued that at the time of their collision a photon transfers to an electron a discrete and uncontrollable amount of momentum (Compton recoil). The more precisely the position of the electron is determined, the greater is the uncertainty of the discontinuous change of the electron's momentum. Yet Compton recoil would not lead to indeterminacy, but to exactly calculable changes. From the conservation laws for energy and momenta, and from the knowledge of the energy and momentum of colliding photon, one can calculate the momentum change of electron exactly. Many of the Copenhagen Gedanken experiments contain mistakes of one kind or another.

Heisenberg uses the same disturbance imagery in order to argue the inapplicability of the concept of path in the interior of an atom, again without recourse to quantum mechanical formalism. One can deduce this fact *"without knowledge of recent theories, simply from experimental possibilities"* (Heisenberg 1927, 1965). Consequently, the impossibility of causal space-time description seems to follow directly and inevitably from the conditions of experience, as Bohr later claimed.

Two voices characterize Heisenberg's discussion on causality, *Anschaulichkeit* and other issues in the uncertainty paper. These conflicting voices persevere in later writings by Göttingen-Copenhagen physicists, often changing their pitch but never completely fading away. One voice is that of "inevitability": through simple analysis of experiments one can see that the state of affairs claimed by quantum theory is unavoidable. The other voice is that of "consistency"—analysis of experiments demonstrate the consistency of quantum theory, because the results of such an analysis do not contradict the quantum mechanical formalism. The inevitability arguments are often mixed with those of consistency. Heisenberg's concluding words in the uncertainty paper are uttered in a more sober, "consistency," voice. The "inevitable" acausality follows from the correctness of quantum mechanics (which covers also the measurement process): *"Because all experiments are subject to the laws of quantum mechanics and therefore to equation $[p_1 q_1 \sim h]$, it follows that quantum mechanics establishes the final failure of causality"* (Heisenberg 1927, 83).

Heisenberg experienced a conflict between his desire to render matrix mechanics intuitively appealing to wide circles and between his infatuation with Dirac's abstract version of quantum theory. This conflict was not resolved: it permeates the uncertainty paper. Heisenberg vacillates between his attempts to "intuit," or "understand," quantum theory through discussion of simple thought-experiments, and between his confident derivation of philosophical consequences from the mathematical formalism of the quantum theory.

Heisenberg was considerably more proficient than Bohr in working with the mathematical tools of the quantum-mechanical formalism. Heisenberg had no reasons therefore to give up the attempts to explore the ontological significance of the quantum formalism, deducing the genuine features of the quantum world from characteristics of the mathematical formalism. The invariant features of this formalism, according to Heisenberg, explain why the uncertainty relations must hold. In order to obtain physical results from Dirac's formulation of quantum theory, one has to associate regular numbers with the q-numbers—matrices, or "tensors" in multidimensional space. A definite experimental setup prescribes a definite direction in this multidimensional space—the question of "value" of the matrix in this direction has a well defined meaning only when the given direction coincides with the direction of the principal axis of the matrix (Heisenberg 1927, 65). When these two directions differ a little from one another, one can still talk about the "value" within a certain statistical error. A definite experiment therefore cannot give exact information about all quantum-theoretical quantities—it rather divides the physical quantities into "known" and "unknown," as is exemplified in the uncertainty formula. In this way Heisenberg "makes sense" out of the abstract quantum formalism.

It was Bohr, of course, who utilized the notion of disturbance to its utmost, until the "disturbance" notion was undermined, and abandoned, due to the EPR challenge. Bohr analyzed the mutual disturbance of the measured object and the measuring device in order to argue the inevitability of acausality (the inevitability of the uncertainty principle) and inevitability of complementarity.

Bohr's arguments of "inevitability" of acausality and complementarity similarly cannot withstand close scrutiny. I have analyzed this issue extensively in (Beller 1999, ch. 9-10). I have argued that Bohr's arguments of inevitability are, at best, disguised arguments of consistency. Here I will present briefly only a few relevant arguments. "The uncontrollable disturbance," or uncontrollable interaction between the measuring and the measured, is the crucial component of all these arguments, and the basic building block on which Bohr's web of arguments for complementarity rests. Bohr employed different strategies for making this "uncontrollable interaction," founded on the "indivisibility of the quantum of action," a fundamental concept in quantum philosophy.

Yet how exactly does the "uncontrollability of the interaction" follow from the "indivisibility of the quantum of action," and what do these terms mean? Their use and their interconnection vary, in fact, significantly, before and after 1935 (the EPR challenge). Before 1935, Bohr's arguments are incorrect, and are based on the discredited later idea of disturbance. After 1935, the arguments are tautological, or circular: textual analysis reveals a web of terms, where "indivisibility of the quantum of action," "individuality of quantum phenomena," "unsurveyability of interaction," or "uncontrollability of interaction" have the same meaning, and are interchangeable.

The following lines are typical in Bohr's pre-1935 writings:

> "Quantum of action means that it is not possible to distinguish between the phenomena and observation... we cannot make the interaction between the measuring instrument and the object as small as we like. Now, that means that phenomena are influenced by observation." (Bohr 1930).

Before 1935 "indivisibility of the quantum of action" simply meant "finitude" of the quantum of action. Measurement "disturbs" the phenomena precisely because of the atomic constitution of measuring devices. The underlying intuition was that because in the quantum domain phenomena and measuring interaction are "of the same order of magnitude," such finite interactions, contrary to the classical case, cannot be neglected or accounted for. Such intuitively appealing argumentation was, of course, incorrect. The concept of disturbance is an inconsistent one: it presupposes the existence of objective exact values that are changed by measurement, contrary to the desired conclusion of their indeterminateness. Moreover, the finitude of interaction does not assure uncontrollability—many classical measurements involve finite interactions that cannot be neglected yet can be calculated.

After 1935, when the EPR challenge undermined the notion of disturbance, Bohr's reasoning underwent an about-face. Bohr's response to EPR constitutes a threshold in respect to this notion. In the paper itself one can discern two conflicting answers, each having a different underlying notion of disturbance (Beller and Fine 1994; Beller 1999). The first (operational) answer dispenses with this notion altogether, while the second answer, which involves the "physical realization of the EPR argument, still implicitly uses the notion of disturbance. In Bohr's post-EPR writings the "uncontrollability" of the interaction follows from its "unsurveyability," which in turn follows from the dogmatically postulated principal difference between the quantum object and classical measuring apparatus. If, in pre-1935 Bohr, the uncontrollability necessitated acausality, in the later writings it is the statistical character of quantum theory that leads to inseparability of the measuring and the measured, and to the "uncontrollability" and "unsurveyability" of the measuring interaction.

There were many attempts, both by physicists and philosophers, to present a coherent and convincing version of Bohr's philosophy. Yet even the most sympathetic authors often either find Bohr's writings hardly comprehensible, or realize the connection between their elaboration of Bohr's ideas and Bohr's own writings remains accidental. Thus, Aage Petersen admitted:

> "As is well known, his style does not readily lend his writings to a review. In Bohr's epistemological writings it seems unusually difficult to separate the content from the style. The following exposition of some of his ideas is therefore, as far as content is concerned, an approximation whose accuracy it is difficult to appraise" (Petersen 1968, 104-105).

And more recently Don Howard proclaimed that *"now there are signs of growing despair ... about our ever being able to make good sense out of his [Bohr's] philosophical view"* (Howard 1994, 201).

I have argued that precisely because Bohr avoids the rigorous analysis of quantum formalism, his philosophy should be understood as merely highly ingenious and seductively appealing metaphor. What looks as strictly "logical" argument under closer scrutiny reveals circularity, at one point or another. Not surprisingly, the reversal of what is to be explained and what is foundational, of explanans and explanandum, is frequent in Bohr's and in Heisenberg's writings. This reversal is an eloquent illustration of circularity and ad hocness of many similar arguments. One of the striking examples is Heisenberg's reversal of the idea of disturbance—if the original idea of disturbance was based on indivisibility, or finitude, of microscopic particles, now the "unavoidable disturbance" is seen as the reason for their indivisibility or atomicity:

> "Nature thus escapes accurate determination... by an unavoidable disturbance which is part of every observation... In atomic physics it is impossible to neglect the changes produced on the observed object by observation... The supposition that electrons, protons and neutrons, according to modern physics the basic properties of matter, are really the final, indivisible particles of matter, is only justified by this fact" (Heisenberg 1979, 73).

Heisenberg's and Bohr's arguments of a foundational, a prioristic status of "uncontrollable interaction," are illegitimate. The uncontrollability of interaction has one, and only one reason—it follows from the mathematical framework of the quantum theory. It is necessary to ensure the consistency of quantum theory, when measuring possibilities are taken into account. David Bohm, after his conversion took place, understood this clearly. Thus, discussing the "uncontrollability" of the interaction in Heisenberg's γ-ray experiment, Bohm pointed out:

> "The disturbance is unpredictable and uncontrollable because *from the existing quantum theory* (italics are his) there is no way of knowing or controlling beforehand what will be the precise angle, with which the light quantum will be scattered into the lens" (Bohm 1957, 82, footnote).

Bohm realized that the indeterminacy relations are *"primarily... a deduction from the quantum theory in its current form,"* and that the attempts to prove its *"absolute and final validity"* are based on nothing else but on an a prioristic and illegitimate insistence on the finality of the quantum theory (ibid.).

Many scientists realized that Heisenberg's analysis of the γ-ray experiment is merely a consistency argument, despite the fact that it is often presented as an "inevitability" one. For example, von Weizsäcker argued that: *"the impossibility of simultaneously observing precise values of position and momentum does not mean the pseudo-positivistic nonsense 'what cannot be observed does not exist,' but the test of consistency: 'what does not exist in the theory cannot be observed in an experiment describable by the same theory"* (von Weizsäcker 1987, 278). In this argument von Weizsäcker was following his teacher Heisenberg who revealed that he considered all of Bohr's complementarity elaborations as demonstrating consistency of the quantum formalism by an analysis of possibilities of measurement. Heisenberg considered all these arguments "trivial":

> "Many experiments were discussed, and Bohr again successfully used the two pictures, wave and particle, in the analysis. The results confirmed the validity of the relations of uncertainty, but in a way this outcome could be considered trivial, because if the process of observation is itself subject to the laws of quantum theory, it must be possible to represent its results in the mathematical scheme of this theory" (Heisenberg 1973).

This position undermines, of course, all of Heisenberg's own "inevitability," or "finality," arguments.

The legitimate function of uncontrollability arguments is to argue consistency, rather than inevitability, of uncertainty and complementarity. Bohr did discuss certain canonical thought experiments along these more legitimate lines (Bohr 1935a). He argued, for example, the impossibility to measure the position of a particle in a set-up measuring its momentum. Using a moveable diaphragm (and by measuring the momentum of a diaphragm—by another test body), we can measure a particle's momentum (by using an appropriate test body) before the passage of a particle through the slit (ibid.). By applying the law of conversation of momenta to a system consisting of a particle and a moving diaphragm we calculate the momentum of a particle passing through the slit. However, because we have exact knowledge of momentum of the diaphragm during, or immediately after, the particle's passage, we block the possibility of also knowing the position of the diaphragm, in accord with uncertainty relations. Consequently, we are denied the knowledge of particle's position which is identical with that of diaphragm location (Beller and Fine 1994). Why does the diaphragm obey the uncertainty relations? Either we just state it (it applies to quantum objects), and then we deduce the uncertainty for the particle (this is, of course, an argument of consistency), or we say that while measuring the momentum of the diaphragm with a test body, the diaphragm undergoes uncontrollable interaction. We deduce consequently the uncontrollable interaction between the particle and the diaphragm—again, these are elementary considerations of consistency.

The crucial point of these thought experiments is that momentum and position cannot be *measured* simultaneously. This is however not satisfactory for Bohr. Why is it not satisfactory? The answer is clear from Bohr's own words. He warns against

> "misunderstanding by expressing the content of uncertainty relations... by a statement as the position and momentum cannot simultaneously be measured with arbitrary accuracy'... such a formulation would not preclude the possibility of a future theory taking both attributes into account" (Bohr 1937, 245).

What is the strategy of precluding a theory, such as Bohm's, where particles have a well-defined positions and momenta, which however cannot be measured simultaneously according to the uncertainty relations? How do we argue that Bohm's theory is impossible in principle? How do we transform consistency arguments into those of inevitability?

With the benefit of hindsight the answer is obvious. Simply postulate that what cannot be measured—does not exist. By defining concepts operationally through procedure of their measurement, and then applying the quantum formalism to the analysis of measurement procedure, we will obtain nothing else but deductions from quantum formalism (such as, for example, uncertainty relations). In this way illusion is created that the features of the theory (such as uncertainty) belong to the very definition of concepts used, and that they inevitably follow from the logical analysis of conditions of experience.

After 1935, when Bohr talked about "wholeness" of experimental arrangement, it is the operational definition of concepts that he had in mind. There is no justification for inflating this lean, positivistic point of view into holism of real properties and entities. When, after 1935, Bohr talked about wholeness, or inseparability, he either uses this operationalistic meaning, or other, more vague and metaphorical images of the idea of wholeness, appealing to our pre-quantum intuitions and imagery. One of these metaphorical images is that of "hiding nature," or of "elusive phenomena"—phenomena that disappear when we try to probe into it more deeply.

Initially, the disturbance (we erase the phenomena while trying to observe them) was the metaphorical tool to express wholeness—Nature hides its detailed working by our interference, though no knowledge without such interference is possible. Later, Bohr distances himself from the disturbance notion and introduced a *"distinction in principle between the objects we want to examine, and the measuring instruments"* (Bohr 1935b, 218).

Yet the idea of disturbance still remained the most vivid and potent metaphorical tool to discuss quantum uncertainty and wholeness without recourse to the mathematical formulas. This is the reason that Bohr and Heisenberg, while denying its viability, often regressed into using the idea of disturbance implicitly. This is clearly apparent in Bohr's notes for the talk at the Institute of Henry Poincaré in Paris in 1937: *"Individual phenomena in quite a new sense in physics. When trying to analyze, phenomena disappear. They appear only under conditions where it is impossible to follow their course"* (AHQP, 18.1.1937).

Bohr's notion of wholeness as eradicating the phenomena if we try to probe into it in more detail, goes back to James' analysis of mental phenomena. Bohr himself connected this notion with James' in an interview with Kuhn: *"If you have some things... they are so connected that if you try to separate them from each other, it just has nothing to do with the actual situation"* (Faye 1991, xvii). James' name in the context of the discussion of wholeness appears in Bohr's notes for lectures at the time. Yet this Jamesian idea of wholeness is poorly fitted to characterize the quantum entanglement or inseparability. Bohr's and James' understanding of wholeness better fits the idea of chemical compound: if we separate its constituent parts (we *can* do it), the phenomenon, the compound, disappears, and *"it... has nothing to do with the actual situation"*.

A chemical compound is radically different from its constituting elements—water is different from hydrogen and oxygen. Such was the understanding of wholeness by Vygotsky, who argued for a holistic approach to thought and language:

> "Two essentially different modes of analysis are possible in the study of psychological structures... The first method analyzes complex psychological wholes into *elements* (italics in the text). It may be compared to the chemical analysis of water into hydrogen and oxygen, neither of which possesses the properties of the whole... Psychology... analyzes verbal thought into is components, thought and word, and studies them in isolation. In the course of analysis, the original properties of verbal thought have disappeared" (Vygotsky 1962, 3).

Jamesian and Bohrian wholeness is more fitted to describe the existence of classical interaction, rather than to provide insight into quantum wholeness which evades classical analogies. Bohr's idea of wholeness has nothing to do either with non-locality or inseparability. Bohr, as Einstein, did not accept non-locality.

Bohr's metaphorical wholeness plays an essential role in his web of seemingly interconnected ideas. Intuitively, but wrongly, Bohr argued for the necessary connection between wholeness and complementarity (impossibility of an unified ontology), and between wholeness and indeterminism.

Yet quantum wholeness does not necessarily imply either complementarity or indeterminism. Bohm's causal ontology of the quantum world stands as an eloquent counterexample to Bohr's categorical assertions. Even though both Bohr and Bohm emphasize the contextual nature of measurement, their notions and uses of wholeness are radically different. Bohr, turning away from mathematics, talked about the "essential unsurveyability" of the measurement interaction. Bohm considered measurement as a special case of the quantum process, and subjected it to detailed mathematical analysis. Bohr, using the notion of wholeness, denied the possibility to provide a space-time description of a two-slit experiment, or a space-time description of an energetic transitions in the atom. Bohm provides a causal realistic space-time description of both (Bohm and Hiley 1993; Cushing et al. 1996). Bohr based his conclusions on the alleged impossibility of coming up with entirely new concepts. Bohm construed a new conceptual tool for describing non-classical space-time behavior – the "quantum potential." Bohm's wholeness is a mathematical tool to explore the quantum world. Bohr's wholeness is tool of legitimization and a weapon of prohibition, which restricts further theoretical exploration.

The connection between wholeness and indeterminism is eminent also in Pauli's writings. As in Bohr's case, so in Pauli's it is often not clear what follows from what, what is the explanandum and what is the explanans. In his later years, Pauli proposed an argument, according to which the inseparability between the observer and the observed phenomena (wholeness) is a direct outcome from the statistical nature of quantum mechanics (acausality): "... *if two observers do the same thing even physically, it is, indeed, really no longer the same: only the statistical averages remain in general the same. The physically unique individual is no longer separable from the observer*" (quoted in Faye 1991, 196). Similarly to the case with Bohr's metaphorical allusions to wholeness, Pauli's argument is intuitively appealing yet conceptually fragile. The notion of the "same experimental arrangement" in which different experimental results are obtained, or of an "observer" that "does the same thing" as another observer, never is adequately defined. This notion presupposes, rather than argues, the impossibility of "hidden variables," and inevitability of indeterminism. Similarly Bohr's arguments, such as the following one, simply beg the question:

> "The very fact that repetition of the same experiment... in general yields different recordings pertaining to the object, immediately implies that a comprehensive account of experience in this field must be expressed by statistical laws" (Bohr 1958, 4).

The post-Bell understanding of quantum inseparability, or wholeness, essentially relies on the mathematical analysis of previously interacting but now separated atomic systems. Bohr's translation of EPR mathematical analysis into his own terms obscured, rather than revealed, the quantum wholeness. Quantum inseparability got buried into the "tranquilizing" metaphorical web supported by positivistic maneuvers, and by repetitive enunciation of "inevitable" complementarity and acausality. Only after mathematical analysis of Bell (Bell 1966) did genuine progress became possible again.

Bohr's intuitions are not congenial neither to non-locality nor to inseparability. Thus, in the case of double-slit experiment Bohr claimed that, if we would be able to give a more detailed deterministic description, we would *"really be lost,"* for *"how could the phenomena depend on whether this hole was open or closed"* (Bohr 1930, 144)—clearly a rejection of non-locality.

The Copenhagen ideology about the supposedly necessary interconnection between the (vague) conception of quantum holism and acausality had far-reaching consequences not only for the philosophy of quantum physics, but for general philosophy of science of the 20th century. The connection between inseparability of measuring and the measured, and between acausality, overthrow of objective reality, etc. was canonized into Heisenberg's and Pauli's notion of closed theories. A closed theory such as quantum mechanics, they argued, is true in a domain of its validity and is not open to modification.

As Heisenberg put it: *"The connection between the different concepts in the system is so close that one could generally not change any one of the concepts without destroying the whole system"* (Heisenberg 1958, 94). The notion of a closed theory implies that the advance to new knowledge demands a jump to a qualitatively new intellectual experience.

If this is reminiscent of Kuhnian paradigms, it is not accidental. I have argued elsewhere that Hanson and Kuhn canonized the Copenhagen ideology into an overarching theory of knowledge (Beller 1997). The alleged impossibility of the gradual modification of a paradigm implied that any genuine change must be revolutionary, disposing the old knowledge wholesale. It also implied the (much criticized) notion of incommensurability between the old and the new.

A closer study of the history of science undermines the notion of the holism of paradigms. Rather than dogmatic commitment to a rigid set of ideas, creative scientific work—whether in the Copernican, the Chemical or the Quantum Revolution—is characterized by the ingenious mingling and selective appropriation of ideas from distinct paradigms. It is this lack of a binding holism that creates room for fruitful dialogue between scientists with different approaches, and consequently for a genuine advance.

In his important Book on Bohmian mechanics (Cushing 1994), Jim Cushing argued that it is a historically contingent fact that the Copenhagen Interpretation, rather than the Bohmian alternative, became the widely accepted one. I would like to alter Cushing's claim—rather, it is a matter of historical contingency that Bohm's version was perceived as an incompatible competing alternative rather than a welcome gradual modification of quantum "paradigm". In which circumstances is a change looked upon as an extension, and in which as a revolutionary break is as much a socio-historical as a philosophical problem.

ACKNOWLEDGEMENTS

I owe a great debt and feel a deep gratitude towards John Stachel for his rare intellectual and personal caring, penetrating scholarly insight, and unfailing support.
This paper is based on my book, *Quantum Dialogue—The Making of a Revolution*, The University of Chicago Press, 1999.

REFERENCES

AHQP: Archive for the History of Quantum Physics, American Philosophical Society, Philadelphia, assembled and edited by T.S. Kuhn, J. Heilbron, P. Forman and Lini Allen. Sources for History of Quantum Physics: An Inventory and Report, Philadelphia: American Philosophical Society.
Bell, J.S. 1966. "On the Problem of Hidden Variables in Quantum Mechanics." Reviews of Modern Physics 38:447-452.
Beller, M. 1993. "Einstein and Bohr's Rhetoric of Complementarity", *Science in Context* 6:241-255.
———. 1997. "Criticism and Revolutions." *Science in Context* 10 (1): 13-37, *Science in Context* 6:241-255.
———. 1999. *Quantum Dialogue—The Making of a Revolution*. Chicago: The University of Chicago Press.
Beller, M. and A. Fine. 1994. "Bohr's Response to EPR." In Faye and Folse: 1-31.
Bohm, D. 1957. *Causality and Chance in Modern Physics*. London: Routledge.
Bohm, D. and B.J. Hiley. 1993. *The Undivided Universe: An Ontological Interpretation of Quantum Mechanics*. London: Routledge.
Bohr, N. 1930. "Philosophical Aspects of Atomic Theory." Address to the Royal Society of Edinburgh (given 26 May). Unpublished manuscript, Bohr Manuscript Collection, AHQP. Sanders: 129-146. [*BCW* has only facsimile of an abstract from *Nature 125*: 958].
———. 1935a. "Can Quantum-Mechanical Description of Physical Reality Be Considered Complete?" *Physical Review 48*: 696-702.
———. 1935b. "Space and Time in Nuclear Physics." Notes from a lecture delivered before the Society for the Advancement of Physical Knowledge, Copenhagen, dated March 21, 1935. Bohr Manuscript Collection, AHQP. Sanders: 205-220.
———. 1937. "Causality and Complementarity." *Philosophy of Science 4*: 289-98. [Originally published in *Erkenntnis* 6:293-303; also in Sanders: 241-252].
———. 1958. "Quantum Physics and Philosophy: Causality and Complementarity." In *Essays*: 1-7. [Originally published in Klibansky].
Cushing, J.T. 1994. *Quantum Mechanics, Historical Contingency and the Copenhagen Hegemony*. Chicago: The University of Chicago Press.
Cushing, J.T., A. Fine and S. Goldstein, eds. 1996. *Bohmian Mechanics and Quantum Theory: An Appraisal*. Dordrecht: Kluwer.

Faye, J. 1991. *Niels Bohr: His Heritage and Legacy. An Antirealism View of Quantum Mechanics*. Dordrecht: Kluwer.

Faye, J. and H. Folse, eds. 1994. *Niels Bohr and Contemporary Philosophy*. Dordrecht: Kluwer.

Heisenberg, W. 1927. "Über den anschaulichen Inhalt der quantentheoretischen Kinematik und Mechanik." *Zeitschrift für Physik 43*:172-198.[Also appears in Wheeler and Zurek: 62-84. Page references are to this English translation.]

———. 1958. *Physics and Philosophy*. New York: Harper and Row.

———. 1973. "Development of Concepts in the History of Quantum Theory." In Mehra: 264-275.

———. 1979. *Philosophical Problems of Quantum Physics*. Woodbridge, CN: Ox Bow Press. [Reprinted from *Philosophical Problems of Nuclear Science*. 1952. New York: Pantheon Books.]

Hermann, A., K. v. Meyenn and V.F. Weisskopf, eds. 1979. *Wolfgang Pauli, Scientific Correspondence with Bohr, Einstein, Heisenberg, A.O., Volume 1*. Berlin: Springer.

Howard, D.A. 1994. "What Makes a Classical Concept Classical? Toward a Reconstruction of Niels Bohr's Philosophy of Physics." In Faye and Folse: 201-230.

Petersen, A. 1968. *Quantum Physics and The Philosophical Tradition*. Cambridge, MA: MIT Press.

Sanders, J.T. 1987. *Niels Bohr: Essays and Papers*. Two volumes of Bohr's writing on epistemological matters, gathered from published papers and the unpublished manuscripts in the Bohr Manuscript Collection, AHQP. Translations from Danish by Else Mogensen. Deposited at the Niels Bohr Archive in Copenhagen, the Niels Bohr Library at the American Institute of Physics and a number of other institutions.

van Fraassen, B.C. 1980. *The Scientific Image*. Oxford:Clarendon.

Vygotsky, L. 1962. *Thought and Language*. Cambridge, MA: MIT. Press.

von Weizsäcker, C.F. 1987. "Heisenberg's Philosophy." In Lahti and Mittelstaedt: 277-293.

Wheeler, J.A., and W.H. Zurek, eds. 1983. *Quantum Theory and Measurement*. Princeton: Princeton University Press.

MICHEL PATY

THE CONCEPT OF QUANTUM STATE: NEW VIEWS ON OLD PHENOMENA

"I think (…) that a theory cannot be produced from results of observations, but only from an invention."

"I am not ashamed to put the concept of «real state of a physical system» ["existing objectively, independently of any observation or measure, and that can in principle be described through the means of expression of physics"] at the very centre of my meditation."
Albert Einstein[1]

"(…) It is interesting to speculate on the possibility that a future theory will not be *intrinsically* ambiguous and approximate. Such a theory could not be fundamentally about «measurements», for that would again imply incompleteness of the system and unanalyzed interventions from outside. Rather, it should again become possible to say of a system not that such and such may be *observed* to be so, but that such and such *be* so. The theory would not be about *observ*ables, but about «*be*ables»."
John S. Bell (Bell 1973; 1987a)

Abstract. Recent developments in the area of quantum systems have led to accept statements, which originally appeared to be mere interpretations, as representing physical facts that appeared formerly to be more related to interpretation with free options. Of such a nature are the statements relating to quantum behavior of individual particles (diffraction, etc.), neutrinos oscillations, distant quantum correlations (local non-separability), Bose-Einstein condensation, cooling isolation of atoms and, recently, decoherence of quantum superposition states interacting with environment measurement apparatus, that allows a better understanding of the transition from the quantum domain to the classical-macroscopic one. The debate on the interpretation of quantum mechanics has imperceptibly changed its nature through these developments, giving higher weight to a "physical interpretation" more clearly distinct from the philosophical one than in the old days of quantum mechanics. In particular, the concept of quantum state has undoubtedly acquired a direct physical meaning, in terms of *properties of a physical system* that is fully represented by a linear superposition of eigenstates, and able to propagate as such in space and time. The price for this new situation is an extension of meaning of the concepts of *physical magnitude* and *physical state* towards ones that do not correspond directly with numerical values.

1. INTRODUCTION:
STATE FUNCTION AND "DIRECT REPRESENTATION"
OF A QUANTUM SYSTEM OR STATE

Quantum physics aims at understanding the deep structure of matter in general, from bodies of our environment and molecular associations of atoms to atomic nuclei and to elementary particles actually or "virtually" contained in the latter, including even cosmic objects as well as the primordial phases of cosmology. In our understanding it underlies the unity of matter in the variety of its organization patterns. The means of the theoretical understanding of this domain of physics is constituted by *quantum mechanics*. This is, in turn, applied to particular (atomic, nuclear) theoretical models, and enlarged, from a more fundamental point of view, to *quantum field theory* ranging from quantum electrodynamics (QED)[2] to electroweak and chromodynamic gauge field theories.[3] These recent theories have been established within the conceptual frame of quantum mechanics and, as an effect, have confirmed its heuristic power and permitted at the same time physicists to get used to work with this tool-for-thought that is indispensable to explore quantum phenomena.

Quantum mechanics, as a theoretical scheme, is *practiced* successfully. Today physicists, while strictly applying the rules that govern the use of quantum magnitudes in the working process of their physical thought on the phenomena being studied, no longer worry very much about the "difficulties of the interpretation" that had heavily preoccupied the founding fathers and their immediate successors. As for interpretation, if they had to propose any, this would be for most of them the following: "what is important is that it works." And indeed, this might be a mark of unconcern or the expression of an immediately pragmatic philosophy that would remain blind to deeper reasons. This attitude comprises, in all events, a part of truth, of the kind—walking before knowing how: they have the theoretical (and even conceptual) tool and know how to handle it before knowing its exact nature and worrying about it.

However, as soon as physicists ask themselves questions concerning the intelligibility of physical phenomena in the quantum area, they again find the terms of the old debate. But, in contrast to their elders, they meet these "from outside," so to speak, in the sense that these questions appear to them as posed only to a "second order" of the understanding: i.e. when they question themselves *not about this understanding*, that is itself provided by theory, a theory they know so well that it has become "second nature" to their thought, but *about the reasons for it*.

To understand "to the first degree," that is at the level of their work in physics itself, is achieved by handling the concepts and magnitudes that represent, reproduce or create the phenomena of interest. When physicists today speak of an "elementary particle" (for example, a proton), they mean, indeed, that it is described by "quantum numbers" or quantities that are "eigenvalues" of operators representing the adequate physical magnitudes. They have abandoned the classical image of a directly visible corpuscle, that no longer belongs to their referential background. *Quarks* themselves are quantum particles considered in this sense. Such *entities* or physical *"systems"* or *quantons*,[4] conceived in the specific way of quantum physics, are implicitly supposed to constitute in one way or another, according to the modalities of their description, objective elements of the real world that manifest themselves to human understanding. Symmetries of quantum particles and fields allow us to *understand* in that way their properties and their arrangements.

Philosophical and epistemological difficulties arise only when one intends to understand *the nature of this understanding*: this is what we mean by "intelligibility to the second degree." This difference of degree with the founding debates comes from the fact that the theoretical tool, "formalism," in the usual expression, is now already justified as a representation of its success. Physics builds its tools in an abstract manner and contents itself with these being well conceived, without trying to naturalize their origin; admittedly they are abstract, symbolic, mathematical, and elaborated by thought from necessities enforced by phenomena (physics has indeed established rules for that purpose such as, for example, statements in the form of principles).

From this one sees clearly and undoubtedly better than in the past when it was necessary to construct these tools, that the problem of interpretation is twofold but separate: physical and philosophical. The physical interpretation deals essentially, as was traditionally the case since the birth of modern physics (in the seventeenth century), with the relationships between mathematically expressed magnitudes and the corresponding physical contents. The difference, from the physical point of view, between theoretical and conceptual elaborations related to quantum phenomena and systems, and those dealing with classical ones, is that the quantum phenomena are farther than the classical ones removed from the processes of observation by which these phenomena reach our senses.

Niels Bohr rightly emphasized this difference of nature between the quantum and the classical. But he formulated it in a manner that abruptly changed a simple physical state of things into a philosophically problematic state of knowledge. There was, according to him, a barrier between the quantum and the classical worlds, a barrier that was due to measurement. It resulted from this state of affairs, in his view, that the knowledge of quantum phenomena cannot grasp these directly, but has always to refer to classical representations.

The disjunction between the representation of quantum phenomena and that of classical ones can be pointed at in a philosophically more neutral manner: while classical physical phenomena and systems are *homogeneous* to the processes of their observation, quantum phenomena and systems are not, since observations and measurements relative to them belong ultimately to the domain of classical physics. But that does not entail any impossibility to represent "directly," that is to say in terms of *properties* and *objects*, quantum phenomena and systems, at least if one refers *intelligibility* not to *perception* but to *understanding*, as it seems logical. If one can conceive objects of a quantum area or "world" in this sense, the questions of *physical interpretation* will be therefore largely independent from those bearing on more general considerations on knowledge, i.e. *philosophical interpretations*. One would then have shifted, so to speak, from a concern for *interpretation in general* to a more precise interest for the *physical meaning of quantum magnitudes* provided by quantum theory itself.

The philosophical aspects of interpretation would then present themselves at a different level, granting a large autonomy to quantum physical thought, sensibly the same as for the other areas of physics: the former would no longer have to sacrifice to a so-called "foundational" need of being based on a peculiar philosophical interpretation, as in the early period of quantum mechanics.[5] And one would have then satisfied, up to some point, the realist demand of the theoretical physicist asking, with John Bell, for a theory that is not fundamentally about measurements, that considers physical systems in their inner completeness, and of which one could tell "not that such and such *may be observed* to be so, but that such and such *be* so." In other words, a theory that "would not be on *observ*ables" [magnitudes able to be *observed*], "but about *be*ables" [magnitudes able to *be*].[6] This is what will be achieved in the following without modifying in any way the form of standard quantum theory, and only by understanding it differently (to an intermediate degree of understanding, involving some kind of *physical* interpretation).

It remains to see how to "interpret physically" the theoretical, conceptual or factual states of things that were problematic in a physico-philosophical mode for the "orthodox" or the "complementarity" "interpretation." We shall restrict ourselves, in what follows, to revisiting some characteristic and relatively simple quantum phenomena, renewed by recent results from high precision experiments, in the light of the proposed perspective on the physical interpretation of magnitudes and of theoretical formalism referred to the description of a world of properly quantum objects or systems. We shall see that they invite us directly to conceive quantum magnitudes in this manner, which entails the need to expand the meaning generally given to the concept of physical magnitude, and especially to the concept of the state function representative of a physical system.

Since the beginnings of quantum mechanics, these phenomena usually served to illustrate the problems of interpretation. By a fair reversal of things, it is today possible to extract directly from them the physical interpretation they are calling for. These phenomena are, first, *local non-separability*, whose epistemological status has undergone changes from a formal feature with optional physical meaning to an established physical fact, corroborated by experiments with distant correlated systems. Then, *diffraction of quantum particles*, no longer performed with many particle beams for statistical results, but with individual quantum systems for probabilities of individual events. Also, *indistinguishability of identical particles*, initially postulated or conceived as a formal property, and thereafter demonstrated by direct physical effects such as Bose-Einstein condensation, where a great number of identical atoms are accumulated in the same fundamental state up to a quasi-macroscopic level. Finally, recent *experiments of "decoherence"* have permitted to visualize superpositions of states in relation with mesoscopic systems when measured by a classical device, in a tiny time interval before the dissipative loss of information occurred from interaction with the environment.

All these results converge towards a specification of the physical meaning of quantum concepts and magnitudes implied by the corresponding phenomena, obliging us to associate factual evidence and physical contents conceived in terms of *properties of systems*, with "formal" properties whose interpretation remained until then optional or problematical. We will analyze some aspects of this new situation, trying to make out in which way they may contribute to deepening, modifying, or finding a foundation for our theoretical comprehension of quantum features, by reducing the latitude of arbitrary choice in the interpretation and by adapting the norms of our intelligibility.

2. LOCAL NON-SEPARABILITY AS A FACT AND AS A PRINCIPLE

The objection opposed by Einstein in 1935 to the claim that quantum mechanics is a fundamental theory that will serve as the *basis* for any further progress in physics raised several questions that overlap with interpretation.[7] It became known as the "EPR argument," and was later reformulated and refined by its main author. The problem was to know whether the theory (quantum mechanics) describes, or not, *real individual physical systems*, and if it describes them *completely*, that is to say adequately to all aspects rightfully attached to their *individuality* and in a one-to-one manner. The orthodox interpretation (in the philosophical sense) challenged the legitimacy of speaking of elements of reality independently of their conditions of observation, and Bohr's reply to Einstein's argument was exactly founded upon this position (Bohr 1935). It had no chance to be listened to by Einstein, who could not accept its principle. Any progress in the debate on this question supposes from then on that one tries to leave aside the philosophical *diktat* of Bohr's reply, and adheres to physical theory and the content of its concepts.

The question, as contemplated by Einstein, was to know whether the theory is complete in the indicated sense (it was to him the minimal requirement for a theory to be considered fundamental).[8] If one can completely characterize an individual quantum system (be its theoretical representation probabilistic or not), its state function has to represent it as such. If that is impossible because of some feature of the theory, then this theory can be only a statistical description (such was, indeed, the conclusion of the "EPR" argument). We will see further that this is also the point at stake with the *interference* of distinctive *individual* quantum systems when put actually in evidence. The EPR argument suppressed in principle the possibility to elude such a question in the name of an operationalist philosophical interpretation, and indeed the construction of rarefied beams later suppressed it effectively. Individualization for a system was indeed usually prevented by the alleged necessity to detect and measure it, if one wanted to know something of it (with a particle counter to know whether there came only one), and this act would destroy it immediately as a quantum system (it would project it on a classical particle state), forbidding all further knowledge of its quantum state (through the manifestation of wave properties).

In the EPR case, the system under study (U) was conceived in correlation to another one (V) while it did not maintain any dynamical interaction with it.[9] The correlation, expressed by the conservation of a magnitude (A) used in the description of these systems, and known for the initial state formed by the two subsystems,[10] allowed the determination of the state of the first without perturbing it by a measurement, by deducing it from the state of the second, measured (supposedly) independently of it.[11] Measuring magnitude (A) of the second destroyed its state at the very moment of its determination, forbidding any meaning to the consideration of an alternative measurement for another magnitude (B) incompatible with the first: the initial system being no longer available, no effective comparison can be made. But it would have nevertheless been logically possible, as a matter of principle, to perform the second measurement instead of the first, and it would have provided another state function for the second system; from it, the first system would have been deduced, *a priori* different from the preceding result.[12] One could therefore have two different state functions to describe one and the same physical system: it would obviously be a theoretical weakness.

But this reasoning was dependent on a statement that did not belong to quantum formalism and that was at this time considered optional: the separability of two far distant systems, that is to say their mutual independence in their respective locations. Einstein gave a precise definition of this *principle of separability*,[13] although recognizing that he added it to quantum theory. Without this principle, he believed, however, one could not characterize separately localized individual systems, unless one admitted a non-physical interaction (instantaneous action at a distance) between them. He concluded from this that quantum mechanics does not describe individual physical systems, but only statistical ensembles of systems, for which the objection does not hold.[14]

Further progress, both theoretical, with the establishment of John Bell's theorem (1964), and experimental, with experiments of correlation from a distance, has essentially consisted in analyzing local separability, a concept identified by Einstein, and in testing it for quantum systems. Bell's theorem on non-locality demonstrated the existence of a contradiction between *local separability* and some predictions of *quantum mechanics* for systems of two correlated particles (*strong correlation* relationships for quantum systems expressed by *equalities* between averages for magnitudes were opposed to *weaker correlations* in the form of inequalities for the local separation case). This theorem provided the sensitive relationships able to discriminate the local separability hypothesis and quantum theory.[15] From then on, experiments have decided in favor of quantum mechanics in a hardly disputable manner, especially that of Freedman and Clauser, realized in 1972, and that of higher precision performed in 1981 by Alain Aspect.[16] *Local non-separability* was henceforth established as a physical *fact*, a *general property* of quantum systems having been put in correlation, well identified from the phenomenal point of view.[17]

This property corresponds to a characteristic feature of the state function in quantum mechanics: the state functions of subsystems that have been once correlated are not factorizable (i.e. independent of each other, i.e. separable). Having been linked together to form, even momentarily, one single system, two quantum (sub-) systems cannot be dissociated: this *"entanglement"*[18] is a fundamental property of quantum formalism, and possesses therefore a direct counterpart in the phenomena it describes, *local non-separability*. This can equally be considered as an aspect of *non-locality* of quantum systems. The general and fundamental character of this property, and its inscription in the formulations that define in the theory the state of a system, suggest at the same time a fact *of experience* and a *principle* for quantum physics.

An important conceptual aspect of local non-separability is its place in the economy of quantum theory. From the point of view of the conceptual and theoretical consistency requirement adopted here (which may also be called of "critical realism"), one can analyze it in the following manner. As it is directly linked to the definition of quantum systems and of the magnitudes by which we represent their states, local non-separability relies only on these and does not have to refer to other magnitudes that would be defined outside this theory. It bears on systems that, spatially speaking, are "extended systems," and for which the space variables have not, as an effect, a direct part in their definition; in this sense, it is not concerned by special relativity. It does not contradict it and has nothing to do with (instantaneous) actions at a distance.[19]

It is fair to say, however, that many physicists and philosophers of science would still disagree with this conclusion, which seems compelling from the point of view adopted here. It is, indeed, difficult to think physically without the help of spatial intuition, and this is probably the main reason for their dissatisfaction with "pure quantum reasoning." But who can say what kind of intuition is adequate for the quantum domain? It seems to me that quantum physicists have developed over the years an adequate intuition in this respect that is basically founded on the quantum formalism as a practiced intellectual tool for exploring and understanding quantum phenomena (the epistemological implications of which we are exploring here, taking a point of view of general consistency). John Bell, who was reluctant to accept the above argument, which he viewed as too formal and even as a "verbal" solution,[20] admitted nevertheless non-separability as a fundamental fact and eventually as a physical principle.[21] But he would have preferred to have a dynamical interpretation of it. It seems to me, on the contrary, that as a principle it definitely does not need an explanation, but stands as one of the primary conceptual references toward which the other quantum concepts must be consistently obliged (in a way similar to the principle of special relativity ruling the transformation laws of the concepts related to the motion of bodies).

As a fundamental quantum fact, one should perhaps consider that local non-separability is to quantum physics, for example, what the principle of equivalence (of inertial and gravitational masses) is to the general relativity theory of gravitation. One can see it as a true *principle*, both a synthetic proposition based on experimental facts and a theoretical statement of a central, and perhaps foundational, importance; it could serve to formulate quantum theory in a less formal manner than the usual presentation, which would make it come closer to the other physical theories from which until now it parted in this respect.

Local non-separability can be seen as an even wider theme of reflection, rejoining a cosmological perspective. One can, indeed, make rapprochements with other features of "disindividualization"[22] or of "desingularization" or, better, of *indifferentiation* such as *indistinguishability* (that partakes as well of the superposition principle), and perhaps as symmetries of matter, that are important features of primordial cosmology (Paty 1999b). One can also see in it, with David Bohm, the mark of a more general indivisible wholeness of material reality (Bohm 1980). With regard to this point, one must nevertheless observe that to grasp an underlying order, thought separations in such a wholeness are needed as a necessary approximation, without which the concept of wholeness would lose all utilizable physical content. Extended in an absolute manner to the whole Universe, the principle of non-separability would present the same kind of disadvantage as the one pointed out by Poincaré[23] regarding the principle of relativity of space if we were to formulate it with respect to all bodies of the Universe: being tautological it would not give us a hold on phenomena. But, at any rate, it might give us some hint on cosmological conditions, of the kind Einstein obtained for a closed and unlimited Universe[24] (for example, in quantum cosmology, some coherence condition for having finite time inside Planck's limit in the primordial Universe).

3. INDIVIDUAL SYSTEMS AND THE TRANSFORMATION OF PHYSICAL PROBABILITY

The phenomenon that is the simplest in its principle to characterize quantum properties is that of *interference*, which confirmed the wave-particle duality of matter and inspired Max Born's idea of the probabilist interpretation of the state function. This archetypal phenomenon illustrates some fundamental aspects of the description of quantum systems and helps, from the physical point of view, to make explicit the interpretation problems that had been raised.

The "orthodox" interpretation of complementarity and observationalism sees in it the necessity of wave-particle duality and the impossibility to go beyond it. The interference pattern (concentric rings, alternately obscure and bright), similar to those of classical waves, is due to the wave property of quantum systems; whilst, on the other hand, the materialization of these varied intensity rings on the screen covered with a sensitized film comes from the corpuscular property of these systems (through their interaction with the grains of photographic emulsion, producing an image). The dual properties, contradictory if they are considered for individual "particles" or systems, can be reconciled as soon as one ceases to be concerned with causality of individual events, and shifts the focus to the statistical aspect of the experiment. If one wanted to examine, in this experiment, the behavior of an individual quantum system, a meaning could not be derived according to the complementarity interpretation, in the name of the very definition of the systems. As a matter of fact, if one wanted to characterize a quantum system as individual, it would be necessary to submit it to a counting experiment, that would indicate which of the slits the quantum system has gone through; by being localized in that way, the system would suffer a perturbation and therefore lose its quantum aspect and its capacity to produce interferences.

Yet, in 1930, Paul Dirac, in his book *The Principles of Quantum Mechanics* (Dirac 1930), already indicated that, according to this theory, one photon interferes with itself and that this is the reason for the interference phenomenon in the case of a single quantum system. This is also the case for any quantum system (particle, atom, etc.). The meaning of it would be that interference is a property of any individual quantum system, and that quantum physics is the theoretical description of such individual systems. The probabilistic turn of this description would not *a priori* be a hindrance for this purpose (after all, statistical mechanics does the same). However, the "complementarity explanation," to which we just referred, blunts and dissolves the force of this statement by making it a mere feature of the formalism in claiming the impossibility by principle to observe it in experiments.

As for the ensemble interpretation of quantum mechanics, according to which the theory is only a statistical one (incomplete for Einstein, complete for others), it only knows averages that have no physical meaning except for an ensemble of systems, and can not pronounce on the significance of individual quantum events.

However, for approximately two decades, experiments have been realized and continuously improved thanks to technical advances with *individual quantum systems* (photons, electrons, neutrons, atoms) that are known to be such without needing to be counted by detection on their path, and therefore without destruction of their quantum state. It has actually been possible to produce beams of such "particles" or quantum systems, extremely rarefied and with a high time definition (better than 0.1 ns), in such a manner that particles get to the interferometer one by one, spread in time, each having got across the experimental arrangement within an interval of time sufficiently small to ensure that the following one has not yet entered.[25] One can then be fairly confident that only one particle at a time has crossed the interference apparatus (and interfered with itself). The detection of impacts on the screen seems in the beginning to be at random. When many single "particles" have gone through the interferometer, the distribution of impacts is seen to obey a law: one obtains, in the end, the same interference pattern as in the traditional experiment with a beam of N identical particles crossing simultaneously the interferometer.

These results require that a physical meaning be attributed to an individual event in an interference experiment. Clearly, the final interference pattern with individual particles can only be obtained statistically, by the realization of a great number of successive one-particle or individual quantum system experiments. The result of N such experiments with single quantum systems gives the same result as a single experiment performed, in the same interferometer, with a beam of N identical systems. But the theoretical inference that one is allowed to make in the two cases is very different. The second case, consisting of the traditional experiment with a great number of simultaneous systems, satisfies a frequentist and purely statistical interpretation of the probability given by the state function.[26]

But the effective occurrence of the first case, N experiments with an individual system identical each time, and represented by the same state function, assures us that each individual phenomenon, independently from the others, contributes to the final interference pattern. One is therefore led to conclude that *it is the individual quantum systems that make the phenomenon* and therefore that, in a way that remains to be specified, each individual phenomenon occurring with each (independent) system potentially constitutes the overall interference phenomenon revealed by the final pattern, obtained statistically. In other words, each phenomenon relative to an individual system is a *quantum phenomenon*, collected on the screen through a *classical measurement* process (the "photon" or quantum particle impact on a grain of silver bromide of the photographic emulsion). One is then inclined to consider that, just before interception on the screen, each of the individual systems having interfered with themselves is in a quantum superposition state. And so, as nothing distinguishes them from each other, all these individual systems in interference are strictly identical. From then on, the only remaining problem would be the measurement process: identical quantum systems provide, after detection, different results, but endowed with probabilities corresponding to the amplitude of probability of their state of superposition.[27]

As a result of what precedes, the ψ state function must be considered as the theoretical representation of an individual particle, which entails the following important consequence of its physical meaning: the *physical probability*, given by the ψ state function[28] (the latter being often named "probability amplitude," in a sense that can only be physical, since nothing of the kind exists in mathematical probabilities), is not liable to be reduced to statistics for ensembles of systems. It has a *theoretical function* from the physical point of view, as it is deduced from a magnitude having a direct physical meaning, the probability amplitude (i.e. the state function itself). One can therefore consider this probability as a *physical magnitude*, which makes it differ from probability in a merely mathematical sense, as well as from probability conceived physically as expressing a frequency.[29]

4. INDISTINGUISHABILITY AND STATE FUNCTION

In quantum physics, the state function that represents a quantum system allows the complete description of all the properties attributed to this system, in such a way that systems represented by the same state function are effectively in the same state and are absolutely indistinguishable. That means that, external to the theory, no other possibility exists to distinguish them. In others words, *a quantum "particle" has no other characteristics than those of its state*, differently from physical systems as described by other theories such as classical mechanics, thermodynamics or relativity theory. These theories describe *what happens* to physical objects that are in other respects defined outside of them. For example, the three-body problem of classical astronomy is about the mechanical processes occurring to celestial objects that are supposedly given. The theory bears not on these objects, but on their interaction properties. The Moon, the Sun and the Earth, for example, possess an identity—and an opacity—defined prior to the laws and equations under study in mechanics and astronomy.

The only theory, except quantum physics, for which the eventuality has been considered that it could be able by itself to describe its object, instead of obtaining it from outside, is the general theory of relativity, at least in a further more elaborated formulation foreseen as a distant purpose (by A. Einstein and J. A. Wheeler notably), where it would be possible to describe in the same system of equations both a field and its source. Such was the "strong" meaning Einstein attached to the notion of theoretical *completeness*.[30] To him, quantum mechanics was not a "complete theory," in this sense obviously, as its status of a framework theory rather than a dynamics suffices to show. But there was another weaker meaning of the same notion, which he considered crucial for the fundamental nature of quantum physics, as we have seen earlier. A theory would be "complete" in a minimal sense if it were able to describe fully its *object*, that is all the properties than can be physically considered about it. It was not the case for Einstein, with quantum mechanics because of EPR type correlations that, invalidating the principle of separability, excluded the description of individual systems.[31]

We do not any more consider this argument in this form, such correlations having proved to be factual and to concern individual correlated systems. On the contrary, actually, completeness at least in the weaker sense would characterize, in principle, the description of *quantum systems* on the background of the physical interpretation envisaged here. The main obstacle to this requirement seems today to remain the "quantum measurement problem." If one sets aside the latter for a moment, one can rightfully be struck by the purpose of quantum mechanics to achieve an *exact covering* of the *described system* by its *state function*, going therefore even, in a way, beyond the restricted completeness requirement.

The most remarkable expression of this covering appears, finally, to be the property of *indistinguishability of identical quantum systems*. But is it a mere feature of the formalism, or a property of physical systems? Both aspects, as always with quantum mechanics interpretation problems, seemed closely connected and not easily disentangled. This property was identified on the eve of the constitution of wave and quantum mechanics, by Satyendra N. Bose and Albert Einstein for quantum systems of null or integer spin (photons and atoms named afterwards "bosons," obeying "Bose-Einstein statistics"), and by Enrico Fermi and Paul Dirac for quantum systems of half-integer spin (electrons, protons, and other "fermions," obeying "Fermi-Dirac statistics"). Indistinguishability of identical bosons (in the case of photons) appeared to be the real underlying reason of the quantification procedure for radiation energy exchanges in black body as performed by Planck in 1900;[32] and indistinguishability of fermions (here, electrons) gave the explanation of the Pauli exclusion principle that accounted for the constitution of atom levels in terms of state occupations by electrons.

This property, corresponding to two types of statistical (or probabilist) processing of quantum systems (the admission of several particles in the same state in the first case or, on the contrary, their mutual exclusion in the second one), opposed to the classical statistical processing *à la* Boltzmann of particles always distinguishable even when occupying the same state (for they possess a proper identity). Indistinguishability therefore limits drastically the possible state occupations. It indicates, actually, that quantum systems *do not occupy* states, but that *they are themselves* states, and are identified with their states.[33] Indistinguishable quantum systems have no other element of identity than those furnished by the theoretical description of their state. The notion of state is identified with that of "particle": a quantum "particle" (or system) *is its state*: it is not "in its state," as a classical system. This situation corresponds to a closer determination of the physical system by the theory. Contrary to the idea that prevailed for quantum physics of a looser determination and a limitation of knowledge because of "indeterminacy" relations.

This formal property, indirectly dictated by factual reasons and finding expression in the principle of superposition,[34] has proven to correspond to fundamental physical properties of quantum systems that could be directly tested and that have implications to the macroscopic level itself.

Supraconductivity and *superfluidity* are such properties directly connected to indistinguishability. Bose-Einstein condensation, already predicted in 1925 by Einstein from the indistinguishability of the identical for some kinds of atoms (it was, actually, the first theoretical description of a phase transition), was long considered as being very far from possibilities of verification. Yet it has recently been experimentally proven thanks to the high technical realization of extreme colds and atoms trapped by laser rays.[35] Tens of thousands of atoms are thus condensed in the lowest energy state (called "of the zero point") with nothing distinguishing them from each other: the superatom they then form corresponds to a fluid in absolute superfluidity state, without viscosity, that can show itself at the macroscopic level (by an effect of *visible non- locality*, the fluid occupying quasi-instantaneously all the space offered to it, rising on the container walls). At this stage, restrictions claimed by the orthodox complementarity interpretation about the direct physical character of the state function appear rather ridiculous, and as an exercise of twisted rhetoric serving only to hide evidence.

One may invoke for the exclusion principle—and therefore for indistinguishability of identical fermions—direct consequences at a highly macroscopic level, concerning cosmic objects corresponding to particular phases of the evolution of stars. "White dwarfs" are compact stars in a state of equilibrium between the gravitational tendency to collapse and the pressure of degeneracy of electrons that cannot fall in the same fundamental state because of the Pauli exclusion principle.[36] "Neutron stars" resist in the same way, they collapse in on themselves because of gravitation due to the degeneracy pressure of the neutrons into which all atomic nuclear constituents have been transformed.

By its directly physical consequences, indistinguishability of identical quantum systems is indeed a physical property of these systems, and not only a feature of the theoretical formalism. It is described precisely by quantum theory in terms of *state function* (submitted to the principle of superposition). There is therefore, as we suggested earlier, a liaison of the property indicated by *indistinguishability* (equivalence of particles of similar characteristics, occupying the same state within a system, that one can count but that nothing distinguishes)[37] and the *theoretical description* by the state function of quantum mechanics (or, at a further stage, of quantum field theory). All this encourages us to see indistinguishability not as a "lack," as would suggest the common intuition of the notion of "particle," taken from the immediate experience of bodies in our environment as well as from the habit of classical physics, but rather as a characteristic and determining physical *property*. For nothing authorizes us, when dealing with such objects, to think of properties that are not pointed out by the theory.

5. REAL PHYSICAL STATE AND SUPERPOSITION, MEASURED STATE AND PROJECTION

The state of a quantum system, as we have tried to characterize it physically, is not identified with that obtained directly by one measurement alone. This last, indeed, is a reduction or at least a projection of the state physically defined by one of its components, according to the choice of the preparation of the system (by a complete set of compatible magnitudes).[38] A measurement device in the usual sense can only measure a classical magnitude. With respect to the state of superposition that represents a system before the operation of measurement, it can only provide one of the components (one of the "eigenstates" of the measured set of magnitudes). One should not be surprised by this as such is its function and its only ability.

The measurement apparatus is, as a matter of fact, by definition a projection device (in the geometrical sense) of the various components of the state of the system. One has claimed that quantum measurement is a non-causal interaction, but this means to pronounce oneself *a priori* on the nature of the interaction between the quantum system and the macroscopic device. If one speaks rightly of a *rule of projection*, or eventually *of reduction*, this rule does not, up to now, mean any directly physical process and nothing allows it to be raised to the status of a physical principle. In the absence of a theory, in the proper sense, of quantum measurement, that would be a general theory of the interaction between quantum system and macroscopic measurement apparatus; one must regard this as merely a practical rule.

Each measurement provides a numerical value for the measured magnitude, one of its possible (classical) values (among the eigenvalues) with some frequency, given by the corresponding probability amplitude (eigenfunction). An experiment with a great number of identical systems, or a great number of independent experiments performed on such systems taken individually, provide the whole spectrum of values of the magnitude with probabilities for each (corresponding to the amplitudes in the superposition). From these results in terms of classical magnitudes, one infers the quantum superposition state that has been submitted to measurement, and of which one can reasonably think that it represents *the quantum system before measurement*, in one of the possible bases; the one chosen by preparation. The state function reconstituted in that way is not a simple catalogue of data, since the system that it represents has the capacity, a clearly physical one, to propagate, to evolve in the course of time, to make interferences or to possibly oscillate between different physical states (on which we shall give more details below). Measurement to determine the state will intervene only after these transformations, which owe nothing to man's hand or thought but everything to nature.

In summary, we propose to consider that the physical quantum states are the states expressed as superpositions themselves, which one can determine from the determination of their components. This reduces to magnitudes endowed with numerical values by classical measurement devices. Actually, this is nothing more than taking von Neumann and Dirac's geometrical vector representations as meaning it: state vectors in Hilbert space are the physical ones, represented by their various possible bases (determined from the preparations according to their possible sets of commuting magnitudes). As a vector, the system state is a basis-free geometrical representation of a physical state, and is more fundamental, because of its invariance, than its "contextual" components.

As an effect, physicists, familiarized by their experience of quantum systems, consider them in this manner: what is important is the representation of these systems' quantum states, i.e. the overall final reconstitution and not the contingent and particular (classical) values obtained by measurement. These values are intermediate entities given by experiment, whose deep physical meaning is obtained only from their immediate translation in quantum terms, necessary in returning to the description of the physical quantum system under study.

6. PHYSICAL PHENOMENA LINKED TO PROPAGATION OF SUPERPOSITION STATES

A physical state, as considered by physicists in their representation of quantum phenomena, and how they *think* about it in their *theoretical work*, is given in an invariant form with respect to its "vector projections," while being generally presented at the same time as a state of superposition on one basis or the other. This is more general than being restricted to the consideration of measurement alone, which after all is nothing else than one of the moments of verification or of experimental test, and is not a purely formal property: *this form rules the physical properties* of quantum systems. We have seen this for the phenomena evoked above, but one can also evoke a number of others of a different nature that show to what extent this is indeed the universal form of the description of all quantum systems. Two examples, both borrowed from elementary particle physics, will show this in a clear and striking manner, all the more as they have no classical analogous: these are the *"mixtures"* of particles states and the *"oscillations"* from one state to another, these mixings and oscillations being expressed directly in terms of state superpositions that propagate.[39]

The neutral "strange pseudoscalar" meson K^0 and its antiparticle, $\overline{K^0}$, are eigenstates of their "mass matrix" (M) and of the *strong interaction* Hamiltonian (H_s) production process (they are physical states in *associated production* conserving the "strangeness" magnitude, $S = +1$ for K^0, $S = -1$ for $\overline{K^0}$ [40] or for any other associated strange particle in the production interaction, for instance the "strange baryon" Λ^0). They behave differently in their *decay through weak interaction*, with the strangeness of non-conserving Hamiltonian H_F.[41] The eigenstates for such processes are the mesons as observed from their decays, characterized by proper lifetimes (τ) and decay modes, the short-lived K^0_S $(\tau = 10^{-10}$ s$)$ and the long-lived K^0_L $(\tau = 10^{-8}$ s$)$. The initial states K^0 (resp. $\overline{K^0}$) are expressed as linear superpositions of K^0_S and K^0_L states, which progressively transform according to the law of exponential decrease in time. If one considers a K^0 meson initially produced (actually, a beam of such mesons, appropriately selected), and one worries about its state at a time t, the superposition initially containing the states K^0_S and K^0_L in equal parts impoverishes in K^0_S, whose time decay is faster, and enriches in K^0_L, that will in the end completely dominate. The then nearly pure beam of K^0_L states can be written as a superposition of the states K^0 and $\overline{K^0}$ in equal proportions. One therefore obtains, in the beam of K^0 mesons, a "regeneration" of $\overline{K^0}$ mesons that were absent in the initial beam. These can be detected through a strong interaction process with respect to which they are well defined, i.e. of which they are eigenstates.

Let us note, incidentally, that the qualification of eigenstate concerns definite states of a Hamiltonian and other physical magnitudes that are not, here, of a classical nature. At this level, the identification of quantum systems in given states does not call for measurement in the classical sense. The latter is needed only at the end of the chain of experimental processes of the detection of "particles" typical of the considered interactions. In a general fashion, an eigenstate given for a set of compatible magnitudes can be projected (in the vector sense) onto another (preparation) basis relative to another set of magnitudes incompatible (non-commuting operators) with the first. This eigenstate of the first set of magnitudes will therefore be written as a superposition of eigenstates of the second set. In others words, the "preparation" of a quantum system concerns proper quantum magnitudes as well as magnitudes submitted to a classical determination by measurement. "Preparation" for measurement is only a particular case of "preparation" in general, that means the choice of a set of physical magnitudes corresponding to a set of eigenstates taken as referential (or as vector basis in the Hilbert space of their eigenfunctions).

One can also consider the behavior of these neutral K particles under the transformation by the CP operator[42] as a product of charge conjugation (C, that changes a particle into its antiparticle) and parity (P, or space symmetry) or, equivalently, by the time reversal (T) operator, the equivalence ($CP = T$) being due to the conservation of the product CPT, following a theorem of the quantum theory of fields.[43] If one represents the eigenstates of the CP magnitude by K_1^0 (with a corresponding eigenvalue $CP = +1$), and K_2^0 ($CP = -1$), and if the operator CP does not commute with the weak interaction Hamiltonian (H_F),[44] the K_L^0 and K_S^0 states are different from the K_1^0 and K_2^0 states and can be considered as linear superpositions of these states. The coefficients in the superpositions are functions of the parameters of CP violation in these weak interaction processes.

Such physical systems propagate with time between the moment of their production and that of their detection. The state that is attributed to them during this course is that given by the state vector (invariant with respect to the basis), that is, for the chosen basis, the linear superposition, whose coefficients vary with time (let the function $\psi_K(t)$ be the representation of this state). That is to say that the *superposition here is the physical state*, without any circumlocution that would bring physical existence only to the state detected after observation or measurement. The quantum system under study (represented by the $\psi_K(t)$ state function) is analyzed by a detector placed on its line of flight that projects it (in the geometrical sense of vector projection) at time t *onto* one of its components chosen by fixing the detection conditions ("preparation"). From the frequencies for each detected state that are a measure of their probabilities, one obtains the coefficients of the superposition or probability amplitudes (probabilities are the absolute squares of the coefficients), as in the usual case. One observes statistically, for K_L^0, a given number of states in the $CP = +1$ mode (for example, $K_L^0 \to 2\pi$) and another one in the $CP = -1$ mode ($K_L^0 \to 3\pi$).

What is interesting for physicists, from a physical point of view, is not so much the final state observed at the detection, which choice is, as a matter of fact, purely contingent, as the indication it provides about the physical state of the K^0 meson *at a time t before its detection*, given by the basis-free or invariant state vector. This state vector is given, for each group of (compatible) magnitudes corresponding to a physical content (either M and H_S, or H_F, or CP), as a superposition of their eigenstates. Conversely, the knowledge of this state permits the characterization of the properties of these magnitudes (for instance, the degree of CP violation in the weak interaction process with a Hamiltonian H_F).[45]

The so-called *"oscillation"* phenomena between quantum particles states are described and thought of in a similar way.[46] Consider *neutrinos*, electrically neutral (fermion) "leptonic" particles existing under the form of three different species,

v_e, v_μ, v_τ, each one endowed with a distinct conservative magnitude, the leptonic, *electronic, muonic, tauic* charges or quantum numbers, shared with the electrically charged corresponding particles, electron, muon, tauon[47] (respectively e^-, μ^-, τ^-), together with which they constitute the three families of *leptons* (the most elementary "particles" of matter with *quarks*). Their mass is very small, possibly null.

If the mass of neutrinos is not exactly zero, one can distinguish three states of mass, v_1, v_2 and v_3, distinct from the states that represent the ("leptonic") neutrinos observed through their "weak interactions" (v_e, v_μ and v_τ and the corresponding antineutrinos). The latter can be described as linear superpositions of the mass states.[48] Neutrinos emitted in nuclear reactions (in β decays of nuclei) are of the type v_e (or $\overline{v_e}$). The evolution with time, during the course of their state function, ψ_v, is given by that of the amplitudes (or coefficients) associated with the states of the superposition. As a consequence, the proportion of the three mass states varies during the propagation. As these mass states can themselves be put in the form of superpositions of the leptonic states, it entails that the initial neutrino (v_e) transforms partly into neutrinos of the other species (v_μ and v_τ), with a given "oscillation length" (or "wave length").[49] Such effects (such *phenomena*) are actively searched for by physicists for the three types of neutrinos.[50]

It is generally considered that physical neutrinos are those characterized by their properties in the (weak) interaction[51] through which they are produced or destroyed (interactions with other particles or eventual decays), that is to say that they are the "leptonic" neutrinos v_e, v_μ and v_τ. Nevertheless, in the propagation of one or the other of these neutrinos, the effective *physical state* would, at any instant of time, under the considered hypothesis (of non-zero masses, and of some degree of leptonic numbers violation), be due to the mentioned transformations, a *linear superposition of these states*, evolving in time in a determined way. The detection by (weak) interaction of one of the states allows, by comparison with the initial state (given by the choice of one of the three types of neutrinos), the physical state at a chosen place on the covered distance to be determined (i.e. at a given time of flight). This detection is based on reactions of interaction where a neutrino transforms into the corresponding charged lepton ($v_e + n \to e^- + p$ and, similarly, $v_\mu + ... \to \mu^- + ...$, $v_\tau + ... \to \tau^- + ...$). For production, these reactions require enough energy to create the mass of the charged leptons.

In the case of neutrinos originated from nuclear reactions, the energies are insufficient to create masses larger than that of the electron. The neutrinos v_e, transformed during their travel into v_μ or in v_τ, will therefore not produce reactions that would detect them and remain sterile. If one finds less v_e than there were at the beginning, it might well be that the pure initial state has been transformed into a superposition of different neutrinos, of which only the projection on the v_e-state is detected. This is, for example, what is supposed to happen with solar neutrinos, whose proportion received on Earth is far less than what is expected if neutrinos continued on their way remaining identical to themselves.[52] We would have there again (actually, the oscillation phenomenon has recently been definitely proven experimentally), an indubitable direct effect of the *physical character of a linear superposition state*.

The example (be it a real phenomenon or a simple possibility) gives indeed also evidence that the thought of such *states of superposition* is hereafter familiar to physicists. A superposition of states has to be understood as a simple change of basis relative to another set of mutually compatible physical magnitudes, corresponding to one of the possible "preparations." The physical state that physicists consider is not restricted *to that after the measurement* (otherwise it would only be the incident deficient neutrino); it is the state that is revealed to them by this measurement, and that also contained another undetected component that can immediately be reconstituted. Recent observations (in 2002) on neutral currents induced by solar neutrinos, which are not dependent on mass threshold effects (as the neutrino is simply scattered by the nucleon target), have yielded the expected rate, confirming that the neutrino beam arriving on Earth is in a superposition state of all the neutrino leptonic states. Of course, all these phenomena are studied with great numbers of "particles," but their description and their explanation must be understood in terms of properties of individual "particles," for the same reasons as those considered previously.

7. BEFORE DECOHERENCE, SUPERPOSITION

It remains to evoke another type of phenomenon of recent production and observation, *"decoherence."* We will not undertake a thorough discussion of its implications and its interpretation here. In particular, we will not pronounce (reserving the discussion for another opportunity) whether this phenomenon gives a solution to the problem of measurement of quantum systems, or whether it brings new views on the relationships between the "classical" and the "quantum." At least it illustrates an important aspect, to my eyes, through its "visualization": it makes us *see* a state of superposition propagating and thus allows us to better *conceive* the possibility and the *physical reality* of such states.[53]

The metastable state of superposition that has been observed recently for "mesoscopic systems"[54] is an *"entangled"* state made by coupling a Rydberg atom in a two-energy states superposition with an electromagnetic field (of few photons) in a two components superposition state. The field is a physical system that plays the role of the Schrödinger's cat of the famous thought experiment (Schrödinger 1935). The overall system is entangled (not factorizable in its various components), and this entanglement (that constitutes the "coherent state") is further multiplied through successive interactions with the various (quantum) elements of the environment (such as those that constitute the observation apparatus), so that in the end the initial coherence does not show anymore, the effect being absorbed rapidly ("decoherence"). In such a production experiment of a coherent entangled state, one can vary the parameters which determine the degree of coherence of the system: these parameters are the number of photons that make the electric field, and the time of propagation of the entangled system (which is the time elapsed between its production and its analysis to determine whether it is still in a coherent state). The coherent state itself manifests as such by some interference which can be observed through a correlation between pairs of the atom-analyzers at detection. Coherence can then be controlled, and the condition and time when coherence ceases marks the shift from quantum to "classical-type" behavior of the system. This shift is attributed to the many interactions occurring between the system and the quantum components of the environment. The simple original entangled system combines itself with the states of the latter (each one being itself in a linear superposition), giving rise to a further entanglement: as the process is going further, it leads in an irreversible way to a many component entangled system. Quantum non-separability forbids going back to the original components simply entangled, and that original entanglement is lost in the end, as it becomes diluted in the multiple entangled overall system, and has become definitely inappreciable. In the end, the quantum character of the state under study has been lost, although the whole process has been considered from a purely quantum point of view. So to speak, a "classical" behavior (a non quantum one) has been generated from quantum states merged inside entangled multiplicities.

It is clear that the process of decoherence is not identifiable with that of measurement, for it happens softly through the quantum interactions themselves, whereas measurement is a process which immediately chooses one of the final states by suppressing the others: the continuous soft (natural) process is (artificially) interrupted by the arrangement of apparatus itself, which favours at random only one of the components of the final state and destroys the superposition. So to speak, measurement is decoherence plus projection (reduction) on only one of the components of the initial state of the physical system under consideration. Nevertheless, decoherence helps to understand the initial stage of such a transition, which seems, in the final stage, to be purely of a statistical mechanics and thermodynamics nature. But I do not want to comment further on this, leaving it for another opportunity.[55] I rather content myself with observing that evidence for the process of coherence to decoherence is *per se* evidence for the physical character of the coherent, entangled, i.e. quantum linear superposition state, shown as propagating in space and time.

8. CONCLUSION

All the physical phenomena examined so far persuade us that the state function ψ represents (or describes) the state of the physical system completely. I mean by "complete representation" adequacy and covering: there is nothing more in the physical system than what is comprised in its theoretical representation by the state function.

If we restrict the question of the theoretical representation of quantum systems to the mere quantum level where these systems exhibit properties and interact with others systems of a similar nature, the concepts of state function, quantum system, quanton, quantized field, with the magnitudes that qualify them, are self-sufficient: for conception and handling in theoretical work, they do not ask for any physical or conceptual underlying classical basis such as that of a undulatory or corpuscular substance, distinguishable and localized. For the quantum physics of atomic and subatomic phenomena and quantized fields the "quantum level" where these concepts operate is the fundamental level, and, in particular, physical systems are effectively represented by their "state functions," and physical magnitudes by their "operators." At this level of representation, it is not necessary to go back, for each magnitude and each state, to the practical circumstances of their determination that refer ultimately to observations with the help of classical apparatuses.

For theoretical thought at the quantum level, the classical systems constituting these apparatuses are only intermediary instances in the process of the constitution of data that are in the end translated into quantum terms. The data being acquired, the quantum domain allowed itself to be conceived and explored in full conceptual and theoretical independence with respect to the classical domain.

This consideration does not diminish the problem of the quantum-to-classical relationship: it simply puts it aside, provisionally, as a fundamental problem. It is an epistemological and philosophical decision, taken in order to give the quantum domain and its theoretical representation the largest autonomy with respect to particular philosophical perspectives on knowledge. It has often been considered that (physical) knowledge is to be referred to observation, in the name of a primacy of perception in characterizing phenomena. However, contemporary reflection on science, and particularly on the various areas of physics, has led to conceive of the relations of concepts and theories to *perception* as most indirect. The demand for *intelligibility* requires, as I suggested in the beginning, a direct and close connection with the *understanding* that undertakes its theoretical elaboration by following a process of *rational construction* that is linked only in an intermediate manner with the forms of *perception*. Regarding the conceptualization and the theoretical insight obtained from them, the phenomena under consideration are first brought to the understanding and secondly to the perception. If we refer these *phenomena* to (quantum) *objects*, that means that the latter are rationally constructed before being secondarily and indirectly perceived.

The question of the *physical meaning of magnitudes*, among which the representative state function of a system is foremost, is henceforth more directly illuminated than by the current ("orthodox") interpretation, conceiving this meaning *through reference to measurement*. The reference, according to the view proposed here, is to quantum phenomena, whose access is indirect but recognizable by a rational and consistent construction, that is supported by data coming in the last instance from the perceptual (observation and experiment). Consequently, there is nothing to oppose considering the state function in the form of a superposition (but basis-invariant) *describing effectively the state* of a physical system evolving in the course of time.

The notion of *quantum physical state* differs from the current idea of a *physical state*, referring generally to magnitudes that are directly observable through instruments ruled by the laws of classical physics. The difference between a physical phenomenon (or system) at the quantum level and a phenomenon (or a system) at the classical level is that *the second is closer (if not homogeneous) to its conditions of observation referred to perception*, whilst the first remains radically *distant from them* and is definitely *heterogeneous* to them.[56] This formulation of the difference between the classical and the quantum domains is free of philosophical bias about knowledge: it has the advantage of not arbitrarily limiting the capacity of the quantum to be intelligible. If they are dissimilar in their relationship with *perception*, their links to *understanding* are not of a different nature: all concepts of physics, classical as well as quantum, are expressed by magnitudes that are constructed (by man) and abstract.[57]

That a *quantum state* be accessible to experiment only indirectly does not affect the possibility *to acquire knowledge of it*. *Magnitudes* that characterize it are also not directly accessible, since they are not endowed with numerical values. To take into account all the elements considered in what precedes, we must therefore conceive an *extension of meaning* to the quantum domain, of the notions of *physical magnitude* and of *physical state*, beyond the meaning usually accepted for them in classical physics (including the theory of relativity). This extension, legitimated by the *phenomena* (with a sense of this term that does not reduce them to mere objects of perception but that conceives them according to their capacity to be brought to *knowledge*), is actually already realized in practice by the main properties of the very formalism of quantum theory.[58]

If we look back to their history, such extensions of meaning have been a common procedure in mathematics as well as in physics: an example among many others in mathematics is the extension of the concept of number from integer to fractional, to irrational and then to imaginary and complex numbers; as for physics, consider only motion, force, energy and also the extension of finite to differential magnitudes. In all cases, such extensions were not the least obvious and led to hard scientific and philosophical debates and controversies.

By proposing this extension of meaning for the concept of physical magnitude to forms that are not endowed with numerical values, to states that are linear superpositions of eigenstates, in order to ensure epistemological *aseity* (self-contentedness) for the quantum domain and its theoretical representation, we give primacy to understanding over perception, which is driven to an ancillary status. This is a pragmatic decision that avoids deciding on the fundamental problem that still remains open to the relationship between the classical and the quantum, but that allows us at the same time to consider with full legitimacy a wide range of phenomena that might well be the base of all others. But, considering the present state of our knowledge, we cannot be sure of this. We can only relate it to the more fundamental and general question, still standing and in evolution, of the unity of physical phenomena and of a unified approach to them. But precisely this kind of approach might still be doomed to remain out of reach for present theories, until a deeper penetration of the unity of physical phenomena is obtained through a sound unification of the fundamental interaction fields of matter.

To find a solution outside this perspective, if proven possible, the "quantum problem of measurement," that is to say the nature of the relationship of the quantum and the classical, would be finally only of limited interest. With the practical rule connecting, through probabilities, quantum magnitudes and their state functions with the corresponding classical entities determined from measurement devices, we have the minimal algorithm needed to place on a pragmatic basis the quasi autonomous existence of two coherent and intelligible domains of physical reality, with reference to their proper and specific phenomena and objects: the classical and the quantum.

ACKNOWLEDGEMENTS

To John Stachel, who contributed so much (and still does) to pave the way to a deeper understanding of the rational exigencies of physics, through his pioneering Einsteinian studies, particularly on the lifelong "Einstein's struggle" with quanta, and his exemplary harmonious practice of physical intuition and elaborated mathematical formalization in considering the physics of today as well as that of the past, always combined with sound epistemological insights. The reflections proposed above are a tribute to his enlightening and stimulating endeavors as well as—I hope—a testimony of intellectual kinship. I would also like to acknowledge the useful comments of the anonymous referee which have helped me to improve this paper.

Centre National de la Recherche Scientifique (CNRS), Paris

NOTES

1. Respectively: Einstein, letter to Karl Popper, 11.9.1935, published as an appendix in (Popper 1959; Einstein 1953, 6-7).
2. On the history of quantum electrodynamics, see (Schweber 1994).
3. Cf., for example, (Bimbot and Paty 1996).
4. This term has been introduced by Mario Bunge (Bunge 1973).
5. See (Paty, 2000a).
6. Bell (1973, 1987a). See the complete quotation in the epigraph.
7. (Einstein, Podolsky, and Rosen 1935; Einstein 1948; 1949). Cf. (Paty 1988a; 1986; and in press).
8. In particular in (Einstein, Podolsky, and Rosen 1935; Einstein 1948; 1949). See (Paty 1986; 1988b; 1995a; and in press).
9. The two systems U and V form at initial time one only system $U \oplus V$ and are allowed thereafter to move away from each other at arbitrary distances (for instance, two photons emitted in correlation by an atom).
10. For instance, the overall momentum, $\vec{P} = \vec{P}_u + \vec{P}_v$, or the overall spin, $\vec{J} = \vec{J}_u + \vec{J}_v$ (its modulus J and one of its components, $J_1 = J_{u1} + J_{v1}$).
11. Measurement of magnitude A for the system V determines its state function, ψ_V^A, and the correlation gives the corresponding magnitude for the system U, and therefore the state function of the latter, ψ_U^A.
12. Let ψ_V^B be determined by the measurement of B, from which we deduce, without measuring, ψ_U^B. *A priori*, ψ_U^A and ψ_U^B are different, although they were not, *in principle*, perturbed, as they have not been directly measured.
13. In particular in (Einstein 1948; 1949). See (Paty 1995; and in press).
14. Ensembles of systems can admit a non-biunivocity of their state function, if the latter is only about mean values.
15. They are called "Bell's inequalities": see (Bell 1964; 1966; 1987b). They are relevant for the property of locality generally speaking, independently of they being or not related to determinist hidden variables, to which they had been linked in a first period. More general relationships have been obtained since then: Bell's theorem for locality without hidden-variables (Bell 1971; Eberhard 1977; Peres 1978; Stapp 1980), and for more than two quantum correlated particles (Greenberger, Horne, and Zeilinger 1989; 1990; Mermin 1990).

16. (Freedman and Clauser 1972; Aspect, Grangier, and Roger 1981; 1982; Aspect, Dalibar, and Roger 1982; Aspect 1983). See the following reviews and analyses of the experimental results: (Bell 1976a; Paty 1977; 1986; Clauser and Shimony 1978).
17. See (Bohm and Hiley 1975, in Lopes and Paty 1977, 222; Paty 1988a, ch. 6, and Paty 1986).
18. The use of this word, coined by Schrödinger in 1935 (Schrödinger 1935; 1984), has been reactivated recently (Shimony 1993; d'Espagnat 1994; Cohen, Horne, and Stachel 1997a, b, etc.).
19. This aspect has been emphasized in the article quoted (Paty 1986), and already in 1980 (Paty 1981; 1982). Bernard d'Espagnat seems to come also to the same conclusion in one of his recent books (d'Espagnat 1994, 430).
20. I remember my discussions with him: we disagreed on this point. This question represented to him an intellectual challenge whose difficulty remained untouched.
21. (Bell 1987b). I have quoted elsewhere (Paty 1988a, 245), a letter in this sense of John Bell to Alain Aspect.
22. This word is inadequate if by individuality one means a unity. Undifferentiated quantum systems can be counted: they keep *cardinality*.
23. Poincaré (1912). See (Paty 1996).
24. See (Paty 1993, ch. 5 and 7).
25. See, in particular (Pflegor and Mandel 1967; Grangier 1986). The concepts of quantum theory of field, that permit the definition of states with a given number of particles, underlie these experiments. It is necessary, for example, to prepare one-photon states of the electromagnetic field (Grangier 1986).
26. $P = |\psi|^2$.
27. Consider an initial individual system crossing a diaphragm with two slit a and b, and whose state is represented by $\varphi(x) = \frac{1}{\sqrt{2}}[\psi_a(x) + \psi_b(x)]$. Let z be the variable corresponding to various localizations on the screen, placed at a distance x from the diaphragm. The state $\varphi(x)$ of the individual interfering system can be considered as a linear superposition of states prepared along the values z_i of the variable (or magnitude) $z : \varphi(x) = \sum_{z_i} \alpha_i \vartheta(z_i)$. The probability of an impact on the screen in z_i is $|\alpha_i|^2$.
28. By the square of its modulus.
29. Cf. (Paty 1990).
30. See (Paty 1988b; 1993, ch. 10; and in press).
31. Cf. (Paty 1995a; in press). See above.
32. Already in 1911-1912, Ladislas Natanson and Paul Ehrenfest had diagnosticated the non-classical character of the statistics corresponding to Planck's radiation law. See, for instance, (Kastler 1981; Darrigol 1988; 1991; Pesic 1991).
33. This includes the invariant characteristics shared by the various possible states of a system, that contribute to define the system and its particular states corresponding to given magnitudes.
34. Consider, in effect, a system of two identical quantum particles *1* and *2*, each in its state, represented by the state functions ψ_1 and ψ_2. The state function of their coupled system is symmetrical for the permutation of the particles in the case of Bose-Einstein statistics, hence: $\psi_{12} = \frac{1}{\sqrt{2}}(\psi_1 \otimes \psi_2 + \psi_2 \otimes \psi_1) = \psi_{21}$. Nothing forbids identical (indistinguishable) particles *1* and *2* from being in the same state inside the system (identical bosons can accumulate in the same state inside a system). For the case of Fermi-Dirac statistics, the coupled state function is antisymmetric: $\psi_{12} = \frac{1}{\sqrt{2}}(\psi_1 \otimes \psi_2 - \psi_2 \otimes \psi_1) = -\psi_{21}$. If the identical fermions *1* and *2* were totally indistinguishable, occupying the same state in the system, then one would have: $\psi_{12} = -\psi_{12} = 0$: two

identical fermions cannot occupy the same state inside a quantum system (exclusion principle).
35. (Cornell and Wiemann 1998). Cf. (Griffin et al. 1995).
36. The mechanism was proposed by R. H. Fowler as soon as he knew the statistics studied by Paul Dirac, who was his student (cf. Doncel et al. 1987, 274).
37. From the point of view of arithmetic, concerning how to count or to identify by a number, such objects are characterized by cardinality, but not by ordinality. It has been proposed from a logic point of view to describe them with a set theory whose elements would possess this property, different from that of Zermelo-Frenkel (cf., for instance, French and Krause 1996).
38. Or, according to the usual terminology, "a complete set of observables that commute."
39. Strictly speaking, the representation of these "particles" makes use of the quantum theory of fields. However, the features of their properties that we discuss here are only those of the basic formalism of quantum mechanics (the definition of a state from physical magnitudes and the principle of superposition for the state functions).
40. The magnitudes (the "observables," in the quantum jargon), H_s, M and S commute between themselves ($[H_s, S] = 0$, etc.) and have the same eigenstates.
41. H_F and S do not commute ($[H_s, S] \neq 0$).
42. Let us recall that in quantum theory the mathematical form of *physical magnitude* is a linear *operator* acting on the state function.
43. Due to Gerhart Lüders, Wolfgang Pauli, and Julian Schwinger, who established it around 1952-1955 (see Lüders 1952 and especially 1954; Pauli 1955; Schwinger 1951-1953). See comments in (Enz 1973; Doncel et al.1987; Yang 1982).
44. In fact, weak interaction does not conserve CP in these processes.
45. The whole thought of "elementary particles" physics is, as quantum physics in general, ruled by the superposition principle. We could have taken other examples of state mixtures as superpositions: the neutral states of "vector mesons" (ω^0, φ^0, with spin-parity $J^P = 1^-$) under the conservation of a given magnitude (for example under SU_2 "isospin" symmetry or SU_3 "unitary spin" symmetry), or the state superpositions of the neutral "intermediate bosons" (γ and B) of the gauge symmetry electroweak theory of A. Salam, S. Weinberg, and S. Glashow (cf., p. ex. Paty 1970; 1985). These bosons, and also the charged "intermediate bosons" ($W^\pm$), are initially supposed to have a vanishing mass as the photon, and their mixture, or superposition, is characterized by a coefficient (ϑ_{W-S}) called "Salam-Weinberg mixing angle," that is the parameter of the theory. The symmetry breaking generates the finite masses of the "physical" "intermediate bosons" ($W^\pm$, W^0), related to the mixing parameter (see, f. ex., H. Pietschmann and D. Haidt, in Gaillard and Nikolic 1977; Paty 1985). All this however is happening inside the limits of the range of weak interaction, that is extremely small. The examples that we have presented in the text are more striking for our purpose, insofar as they correspond to phenomena that are manifested on large spatial distances, covered during the propagation, and for which one hardly could refrain to speak of *physical states*, beyond the mere mathematical formalism of the theory.
46. One example, hypothetical but theoretically founded, would be eventual oscillations of neutrons into antineutrons ($n \rightarrow \bar{n}$), through an interaction field violating baryonic number (such as required by the "Grand Unification" theories).
47. Or "heavy lepton" (with mass 1777 MeV, the muon mass being 106 MeV, and the electron mass 0,5 MeV; the mass unit is MeV, million of electron-volts, in the appropriated unit system commonly used in subatomic physics).
48. See, f. ex., (Paty 1995b); Alexei Smirnov in (Nguyen-Khac and Lutz 1994). Leptonic numbers are no more completely conserved, and the heavier neutrinos can decay into a lighter neutrino together with other particles (a different process than "oscillations" considered here).
49. "Oscillations" are a function of neutrino mass differences, energies and covered distances.

50. These experiments concern, besides nuclear reactor or solar neutrinos (essentially $\bar{\nu}_e$), atmospheric neutrinos (and antineutrinos) (mainly ν_μ) and those produced at particle accelerators (ν_μ and ν_τ).
51. Neutrinos interact only through "weak interactions."
52. ν_e neutrinos are detected by their capture by a nucleus with emission of an electron (or of a position in the case of $\bar{\nu}_e$ antineutrinos). Neutrinos of other kinds resulting from oscillation are sterile for this type of reaction, and escape detection. But they are indeed part of the incident flux.
53. On the theoretical interpretations of the phenomena and of the experiments, see notably (Zurek 1982; 1991; d'Espagnat 1994; Omnès 1994a and b).
54. In the experiment performed at the Laboratoire de physique de l'École Normale Supérieure, Paris: (Haroche, Brune and Raimond 1997).
55. For a reflection on this state of things, see (Paty, in 2000a).
56. There still remains, anyhow, between a *physical system* qualified as such, be it a classical or a quantum one, and its *conditions of observation*, a difference of nature. I want only to underline here that the working modes of measurement devices are referred to classical phenomena.
57. Cf. (Paty 1988a, and 2000a).
58. Intuitively perceived by such theoreticians as Dirac, who extended the notion of commutative magnitudes expressed by ordinary numbers (*c*-numbers), to non-commutative ones (*q*-numbers) (Dirac 1926a and b; 1928). Cf. (Mehra and Rechenberg 1982, vol. 4, 162 sq.; Darrigol 1992), it has not, however, been explicitly legitimated as such, which ensured the permanence of the dominant philosophical interpretation (cf. Paty, 2000a).

REFERENCES

Agazzi, Evandro, ed. 1997. *Realism and quantum physics*. Amsterdam: Rodopi.

Aspect, Alain, Philippe Grangier, and Gérard Roger. 1981. "Experimental tests of realistic local theories via Bell's theorem." *Physical Review Letters* 47: 460-463.

———. 1982. "Experimental realization of Einstein-Podolsky-Rosen-Bohm *Gedankenexperiment*: a new violation of Bell's inequalities." *Physical Review Letters* 49: 91-94.

Aspect, Alain. 1983. *Trois tests expérimentaux des inégalités de Bell par mesure de polarisation de photons*. Ph.D. Thesis, doctorat ès-sciences physiques. Orsay: Université Paris-Sud.

Aspect, Alain, Jean Dalibar, and Gérard Roger. 1982. "Experimental tests of Bell's inequalities using time-varying analyzers." *Physical Review Letters* 49: 1804-1807.

Bell, John S. 1964. "On the Einstein-Podolsky-Rosen paradox." *Physics* 1: 195-200. Reprinted in (Bell 1987b, 14-21).

———. 1966. "On the problem of hidden variables in quantum mechanics." *Review of Modern Physics* 38: 447-452. Also in (Bell 1987b, 1-13).

———. 1971. "Introduction to the hidden-variable question." Pp. 171-181 in (d'Espagnat 1971); reprinted in (Bell 1987b).

———. 1973. "Subject and object." Pp. 687-690 in (Mehra 1973).

———. 1976a. "Einstein-Podolsky-Rosen experiments" in *Proceedings of the Symposium on Frontier problems in high energy physics, in honour of Gilberto Bernardini on his 70th birthday*. Pisa: Scuola, pp. 33-45. Normale Superiore. Also in (Bell 1987b, 81-92).

———. 1976b. "The measurement theory of Everett and de Broglie's pilot wave." Pp. 11-17 in *Quantum mechanics, determinism, causality, and particles*, eds. M. Flato et al. Dordrecht: Reidel. Also in (Bell 1987b, 93-99).

———. 1982. "On the impossible pilot wave." *Foundations of Physics* 12, No. 10: 989-999. Also in (Bell 1987b, 159-168).

———. 1987a. "Beables for Quantum Field Theory (1987)." In (Bell 1987b).

———. 1987b. *Speakable and unspeakable in quantum mechanics*. Cambridge: Cambridge University Press.

———. 1989. "Against measurement." Pp. 17-32 in (Miller 1989).

Ben-Dov, Yoav. 1987. "Bell's version of the 'pilot-wave' theory." *Fundamenta Scientiae* 8: 331-343.
———. 1988. *Versions de la mécanique quantique sans réduction de la fonction d'onde.* Ph.D. Thesis. Université Paris-13.
———. 1989. "De Broglie's causal interpretation of quantum mechanics." *Annales de la Fondation Louis de Broglie* 14: 349-360.
Bimbot, René, and Michel Paty. 1996. "Vingt cinq années d'évolution de la physique nucléaire et des particules." Pp. 12-99 in (Yoccoz 1996).
Bohm, David. 1980. *Wholeness and the implicate order.* London: Routledge and Kegan Paul.
Bohm, David, and David Hiley. 1975. "On the Intuitive Understanding of Non-Locality as Implied by Quantum Theory." *Foundations of Physics* 5: 93-109; also in (Leite Lopes and Paty 1977, 206-225).
Bohr, Niels. 1935. "Can quantum-mechanical description of physical reality be considered complete?" *Physical Review* 48: 696-702.
———. 1958. *Atomic physics and human knowledge.* New York: Wiley. French translated by Edmond Bauer, and Roland Omnès. *Physique atomique et connaissance humaine.* Paris: Gauthier-Villars; new ed. established by Catherine Chevalley, Paris: Gallimard, 1991.
Broglie, Louis de. 1953. *Louis de Broglie, physicien et penseur.* Paris: Albin Michel.
Brown, Laurie M., Abraham Pais, and Brian Pippard, eds. 1995. *Twentieth century physics,* 3 vols. New York: Philadephia Institute of Physics.
Bunge, Mario. 1973. *Philosophy of physics.* Dordrecht: Reidel.
Clauser, John F., and Abner Shimony. 1978. "Bell's theorem: experimental tests and implications." *Reports on Progress in Physics* 41: 1881-1927.
Cohen, Robert S., Michael Horne, and John Stachel, eds. 1997a. *Experimental Metaphysics. Quantum Mechanical Studies for Abner Shimony. Vol. One.* Dordrecht: Kluwer.
———. eds. 1997b. *Potentiality, Entanglement and passion-at-a-distance. Quantum Mechanical Studies for Abner Shimony. Vol. Two.* Dordrecht: Kluwer.
Cornell, Eric, and Carl Wiemann. 1998. "Bose-Einstein condensation." *Scientific American* 1997,—French translation, "La condensation de Bose-Einstein." *Pour la science,* No. 247, May 1998: 92-97.
Costa, Newton da. 1997. *Logiques classiques et non classiques.* French translation from Portuguese (Brazil) and completed by Jean-Yves Béziau. Paris: Masson.
Darrigol, Olivier. 1988. "Statistics and combinatorics in early quantum theory." *Historical Studies in the Physical Science* 19: 17-80.
———. 1991. "Statistics and combinatorics in early quantum theory, II: Early symptoms of indistinguishability and holism." *Historical Studies in the Physical Sciences* 21 (2): 237-298.
———. 1992. *From c-Numbers to q-Numbers. The classical Analogy in the History of Quantum Theory.* Berkeley: University of California Press.
Dirac, Paul A. M. 1926a. On the theory of quantum mechanics. *Proceedings of the Royal Society of London* A 112: 661-677.
———. 1926b. "The physical interpretation of the quantum dynamics." *Proceedings of the Royal Society of London* A 113: 621-641.
———. 1930. *"The principles of quantum mechanics."* Oxford: Clarendon Press, 4th ed.
Doncel, Manuel, Armin Hermann, Louis Michel, and Abraham Pais. 1987. *Symmetries in physics (1600-1980).* Bellaterra (Barcelona): Universitat Autônoma de Barcelona.
Eberhard, P. H. 1977. "Bell's theorem without hidden variables." *Nuovo Cimento* 38B: 75-80.
Einstein, Albert. 1948. "Quantenmechanik und Wirklichkeit." *Dialectica* 2: 35-39.
———. 1949. "Reply to criticism. Remarks concerning the essays brought together in this cooperative volume." Pp. 663-693 in (Schilpp 1949). (English translation by P. A. Schilpp on the original in German, "Bemerkungen zu den in diesen Bände Vereinigten Arbeiten." Pp. 493-511 in (Schilpp 1949) German edition).
———. 1953. "Einleitende Bemerkungen über Grundbegriffe. Remarques préliminaires sur les principes fondamentaux." (French transl. by Marie-Antoinette Tonnelat). Pp. 4-15 in (de Broglie 1953).
———. 1989-1993. *Oeuvres choisies.* French transl. by the transl. group of ENS Fontenay-St-Cloud et al., ed. by Françoise Balibar et al. Paris: Seuil/ed. du CNRS, 6 vols.
Einstein, Albert, and Max Born. 1969. *Briefwechsel 1916-1955.* München: Nymphenburger Verlagshandlung.

Einstein, Albert, Boris Podolsky, and Nathan Rosen. 1935. Can quantum-mechanical description of physical reality be considered complete? *Physical Review* ser. 2, XLVII: 777-780.

Enz, Charles P. 1973. "W. Pauli's Scientific Work." Pp. 766-799 in *The Physicist's Conception of Nature*, ed. J. Mehra. Dordrecht: Reidel.

d'Espagnat, Bernard, ed. 1971. *Foundations of Quantum Mechanics. Proceedings of the International School of Physics Enrico Fermi, Course xlix.* New York: Academic Press.

d'Espagnat, Bernard. 1984. "Nonseparability and the Tentative Descriptions of Reality." *Physics Report* 110: 201-264.

———. 1994. *Le réel voilé. Analyse des concepts quantiques.* Paris: Fayard.

Flato, Moshe et al., eds. *Quantum mechanics, determinism, causality, and particles.* Dordrecht: Reidel.

Freedman, Stuart J., and John F. Clauser. 1972. "Experimental test of local-hidden variable theories." *Physical Review Letters* 28: 938-941.

Freire Jr., Olival. 1995. *A emergencia da totalidade. David Bohm e a controversia dos quanta.* Ph.D Thesis. Dpt. de História, Universidade de São Paulo.

Freire Jr., Olival, Michel Paty, and Alberto Rocha Barros. 2000. "Physique quantique et causalité selon Bohm. Un cas d'accueil défavorable." in *History of Modern Physics*, eds. H. Kragh, P. Marage, and G. Vanpaemel. Archives de l'Academie internationale d'histoire des sciences, *XXè Congrès International d'Histoire des Sciences, Liège, 1997, Actes,* Collection de Travaux de l'Académie Internationale d'Histoire des Sciences. Liège: Brepols, 2002.

French, Steven, and Decio Krause. 1999. "The Logic of Quanta." Pp. 324-342 in *Conceptual Foundations of Quantum Field Theory*, ed. Tian Yu Cao. Cambridge: Cambridge University Press.

Ghirardi, Gian Carlo, and T. Weber. 1997. "An interpretation which is appropriate for dynamical reduction theories." Pp. 89-104 in (Cohen, Horne and Stachel 1997b).

Grangier, Philippe. 1986. *Étude expérimentale de propriétés non-classiques de la lumière; interférences à un seul photon.* Ph.D. Thesis, doctorat ès-sciences physiques. Orsay: Université Paris-Sud.

Greenberger, D. M., M. A. Horne, and A. Zeilinger. 1989. "Going beyond Bell's theorem." P. 73 sq. in *Bell's Theorem, Quantum Theory and Conceptions of the Universe*, ed. M. Kafatos. Dordrecht: Kluwer.

———. 1990. "Bell's theorem without inequalities." *American Journal of Physics* 58: 1131 sq.

Griffin, A, D. W. Snoke, and S. Stringari, eds. 1995. *Bose-Einstein condensation.* Cambridge: Cambridge University Press.

Haroche, Serge, M. Brune, and Jean-Michel Raimond. 1997. "Experiments with single atoms in a cavity: entanglement, Schrödinger's cats and decoherence." *Philosophical Transactions of the Royal Society of London* A: 355, 2367-2380.

Jammer, Max. 1974. *The Philosophy of Quantum Mechanics. The Interpretations of Quantum Mechanics in Historical Perspective.* New York: Wiley.

Jauch, José-Maria. 1968. *Foundations of Quantum Mechanics.* Reading, MA: Addison-Wesley.

Kastler, Alfred. 1981. "On the historical development of the indistinguishability concept for microparticles." Pp. 607-623 in (Merwe 1981).

Leite Lopes, José, and Michel Paty, eds. 1977. *Quantum mechanics, a half century later.* Dordrecht: Reidel.

Lüders, Gerhart. 1952. "Zum Bewegungsumkehr in quantisierten Feldtheorien." *Zeitschrift für Physik* 133: 325-339.

———. 1954. *Kgl. Danske Videnskab. Selskab, Mat.-Fys. Medd.* 28, No. 5.

Mehra, Jagdish, ed. 1973. *The Physicist's Conception of Nature.* Dordrecht: Reidel.

Mehra, Jagdish, and Helmut Rechenberg. 1982. *The historical development of quantum theory,* 7 vols. New York: Springer.

Mermin, N. D. 1990. Quantum Mystery Revisited. *American Journal of Physics* 58: 731 sq.

Merwe, Alwyn van der, ed.1981. *Old and new questions in physics, cosmology, philosophy and theoretical biology. Essays in honour of Wolfgang Yourgrau.* New York: Plenum Press.

Miller, Arthur, ed. 1989. *Sixty-two years of uncertainty. Historical, philosophical and physical inquiries into the Foundations of Quantum mechanics.* (Proceedings of the International School of History of Science, Majorana Centre for Scientific Culture, Erice-Sicilia, 5-15 Aug. 1989). New York: Plenum Press.

Neumann, John von. 1932. *Mathematische Grundlagen der Quantenmechanik*. Berlin: Springer, 1932; 1969; New York: Dover, 1943. English translation, *The Mathematical Foundations of Quantum Mechanics*. Princeton: Princeton University Press, 1955.

Nguyen-Khac, Ung and Anne-Maris Lutz, eds. 1994. *Neutral currents twenty years later*. Proceedings of the international conference, Paris, July 6-9, 1993. Singapore et al.: World Scientific Publishing Company.

Omnès, Roland. 1994a. *Philosophie de la science contemporaine*. Paris: Gallimard.

———. 1994b. *The interpretation of Quantum Mechanics*. Princeton: Princeton University Press.

Paty, Michel. 1970. "Désintégrations électromagnétiques des bosons." *Ecole internationale de physique des particules élémentaires, Herceg-Novi*, 80 p.

———. 1977. "The recent attempts to verify quantum mechanics." Pp. 261-289 in (Leite Lopes and Paty 1977).

———. 1981. "Interventions sur les *Implications conceptuelles de la physique quantique, Colloque de la Foundation Hugot du Collègue de France, juin 1980*." *Journal de Physique*, Supplément, fasc. 3, Colloque No. 2: C2 37-40, 78-80, 96-98, 109-111, 116-118.

———. 1982. "L'inséparabilité quantique en perspective, ou: Popper, Einstein et le débat quantique aujourd'hui." *Fundamenta Scientiae* 3: 79-92.

———. 1985. "Symétrie et groupes de transformation dans les théories contemporaines de la matière: jalons épistémologiques." *Colloque Abel-Galois, Lille, 21-25 février 1983, Première partie*. Institut de Recherches de Mathématiques Avancées (IRMA). Lille, fasc 5, 54 p.

———. 1986. "La non-séparabilité locale et l'objet de la théorie physique." *Fundamenta Scientiae* 7: 47-87.

———. 1988a. *La matière dérobée. L'appropriation critique de l'objet de la physique contemporaine*. Archives contemporaines. Paris.

———. 1988b. "Sur la notion de complétude d'une théorie physique." Pp. 143-164 in *Leite Lopes Festchrift. A pioneer physicist in the third world*, eds. Norbert Fleury, Sergio Joffily, José A. Martins Simões, and A. Troper. (Dedicated to J. Leite Lopes on the occasion of his seventieth birthday). Singapore: World Scientific Publishers.

———. 1990. "Reality and Probability in Mario Bunge's *Treatise*." Pp. 301-322 in *Studies on Mario Bunge's Treatise*, eds. Georg Dorn, and Paul Weingartner. Poznan studies in the philosophy of the sciences and humanities. Amsterdam-Atlanta: Rodopi.

———. 1993. *Einstein philosophe. La physique comme pratique philosophique*. Paris: Presses Universitaires de France.

———. 1995a. "The nature of Einstein's objections to the Copenhagen interpretation of quantum mechanics." *Foundations of Physics* 25, No. 1 (January): 183-204.

———. 1995b. "Les neutrinos." Pp. 294-300 in *Encyclopaedia Universalis*, vol.12. Paris.

———. 1997. "Predicate of existence and predictability for a theoretical object in physics." Pp. 97-130 in (Agazzi 1997).

———. 2000a. "Interprétations et significations en physique quantique." *Revue Internationale de philosophie*. Brussels, n° 212, 2 (June), 2000, 199-242. (Interpretations and significations in quantum physics.)

———. 2000b. "The quantum and the classical domains as provisional parallel coexistents." In honor of *Newton da Costa*, eds. Steven French, Décio Krause and Francisco Doria. *Synthese*: 125:1-2. Dordrecht: Boston.

———. In press. *Einstein, les quanta et le réel (critique et construction théorique)*.

———. 2001. "Physical quantum states and the meaning of probability." Pp. 235-255 in *Stochastic Causality*, eds. Maria Carla Galavotto, Patrick Suppes and Domenico Costantini. Stanford: CSLI Publications.

Pauli, Wolfgang 1955. "Exclusion Principle, Lorentz Group and Reflection of Space-Time and Charge." Pp. 30-51 in *Niels Bohr and the Development of Physics*, eds. W. Pauli, Léon Rosenfeld, and Victor Weisskopf. London: Pergamon.

Peres, Asher 1978. "Unperformed experiments have no results." *American Journal of Physics* 46: 745-747.

Pesic, Peter D. 1991. "The principle of identicality and the foundation of quantum theory. 1. The Gibbs paradox;" "2. The role of identicality in the formation of quantum theory." *American Journal of Physics* 59 (11): 971-974, 975-979.

Pflegor, R. L., and L. Mandel. 1967. "Interference of independent photon beam." *Physical Review* 159: 1084-1088.
Popper, Karl R. 1959. *The Logic of Scientific Discovery*. London: Hutchinson, 1959; 1968. (English translation modified and augmented from the original in German, *Logik der Forschung*, Springer, 1935).
Schilpp, Paul-Arthur, ed. 1949. *Albert Einstein: philosopher—scientist*. The library of living philosophers. Lassalle. IL: Open Court, 1949. Ré-ed. 1970. Translated in German. *Albert Einstein als Philosoph und Naturforscher*. Stuttgart: Kohlhammer Verlag, 1955 (original texts of Einstein and Pauli).
Schrödinger, Erwin. 1935. "Die gegenwärtige Situation in der Quantenmechanik." *Die Naturwissenschaften* 23: 807-812, 824-828, 844-849. Also in (Schrödinger 1984, vol. 3, 484-501.
———. 1984. *Gesammelte Abhandlungen. Collected papers,* 4 vols. Wien: Verlag der Oesterreichischen Akademie der Wissenschaften; Braunschweig: Vieweg und Sohn.
Schweber, Sam S. 1994. *QED and the men who made it*. Princeton: Princeton University Press.
Schwinger, Julian. 1951-1953. "The theory of quantized fields, I." *Physical Review* 82, 1951: 914-927; II, *ibid.* 91, 1953: 713-728 (reprinted in J. Schwinger, ed., *Selected papers on electrodynamics*. New York: Dover, 1958: 342-355, 356-371); III, *ibid.*, 91, 1953: 728-740; IV, *ibid.*, 92, 1953: 1283-1299 (reprinted in J. Schwinger. *Selected papers (1937-1976)*, ed. M. Flato, C. Frondsal, and K. A. Milton. Dordrecht: Reidel, 1979: 43-55, 56-72).
Seidengart, Jean, and Jean-Jacques Sczeciniarz, eds. 2000. "Cosmologie et Philosophie. En hommage à Jacques Merleau-Ponty," numéro spécial de *Épistémologiques. Philosophie, sciences, histoire. Philosophy, Science, History* 1, No.1-2 (Jan. – June). Paris, São Paulo.
Shimony, Abner. 1993. *Search for a naturalistic world view*. Cambridge: Cambridge University Press, 2 vols.
Stapp, Henry. 1980. "Locality and reality." *Foundations of Physics* 10: 767-795.
Weyl, Hermann. 1931. *Gruppentheorie und Quantenmechanik*, english translation (on the 2nd rev. German ed.) by H. P. Robertson, *The Theory of Groups and Quantum Mechanics*. New York: Dover, 1950.
Wheeler, John A., and Wojcieh H. Zurek, eds. 1983. *Quantum theory of measurement*. Princeton: Princeton University Press.
Wigner, Eugen P. 1967. *Symmetries and reflections*. Bloomington: Indiana University Press.
Yang, C. N. 1982. "The discrete symmetries P, T and C." Pp. 439-449 in *Colloque International sur l'histoire de la physique des particules (Paris, 21-23 July 1982). Journal de Physique,* Colloque C8, Supplt au No. 12, Tome 43, December 1982.
Yoccoz, Jean, ed. 1996. *Physique subatomique: 25 ans de recherche à l'IN2P3, la science, les structures, les hommes*. Gif-sur-Yvette: Editions Frontières.
Zurek, Wojcieh H. 1982. "Environment-induced superselection rules." *Physical Review* D 24: 1516-1525.
———. 1991. "Decoherence and the Transition from Quantum to Classical." *Physics Today* 44: 36-44.

DAVID RITZ FINKELSTEIN

ELEMENTARY PROCESSES

Abstract. Since non-semisimple groups are unstable, every non-semisimple group in physics will probably be the locus of a major conceptual reformation. We carry out such reformations of the canonical commutation relations, the space-time continuum, and the interface between experimenter and system. The canonical reformation introduces a large orthogonal group and its Clifford algebra and spinors. The space-time reformation eliminates the differential calculus from basic physical laws, introducing a generic elementary quantum process and a fundamental time χ. The interface reformation leads to a non-operational cosmic quantum theory that treats the physical experimenter as a quantum system, recovering the operational quantum theory by a systematic quotient process. We propose a real form of Fermi statistics for the elementary processes, in a Clifford algebra generated from the real line (which represents the empty set) by iterated quantification. We use this quantum set language to program the spin and space-time variables and equations of motion of a Dirac particle.

1. REFORMATIONS

The three main evolutionary leaps of twentieth century physics have a suggestive family resemblance. Special relativity introduced a non-commutativity of boosts. General relativity and gauge theory introduced a non-commutativity of infinitesimal translations. Quantum theory introduced a non-commutativity of observations.

Do all historic radical changes in the foundations of physics introduce non-commutativities? Is this the form the next such change will likely take?

The analyses of Segal (1951) and Inönü and Wigner (1953) suggest that the answer to these questions is yes. Groups and Lie algebras that are not semisimple we call *compound*. A compound group is unstable with respect to small changes in its structure and is almost certainly but an approximation to a more physical stable semisimple Lie algebra, subject to verification when measurements become more accurate. Non-commutativity is what stabilizes and simplifies an unstable compound group, as curvature stabilizes a sail.

The compound group generally respects an absolute A that is presumably false: an *idol* in the sense of Francis Bacon. Dethroning such an idol is a radical conceptual change.

What shall we call this inverse process to group contraction? "Revolution" is the common name but is inappropriate, since the deformed theory has the utmost respect for the old, incorporates it as a limiting case, stays as close to it as possible, and never supersedes it entirely. (Centuries after Copernicus we still speak of sunrise.) "Deformation" is too general; this deformation must begin from a singular limiting case and remove its instability.

Since the process is a de*formation* that *re*pairs the theory let us call it *reformation*.

The general prescription for reformation is clear.

1. Select a compound group of the existing theory. The compound group generally has an idol.

2. Find the physical effects that allegedly demonstrate the existence of the idol. They show what the idol couples into.f

3. If A couples into B, couple B into A reciprocally, thereby introducing a small constant coefficient, the *reformation parameter*. If A couples to nothing, eliminate it. It was unnecessary.

A is then no longer an absolute of the theory.

Reformation always inaugurates such a reciprocity. This implies that physical groups ultimately couple their variables reciprocally. Such a "reciprocity principle" was applied to the coupling between space-time geometry and matter by Mach and Einstein. Since some of our habits of thought still harbor idols that violate reciprocity, we must expect future reformations as radical as the reformations that introduced quantum theory, special relativity, and general relativity.

This principle does not tell us which reformation is next, or predict the size of a reformation parameter, or tell us which of several possible reformations is most relevant to current experiment. Since there are many idols in circulation today, the probability of a proposed reformation is small. On the other hand the probability of the unreformed theory is 0.

The standard example of reformation for Segal and for Inönü and Wigner is Einstein's replacement of the compound Galilean group of rotations and boosts by the simple Lorentz group. This historic process eliminated the idol of absolute time. When Galileo transforms from dock to ship, he couples time into space ($x' = x - Vt$) but not space into time ($t' = t$). The Galileo space/time algebra is not stable/semi-simple because there is this one-way time-to-space coupling under transformations from one observer to another.

The stability principle is not devoid of discrimination. For example, it does not lead to string theory which leaves the all compound groups of present physics still compound. Indeed, if the string moves in a flat space-time, perhaps one of many dimensions, as originally proposed, then string theory is a counter-reformation, re-establishing an idol that general relativity dethroned.

Special relativity, general relativity, gauge theory, and quantum theory are all reformations. These retrodictive successes suggest that we test the predictive consequences of Segal's principle.

Any theory starts by assuming some physical invariants that couple to physical variables without back-coupling. Therefore the reformation strategy promises a never-ending succession of reformations, each introducing a new fundamental constant as the value of a reformation parameter after the examples of c, $\hbar$ and G.

2. THE CANONICAL REFORMATION

The first stage of reformation is idol-detection.

The most conspicuous current idol is the quantum imaginary $\hbar i$ of the canonical commutation relations

$$[q, p] = \hbar i, \quad [p, i] = 0, \quad [q, i] = 0 \tag{1}$$

(Segal 1951). The algebra of (p, q, i) is isomorphic to a subalgebra $(\partial_x, b_x, \partial_t)$ of the Galileo algebra where ∂_x and b_x respectively translate and boost infinitesimally in the x direction. Heisenberg's group is just as likely to be wrong as Galileo's, being just as unstable. It would be a miracle — more precisely, an event of probability 0 — if the canonical commutation relations were exact.

The reformed canonical commutation relations are

$$[q, p] = \hbar i, \quad [p, i] = \lambda q, \quad [q, i] = \mu p \tag{2}$$

with small new fundamental *Segal constants* $\lambda, \mu \neq 0$ supplementing Planck's constant. We can make a dimensionless combination $N = \lambda\mu$ from them, and a time χ defined below.

This reformation relativizes Heisenberg's $\hbar i$ and suspends its commutativity with p and q.

Relativizing Heisenberg's absolute i is isomorphic to relativizing Galileo's absolute time.

This reformation, however, triggers an avalanche of many secondary ones, far more than its Einsteinian prototype. Different particles have different space-time coordinates, momenta, and boosts, but the same $\hbar i$ couples to all canonical pairs. Unifying each pair with $\hbar i$ unifies them all with each other. This fuses a great many groups SO(3) into a vast SO(N) group and replaces the p's, q's and $\hbar i$ by spin operators of that group.

Then the same instability attacks the differential calculus relation $[d/dx, x] = 1$. Its reformation eliminates the differential calculus from all basic laws, replacing differential operators too by spin operators. This incidentally makes the theory finite.

Nevertheless, this is a conservative reconstruction of physics. Any reformation makes changes in the predictions of today's working physics that can be as small as we like. We can always return as closely as we like to the predictions of the old theory by choosing the reformation parameter small enough. The prediction, however, is that eventually experiment will favor the reformed theory over the unstable one and provide a non-zero value for the reformation parameter. Until that happens, we might think that this reformation is merely a way of regularizing the present singular theory.

There is a note out of tune. The isomorphism between Segal's reformation of Heisenberg and Einstein's reformation of Galileo seems imperfect: Einstein used only one constant c to reach special relativity from Galileo relativity. Why does Segal need two?

The difference is that Segal set out to simplify a group and Einstein did not. The Poincaré group of special relativity still is not simple, thanks to its translation subgroup. Had he set out to simplify the Galileo group, Einstein would have gone all the way to the de Sitter group in the first step, introducing a second constant, the radius of the universe, besides lightspeed c. The Galilean reformation leads at once not only to special relativity but also to the expanding universe.

After the canonical reformation i is no longer a c number but a dynamical variable. Therefore we start from a quantum kinematics that lacks it, a real quantum theory. A variable i once provided a Higgs field (in quaternionic gauge theory, (Finkelstein et al. 1963)). It seems likely to do so again.

The orthogonal group of the reformed Heisenberg algebra has a single finite-dimensional defining spin representation from which all the physical representations can be built algebraically. The usual infinite-dimensional representation of momentum-energy by differentiation and position by multiplication gives way to a finite-dimensional representation of all the components of space-time-momentum-energy-$\hbar i$ by sums of products of spin operators. The ultimate factors correspond to finite elementary quantum processes, chronons.

3. THE INTERFACE REFORMATION

The classic prototype of a theory ripe for radical change, Galileo's space-time, is a bundle with time as base and space as fiber. The reformation unifies fiber and base.

All bundle theories share the instability of Galilean space-time and are as much in need of reformation. The group of the bundle couples the base into the fiber but not reciprocally. We must eventually unify the fibers and the base of the gauge theories. The unification of p and x that Segal proposed is another special case, the cotangent bundle of space-time.

In modern field theory, however, the space-time base is classical, part of the exosystem, and the fiber is quantum. To unify them we must also unify experimenter and quantum system. We cannot eliminate experimenters but we can relativize the absolute experimenter of operational quantum theory, thus eliminating another idol.

One does this most often by imaging a cosmic viewpoint that covers all possible divisions into experimenter and system. This results in what claims to be a quantum theory of the cosmos, a concept that Heisenberg and Bohr resisted strongly at first, but which was finally accepted by Bohr as a necessary step. Unity and operationality ultimately part, and we pursue unity. We assume – like Laplace — a metaphoric or virtual cosmic experimenter (CE) who can input, transduce, and outtake the entire cosmos including all the experimenters in it. Quantum field theory has implicitly taken the cosmic perspective from the start, since the actual experimenter is made of the same field as the system under experiment.

A maximally informative experiment has three stages corresponding to the factors in the transition amplitude $\langle 3|2|1\rangle$. Therefore the experiments of the CE also have three parts. The CE inputs the cosmos in the remote past in an initial mode $|1\rangle$, sets up a medial dynamical development 2, and outtakes the cosmos in the distant future with a final dual vector $\langle 3|$, experimentally testing the cosmic transition amplitude $\langle 3|2|1\rangle$.

One should not be misled by the revealing miscount of two experimental interventions given in (von Neumann 1932). One might make the same miscount today by counting as interventions, say, only what goes on at CERN in the target and detector buildings, overlooking the accelerator facility that produces the input beam.

Some do not express quantum theory operationally but in terms of collapsible states; others speak of actual operations and actions. We call these formulations ontic and praxic respectively. Ontists naively imagine that a unitary operator represents an individual action faithfully, like a classical rotation matrix. To give the operator something definite to act on they count the input mode vector $|1\rangle$ as a state. Then they count as interventions only the dynamical process 2 and the outtake action $\langle 3|$. Praxists recognize that an operator representing an action is a matrix of transition amplitudes referring to ensembles of many experiments. They count all three actual stages of an experiment.

The ontic formulation of the quantum theory is an atavistic vestige of the discredited wave theory of quantum phenomena, but it gets to the same physical results as the more accurate praxic formulation by making two errors that cancel. The first error is calling the first action a state of the system. Actually we cannot extract information specifying the first action from the system, according to the amplitude formula. It may be called a state, but it is not of the system but of the exosystem, just as a probability distribution is not of one case but of its context, a source or an ensemble of cases. Calling the last action a collapse to a new state is a similar error that compensates for the first. The same artifice works just as well for cosmic quantum theory.

Any operational theory, classical or quantum, divides the cosmos into a system that is treated dynamically and an exosystem that is not. The exosystem always includes the experimenter. We call this division the (system-exosystem) interface. The two vertical bars in the amplitude formula $\langle 3|2|1\rangle$ may be used to represent the interface, since the dynamics operates only inside them. The interface is sometimes called the quantum or Heisenberg cut but it is not peculiar to quantum theory, accept insofar as quantum theory dispelled some of the naivety that permitted most people most of the time to overlook the interface in pre-quantum physics.

Introducing a cosmic experimenter and a cosmic dynamical law does not eliminate the interface but shifts it from between system and exosystem to between cosmos and the CE. It does not reduce the number of interventions but shifts them from the actual experimenter to the virtual CE.

4. THE DYNAMICAL REFORMATION

The Hamiltonian giving the dynamical law of the system under study is an idol of the current quantum theory, acting on the system with no reciprocal action. As step 2 of the reformation process we do thought experiments in which we determine the dynamical law for a system under maximal quantum resolution.

In the quantum theory of Heisenberg and Bohr, one postulates the transition amplitude formula $A = \langle 3|2|1\rangle$. This familiar assumption of quantum theory is approximate on several counts. It splits the experiment into three stages represented by the three factors in A. In stages $|1\rangle$ and $\langle 3|$ the system development is totally dominated by its interaction with the exosystem and in stage 2 the system is totally decoupled from the exosystem and develops according to a known Hamiltonian operator H. The expression of H in terms of primitive system variables is idealized as completely knowable, hence classical, even though it governs a quantum system, because it summarizes an infinite number of quantum experiments, not one. The three stages are assumed to be independently variable.

Our experiments, however, are not infinitesimal in size and infinite in number, but necessarily finite in size and number, for several reasons, including their quantum structure, their gravitational fields, and their cosmological limits. Therefore our knowledge of H is incomplete in principle.

Since we measure H on an ensemble, we should treat H as a quantum variable of the ensemble, with maximal descriptions combining by quantum superposition as usual.

Furthermore, if we make the dynamics a quantum variable then we probably need no others. Newton and Mach already raised the possibility that the dynamical law is variable. Here we further propose that it is the sole variable. To be sure it is typically a highly composite variable, but its parts too are dynamical processes This corresponds to taking the variable metric of general relativity to be the sole dynamical variable, as in Einstein's unified field theory.

In some earlier efforts we took the causal relation between events as the prime variable, and sought the dynamics that governs it. Now the dynamics is the sole independent quantum variable, with its own algebra. We are to find all other physical variables within that algebra. The causal relation and the metric tensor give a reduced statistical description of the dynamics, adequate for a planet but not for a proton.

A sharp description of the dynamics is a spinor of $SO(N)$ that we call a *dynamics mode*. The relation between spinor and dynamics is like that between a wave function and an atom. The spinor describes the dynamics sharply but is not the dynamics. The addition of operators to form a Hamiltonian is a classical process distinct from the quantum superposition of spinors.

The high constancy of the dynamics over the sidereal cosmos suggests that it is all one order-domain of the dynamics, and that some basic cell serves as order parameter, repeating its structure throughout our ambient vacuum and defining the dynamics of our cosmos. This unit cell is then the molecule of dynamics, composed of some number of chronons that remains fixed as the total number N varies.

The space-time metric tensor is a partial description of dynamics. As the Lagrangian of a massive point particle, it defines, and is operationally defined by, how a test body — a smallish system whose tides hardly effect its orbit — moves under gravity. The theory of the space-time metric can therefore give useful hints toward a theory of the dynamics in general. The usual quantum assumption of a fixed dynamics for arbitrary initial and final actions is a quantum analogue of the pre-relativity assumption of a flat space-time for arbitrary space-time content.

In classical thought the path description varies from experiment to experiment on the same system, while the action function is fixed. The specification of the dynamics leaves a set of possible paths, not a sharp description of the path. The initial data then complete the determination of the path.

In the quantum theory, however, the dynamics assigns a probability amplitude to each path, and is therefore identical with a superposition of path descriptions, up to a duality; or to a highly entangled description of one q path. Such a path description is complementary to the specification of the end-points or for that matter of any point on the path.

In the c theory the endpoint specification is *supplementary* to the dynamics description, and completes it. In the q theory the endpoint data is *complementary* to the dynamics description and violates it. The classical theory describes its dynamics only crisply, as a set of many possible paths. The existing quantum theory describes its dynamics sharply, as a single entangled path. We follow the quantum model rather than the classical in setting up a chronon dynamics.

Smolin (1992, 1997) has suggested that the dynamics not only varies but undergoes Darwinian evolution, like genes, without proposing a microscopic locus for the dynamics. The development of the dynamics may be like crystal growth, guided by the entire unit cells, or like cell multiplication, guided by specialized code structures buried within each cell. In either case we require a higher-level dynamics to control the lower-level dynamics under consideration, as Einstein's gravitational equations control the metric.

We do not yet understand well enough how the approximation of a fixed dynamics can work so well if the dynamics is as highly variable as we claim. This dynamical fact corresponds to the chronometrical fact that the flat space-time metric is a good approximation for much of physics, despite the variability of the space-time metric.

The chronon dynamics must contract both to a standard q field theory and to general relativity, in two appropriate contractions with $\chi \to 0$ and $N \to \infty$. It would suffice, for example, if in the limit $\hbar \to 0, \chi \to 0, N\chi \to \infty$ we recover classical space-time and general relativity; and in the limit $\hbar =$ constant, $\chi \to 0$, $N \to \infty$, $N\chi = t =$ constant, we return to the continuum-based quantum field theory.

Our main variable is no longer a causal relation as in our first efforts (Finkelstein 1969). A causal relation is a weak statement about what dynamics is possible. It is stronger and sharper, hence more vulnerable, to give the actual dynamical-development operator D for the universe. This contains the causal connections and more. The space-time metric is a reduced statistical description of D.

5. THE SPACE-TIME REFORMATION

The usual quantum theory assumes an underlying fundamental flat space-time, ignoring Einstein locality. But the space-time in the experimental chamber can be determined operationally from the dynamics, and so should not be postulated. We assume neither flatness nor even any definite dimension or signature under high resolution. These all come from how the atomic elements of the dynamics are interconnected in a particular mode.

We consider the simplest possibility only: that the atomic elements of the dynamics — chronons henceforth — have no parts, cannot be factored. We expect that the standard classical space-time points are the correspondence limit of many such quantum events.

All field theories are ripe for reformation. Field theories, including string and membrane theories, exhibit the same symptom of impending failure as Galilean classical mechanics, a basic one-way coupling, formerly from time to space, now from space-time to field, resulting in a non-semisimple transformation group, presumptive evidence of an algebra contraction in the Segal-Inönü-Wigner sense (Marks et al. 1999). If such a theory is valid in some domain, it is expected to be a contraction of a simpler theory of greater domain.

The bundle basic to all field physics is the tangent bundle of space-time. The symmetry between p and x that the canonical commutation relations imply, even before their reformation by Segal, is violated by any bundle over space-time. Simplicity requires us to unify the field variables and the space-time variables.

The differential-calculus limit $\Delta t \to 0$ results in the canonical commutation relations $[\partial_x, x] = 1$, which have a non-semisimple group. To deform this group to SO(3) (or SO(2, 1)) we cut off the limit $\Delta t \to 0$ at a finite value $\Delta t = \chi$, introducing a finite elementary dynamical process — the chronon — and a fundamental constant χ giving the time-scale of the chronon.

We suppose that the dynamical process is an aggregate of finite elementary quantum processes, chronons. We will specify the structure of this aggregate after we develop necessary concepts of Clifford-algebraic quantum theory and quantum statistics.

The usual operator-algebraic quantum kinematics lacks space-time concepts. One appends these as necessary. Frequently this disturbs simplicity by creating a fiber bundle. and a continuous time coordinate. Then one can analyze the dynamics into ever finer transformations and never reach elementary or atomic transformations. This assumption leads to infinities and is likely wrong. The atomism principle cuts this process off, and yet does not conflict with exact Lorentz invariance on sub-cosmological scales of time and length. Snyder space-time is an early example of such a Lorentz-invariant quantum space-time with discrete space, though it still has continuous time.

A crystal has many characteristic times with different physical meanings. So does the vacuum, experimentally speaking.

The space-time analogue of the crystal cell-size l is the fundamental time-scale χ of the chronon. The cell of the q space-time is an assembly of several chronons.

The vacuum must also have a *coherence number* N_c, analogous to the pure number $(\lambda_c/l)^D$ for a superconductor, where λ_c is the Ginzburg-Landau coherence length, l is the the cell size of the crystal, and $D = 2$ or 3 is the dimensionality of the superconductor.

Further, each mode that can not propagate in a crystal has a penetration depth l_P. Similarly each mode to which the space-time q crystal is not transparent has a penetration depth l_P. Penetration depths may manifest experimentally as Compton wavelengths defining masses, for example of gauge fields.

A well-known elementary qualitative argument equates the Planck mass to a quantum black-hole mass. Then the Planck length is not the cell size but the penetration length of a high mode of the network.

Standard relativistic quantum field physics incorporates an absolute locality principle that has been enormously fertile, producing electromagnetic field theory, general relativity, and the gauge revolution of recent decades. The chronon replaces infinitesimal locality by a finite locality. This act of desperation calls for several excuses and a promise to make restitution. We must restore infinitesimal locality by a suitable contraction of the finite locality.

Originally we dropped infinitesimal locality just for the sake of finiteness. Present continuum-based field theory is a kind of science-fiction, postulating acts that are impossible in principle, exploring some consequences of this fantasy, and carefully ignoring others. A better quantum theory would postulate only acts that are possible in principle. Such acts cannot be infinitesimals. We retain a finite locality, in which elementary interactions couple events separated by $\chi > 0$.

Nowadays we further justify this renunciation on Segal's grounds of stability and algebra simplicity. They couple x and p into one simple orthogonal-group algebra. Since x has local matrix elements $\sim \delta(x - x')$ and p has slightly nonlocal ones $\sim \delta'(x - x')$, mixing x and p breaks locality.

To put it bluntly: One measures x with a small pinhole and p with a large diffraction grating. The reformed theory cannot tell its diffraction grating from its pinhole.

The standard quantum theory is local but not simple; we seek a quantum theory that is simple, hence not local. The c correspondents of our q events are now points of at least an 8-dimensional space, combining x and p, not a 4-dimensional one. The 8-dimensional quadratic space $8\mathbb{R} = \{(x^\mu, p_\mu)\} = 4\mathbb{R} \oplus 4\mathbb{R}$ has possible signatures $(6 - 2)$, $(4 - 4)$, and $(2 - 6)$, based on the Minkowski signature $(3 - 1)$. In section 10 we opt for the neutral quadratic form. Then time t and momentum are composed of Clifford units with signature $+1$ and space x and energy E with signature -1.

6. CLIFFORD † ALGEBRA

A roman capital designates a quantum entity: X stands for the chronon and D for the dynamics much as H stands for hydrogen. Each system S has a quadratic vector space $V = V_S$ of quantum modes or channels and an operator algebra $A = A_S = \text{Endo}\, V$ representing possible actions upon the system. $\text{Cliff}(N_+, N_-)$ and CLIFF V are real Clifford algebras with an adjoint † taken to be the main anti-automorphism. CLIFF is the Clifford functor from real † spaces to their free Clifford algebras. Cliff is the Clifford map $V \to C = \text{CLIFF}\, V$, linearly mapping each vector $v \in V$ to a Clifford element $\text{Cliff}\, v \in C$. By a real Clifford † algebra over a quadratic space V with quadratic form g_{ab} — possibly indefinite or singular — we mean a real † algebra C generated by $\mathbb{R}$ and an isomorph of V in which the *Clifford relations* hold in the form

$$(\forall v \in V)(\text{Cliff}\, v)^2 = -\|v\|. \tag{3}$$

The Clifford algebra is *free* (or universal) if all its relations follow from these postulates. We define the anti-automorphism † $: C \to C$ by postulating that $(\text{Cliff}\, v)^\dagger = -\text{Cliff}\, v$ for all $v \in V$. The *top unit* of a basis $\{i_a\}$ of V is

$$\chi_\uparrow := \overleftarrow{\prod_a} \chi_a = \chi_N \cdots \chi_1. \tag{4}$$

If $A = \text{CLIFF}\, V$ is a Clifford algebra and $A \sim \text{Endo}\, V'$ then V' is a *free spinor space* for A and we write $V' = \sqrt{A}$. $\sqrt{}$ is a kind of square root, extracting a vector space of dimension D from an algebra of dimension D^2. Dimension grows exponentially from V to A: $\text{Dim}\, A = 2^{\text{Dim}\, V}$.

7. QUANTIFICATION

What is the statistics of the chronon?

Before we describe our latest model, we recapitulate the concept of statistics in general.

One usually defines a statistics for indistinguishable particles by representing the permutations of individuals in the aggregate algebra. This creates a semantic problem: What physical operations can be meant by an exchange of indistinguishable objects?

We evade this problem if we consider a unified system of a variable number of quantum entities, with number ranging from none to very many. Then an exchange of a pair may act trivially on the pair but non-trivially on each of its members by itself. We therefore define an exchange as an automorphism of the many-body algebra that exchanges specified elements of the one-body algebra. We define a statistics so that it provides a Lie homomorphism $\iota : A \to A'$ of the 1-quantum endomorphism Lie algebra A into the many-quantum algebra A', describing how aggregates transform under transformations of the individual.

For a quantum statistics we further impose a *coherence condition*: that ι be induced by an operator-valued form $Q^\dagger : V \to A'$ here called the *quantification form* so that for all $a \in A$, $\iota(a) = Q^\dagger a Q$.

For example if ϕ is any boson state vector and $Q^\dagger = Q_2^\dagger$ is the boson quantification form then $Q_2^\dagger \phi$ is the boson annihilation operator associated with mode ϕ. For complex Fermi statistics, the quantification form $\iota = \iota_1$ maps a state vector ψ to a fermion creation operator.

Any quantification functor Q fits into a commutative diagram

$$\begin{array}{ccc} & \text{ENDO} & \\ \text{VEC}_0 & \to & \text{ALG}_0 \\ & & \\ Q \downarrow & \searrow \quad \downarrow & \\ & & \\ \text{VEC}_0 & \to & \text{ALG}_0 \end{array} \qquad (5)$$

VEC_0 designates the category of neutral mode vector spaces, ALG_0 the category of endomorphism algebras of such spaces. The horizontal arrows ENDO lead from † spaces describing quanta to their endomorphism algebras. The vertical arrows Q lead from individuals to collectives. The resultant morphism Q ENDO = ENDO Q transforms the individual chronon state-vector space to the aggregate dynamics algebra. Associated with each mode of quantification Q and each † space V there is an operator-valued form $Q^\dagger : V \to A' = QV$ that we call the quantification form. Its dual is an operator-valued vector Q. Since $A = V \otimes \text{Co}\, V$, Q induces a mapping $\iota : A \to QA$, $a \mapsto Q^\dagger a Q$ that we require to be a Lie homomorphism, respecting commutation relations $[a, b] = c$. The mapping ι generalizes the braces $\{\ldots\}$ of set theory and the successor function ι of Peano.

We tag the real Fermi, complex Fermi, and Bose quantification operators and unitizers with a subscript 0, 1, 2 counting the independent imaginaries in the coefficient number system of the associated classical group.

We may imbed permutations in the orthogonal group by applying them to the axes of a frame in the input vector space of the individual system. Then relative to any frame in $V\mathrm{I}$, Q also defines a representation of the permutation group $S_N \subset O(N)$, thus subsuming the usual concept of statistics.

In the standard quantum theory the elements of a time-translation process are infinitesimal, distinguishable, and obey Maxwell-Boltzmann statistics. The true chronons must have an i-less irreducible (simple) statistics. Maxwell statistics is not simple, Bose is neither real nor stable. Neither will do. This seems to favor Fermi-Dirac statistics. Finkelstein (1996) was based on Fermi statistics.

But Fermi statistics is usually expressed in a Grassmann algebra. Grassmann algebra is unstable; its reformation is a Clifford algebra. Finkelstein and Rodriguez (1981, 1984) proposed such a reformed Fermi statistics. Here we use a Fermi statistics formulated from the start within a stable Clifford algebra (Plymen and Robinson 1995).

The global dynamics $D = Q_0 X$ is a q set of microscopic elementary dynamical processes or chronons X. Q_0 is a special quantification functor defining the real Fermi-Dirac statistics of the chronon, discussed in the next section.

The number N of X's in D reflects the space-time measure of the process we choose to study. We take N to be finite, leaving any limit $N \to \infty$ for last. In ordinary quantum experiments $N \gg 1$.

Since we seek an analysis into elementary processes, we suppose that the basic dynamics describes only one step in history, from one moment to the next, a step of χ in cosmic time. For autonomous systems, the whole story is constructed by iterating D.

To describe Fermi-Dirac spin 1/2 quanta we assume that the spinor space V too is a real quadratic space with a neutral bilinear form †. Then its automorphism group is an orthogonal group $O(N, N)$; its Clifford algebra $\mathrm{CLIFF}\, V$ is a fermionic algebra generated by creators and annihilators of spin 1/2 excitations; and the spinors of V represent assemblies of such processes.

All actions on a set of chronons can be expressed algebraically in terms of chronon permutations. Our real Fermi statistics is thus more expressive than the usual, where *no* (nontrivial) system variables can be expressed in terms of quantum permutations $(= \pm 1)$.

It is well-known in principle how to construct approximate bosons out of many fermions.

In the early days of quantum physics, the 2-valued representations of the rotation group were overlooked for a time because they do not occur in a tensor product of spinless quanta. We overlooked the much-studied 2-valued representations of the permutation group and 2-valued statistics until recently for much the same reason: The spinor space does not occur in a tensor product of the vector spaces of individual quanta.

Our main constitutive assumption about the dynamics can now be expressed:

The dynamical process D is a real Fermi-Dirac aggregate of finitely many chronons.

A spinor D of $\text{Cliff}(N_+, N_-)$ maximally describes the unified quantum space-time-matter-dynamics D. We expect a suitable D to define both the action functionals of q field theory and gravity in appropriate contraction limits.

Quantum theories that rest on Hilbert space, with its definite sesquilinear form, are not sufficient for our purpose. Our spinors have real neutral quadratic forms. They are neutral so that they can split relativistically into space-time and energy-momentum forms, which in turn are indefinite so that they can split relativistically into spacelike and timelike parts..

8. MULTIPLE QUANTIFICATION

Standard quantum physics uses "$\infty + 1$" levels of quantification: the ∞ classical quantifications of classical set theory to produce the underlying classical space-time continuum from the empty set, and one quantification in the quantum theory to produce fermionic or bosonic algebras from a 1-quantum theory.

For example to build up the hierarchy of points, lines, surfaces, volumes, ... of the space-time manifold. Two levels that occur in the quantum theory are those of the space of one-fermion wave functions of Dirac's first electron theory, and its quantification, the many-fermion field theory.

Since we are unifying the experimenter with the system, we must unify their logics, including their quantification processes.

To constrain the problem we suppose that all quantifications are of one kind. This is clearly a vestige of set theory but it might still be right. We seek one quantification whose iteration successively produces space-time, a one-fermion system, and the many-fermion system.

We have defined the quantification functor so that it can be iterated. We may iterate Q and ENDO and extend the diagram (5) downward and to the right as necessary, just as we iterate the power set functor in set theory. For example, $Q^2 = QQ$ produces an aggregate of aggregates with a well-defined algebra and vector space. These have internal structures resulting from the preceding aggregation.

9. SPINS AS AGGREGATES

The most appropriate statistics for the chronon, we propose, is a real generalization of Fermi statistics that describes the many-fermion system by a spinor space of state-vectors and a Clifford † algebra of observables.

It is already well-known that spinors arise from statistics without reference to spin (Wiman 1898; Schur 1911; Hoffman and Humphreys 1992; Nayak and Wilczek 1996; Wilczek 1997; Wilczek 1998; Wilczek 1999) and can be used to describe complex Fermi ensembles (Plymen and Robinson 1995). We give such a spinorial description for real quantum statistics.

We define the real *Fermi quantification* functor Q_0 (and a real Fermi statistics) by

$$V' = Q_0 V = \sqrt{\text{CLIFF}(V \oplus \text{Co } V)} \tag{6}$$

Here $\mathrm{Co}\, V$ is the † space of real forms on V.

This spinorial representation for Fermi-Dirac statistics is essentially that of Dirac (1974) and Plymen and Robinson (1995). It supersedes Finkelstein (1996) where Fermi quantification was represented by a different functor

$$V' = Q_G V = \text{VEC GRASS}\, V, \qquad (7)$$

in which $\text{GRASS}\, V$ is the Grassmann algebra over V and VEC is the forgetful functor that remembers only the † vector space of a † algebra. It happens that both quantifications produce vector spaces V' of the same dimension 2^D, where D is the dimension of V. But Q_0 has a larger invariance group. Finkelstein (1996) was written before we encountered Segal (1951).

To verify that the new formulation captures the essential features of the usual Fermi quantification, consider orthonormal basis vectors $\psi_n \in V$ and dual basis vectors $\psi^{n\dagger} \in \mathrm{Co}\, V$ with $\{\psi_n, \psi^{m\dagger}\} = \delta_n^m$. Because the natural metric on $W = V \oplus \mathrm{Co}\, V$ is neutral and has V and $\mathrm{Co}\, V$ as maximal isotropic subspaces, the corresponding Clifford generators in $\text{CLIFF}\, W$ indeed obey Fermi-Dirac relations

$$\{\psi_n, \psi_m\} = 0, \quad \{\psi_n, \psi^{m\dagger}\} = \delta_n^m. \qquad (8)$$

The exchange $X : \psi_1 \leftrightarrow \psi_2$ of two orthogonal unit vectors in V induces an exchange pair $\psi_1 \leftrightarrow \psi_2, \psi_1{}^\dagger \leftrightarrow \psi_2{}^\dagger$ in $\text{CLIFF}(V \oplus \mathrm{Co}\, V)$.

Thus the real Fermi statistics can represent one swap $\chi_1 \leftrightarrow \chi_2) = (12)$ projectively in several ways: by the reflection $\chi_1 - \chi_2$, the reflection $\chi_1 + \chi_2$, or the π rotation $1 \pm \chi_1\chi_2$. Read (2002) uses some of these in a comparison of several statistics, including braid statistics. These operators agree on the rays of χ_1 and χ_2 but not on other rays in the $\chi_1\chi_2$ plane. To swap two fermions, moreover, we must swap both the hermitian parts $\chi_1^+ \leftrightarrow \chi_2^+$ and the anti-hermitian parts $\chi_1^- \leftrightarrow \chi_2^-$ of their creation operators.

We shall assign positive indices n to the real parts ψ_n of the complex vectors $|n\rangle$ and negative indices $-n$ to the corresponding imaginary parts ψ_{-n}. To avoid ambiguity we do not use the index $n = 0$. Then the creation and annihilation operators for the complex mode ψ_n with $n > 0$ are (up to a possible normalization factor) respectively

$$\begin{aligned} \iota^\dagger(\psi_n + \psi_{-n}) &= \chi_n + \chi_{-n}, \\ (\psi_n - \psi_{-n})^\dagger \iota &= \chi_n - \chi_{-n} \end{aligned} \qquad (9)$$

where $\chi_n = \text{Cliff}\,\psi_n \in \text{CLIFF}\, W$. The fermion swap too can be done in several ways too, such as $(\chi_1^+ - \chi_2^+)(\chi_1^- - \chi_2^-)$.

Since all of these representations are in the same algebra with the same physical interpretation, they belong to one physical theory and we propose to regard them as consequences of one statistics, the real Fermi statistics. We assign the following names:

Let $m, n > 0$. A *semi-swap* projectively represents the finite transformation

$$(mn) : V \to V,$$
$$\psi_m \mapsto \psi_n,$$
$$\psi_n \mapsto \psi_m;$$
$$(\forall k \neq m, n) \quad (mn) : \psi_k \mapsto \psi_k, \qquad (10)$$

by the first-grade Clifford operator $\chi_m - \chi_n$.

A swap projectively represents the finite transformation (mn) by the second-grade operator $(\chi_m - \chi_n)(\chi_{-m} - \chi_{-n})$

Spin represents the infinitesimal orthogonal transformation

$$L_{mn} : \psi_m \mapsto \psi_n,$$
$$\psi_n \mapsto -\psi_m;$$
$$(\forall k \neq m, n) \quad L_{mn} : \psi_k \mapsto 0, \qquad (11)$$

by the Clifford operator $\frac{1}{2}(\chi_m \chi_n + \chi_{-m} \chi_{-n})$.

Swap is conceptually more primitive than (Lorentz) spin as permutation is more primitive than rotation, invoking no spatial concepts of length or angle but only identity. We propose to reduce all physical actions and gauge transformations, including the dynamical development, to swaps of elementary events, as figures in a tapestry may be created, moved, and annihilated by interweaving its threads.

The Grassmann algebra of Q_G is the standard representation of the spinors of $V \oplus \text{Co}\, V$ in the form given by Chevalley (1954). The spinor representation has the invariance group SO(W) of dimension $D(2D + 1)$, mixing creators and annihilators, while the Grassmann representation reduces that to the group GL(D) of dimension only D^2.

Furthermore the spinor spaces $V' = QV$ resulting from Q_0 have a natural neutral metric. This enables us to iterate the functor Q_0 to make a hierarchy of nested structures.

We shall call the Grassmann theory the restricted Fermi statistics and the spinor theory the real Fermi statistics.

Since the spinors that describe spins maximally are isomorphic to the spinors that describe real Fermi-Dirac aggregates maximally, we shall identify spins of SO(N, N) with real Fermi-Dirac ensembles of $2N$ elements.

Because intrinsic spin is so much simpler than orbital angular momentum and the other variables of quantum field theory, many have considered spinors to be more fundamental than space-time vectors. Schur's theory of spinors — preceding Cartan's — shows that this is wrong. Spinors over W clearly describe sets of quanta described individually by vectors in W.

This is consistent with the fact that spinors are an irreducible representation of the Lorentz group. Mode vectors and the entities they describe are categorically dual, and behave dually under decomposition. An irreducible tensor may describe a composite object. That vectors are quadratic in spinors is actually primafacie evidence that the spinor describes a larger aggregate than a vector describes. The spinor provides enough information to define a vector, but not conversely. The spinor therefore describes a more complex structure than the vector does.

When the space-time melts down into its quantum elements and the classical Lorentz group loses meaning, spin loses physical meaning, but swap has meaning as long as identity does.

It is natural to suppose that all spin derives from swap, the more primitive concept of the two. Then every 2-valued spin representation signals a deeper 2-valued statistics. Quanta of spin 1/2 are likely aggregates of at least four fermionic entities.

Empirically, the standard spin-statistics linkage applies to quanta and not to space-time points. For example the classical space-time points of standard quantum theory have M-B statistics, not Bose or Fermi. In chronon dynamics too the spin-statistics connection must hold for excitations on the top level, that of dynamics, and not on the lower level of space-time.

Earlier we proposed that the spin-statistics correlation arose from the fact that processes of 2π rotation and exchange are homotopic for skyrmeons (Finkelstein and Misner 1959). It seems now that its source is still deeper; that spin is correlated with statistics, and 2π rotation with pair exchange, because fundamentally rotation *is* exchange, a permutation operation.

10. THE DIRAC REFORMATION

In a theory of space-time quanta, the dimension and signature of space-time are order parameters of the vacuum condensate. Algebra contraction must produce the usual Maxwell-Boltzmann aggregate of dimension 4 and signature 2 or -2, and a complex quantum theory.

In Fermi statistics, squads of eight, *octads*, are special, thanks to the octal periodicities of the spinorial clock (Atiyah, Bott, and Shapiro 1964) and the spinorial chessboard (Budinich and Trautman 1988). A set (Fermi-Dirac aggregate) of $8N$ fermions of neutral signature is algebraically isomorphic to a sequence (Maxwell-Boltzmann aggregate) of N octads of neutral signature:

Octad lemma

$$\mathrm{Cliff}(4N, 4N) \cong \mathrm{Cliff}(4, 4) \otimes \ldots \otimes \mathrm{Cliff}(4, 4) \otimes = [\mathrm{Cliff}(4, 4) \otimes]^N. \tag{12}$$

This lemma would not hold with any squad smaller than the octad. It would not hold with signature $\pm(6-2)$. Therefore we choose the neutral signature for the octad. This extracts the Maxwell-Boltzmann statistics that we need for space-time from Fermi statistics.

Therefore we suppose that the formation of space-time spontaneously reduces symmetry from $\mathrm{SO}(4N, 4N)$ to $\mathrm{SO}(4, 4)^N$ in the way described by the octad lemma. Fermionic chronons assemble into maxwellonic octads. The resulting quantum M-B space includes a Bose subspace in which further Bose condensation can proceed.

If the chronon is the atom of time, the octad might be the molecule or unit cell.

Two classical 8-dimensional spaces vie for election to be the correspondent of the $\mathbb{R}(4, 4)$ underlying each quantum octad: the complex Minkowski space-time $\mathbb{C} \otimes \mathcal{M}^4$ and the symplectic Minkowski tangent-cotangent space $\mathcal{M} \oplus \mathcal{M}^\dagger$. Conveniently, we have already inferred from Segal simplicity that under higher resolution the two become the same.

We can now obtain the infinite-dimensional algebra $\mathcal{A} = \mathcal{A}(x^\mu, \partial_m)$ of the space-time manifold, generated by the variables x^μ, ∂_m, by contracting an atomistic Clifford algebra, as required by Fermi statistics.

Feynman (1971) proposed that the space-time coordinate-difference operator is the sum of many mutually commuting tetrads of Dirac vectors:

$$\Delta x^\mu = \text{Const} \sum \gamma^\mu(n) \tag{13}$$

It is a quantum form of the proper-time Heisenberg-Dirac equation $dx^\mu/d\tau = \gamma^\mu$. We incorporate and flesh out this proposal in the chronon dynamics that follows.

Our reformation requires a larger Clifford algebra than Feynman's, because we reform the canonical commutation relations. For the present we use the Clifford algebra $C := \text{Cliff}(3N, 3N)$ with defining relations

$$\{\chi_{\nu',\tau'}, \chi_{\nu,\tau}\} = 2g_{\nu'\nu}\delta_{\tau'\tau}, \tag{14}$$

where $\mu = 1, 2, 3, 4$ indexes 4 chronons in a Dirac tetrad, $\mu = 5, 6$ indexes two chronons in a reformed complex plane, and $\tau = 0, 1, \ldots, N-1$ indexes N neutral hexads.

We write the top Clifford element of each hexad in C as

$$\chi_{\uparrow,\tau} := \chi_{6,\tau}\chi_{5,\tau}\cdots\chi_{2,\tau}\chi_{1,\tau} \tag{15}$$

To reform the canonical commutation relations we suppose that each hexad contributes

$$\Delta x^\mu = \chi^{\mu 5} \tag{16}$$

to the space-time coordinate,

$$\Delta p_\mu = \chi_{\mu 6} \tag{17}$$

to momentum-energy, and

$$\Delta i = \chi^{56} \tag{18}$$

to the complex i; we drop the hexad index τ for the moment and use $\frac{1}{2}\hbar = c = \chi = 1$ (chronon units). These are the antihermitian generators associated with the usual observables. At the end we must supply a factor i to make a hermitian observable.

By duality, when we sum contributions to the total space-time coordinate we average contributions to the momentum-energy. To approximate a square root of -1 we also average the contributions Δi, supposing that they all align in the vacuum.

Then in this quantum space-time the reformed manifold-algebra commutation relations are for each hexad:

$$\begin{aligned}\frac{1}{2}[\Delta x^\nu, \Delta p_\mu] &= \Delta i\, \delta^\nu_\mu, \\ \frac{1}{2}[\Delta i, \Delta x^\mu] &= \Delta p^\mu, \\ \frac{1}{2}[\Delta p_\mu, \Delta i] &= \Delta x_\mu.\end{aligned} \tag{19}$$

The reformed i generates the symplectic symmetry between x and p of classical mechanics, as Segal (1951) proposed.

The sum over the hexads is

$$\begin{aligned} x^\mu &= \chi \sum \Delta x^\mu \\ p_\mu &= \frac{\hbar}{2N\chi} \sum \Delta p_\mu \end{aligned} \qquad (20)$$

Here we restore x to seconds and p to joules, leaving $c = 1$.

The time unit χ implies discreteness for space-time-coordinates resulting from single measurements, but not for expectation values like centroids. Averages over sufficiently many measurements may still be arbitrarily small.

There are far too many i's for comfort in this Clifford algebra. There is only one i in the canonical algebra, but there is a Δi^τ for each of the $2N$ octads in C. Within this Maxwell-Boltzmann algebra of hexads there is however a Bose subalgebra, and in this Bose subalgebra there is a projector describing a condensation of all the local i's into one global average i.

The leading 4×4 components of

$$L_{\mu\nu} = \frac{\hbar}{4} \sum_\tau [\chi_{\mu\tau}, \chi_{\nu\tau}] \qquad (21)$$

are infinitesimal generators of the Lorentz group. The 6×6 totality of the L's infinitesimally generate the reformed Poincaré group, which is the de Sitter group.

To produce locally commuting squads of fermions from globally anticommuting ones, we have linearly ordered them, by a kind of cosmic-time parameter τ. This cosmic time seems have no correspondent in field theory. It is analogous to proper-time parameters of Feynman and Schwinger and is similarly helpful. It permits one to formulate a Lorentz-invariant concept of dynamics operator, as an operator that generates a unit increase of τ. It permits us to identify a canonical conjugate to τ with rest mass $M \sim \hbar i \partial_\tau$ (Schwinger 1951). It makes possible and natural a space-time quantization much like Feynman's (13)

and extends it to momentum-energy space.

It is straightforward to reform the one-particle Dirac equation within this Clifford algebra (Galiautdinov and Finkelstein 2002).

The reformation of the usual quantum field theory of a spinor field requires another stage of quantification. The reformation of the one-graviton theory of spin 2 resembles that of four fermions of spin 1/2. The reformation of gravitation theory combines these two extensions. These are now under study. The reformation of the Dirac equation required only two new reformation parameters, a chronon time χ and a chronon number N. More will likely be needed.

We have reconsidered yet again the statistics of these finite elementary processes. Real Fermi statistics is the most plausible we have tried so far. It produces 2-valued spinor representations of rotations, the Dirac spin operators and the fermionic creation operators of the Dirac equation. It leads to an Maxwell-Boltzmann sequence of octads like the M-B assembly of the usual space-time points, with four space-time dimensions. It lends itself to a hierarchy of multiple quantifications that might unify space-time and fields in one algebraic theory.

ACKNOWLEDGEMENTS

This contribution includes work done in collaboration with James Baugh, Andrei Galiautdinov, Michael Gibbs, William Kallfelz, Dennis Marks, Tony Smith and Zhong Tang. I am grateful for stimulating discussions with Giuseppe Castagnoli, Shlomit Ritz Finkelstein, Raphael Sorkin, and Frank Wilczek. I thank the state of Georgia, the Institute for Scientific Interchange, the Elsag-Bailey Corporation, and the M. and H. Ferst Foundation for their generous support.

Georgia Institute of Technology

REFERENCES

Atiyah, M. F., R. Bott, and A. Shapiro. 1964. "Clifford modules." *Topology 3 Supplement* 1:3–38.
Baugh, J., D. Finkelstein, H. Saller, and Z. Tang. 1998. "General covariance is Bose statistics." In *On Einstein's Path*, ed. H. Allen. New York: Springer.
Budinich, P., and A. Trautman. 1988. *The Spinorial Chessboard*. Berlin: Springer.
Chevalley, C. 1954. *Algebraic Theory of Spinors*. New York: Columbia University Press.
Dirac, P. A. M. 1974. *Spinors in Hilbert Space*. New York: Plenum.
Feynman, R. P. 1971. Private communication. Feynman studied this space-time quantization shortly before his work on the Lamb shift.
Finkelstein, D. 1969. "Space-time code." *Physical Review* 184:1261–1271.
———. 1996. *Quantum Relativity*. New York: Springer Verlag.
———. 1999. "Quantum Relativity." In *Frontiers of Fundamental Physics*, eds. B. G. Sidharth and A. Burinski. Hyderabad: Universities Press (India).
Finkelstein, D., and J. M. Gibbs. 1993. "Quantum relativity." *International Journal of Theoretical Physics* 1801:32.
Finkelstein, D., J. M. Jauch, S. Schiminovich, and D. Speiser. 1962. "Foundations of quaternion quantum mechanics." *Journal of Mathematical Physics* 3:270.
———. 1963. "Principle of General Q-Covariance." *Journal of Mathematical Physics* 4:788–796.
Finkelstein, D., and C. A. Misner. 1959. "Some new conservation laws." *Annals of Physics* 6:230–243.
Finkelstein, D., and E. Rodriguez. 1981. "Applications of quantum set theory." Pp. 70–82 in *Quantum Theory and the Structures of Time and Space*, vol. 6, eds. L. Castell and C. F. v. Weizsäcker. Hanser.
———. 1984. "The quantum pentacle." *International Journal of Theoretical Physics* 23:887–894.
Finkelstein, D., H. Saller, and Z. Tang. 1998. "General covariance is Bose-Einstein statistics." In *On Einstein's Path*, ed. Alex Harvey. New York: Springer.
Galiautdinov, A., and D. Finkelstein. 2002. "Chronon Corrections in the Dirac Equation." *Journal of Mathematical Physics* 43:4741–4752.
Hamermesh, M. 1962. *Group Theory and its Applications to Physical Problems*. Reading: Addison-Wesley.
Hestenes, D. 1966. *Space Time Algebra*. Gordon and Breach.
Hoffman, P. N., and J. F. Humphreys. 1992. *Projective Representations of the Symmetric Groups: Q-functions and Shifted Tableaux*. Oxford Mathematical Monographs.
Inönü, E., and E. P. Wigner. 1953. "On the contraction of groups and their representations." Pp. 510–524 in *Proceedings of the National Academy of Sciences*, vol. 39.
Karpilovsky, G. 1985. *Projective Representations of Finite Groups*. New York: M. Dekker.
Marks, D. W., A. Galiautdinov, M. Shiri, J. R. Baugh, D. R. Finkelstein, W. Kallfelz and Z. Tang. 1999, April. "Quantum Network Dynamics 1-7." In *Bulletin of the American Physical Society (April)*. Seven consecutive abstracts on chronon dynamics.
Nayak, C., and F. Wilczek. 1996. *Nuclear Physics* B479:529.

Nazarov, M. 1997. "Young's symmetrizers for projective representations of the symmetric group." *Advances in Mathematics* 127:190–257.

Penrose, R. 1971. "Angular momentum: an approach to combinatorial space-time." Pp. 151–180 in *Quantum Theory and Beyond*, ed. T. Bastin. Cambridge: Cambridge University Press.

Plymen, R., and P. L. Robinson. 1995. *Spinors in Hilbert Space*. Cambridge Tracts in Mathematics 114. Cambridge: Cambridge Univ. Press.

Porteous, I. R. 1995. *Clifford Algebras and the Classical Groups*. Cambridge: Cambridge University Press.

Read, N. 2002. "Nonabelian braid statistics versus projective permutation statistics." *Journal of Mathematical Physics (to appear)*. hep-th/0201240.

Rehren, K. H. 1991. "Braid group statistics." In , eds. J. Debrus and A. C. Hirshfeld. Heidelberg: Springer.

Schur, I. 1911. "Über die Darstellung der symmetrischen und der alternierenden Gruppen durch gebrochene lineare substitutionen." *Journal für die reine und angewandte Mathematik* 139:155–250.

Schwinger, J. 1951. "On gauge invariance and vacuum polarization." *Physical Review* 82:664.

Segal, I. E. 1951. "A class of operator algebras which are determined by groups." *Duke Mathematical Journal* 18:221.

Smolin, Lee. 1992. "Did the universe evolve?" *Classical and Quantum Gravity* 9:173–191. Reprinted in (Smolin 1997).

———. 1997. *The Life of the Cosmos*. Oxford.

Stembridge, J. R. 1989. "Shifted tableaux and the projective representations of symmetric groups." *Advances in Mathematics* 74:87–134.

von Neumann, J. 1932. *Mathematische Grundlagen der Quantenmechanik*. Berlin: Springer. Translated by R. T. Beyer, *Mathematical Foundations of Quantum Mechanics*. Princeton, 1955.

Wilczek, F. 1982. *Phys. Rev. Lett.* 48:1144.

———. 1997. "Some examples in the realization of symmetry." Talk given at Strings '97, 18–21 June 1997, Amsterdam, The Netherlands. *Nuclear Physics* Proceedings Supplement. Also hep-th/9710135.

———. 1998. Private communication.

———. 1999. "Projective Statistics and Spinors in Hilbert Space." Submitted for publication. Also hep-th/9806228.

Wiman, A. 1898. *Mathematische Annalen*. 47:531, 52:243.

ABNER SHIMONY AND HOWARD STEIN

ON QUANTUM NON-LOCALITY, SPECIAL RELATIVITY, AND COUNTERFACTUAL REASONING

> "Contrariwise," [said] Tweedledee, "if it was so, it might be; and if it were so, it would be: but as it isn't, it ain't. That's logic."

1. INTRODUCTION

John Bell (1964), and others following him (Clauser et al. 1969; Wigner 1970; Belinfante 1973), have shown that some of the predictions of quantum mechanics conflict with "local hidden variables theories" (also called "local realistic theories"). Henry Stapp[1] has worked over many years to strengthen these results by showing, without recourse to any assumption of "realism" or "hidden variables" or (following Einstein, Podolsky, and Rosen (1935)—henceforth referred to as "EPR") "elements of reality," that quantum mechanics contradicts the locality demanded by the special theory of relativity. In a recent article (Stapp 1997), which he regards as perhaps simpler, and technically better, than his earlier ones, he argues that there is inconsistency among the following: "(1) the validity of some simple predictions of quantum theory; (2) the explicitly stated locality conditions [that is: three principles—labeled "LOC1," "LOC2," and "LOC3"—described by Stapp as aspects or forms of "the crucial locality assertion suggested by the theory of relativity"]; (3) the general idea that physical theories can cover a variety of special instances that can be imagined to be created by free choices of experimenters; and (4) the *general* principles of modal logic."[2] Of the last category of premises—the general principles of modal logic—Stapp says that they "merely formalize what we mean by modal language;" and he argues that modal, and in particular counterfactual, statements can play a legitimate role in scientific reasoning. This last claim, put so generally, seems to us quite reasonable; but we wish to insert a caveat at the outset as to the statement that the principles of modal logic "merely formalize what we mean by modal language": modal language and modal argumentation are tricky; and "merely what we mean" may conceal presuppositions legitimately made in the context of ordinary language, but challengeable in the light of fundamental scientific theories. (As an analogy: it might have been objected, indeed, it *was* objected, to the special theory of relativity, that asserting the existence of an equivalence-relation, "simultaneity," among events in the world "merely formulates [part of] what we mean by temporal language." And such a statement was literally correct—except for the connotations of the word "merely"!—since what "we" ordinarily "mean" by temporal language conceals a presupposition that is substantively *false* about the actual fundamental structure of the physical world; more precisely, a presupposition that holds to an approximation that is more than adequate to the concerns of everyday life, but that breaks down very seriously in more refined contexts.)

David Mermin (1998) and Stapp (1998) have recently published an interesting debate on Stapp (1997). Their discussion, however, seems to us inconclusive, and we wish to present a new critique that we believe will help to clarify the main issues.

2. SOME MODAL LOGIC

The argument presented by (Stapp 1997) has the guise of a formal logical derivation, each step being justified by appeal either to "LOGIC," to one of the three locality conditions "LOC1," "LOC2," "LOC3," or to quantum-mechanical predictions ("QM"). It uses only elementary *propositional* logic; but besides the usual (Boolean) propositional connectives—including the material conditional, $p \to q$, which is equivalent to "either not-p or q"—it employs two modal conditional connectives: *the strict conditional or entailment, $p \Rightarrow q$*, understood as "p implies q *with necessity*"; and the so-called "counterfactual conditional, symbolized as $p \square \to q$, taken to mean: "whether or not p is true, if p *were* true, then q *would be* true."

These informal explanations obviously require some filling-out. A clear semantical account of the intended meanings of the modal connectives is particularly important in view of two facts: first, the debate between Stapp and Mermin turns upon the question whether a certain counterfactual assertion that occurs in Stapp's formal argument *has any meaning* in the context in which it occurs; second—and still more critical—although, as has been said, Stapp's argument is presented as justified by certain purely *formal* principles of logic (including modal logic), we shall presently point out that the principles he appeals to are *unjustified* by the *semantics* of modalities upon which he in fact takes his stand.—In view of this situation, we are especially grateful to Stapp himself for a clarification of his semantic interpretation of the modalities, which he made in a private communication to one of us, commenting on a draft essay that is an ancestor of the present paper.

The semantics in question is based upon that which has been standard in modal logic since the pioneering work of Saul Kripke in the early 1960's (Kripke 1963). The fundamental notion is that of *a possible world*. It is presupposed here that one has a well-constituted *language*, within which ordinary (non-modal) sentences are formulable. A *world* is then understood to be constituted by any specification that determines a definite truth-value—*true* or *false*—for every conceivable sentence of that language, in a way that respects the usual conditions on the relation of the truth-values of complex sentences to their constituents. (This formulation is technically oversimplified. For instance, "conceivable sentence of that language" has to be construed in a special way, to take account of the fact that what we aim for as the set of "conceivable sentences" will in general be larger than the set of all sentences *actually* formulable in the language: e.g., we cannot formulate sentences referring, severally, to *each moment of time,* since there are uncountably many of the latter, and only countably many sentences in a language as ordinarily envisaged. The method of handling such problems is well understood in logic; and the matter has no bearing upon the present topic.) What, then, is to be understood by a *possible* world? — In the Kripkean analysis, "possibility" is a *relative* notion: certain worlds are, and others are not, possible *relative to* a given world; and, indeed — in the context of this theory — our statement above of what specifications *constitute* a world is deficient: one needs, further, a specification of which *other* worlds *count as possible* relative to the one being "constituted." But this complication may be finessed here; for Stapp's argument is concerned with *physical* possibility, and *this* may be construed — when we are dealing with a particular formulation of physics — in a *uniform* way, the same for all the worlds in question. Namely, we presuppose that, for any given "world" of the kind we are considering, the specification of that world — which, it should be noted, means a specification *as complete as the vocabulary and the laws of our physics allow* of *everything* about that world *through all time* (a "complete history" of the world) — determines the truth-value of every proposition, and thus determines the truth-value of every "law" of our physics. (In the quantum-mechanical case, the "propositions" in question should be understood to be, not "observables associated with projection-operators," but propositions about (a) quantum states and their evolution, and (b) experimental arrangements and experimental outcomes — it is such things as these that Stapp's discussion involves. To suppose that a language is available in which such specifications — and *maximal* ones — can be made, is actually to idealize the present situation in physics; but such an idealization seems to be required as the necessary background for a discussion like Stapp's.) The world, then, is *physically possible* if and only if every law of physics is *true* in it (note that we have built the truth of the laws of *logic* into the very notion of a "world;" it would be possible to handle this otherwise, but that seemed simplest). In short, we have the following very important principles: (a) A possible world is a *physically* possible one; i.e., one that conforms to the laws of physics. (b) This property is uniform — that is, independent of the world "from which" the possibility of some other world is being assessed. (c) That a world is or is not possible is determined by its "internal" specification and by the laws of physics, *which do not themselves involve any modal considerations:* the semantics of modality are *defined by* the conception of a possible world — the latter is constitutive of, not based upon, the former. We now define a sentence to be (physically) *necessary* if it is *true in every possible world;* and it is only upon the basis of this definition that we *now* say that "the laws of physics are physically necessary."

That semantics suffices to characterize the strict conditional quite simply: a sentence of the form $p \Rightarrow q$ is *true* if and only if the corresponding material conditional, $p \rightarrow q$, is *necessary*.

The counterfactual conditional is a little more subtle; its explication, due to David Lewis,[3] involves a somewhat troublesome element, which Stapp has for his purposes ingeniously managed to avoid. Lewis introduces into the "space of possible worlds" a notion (in effect) of "comparative closeness," of the form "world w'' is at least as close to world w as world w' is"; and he defines a sentence of the form $p \square \rightarrow q$ to be *true in* (or, as Lewis says, "at") *the world w* if and only if, for some possible world

w', p is true in w', and for *every* possible world w'' that is at least as close to w as is w', $p \to q$ is true in w''.[4] Lewis's theory lays down conditions that the relation of comparative closeness (or an equivalent conception that he elects to use as the foundation) has to satisfy, in order for this notion of the counterfactual conditional to be reasonable. But apart from such formal constraints, his theory does not determine what the relation in question *is*. Of course, the general theory of possible worlds also fails to define the notion of a possible world; but for that, at least for the centrally important notion of physical possibility, we have recourse to physics itself—and presumably for other notions of possibility, we should have recourse to other non-logical criteria; whereas with regard to the notion of comparative closeness of possible worlds, we are quite seriously in the dark.

It is on just this point that Stapp makes an admirably simplifying move. He considers a situation in which the "counterfactual" condition p, in the conditional $p \square \to q$, describes an action, taken by an experimenter (cf. assumption (3) in the first paragraph of section 1 above), localized to a certain set, S, of space-time points. In these conditions, he takes that counterfactual conditional to be true in a world w if and only if q is true in every world w' that differs from w *only by the consequences of doing the action described by p*; and in effect this means: every world w' in which p is true, and which agrees with w everywhere outside the future light-cone of the set S. (Here by the future light-cone of a point x is meant the result of deleting the point x itself from the closure of the set of all points y such that the vector xy is time-like and future-pointing; and by the future light-cone of a set S of points is meant the set-theoretic union of the future light-cones of all the points of S.) This criterion is entirely explicit: it fixes unambiguously the meaning of a counterfactual conditional of the type Stapp is concerned with, and therefore provides a firm basis for the critical evaluation of his argument. (The question whether Stapp's theory of the counterfactual is a special case of Lewis's, or a departure from it, we shall not discuss here; but since Stapp's theory is in any case like Lewis's, and inspired by it, we shall refer to it as the Lewis-Stapp counterfactual semantics.)

3. THE LOGIC OF STAPP'S ARGUMENT

Stapp considers a single physical system with two components, described by an entangled quantum state Ψ that was invented by Lucien Hardy (1992). One of the two components is analyzed in space-time region **L** and the other in space-time region **R**, the two regions having spacelike separation. In region **L** an experimenter may freely choose to perform one of two experiments $L1$ or $L2$, each with two possible outcomes, denoted "+" and "-"; and likewise in **R** an experimenter may freely choose to perform one of two experiments $R1$ and $R2$, each with two possible outcomes + and -. We also let "$L1$" denote the proposition that the experiment $L1$ is performed, and permit a similar equivocal (but unconfusing) usage of "$L2$", "$R1$", and "$R2$". We use the notation "$L1+$", "$L1-$", "$L2+$", "$L2-$", "$R1+$", "$R1-$", "$R2+$", "$R2-$" to denote statements, "$L1+$" being the statement that experiment $L1$ is performed in region **L** with outcome +, "$R2-$" being the statement that experiment $R2$ is performed in region **R** with outcome -, etc.

Among the quantum mechanical predictions based on state Ψ are the following (which are represented in fig. 1):

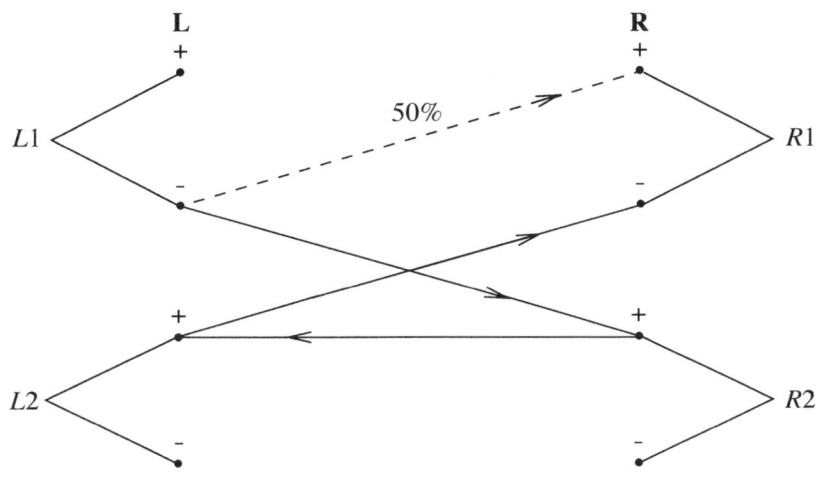

Figure 1. Some quantum predictions of the Hardy state.

(i) If the experiments $L1, R2$ are performed and the result of the $L1$ experiment is -, then the outcome of the $R2$ experiment is certain to be +.
(ii) If the experiments $L2, R2$ are performed and the result of the $R2$ experiment is +, the outcome of the $L2$ experiment is certain to be +.
(iii) If the experiments $L2, R1$ are performed and the result of the $L2$ experiment is +, then the outcome of the $R1$ experiment is certain to be -.
(iv) For i, j, = 1 or 2 (in any combination), the fact of the performance of the pair of experiments Li, Rj, is compatible with either outcome (+ or -) in **L**; and also compatible with either outcome in **R** (not, of course, compatible with these outcomes *jointly* in **L** and **R**, as (i), (ii), and (iii) make clear).
(v) If the experiments $L1$ and $R1$ are performed and the result of the $L1$ experiment is -, then there is 50% probability that the outcome of the $R1$ experiment will be + and 50% probability that it will be -.
(Stapp assumes that these probability evaluations are meaningful for an individual system, even though an ensemble of systems prepared in the same quantum state may be required for an empirical test of the evaluations.)

Propositions (i), (ii), (iii), (iv), and (v) are the quantum-mechanical predictions referred to in (1) of our Introduction. Stapp claims to exhibit, in Sect. III of (Stapp 1997), the inconsistency of the conjunction of these propositions with his assumptions (2), (3), and (4) (for which, again, see our Introduction).

As we have remarked in Sect. II above, the formal derivation given by Stapp is defective. More precisely, there occur in that derivation, first, a (tacit) invocation, as a *formal inference-rule of modal logic,* of something that is in fact not a valid rule—

this mistake, as we shall show, is rooted in an oversight on Stapp's part concerning *the meaning* (and *therefore* the "logic") *of the strict conditional;* and, second, three *substantive* errors in the formulation of propositions—errors rooted in essentially that same oversight in construing the strict conditional. (One other point should perhaps be made about the formal derivation: It should really be regarded as elliptical, containing throughout a *tacit* premise that posits the Hardy state and the experimental arrangements allowing the choice between $L1$ and $L2$ in the one region, and that between $R1$ and $R2$ in the other. To treat this premise correctly in the modal context might require a little care; but it is pretty clearly manageable, and it is therefore unnecessary to dwell upon the matter.)

The first of the mistakes is the following: Stapp uses the rubric "LOGIC"—i.e., he invokes *general logical principles*—as warrant for the inference from a proposition of the form $(p \wedge q) \Rightarrow r$ to the corresponding proposition $p \Rightarrow (q \Rightarrow r)$. Analogy to the material conditional certainly makes this tempting; but it is a blunder in modal logic. To make the point entirely clear, we give first an example that should convince the reader that—quite apart from the technicalities of standard modal semantics—to allow this inference would be catastrophic for modal logic; then we shall give the relevant semantical analysis.

The example: For p, in the above propositions, take the material conditional $q \to r$. We have, of course, that $((q \to r) \wedge q) \Rightarrow r$ is true (indeed, it is a truth *of logic*). Therefore, if the inference in question were valid, we should be able to infer the truth of the proposition $(q \to r) \Rightarrow (q \Rightarrow r)$: for *any* propositions q, r, the *material* conditional $q \to r$ would[5] *entail* the *strict* conditional $q \Rightarrow r$. Since the converse entailment is clear, the material and strict conditionals would have turned out to be—strictly!—equivalent; the distinction between the modal and the simple conditional would have been effectively erased. And this clearly is a disaster. (We have already remarked that modal language and modal argumentation are tricky.)

As to the semantical analysis: A proposition of the form $q \Rightarrow r$ is—given the "absolute" character of the notions of possibility and necessity that we are working with (i.e., the fact that these do not vary from world to world among our "possible" worlds)—either *true in every possible world*—i.e., *necessary*—or *false in every possible world*—i.e., *impossible*. If the former holds, then of course $p \Rightarrow (q \Rightarrow r)$ will also be true (in every possible world); but if the latter holds—that is, if $q \Rightarrow r$ *fails* to be necessary—then unless p is itself *impossible*, $p \Rightarrow (q \Rightarrow r)$ will be *false* (again: in every possible world!). In short, in the standard semantics (but with an "absolute" notion of possibility), the *meaning* of $p \Rightarrow (q \Rightarrow r)$ is that *either p is impossible, or $q \Rightarrow r$ is necessary*. And this should make plain to the reader the reason for the invalidity of the inference under discussion.

A natural emendation of this defective rule of inference suggests itself: namely, from (a): $(p \wedge q) \Rightarrow r$ to infer (b): $p \Rightarrow (q \to r)$; for not only is this inference valid, but it is as strong as possible, in the sense that the formulas we have labeled "(a)" and "(b)" are in fact *logically equivalent*. We shall consider presently whether this change allows Stapp's argument to be rescued. But before we can do so, it is necessary to discuss the other mistakes, because if the argument is to be saved these too must be repaired.

Of the substantive errors in formulation, two concern principles of locality. The simpler of these is the one Stapp calls "LOC3." Here is his formulation of this principle:

> The final locality condition, LOC3, asserts the commutability of $L1\text{-}\Rightarrow$ and $R1\square\rightarrow$. This result follows from the fact that, according to the putative locality property, the truth of $L1\text{-}$ is unaffected by changing the choice in **R** from $R2$ to $R1$: $L1\text{-}$ is the same condition whether asserted under condition $R2$ or $R1$.

The claim, in other words, is that, on the general locality grounds of Stapp's conception of the counterfactual conditional, a proposition of the form $L1\text{-}\Rightarrow(R1\square\rightarrow p)$ and the corresponding proposition $R1\square\rightarrow(L1\text{-}\Rightarrow p)$ are equivalent. From our discussion of the first error, it should be clear immediately that something must be wrong with this; for the first of the two propositions, a strict conditional, is true *only* if it is *necessary*, whereas to say that the second of them is necessary is in effect to assert the proposition $R1\Rightarrow(L1\text{-}\Rightarrow p)$. What the grounds that Stapp cites genuinely establish is the equivalence (given his semantics) of *two weaker propositions:* namely, $L1\text{-}\rightarrow(R1\square\rightarrow p)$ and $R1\square\rightarrow(L1\text{-}\rightarrow p)$; they therefore justify the inference from $L1\text{-}\Rightarrow(R1\square\rightarrow p)$ to $R1\square\rightarrow(L1\text{-}\rightarrow p)$. We may call this modification of LOC3—allowing the inference just described—"LOC3´".

Stapp's principle LOC2 is more complex; and we shall see later that the issues raised by it are also more complex. LOC2 licenses the inference from the first to the second of the following displayed propositions:

$$L2\Rightarrow(R2+\Rightarrow(R1\square\rightarrow R1\text{-})); \qquad (1)$$

$$L1\Rightarrow(R2+\Rightarrow(R1\square\rightarrow R1\text{-})). \qquad (2)$$

We defer quoting Stapp's justification for this principle until we come to discuss the more complex issues just mentioned; but for the present, it is to be remarked that the criticism already made of Stapp's erroneous use of *nested* strict conditionals is clearly relevant here—so that what should really be at issue is a justification of inference from the first to the second of the following pair of *weaker* propositions:

$$L2\Rightarrow(R2+\rightarrow(R1\square\rightarrow R1\text{-})); \qquad (1´)$$

$$L1\Rightarrow(R2+\rightarrow(R1\square\rightarrow R1\text{-})). \qquad (2´)$$

We shall call this amended inference-rule "LOC2´."

The last mistake is in the formulation of one of the quantum-mechanical predictions—namely, (i) above. Stapp puts this—in step (8) of his proof—as follows: $(L1\wedge R2)\Rightarrow(L1\text{-}\Rightarrow R2+)$. Once again, this rests on the same mistake about nested strict conditionals; the correct formulation is: $(L1\wedge R2)\Rightarrow(L1\text{-}\rightarrow R2+)$.

In the appendix at the end of this paper we reproduce the details of Stapp's proof (with slight modifications of his notation: namely, we use the notation of the main body of Stapp's paper, which is the same as that introduced above; whereas he, in his formal statement of his proof, uses some modifications of that notation, which seem to us actually to make the steps a little harder to follow; but the difference is in any case really trivial). We note there the specific places where the mistakes we have discussed occur; and then present the result of modifying the proof in accordance with the suggestions we have made. The net result of this is the following:

Stapp, at the end of his proof, says: "The conjunction of [steps] 11 and 14 contradicts the assumption that the experimenters in regions **R** and **L** are free to choose which experiments they will perform. Thus the incompatibility of the assumptions of the proof is established." Now, steps 11 and 14 of Stapp's proof, slightly reformulated in the interest of perspicuity, are:

$$(R2 \wedge L1) \Rightarrow (R1\square \to (L1\text{-} \Rightarrow R1\text{-})); \tag{XI}$$

$$(R2 \wedge L1) \Rightarrow (R1\square \to \neg(L1\text{-} \Rightarrow R1\text{-})); \tag{XIV}$$

If these both hold, then for any possible world w in which the experiments $L1$ and $R2$ are both performed, it must be the case that *were* $R1$ to be (or to have been) performed (instead of $R2$), a *contradiction* would result (both $(L1\text{-} \Rightarrow R1\text{-})$ and its negation, $\neg(L1\text{-} \Rightarrow R1\text{-})$); and this *seems* to say that whenever $L1$ and $R2$ are both performed, $R1$ "could not have been" performed "instead of" $R2$ —indeed contradicting the assumption of the independent freedom of choice of experiments in the regions **L** and **R**. There is in fact, however, something problematic about this reading of the result. For (XI) and (XIV) together do *not* imply that either there is no possible world in which $L1$ and $R2$ are both performed or there is none in which $L1$ and $R1$ are both performed; they *only* imply—given Stapp's interpretation of "$\square \to$", that any two such possible worlds *must differ somewhere outside the future light-cone of the region* **R**. Stapp would doubtless hold that this result is in conflict with the same principle—or "intuition," as philosophers unfortunately like to say—about locality that grounds his explicit locality assumptions and his interpretation of "$\square \to$" itself; but to say this is different from the claim that a formal contradiction has been derived—it means that Stapp, even if his proof were otherwise correct, has failed in his technical aim, and that some further *formal* "locality principle" would be required to render his proof complete. (For further discussion of this somewhat bizarre aspect of the theory of counterfactuals employed by Stapp, see section 5 below.)

But all this is rendered otiose by the fact that Stapp's proof is *not* otherwise correct. If one substitutes our revised formal argument, instead of (11) or (XI), one obtains:

$$(R2 \wedge L1) \Rightarrow (R1\square \to (L1\text{-} \to R1\text{-})). \tag{11$'$}$$

Proposition (XIV), it turns out, is valid. But the pair of results (11$'$) and (XIV) together have no smack of contradiction with freedom of choice. Their conjunction only says that if you perform $R2$ and $L1$, then if you had performed $R1$ instead of $R2$, the results $L1\text{-}$ and $R1+$ "would not both have occurred;" but that it is *not* the case that the performance of $L1$ with result $L1-$ entails—in *all* possible worlds[6]—the performance of $R1$ with occurrence of $R1-$. This statement about what is *not entailed* is ludicrously weak, and contradicts nothing in sight; indeed, if the entailment in question *did* hold, *that* would deny the freedom of choice of the experimenter in **R**!

But this is not the end of the story. Although we believe our discussion reveals irreparable weaknesses in Stapp's *formal* argument, that does not settle the question of the actual compatibility of the assumptions from which he starts. In the next section, we shall show that these assumptions—suitably revised in the light of the foregoing critique—are indeed *not* compatible; so on this score, Stapp is right. We shall show, too, that the incompatibility holds without any need to make use of the two locality principles LOC1 and LOC3´. This result serves to focus attention upon the remaining (revised) locality principle, LOC2´. Since this (in its original form, LOC2; but we shall take for granted the emendation to LOC2´) is also the focus of the debate between Mermin and Stapp in (Mermin 1998; Stapp 1998), we shall have occasion there to comment on that debate as well.

4. FURTHER DISCUSSION, AND THE MERMIN-STAPP DEBATE

LOC2´, it will be recalled, licenses the inference from:

$$L2 \Rightarrow (R2 + \rightarrow (R1\square \rightarrow R1-)) \qquad (1´)$$

to:

$$L1 \Rightarrow (R2 + \rightarrow (R1\square \rightarrow R1-)). \qquad (2´)$$

Now, (1´) is certainly *true*, according to Stapp's specialization of Lewis's semantics for the counterfactual conditional. For consider any possible world, w, in which our quantum system is in the Hardy state, and in which the experiments $L2$ and $R2$ are performed, the latter with outcome +. According to the quantum-mechanical prediction (ii), the outcome of $L2$ in w will be +. Applying the semantics of the counterfactual conditional to determine the truth-value, in w, of the counterfactual clause in (1´), we must next consider any possible world w' that agrees with w everywhere outside the forward light-cone of **R**. Since **L** is outside the forward light-cone of **R**, in w' as in w $L2+$ will be true. Therefore, by quantum-mechanical prediction (iii), the outcome of $R1$ in w' will be -. This establishes the truth of the counterfactual clause in w, and thereby establishes the truth of (1´) (everywhere!).

On the other hand, a similar analysis shows that (2´) is false—and therefore, again by the nature of the strict conditional, *impossible*: false in every possible world. For consider a world w in which $L1$ is performed with the outcome -, and in which $R1$ is performed with the outcome + (that such possible worlds exist is implied by the quantum-mechanical proposition (v), on any reasonable understanding of the concept of probability). Next consider a world w' that agrees with w everywhere outside the future light-cone of **R**, but in which $R2$ is performed rather than $R1$ (that this is possible—i.e., that there is such a world w' —is implied by Stapp's "local freedom-of-choice" assumption). The outcome of $R2$ in w' will be +, by quantum-mechanical proposition (i). Since the world w agrees with w' everywhere outside the future light-cone of **R**, whereas in w $R1+$ is true—i.e., $R1$ is true but $R1-$ is not—the counterfactual $R1\square \to R1-$ is false in w''. On the other hand, $L1$ is true in w'; and this demonstrates the falsity of $(2')$—everywhere. Therefore LOC2´ is, as we have said, incompatible with Stapp's assumptions (1), (3), and (4), without reference to the other two locality principles.

This result confirms a part of the views expressed both by Mermin and by Stapp; for Mermin, in (Mermin 1998), sharply attacks the principle LOC2; and Stapp, in (Stapp 1998), although he defends his *argument* for LOC2 (as a principle warranted by the space-time assumptions of the theory of relativity), *agrees* with Mermin to the extent of saying: "If one adds LOC2 [to a certain set of assumptions] then one can make [certain] inferences. If [Mermin's] intent is to say that any such inference is false, and hence that LOC2 is false, then I would agree that this is probably correct: I agree that the (trial) assumption LOC2 is probably false."

But there is an essential disagreement between Mermin and Stapp: namely, the former rejects Stapp's view that LOC2 is a consequence of relativity theory. And on this Stapp's central contention hangs; for his intention is to show that quantum mechanics demands some *alteration* in—or supplementation of—the space-time structure of relativity theory if contradiction is to be avoided. In order to assess this claim, we must consider Stapp's argument in justification of LOC2 on the basis of the special theory of relativity; or rather, we must consider the natural modification of that argument when LOC2 is replaced by LOC2´.

Here, then, is Stapp's argument for LOC2 (unmodified): He abbreviates the proposition $R2+ \Rightarrow (R1\square \to R1-)$ by *"SR"*; and says:

> Notice that everything mentioned in SR is an observable phenomenon in region **R**. ... The assumption in LOC2 is that statement SR has been proved under the condition that $L2$ was performed in the far-away region **L** at an earlier time.[7] Now special relativity comes in. According to the ideas of special relativity it should not matter which frame is used to specify "earlier" and "later." So let's use a frame in which the choice to perform $L2$ in **L** is made *later* than all the phenomena referred to in SR. But a statement about what would happen in **R**, under certain conditions defined there, that is known to be true under the condition that the later free choice in **L** is $L2$, cannot be rendered false by changing that later free choice, provided the idea is upheld that observable effects can propagate only into the future (light-cone). [At this point, Stapp first *formulates* LOC2, as justified by the foregoing reflection, as follows:] So LOC2 asserts this: "If SR is proved to be true under the condition that $L2$ is freely chosen in **L**, the SR must be true also under the condition that $L1$ is freely chosen there instead." (Stapp 1997, 301-302)

Let us suppose, then, that in the above argument reference to LOC2 is replaced by reference to LOC2´, and that reference to SR is likewise replaced by reference to SR' —where the latter is obtained from the former by substituting the material for the strict conditional. We contend that the argument—in either case: modified or unmodified—is unjustified, in the first place for the following reason: Stapp has asked us to "notice" that everything mentioned in SR (or in SR') is an observable phenomenon in region **R**. But *SR* and *SR'* contain the clause "$R1\square \to R1\text{-}$". To what does this clause "refer"?—Certainly to the observable phenomenon "experiment $R1$ is performed, in **R**, with outcome -." But is this all? According to the semantics of Lewis as specialized by Stapp, what this clause asserts—in whatever world w it is asserted—is that in every possible world w' that agrees with w *everywhere outside the forward light-cone of* **R**, and in which experiment $R1$ is performed, its outcome is -. How can this statement be construed not to refer to anything but observable phenomena in **R**? To maintain the latter would seem to be to ignore what is *behind*—that is, what is *tacitly* "referred to" by—the contrary-to-fact connective "$\square \to$" itself.—This criticism applies even if we relax Stapp's claim about everything mentioned in *SR*, replacing "is an observable phenomenon in region **R**" by "is a reality confined to region **R**."

This point can be regarded as an unpacking of what may be involved in a claim put forth by Mermin: namely, that *"A statement about the specific outcome of an experiment at a given time that was not actually performed, may derive its truth or falsity or its very meaning from events in the future that had not yet happened at that given time"*—and that this is in fact the case of the clause in question in *SR* and *SR'* (Mermin 1998, 922-923).[8] Stapp contests this claim—which, as Mermin asserts it, is based in effect upon Mermin's impression of the "meaning" of that clause, but not upon an explicit analysis of that "meaning." It is thus our contention that when Stapp's own analysis of the meaning of the counterfactual conditional is applied, Mermin's impression is borne out as correct.

We shall not consider explicitly Stapp's argument against Mermin (Stapp 1998), but shall only say that nothing in that reply deals with the point we have raised: that is, Mermin having rested his case on an *informal* "sense" of what the counterfactual statement means, Stapp claims that his own stand on that meaning is not vulnerable to Mermin's strictures—but does so without himself even stating explicitly the semantics of the counterfactual upon which his considerations do rest.

It is quite possible that Stapp may have objections to our conclusions from his semantics. But there is a logical consideration of a quite different sort, that we think is independently decisive against the *coherence* of any argument of the kind Stapp has advanced.

First, however, a preliminary point of clarification: We are certainly not in a position to *prove* the consistency of quantum mechanics and of the special theory of relativity jointly. Indeed, two well known foundational problems of physics—(a) the measurement problem in quantum mechanics, in view of which it cannot be said that there is a generally agreed upon *systematic, logically formalizable* framework within which the principles of quantum mechanics and an account of actual *experiments* and *observations* ("phenomena") can be formulated together in a consistent way; and (b) the problem of combining quantum mechanics and special relativity within an explicit axiomatic framework—make any statement of consistency here a *hypothesis* about how the *future* development of this vast subject may turn out. On the other hand, we note that Stapp is claiming to derive a contradiction, not from the ordinary—non-modal—statements of these physical theories and formulations of observed phenomena, but specifically from the application *to* these statements of principles *about* modal connections and modal reasoning (together with the assumption of free choice of experiments by experimenters). It therefore seems appropriate, in the context of the issues he has raised, to make the hypothesis in question. Indeed, without it, the grounds that underlie Stapp's own modal arguments would be removed. For the semantics of the modal propositions on which he bases his considerations require reference to the population of "possible worlds;" and these "possible worlds," taken each by itself, are—as we have explained earlier—characterized as complete "world-histories" *that are compatible with the laws of physics* (expressed in non-modal language). If those (alleged) laws are themselves contradictory, no such "possible worlds" exist at all.—Again, supposing this hypothesis, of (let us say) the compatibility with one another of *the constraints upon phenomena*—in *one* world—implied by special relativity and quantum mechanics, to be correct, we also do not know whether the further assumption of the compatibility of these assumptions with that of *local freedom of choice of experiments by experimenters* is (or "would be") true. (Note that this issue involves the age-old problem of how to *understand* the notion of "free choice"—in relation to physics, or simply in its own right. Perhaps it should also be noted, not directly relevant to the present discussion, that in the context of *classical* physics—or of any *deterministic* physics—the principle of free choice would indeed have introduced a contradiction; and yet this was never taken as a fatal objection to classical physics.) But—again—Stapp's claim seems to *grant* the notion of a population of possible worlds, in each of which, separately, the assumptions in question do hold; for his argument from *counterfactual* reasoning claims to show, in effect, that the possible worlds that accord with the predictions of quantum mechanics in the situation of the Hardy state and of the posited experiments *cannot have the relations to one another* that are demanded by the "LOC" principles he sees as derived from special relativity. It is the coherence of *this* strategy, as a way to make a case for the need to revise special-relativistic space-time, that we mean to call into question.

The point is simple: The population of possible worlds is determined, in this setting, by the condition that the "complete history" characterizing the world is compatible, jointly, with special relativity theory and quantum mechanics (formulated without the help of modal-logical notions); and the *content* of these theories, jointly—exactly what they together *assert*—is exhausted by that condition upon the possible worlds. What modal statements are then true—in consequence of these theories—is fully determined by the modal semantics. Or, rather: *ordinary* modal statements (involving possibility and necessity—and, thus, including strict conditionals) have their truth-values so determined (for instance, the laws of physics themselves, which have been formulated without any use of modalities, will of course be seen as *necessary*—because conformity to them is the defining standard of the notion of "possible world," and they are therefore *ipso facto* "true in all possible worlds"). As for statements involving counterfactual conditionals, these will have their truth-values (in any given possible world) determined, according to Lewis's semantics, by the further choice of a notion of comparative similarity among worlds. Since Stapp has made such a choice, the truth-values of such statements are also, for his system, fully determined. It may be objected here that in the case of quantum theory there is a modal principle implicit in the theory itself, since a quantum state entails the probabilities of true and false outcomes of the testing of any proposition in the lattice associated with the system, not just of a proposition that is in fact tested. The probabilities of quantum theory are thus *conditional,* leading to empirical predictions only when a condition (*viz.,* the testing of a specific proposition) is satisfied in a sufficiently large class of instances. Conditionality, however, is conceptually different from counterfactuality. In particular, there seems to be no need to estimate closeness of possible worlds — as required for counterfactual statements — in order to judge the truth value of a statement about conditional probability in quantum theory; and surely the application of the notion of conditional probabilities in non-relativistic classical physics *and* quantum mechanics shows that this notion is manageable in practice with no appeal to Stapp's criterion of counterfactuality. There is therefore *in principle* no room for *further* modal, or counterfactual, propositions, as *further* principles of the physical theory; and the modal and counterfactual principles that in fact follow from the physical theory and the indicated semantics *cannot* lead to contradiction — so long as the physical theory is, as we here assume, consistent in its non-modal form.

But perhaps this seems *too* simple — too flat — to be quite convincing. Let us pursue the issue just a little further. Stapp's argument does introduce one sort of supplementary modal assumption for which a justification can be adduced: namely, the assumption of free choice of experiments; for this says, in effect, that, given a possible world w in which (say) $R1$ has been performed, there exists another possible world w' that agrees with w in the entire *past* light-cone of **R** (understanding **R** itself to be excluded from this region), and in which $R2$ is performed, not $R1$. *This* assumption clearly and explicitly goes beyond physics. As we have remarked above in passing, this is also a *problematic* assumption. But it is introduced into the argument as plausibly connected with ordinary ways of thinking about experiments, and it seems to us that — if regarded with proper caution — it is indeed allowable for argument's sake. But the LOC assumptions of Stapp are grounded in a quite different way — namely, as assumptions warranted *by* physics. Our argument in the preceding paragraph has claimed to show, somewhat abstractly, that this warrant cannot obtain, because the physics has already done its work in singling out the possible worlds. We now remark (and it is our last point on this question) that to *deny* that statement — to maintain, in particular, that the counterfactual aspect of LOC2 or LOC2´ ought to be accepted as a further part of what the physical theory says — is indefensible on grounds of the methodology of physics itself. For physics has been created by study of phenomena in just one world. Whatever conceptions we have formed, on the basis of that study, about "what is possible" (and thus about the whole population of "possible worlds"), we have arrived at entirely by the consideration of just that one world. We have no independent access to the planetary system, or the galaxy, of possible worlds surrounding our own. Therefore, it cannot be an objection to a physical theory to say that it gives a consistent and correct account of phenomena in a given world, but that it fails to give a consistent and correct account of the *relations* to some given world of *other* possible worlds. The truth or falsity of all the modal statements, and all the counterfactual statements, and all combinations thereof, pertaining to a physical theory, must be — as they are, on the Lewis-Stapp semantics — *determined* by the construction of "possibilities" *as constrained by nothing but the non-modal propositions of the theory.*

5. A RECONSIDERATION OF STAPP'S "LOCAL SEMANTICS" OF THE COUNTERFACTUAL; COMPARISON WITH EPR'S "ELEMENTS OF PHYSICAL REALITY"

Let us first reconsider the point made in section 3 about Stapp's propositions (11) and (14): that together they *don't* show that if $L1$ and $R2$ are jointly possible, $L1$ and $R1$ are not jointly possible—*despite* the fact that they say that if $L1$ and $R2$ are both performed, the performance of $R1$ instead of $R2$ "would have entailed" a contradiction. This situation, it can be argued, represents an anomaly in the logic of counterfactuals itself. How does it arise?—The answer is that in the Lewis semantics, it is *not* assumed that when one establishes the "grading" of "comparative similarity" to one possible world of other possible worlds, there is always a "lowest degree of similarity" such that *every world that is possible at all* has *at least* this lowest degree of similarity to a given world. Lewis considers the possibility of making this assumption, and rejects it—he does not *rule out* relations of comparative similarity for which this holds; but he doesn't *require* it to hold. Now, *if* that assumption *did* hold, the anomaly in question could not arise, as we shall see in a moment; so the fact that the anomaly does exist on Stapp's specialization of Lewis's semantics can be regarded as a problematic feature of that specialization.

To establish the claim just made, let us consider a proposition of the form $(p \wedge q) \Rightarrow ((p \wedge \neg q) \square \rightarrow (r \wedge \neg r))$. Suppose this proposition is true, and suppose there is a possible world w in which $p \wedge q$ holds. Then $(p \wedge \neg q) \square \rightarrow (r \wedge \neg r)$ must also hold in w. This means, according to Lewis's general semantics of the counterfactual, that *either* there is no world with any "degree of similarity" at all to w in which $p \wedge \neg q$ holds, *or* there is a world w' such that (a) $p \wedge \neg q$ holds in w', and (b) $r \wedge \neg r$ holds in every world whose "similarity" to w is at least that of w'. But clauses (a) and (b) together imply in particular that $r \wedge \neg r$ holds in w'. This (of course) cannot be—there *is* no possible world in which $r \wedge \neg r$ holds. Therefore, if $(p \wedge \neg q) \square \rightarrow (r \wedge \neg r)$ holds in w, there is no world w' with any degree of similarity to w such that $p \wedge \neg q$ holds in w'. Thus if we postulate that there is, for each w, a "lowest degree of similarity," such that *every* possible world has "at least this lowest degree of similarity" to w, it will follow that there is no possible world whatever in which $p \wedge \neg q$ holds.[9] But this conclusion has been drawn from the premises (i) that $(p \wedge q) \Rightarrow ((p \wedge \neg q) \square \rightarrow (r \wedge \neg r))$ is true, and (ii) that there is a world in which $p \wedge q$ is true. Therefore from (i) *alone* it follows that *either* there is no possible world in which $p \wedge q$ is true, *or* there is none in which $p \wedge \neg q$ is true. Another way of putting this is that either $p \Rightarrow q$ or $p \Rightarrow \neg q$ must be true. Thus we have derived—*on the assumption* of a "minimal degree of closeness," in which every possible world stands to any given one—a rule of modal logic stating that from a proposition of the form $(p \wedge q) \Rightarrow ((p \wedge \neg q) \square \rightarrow (r \wedge \neg r))$ one can validly infer the corresponding proposition $(p \Rightarrow q) \vee (p \Rightarrow \neg q)$. On the other hand, without that assumption, such an inference is *not* valid; and on Stapp's prescription, the assumption is false.

Now, *Lewis* clearly does not consider the resulting situation problematic, since he refrains from making the assumption that would avoid it; but since the unavailability of the inference in question subverts *Stapp's* proffered argument, it seems fair to say that it contradicts Stapp's own counterfactual "intuitions". This is another indication of the propriety of the caveat we inserted near the beginning of this paper—that modal language and modal argumentation are tricky, and that "merely what we mean" in such reasoning may conceal presuppositions whose implications may not be altogether innocent. But can we understand better what is going on here?

The view that a certain world w', although possible in itself, is so dissimilar to a given possible world w as to be *ineligible as a candidate for alternative in any counterfactual statement* has the flavor of (something like) a quantum-mechanical "selection rule," forbidding certain transitions; as if to say, "given that the world is w, although w' is another possibility *simpliciter*, it cannot be regarded as an alternative "that would have happened," from the perspective of w, "if something had occurred differently" (w' is "inaccessible from" w). Stapp's light-cone criterion has the effect, *even without his further "locality" principles* (in particular, LOC2—or LOC2´), of imposing such a "selection rule."

But this—on reconsideration—already seems arbitrary and unjustified, on the grounds of the points made at the end of the preceding section: namely, such "transitions" are *mythical*—we do not observe alternative possibilities as "transitions from one world to another"—and therefore the *locality* restrictions that we are led to postulate from our observations, and our observation-grounded theories, in *this* world, don't really have any claim to extend to something like "possible transitions *between worlds*."

A quite similar consideration clearly applies to the *prima facie* plausibility of LOC2´. *Nevertheless* it is of some interest to ask whether the introduction of an EPR-type assumption about "elements of reality" might serve to reinstate LOC2´, or something analogous—always bearing in mind that this question is of interest at most as possibly clarifying the *prima facie* appeal of Stapp's proposed counterfactual principles: we already know that LOC2´ contradicts Stapp's other assumptions, and by the argument of the last section, if a proposed principle like that of EPR supports LOC2´—or some surrogate that does the same effective work in the argument—then that principle must also contradict those assumptions *without* any appeal to counterfactual reasoning to establish the contradiction.[10] And this is indeed the case, as we shall now show first (and then proceed to consider how far EPR may support LOC2´).

EPR (Einstein, Podolsky, and Rosen 1935, 777) propose the following sufficient (clearly not *necessary*) condition for the existence of an element of physical reality:

> If, without in any way disturbing a system, we can predict with certainty (i.e., with probability equal to unity) the value of a physical quantity, then there exists an element of physical reality corresponding to this physical quantity.

We assume, further, in accord with Einstein's general commitment to the *locality* of physical determination, that the "element of reality" in question is located where the "system" whose property is in question is located.

EPR's criterion will now be applied in three steps to a specific system governed by the correlations that are implied by the Hardy state and represented in fig. 1. One part of this specific system is subjected to observation in region **L**, and it is assumed that experiment $L1$ is chosen with outcome $L1-$ (an assumption which is certainly true in some possible world), and the other part is subjected to observation in region **R**.

The quantum mechanical prediction (i) given in Sect. III is a conditional prediction made with certainty: that if experiment $R2$ were to be performed in **R**, the outcome would be $R2+$. Since this prediction is made without in any way disturbing the part of the system in **R**, EPR's criterion implies that there exists in **R** an element of physical reality **r** determining the outcome $R2+$ if experiment $R2$ is performed; this element of reality **r** may be characterized as "an instruction sheet," in Mermin's (Mermin 1994) graphic terminology. Note that the existence of the instruction sheet **r**, on the EPR assumption, is independent of whether $R2$ is actually performed in **R**, or $R1$ is performed, or indeed neither; that is the whole point of their assumption. Now, the term "instruction sheet" is really to be taken as presupposing—again, this is the intention of EPR—the existence of a *theory* in accordance with which the "instructions" **r** imply that if $R2$ is performed the outcome will be $R2+$. Therefore it is presupposed that any consequence of the outcome $R2+$ is also to be a consequence of **r**'s presence in **R** conjoined with the performance of $R2$; but then, by locality, any consequence in **L** of the outcome $R2+$ is simply a consequence of **r**'s presence in **R**.

Next, if **r** is present in the region **R**, there must exist in **L** an "instruction sheet" **r**′ determining that if $L2$ is performed there, the outcome will be $L2+$; for quantum mechanics tells us that if $R2$ is performed with outcome $R2+$, then we can predict that if $L2$ is performed, the outcome will be $+$; our hypothesized theory and locality tell us that any consequence in **L** of $R2+$ is a consequence of **r**; and EPR tell us that if the outcome of $L2$ is predictable without operating upon **L**, there must be an element of reality there carrying the "instructions" that determine the outcome.

In exactly the same way, we conclude (since, according to quantum mechanics, $L2+$ and the performance of the experiment $R1$ imply the outcome $R1-$) that the existence in **L** of **r**′ implies the existence in **R** of an element of reality **r**″, an "instruction sheet" determining the outcome of $R1$ to be $R1-$.

But this sequence of inferences leads to the conclusion that in any possible world in which $L1$ is performed with the outcome $L1-$, the region **L** must contain the element of reality **r**′, and the region **R** must contain *both* the element of reality **r** *and* the element of reality **r**″; in particular, then, in any such world, if $R1$ is performed, the outcome must be $R1-$. And this contradicts the prediction of quantum mechanics.

This conclusion does, in the end, show that LOC2′ (and even LOC2) is a consequence of all the assumptions used—including that of EPR—since *everything* follows from contradictory premises; but that is of course of no interest at all. However, it is still worth considering what relation might hold between the EPR assumption—or something like it—and the basis alleged by Stapp for LOC2′. It will be recalled that Stapp argues for LOC2 on the ground of his claim that "everything mentioned in SR is an observable phenomenon in region **R**"—where SR is the proposition $R2+ \Rightarrow (R1\square \rightarrow R1-)$ that stands as the consequent in each of the statements connected by the rule of inference LOC2. We are concerned instead with LOC2′ and SR', where SR' is obtained from SR by the substitution of the material conditional for the strict conditional, and where LOC2´ is obtained from LOC2 by replacing SR by SR'; moreover it would suffice, for Stapp's purposes, to make the weaker claim that everything mentioned in SR — or SR' — is a reality confined to **R**. On the other hand, the argument of Sect. IV above has shown the unacceptability of this claim. The question we now raise, therefore, is whether the introduction of the elements of reality **r** and **r″** — both contained in **R** — by the reasoning we have here described, would somehow save LOC2´.

Our answer is that such a strategy fails, if all we have to appeal to is the carefully formulated criterion of EPR. The proposition that the element of reality **r″** exists in **R** — let us write for it $\mathbf{r}'' \in \mathbf{R}$ — certainly does mention nothing but realities confined to **R**; and so does the proposition $R2 \to \mathbf{r}'' \in \mathbf{R}$; so if we had a way of inferring *this* from $L2$, we should have a premise, $L2 \Rightarrow (R2 \to \mathbf{r}'' \in \mathbf{R})$, analogous to the premise in the rule LOC2 or LOC2´; and (the analogue of) Stapp's claim would be true of the clause that replaces SR' in this premise. However, we do *not*, in any straightforward way, have the required inference from $L2$ to $R2 \to \mathbf{r}'' \in \mathbf{R}$ on the basis of the EPR criterion (as a review of the three-step argument above should make clear).

But let us consider further the full implications of Stapp's claim about SR, as applied instead to SR', in the following way: Suppose there were available a surrogate for SR', of which it could correctly be maintained that it mentions only (let us say) *facts* based in the region **R**. Since SR' contains the clause $R1\square \to R1\text{-}$, there must be a proposition equivalent to that clause that itself mentions only facts based in the region **R**. Let us take this to mean a proposition, strictly equivalent to the counterfactual clause, which explicitly "asserts something about **R**" — so that it may be appropriately symbolized by "S(**R**)", and so that its semantical truth-conditions can be assumed to make no reference to other possible worlds, and in an appropriate sense to "depend only upon the state of affairs within **R**." Then the proposition $L2 \Rightarrow SR'$ will be equivalent to the proposition $L2 \Rightarrow (R2+ \to S(\mathbf{R}))$. But now — since by assumption the choice whether or not to perform the experiment $L2$ is entirely independent of anything in region **R** — a bona fide locality argument will allow the conclusion, not merely that $R2+ \to S(\mathbf{R})$ must also be entailed by $L1$, but quite simply that $R2+ \to S(\mathbf{R})$ must be *true in every possible world*. Thus, *a fortiori*, the analogue of LOC2´ — the rule permitting inference from $L2 \Rightarrow (R2+ \to S(\mathbf{R}))$ to $L1 \Rightarrow (R2+ \to S(\mathbf{R}))$ — would be valid, and a precise analogue of Stapp's derivation of a (quasi-)contradiction would be correct.

Of course, to hold that there is available a proposition S(**R**), expressing a state of affairs localized within the region **R** and *equivalent* to the counterfactual $R1\square \to R1\text{-}$, is in effect to hold that the counterfactual can only be true in virtue of "something in **R** that determines the result $R1\text{-}$ if $R1$ is performed" — i.e., by virtue of an "instruction sheet" in Mermin's sense, or an "element of reality" in the sense of EPR. We see, therefore, (a) that the claim on which Stapp bases his argument for LOC2 (or its amended version LOC2´) is not "heuristically" underwritten by EPR's principle; but (b) that what one might call a more *aggressively* "realistic" assumption than that of EPR not only would support the claim, but appears also to be *entailed* by it. Stapp himself intended, through his recourse to counterfactual reasoning, to avoid any such assumption; in fact, however, the particular mode of counterfactual reasoning he advocates—in the light of the grounds he puts forward to justify it—goes strictly beyond what the EPR criterion for "elements of reality" would support.

6. CONCLUSIONS

Our analysis, then, has led to the following conclusions:

(1) There is a contradiction in the joint assumption of the correctness of the quantum-mechanical predictions concerning a system in the Hardy state, of special-relativistic locality, and of the EPR principle concerning elements of reality. (That quantum mechanics, locality, and EPR jointly are contradictory is of course not news—that follows easily from the results of Bell.)

(2) Stapp—who undoubtedly was aware of the easy argument leading to the foregoing conclusion, but who wished to avoid the assumption of EPR—has not succeeded in his attempt to derive an analogous contradiction, without the EPR assumption, by a simple appeal to counterfactual reasoning; and we have given reasons for the claim that no such attempt can succeed, unless a contradiction already exists *without* the use of counterfactual reasoning.

Now, it is worth noting that in classical physics and in situations of ordinary life counterfactual propositions are routinely based upon assumptions of the existence of "elements of physical reality." For example, "If a catalyst c were added to non-reacting substances s_1, s_2, and s_3, the reaction r would have occurred" is judged to be true because of the constitutions of c, s_1, s_2, and s_3 (all of which are elements of physical reality), and the laws governing them. In view of the great success of this type of realistic reasoning in pre-quantum reasoning, it is tempting to maintain a residue of it even after the acceptance of quantum theory; but what we have seen demonstrates how misleading this can be, and how much care is required to avoid fallacy in the non-deterministic context of quantum mechanics.

The correlations-at-a-distance predicted by quantum theory, and strongly confirmed by experiment, for pairs and multiples of systems in entangled states are unprecedented in physics. *Prima facie,* these correlations are either bizarre coincidences or violations of relativistic restrictions on causal connection. But this statement of alternatives needs to be treated with reserve. The sense in which relativity theory forbids "causal connection" at a distance perhaps deserves more careful critical discussion than it has received—or than can be offered in this place; but it would appear that the chief point is this: it must, if special relativity holds, be impossible to *transmit energy* at a rate faster than the speed of light. Now, there are well-known theorems (Eberhard 1978; Ghirardi, Rimini, and Weber 1980; Page 1982) to the effect that quantum correlations-at-a-distance cannot be used to send messages superluminally, and therefore that they do not involve "causal connection" of the prohibited kind.

The results of Bell, and the rather strange interrelationships that characterize the Hardy state, appear to demonstrate a feature of the world that could hardly have been anticipated in earlier stages of physics. To put the point in terms of the famous vivid metaphor of Einstein—a point, however, contrary to what Einstein maintained—God (or *nature*—Spinoza's *Deus sive natura*) *does* "play dice" with the world; and moreover, does so at spatially separated points, with dice that are neither loaded nor interacting, but are nevertheless *correlated!*—How does he do this?—Well, when it comes to that, we don't know "how" nature "does" *anything*. At the beginning of modern science, in the seventeenth century, it was thought that the only "conceivable" way that bodies could *interact* was by pressure or impact; and Newton's introduction of forces other than those of pressure or percussion was considered by many to be a violation of intuitively evident principles. Even then, the question was raised by some acute philosophers whether interaction by impact really was intrinsically intelligible (Locke 1974, Book II, ch. xxiii, §28) and indeed whether what was taken for "intuitive evidence" was anything more than a *sense of familiarity* with such modes of action—familiarity based upon frequent experience (Hume 1993, §7, Part II, third paragraph). In later physics, *neither* Newtonian action at a distance *nor* interaction by impact came to be seen as the norm for the actions of bodies upon one another—rather, one was led to suppose local action between particles and fields, and locally determined propagation of fields. In particular, our understanding of impact today makes this a very complicated process—anything but an "intuitively evident" mode of interaction. All this suggests another cautionary remark: that in domains very remote from our ordinary experience, we have to be prepared for modes of lawful connection that may not conform to our prior notions of intelligibility.

But the sense of something strange in this notion of non-interacting correlated dice persists; and one must recall that—as we have remarked earlier—the present state of physics does not even give us a *language* in which the description of a "possible world" in the sense of the logicians can be formulated. This fact, and the well known problems of integrating, not merely special, but *general* relativity with quantum mechanics, lead us to consider a note of hesitation to be the right one on which to end our discussion. It is also, we think, an appropriate note on which to end a paper dedicated, with affection and admiration, to John Stachel—one of the last people in the world to be satisfied with a glib or premature answer to a hard problem.

APPENDIX: STAPP'S FORMAL ARGUMENT (AND ITS CORRECTED COUNTERPART)

In the following representation of Stapp's formal argument, we have, as already remarked made some small notational changes; otherwise the steps of the argument are exactly as in (Stapp 1997); and we use Stapp's numbering of the steps. The capitalized labels prefixed to the steps, taken over from (Stapp 1997), are not *names of the propositions* that follow them, but indications of the *grounds appealed to as justifying* the assertion of those propositions. (The first step, labeled "LOC1," is in fact the symbolic expression of LOC1 itself; we have not discussed this principle, but we have no quarrel with it.) Steps to whose number we prefix an asterisk are ones whose legitimacy we have questioned in the discussion above. We place an asterisk before steps (6) and (10), although the propositions that occur there do follow correctly from premises laid down by Stapp, because, as we have pointed out, the *formulation* of those premises is marred by a logical mistake—namely, one concerning the semantical force of an "imbedded" strict conditional clause. Some steps are in themselves "correct," in that they indeed follow from what precedes them by the principles named, but still need to be corrected because what precedes them is *itself* defective. Such steps we do not mark with an asterisk—but of course we do correct them in our revised derivation below.)—Bracketed comments are Stapp's, directly from the original; parenthetical comments following are ours.

$$\text{LOC1: } (L2+ \wedge R2) \Rightarrow [R1\square \rightarrow (L2+ \wedge R1)]. \tag{1}$$

$$\text{QM: } (L2 \wedge R2+) \Rightarrow (L2+ \wedge R2). \tag{2}$$

$$\text{QM: } (L2 \wedge R1) \Rightarrow (L2 \wedge R1-). \tag{3}$$

$$\text{LOGIC: } (L2 \wedge R2+) \Rightarrow [R1\square \rightarrow (L2 \wedge R1-)]. \text{ [From 1,2,3]} \tag{4}$$

$$\text{LOGIC: } L2 \Rightarrow [R2+ \Rightarrow (R1\square \rightarrow R1-)]. \text{ [Logically equivalent to 4.]} \quad *(5)$$
(No!)

$$\text{LOC2: } L1 \Rightarrow [R2+ \Rightarrow (R1\square \rightarrow R1-)]. \quad *(6)$$

$$\text{LOGIC: } (L1 \wedge R2) \Rightarrow [R2+ \Rightarrow (R1\square \rightarrow R1-)]. \text{ [Equivalent to 6.]} \tag{7}$$

$$\text{QM: } (L1 \wedge R2) \Rightarrow (L1- \Rightarrow R2+). \quad *(8)$$

$$\text{LOGIC: } (L1 \wedge R2) \Rightarrow [L1- \Rightarrow (R1\square \rightarrow R1-)]. \text{ [From 7 and 8.]} \tag{9}$$

$$\text{LOC3: } (L1 \wedge R2) \Rightarrow [R1\square \rightarrow (L1- \Rightarrow R1-)]. \quad *(10)$$

$$\text{LOGIC: } R2 \Rightarrow [L1 \Rightarrow (R1\square \rightarrow [L1- \Rightarrow R1-])]. \text{ [Equivalent to (10).]} \quad *(11)$$
(No!—although the point is minor.)

$$\text{QM: } L1 \Rightarrow [R1 \Rightarrow \neg(L1- \Rightarrow R1+)]. \text{ (Note: this is true; but rather weak,} \tag{12}$$
since $\neg(L1 \Rightarrow R1+)$ is *itself (nessessarily) true*, by QM.)

$$\text{LOGIC: } L1 \Rightarrow [R1\square \rightarrow \neg(L1- \Rightarrow R1+)]. \text{ [Entailed by 12.]} \tag{13}$$

$$\text{LOGIC: } R2 \Rightarrow [L1 \Rightarrow (R1\square \rightarrow \neg[L1- \Rightarrow R1+])]. \text{ [Entailed by 13.]} \tag{14}$$

And now the corrected derivation:

$$\text{LOC1: } (L2+ \wedge R2) \Rightarrow [R1\square \rightarrow (L2+ \wedge R1)]. \tag{1}$$

QM: $(L2 \wedge R2+) \Rightarrow (L2+ \wedge R2)$. (2)

QM: $(L2+ \wedge R1) \Rightarrow (L2 \wedge R1-)$. (3)

LOGIC: $(L2 \wedge R2+) \Rightarrow [R1\square \rightarrow (L2 \wedge R1-)]$. (From 1,2,3.) (4)

LOGIC: $L2 \Rightarrow [R2+ \rightarrow (R1\square \rightarrow R1-)]$. (This *is* logically equivalent to 4.) (5´)

LOC2´: $L1 \Rightarrow [R2+ \rightarrow (R1\square \rightarrow R1-)]$. (6´)

LOGIC: $(L1 \wedge R2) \Rightarrow [R2+ \rightarrow (R1\square \rightarrow R1-)]$. (Equivalent to 6´.) (7´)

QM: $(L1 \wedge R2) \Rightarrow (L1- \rightarrow R2+)$. (8´)

LOGIC: $(L1 \wedge R2) \Rightarrow [L1- \rightarrow (R1\square \rightarrow R1-)]$. (From 7´ and 8´.) (9´)

LOC3´: $(L1 \wedge R2) \Rightarrow [R1\square \rightarrow (L1- \rightarrow R1-)]$. (10´)

LOGIC: $R2 \Rightarrow [L1 \rightarrow (R1\square \rightarrow [L1- \rightarrow R1-])]$. (Equivalent to (10´).) (11´)

QM: $L1 \Rightarrow [R1 \Rightarrow \neg(L1- \Rightarrow R1+)]$. (Cf. Note on (12) above.) (12)

LOGIC: $L1 \Rightarrow [R1\square \rightarrow \neg(L1- \Rightarrow R1+)]$. (Entailed by 12; but, again a simpler and stronger statement is possible: just $\neg(L1- \Rightarrow R1+)$.) (13)

LOGIC: $R2 \Rightarrow [L1 \Rightarrow (R1\square \rightarrow \neg[L1- \Rightarrow R1+])]$. (Entailed by 13; but the some comment is in order. However, the strengthened statement does not save the argument.) (14)

ACKNOWLEDGEMENTS

We wish to thank Henry Stapp and David Mermin for penetrating comments on an earlier draft of this paper, and William Harper for stimulating questions about the implications of stochastic physics for possible world semantics. One of us (A. S.) acknowledges with gratitude a visit to the Institute for Advanced Studies of the Hebrew University, where part of this work was done.

Boston University
University of Chicago

NOTES

1. See (Stapp 1971, 1975, 1979, 1994a, 1994b) and other articles referred to in these.
2. See (Stapp 1997, 302-303; emphasis in the original).
3. See, for example, (Lewis 1973).
4. Note that this entails—given that it is always the case that w' is at least as close to w as is w' itself—that p and q are both true in the world w'.
5. Note, by the way, a use of the contrary-to-fact conditional that we do not attempt to subsume under possible-worlds semantics!
6. We remind the reader that a strict conditional—even if it occurs as the consequent clause in a conditional, whether material, strict, or counterfactual—is itself either *true in all possible worlds, or false in all possible worlds*.
7. Note that **L** is a space-time region *in space-like relation* to **R**; "at an earlier time" is a *frame-dependent* characterization, having no *objective* significance within the special theory of relativity—and this is taken account of by Stapp in his next sentence.
8. We are unsure, when Mermin says that a statement "may derive ... its very meaning" from events that were still future, whether he means that the *specific meaning* of the statement may involve such events, or that the very *existence* of a meaning may depend on the character of those events. The former seems to us justified, on our analysis of the consequences of Stapp's semantics; the latter not, in that Stapp's semantics ascribes a well-defined meaning to the statement in question in all circumstances.
9. Note again that *without* the postulate of that "least degree of similarity," it would only follow that there is no possible world that is "similar to w in any degree allowed by the theory" and in which $p \wedge \neg q$ holds.
10. More exactly: The *heuristic* reasoning of EPR *does* involve counterfactuals "One *could have* predicted A or B *if one had measured* [etc.]"; but such heuristic considerations *motivate* the postulation of theoretical laws, which themselves have simple (non-modal) universal form.

REFERENCES

Belinfante, F. J. 1973. *A Survey of Hidden-Variable Theories*. Oxford: Pergamon.
Bell, J. S. 1964. "On the Einstein-Podolsky-Rosen Paradox." *Physics* 1:195-200.
Clauser, J. F., M. A. Horne, A. Shimony, and R. A. Holt. 1969. "Proposed Experiment to Test Local Hidden-Variable Theories." *Physical Review Letters* 23:880-884.
Eberhard, P. H. 1978. "Bell's Theorem and the Different Concepts of Locality." *Nuovo Cimento* 46 B, series 2:392-419.
Einstein, A., B. Podolsky, and N. Rosen. 1935. "Can Quantum-Mechanical Description of Physical Reality be Considered Complete?" *Physical Review* 47:777-780.
Ghirardi, G. C., A. Rimini, and T. Weber. 1980. "A General Argument against Superluminal Transmission through the Quantum Mechanical Measurement Process." *Lettere al Nuovo Cimento* 27, series 2:293-298.
Hardy, L. 1992. "Quantum Mechanics, Local Realistic Theories, and Lorentz-Invariant Realistic Theories." *Physical Review Letters* 68:2981-2984.
Hume, D. 1993. *An Enquiry Concerning Human Understanding*. Indianapolis: Hacket.
Kripke, S. 1963. "Semantical Considerations on Modal Logic." *Acta Philosophica Fennica* 16:83-94.
Lewis, D. 1973. *Counterfactuals*. Cambridge, MA: Harvard University Press.
Locke, J. 1974. *An Essay Concerning Human Understanding*. New York: New American Library.
Mermin, N. D. 1994. "Quantum Mysteries Refined." *American Journal of Physics* 62:880-887.
———. 1998. "Nonlocal Character of Quantum Theory?" *American Journal of Physics* 66:920-924.
Page, D. N. 1982. "The Einstein-Podolsky-Rosen Physical Reality is Completely Described by Quantum Mechanics." *Physics Letters* 91 A:57-60.
Stapp, H. P. 1971. "S-Matrix Interpretation of Quantum Theory." *Physical Review* D 3:1303-1320.
———. 1975. "Bell's Theorem and World Process." *Nuovo Cimento* 29:270-276.
———. 1979. "Whiteheadian Approach to Quantum Theory and the Generalized Bell's Theorem." *Foun-

dations of Physics 9:1-25.

———. 1994a. "Strong Versions of Bell's Theorem." *Physical Review* A 49:3182-3187.

———. 1994b. "Reply to 'Stapp's Algebraic Argument for Nonlocality.'" *Physical Review* A 49:4257-4260.

———. 1997. "Nonlocal Character of Quantum Theory." *American Journal of Physics* 65:300-304.

———. 1998. "Meaning of Counterfactual Statements in Quantum Physics." *American Journal of Physics* 66:924-926.

Wigner, E. P. 1970. "On Hidden Variables and Quantum Mechanical Probabilities." *American Journal of Physics* 38:1005-1009.

GREGG JAEGER AND SAHOTRA SARKAR

COHERENCE, ENTANGLEMENT, AND REDUCTIONIST EXPLANATION IN QUANTUM PHYSICS

Abstract. The scope and nature of reductionist explanation in physics is analyzed, with special attention being paid to the situation in quantum mechanics. Five different senses of "reduction" are identified. The strongest of these, called "strong reduction," is the one that purports to capture the relations between macroscopic and microscopic physics. It is shown that the criteria for strong reduction are violated by explanations in quantum mechanics which involve "entangled states." The notion of "coherence" in physical systems is also defined. It is shown that, contrary to many current views, the invocation of coherence does not necessarily lead to the violation of strong reduction. However, entangled systems also exhibit coherence. Therefore, the subclass of coherent systems that are entangled presents problems for strong reduction.

1. INTRODUCTION

In recent years, the problems surrounding the role of reductionism in science have been extensively studied, mainly in the contexts of biology and psychology.[1] Philosophers seem to have generally assumed that (i) instances of reduction in the physical sciences are ubiquitous; and (ii) these instances are straightforward in the sense that they are trivially captured by at least one of the alternative models of reduction that have been developed by philosophers of science.[2] These assumptions seem to be based on intuitive analyses of putative instances of reduction such as the following:

(i) there is a straightforward sense in which Newtonian mechanics can be obtained from the special theory of relativity (for instance, by taking the $c \to \infty$ limit, where c is the speed of light in vacuum). Thus, Newtonian mechanics is reduced to special relativity;
(ii) similarly, the so-called "Newtonian" limit of general relativity yields Newton's theory of gravitation. Once again, the latter is thus reduced to the former;
(iii) geometrical optics is reducible to physical optics through Maxwell's electromagnetic theory in the sense that the behavior of light waves, as predicted by Maxwell's theory, shows why geometrical optics is correct to a good approximation;
(iv) classical thermodynamics can be derived from, or reduced to, statistical mechanics by the construction of kinetic models of gases. This putative reduction is particularly fascinating because mechanics, the laws of which are invariant under time reversal, is supposed to give rise to the second law of thermodynamics which embodies a direction of time;

(v) quantum mechanics reduces classical chemical bonding theory, and accounts for the various valency rules of chemistry that have been used since at least the early 19th century

It is quite possible that a half-century ago most, though not all, physicists and chemists would have accepted these cases as successful reductions. Most of them would also have endorsed a positive and unproblematic assessment of the status of reductionism in the physical sciences. The developments in physics during the last fifty years, however, have called these judgments into question. Among the most startling have been scaling theory and renormalization "group" techniques in condensed matter physics. A perusal of these developments has even led Leggett (1987) to suggest that the appropriate relationship between macroscopic and microscopic physics is only one of consistency: microscopic physics has no significant role in the explanation of macroscopic phenomena.[3] Leggett's views are by no means idiosyncratic. For instance, Fisher (1988), one of the founders of scaling theory, has argued that there are aspects of condensed matter physics for which the underlying microphysics (viz., quantum mechanics) is explanatorily irrelevant.

Meanwhile, Fröhlich (e.g. 1968, 1969) initiated a research program that aims to prove that in several instances biological entities must be considered as quantum mechanical systems that exhibit "coherence" or phase correlations in their dynamics. Fröhlich (1973) has claimed that biological explanations that invoke such phenomena are not "mechanistic," that is, reductionist in the usual sense of that term (see section 2). This claim has also been endorsed by Ho (1989).[4] What is particularly philosophically interesting about Fröhlich's conjecture is that, if it turns out to be correct, these attempts at physical explanation in biology would fail to satisfy the strictures of reduction not because of any peculiarity of biological systems but because of a failure of reduction (e.g., as it will be construed in section 2) in physics (more specifically, quantum mechanics) itself.

Finally, it has become apparent that at least some of the ostensibly straightforward cases of reduction, such as those mentioned above, are not quite as unproblematic as they customarily have been taken to be. Returning to the cases mentioned above:

(i) Though, from a strictly mathematical point of view, Newtonian mechanics does emerge from special relativity in the $c \to \infty$ limit, that limit is *counterfactual*: the speed of light is finite;

(ii) For the Newtonian limit of general relativity to exist and yield Newton's theory of gravitation, constraints have to be imposed on general relativity (Ehlers 1981; Malament 1986). In particular, the spatial part of space-time must be flat;

(iii) A series of approximations is required to obtain geometrical optics from Maxwell's laws.

(iv) The relation of thermodynamics to statistical mechanics has turned out to be even more complicated. It should consist of the derivation of the thermodynamic laws from statistical mechanics in the so-called "thermodynamic limit," the one in which the number of particles (N) and the volume (V) of the system both go to ∞ while the density (N/V) remains constant. All thermodynamic parameters must approach well-defined values in this limit. In the classical (non-quantum) realm, the existence of the thermodynamic limit so far has been rigorously proved for only very contrived and simple systems (Thompson 1972). For classical systems with only electrostatic interactions, it is even easy to show that the thermodynamic limit does not exist. If quantum mechanics is invoked, however, the limit is defined (Ruelle 1969, 60-68). Perhaps there is even some unexpected insight to be gleaned from this situation: that classical thermodynamics is reducible to quantum statistical mechanics, but not to classical statistical mechanics;

(v) To obtain the classical rules of valency from quantum mechanics, approximations about the nature of the wave-functions, about the Hamiltonian for complex atoms, and assumptions about the convergence of solutions must all be brought into play (Pauling 1960).

These observations should indicate that the issues surrounding reduction are far from settled in contemporary physics. Our immediate purpose in this paper is to examine whether explanations invoking the concept of "coherence" (including the claims of Fröhlich) really do violate reductionism. We conclude that the use of "coherence" does not necessarily preclude an explanation from being reductionist. Nevertheless, there exists a class of quantum-mechanical systems exhibiting coherence, those having "entangled states," for which explanations are no longer clearly reductionist. This class of systems has long been of philosophical interest because one of its subclasses, that of entangled two-particle systems each with an associated two-dimensional Hilbert space, includes the systems that violate Bell's inequality. It appears likely that the results establishing the relationship between entanglement and Bell-inequality violations can be generalized to the result that all entangled systems are capable of violating Bell-type inequalities. Furthermore, it turns out that though the type of model that Fröhlich has invoked involves coherence, any failure of reduction will be due to *entanglement* rather than coherence. In the long run, we hope that our analysis will help reintroduce detailed discussions of the place of reductionist explanation in physics.

In section 2 we discuss what we mean by a "reductionist explanation." In section 3 we give several examples of classical and quantum systems that exhibit "coherence." We then attempt a general definition of that and related concepts, which is necessary because a sufficiently general definition has proved to be elusive in the past. We are not fully satisfied with our definition; we hope, however, that it will spur further discussion by others. In section 4 we show that all entangled (quantum) states exhibit coherence, and that the coherent states invoked by Fröhlich are entangled. We observe that the quantum states that violate Bell's inequality are also entangled. In both sections 3 and 4 our more technical conclusions are presented as relatively precise theorems. We give proofs of these theorems in those cases in which they are not explicitly available in the extant literature. In section 5 we show that, while the use of coherence does not necessarily lead to an explanation failing to be reductionist, entanglement leads to such a failure. We note some of the implications of our analysis in the concluding section (section 6).

2. ON REDUCTION

Systematic analysis of the concept of reduction in the natural sciences began with the pioneering efforts of Nagel (1949, 1961) and Woodger (1952), who viewed reduction as a type of inter-theoretic explanation. This approach has since been significantly extended, especially by Schaffner (1967, 1994). All these approaches view explanation as deductive-nomological. Alternative accounts that view reduction as a relation between theories, though not necessarily one of explanation, have been developed by Kemeny and Oppenheim (1956), Suppes (1957) and, more recently, Balzer and Dawe (1986, 1986a). Meanwhile analyses of reduction that view it as a form of explanation but not necessarily as a relation between theories have also been developed (e.g., Kauffman 1971; Wimsatt 1976; Sarkar 1989, 1992, 1998). These analyses have been motivated primarily by the situation in molecular biology, where reductionist explanation seems to be rampant, but the explanations refer to a variety of mechanisms rather than to theories. Using a set of distinctions introduced by Mayr (1982), Sarkar (1992) classified these models of reduction into three categories: (i) *theory reductionism* which consists of those models that view reduction as a relation between theories; (ii) *explanatory reductionism* which consists of those models that view it as explanation, but not as a relation between theories; and (iii) *constitutive reductionism* which consists of those models such as the various types of supervenience that eschew both theories and explanation.[5]

Accounts in all three categories make both epistemological and ontological claims. In particular, all models of reduction share the rather innocuous ontological claim that what happens at the level of the reduced entities (theories or not) is not novel in the sense of being inconsistent with what happens at the level of the reducing entities — otherwise these would not be models of *reduction*. Throughout our discussion we will continue to make this assumption. However, we will ignore other ontological issues that have been controversial: whether the terms invoked in a reduction refer to "natural kinds"; whether reductions establish relations between "types" at the two levels or between "types" and "tokens", etc. These issues, though often regarded as philosophically important, are orthogonal to our present purpose.[6] We will also ignore those formal epistemological issues that have persistently been the focus of dispute: (i) whether the factors involved in a reductionist explanation are codified into theories; (ii) whether the structure of the explanation is basically deductive-nomological (Schaffner 1994) or statistical (Wimsatt 1976), *etc*. Thus our present analysis will be consistent with any of the models of reduction that view it as a type of explanation.[7] Meanwhile, we will focus on three substantive claims that have often been implicit in models of reduction but have rarely been discussed in detail. Our analysis will reveal some rather surprising subtleties about reductionist explanation.

We assume that what we have at hand is an explanation, i.e., it satisfies whatever strictures that one chooses to put on "explanation." Our problem is solely to specify additional criteria by which we can decide whether that explanation is reductionist. The reasons for this move are to avoid disputes about the explication of "explanation" (on which there is no consensus), and to focus attention precisely on *those factors that make an explanation reductionist*.[8]

We suggest that, at the substantive level, the three most important criteria are:

(i) *Fundamentalism:* the factors invoked in an explanation are warranted by what is known, either from theoretical considerations (characteristically involving approximations) or only from experiments entirely at the level of the reducing theories or mechanisms which are more "fundamental" than those at the level of the reduced entities in the sense that their presumed domain of applicability is greater.[9] Satisfaction of this criterion is a matter of degree. If the demonstration of such a warrant involves only theoretical derivation, from first principles, with no approximation, and so on, then its satisfaction is most complete. Approximations, especially counterfactual or mathematically questionable approximations, hurt the degree to which this criterion is satisfied;

(ii) *Abstract hierarchy:* the complex entity whose behavior is being explained is represented as having a hierarchical structure (with identifiable levels and an ancestral relation between levels) in which only the properties of entities at lower levels (of the hierarchy) are used to explain the behavior of the complex entity.[10] Such an abstract hierarchical representation can involve any space, not necessarily physical space, that is used to model a system. For instance, it can be a hierarchy in any configuration or phase space in (classical) analytical mechanics or a Hilbert or Fock space in quantum mechanics;

(iii) *Spatial hierarchy:* the hierarchical structure of the entity (that is invoked in the explanation) must be realized in physical space, that is, entities at lower levels of the hierarchy must be spatial parts of entities at higher levels of organization.

These criteria are not all independent of one another: (iii) can only be satisfied provided that (ii) is. Moreover, if (i) is not satisfied at all, it is doubtful (at least) that an explanation should be considered a reduction. It will be assumed here that, for all reductionist explanations, (i) is at least approximately satisfied. What will distinguish the different types of reduction are the questions whether (i) is fully satisfied and which of the other two criteria, if either, is also satisfied. With this in mind, five different senses of "reduction" based on these criteria can be distinguished. For each of these senses, several illuminative biological examples exist—these are discussed in Sarkar (1996). Here, we only mention putative examples from physics:

(a) criterion (i) is (fully) satisfied while none of the others are: examples include the reduction of Newtonian mechanics to special relativity, and of Newtonian gravity to general relativity. More controversially, the reduction of geometrical optics to physical optics is a reduction of this sort.[11]

(b) criterion (ii) is fully satisfied while (i) is approximately satisfied: this is clearly a very weak sense of reduction. Little more than a hierarchical structure is assumed. In the study of critical phenomena, explanations involving renormalization in parameter space are of this type: the system (say, a ferromagnet) is given a hierarchical representation in parameter space (though not in physical space) but the interactions posited at each stage as the number of units of which the system is composed (in our example, magnetic spins) is iteratively decreased have *at best* only approximate warrants from the underlying mechanisms;[12]

(c) criteria (i) and (ii) are fully satisfied but (iii) is not: in these reductions, the organizational hierarchy does not correspond to a hierarchy in the usual physical space. If quarks really are confined, in the sense that free quarks do not exist, explaining the properties of hadrons on the basis of properties of quarks would be an explanation of this type. There is a hierarchical structure: hadrons consist of quarks but this is clearly not a hierarchy in physical space. Examples of this sort abound in particle physics;
(d) criteria (ii) and (iii) are (fully) satisfied while (i) is approximately satisfied. Once again, in the study of critical phenomena, real-space renormalization, that is, renormalization with a representation in physical space involves a reduction of this type;
(e) all three criteria are satisfied: this is obviously the strongest sense of reduction. It is the one that is invoked in the putative reduction of thermodynamics to statistical mechanics. It is also the one that will be most relevant to the discussion in section 5. We will refer to this sense as "strong reduction."

Criteria (ii) and (iii) together capture the intuition behind those types of reduction which refer to a spatial whole being made up of identifiable constituent parts. When (i) is also satisfied, the explanatory force in a reduction, which comes from the properties of the parts, is supposed to provide a *deeper* explanation entirely from the lower level. In general, this is the one which is usually assumed to capture the relation between macroscopic and microscopic physics. In the particular case of explanations involving the notion of coherence, this is the intuition that we will explore in this paper. Of course, if what we have already said about explanations involving renormalization theory in condensed matter physics is correct, there are other reasons for doubting that strong reduction describes the relation between microscopic and macroscopic physics—we leave a full discussion of renormalization for another occasion.

3. COHERENCE

We turn, now, to the concept(s) of coherence. Any attempt at an explication of this concept faces a peculiar quandary: though it is quite routinely used in a variety of areas within physics, it is almost never defined outside optics, where it has become associated with various measures of correlation between field variables at two space-time points (cf. Mandel and Wolf 1995, chs. 4-8, 11). Usage in physics and biophysics is so varied that it is open to question whether there are non-trivial criteria that all uses share. This is so despite 75 years of very gradual development and several attempts to broaden the definition beyond quantum optics, where there is a fairly consistent pattern of use (Mandel and Wolf 1970, 1995). Because of this, we start with some remarks in the quantum optics literature regarding this definition. We then attempt to generalize it to be generally applicable across physics. But we are less that sure that the result is non-trivial. So we proceed to avoid triviality by relativizing our definition to an independently characterized class of physical systems. At the very least, we hope that our attempt will provoke not just criticism but other attempts at providing a general definition of coherence.

Section 3.1 discusses examples of coherence across physics. Section 3.2 presents some illuminating comments by others regarding usage of the term, as well as our explication of one trivial and one non-trivial concept of coherence; we demonstrate the triviality of the former by an explicit (and itself somewhat trivial) theorem.

3.1 Examples

Skepticism about the possibility of reductionist explanation in physics has often been based on the belief that the components of composite systems exhibit collective behavior that cannot be accounted for by a straightforward examination of the properties of these parts taken in isolation. In physics, descriptions of such behavior often invoke the notion of "coherence." As we noted above, this notion is almost never given a definition when it is used outside optics. To motivate our suggested definitions to be given at in section 3.2, consider the following four examples. The first is from classical physics. The other three involve quantum concepts:

(i) *Coupled harmonic oscillators:* Consider a pair of identical oscillators, having the same spring constant k and mass m, coupled by a third spring of spring constant k'. Such a pair of oscillators are capable of moving in a "high-frequency normal mode" of vibration in which the motion of each oscillator is of constant amplitude and the same frequency ω higher than their common "natural frequency," $\omega_0 = (k/m)^{1/2}$. The equations of motion for the two oscillators are:

$$x_1 = A\cos\omega t, \; x_2 = A\cos\omega t, \tag{3.1}$$

where x_1 and x_2 are the displacements of each of the oscillators from its equilibrium position, t is the time, and A and B are the (constant) amplitudes of oscillation. There is a correlation in the position of the pair of oscillators because both oscillators vibrate with the same frequency. Note that, given t and knowing $\omega = \sqrt{\omega_0^2 + k'/m}$, we can infer x_2 from x_1 and vice versa;

(ii) *Rabi oscillators:* Consider an electron capable of being in one of two coupled states $|\phi_1\rangle$, and $|\phi_2\rangle$ with energies E_1 and E_2 (respectively), of an atom, ion, or molecule, so that its quantum state, which can be written

$$|\psi(t)\rangle = a_1(t)|\phi_1\rangle + a_2(t)|\phi_2\rangle, \tag{3.2}$$

where $a_i(t)$ $(i = 1, 2)$ are state probability amplitudes and t is the time, obeys the standard time-dependent Schrödinger equation.[13] It is convenient to decompose $|\psi(t)\rangle$ using the eigenvectors, $|\psi_+\rangle$ and $|\psi_-\rangle$, of the Hamiltonian, $H = H_0 + W_{12}$, where W_{12} is the operator representing the "energy" of the "force" coupling the two states. Let $|\psi_+\rangle$ and $|\psi_-\rangle$ have the energies E_+, E_- (respectively). Taking $|\psi(0)\rangle = |\phi_1\rangle$, we have $|\psi(t)\rangle = \lambda e^{-iE_+ t/\hbar}|\psi_+\rangle + \mu e^{-iE_- t/\hbar}|\psi_-\rangle$, where λ and μ are constants. The probability of the electron being in either of the two states (or in the other) varies periodically with a frequency that depends on the strength of the coupling $|W_{12}|$ between them: assuming W_{12} to be purely non-diagonal, the probabilities $P_2(t)$ and $P_1(t)$ of respectively being in states $|\phi_2\rangle$ and $|\phi_1\rangle$ are

$$P_2(t) = \sin^2\theta \sin^2\left(\frac{E_+ - E_-}{2\hbar}t\right) \tag{3.3a}$$

$$P_1(t) = 1 - P_2(t) = 1 - \sin^2\theta \sin^2\left(\frac{E_+ - E_-}{2\hbar}t\right) \tag{3.3b}$$

where $\theta = \left(\frac{2|W_{12}|}{E_1 - E_2}\right)$ is the oscillation period and $\hbar$ is $h/(2\pi)$ (with h being Planck's constant).

There is a strict correlation between the *probabilities* of the electron occupying the two states, thus of the atom, ion or molecule having the feature of one energy level or its alternative energy level occupied and the probabilities can be inferred from one another using eq. (3.3b). Note that this discussion can be extended beyond two-state systems to cases with several states and different associated energy levels. The correlations are then less trivial;

(iii) *Fröhlich systems*: (Fröhlich 1968, 1975) claims that quantum states of a macroscopic system described by "macro wave functions" are needed to explain a wide range of biological phenomena. These wave functions are supposed to exist when there is "off-diagonal long-range order" (ODLRO) in a system. Yang's (1962) definition of ODLRO, which is a property of a "reduced density matrix," will now be used to show how coherence is present in such a system.[14]

Let a large system Σ containing a subsystem σ be described by the density matrix ρ. The "reduced density matrix" representing subsystem σ is first obtained by taking a weighted average of the parameters specifying the portion of Σ not including σ. This averaging is achieved by "tracing out" the parameters from the density matrix ρ representing Σ, yielding a reduced density matrix ρ_σ for σ. ODLRO is a property of ρ_σ, illustrated here by the following two examples.

The simplest and most interesting example in our context is the one-particle reduced density matrix ρ_1 describing single particles. This matrix has the elements

$$\langle j|\rho_1|i\rangle = Tr(a_j \rho a_i^\dagger), \tag{3.4}$$

where i and j are the labels of the possible states for the individual particles of the system, ρ is the density matrix for the entire system, $Tr(\cdot)$ is the trace of a matrix "$\cdot$", the a_i's are the annihilation operators for single-particle states and the $a_i^\dagger$'s are the corresponding creation operators. The second example is the two-particle reduced density matrix describing pairs of particles, having the elements

$$\langle kl|\rho_2|ij\rangle = Tr(a_k a_l \rho a_i^\dagger a_j^\dagger), \qquad (3.5)$$

where the new labels k and l refer to particle states just as i and j do. (Note that elements of this reduced density matrix are labelled by four indices.) Other reduced density matrices for yet larger subsystems of particles, ρ_n, for n-particle states can be similarly defined. The pertinent subsystems of many-particle systems, which we will call "Fröhlich states (F-states)" (following Pokorny 1982), are those whose single-particle reduced density matrices exhibit ODLRO. These consist of a large number, N, of bosons (either simple bosons or bosonic quasiparticles formed from fermions) that are said to exhibit ODLRO when they can be represented by single-particle reduced density matrices of the form

$$\rho_1(r', r'') = \alpha N \Phi(r') \Phi^*(r'') + \chi(r', r''), \qquad (3.6)$$

where $\Phi(r)$ is the quantum "macro wavefunction" attributed to the subsystem, $\Phi^*(r)$ is its complex conjugate, $\chi(r', r'')$ is a positive operator, $0 \le \alpha \le 1$ and r' and r'' represent two (spatial) positions of the subsystem.

The reduced density matrix for the subsystem has the spectral resolution

$$\rho_1(r', r'') = \sum_{i=1}^{\infty} \mu_i \xi_i(r') \xi_i^*(r''), \qquad (3.7)$$

where the $\xi(r)$ are energy eigenstates and μ_i are weights. Most of the N particles in our system lie in the same state, $\xi_n(r) = \Phi(r)$, for which $\mu_n = \alpha N$ (for some value n of i and α is a constant in [0,1] near 1) and the weights $\mu_i \ll 1$ for all $i \ne n$. Thus

$$\rho_1(r', r'') = \alpha N \Phi(r') \Phi^*(r'') + \sum_{\substack{i=1 \\ i \ne n}}^{\infty} \mu_i \xi_i(r') \xi_i^*(r''), \qquad (3.8)$$

where $\sum_{\substack{i=1 \\ i \ne n}}^{\infty} \mu_i = (1-\alpha)N$. With $\sum_{\substack{i=1 \\ i \ne n}}^{\infty} \mu_i \xi_i(r') \xi_i^*(r'') \equiv \chi(r', r'')$, which describes that portion of ρ_1 describing individual particles of our subsystem not in the state Φ, we have

$$\rho_1(r', r'') = \alpha N \Phi(r') \Phi^*(r'') + \chi(r', r''), \qquad (3.9)$$

where $\chi(r', r'')$ is small compared to αN except when $r' \approx r''$. As a result of the presence of the first term, $\rho_1(r', r'') \not\to 0$ even as $|r' - r''| \to \infty$. Thus, for F-states, an identity of quantum state—and, therefore, perfect correlation—will persist over large spatial distances. This is the sense in which ODLRO is a form of long-range order.

(iv) *EPR states*: An infamous set of quantum-mechanical states exhibiting coherence are the so-called "Einstein-Podolsky-Rosen (EPR) states." An example of an EPR state is singlet state of the composite system formed by a pair of two spin -1/2 particles, such as an electron (e^-) and a positron (e^+). This system is described by the wavefunction,

$$|\Psi\rangle = \frac{1}{\sqrt{2}}[|+\rangle_{e^-}|-\rangle_{e^+} - |-\rangle_{e^-}|+\rangle_{e^+}], \qquad (3.10)$$

where, in each term, the subscripts refer to the Hilbert space of the corresponding particle, $|+\rangle$ is the single-particle state "spin up," and $|-\rangle$ is the single-particle state "spin down."

The system is correlated in two ways: (a) there is a correlation of both quantum phases, the angles θ of the complex exponential part of each complex probability amplitude, for the combined system (as explained in theorem 3.1 below); and (b) the spins are strictly anticorrelated, that is the spin states of the electron and positron are such that if one has spin "up" then the other has spin "down"; knowing the spin state of either particle allows one to infer the spin state of the other. The example will be relevant to section 4.

Such a system is traditionally referred to as being in a "coherent" superposition of states. Here this involves two quantum-mechanical states: one in which the electron has the relevant component of its spin in the "up" state and the positron has the same spin component "down" (i.e. it is described by $|+\rangle_{e^-}|-\rangle_{e^+}$), and one in which the electron has spin "down" and the positron has spin "up" (described by $|-\rangle_{e^-}|+\rangle_{e^+}$).

3.2 Tentative Explication

The notion of coherence is often clear enough in each case, given its context, but there seems to have been no obvious context-independent physical definition of the concept, beyond reference to increasingly complex sets of correlation functions, despite its rather routine use. This point has sometimes been explicitly recognized. As Pippard (1956, 765) puts it: "It is not possible at present to define precisely, in microscopic terms, what meaning is to be attached to the term 'coherence'... since it is still a somewhat vague empirical concept." Somewhat later, Senitzky (1962, 2864) remarked in the context of quantum optics: "[An] unsatisfactory situation exists with respect to the concept of coherence ... because of the various different meanings attached to the word 'coherent.'" In particular, he pointed out that:

> As mentioned above, the word 'coherent' is used with various meanings. One hears the expression 'coherent oscillator,' denoting an oscillator the output of which is a sine wave; here coherence means monotonicity. The expression 'coherent signal' is often used to distinguish information from noise, and coherence implies, in this connection, non-random variation with time but not necessarily monochromatic variation. The word

'coherent' is also widely used in physics to indicate correlation between two or more functions of either space or time, (such as its use in the description of two light beams obtained by the splitting of a single beam), although the functions themselves may have some random properties.

Klauder and Sudarshan (1968, 56) echo these sentiments in their own attempt to characterize "coherence" broadly:

... we must come to terms with the vague concept of 'coherence.' Being purposely general, let us say that a 'coherent feature' of a statistical ensemble is an observable aspect held in common by each member of the ensemble. Different ensembles will in general have different coherent features depending on what collection of quantities is deemed 'observable.' This suggests that we call a 'relative coherent feature' one which fulfills the criteria for a coherent feature for a subset of the observables. That is, the criteria for a relative coherent feature are necessary but not sufficient for a coherent feature. ... 'Full coherence' can be said to exist if the members of the ensemble are identical in all their observable aspects.

This definition, which is explicitly intended as a general one (though motivated by Klauder and Sudarshan's interest in optics), is unnecessarily restricted to situations where a statistical ensemble exists. One should be able to speak of "coherence" in systems that are not normally regarded as ensembles. This point has been long recognized (e.g by Hopkins 1952, 263; and Senitzky 1962, 2865). It is also suggested by our examples, all of which are systems of this sort. Furthermore even in the context of ensembles, it seems unreasonable to require that *all* properties of a system be correlated for a system to be called "fully coherent." Similarly, in one of the most highly regarded definitions requires that a fully coherent field be "defined as one whose correlation functions satisfy an infinite succession of stated conditions" (Glauber 1963), which also seems unreasonably strict.

However, Klauder and Sudarshan are correct to indicate that even a context-dependent definition of coherence can invoke nothing more than the existence of a correlation. Ho and Popp (1993) also come to the same conclusion.[15] We might, therefore, attempt modify and extend such a putative definition in the following way: a system is *coherent* if and only if, for at least two of its features, A and B, there exists a correlation between their values.[16] Then, either the value of B can be estimated from that of A or the value of A from that of B.

The trouble with this definition of coherence is its *weakness*: almost all systems exhibit coherence.[17] For the purposes of discussion we will instead call this property "trivial coherence." In classical physics almost all systems exhibit such coherence because of the fully general laws such as Newton's third law, which routinely provide the framework in which all models are constructed; each such model will have coherence automatically built into it. In the case of Newton's third law, $\vec{F}_{ij} = -\vec{F}_{ji}$ (where $\vec{F}_{ij}$ is the force of one object, i, on another, j), each time two objects interact, the forces involved will correlate their motions in accordance with Newton's second law, $\vec{F}_{ij} = m\vec{a}_{ij}$, where $\vec{a}_{ij}$ is the contribution to the acceleration of each object due to the force $\vec{F}_{ij}$ of one object, i, on the other j. In quantum mechanics, for all pure states a similar situation arises because all systems are represented by a ray in an associated Hilbert space. The following trivial theorem can be proved:

Theorem 3.1: *All isolated nonstatistical (pure) quantum states exhibit trivial coherence.*

Proof. Consider the case when the Hilbert space of the quantum-mechanical system in question is countable. Every isolated such a system can be written as a pure state of the following form: $|\psi\rangle = \sum_{j=1}^{n} \lambda_j |\gamma_j\rangle$, where $\lambda_j \equiv r_j e^{i\xi_j}$, $\xi_j = -\frac{1}{\hbar} E_j t$ (E_j being the energy of the system in state $|\gamma_j\rangle$, t the time, r_j real constants) and the eigenvectors $|\gamma_j\rangle$ form an orthonormal basis for the Hilbert space of the system. A sustained correlation will exist between the properties corresponding to the eigenvectors $|\gamma_j\rangle$ since the quantity $\Delta_{kl} \equiv (1/t)(\xi_k - \xi_l) = -\frac{1}{\hbar}(E_k - E_l)$ is constant for any k and l, for all times t; since, for isolated systems, the energies E_j are constant in time, a correlation exists, represented by the energy difference Δ_{kl} between the values of ξ_k and ξ_l at any time t. This correlation is observable in the form of quantum interference. (The proof in the case of uncountably infinite Hilbert spaces is identical but with the sums replaced by integrals and discrete indices replaced continuous ones.)

Every physical system will usually exhibit some such trivial coherence which makes it, therefore, not a very satisfying notion. We will, therefore, introduce the notion of *(non-trivial) coherence* and relativize it to a class K of physical systems which forms the background against which new correlations become interesting. A system is *(non-trivially) coherent with respect to K* if and only if it exhibits a type of correlation between two features that is not exhibited by all members of K. This definition is context-dependent since it is relativized to the class K. Nevertheless it remains, in a sense, largely context independent since it is applicable to a wide variety of situations. Non-trivial coherence is most interesting when class K is almost universal, for instance, the class of all Newtonian systems (N) or the class of all quantum-mechanical systems (Q). Returning to our examples, it is obvious that the coupled oscillators exhibit non-trivial coherence with respect to N and the other three cases exhibit non-trivial coherence with respect to Q.

4. ENTANGLEMENT AND COHERENCE IN QUANTUM SYSTEMS

In the quantum mechanics of composite systems, *entangled states* are those states system that cannot be expressed as a product of states of its individual subsystems. For simplicity, consider a system involving just two subsystems, each representable by a finite-dimensional Hilbert space. In this case, each possible state of the composite system is represented by a vector in a Hilbert space $\mathbf{H}_1 \otimes \mathbf{H}_2$, where $\mathbf{H}_1$ and $\mathbf{H}_2$ are the Hilbert spaces of each of the two subsystems in isolation. The *product states* of such a system are those represented by vectors $|\psi\rangle_{1+2} \in \mathbf{H}_1 \otimes \mathbf{H}_2$ such that $|\psi\rangle_{1+2} = |\eta\rangle_1 |\theta\rangle_2$, for a pairs of states, $|\eta\rangle_1 \in \mathbf{H}_1$ and $|\theta\rangle_2 \in \mathbf{H}_2$. The *entangled states* of such a system are those represented by vectors $|\psi\rangle_{1+2} \in \mathbf{H}_1 \otimes \mathbf{H}_2$ such that $|\psi\rangle_{1+2} \neq |\eta\rangle_1|\theta\rangle_2$ for any of the possible pairs of states $|\eta\rangle_1 \in \mathbf{H}_1$ and $|\theta\rangle_2 \in \mathbf{H}_2$. Note that the state vector of any composite system of two objects (each having a finite-dimensional Hilbert space) can be written as a linear combination of unit vectors $|\alpha_i\rangle|\beta_i\rangle$, $i = 1, \ldots, n$ forming an orthonormal basis for $\mathbf{H}_1 \otimes \mathbf{H}_2$. This basis is known as the Schmidt basis and the linear combination is called the "Schmidt decomposition." This basis makes manifest the character of a quantum state: if only one of these components is non-zero then the state is a *product state;* otherwise it is an *entangled state.* (Note that the Schmidt decomposition cannot, in general, be achieved for systems with more than two parts. Nevertheless, in such cases, any n-dimensional system can still be rather artificially decomposed into two subsystems, one with dimension $k (0 < k < n)$ and the other with dimension $n - k$.)

It is hard to over-emphasize the importance of entanglement for understanding quantum systems: as Schrödinger (1935, 555), who introduced the concept, put it: "I would not call [entanglement] *one* but rather *the* characteristic trait of quantum mechanics, the one that enforces its entire departure from classical lines of thought." However, what are most interesting here are the following results.

Theorem 4.1 (a): *All isolated composite systems (having finite-dimensional Hilbert spaces) in entangled states are coherent with respect to Q;* **(b):** *However, not all quantum-mechanical systems that are coherent with respect to Q are entangled.*

Proof (a): Any entangled state for which the Schmidt decomposition can be performed will exhibit correlations between at least two pairs of properties of its *component systems*, namely whose eigenvectors form the Schmidt basis. Such an entangled state may be written as: $|\psi\rangle = \sum_{i=1}^{n} \mu_i(t)|\alpha_i\rangle_1|\beta_i\rangle_2$, where the $\mu_i(t)$ are complex numbers and $|\alpha_i\rangle$ and $|\beta_i\rangle$ are the basis vectors. For that basis, there are clearly n ordered pairs (α_i, β_i) ($i = 1, \ldots, n$, where $n > 1$) of values of the observables A (of system 1) and B (of system 2). These n ordered pairs of properties are perfectly correlated. Such a correlation is not found in all quantum systems. Thus a system in an entangled state will be non-trivially coherent with respect to Q;

Proof (b): This is shown by exhibiting a counterexample: the system above exhibiting Rabi oscillations (see section 3) exhibits coherence but is not entangled. There are many other counterexamples that could be supplied here as well.

The discussion of entanglement that we have given above refers only to systems with two subsystems. This definition of "entanglement" and theorem 4.1, can be straightforwardly extended to systems with countably infinite subsystems. However, entangled states of two-particle systems have long been of interest to philosophers because, to begin with:

Theorem 4.2 (a): *States of two-particle (quantum) systems each having Hilbert spaces of dimension 2 that violate the Bell inequality are entangled.* **(b):** *Moreover, states of two-particle (quantum) systems that violate the Bell inequality are non-trivially coherent with respect to Q.*

Proof (a): This has long been part of the folklore of foundations of quantum mechanics.[18] **(b):** This now follows directly from theorem 4.1(a).

Furthermore, these results are very likely extendable to apply to larger systems and systems of higher dimensionality. Finally, returning to F-states, the coherence of which has been a source of disquiet about reductionist explanation in biology, it is easy to prove that:

Theorem 4.3 (a): *Some F-states exhibit entanglement;* **(b):** *Not all F-states exhibit entanglement.*

Proof (a): Yang (1962) has demonstrated that ODLRO is present only in density matrices of systems of bosons or fermion pairs forming bosonic quasiparticles (via pair occupation of single-particle states by fermions). Being composed entirely of bosons, Fröhlich systems must have wave functions $\Psi(1, 2, ..., N)$ that are symmetric under exchange of particle labels. Whenever the particles are not in the same single-particle state, Φ, the overall system state will exhibit entanglement because this symmetry requirement yields a state of the system that is unfactorizable, i.e. $\Psi(1, 2, ..., N) \neq \Phi(1)\Phi(2)...\Phi(N)$ where $\Phi(i)$ are wavefunctions describing individual particles. For example, if one particle is in a state $\Xi \neq \Phi$, the state of the other $N - 1$ particles, Ψ is be a sum of N different terms each containing one factor $\Xi(i)$, for exactly one value of $i \in Z_n^+$, and $N - 1$ factors $\Phi(j)$, $j \neq i$.

Proof (b): F-states that have reduced density matrices for which $\chi(r', r'') = 0$, in which case all particles are in the same single-particle state, are not entangled. For them, $\rho_1(r', r'') = N\Phi(r')\Phi^*(r'')$ and $\Psi(1, 2, ..., N) = \prod_{i=1}^{N} \Phi(i)$, where $\Phi(i)$ is the wavefunction attributed to the ith particle (for example, in the case of the ground state of the entire system, $\Psi = \phi_0(1)\phi_0(2)\phi_0(3)...\phi_0(N)$, so that for all i, $\Phi(i) = \phi_0(i)$, the single particle ground state). Exchanging particle labels in Ψ leaves the state unchanged in compliance with the symmetry requirement for bosons. From the definition of entanglement, such product states, which are F-states, are not entangled.

5. THE QUESTION OF REDUCTION

In the explanation of the coherent behavior of all the systems that we have discussed, our first criterion of reduction (fundamentalism) is easily satisfied. Thus, *none of our examples fail to satisfy the condition for the first sense* (a) *of reduction*. It is important to note that this is a non-trivial claim and should not be confused with the ontological question whether any new process exists at the higher level. Explanations from the more fundamental assumptions routinely involve making approximations or taking limits and this can be philosophically problematic. The relevant procedures might well be *contrived* in the sense that the particular way in which a limit is taken or an approximation made may be motivated specifically by the desired result (Sarkar 1996, 1998). In such a situation, the fundamentalist criterion is violated. It is an open question how often this happens in physics.

Turning now to strong reduction, the analysis in section 4 permits the following three conclusions to be drawn:

(i) *The existence of coherence does not always imply a failure of strong reduction for either classical or quantum systems.* The coherent motions of the (classical) coupled oscillators system can be simply explained by describing each oscillator separately and noting the correlations between their positions. The coherent features of the Rabi oscillations are explained by the coupling between the states, an interaction between parts of a hierarchy that can be spatially instantiated (though, as in most quantum systems, it cannot be easily and accurately visualized);

(ii) *Explanations of coherence in quantum systems that involve entangled states violate strong reductionism because they violate the second condition (hierarchical structure) and, ipso facto, also the third condition (spatial instantiation).* If a composite system is described by an entangled state, definite states in general cannot be attributed to its individual subsystems. As Schrödinger (1935) pointed out, in such a state the subsystems cannot be in general be individuated, and locutions such as "subsystem A" and "subsystem B" do not refer to any precise entity within the entire state. Consequently, often no hierarchical relation between the subsystems' states and that of the composite system exists, let alone is instantiated in physical space. This conclusion has some serious consequences. Consider, for example, a hydrogen atom which consists of an electron and a proton interacting with each other. In a *fully* quantum description of this atom, it would be represented by an entangled state. Therefore it cannot be represented as a hierarchical structure with identifiable individual states for the proton and electron. The situation is the same for atoms other than hydrogen (Jaeger 2000). What this means is that once entanglement becomes involved, the usual hierarchical picture of the composition of matter breaks down in quantum mechanical explanations.

(iii) *Such explanations invoking F-states can violate strong reduction.* This is trivially a consequence of the last point. Note, however, that by theorem 4.1b the F-state may exhibit no entanglement, in which case all three criteria for strong reduction can be satisfied.[19] However, composite systems that are represented by product states and, therefore, do not pose problems for (at least abstract) reductionist explanation can yet exhibit unusual properties, such as superfluidity and superconductivity: fluid motion with zero viscosity and electrical conduction with zero resistance, respectively. Superfluids or superconductors in their ground states are such systems.[20] Nevertheless, the point is that if there is a the real culprit is entanglement, not coherence. Unusual behavior is no guarantee that the hierarchy criterion is violated,

6. CONCLUSIONS

We only have two major conclusions. *First*, we have shown that the invocation of quantum-mechanical entanglement violates the conditions for strong reduction within physics though coherence alone is not sufficient for that purpose. Entanglement destroys the possibility of strong reduction because an entangled system cannot be described as being hierarchically organized. This does not, of course, imply that this is the only way in which strong reduction can fail in physics. Leggett (1987) and Fisher (1988), for instance, have argued that the fundamentalist assumption itself fails in some types of explanation in condensed matter physics. In our discussion of examples of types (b) and (d) of reduction we, too, have indicated that strong reduction fails in those explanations of critical phenomena in condensed matter physics that involve renormalization. Whatever be the merit of those claims, we hope that our analysis has drawn attention to the fact that not all the issues surrounding the role of reductionist explanation in physics have been resolved.

Second, physicalist explanations in molecular biology have so far satisfied the conditions for strong reduction (Sarkar 1989, 1992, 1996, 1998). However, if entangled F-states have to be invoked in such explanations, strong reduction will fail because of the failure of the hierarchy assumption. Whether F-states have any explanatory role in molecular biology remains highly controversial.[20] It suffices here simply to note that there is at present no plausible candidate for a biological mechanism involving an F-state. It remains an empirical question whether, eventually, such a mechanism will be discovered, and lead to a failure strong reduction in biology. Should that happen, what will perhaps be most ironic is that it will fail not because of any special property of biological systems (as anti-reductionists have usually held), but because of the nature of physics itself. Thus, if the reduction of biology to "fundamental" physics is a problem, then what is problematic is not the relation of biology to physics, but the relation of macroscopic physics to "fundamental," that is microscopic physics. The full oddness of this situation seems not to have been fully appreciated, despite its having been occasionally pointed out at times before (e.g. Shimony 1978).

Boston University
University of Texas at Austin

ACKNOWLEDGEMENTS

We would like to thank Abner Shimony for initiating our interest in the problems of reductionism in quantum mechanics. Besides Abner, we would like to thank Tian Yu Cao, Don Howard, Simon Saunders, John Stachel and Chuck Willis for comments and criticism of an earlier version of this paper. This research was partly supported by an NIH grant (No. 7 ROl HG00912-02) to SS.

NOTES

1. See Wimsatt (1979), Sarkar (1992, 1998), and Schaffner (1994) for reviews.
2. See, for example, Nagel (1961). However, Specter (1978) and Shimony (1987) are notable exceptions.
3. See Leggett (1987, 111-143).
4. See also Ho and Popp (1993).
5. For models of supervenience see, e.g., (Davidson 1970; Rosenberg 1984) and, especially, Kim (1993).
6. A detailed discussion of ontological issues can be found in Wimsatt (1995). Arguments against the philosophical importance of these issues are developed by Sarkar (1998).
7. This analysis is not straightforwardly compatible with models of reduction which do not view it as a form of explanation — e.g. Balzer and Dawe (1986, 1986a) and Ramsey (1995). However, it is unclear that these models should at all be regarded as those of "reduction" in any usual philosophical sense, given that this term has almost always been used to refer to a particular mode of explanation.
8. The approach that is summarized here is developed in Sarkar (1998). That work includes a discussion of the epistemological problems posed by approximations, which is glossed over in this paper.
9. The provision of a "warrant" may be weaker than logical deduction or even derivation. A warrant can potentially be a theoretical claim that is made on the basis of experimental facts known at the reducing level (see Sarkar 1989). Note that the reducing level need not be a lower level of any hierarchy as, e.g., in the reduction of Newtonian gravitation to general relativity or in the reduction of Newtonian mechanics to special relativity.
10. The simplest kind of such a hierarchical structure is a (graph-theoretical) tree. More complicated structures would only require a directed graph with an identifiable root and no cycles.
11. That one must distinguish between reductions only satisfying (i) and those that satisfy the other criteria was pointed out by Nickles (1973) though, in his usage, the reductions occurred in the opposite direction than the one discussed here (Wimsatt 1976).
12. For a particularly clear exposition, see Binney, Downick, Fisher and Newman (1993). Renormalization theory in condensed matter physics has not received nearly the kind of philosophical attention that it deserves. (For discussions of renormalization theory in the context of quantum field theory see (Brown 1993).)
13. Two quantum states of the system are said to be "coupled" if the system is subject to a force, with corresponding energy (here W_{12}), capable of causing the system to change from one state to the other.
14. See (Yang 1962; Penrose and Onsager 1956; Penrose 1951; and Ginzburg and Landau 1950).
15. Ho (1993, 141) admits that "[c]oherence in ordinary language means correlation, a sticking together, or connectedness; also, a consistency in the system." He eschews any attempt at explicit definition and later (pp. 150-151) only provides two jointly sufficient conditions for coherence, but even these are restricted to a quantum context.
16. We mean "feature" quite generally in the sense that even possible states of a system will be regarded as "features."
17. For instance, a classical system consisting of a free particle with mass m, energy E, and momentum $\vec{p}$ is coherent because E can be estimated from $\vec{p}\left(E = \frac{\vec{p}\cdot\vec{p}}{2m}\right)$ though $\vec{p}$ cannot be estimated from E. Therefore this system will not be fully coherent.
18. See, for example, the proof of Gisin (1991). See also (Capasso et al. 1973; Werner [unpublished, cited in Popescu and Rohrlich 1992]; Gisin and Peres 1992; and Popescu and Rohrlich 1992).
19. This does not, of course, mean that such explanations necessarily satisfy the criteria of strong reduction — that will depend on whether the other explanatory factors involved also satisfy these criteria.
20. T. Y. Cao (personal communication) has argued that the proper response to this situation is to abandon the characterization of "hierarchy" that we use, and adopt a less restrictive one and thus save the usual hierarchical picture of matter. We find this unpalatable for two reasons: (i) the notion of hierarchy we use is the standard one appropriate not only for biology and the social sciences, but also for classical physics. It would be odd philosophical strategy to weaken a generally useful notion simply to include

one special case; (ii) such a strategy would prevent a recognition of yet another way in which quantum mechanics undermines our classical intuitions and this is precisely what we think is of most interest.
21. See, for example, (Cooper 1978; Yushina 1982; Fröhlich 1983; Mishra and Bhowmik 1983; Ho and Popp 1993).

REFERENCES

Balzer, W. and C. M. Dawe, 1986. "Structure and Comparison of Genetic Theories (1): Character-factor Genetics." *British Journal for the Philosophy of Science* 37:55-69.
———. 1986a. "Structure and Comparison of Genetic Theories (2): The Reduction of Character-factor Genetics to Molecular Genetics." *British Journal for the Philosophy of Science* 37: 117-191.
Binney, J. J., N. J. Downick, A. J. Fisher, and M. E. J. Newman. 1993. *The Theory of Critical Phenomena: An Introduction to the Renormalisation Group*. Oxford: Oxford University Press.
Brown, L., ed. 1993. *Renormalization*. Berlin: Springer-Verlag.
Capasso, V., D. Fortunato, and F. Selleri. 1973. *International Journal of Modern Physics* 7:319.
Cooper, M. S. 1978. "Long-Range Coherence and Cancer." *Physics Letters* A 65:71-73.
Davidson, D. 1970. "Mental Events." Pp. 79-101 in *Experience and Theory*, eds. S. Foster et al. Amherst: University of Massachussetts.
Ehlers, J. 1981. "Über den Newtonschen Grenzwert der Einsteinschen Gravitationstheorie." Pp. 65-84 in *Grundlagen probleme der modernen Physik*, eds. J. Nitsch, J. Pfarr, and E. W. Stachow. Mannheim: Bibliographisches Institut.
Fisher, M. E. 1988. "Condensed Matter Physics: Does Quantum Mechanics Matter?" Pp. 65-115 in *Niels Bohr: Physics and the World*, eds. H. Feshbach, T Matsui, and A. Oleson. Chur: Harwood.
Fröhlich, H. 1968. "Long-Range Coherence and Energy Storage in Biological Systems." *International Journal of Quantum Chemistry* 2:641-649.
———. 1969. "Quantum Mechanical Concepts in Biology." Pp. 13-22 in *Theoretical Physics and Biology*, ed. M. Marois. Amsterdam: North-Holland.
———. 1973. "Life as a Collective Phenomenon." Pp. vii-xii in *Cooperative Phenomena*, eds. H. Haken, and M. Wagner. New York: Springer.
———. 1975. "Evidence for Bose Condensation-like Excitation of Coherent Modes in Biological Systems." *Physics Letters* A51:21-22.
———. 1977. "Long-Range Coherence in Biological Systems." *Rivista del Nuovo Cimento* 7:399-418.
———. 1983. "Evidence for Coherent Excitation in Biological Systems." *International Journal of Quantum Chemistry* 23:1589-1595.
Ginzburg, V. L. and L. D. Landau. 1950. "On the Theory of Superconductors." *Zhurnal Eksperimental'noi i Teorechiskoi Fiziki* 20:1064-1082.
Gisin, N. 1991. "Bell's Inequality Holds for All Non-Product States." *Physics Letters* A 154:201-202.
Gisin, N. and Peres, A. 1992. "Maximal Violations of Bell's Inequalities for Arbitrarily Large Spin." *Physics Letters* A 162:15-17.
Glauber, R. J. 1963. "The Quantum Theory of Optical Coherence." *Physical Review* 130:2529-2539.
Ho, M. W. 1989. "Coherent Excitations and the Physical Foundations of Life." Pp. 162-176 in *Theoretical Biology*, eds. B. Goodwin, and P. Saunders. Baltimore: Johns Hopkins.
———. 1993. *The Rainbow and the Worm*. Singapore: World-Scientific.
Ho, M. W,. and F. A. Popp. 1993. "Biological Organization, Coherence, and Light Emission from Living Organisms." Pp. 183-213 in *Thinking About Biology*, eds. W. Stein and F. J. Varela. Reading: Addison-Wesley.
Hopkins, H. H. 1951. "The Concept of Partial Coherence in Optics." *Proceedings of the Royal Society Series* A 208:263-277.
Jaeger, G. 2000. *The Reduction of Structural Chemistry to Quantum Mechanics*. Unpublished manuscript.
Kauffman, S. A. 1971. "Articulation of Parts Explanation in Biology and the Rational Search for Them." *Boston Studies in the Philosophy of Science* 8:257-72.
Kemeny, J. and P. Oppenheim. 1956. "On Reduction." *Philosophical Studies* 7:6-19.

Kim, J. 1993. *Supervenience and Mind: Selected Philosophical Essays*. New York: Cambridge University Press.
Klauder, J. R. and E. C. G. Sudarshan. 1968. *Foundations of Quantum Optics*. New York: Benjamin.
Leggett, A. J. 1987. *The Problems of Physics*. Oxford: Oxford University Press.
Malament, D. 1986. "Newtonian Gravity, Limits, and the Geometry of Space." Pp. 181-201 in *From Quarks to Quasars*, ed. R. G. Colodny. Pittsburgh: University of Pittsburgh Press.
Mandel, L. and E. Wolf. 1970. *Selected Papers on Coherence and Fluctuations of Light* (2 vols.) New York: Dover.
———. 1995. *Optical Coherence and Quantum Optics*. Cambridge: Cambridge University Press.
Mayr, E. 1982. *The Growth of Biological Thought*. Cambridge: Harvard University Press.
Mishra, R. K. and K. Bhowmik, "Theory of Living State. VII. Bose-Einstein-Like Ordering in Temperature and Time Domain." *International Journal of Quantum Chemistry* 23:1579-87.
Nagel, E. 1949. "The Meaning of Reduction in the Natural Sciences." Pp. 99-135 in *Science and Civilization*, ed. R. C. Stauffer. Madison: University of Wisconsin Press.
———. 1961. *The Structure of Science*. New York: Harcourt, Brace and World.
Nickles, T. 1973. "Two Concepts of Inter-theoretic Reduction." *Journal of Philosophy* 70:181-201.
Pauling, L. 1960. *The Nature of the Chemical Bond* [3rd ed.]. Ithaca: Cornell University Press.
Penrose, O. 1951. "On the Quantum Mechanics of Helium II." *Philosophical Magazine* 42:1373-1377.
Penrose, O. and L. Onsager. 1956. "Bose-Einstein Condensation and Liquid Helium." *Physical Review* 104:576-584.
Pippard, A. B. 1953. "The Coherence Concept in Superconductivity." *Physica* 19: 765-774.
Pokorny, J. 1982. "Multiple Fröhlich Computer States in Biological Systems: Computer Simulation." *Journal of Theoretical Biology* 98:21-27.
Popescu, S. and D. Rohrlich 1992. "Generic Quantum Nonlocality." *Physics Letters* A 166:293-297.
Ramsey, J. L. 1995. "Reduction by Construction." *Philosophy of Science* 62:1-20.
Rosenberg, A. 1984. "The Supervenience of Biological Concepts." Pp. 99-115 in *Conceptual Issues in Evolutionary Biology*, ed. E. Sober. Cambridge: MIT Press.
Ruelle, D. 1989. *Statistical Mechanics*. Redwood City, CA: Addison-Wesley.
Sarkar, S. 1989. "Reductionism in Molecular Biology: A Reappraisal." Ph.D. Dissertation, University of Chicago.
———. 1992. "Models of Reduction and Categories of Reductionism." *Synthese* 91:167-194.
———. 1996. "Biological Information: A Skeptical Look at Some Central Dogmas of Molecular Biology." *Boston Studies in the Philosophy of Science* 183:187-232.
———. 1998. *Genetics and Reductionism*. Cambridge: Cambridge University Press.
Schaffner, K. 1967. "Approaches to Reduction." *Philosophy of Science* 34:137-47.
———. 1994. *Discovery and Explanation in the Biomedical Sciences*. Chicago: University of Chicago Press.
Schrödinger, E. 1935. "Discussion of Probability Relations Between Separated Systems." *Proceedings of the Cambridge Philosophical Society* 31:555.
Senitzky, I. R. 1962. "Incoherence, Quantum Fluctuations, and Noise." *Physical Review* 128:2864-2870.
Shimony, A. 1987. "Metaphysical Problems in the Foundations of Quantum Mechanics." *International Philosophical Quarterly* 18:3-17.
———. 1987. "The Methodology of Synthesis: Parts and Wholes in Low-Energy Physics." Pp. 399-423 in *Kelvin's Baltimore Lectures and Modern Theoretical Physics*, eds. R. Kargon, and P. Achinstein. Cambridge: MIT Press.
———. 1989. "Conceptual Foundations of Quantum Mechanics." Pp. 373-395 in *The New Physics*, ed. P. Davies. Cambridge: Cambridge University Press.
Specter, M. 1978. *Concepts of Reduction in the Physical Sciences*. Philadelphia: Temple University Press.
Suppes, P. 1957. *Introduction to Logic*. New York: Van Nostrand.
Thompson, C. J. 1972. *Mathematical Statistical Mechanics*. New York: Macmillan.
Werner, R. F. 1989. "Quantum States with Einstein-Podolosky-Rosen Correlations Admitting a Hidden-Variable Model." *Physical Review* A 40:4277.
Wimsatt, W. C. 1976. "Reductive Explanation: A Functional Account." *Boston Studies in the Philosophy of Science* 32:671-710.

———. 1979. "Reduction and Reductionism." Pp. 352-377 in *Current Research in the Philosophy of Science*, eds. P. D. Asquith and H. Kyburg. East Lansing: Philosophy of Science Association.

———. 1995. "The Ontology of Complex Systems: Levels of Organization, Perspectives, and Causal Thickets." *Canadian Journal of Philosophy.* 20:207-274.

Woodger, J. H. 1952. *Biology and Language*. Cambridge: Cambridge University Press.

Yang, C. N. 1962. "Concept of Off-Diagonal Long-Range Order and the Quantum Phases of Liquid He and of Superconductors." *Reviews of Modern Physics* 34:694-704.

Yushina, M. Y. 1982. "Derivation of Kinetic Equation in Fröhlich Anharmonic Model." *Physics Letters* 91A:372-374.

IV

Science, History, and the Challenges of Progress

ALLEN I. JANIS

PHYSICS AND SCIENCE FICTION

1. INTRODUCTION AND DEDICATION

John Stachel is a person with a wide range of interests and a broad sense of humor. In thinking about a suitable contribution for this volume, I recalled that about a decade ago I was asked to contribute an essay (Janis 1989) on astronomy and science fiction to a volume in memory of the astronomer M.K.V. Bappu. (This invitation came because, for many years, I taught a course for non-science students in which I used science fiction as the setting for teaching many aspects of physics and astronomy.) It occurred to me that an essay on physics and science fiction, although undoubtedly rather different from the usual sort of *festschrift* contribution, might be a pleasant change of pace. My earlier essay on astronomy and science fiction assumed little, if any, scientific background. For the present volume, given its more technical but still interdisciplinary character, I have tried to write at a level that assumes some technical background but is still comprehensible to non-specialists. At any rate, it has been fun for me to write this essay, and I hope that John, as well as other readers, will have fun reading it.

Since a comprehensive treatment of physics and science fiction would fill (at least) an entire book, it is necessary to limit the scope of this essay. After some brief remarks (in section 2) about the different ways in which science appears in science fiction, I shall (in section 3) describe the use (and sometimes misuse) of physics in a small selection of science fiction stories; they are all stories that I have used in teaching. Finally (in section 4) I shall make some further comments on the use of science fiction as a teaching tool.

It is a pleasure for me to offer this essay for John Stachel's enjoyment as part of this celebration of his 70th birthday.

2. THE SCIENCE IN SCIENCE FICTION

In terms of their use of science, science fiction stories can differ widely. The stories that are usually the most fun for scientists to read are those that use sound scientific ideas in imaginative ways. Sometimes known scientific principles will be deliberately ignored in order to make a story possible (for example, by postulating various ways of traveling faster than light), but the other aspects of the story will strictly respect the known laws of nature. Sometimes a story that is almost entirely good science will slip up at some point and allow an error to slip in. Other stories are so full of errors that one can believe that they are called science fiction only because the science in them is fictitious. Even such a story, within a classroom setting, can sometimes be a useful way to introduce students to scientific ideas.

Some writers of science fiction are excellent scientists themselves. The best of such authors will sometimes use well established scientific principles in surprising and delightfully imaginative ways, and sometimes go beyond the bounds of known science to speculate in ways that nevertheless have sound underpinnings. Some have even been known to put in a plug for a favorite, but controversial, theory (see the next section).

Science fiction, of course, is not written solely as a showcase for scientific ideas. Even those stories with the best use of science have, like all literature, a variety of aims, some light-hearted (e.g., humor), others serious (e.g., social or political commentary). Some of my comments on specific stories will include remarks along these lines.

3. SOME STORIES (AND SOME ERRORS)

Larry Niven's short story *Neutron Star* (1966) contains much good science. In brief, the protagonist, Beowulf Schaeffer (a character appearing in a number of Niven's stories, and one whose personal characteristics suggest that his initials may not have been chosen entirely at random), has been blackmailed into piloting a spaceship to the vicinity of a neutron star in order to discover what killed Sonya and Peter Laskin, who had piloted a similar ship there. The makers of the ship had a reputation to maintain: their ships were guaranteed to be impervious to everything except visible light, but something had wrecked much of the ship's interior and reduced the Laskins to a collection of bloody smears. Schaeffer discovers the answer: intense tidal forces near the neutron star ripped things apart, throwing virtually everything inside the ship into one end or the other with great violence. Schaeffer escapes with his life by staying at the center of mass.

Niven correctly describes many aspects of the physics of strong gravitational fields: gravitational bending of light, including Einstein rings; the gravitational frequency shift, which in the case of Schaeffer observing distant stars from the vicinity of a neutron star is a blue shift; and, of course, strong tidal effects, including the tidal locking of the ship into a position with its axis pointing toward the star. However, Schaeffer would not have survived the trip.

I have been told that Niven later found a problem with the story. Since the tidal forces caused the ship's axis to line up with the neutron star, the ship's rapid passage by the star would have imparted such a high angular momentum to the ship that it would not have been safe for Schaeffer to leave the center of mass—had he done so, the rapid spin would have smashed him into one of the ends of the ship. But there is another calculation that Niven apparently never did.

Let us try to estimate the tidal gradients that Schaeffer encountered. According to data given in the story, the star had a mass of 1.3 solar masses and a diameter of about 12 miles, and the ship's closest approach to the surface was about one mile. Since this means that the ship came within 3 Schwarzschild radii of the star's center, the effects of general relativity would certainly not be negligible. Nevertheless, just to get a very rough approximation, let us see what Newtonian physics would say.

From the Newtonian expression GM/r^2 for the strength of the gravitational field (or, equivalently, the acceleration of a freely falling test mass) produced by a star of mass M at a distance r from its center, it follows that the difference in fields at two points Δr apart (where Δr is small compared to r) has the magnitude $(2GM/r^3)\Delta r$. If we evaluate this at periastron and express the answer in terms of g's (one g being the acceleration of gravity at the Earth's surface), we get approximately $2.5 \times 10^7 \times \Delta r$ g's, where Δr is measured in meters. This means that if Schaeffer could curl up into a diameter of one meter, the difference in the gravitational field strength across his body would be 25 million g's (curling up a bit smaller would not help much). Even allowing for the poor approximation, it seems likely that tidal forces would have ripped Schaeffer apart.[1]

The title 'character' in Fred Hoyle's novel *The Black Cloud* (1957) is a large (big enough to enclose the Sun and cause catastrophic cooling on Earth) cloud possessing superhuman intelligence. This is certainly one of the most creative conceptions of extraterrestrial life to be found. Hoyle goes into some detail as to how the cloud might be organized. Basically, different parts of the cloud are linked by radio waves into a single neurological entity. Electrical discharges in its outer parts cause sufficient ionization to prevent disruption by unwanted external signals, much as our skulls protect our brains. An electromagnetic pump maintains a flow of gas that acts analogously to a blood supply and passes through a filter that acts analogously to kidneys. Overall, the cloud controls vast amounts of energy, but sometimes needs to be recharged by absorbing energy from a star; hence its excursion into our solar system. When communication is established between the cloud and a group of terrestrial scientists, the cloud expresses surprise (p. 149):

> for it is most unusual to find animals with technical skills inhabiting planets, which are in the nature of extreme outposts of life.

Hoyle's treatment of the effects of the cloud's blocking of the Sun makes good use of thermodynamical and statistical mechanical concepts. Much of what he foresees reminds one of the discussions of 'nuclear winter' that came much later, but the 'cloud's winter' was even more extreme. The book also makes good use of a number of other concepts of physics and astronomy, making this an enjoyable novel for readers that like science fiction containing lots of good science imbedded in a good story. One of Hoyle's predictions in this novel, however, did not turn out to be right.

Prior to close-range studies of the Moon's surface, there was considerable debate as to whether the Moon was covered by dust. The arguments were fueled by observations of the Moon's rate of cooling during lunar eclipses. There was even speculation that the dust would be deep enough to engulf any ship that landed on the Moon. Hoyle's views on the subject are made clear in *The Black Cloud* (p. 105):

> The existence of vast drifts of dust on the Moon was confirmed in dramatic fashion.

As the cloud approached the Sun, it slowed down by ejecting gas at high speed.

Some of this gas hit the Moon, dramatically changing its appearance as seen through a telescope. Hoyle has one of the astronomers explain (p. 106):

> Those dark areas are gigantic drifts of dust, drifts perhaps two or three miles deep. What is happening is that the high speed gas is causing the dust to be squirted hundreds of miles upwards from the surface of the Moon.

Hoyle also includes a plug for his favored, steady-state cosmology. This comes during a discussion between a human and the cloud about the cloud's reproductive abilities. The cloud explains (p. 156):

> If I, for instance, were to find a suitable cloud not already endowed with life I would plant a comparatively simple neurological structure within it. This would be a structure that I myself had built, a part of myself.

The cloud further explains that it would endow its 'infant' with an electromagnetic screen to prevent radioactive materials from penetrating its neurological regions. It continues:

> The point of this example is that we can provide our 'infants' both with screens and with the intelligence to operate them, whereas it would be most improbable that such screens would develop in the course of a spontaneous origin of life.

When the human interposes:

> But it must have happened when the first member of your species arose

...the cloud replies...

> I would not agree that there ever was a 'first' member.

At this point, two of the astronomers listening in

> ...exchanged a glance as if to say: "Oh-ho, there we go. That's one in the eyes for the exploding-universe boys."

A short story that makes clever use of standard physics is David Brin's *Tank-Farm Dynamo* (1983). The tank farm of the title consists of two parallel flat 'decks' connected by six very long, parallel cables. The decks are composed of rows of giant cylinders covered by aluminum plating on the sides facing each other. The farm is in Earth orbit, and tidal effects create a small amount of 'artificial gravity' at each end, directed away from the center of mass (as in *Neutron Star*, but of course smaller by an enormous factor). An elevator connects the decks, and the apparent gravitational field is enough to allow farming to take place in the tanks and to allow the decks to be used for spacecraft landings and takeoffs. In fact, the farm is used to give a boost to spacecraft that land on the deck closer to Earth, are carried to the farther deck, and then slip off into higher orbits. There is a small engine to enable the farm to counteract the inevitable orbital decay. The tanks that make up the ever growing farm are cast-off fuel tanks from spacecraft. The residual fuel in these tanks—oxygen and hydrogen, water in effect—provides life support and fuel for the farm's engine. And now the government is threatening to cut back severely on this supply of water. While a delegation from Earth is at the farm to deliver an ultimatum, the farm's scientists come up with the solution to their problem. They use solar power to force a current through the tethering cables in the direction that, in the Earth's magnetic field, will feed energy into their orbit. The bad guys, who wanted to drive the farm out of business, are left sputtering.

Earlier in the story, the farm's scientists (or, more accurately, Brin, who has a Ph.D. in Applied Physics and Space Science) realized that although their motion through the Earth's magnetic field would generate an electromotive force across the cables, they could not draw power from this without further degrading their orbit. This follows from Lenz's Law, which says that the induced current would flow in the direction that would cause magnetic drag. Whether this was understood in real life by the relevant scientists at the American space agency, NASA, has been questioned. As described by Robert Park (1998, 5):

> [I]n 1992, NASA attempted to deploy a small satellite from the shuttle Atlantis tethered by a 20-km wire. The plan was that the conductor moving through Earth's magnetic field would generate electric power for the spacecraft. The mission manager described this as 'a free lunch'.

One of the tank farm's scientists comments when a similar scheme is proposed in the story (Brin 1983, 111):

> You know there ain't no such thing as a free lunch.

After mentioning Lenz's Law, Park continues:

> To maintain its orbit the spacecraft would have to fire its rockets. In effect, the electricity would be generated by the rockets—and not very efficiently. In any case, the reel jammed...Incredibly, NASA tried to refly the $1B mission four years later. This time the tether broke. Fortunately, NASA seems to have given up.

However, if NASA's intention had been to convert orbital energy to electrical energy, as suggested by Geoffrey Landis (1999, 2) in a reply to Park, then the criticism would be unjustified.

Tank-Farm Dynamo reminds me of Isaac Asimov's story *The Martian Way* (1952), in which a demagogue on Earth stirs up the people to support cutting off the supply of water to a colony on Mars. This, too, is a story making good use of physics; conservation of momentum, for example, is clearly explained and illustrated. In this story, a heroic expedition to bring ice back from the rings of Saturn saves the people on Mars and leaves the demagogue looking foolish. Asimov (1973, 11) has explained that this story represented his reaction to the McCarthy era in American politics.

Ross Rocklynne's story *The Men and the Mirror* (1938) is an interstellar cop-and-robber chase. It is one of a series in which the officer and the outlaw (the cleverer of the two) find that the chase has put them both in a dangerous situation from which they ultimately escape by clever use of the laws of physics. In *The Men and the Mirror*, they have landed on a planet in which is embedded a huge concave mirror, approximately a thousand miles in diameter with a depth of about three hundred miles, and they have, unfortunately, fallen into it. The mirror's surface is almost frictionless, but 'almost' is part of their problem: There is just enough friction to keep them from rising back to the rim and effect their escape. Their problem is two-fold: not only how to get out but, if they succeed, to do so somewhere in the vicinity of their ships. The answer to the first part comes from use of the conservation laws for angular momentum and energy. The two men are connected by a rope. Having acquired some angular momentum about their common center of mass, they haul in on the rope, decreasing their moment of inertia about their center of mass. Conservation of angular momentum then dictates that their angular velocity must increase. They shorten their separation until they have acquired sufficient kinetic energy for escape. At a moment when they are at the apex of one of their oscillations across the mirror and when in the course of their rotation one of them is aimed directly at the closest point of the rim, they cut the rope. The one aimed at the closest rim shoots over it and out of the mirror. Conservation of energy guarantees that, if the one escapes in this fashion, the other will similarly escape on the other side (although, due to the planet's rotation, not directly opposite) provided that the kinetic energy of each when the rope is cut exceeds that needed for escape by at least the amount that each loses to friction. The solution to making one of them come out near their ships comes from an analysis of their oscillatory motion in relation to the rotation of the planet. Careful observations allow them to time their escape so that the second one out is near their ships. (There was good reason for making the second one out the one to be near the ships. The first one out would have lost less kinetic energy to friction since the cutting of the rope, and thus would be more likely to suffer injury when landing. In the story, the first one out breaks a leg, while the second one emerges unscathed and brings his ship to the rescue of the first.) Unfortunately, the solutions to both parts of the problem seem to be flawed.

The problem with the first part is the same conservation of angular momentum that the men used to escape from the mirror. The story seems to make clear that they had little, if any, angular momentum to begin with. It is by flailing about that they build up their rotational motion. In the absence of friction, conservation of angular momentum would forbid this. It is true that the presence of friction is important in the story. It makes their rise less with each oscillation, and provides a time constraint: If they do not effect their escape soon enough, they will no longer rise high enough (with correspondingly low enough escape velocity from that height) to be able to build up escape velocity as they did. But the smallness of the friction seems incompatible with the rapidity with which they build up their angular momentum.

As for the timing of their escape, the analysis presented in the story has them moving like a plane pendulum while the planet rotates under them (the axis of the mirror coincides with the rotational axis of the planet); in effect, they are a Foucault pendulum at one of the planet's poles. However, when they fell into the mirror, they were at rest with respect to the planet (at a point off the rotational axis, of course, since they were at the mirror's rim), not at rest in the non-rotating frame of reference. Their motion would not be that of a plane pendulum.

Arthur C. Clarke's novel *Rendezvous with Rama* (1973) is a good story with lots of good physics. Rama is an enormous, mostly hollow cylinder, 50 km long and 20 km in diameter, rotating at just the right speed to produce an apparent gravitational field of about 1 g for someone standing on the inside surface. Rama has penetrated the solar system, and has been spotted by a monitoring program that was set up after the catastrophic collision of a thousand-ton meteorite with Earth in 2077; it is now

2131. Rama is clearly a manufactured object, the first indication of intelligence elsewhere in the Universe, but it shows no signs of activity. An exploratory mission is dispatched, which lands on one of the end faces and enters the cylinder along its axis. Although there are structures inside, mostly along the inside cylindrical surface, the explorers find no signs of activity nor sources of light (other than the ones they bring in themselves). But Rama warms up as it gets closer to the Sun, and things begin to happen. Lights come on, a circular band of ice going around the inside of the cylinder melts (this encircling body of water is called the Cylindrical Sea), and robotic creatures appear.

Rama provides an excellent setting for exploring the physics of rotating reference frames. The apparent gravitational field, already mentioned above, is one example of this. The terrestrial explorers find they can get 'down' to what they call the Raman 'plain' (in spite of its curvature) by means of huge staircases built into the end face where they entered. These staircases change gradually from ladder-like to stair-like as they descend and the apparent gravitational field increases. Coriolis force gets its share of attention also. As Rama warms up, atmospheric currents are created and are affected by Coriolis forces, ultimately resulting in hurricanes that cause the explorers to withdraw until a new equilibrium has been established. There is also a waterfall cascading from one of the end faces that, due to the Coriolis effect, lands many kilometers to the side of the point directly below its source.

Another interesting point of physics relates the heights of the cliffs on either side of the Cylindrical Sea to Rama's acceleration. On one side the cliff is only 50 m high, but on the other it is ten times that. Assuming that the cliffs are meant to keep the sea from sloshing over onto the plain, the larger cliff height and the width of the Sea are used to calculate the maximum acceleration that Rama can sustain (and, of course, its direction). At the end of the story, Rama is found to leave with an acceleration that is 75% of this calculated maximum.

There are many other examples of the good use of a variety of physical concepts in the story, and one intentional break from known physics: Rama's propulsion mechanism. At the end of the story, after the explorers have left Rama for good, they find their ship rolling in Rama's wake. They can tell they are rolling by looking at the stars, but they feel nothing (they are rolling fast enough that they would expect to feel it) and their instruments show no rotation. They conclude that only a powerful gravitational field, among the known fields, could produce such an effect. There is no evidence of any exhaust—Newton's third law seems not to have anything to do with Rama's acceleration.

Jerome Bixby's story *The Holes Around Mars* (1954) provides a stark contrast to *Rendezvous with Rama*, at least insofar as the correct use of physics is concerned. Asimov (1971, 235) introduces his own comments about this story with the words:

> Jerome Bixby, although an excellent piano player, is not an author usually associated with 'scientific' s.f. *The Holes Around Mars* is entertaining and fascinating, but it has many scientific flaws.

A terrestrial expedition to explore Mars discovers a series of holes, apparently all at the same altitude (although one of the scientific confusions in the story is the suggestion that the holes are both in a perfect circle around the center of Mars and in a straight line), in anything on the Martian surface that is higher than that altitude and lying along a 'straight line' (presumably, a great circle). By the end of the story, they discover that the holes have been formed by a hitherto unknown, very small (about four inches in diameter) moon, which is christened Bottomos by the expedition's leader, who has a penchant for puns (much to the disgust of his companions). How this moon could punch its way through, among other things, solid rock is not satisfactorily explained. Even stranger, doing so does not seem to affect its orbit, for it goes through the same holes on each passage. That it frequently encounters fresh obstacles is made clear in the story, for one of the holes discovered by the explorers is in a cactus-like plant that is still oozing from the impact; but Bottomos' path is seemingly unaltered by the encounter. At the end of the story, it is suggested that Bottomos initially was traveling fast enough to punch its way through such things as rocks, but that these encounters slowed it until now it can only penetrate softer materials—and this slowing apparently took place without any change of orbit! Another problem is that, since the orbit does not change relative to the Martian surface, it must be equatorial, but internal evidence in the story shows that at least some of the holes are not at the equator. Near the climax of the story, the explorers come upon a Martian village (this story was written before such a thing became totally unthinkable). Bottomos' orbit takes it down the main street of this village a couple of feet above the ground, so the inhabitants have a calendar posted by the street on which they keep track of the days when Bottomos will arrive; they apparently keep track of the actual time of arrival by checking the lengths of their shadows. But let us take a closer look at Bottomos' orbital period.

Kepler's third law tells us that, for objects orbiting the same central body, the squares of the periods are proportional to the cubes of the semi-major axes. Using orbital data from the known moons of Mars, Deimos and Phobos, and taking the radius of Bottomos' orbit to be approximately the equatorial radius of Mars, we find that Bottomos' period would be about 100 minutes. The Martian villagers would hardly need a calendar to keep track of Bottomos' arrival.

Having been reminded of puns by thinking about *The Holes Around Mars*, I shall conclude this section by describing Larry Niven's *very* short (fewer than 200 words) story *Unfinished Story #1* (1970), which is essentially nothing but a pun (but, still, the story contains some good physics). A certain warlock keeps his cave cool in the summer by using a demon who sits at the entrance and keeps the fast-moving molecules of air from entering and the slow-moving ones from leaving; the rest are allowed to pass. Naturally, the process is reversed in the winter. When a visiting sorcerer compliments the warlock on his ingenuity, the warlock is quick to place the credit where it belongs. The idea came from his clerk, Maxwell.[2]

4. SCIENCE FICTION AS A TEACHING TOOL

My experience in teaching our department's course in Physics and Science Fiction has shown me that science fiction can be used as an effective tool to teach the basic ideas of both classical and modern physics (I also included some basic ideas of astronomy and cosmology) to students with little background in science and mathematics. For the most part, the students seemed to learn a good bit and to enjoy doing so, and I found the course enjoyable to teach.

The science fiction readings, both short stories and novels, were chosen primarily because of their use (or misuse) of physics, although I certainly tried to choose stories that the students would enjoy reading. My lectures were about physics and included many demonstrations to illustrate the ideas, but I would also illustrate them by discussing how they fit into the assigned science fiction readings. I used *Unfinished Story #1*, for example, to illustrate the concept that the Kelvin temperature of a gas is proportional to the mean kinetic energy of the random translational motions of the molecules (and I explained the story's pun). The examinations tested their knowledge of physics, but I often worded the questions in ways that referred to the stories; doing so provided added motivation for the students to actually do the readings without actually testing them on things unrelated to physics.

Although the emphasis was on understanding the ideas, the course was not wholly qualitative. For example, the students had to understand quantitatively what it means for something to be proportional to the inverse square of the distance, and they had to use conservation laws quantitatively to find how one thing changes given the change of something else. They could calculate the gravitational acceleration on the surface of another planet knowing the factors by which its mass and radius differed from the Earth's, and by what factor the rotating men in *The Men and the Mirror* would increase their angular velocity given the factor by which they decreased their moment of inertia as they pulled in on the rope joining them. They could also use graphs of gravitational potential energy as a function of altitude to predict (quantitatively) the behavior of objects launched from various altitudes with various kinetic energies. Still, the main emphasis was on having the students achieve a clear, qualitative understanding of the basic concepts.

Perhaps some readers connected with academic institutions feel motivated to try teaching a course of this nature themselves. If so, and if they would like more information about the course that I taught, they should feel free to contact me.

University of Pittsburgh

NOTES

1. A more accurate calculation would use the equation of geodesic deviation, and would require knowing the star's angular momentum and further details of the ship's trajectory. Most neutron stars are rotating. Niven indicates that the one in the story does rotate, but he doesn't give any indication of how rapidly. An estimate using the equation of geodesic deviation, neglecting rotation and estimating the relevant trajectory data, indicates that tidal forces would still destroy Schaeffer.
2. For the sake of any reader unfamiliar with both Maxwell and his demon, let me point out that the full name of the person whose idea Niven borrowed was James Clerk Maxwell, and that he is sometimes referred to as Clerk Maxwell.

REFERENCES

Asimov, Isaac. 1952. *The Martian Way*. Reprinted 1973 in *The Best of Isaac Asimov*, Isaac Asimov.

———. ed. 1971. *Where Do We Go From Here?* Greenwich, Connecticut: Fawcett.

———. 1973. *The Best of Isaac Asimov*. New York: Fawcett Crest.

———. ed. 1974. *Before the Golden Age*. Garden City, New York: Doubleday.

Bixby, Jerome. 1954. *The Holes Around Mars*. Reprinted 1971 in *Where Do We Go From Here?* ed. Isaac Asimov.

Brin, D. 1983. "Tank-Farm Dynamo." *Analog: Science Fiction/Science Fact*. New York: Davis Publications, (November):104-118.

Clarke, A. 1973. *Rendezvous with Rama*. New York: Ballantine.

Hoyle, F. 1957. *The Black Cloud*. New York: New American Library.

Janis, A. 1989. "Astronomy and Science Fiction." Pp. 233-249 in *Cosmic Perspectives*, ed. S.K. Biswas, D.C.V. Mallik, and C.V. Vishveshwara. Cambridge: Cambridge University Press.

Landis, G. 1999. "Tethered Satellites as 'Voodoo Science'." *Physics & Society* 28/1:2, 14.

Niven, Larry. 1966. *Neutron Star*. Reprinted 1971 in *Where Do We Go From Here?* ed. I. Asimov.

———. 1970. *Unfinished Story #1*. Reprinted 1971 in *All the Myriad Ways*, L. Niven.

———. 1971. *All the Myriad Ways*, New York: Ballantine.

Park, R. 1998. "Voodoo Science: Perpetuum Mobile." *Physics & Society* 27/4:4-5.

Rocklynne, Ross. 1938. *The Men and the Mirror*. Reprinted 1974 in *Before the Golden Age*, ed. Isaac Asimov.

LASZLO TISZA

CAN WE LEARN FROM HISTORY? DO WE WANT TO?*

1. INTRODUCTION

Amicus Plato amicus Aristoteles magis amica veritas. Plato is a friend, Aristotle is a friend, but a greater friend is Truth. This "slogan" was found by R.S.Westfall in the earliest student notebook of Newton. Henry Guerlac (1981, chap.1) tracked down its history and found it quite common in the later Middle Ages and the Renaissance. With some variation it even goes back to antiquity.

This is a sobering rebuttal of our modern day hubris that critical evaluation of authority is an invention of the Scientific Revolution. I'll go one step further and suggest that Einstein is a friend, Bohr is a friend, but a greater friend is Truth. The young Einstein and the young Bohr made momentous discoveries for which they received the early recognition of the community, yet they could not resolve their lifelong debate on quantum mechanics (QM). However now, after all these years, and after they have passed from the scene, I like to picture them as waiting impatiently in purgatory that their discoveries in the public domain be freed from imperfections that thus far prevented the integration of their partial truths into the permanent acquisition. Let them grant the peace of mind to enter the Elysian Fields.

The scientific enterprise is a balancing act of opposite guidelines. The free creative imagination in the course of the heuristic process has usually flaws even when it is basically successful. These flaws can be pinpointed after the situation is clarified, and must be removed to ensure progress. Thus the heuristic foundation is only a temporary scaffolding that ceases to carry a logical weight as a more permanent foundation is taking shape; at this point it only hides the beauty of the façade. This contrasting appraisal of the scaffolding is possible only if we consider science as a two-stage process. In the heuristic stage the prime concern is to extend the frontier even at the price of contradiction and paradox. In the stage of consolidation, or rational reconstruction all paradoxes must be removed it appears that the cathedral stands even as the scaffolding is removed. The logical standards at the two stages are very different.

The characteristic feature of this century is the enormous success of imaginative invention. An important measure of consolidation was achieved as the transition from the old quantum theory to quantum mechanics (QM) eliminated the mathematical inconsistencies. Consolidation remained incomplete, however, since conceptual paradoxes were accepted as fixtures of epistemology. Paradoxes generate controversies, and this gave the opportunity for post-modern sociologists of science to suggest that on confronting tradition and innovation physicists are not different from everyone else.

Physicists reject this view, as they see no need to improve the way physics is currently practiced. This is the setting of the hotly debated "science war" of recent years. By pitting opinion against opinion, this war is currently fought by the rules of the sociologists. The purpose of this paper is to suggest, that the exact scientists ought to feature the credentials that sets them apart from other scholars, namely the combination of quantitative experiments with rigorous mathematics. This coordination is not unique, however, and calls for additional restrictions. It is usually required that the formalism be simple and beautiful. This rudimentary criterion was sufficient to generate the efficient cluster of classical theories which are as valid as ever when restricted to macroscopic phenomena. Unfortunately, "simplicity" and "beauty" blend objective logical and subjective psychological elements. Their separation is essential for handling the transition from macro- to microphysics. It is suggested in (Tisza 1997, 213), referred to as l.c., to use the contrasting terms *optimal* vs. *familiar*. I claim that paradoxes arise if the familiar is favored over the optimal. Let us illuminate this contrast in the context of QM.

The central paradox of QM is that its entities are both undulatory and corpuscular, attributes which in macrophysics are incompatible with each other. Since experiment cannot be internally inconsistent, it follows that we inadvertently change the meaning of the terms "wave" and "particle" as we cross over from macro- to microphysics. Could we purposefully create such flexible concepts beyond our familiar experience? Niels Bohr held that the only thing we can do is to use the familiar terms even if contradictory. He developed the ingenious philosophy of "complementarity" that justifies this usage and assures that we do not arrive at contradictory predictions with respect to measurable entities. This was clearly a useful scaffolding and continues to be pragmatically sufficient in atomic and nuclear physics. However, a logically sound definition for the "particle" would be preferable for understanding what we are doing.

The mathematicians handle abstract concepts with precision in terms of "definition by postulation" in the context of postulational or formal systems, which express precise truths about a limited domain of entities. They learned this only as the discovery of non-Euclidean geometry replaced the dogmatic view that Euclidean geometry is the universe of discourse for all of mathematics.

The purpose of this paper is to show that in a late stage of consolidation such a use of formal systems can be adapted to the empirical character of physics. This proves to be an eminently labor-saving device that lets us review the turning points of classical physics in a short paper. While mathematical disciplines may be handled independently, disciplines of physics must be mutually consistent because they deal with different aspects of the same reality. Einstein used this idea in his foundation of special relativity (SR). His program of bringing quantum theory to the same level of logical standards is becoming ready for realization.

2. FITTING FORMAL SYSTEMS FOR EMPIRICAL USE

A. The Euclidean legacy

Euclid has a unique role in the history of science. He is credited with the first instance of the axiomatic method presented in terms of a deductive system, a device in active use in all of the exact sciences after twenty three centuries. This is satisfactory only to the extent that the Platonic idealistic features of the original Euclid have been tempered to accommodate empirical elements. Here is a brief score card of this modernization process in physics and in mathematics. These efforts had their independent origins and were not coordinated with sufficient care. This is unfortunate, since the deductive system in the exact sciences is the connective tissue between mathematics and experience.

B. Newton invokes Euclid to account for motion

As a student in Cambridge Newton was exposed to a heavy dose of Aristotelian and late Peripatetic doctrine. According to Aristotle the intelligent man does not apply mathematics where it does not belong. This was a discreet stab at his master Plato for whom reliable knowledge was anchored in mathematics (geometry at the time) with insufficient attention to experience. The "slogan" quoted at the beginning of this paper suggests that Newton may have decided to overcome the blind spots of both authorities yet learn from both. It seemed indeed a good way to revitalize the stale blend of late Peripatetic empiricism and logic chopping with an injection of visionary Platonism, particularly Euclidean geometry.

When Newton decided to use the Euclidean method in natural philosophy, he had to clean it from the Platonic antiempirical bias that discouraged such an application. This adjustment would have called for several major steps, of which Newton performed only the first. He abrogated the Platonic thesis that knowledge is created only from *a priori* mathematical insight to be imposed on common experience. The construction of a deductive system is a rational effort involving imagination. Yet there is a subtle distinction: Is the intuition generating the primitive concepts entirely nonempirical as presented by Euclid in the spirit of Plato, or does it start by recognizing the potential of a hidden mathematical structure in a well-chosen range of experience? Newton's genius was to demonstrate the latter alternative in terms of an actual construction. He expected (see his Preface to the first edition of the *Principia*) that a rational mechanics would address "motion" with the same precision as geometry handles the size and shape of idealized objects. He started from observational facts and got rid of their fuzziness by abstraction. I believe that this alternative to arrive at the basis of a formal system may be what Einstein meant by "theory of principle" (TOP), see (Pais 1982, 27, 31) also what Abner Shimony means by "experimental metaphysics."

Newton did not prove this thesis in all generality, but supported it by the foundation of a unified mechanics for planets and projectiles. He was bridging a gulf which millennial tradition saw as unbridgeable. The result was duly celebrated but the expansion to new areas of experience was obscured by ambiguities.

The concept of "motion" has many aspects and the field had to be narrowed to make it amenable to the then available mathematics. Newton did not spell out his simplifying steps, but just as David Hilbert identified the tacit assumptions of Euclid, we ought to do likewise for Newton.

The types of motion can be classified into (i) rigid translations, (ii) spinning and (iii) undulation. By confining himself to class (i) Newton achieved a huge simplification: an adequate representation of the translation of a rigid body is the concentration if its mass in the center of mass. The resulting orbit of a mass point can be associated with a curve traced by an Euclidean point.

The next step was to make this simple model rich enough for significant applications. A key idea was to consider nonuniform translation whereby nonuniformity, i.e. acceleration, was associated with the newly created concept *force*. Specifically, universal gravitation proved to be an extraordinary discovery. It was also easy to arrive from Euclidean space at the vector spaces of displacement, of velocity and of momentum. Since the momentum vector can change either its direction or its magnitude, the unification of celestial central motion and terrestrial free fall could be achieved in phase space with the stroke of the pen. (While the term "phase space" is of later origin, the concept is implicit in Newtonian mechanics.) All this was within the context of translational motion. Such a confinement of the argument was justified by short-term results, but having passed over in silence spinning and undulation was going to cloud the long-term prospect.

C. Analytical mechanics, the upgrading of the geometrical language

The first major development beyond the *Principia* was the work of mathematicians who improved the mathematics to fit the Newtonian model. The scene shifted to the Continent where the mathematicians discovered in Newton's model the proving ground on which to hone the evolving infinitesimal calculus. Within a few years Newton's geometrical language was replaced by infinitesimal analysis. Lagrange boasted that his *Mécanique Analytique* contained not a single figure. Back to the British Isles the Hamiltonian form of Newtonian mechanics marked a measure of perfection that was acknowledged by the "canonical" designation. This was an overstatement leading to the expectation that the joint use of position and momentum must have a universal significance although the method was established only for translation.

It would have been more proper to say that Hamiltonian mechanics is the *optimal* formalization of the Newtonian model. Whereas Newton's geometrical language stemmed from his subjective experience, the Hamiltonian and Hamilton-Jacobian methods marked an objective improvement of the formalization of the model. I suggest that "optimal" is a more precise formulation of what is usually called the "beauty" of a theory. Note that "optimal" is model dependent and does not foreclose the use of other models as suggested by the dogmatic "canonical" designation. Gauging the scope and limit of the model is an entirely different question that involves empirical adequacy. For an illustration we turn to Newton's *Optics*.

D. Newton's Optics and the tortuous history of the photon

The *Optics* has been justly celebrated for the deep insights Newton extracted from his own experiments. Whereas his corpuscular conception fell in disfavor in the middle of the 19th century, it was revived after the emergence of the light-quantum hypothesis. Newton was credited by an anticipation of wave-particle duality. Yet at the same time he planted the seeds for our paradoxical view of this experimental phenomenon. In fact, surveying the history of the photon from the vision of hindsight, we note a curious blend of deep insight with prejudice. Clarification calls for nothing more than the purging of logical contradictions. The following quotations are culled from the queries in the Dover reprint of the Fourth Edition.

> Qu. 29. Are not the Rays of Light very small Bodies emitted from shining Substances?
>
> Qu. 26. Have not the rays of Light several sides, endued with several original Properties.
>
> Qu. 30. Are not gross Bodies and Light convertible into one another? ...
> The changing of Bodies into Light, and Light into Bodies, is very conformable to the Course of Nature, which seems delighted with Transmutations.
>
> Qu. 13. Do not several sorts of Rays make Vibrations of several bignesses, which according to their bignesses excite Sensations of several Colours, much after the manner that the Vibrations of the Air, according to their several bignesses excite Sensations of several Sounds?

Is the "small body" in Qu. 29 identical to the point mass of Newtonian mechanics? This is a key question that Newton does not address explicitly but he does *implicitly* identify the two concepts. The concluding Propositions XCV – XCVIII in BOOK I of the *Principia* use the mechanical context to derive the Snell – Descartes sine law of refraction. In addition it followed also that light velocity in a medium is larger than in vacuum.

When experiment decided otherwise, the corpuscular theory was rejected. Later, in 1905 and 1909 Einstein revived the theory, without explaining whether and how the earlier objection could be met. It is hard to pass judgement on the adequacy of a concept defined in terms of the logically ambiguous language of a heuristic setting. Eliminating such fuzziness and separating the perennial from the obsolescent is what the age of consolidation is about.

By strict logic the above propositions are invalid, even though the sine law is correct. The Newtonian particle is conserved by definition: a mechanical process is an orbit in phase space. Creation or destruction of a particle involves the replacement of one phase space with another. This is not an orbit, hence not a mechanical process. The photon is not conserved as correctly implied by Qu. 30. The use of the term "small body," or in modern terms "particle," both for conserved and for nonconserved entities is a semantic sleight of hand. In order to return to the straight path, we have to keep the two concepts apart.

I will call the conflict of the two particle definitions the *Newtonian ambiguity*. It appears from the foregoing analysis that Newton was not aware of any difficulty. Yet glossing over it condemned the particle concept to a paradoxical status and nipped in the bud the articulation of a logically sound particle concept in harmony with particle physics.

It might have been within Newton's vision that particles subject to creation and destruction ought to be within the purview of chemistry rather than mechanics. However, the time was not ripe for the solution of this fundamental problem. The extent of Newton's walking over thin ice without displaying any hesitation is evident from the last query in the Optics. After an extensive discussion of chemical reactions he concludes:

> Qu 31. ... it seems probable to me that God in the Beginning form'd Matter in solid, massy, hard, impenetrable moveable Particles...that these primitive Particles being Solids; ...even so very hard, as never to wear or break in pieces. ...While the particles continue entire, they may compose Bodies of one and the same Nature and Texture in all Ages: But should they wear away, or break in pieces, the Nature of Things depending on them would be changed.

What is a remarkably positive feature in these passages is the preoccupation with "chemical identity." Although this is a problem both puzzling and important, It is seldom invoked with such an emphasis. A noteworthy exception is Maxwell (1965, 361). What Newton and Maxwell have in common is their spelling out their concern about chemical identity. Whereas Maxwell grants that the state of the art is insufficient to explain this striking phenomenon, Newton's explanation in terms of the absolute permanence of the chemical particles reveals a lack of appreciation of how rudimentary the chemistry of his day was to reconcile transmutation and invariance.

The two particle concepts differ not only with respect to conservation laws but are associated also with different types of motion: Qu.13 suggests undulation. The "sides" in Qu. 26 were always interpreted as "polarization," in modern terms internal angular momentum, hence spinning. Certainly different from translations that admit the point representation of the Newton model. Yet, we should not condescend to Newton for not "noticing" a difference. Only few of the obvious features in nature were ready to be accounted for mathematically at the time. He had reasons to doubt that his post-Euclidean formalism could cope with diversity. This tension led to a characteristic fault line in the rich body of 19th century physics.

3. COPING WITH DIVERSITY

A. Non-Euclidean geometries and the path to diversity

Newton may have had his subjective reasons for not realizing the difference between the particles of light and particles of his mechanics, but there was also an objective reason in the residual dogmatic features of his Euclidean heritage, called the Euclid myth by Davis & Hersh (1981).

Not only do the *Elements* deal with a single geometry, they deal with problems which are assigned today to arithmetic, algebra, number theory within a single deductive system expressed in geometrical terms. It is fitting to call the context the *universe of discourse* for the mathematics of the times.

The discovery of non-Euclidean geometries marked an end to this overestimation of early achievement. It generated a vast amount of literature and, of course, Einstein took a keen interest in the subject but only for use in theories of spacetime. Actually, this discovery has a broader methodological impact. See l.c. The loss of Euclidean certainty had at first a shocking impact on mathematicians and philosophers as vividly described by Richards (1988). It became gradually apparent that nothing was lost except the illusion of being already in possession of the whole truth and new ways were opening up towards acquiring more truths. It is a simple routine to start from Euclid, designate some constructs as primitives and arrive at a new geometry which is internally no less consistent than that of Euclid. It is equally easy to construct vector spaces and one has an unlimited opening into different parts of linear algebra and group theory. I proposed to call such intersystem relations *logical continuation* see l.c.

This step utterly changed the meaning of the term "deductive system." Even the name became the more neutral "formal system." Instead of a single *universe of discourse*, these systems became the *elements of discourse* in the vast diversity of mathematical disciplines, leading to an evolving hierarchy of reliable knowledge. The significance of formal systems was to provide precise "definitions by postulation" for new concepts, to allow the generation of a plurality of mathematical theories centered on precise mathematical entities. It became the prerequisite for the hierarchic organization of a vast amount of information.

Within a few decades, by the late 19th century the dogmatic, fundamentalist view of the foundations of geometry was abandoned. The prevailing method is to structure a vast complexity of mathematical entities into a plurality of intelligible formal systems. These form the elements of a hierarchy.

In sum, in its revised role Euclid is to act as an "ancestor" which spawns an unlimited number of new formal systems. It is an open-ended process and *not* a closed universe of discourse, the methodological counterpart to the Copernican transition from the closed to the open universe.

B. The two branches of classical physics

Since the non-Euclidean developments lead to the separation of Euclidean creativity from dogma, it is plausible to apply the insights gained to a similar improvement within Newtonian science as a Euclidean descendant.

Although in his theory of fluxions Newton did not hesitate to transcend the Euclidean framework, he was probably aware that the rigor of the new mathematics does not compare with that of Euclid. In the context of the *Principia* he was very much Euclidean and considered his mechanics as the universe of discourse for all of physics. Specifically, he was influenced by his success in unifying two layers which could not have seemed more diverse, namely celestial and terrestrial mechanics. Diversity that seemed overwhelming for the casual observer turned out to be only apparent. Newton expected or hoped that such a scenario might occur again. He formulated general rules to this effect under the heading: Rules of Reasoning in Philosophy at the beginning of BOOK III of the *Principia*, Newton (1687, 1999). These rules insist on simplicity and uniformity of Nature, their principal target is to support the universal principle of gravitation. The Third Rule is more specific and deserves scrutiny. Newton explicitly postulated the validity of his mechanics for the smallest constituents of matter. It means that the *interface* associated with Newtonian model ought to assume a *canonical* status for all of physics. Prescribing the conceptual beginnings of an existing discipline for all future applications is contrary to the restraint of the empiricist. Newton hedged in his Fourth Rule: the final word is up to experiment. Also opinions differ whether he admitted the possibility that atoms might obey different rules from the mechanics of rigid bodies. See (McGuire 1970). Regardless of what Newton had really meant, the important point is that classical physics consists of a cluster of theories and these can be divided into two branches depending on whether uniformity between macro- and microphysics is axiomatic, or whether room is left for evolutionary developments.

In order to size up the nature of this schism in the classical theory, we have to realize that the quasi – Euclidean geometrical style of the *Principia* became obsolete within a century. By contrast, the physical model is alive as reformulated in terms of the infinitesimal calculus and preferably the canonical formalism.

The ontological preference that Newton claimed for the mechanics of the *Principia* has been transferred to the canonical mechanics of point masses (CMP). Theories generated by such a reinterpretation of the Third Rule can be classified as belonging in the *canonical program* within classical physics.

The canonical injunction became significant only when it was applied to phenomena beyond mechanics addressing the problems of heat, light, electricity and magnetism. Within the *canonical program* the tendency was to adapt the treatment of gravitational interaction to the new situations by admitting different charges and forces into the canonical framework. The use of electric and magnetic charges led the action at a distance electrodynamics. Ampère became the "Newton of electricity." This, however, was not the end of the classical theory electricity. It took the chemist Faraday to recognize that the canonical dogma is too restrictive to do justice to the phenomena. Maxwell translated Faraday's fields into a mathematical form that was no less elegant that the canonical formalism, but markedly different. At this point the bifurcation of classical physics became complete. Either follow the canonical orthodoxy, or else go to a higher evolutionary stage or optimize the Newtonian harmony between experiment and mathematics. Maxwell's CED is widely considered to be a higher evolutionary stage than the action at a distance theory. In mathematical elegance it is of equal rank as the canonical formalism. It anticipates the algebraic topology developed within mathematics only in the twentieth century. In their highest developments, and not confined by the canonical injunction all the classical theories excel in having selected or produced excellent mathematical theories.

The same situation is repeated within thermodynamics (TD). The canonical form of the theory, namely the kinetic theory of gases yields the kinetic coefficients, such as viscosity, which is beyond TD. However, the entropy of the kinetic theory is in error, because it fails to account for the identity of the gas molecules. Also the equipartition theorem fails at low temperatures. By contrast, TD is reliable within its stated limits. This state of matters refutes the alleged ontological priority of the canonical program. It is also evident that there is an evolutionary sequence of phenomenological classical theories that satisfies the high Newtonian standards each within its specified limits of validity.

In sum, the classical theory consists of an evolutionary sequence of theories that has the remarkable property of accounting for all laboratory scale phenomena. These phenomena are diverse, hence we have a plurality of theories: canonical mechanics of point-masses CMP, classical electrodynamics CED, chemical thermodynamics ChTD, hydrodynamics and gyroscopic rotation. The last two were advanced already by Leonard Euler in a deliberate effort to go beyond the Newtonian model. Canonical mechanics is part of the cluster but it is not capable of reducing the other theories.

This pluralistic cluster of theories is remarkable with respect to *completeness,* but it lost the unification promised by the *canonical program.*

Is there a replacement for the failed canonical reduction? This is a methodological challenge. Dealing with a cluster of theories calls for the handling of intertheory relations which cannot even be formulated if a single formal system is the universe of discourse for all of physics. This is a problem of metatheory which has been examined, l.c.

It is remarkable that the problem was solved by Einstein in a special case which is an opportunity for demonstrating metatheory or metalanguage, just by speaking it.

4. THE EINSTEIN PHENOMENON

A. Einstein discovers transclassical unification

The unification of mechanics and CED was a recurrent problem in the last century. Maxwell tried and failed to reduce the EM field to mechanical models and Poincaré advanced an unconvincing electrodynamic structure for the electron.

Einstein advanced a totally new idea. He asserted in his June 1905 paper that inertial relativity is an essential feature of CMP and argued that CED ought to satisfy inertial relativity as well. Yet the absolute light velocity c inherent in this theory is inconsistent with relativity, unless Newtonian absolute space and time are abandoned. Einstein has set up two hard and fast alternatives: (A) accept a paradoxical conflict between CED and mechanics or (B) abandon the absolute character of Newtonian space and time.

He did abandon this absolute feature and arrived at an unambiguous derivation of the Lorentz group of transformation. The derivations available up to that time failed to reveal the conflict with Newtonian space and time. The rejection of these concepts was a daring idea but the sharp logic ensured ready acceptance within a few years. I suggest that the formalism of special relativity (SR) be named *transclassical* because it subsumes two classical theories from a higher logical level.

Within standard methods it is next to impossible to build a convincing case against a well-entrenched concept. Therefore the landmark status given to Einstein's June paper is even more justified than usually credited for, although his foundation of SR is unnecessarily restricted to macroscopic applications. This comes about because of his use of the Maxwell equations which is invalid under microscopic conditions, as he argued only three months earlier in March. This apparent limitation of SR to macroscopic applications is not germane to the actual scope of the theory: its simplest seminal application is the Compton effect which Einstein could have predicted but did not. See (Stuewer 1975). This does not affect his main achievement, the limitation of Newtonian space and time.

Einstein's argument is an exemplar of the rigorous logic of consolidation. More specifically, it belongs to metatheory as it interrelates theories described as formal systems, in turn compactly represented by their primitive concepts. He, of course, did not mention metatheory, which was not on the books. He seems to have attributed his successful logic to the special choice of spacetime structure as subject, later deepened in general relativity. By contrast, in his March 1905 paper, he advanced the light-quantum in the context of the canonical reduction program.

From the point of view of metatheory the tasks of the March and of the June papers are analogous. We have alternatives: (A') Accept that the light particle has paradoxical wave-particle properties, or (B') abandon both the canonical Newtonian model for the particle and the macroscopic differential equation for undulation.

In this instance Einstein resigned himself to accept the paradox rather than exploring alternative (B'). I suggested l.c. that the alternative to the canonical particle is the chemical particle as summarized below in section 4. The alternative to the wave equation is the algebraic theory of undulation, see (Tisza 1989). These are major steps that would have added an excessive burden to the vast production of his *annus mirabilis*, particularly at that early stage. Be it as it may, the acceptance of the paradox became an incessant irritant and the skeptical reception of the light-quantum compared unfavorably with the success of SR.

The skepticism of the world did not shake Einstein's conviction in the existence of the quantum, but when the world became satisfied by the observation of the photoelectric effect, Einstein did not stop deploring his own failure to understand its meaning. The tension between the March and the June papers developed into a life-long search, of which one aspect was the critique of QM for failure to conform to the standards of the relativity theories. The continued soul-searching ended with his last published statement in the posthumous work (Einstein 1956, 165). A completely revised unified field theory is developed in Appendix II but the last paragraph is an abrupt turnabout:

> One can give good reasons why reality cannot at all be represented by a continuous field. From the quantum phenomena it appears to follow with certainty that a finite system of finite energy can be completely described by a finite set of numbers (quantum numbers). This does not seem to be in accordance with a continuum theory, and must lead to an attempt to find a purely algebraic theory for the description of reality. But nobody knows how to obtain the basis of such a theory.

In the last act the odds are against the favorite field theory, but the new algebraic theory remained in the shadows. As indicated above, I attribute the flaw of the March paper to Einstein's unwillingness to exchange the canonical particle for the chemical particle, and infinitesimal analysis for algebra. On reading Einstein's ultimate message, I believed that he reviewed both of his earlier stands. Alas, this belief is not quite true in view of Stachel (1993, 275). In this brilliant essay: *The other Einstein: Einstein Contra Field Theory,* Stachel documents the evolution of Einstein's thinking over a lifetime. The leitmotif is an alternation of his well-known emphasis on unified fields with a discontinuous algebraic ontology. The (reluctant) willingness to abandon continuous field seems to be partly due to the divergences of field theory and partly to the pragmatic success of the quantum discontinuity. Particularly revealing of Einstein's thinking is his discussion with the mathematician Abraham Fraenkel (p. 287-88) about the remote possibility of a discrete spacetime structure. Einstein did not waver in his focusing on spacetime, rather than matter, but he was willing to give it a new twist. He hoped to derive the quantum discontinuity from a revolutionary discrete mathematics of spacetime, although he was daunted by the incongruity of a discrete spacetime.

There is no hint of the mechanical – chemical dichotomy, which I would like to stress. May we ask: Why not? But before attempting an answer, I will briefly argue why chemistry is not just a footnote among the applications of mechanics.

5. THE CHEMICAL CONNECTION

The discontinuity of natural processes is trivially evident in any chemical reaction. However, this empirical discontinuity may seem just as far from mathematical discontinuity as terrestrial and celestial mechanics seemed before Newton.

How could we build a mathematical connection? There was a major mathematical switch from the Euclid-Newtonian geometry to infinitesimal analysis centering on continuity. In order to cope with atomicity we need another switch to algebra and combinatorics. Formal systems do have a combinatorial nature as the primitive concepts are combined into compound concepts.

The combinatorial nature of chemistry was first revealed by the chemical revolution of Lavoisier. According to this classical theory the elements and their constituent atoms are the stable blocks for the observable world. However, the large and increasing number of elements could not have been mathematically interpreted. Fortunately for the simplifying program, the elements are not absolute invariants, yet this is no "breakdown of classical chemistry." Chemistry has a layered structure and chemists are adept in passing between layers without confusion. The elements in one layer are compounds in a deeper layer, affording an exquisite method of reconciling unity and diversity with a great potential of growth. This escaped the French philosopher August Comte who translated the temporary backwardness of chemistry into a perennial hierarchy of sciences in which chemistry was two notches below astronomy. He also declared in 1835 that knowledge of the chemical composition of stars would forever be denied to man. Quoted by Pais (1988,165).

It is well known that Kirchhoff discovered the sodium spectral lines in the sun already in 1859. However, instead of smiling at Comte's misjudgment, one should recognize it as the consequence of his rigid hierarchy of the sciences which is still a widely shared hubris of the physics community. The observation of the chemical composition of the stars became possible owing to the joint use of chemistry, physics, astronomy and, eventually, mathematics. The nature of this convergence, the great achievement of the last two centuries, calls for elucidation.

Classical physics deals, in addition to mechanics, with electricity, magnetism, heat and light. All of these effects manifest themselves as by-products of chemical structure or of chemical reactions. The first major event was when the "perpetual motion of electricity" revealed by the Volta pile was traced to chemical reactions. While the individual electrolytic reactions belong in the purview of chemists, Faraday's electrochemical equivalent charge is a universal entity which deserves the physicists' attention. Divided by Avogadro's constant it yields the elementary charge, as first pointed out by Helmholtz in 1881, quoted by Pais (1986, 1988, 73).

Avogadro's constant first emerged within chemistry, but chemists were concerned more with moles than with molecules. Its meaning is most intuitive in the relation

$$PV = NkT$$

of the kinetic theory. This harmony between the chemical and the physical molecule is deceptive. Boltzmann's counting of complexions in the kinetic theory has to be corrected by a factor N! in order to yield an entropy proportional to N, as required by ChTD. This is a symptom of an important difference between the individual canonical particles moving along their distinct orbits and the chemical particles that form classes of indistinguishable particles, and classes are distinguished from each other by their intrinsic properties.

With respect to distinguishability QM agrees with the chemical experience, but in a refined form. There are particles obeying Fermi-Dirac counting while others follow Bose-Einstein statistics. Recently well - deserved credit was given to Einstein's imagination as the Bose-Einstein condensation (BEC) is verified in terms of spectacular experiments. Note that Einstein established BEC six months before the beginning of QM. It is unfortunate, however, that Einstein did not spell out the meaning of his new statistics as a transition from the canonical to the chemical particle identity. See sections 5 and 6 in (Tisza 1997a).

To shed light on the relation of the two sorts of particles let us examine the discovery of spectroscopy. The initiative was Bunsen's, who worked on problems of qualitative analysis. The epistemological significance of this procedure has not received the attention it deserves. Note the chemists' ability to purify substances and to establish their identity with precision. The first prerequisite is the concept of "chemical identity" which aroused the curiosity of Newton and Maxwell, see section 2.C. There appears to be a discrete set of potential compounds. The second prerequisite is that the reactions of an unknown sample with standard reagents let us establish the identity of the sample. The third point was Kirchhoff's discovery of the line spectrum as a superior tool for qualitative analysis. The experimental facts that emerged within a few years are so specific that one can marshal the facts to guide the construction of a theory. Since the line spectrum is produced by a diffraction grating, the light beam is undulatory. Since we know that matter is atomic, the light beam must originate in individual atoms in discrete batches of conserved properties. This is just what we express in the quantum condition for the photon and there is no empirical justification for considering the photon as a canonical point mass.

We have also the antithetical definitions of the chemical and the canonical particles: Canonical particles have precise location but no intrinsic properties, while chemical particles have intrinsic properties and their location is irrelevant.

QM subsumes both types of particles, it is a transclassical theory, just as SR is in its own context. Einstein criticized QM for the weakening of the canonical description due to uncertainty, but he did not credit it for doing justice to the chemical description. Yet the qualitative analysis resulting from the junction of chemical and physical procedures lets us structure a set of statistical measurements into ontological information.

For the beginning of a quantitative foundation of QM based on optimal rather than canonical procedures see (Tisza 1989). I consider this paper still up to date, except the conjecture in the last paragraph. The spinorial representation of the Lorentz group that is postulated at the beginning can be shown to be "optimal."

6. CONCLUSIONS

It is an experimental fact that the entities of microphysics are different from macroscopic bodies. There is nothing paradoxical about the fact that bricks are different from buildings. The pioneers of quantum physics were overwhelmed by this difference partly because they narrowed the meaning of "classical" to "canonical," and partly because the philosophical underpinning of the Newtonian enterprise had no provision for diversity.

The Euclidean deductive system was supposed to determine the universe of discourse for all the mathematics of the times. Newton himself considered his mechanics the universe of discourse for all of natural philosophy. Mathematicians discovered the narrowness of Euclidean color-blindness to diversity in the context of non-Euclidean geometries. The "deductive, or better, formal system" turned from *universe of discourse* into an *element of discourse*. The former set dogmatic limits, whereas the latter opened the way for the precise definition of new concepts. It is shown in (Tisza 1997, 213), and more compactly in this paper, how this progressive interpretation can be adjusted to empirical requirements.

This philosophical preconception against diversity was too narrow even for Newton's own oeuvre. As discussed in section 2.D., his light corpuscles were strikingly different from the rigid bodies of mechanics. Glossing over this difference caused the false prediction that light velocity in water would be higher than in vacuum. This difference between particles was confirmed by Kirchhoff's spectroscopy, since light was wave-like, yet presumably emitted by atoms in pulses. The wave-particle duality was implicit, although was not spelled out. There was no paradox, because the particle was not supposed to be point-like. Why did no one think of celebrating the birth of a new discipline: "*opticochemistry*"?

Einstein and Bohr could not reconcile their disagreement because they shared the subjective preconception for the canonical particle that was the origin of the paradoxes.

The physicists ability to overcome subjective feelings was colorfully expressed by Robert J. Oppenheimer as he wrote to his physicist brother Frank:

> ...physics has a beauty which no other science can match, a rigor, an austerity and depth we come a little to see the world without the gross distortion of personal desire (Holton 1984, 155).

Is this exalted picture of physics deserved? Does it hold for the individual physicist during the excitement of discovery? I suggest "yes" for the first question and "no" for the second. We should indeed have the perspective to free the creations of the great pioneers from the distortions of their desire. For the beginning of a paradox-free path to QM see (Tisza 1989).

Massachusetts Institute of Technology

ACKNOWLEDGEMENTS

I am pleased to acknowledge stimulating discussions with Abner Shimony and Gennady Gorelik.

NOTE

* This paper is dedicated to John Stachel on his 70th birthday. As an Editor of the Einstein Papers he helped us to understand Einstein's legacy. In addition he also showed us an Einstein perplexed and unable to make up his mind. He thus enables us to see further by standing on his shoulders.

REFERENCES

Davis, Philip J., and Reuben Hersh. 1981. *The mathematical experience*. Boston: Birkhauser.
Einstein, Albert. 1956. *The Meaning of Relativity*. Fifth ed. Princeton: Princeton University Press. [Ms. finished December 1954.]
Guerlac, Henry. 1981. *Newton on the Continent*. Cornell University Press.
Holton, Gerald. 1984. "Success Sanctifies the Means." In *Transformation and Tradition in the Sciences*, ed. Everett Mendelsohn. New York: Cambridge University Press.
Maxwell, J. C. 1965 [1890]. *The Scientific Papers of J. C. Maxwell*, ed. W.O. Niven. Dover Reprint. Vol. II. *Molecules*.
McGuire, J. E. 1970. McGuire, Studies in History and Philosophy of Science, 1, 3
Newton, Isaac. 1687. *Philosophiae Naturalis Principia Mathematica*. London.
——. 1999. *New translation of the Principia by I. B. Cohen and Anne Whitman*.
University of California Press.
Pais, A. 1982. *Subtle is the Lord*. Oxford, New York: Clarendon Press.
——. 1988. *Inward Bound*. Oxford: Oxford University Press.
Stachel, J. 1993. "*The Other Einstein: Einstein Contra Field Theory.*" In *Science in Context*, 6, 1.
Stuewer, Roger H. 1975. *The Compton Effect*. New York: Science History Publications.
Tisza, L. 1989. "Integration of Classical and Quantum Physics." *Phys. Rev.* A, 40:6781-6790.
——. 1997. "The reasonable effectiveness of mathematics in the natural sciences." In *Experimental Metaphysics*, eds. R. S. Cohen et al. Dordrecht: Kluwer Academic Publishers. Referred to as l.c.
——. 1997a. *End of Century Reflections on Planck's Quantum Theory and Philosophy*. Preprint #108 of the Max Planck Institute for the History of Science.

KOSTAS GAVROGLU AND MANOLIS PATINIOTIS

PATTERNS OF APPROPRIATION IN THE GREEK INTELLECTUAL LIFE OF THE 18TH CENTURY

*A Case Study on The Notion of Time**

INTRODUCTION

Reception or transmission studies are not, of course, something new. There have been studies discussing the diffusion of the new ideas about nature in England, Scotland, France, the Low Countries and Germany during the seventeenth and eighteenth centuries. Many problems related to the reforms by Peter the Great in Russia have also been analyzed. There have been studies on the introduction of the new scientific ideas in Latin America. So is the case for many aspects of science in the Scandinavian countries. Furthermore, there have been many studies on the question of science, technology and imperialism. There have also been accounts of the establishment of university chairs in many countries. The introduction of modern physics in a number of countries is also well documented. The reactions to the Darwinian theory have been the subject of serious scholarship. Nevertheless, studies in languages other than the local languages for the Balkans, the Ottoman Empire, the Central European countries, the Baltic countries, Portugal, but also Spain have been very few and mostly from a philological point of view. The lack of studies for any subject by itself does not, of course, constitute a legitimate reason for starting to work on it; nevertheless, recent developments in the history of science raised many interesting historical questions to warrant an analytical discussion of these issues (Gavroglu 1999, Abattouy et al. 2001).

Although a simple bipolar distinction between center and periphery is useful for broadly delineating the situation, it is incapable of capturing many salient details. There are first of all many centers and many peripheries. Moreover, and depending on the subject one is discussing, a place may be both center and periphery. A center may, over time, change into a periphery, and vice-versa. And a single country may contain both centers and peripheries, thereby making purely national distinctions problematic. Nevertheless, in the following we shall use the term center-periphery to denote the dynamics of the transmission and appropriation of the new scientific ideas from the region broadly defined by the British Isles, France, Switzerland, Germany, and the Low Countries to the rest of Europe during the eighteenth century.

1. TRANSMISSION VERSUS APPROPRIATION

The concept of the "transfer" of ideas, used extensively by those who have discussed these issues, is found to be ultimately inadequate in contextualizing the dissemination of the new sciences in the societies of the European periphery. We shall argue that "appropriation" can be a more coherent and fruitful analytic concept. Appropriation directs attention to the measures devised *within the appropriating culture* to shape the new ideas within the local traditions which form the framework of local constraints—political, ideological as well as intellectual constraints. To examine such issues requires discussing the ways in which ideas that originate in a specific cultural and historical setting are introduced into a different milieu with its own intellectual traditions as well as political and educational institutions.

A historiography based on the concept of transfer can easily degenerate into an algorithm for keeping tabs on what is and what is not "successfully" transmitted. A historiography built around the concept of appropriation is more comparable to the procedures of cultural history; acceptance or rejection, reception or opposition are intrinsically cultural processes. Such an approach also permits the newly introduced scientific ideas to be treated *not* as the sum total of discrete units of knowledge but as a network of interconnected concepts. The practical outcome of a historiography based on the notion of appropriation is to articulate the particularities of a discourse that is developed and eventually adopted within the appropriating culture.

Undoubtedly the concept of transmission of ideas is of some use to the historian of ideas. This, however, is apparent only in the case of comparative studies, when the historical inquiry focuses on the differential reception of a certain system of scientific ideas in a variety of cultural contexts. Nevertheless, even in such cases one must always recognize that ideas are not simply transferred as if they were material commodities. They are always transformed in unexpected and sometimes startling ways as they are appropriated within the multiple cultural traditions of a specific society during a particular period of its history. Indeed, a major challenge for historians who examine processes of appropriation across boundaries is precisely to transcend the merely geographical reference, and to understand the character of what one might call *the receptive modes and devises of the receiving cultures.*

Adopting the notion of appropriation directs attention to the production of a distinctive scientific and philosophical discourse through the reception of the new scientific ideas. This is a crucial point and misconceptions abound. Many historians assume that the scholars of the periphery introduce the new scientific ideas having already adopted the same constitutive principles of the new discourse as those adopted by the scholars at the center. But this is hardly the case; rather, one should adopt the view that the whole enterprise of appropriating the new ideas during the eighteenth century could only be achieved through the formation of a *new* discourse as the optimum way of overcoming the local constraints. Ideas, techniques and practices are not simply transferred; they are being appropriated in order to form a discourse adapted to local intellectual traditions, educational strategies and ideological commitments. In this sense, what is to be systematically studied are the metamorphoses the new ideas underwent through the various stages of assimilation and the kinds of attempts by "local" scholars to incorporate them into existing traditions. For it appears that at the initial stages of the attempts to introduce new ideas, these scholars were able to choose from a host of many different alternatives for developing a proper discourse, and their works expressed different intellectual and social prerogatives. The detailed study we will later present of the way Eugenios Voulgaris appropriated the concept of time by "intervening," in a way, to the dispute between Newton and Leibniz, will clarify these considerations. Time was always a disputable notion and many contemporary historians and philosophers have attempted to trace the various transformations this notion underwent in the past few centuries. What we shall try to show here is how a scholar of the European periphery elaborated this fundamental notion of science and philosophy while the dust from the famous Newton-Leibniz debate had not yet settled down.

One of the main aspects of such an approach is to understand the dynamics and the conditions under which the creation of legitimizing space for the new ideas becomes possible. The problem is relatively simple in those cases where we are confronted with well discerned and clearly defined spaces such as universities and academies. But in many instances at the countries of the periphery one may not be able to even find such spaces. In this case one will have to understand the role of many priests who have written extensively on the subjects we are interested in and have spent all their lives teaching at schools in remote agricultural regions. Likewise one should explain the many cases of lay people who had written philosophical and scientific works and never had the opportunity to communicate them through the standard institutional settings. So where shall we direct our attention to find the legitimizing spaces? Travel itineraries, publishing programs by authors, editors or publishers, lists of subscribers at the end of books, may be some alternative indications. Disputes among scholars have also been a particularly advantageous method for understanding the dynamics of legitimizing spaces. But somehow in the more standard accounts, disputes presuppose an audience with an inclination or at least a potential interest to engage in the issues involved in the dispute. It has quite often been the case that those who are directly involved in a dispute are preoccupied almost exclusively with the audience rather than the adversary. But what about public disputes before an audience totally ignorant of the issues involved but supportive of the overall agenda of particular scholars? Can under such circumstances our studies concentrate in understanding the cognitive content of disputes? Our answer is yes, it is possible to deal with the cognitive content, but only if one stops looking at disputes as intricate scientific rituals and analyze them as *alternative cultural processes*. In this sense, understanding the creation of legitimizing spaces for the new ideas presupposes the comprehension of the nature and features of resistance to these ideas. *Resistance is usually expressed because when new ideas are introduced, they provide alternative methods and answers to questions for which peoples and cultures have already adequate answers. In other words, new ideas are not introduced to be placed in any kind of void, but they are asked to displace other, usually strongly entrenched ideas.* Therefore, understanding the creation of legitimizing spaces for the new ideas cannot be achieved independently from the understanding of the ways resistance is expressed against these ideas.

2. FRENCH ENLIGHTENMENT AND NEWTONIANISM

Let us now discuss a problem which, we feel, has undermined a large number of studies on these issues. It seems that many historical works imply a dependence on a double equation. Enlightenment equals French Enlightenment and the introduction of the new scientific ideas during the eighteenth century equals the reception of Newtonianism.

The first problem is the almost exclusive attention being given to the French Enlightenment. The French Enlightenment is taken as the paradigmatic expression of the Enlightenment, and all other expressions of the Enlightenment are considered as being either unfulfilled versions of the French case or cases which tended to the ideal and pure program which was expressed by the French *lumieres* and *philosophes*. The French Enlightenment has been particularly dear to the heart of a number of historians at the countries of the periphery and especially of philologists, whose studies concentrated on scholars with a social and political agenda that was a significant part of their life and work.

If one looks, however, at the German case and studies a man like Christian Wolff, his followers and other rationalists of their time, one realizes that they did not enter into a confrontation with either the political or the religious establishment, though they were definitely unwilling to accept their all-pervasive power. In fact, this contradictory attitude, this practice of not wanting to come into a conflict, yet questioning the authority of the state and ecclesiastical powers, characterised this practice and set it apart from that of the French *lumieres*. It was not an antagonistic view of the Enlightenment, but rather a complementary one. Referring to Frederick II, Venturi notes:

> But his limitation, his desire never to go beyond certain definite barriers finally elicited from the man who set himself up as the protector of philosophers, Frederick II, a statement which defined with the utmost clarity that detachment, that division of labour between men of culture and statesmen which was only to be overcome with very great difficulty later in Germany. This is precisely the limitation of the greater part of the *Aufklärung* as opposed to the lumieres. Frederick II would proceed to write that the philosophers 'instruct the world through their reasoning we through exemplary practice.' It was a division of labour which also meant putting the philosophers in their place, a definition of enlightened absolutism (Venturi 1972, 21).

The point is clear: There have been many societies where it was often the case for persons holding high offices to be consciously initiating elements which in the local context constituted Enlightenment policies. To study these cases—especially for the societies at the European periphery—though an almost exclusive reference to the French case would surely lead to deadlocks. In this respect let us make three points:
- The first point is almost trivial. Enlightenment was not a homogeneous and uniform movement. There are no more sanitized and less sanitized versions of Enlightenment. They are all equally legitimate, and it is wrong to look at the French version as the more advanced and radical if we want to see how the movement in Europe as a whole influenced the rest of the regions. Exclusive attention to the French Enlightenment when studying the reception of the new ideas at the societies of the European periphery during the eighteenth century is, we feel, a methodological choice which is historically sterile.

- The second point is that we should look at the French Enlightenment and the German Enlightenment in their complementary aspects as well as in their contradictory aspects, and emphasis should be placed on the merging and the confluence of traditions. Let us be reminded that the Balkans turned out to be particularly receptive to the practice of enlightened despotism of Germany, Poland and Russia.
- And thirdly, we should deal with the scholars of the periphery as a group of people who turned what appeared a liability into an asset. These scholars functioned within a framework formed not by any paradigmatic case they may have perceived, but by their "eclecticism" among a number of alternatives. The scholars of the periphery became rather assertive and acquired a rather creative freedom when they realized that there was much to be gained by looking at the cracks of the various manifestations of the Enlightenment, by concentrating on its unfinished business, its weaknesses, failures or even exaggerations. In other words, we should look at the scholars of the periphery not as passive agents whose only function was to distribute locally the well-packaged goods delivered to them from the centers of Europe, but rather as active subjects who received many goods with no particularly clear directions on how to dispose of them locally. The French Enlightenment as the paradigmatic case of Enlightenment, apart from being a historiographical construct much in demand in the twentieth century, is also a notion that reduces the local scholars to passive carriers of this otherwise "perfect" program.

Let us now come to Newtonianism. Almost all of the works discussing these issues take for granted that the developments in natural philosophy during the eighteenth century were simply the unfolding of the *Principia*. At best they consider the eighteenth century as the algebraization of the geometrical *Principia*. Nearly no one takes into consideration the deeply diverging opinions on the future, as it were, of mechanics. And even fewer people note that Newtonianism was in a state of flux[1] and that such a state of affairs provided a *much less constraining context* to a lot of scholars of the periphery in their attempt to formulate a new discourse. Such considerations are rather significant for us since what we would be mainly concerned with is the understanding of the attempts to appropriate the new ideas through the formation of a *new* discourse. When we talk of the influence of Newtonianism, or still, when people talk of the ways Newtonianism was introduced at the periphery, the tendency is to see how the local scholars were influenced by the *Principia*, how faithful they were to the particular work or how much influenced they were by those who tried to either "popularize" Newton's work or write simpler scholarly treatises about it. If one deals with at least the first half of the eighteenth century, this is a misguided effort. For it was the period when the notion of force was still an open question, while the precarious procedure of separation of rational mechanics from natural philosophy was still in progress.[2] To understand what came to be known as Newtonianism in the countries of the European periphery will greatly help to comprehend the multiple aspect of the Newtonian program. The formation of different local discourses, namely the procedures of appropriation, involved selections and decisions on the part of the scholars, concerning the synthesis of ongoing programs like "rational mechanics," "experimental philosophy" or "vis viva conservation physics" with local intellectual and, more specifically, theological traditions.

So, what kinds of themes are amenable in such a discussion and what kinds of questions could be raised? Here are some examples among the many themes which suggest themselves: What were the particular expressions of the new ideas in each place? What were the specific forms of resistance encountered by these new ideas? To what extent such expressions and resistances displayed national characteristics? What were the commonalities and the differences between the methods developed by scholars at the "periphery" for handling scientific issues and those of their colleagues in the center? What was the role of the new scientific ideas, texts, and popular scientific writings in forming the rhetoric concerning modernization and national identity? What scientific institutions were slowly consolidating their presence and what had been the opposition by the local scholars? What were the characteristic features of the scientific discourse formed by the local scholars? What were the particular expressions of the relation between political power and scientific culture in the societies of the "periphery"? What were the social agendas, educational policies and (in certain places) the research policies of scientists and scholars? What shifts in ideological and political allegiances were brought about as the landscape of social hierarchy changed? What consensus and tensions appeared as disciplinary boundaries were formed, especially as those were reflected in the establishment of new University chairs? What ideological undertones characterized the disputes, and what was their cognitive content? What was the significance of the disputes for the "becoming" and/or the "emergence" of the respective audiences? What was the character of the institutions and other intellectual spaces legitimizing the newly emerging community?

Before discussing the Greek case let us make a short comment on the ways we approach the individual scholar of the eighteenth century. We will follow Peter Gay in talking about the sub-worlds and mental universes of the scholars which normally reinforce each other but often are in conflict with another (Gay 1972).

The first such sub-world is the world of cultural atmosphere of the age, the environment that assigns positive or negative values to ideas passions and actions, pronouncing some exemplary, others unthinkable. This is the comprehensive world that sets the rules governing the way of living. To quote Peter Gay "Even rebels acknowledge its power…one leaps out of the magic circle of one's culture only so far."

The individual's relations with his or her culture are mediated through his or her social environment: Class, gender, ethnic and religious loyalties, regional affiliations and family ties put strong constraints on the meaning of the words used and the ideals followed, define what aspirations are legitimate and what limits are inescapable. Gay notes that the most interesting ideas "emerge from a position on the margin of defined groups. But whenever ideas stand they stand somewhere."

The interplay of cultural and social environments is not sufficient to account for the emergence of ideas. What we usually call tradition, defines a relatively autonomous network of ideas, skills and values which constitutes another constraining framework.

> These are the three collective pressures –culture, society, tradition. They press on what is the ultimate shaper, the only carrier, of ideas: the individual. This makes the fourth sub-world, the self, so critically important. By "self" I mean the uneasy collaboration between genetic endowment and acquired habits, affection and neuroses, conscious pur-

poses and unconscious wishes, skills and stratagems. Whenever a scholar is seriously engaged with his work, the latter offers substantial evidence of his encounter between his private world and those three other worlds, which he reflects in his distorting mirror, relates to his needs and urges, and reproduces in his own way (ibid., 68).

Let us now discuss a number of points related to the Greek case.[3]

3. GREEK INTELLECTUAL LIFE AFTER THE FALL OF CONSTANTINOPLE

In the present paper we shall mainly be concerned with the regions where Greek-speaking scholars appropriated the new scientific ideas during the Enlightenment. These regions were on the whole part of the Ottoman Empire until the beginning of the nineteenth century, and the Christian Orthodox Church played a dominant role there, through its highest institution, the Ecumenical Patriarchate at Constantinople. The schism between Rome and Constantinople has had a very complicated history. A number of theological and political differences precipitated a crisis in 1054 when the representative of Pope Leo IX, Cardinal Umberto, walked into Saint Sophia and left a letter excommunicating the Patriarch Mihail Kiroularios. Ostensibly the disagreement was over the question of *filioque*—that is on the insistence of the eastern Church that the holy spirit originates from both the father and the son, whereas Rome insisted that it originated only from the father. The enmity between the two Churches grew to such an extent that during the siege of Constantinople there were many people in the city wishing an Ottoman occupation over the rumored salvation by the Catholic fleet.

Immediately after the fall of Constantinople, in 1453, the Sultan Mohammed II not only allowed the Patriarchate to continue its function but also provided it with a written "privilege" that granted the Christian authorities jurisdiction over many aspects of the religious and civil life of the Orthodox populations. One of the most important consequences of this arrangement was that it allowed the Patriarchate to gain full control on the educational procedures and the respective intellectual activities of these populations. This was the situation when, in the late seventeenth century, a period of educational and economic rejuvenation of many Christian sectors of the Ottoman Empire was initiated. By referring to Christian sectors, we mean the Greeks, the Armenians, the Catholics who were mostly the descendants of the Venetians and Genoans and all kinds of small and sometimes not so small groups, especially in Constantinople. Among all these, there was a social group which would play a rather significant role intellectually, politically and educationally. These were the Fanariots, who took their name after Fanari, the neighborhood of Constantinople where the Orthodox Patriarchate was located. From the end of the seventeenth century, the Fanariots acquired an increasingly important role in the administration of the Ottoman state. At the outset of the eighteenth century representatives of the Fanariots were appointed by the Sultan governors and hospodars in Wallachia and Moldavia. The Fanariots would soon take the lead among all the other Orthodox groups dispersed in the Balkans; their political dominance would reinforce the already strong influence of the Greeks in the economic as well as cultural spheres in these regions, while at the same time as administrators and as diplomats they would adopt the line of the enlightened despotism.

This period is characterised by three interdependent developments. The first is that the increasing involvement of the Fanariots in the administrative affairs of the Ottoman Empire undermined the almost exclusive role of the clergy in mediating the relations of the Christians with the Ottoman Court. The second is the increase of the receptivity of the Fanariots for the new ideas coming from Europe. The third characteristic is related to the rise of a new social group. In addition to the Fanariots, the merchants started to assert themselves socially and played a rather significant role in the intellectual orientations of the period. The symbiotic relationship between the merchants and the quasi-administrative group of the Fanariots was not always without conflict. The point is, however, that the social and economic prominence of these two groups slowly led to the weakening of the absolute control the Church had on the schools and on their curricula.

By the early eighteenth century, Greek-speaking scholars started moving all over Europe. Italy ceased to be the almost exclusive place they would go to study. They also started travelling to the Germanic states, the Low countries, Russia, France and elsewhere. They were thus acquainted with a multitude of intellectual traditions and schools. Contrary to the previous generations of Greek-speaking scholars who pursued their careers mainly in the Italian courts, many scholars of the eighteenth century started returning home after the completion of their studies abroad. There were, basically, two reasons favouring this repatriation. The first was the growing need for teachers in the schools that were being founded as a result of the economically thriving Greek communities dispersed in various regions of the Ottoman Empire. From the early eighteenth century, the economic well-being of the Greek communities within the Ottoman Empire with the accompanying social transformations brought about a number of changes in the educational system. There was a gradual redefinition of the teachers' role. The image of the teacher-priest whose work was a religious mission gave way to another kind of scholar: although the great majority of those teachers were still priests, their educational agenda became more secular and their actual work tended to be more "professional." The scholastic teaching of the works of the Fathers of the Orthodox Church, as well as of ancient Greek literature and Aristotle gave way to a curriculum determined through negotiations with the communities which had established and catered for the schools. Teaching began to reflect the social, political and ideological priorities of these communities. These changes strengthened the relative autonomy of the scholars from the Patriarchate and reinforced their role as independent intellectuals. In many schools the curriculum was no longer determined exclusively by the Church. It was, rather, a compromise between the largely similar but at times conflicting aims of the religious hierarchy, of the social groups with significant economic activity and of the scholars themselves.

The second reason for the return of the scholars had to do with the marginalization of their intellectual enterprise with respect to the established community of natural philosophers in Europe. Almost all of the scholars who went to Europe were churchmen having the blessings of the Patriarchate. They were among the best who had mastered the amalgamation of ancient thought together with the teachings of the Church. In their travels to Europe, however, they found a Europe quite different from what the narratives and experiences of the scholars of the preceding generation had led them to expect. By the middle of the eighteenth century they found a Europe dominated by the ideas of the Scientific Revolution, with flourishing scientific communities involved in the production of original scientific work. The institutions where the Greek-speaking scholars could indulge in the all-embracing studies of philosophy, continuing the kind of education they had already acquired, were progressively decreasing. The scholars were faced with a paralyzing dilemma: if they were to become part of the community of the natural philosophers in the places where they were studying, they would have to abandon their own intellectual traditions and probably question the doctrines of their Christian Orthodox faith. *Being ideologically unwilling and intellectually unable to proceed to such a break, they immersed themselves in the study of the new sciences with a view to returning home and assimilating them into their familiar intellectual milieu.* A characteristic consequence of this attitude was the increasing desire to teach the new sciences in a manner that harmonized with the conceptions of the ancients. No wonder that almost all the Greek scholars explicitly expressed in their books their "debt" to their ancient predecessors independently of the subject they were writing about; they almost always included a first chapter where they made sure to state that what would follow in the book is in perfect harmony with the teachings of the ancients. This conception of an uninterrupted continuity and the perfection of ancient knowledge—a conception that was gladly adopted and promoted by the Church—became one of the basic characteristics of the Greek scientific culture during the Enlightenment.

One of the difficulties in trying to analyze the newly emerging community of Greek-speaking scholars has to do with the relative lack of consensus among the scholars as to the *constitutive discourse* of the community. The study of the emergence of the scientific community in the various countries of Western Europe deals with the ways a group of people managed to reach a *consensus* as to the discourse they were to use in discussing, disputing, agreeing and communicating their results in the new field. From the first decades of the eighteenth century until well into the nineteenth century, the discourse that the Greek-speaking scholars developed was a predominantly philosophical discourse. Two reasons, among the many, which favoured the development of such a discourse are the following. Firstly, there were neither internal nor external factors to precipitate a crisis with Aristotelianism and, therefore, no need to reformulate Aristotelianism let alone initiate a break with it. Secondly, although these scholars appeared quite sympathetic to experimental philosophy, what they considered to be experiments was hardly different from verbal descriptions of experimental demonstrations. The emphasis, usually indirect but often explicit, was about the use of the new material for (re)shaping philosophical arguments. From this point of view it is quite remarkable that in almost all the books where mention of experiments is being made, the emphasis is on the confirmation of already known results, rather than on the process of measurement and the heuristic function of the experiment. In this sense, it is quite typical that, in more than one place, one finds passages stating that "rational thought is not less effective than experimental results."

4. CONTEMPLATING TIME: THE NOTION OF TIME IN THE WORK OF EUGENIOS VOULGARIS

4.1 Eugenios Voulgaris and the intellectual life of his time

Eugenios Voulgaris (1716-1806) was probably the most "representative" figure of what G.P. Henderson called "the revival of Greek thought." For twenty years (1742-1762) he was a renowned professor of philosophy in the most important Greek schools of the southern Balkans and a protagonist in the attempts of the Ecumenical Patriarchate and the Fanariots to reform the higher education of the period. Although his biographical details do not form part of the present study, it is important to keep in mind that after his educational career in the Greek-speaking regions of the Balkans he continued his intellectual activities for some years in Leipzig, where he also became personally acquainted with several members of Saxony's philosophical community. Subsequently he set himself under the patronage of Catherine the Great, became a courtier in Saint Petersburg and culminated his career as Archbishop of Slavensk and Cherson—a new diocesan seat created by the Russian Orthodox Patriarchate especially for him.

Voulgaris was a typical man of letters. His contributions were in the fields of theology (like most of his contemporary scholars, he was an ordained clergyman), metaphysics, literature, political philosophy and the "sciences." Especially the latter occupied a central position in his interests throughout his life. He was the first to introduce into the Greek education the philosophy of Descartes (1596-1650), Leibniz (1646-1716), Newton (1642-1727), and Wolff (1679-1754). He was also well acquainted with the works of natural philosophers like Samuel Clarke (1675-1729), Jacob van 'sGravesande (1688-1742), Petrus van Musschenbroek (1692-1761), and Madame du Châtelet (1706-1749) and he incorporated many elements from their textbooks in his teachings and writings. And he translated into Greek many treatises like Voltaire's *Essai historique et critique sur les dissensions des églises de Pologne* (though accompanied by a commentary that questioned the central thesis of the original work), 'sGravesande's *Introductio ad philosophiam,* Antonio Genovesi's, *Elementa metaphysicae mathematicum in morem adornata* and John Locke's *Essay.* In this respect, Voulgaris was an "enlightened" person and this is, *grosso modo,* the way current Greek historiography perceives him: He was the first to import the ideas of the new natural philosophy in the Greek intellectual life; and because of this he encountered the hostility of many contemporary scholars, who were suspicious about the new intellectual trends; and this was one of the main reasons why he failed to fulfill most of his pursuits. What interests us here, however, is *the specific way he became involved with natural philosophy.* What were his intellectual motives and constraints while doing so? What aspects of his cultural and social environment did they reflect? And how they affected not simply his ability to perceive "correctly" the new scientific ideas but the very discourse he *produced* in order to account for nature in consonance with his contemporaneous natural philosophy? Such questions are important in the sense that they may help us not only throw light on the way Eugenios Voulgaris practiced science but also to bring forth one of the many ways of *doing science* in eighteenth-century Europe.

Eugenios Voulgaris was one of the first—if not the first—who became consciously involved with the enterprise of synchronization of the Greek intellectual life with the attainments of European thought. In this capacity he found himself in the midst of multiple diverging traditions. Being, on the one hand, an agent of "modernization" he felt obliged, on the other, to secure the specific intellectual identity of his audience. As a result, the theological particularities of Eastern Christendom and the neoaristotelian tradition maintained a central position in his philosophical endeavors. It is important to stress, however, that the function of this dipole in the Orthodox East was fairly different from the function of the dipole Catholicism-Aristotelianism in the Latin West. The fact that *neoaristotelianism* was perceived as an anti-Catholic trend within the general context of Aristotelianism[4] made it quite attractive for the Orthodox Christians; but there were also other historical circumstances which determined the character of the co-existence of these two traditions. This is not, of course, the place to discuss analytically these circumstances, but it must be stressed that the doctrinal integration of Aristotelianism with Christian faith, which was peculiar to the philosophical synthesis of Thomas Aquinas, never occurred in the Greek intellectual life of the early modern period. Orthodoxy managed to coexist for almost two centuries with a hard core materialist interpretation of the Aristotelian philosophy without being fused with it and, strangely enough, without raising a major dispute against it. At the same time, it was this political "moratorium" along with the profundity of the philosophical teachings of Theophilos Korydaleas (c. 1566-1646), the man who founded Greek neoaristotelianism in the early seventeenth century, that ascribed the latter a legitimate status, and allowed it to dominate in Greek-speaking education throughout the whole period.

The most important trait of Korydalean philosophy was the emphasis it placed upon Aristotle's natural philosophy.[5] Voulgaris emerged from a culture which not only accounted consistently for the whole range of the known natural phenomena, but also had put this concern into the center of its investigations. Voulgaris could not ignore neither overcome easily this cultural state: Korydalean neoaristotelianism provided the conceptual *armamentarium* for the understanding of the natural world, and framed the conceptual context within which the respective knowledge ought to be placed. The structure of Greek intellectual life did not encourage the emergence of major philosophical disputes on such issues like those that shaped the philosophical and scientific controversies in the Western societies of the seventeenth and eighteenth centuries. As a result, Voulgaris represented a philosophy which seemed not to display important inconsistencies and which did not find itself against the problem of reinterpreting natural world, as was the case with other European philosophical and religious traditions.

But Voulgaris was also a learned man of his age who was well aware of the changes that took place in the European intellectual landscape. In this respect, he understood that no version of the Aristotelian philosophy could keep up with these changes. But, although he considered Aristotelianism a problematic philosophical interpretation of nature, which needed to be integrated with the attainments of new philosophy, he did not feel that it was a tradition that ought to be eliminated from the intellectual horizon. Thus, even if the result of such a reconsideration was to lead to a radically new philosophy of nature, Voulgaris' adherence to Korydalean neoaristotelianism was so strong that he could not refrain from incorporating its fundamental principles in the new philosophical synthesis he produced.

Orthodoxy comprised the other cornerstone of his intellectual edifice. Voulgaris was one of the most eminent Greek theologians of the eighteenth century and the author of a great number of relevant treatises. What is important here, however, is not the relationship between his religious and philosophical considerations *per se*, but his very attempt to revive the link between Orthodox Christian religion and philosophy. After more than one and a half centuries of "political" coexistence of Orthodoxy with neoaristotelian philosophy, he is the first who aimed consciously at producing a more tight epistemological fusion between the two. The mid-eighteenth century was a period during which Christian Orthodoxy occupied a significant position in the discussions concerning the character and the future of the newly emerging society. Voulgaris participated actively in these discussions and supported the prospect of a Great Orthodox Empire under Russian domination. Under these circumstances the incorporation of religious elements in his philosophical endeavors becomes a decisive task. And, strangely enough, this is also an important reason why Voulgaris honored to such an extent "Newtonian" philosophy. Although most historians tend to perceive his preference towards Newton as a self-explanatory result of the epistemological superiority of new physics, an equally strong reason for this preference seems to be the fact that Newton and the natural theology of his age brought anew in the foreground the notions of divine intervention and of miracle. And this approach enabled Voulgaris and other contemporary Greek-speaking scholars to develop a religiously oriented natural philosophy keeping distances from both disturbing extremes, namely the potential atheism of Korydalean neoaristotelianism on the one hand, and the fusion of Aristotelian natural philosophy with Catholicism on the other.

Voulgaris' dialogue with his contemporaneous natural philosophy, therefore, was mediated, to a great measure, by the particularities of his own intellectual and cultural milieux. He did not simply "transfer" nor "translate" nor "canalize" the scientific attainments of the Enlightenment into the Greek intellectual life; he attempted to *produce* a new philosophical synthesis, which reflected his intellectual and social pursuits. Besides, what exactly Voulgaris could have "transferred" or "translated" or "canalized"? "Science"; yes, but *what* "science"? "Natural philosophy"; yes, but *what* "natural philosophy"? "Newtonianism"; yes, but *what* "Newtonianism"? As we have already mentioned above, what we nowadays tend to perceive as an integrated and homogeneous pattern of scientific activity had not yet been implemented in the mid-eighteenth century. There was an extremely wide spectre of philosophical interpretations and research directions concerning the understanding of natural phenomena. Even Newton's own works seemed to indicate diverging directions and to accommodate different patterns of natural investigation. Thus, the further elaboration of mathematical principles of motion developed in the *Principia*, offered a sound foundation for rational mechanics (which, must be noted, was a branch of *mathematics*), while, at the same time, his experiments in *Opticks* and his concern about the theoretical foundation of experimental induction contributed to the "advancement" of experimental philosophy. On the other hand, Newton displayed a strong interest in exploring the nature of matter and force, and his respective metaphysical contemplations were incorporated in a long series of contemporary philosophical discourses, frequently juxtaposed with the ideas of his major philosophical opponent, Gottfried Wilhelm Leibniz. In this sense, Voulgaris' philosophical enterprise took place within an intellectual space defined by a multiplicity of philosophical approaches, which had their roots chiefly in Newton's works (or referred to them), but which also differed considerably from one another, reflecting the specific intellectual origins and pursuits of their agents.

It seems that Voulgaris was well acquainted with the respective literature. He was at home with a broad range of works, which represented the various aspects of "Newtonian" philosophy. And, most importantly, he seemed to understand that, above all, the eighteenth century was a period of *trial* of the various philosophical discourses about nature. It was a period, during which a great number of open problems came into consideration and formed the object of fervent discussions throughout the centers of European philosophy. Many of these problems were related to those principles of natural philosophy which comprised its constitutive elements—as became evident from the debates among the representatives of the Cartesian, Leibnizian and Newtonian traditions. At the same time, many fundamental concepts of natural philosophy —like those of *matter, vis inertiae, force* and *attraction*—seemed to bear multiple and not necessarily well-defined contents. In fact, the codified way we perceive nowadays Newtonian physics is a result of the developments that took place during the nineteenth century. In the eighteenth century, however, the interpretation that was eventually to prevail had not become yet clear, and the resolution of the various pending problems was going to determine to a great extent the outcome of this procedure.

As a matter of fact, in order to fulfill reliably his philosophical undertaking Voulgaris was obliged to take sides in the various open disputes concerning either the metaphysical foundations of natural philosophy or the physiognomy and the patterns of natural investigation. *In other words, he had to function as a genuine natural philosopher and not as a mere intermediary between different cultural milieux.* In this respect, he would not be preoccupied with the "transfer" of a definite set of scientific theories and practices, but with the articulation of his own philosophical discourse where the philosophical and scientific attainments of Enlightenment would be fused with the intellectual traditions and the social pursuits of his own cultural milieu.

4.2 Debating Time

Time was one of the central issues of the "Leibniz-Clarke correspondence." This dispute took place between 1715 and 1716 and, as is well known, was in fact a dispute between Newton and Leibniz about a number of metaphysical considerations about the foundations of natural philosophy (Cohen and Koyré 1962). Voulgaris was well aware of the dispute between the two philosophers. As is evident from the footnotes and the cross-references that occur in his philosophical works, he had studied thoroughly the correspondence between Leibniz and Clarke, which was published by Clarke in 1717. Indeed, in many cases he referred to the specific edition in order to draw information about Newton's and Leibniz's views on various subjects of natural philosophy. The death of Leibniz put an end to the debate, but the issues it raised maintained their significance for many European philosophers for more than forty years afterwards. What was at stake as far as *time was concerned?*

In the first "Scholium" of *Principia,* with which the introductory "Definitions" of the first book conclude, Newton clarifies the meaning of some terms, which are in common use, but their empirical context differs from the notion he is going to ascribe them. These terms are *time, space* and *motion*. He writes about time: "Absolute, true, and mathematical time, in and of itself and of its own nature, without reference to anything external, flows uniformly and by another name is called duration. Relative, apparent, and common time is any sensible and external measure (precise or imprecise) of duration by means of motion; such a measure —for example, an hour, a day, a month, a year—is commonly used instead of true time" (Cohen and Whitman 1999, 408).

The *sensible* time, therefore, is always relative. In astronomy, however, Newton remarks, astronomers are obliged to correct the apparent durations by reducing them to an absolute measure of time so that they can calculate the celestial motions more precisely ("on the basis of a truer time"). This measure represents the equable flow of the absolute time and is defined on the basis of a uniform motion. But no such motion really exists, since all real motions are either accelerated or retarded (ibid., 409). As a matter of fact, the need for a measure of absolute time arises from the practice of astronomers and experimentalists who seek accurate quantitative results, *but* it cannot be fulfilled on the basis of the real motions of the sensible world these groups study. In fact, the quest for the definition of such a measure is an implication of the *metaphysical* assumption that the duration of the things *does not* depend on the things themselves and the transformations they undergo. And, in this sense, it comprises one of the most fundamental presuppositions of a crucial transition: Without the "absolute, true and *mathematical"* time, the study of motion would never be able to overcome the fragmentary character of geometry and enter the world of algebraic relations.[6]

According to Newton's conception, therefore, time is an infinitely extended substance independent of matter and its manifestations. Time would continue existing even if matter ceased to exist. On the other hand, matter exists only within time and all natural phenomena have an absolute temporal duration. Newton's epistemology is permeated by a fundamental distinction between *apparent* nature, which is being perceived through the senses, and *true* nature, which can be perceived only through abstraction and the proper mathematical processes. Absolute time forms part of the latter. And this is the reason why Newton draws the attention of his readers to "certain preconceptions" which may lead to the confusion of mathematical time, space and motion with the sensible ones. And that is why there is a kind of an (implicitly evaluative) asymmetry in the expressions he uses: "absolute, true and mathematical" for the former, "relative, apparent, and common" for the latter.[7]

The dispute between Newton and Leibniz had its roots in the ideas of the two philosophers about the relation of matter with space and time. According to Newton a material particle considered from the point of view of mathematical physics occupies, by definition, a certain position in space and time. It is possible, then, for identical material particles to be distinguished only on the grounds of their differential positioning in space and time. But this is exactly the point where Newton's ideas clashed with the two fundamental principles of Leibniz's philosophy: The principle of sufficient reason and the principle of the identity of indiscernibles.

The principle of sufficient reason is the cornerstone of Leibnizian metaphysics. According to it there must be always a specific and well-defined reason for whichever action in the universe; nothing occurs by itself or, better, nothing occurs without a reason; and this holds even for God.[8] The Newtonian perception of time clashed with this principle, since in a continuous and uniform flow of time there cannot be found a reasonable cause for the location of an event in a certain position instead of anywhere else in this flow. The principle of the identity of indiscernibles, on the other hand, states the impossibility of the simultaneous existence of two different entities which are absolutely identical; if so, they should be a unique entity. The identical material particles of Newtonian atomism which are distinguished only by their differential positioning in space and time are apparently in conflict with this principle. But there is, also, an even more profound conflict between the two approaches: The very same points of mathematical space and time, deprived from any quality that might distinguish them from each other, could not but be reduced to a unique point; in other words, Leibnizian philosophy contests the possibility itself for absolute space and time to be substantiated. According to Leibniz, therefore, time and space are only relative: Space is the order of coexistent phenomena and time is the order of successive phenomena. Or, at least, this was the way his contemporaries perceived his ideas about time and space. Today we know that his views on the subject were not reduced to this statement. Ernst Cassirer in his systematic discussion of 1943 suggests that space and time for Leibniz were not only a result of the sensible relations between things but they also comprised the totality of the relations among the terms of *every possible* experience. Leibniz himself remarks:

> Space and time together are the order of possibilities of the entire universe, such that they order not only that which actually is, but also that which could be put in its place, just as numbers are indifferent to that which is numbered.

> It is as I said that time and space mark the possible apart from the supposition of existence. Time and space are of the nature of eternal truths which obtain equally in the possible and the existent.[9]

This view seems to contradict the idea of relativity of space and time which Leibniz admits in other places. What he claims here is that space and time belong to the realm of *eternal truths,* while elsewhere he insists that they are only apparent impressions derived from the sensory perceptions. The fact, however, is that the point he tries to make during his controversy with Clarke, as well as elsewhere in his works, is that space and time are *ideal* conditions through which every possible world can be substantiated. This means that the *knowledge of space and time* is distinguished from the *knowledge of their empirical terms*. Although the *notions* of space and time are always a result of the relations between the various bodies and events which are being perceived through our senses, space and time themselves are not sensible. The concepts of spatial order and temporal succession do not derive directly from sensory perception, since they presuppose the processing of this perception by means of reason. As a result, the knowledge of space and time does not consist of the attributes of the apparent world we perceive through our senses, but originates from the eternal truths, the validity of which is founded on the law of contradiction. Thus, Leibniz agreed with Newton as far as the *ideal* notion of space and time was concerned but disagreed with him on the grounding of these notions in the real world.

Voulgaris had not studied the original works of Leibniz; he got acquainted with them through the textbooks of Christian Wolff and, especially, of Madame du Châtelet. The latter conceives Newtonian time in the way most European philosophers read Newton's "Scholium" on space, time and motion:

> Ansi, on se le figure comme un Etre composé de parties continues, successives, qui coule uniformément, *qui subsiste indépendamment des choses qui éxistent dans le Tems,* qui a étédans un flux continuel de toute éternit, & qui continuera de même (Madame du Châtelet 1742, 119 [our emphasis]).

Talking about the Leibnizian view on the subject which she herself chose to adopt she says:

> Le Tems n' est donc réellement autre chose que l' ordre des Etres successifs; & on s' en forme l' idée, entant qu' on ne considère que l' ordre de leur succession. Ansi, *il n'y a point de Tems sans des Etres successifs* rangés dans un suite continue; & il y a du Tems aussi-tôt qu'il existe de *tels Etres* (ibid. 1742, 124 [our emphasis]).[10]

Along the same lines, but speaking as an opponent of Leibniz's view, Voulgaris observes in his *Metaphysics*:

> Wolff's followers do not seem to me to have been dealing successfully with the problem of time's nature. For they consider time to be the order of successive events that occur in a continuous manner (*Metaphysics*, part II, "namely *Cosmology*," 130).

4.3 The multiple aspects of temporality

Voulgaris' views on time are elaborated in two of his major works. The first is the *Elements of Metaphysics, written by Deacon Eugenios Voulgaris ... containing his teachings before his past students,* namely the philosophical lectures he delivered between 1742 and 1762. The work was published in 1805 thanks to the patronage of the Zosimas family, a family who sponsored the edition of a great number of Voulgaris' works that remained unpublished until the last years of his life. The other work took its title from pseudo-Plutarch's *Placita Philosophorum* (= Philosopher's Favorites) but it is, in fact, a work on the natural philosophy of his times. It was also published thanks to Zosimas' funding in order to be delivered to the Greek schools of the period. It came out in 1805 as well, but it was printed in Vienna while *Metaphysics* was printed in Venice. It is estimated that *Placita* must have been written between 1763 and 1771, while Voulgaris was living in Leipzig. This book does not contain his past teachings but it is a programmatic work on various subjects of natural philosophy.

The chapter "About time and space, as well as about Vacuum" of *Placita* is introduced with a bold claim: "Time is not a real being, neither is it a substance of the temporal beings that can be divided"[11] (*Placita*, 73). This claim brought apparently in the foreground the central issue of the controversy between Leibniz and Newton about time: What is the *nature* of time? Voulgaris' most complete answer did not occur in the *Placita* but in his *Elements of Metaphysics*. There, he devotes to the subject an extended section of the second chapter of book two, which is entitled *Cosmology*.[12] The section is entitled "About Continuance as well as about Time" and predisposes the reader for a significant ambivalence even in the definitions themselves: Voulgaris is going to deal not with one, but with two concepts, "time" and "continuance."[13] The fundamental one is the latter. "Continuance is the continuous extension of a being's existence"[14] (*Cosmology*, 109). But continuance has a double meaning as well. Insofar as it represents the duration of a being's existence *per se* it is called "irrelative,"[15] while when it is being measured on the basis of the events that take place in other beings it is called "relative." Voulgaris evokes Newton's "Scholium," although without making an explicit reference to it: "And the irrelative [continuance] is also called true and mathematical" and resembles the equable motion of a point that describes an infinite straight line; the "relative," on the other hand, is also called "apparent" and is being affected by the changes in the rate of the external events, "being, thus, particularly irregular" (*Cosmology*, 110). Though one can easily recognize Newton's presence in this passage, it is still extremely difficult to recognize Newtonian time. What is the relation between "continuance" and "time"?

As is the case with many other concepts of traditional Metaphysics, this relation seems to be an *hierarchical* one: Continuance has three ontologically distinguished states, according to Voulgaris: The superior state is "eternity," namely the duration that has neither a beginning nor an end and is peculiar only to God. Second comes "perpetuity," which is connected to the existence of *pure forms*.[16] Since these forms are created entities, they cannot be considered as eternal beings; although their existence never comes to an end, it certainly has as starting point the act of Creation. That is why this kind of continuance, which is open at the one end, is called "perpetuity" and —contrary to the common beliefs— it is not identical to "eternity." "Time" lies on the lowest level of temporal hierarchy; it is a duration closed at both ends, since it always has a starting point and an end, and corresponds to the duration of the natural bodies that belong to the world of creation and corruption (*Cosmology*, 110).

Voulgaris feels obliged to distinguish the thing *per se* from its conceptual representation. Notwithstanding the differentiation of continuance, he remarks, the ability of human beings to perceive temporal relations is limited. In fact, the only way we perceive a temporal relation is by comparing the duration of a being with a sequence of events that take place in our intellect. For, as Locke observed, *but also Aristotle much earlier than him had realized*,[17] if we are unable to take notice of the changes that occur in our mind, it is impossible to perceive time. This is so because we attach what precedes to what follows and we miss what happened in the meantime.[18] Hence, if we concentrate on a unique thing in such a manner as to follow its changes without taking notice of any external events, we will get the impression of a continuous present and, thus, we will miss the time that elapsed. Fortunately, this is impossible because our mind tends to be distracted by the multiple events that take place in the world of natural beings and to order them according to their temporal succession. As a result, human beings have at every moment a perception of time. But what does this manner of fashioning the notion of temporality mean for our ability to perceive "continuance"? Since human beings can only take notice of the changes that occur in the world of creation and corruption, the only dimension of continuance they actually perceive is "time"; and insofar as they take notice of the time elapsed only through the comparison of different events, the only "time" they can really perceive is *relative* time. The two other aspects of continuance are intelligible but not perceptible. Hence, the fact that the human beings perceive the various aspects of continuance in the same way—that is, as "time"—reflects the limitation of human intellect. Continuance, however, as a general concept for temporality is actually divided into three different levels, each of which corresponds to a different group of beings. What is the reason for this "idiosyncratic" manipulation?

After a general introduction to the subject, Voulgaris turns to a series of theorems that refer to the various features of continuance. The first theorem is of a particular philosophical significance: "Continuance cannot be distinguished from the continuate."[19] Voulgaris proves the proposition in three different ways, all of which are based on the presumption that the dissociation of *existence* from its duration either renders the former unintelligible or leads to logical absurdities. But the purpose and the meaning of the theorem are better explained in the ensuing "Corollary" and "Scholium." The corollary is the positive rephrasing of the theorem: Consequently, neither "eternity" is distinguished from eternal God nor "perpetuity" from "perpetual natures" nor "time" from natural bodies; on the contrary, "eternity" is *identical* with God,[20] perpetuity is *identical* with those natures that remain immutable since the moment of their creation, and time is *identical* with natural bodies that are in a process of continuous change, governed by the necessity of creation and corruption. As a matter of fact, "most wisely Aristotle proved that time is nothing *per se*." The only way we have to distinguish the various aspects of continuance from the respective beings is through the help of reason. Voulgaris here reproduces a description that is found in many textbooks of natural philosophy of his time: As we distinguish number from the objects it counts and dimension from extended bodies, so we can also distinguish continuance from the beings whose duration it represents: That is, by *abstraction*. But virtually, continuance is not a self-existent being and, most importantly, its various aspects do not exist *prior to* the respective beings (*Cosmology*, 113-114).

So, this is the answer Voulgaris gives to the problem of time: *Continuance is an existential condition of the beings*. And this is the reason why he introduces the section on time of *Placita* with the programmatic declaration that "time is not a real being, neither is it a substance of the temporal beings that can be divided"; and that is why, in what follows he repeats that "most wisely Aristotle proved that time is nothing real";[21] and that is also why Voulgaris stresses that "time is the continuance and duration of existents, namely the progress, and the advancement, and the prolongation of their existence"[22] (*Placita*, 73). Voulgaris makes a philosophical as well as a theological point: Aristotle teaches us that time is an existential condition of the beings and thus it cannot subsist independently of, prior to, or after the end of them. On the other hand, the nature of God differs from the nature of humans and the nature of the eternal beings from the nature of God. Consequently, the times that represent the durations of these beings should be qualitatively different. "Time" as an existential condition of natural beings cannot account for the existence of eternal God; the same holds for "perpetuity," as well. "Eternity" and "perpetuity," on the other hand, due to their immesurability,[23] are inadequate for the estimation of the duration of natural bodies, which observe the necessity of creation and corruption. Thus Voulgaris' view on temporality is articulated as an alternative answer to the problem of time when compared to Newton's and Leibniz's conceptions. Time is neither a *self-existent entity*, which flows *independently* from the presence and the transformations of matter (Newton) nor an *apparent notion* that ensues from the comparison of the changes that take place in the natural world (Leibniz). Time —or, more precisely, "continuance"— represents an existential condition for the various beings both of the natural as well as of the transcendental world.

There is, however, a difficulty: Voulgaris displays undoubtedly a special preference for Newton's philosophy; he is, indeed, the first who introduced Newtonian ideas in the Greek intellectual life. Thus in two points, one in *Cosmology* and one in *Placita*, he reproduces faithfully the distinction made by Newton between absolute and relative time. In *Cosmology*, as we have already mentioned, he writes that "continuance" is divided into "irrelative" and "relative." "And the irrelative one is also called true and mathematical" and resembles to the continuous and uniform motion of a point that describes an infinite straight line. The "relative" one, on the other hand, is also called "apparent" and since it follows the rate of external events, its flow is irregular (*Cosmology*, 110). Along the same line in *Placita* he remarks that "absolute time, which is also called true and mathematical" is an endless uniform flow, without accelerations, retardations or interruptions which helps us "apprehend the endurance itself of existence." And "relative" time is the measurable duration of a change, which can be perceived through our senses as a result of the comparison with other changes (*Placita,* 74). Thus the difficulty we have is the following: How can one relate the "*true* continuance" and the "*true* time" which occur in the above statements with Voulgaris' programmatic declaration that "time is nothing real"? How can these views be reconciled? What does it mean that Voulgaris seems to adopt the "true time" of *Principia,* while at the same time he praises Aristotle because he proved that time is not real?

The answer lies in the distinction Voulgaris makes between the terms "true" and "real." For Newton the two terms are identical, insofar as time is ontologically self-existent and independent from matter. The "absolute, true and mathematical" time of *Principia* does not *correspond* to a natural entity, *it is* a natural entity. For Voulgaris this presumption is not valid. "Continuance" is being substantiated, only to the extent that material or immaterial entities exist, as a condition that represents the duration of their existence; as a result, "continuance" cannot not be real *per se*. But, it *can* be "true": If it becomes intelligible only in respect to the being whose duration it counts, it is "true" in the sense that it informs us about the real conditions of the existence of a being. On the contrary, when it is measured on the basis of the changes that take place in other beings it is only "apparent," because it does not inform us about what actually occurs to the being itself. This semantic manipulation makes it possible for Voulgaris to accommodate Newton and Aristotle under the same theoretical synthesis; or, to be more precise, to eliminate the potential contradiction between the two approaches and to construct a philosophical context within which the Aristotelian view on time becomes compatible with the "absolute, true and mathematical" time of *Principia. Reality* and *truth* are two distinguished states upon which the co-existence of the two different approaches is firmly founded.

5. CONCLUSIONS

Undoubtedly Voulgaris' elaboration of scientific ideas formed part of a *legitimate* cognitive enterprise. The fact that this enterprise was not able to be fully integrated within the broader stream of the emerging scientific thought was, of course, a result of the particular historical and cultural circumstances under which it was shaped.

The introduction of the new scientific ideas by the Greek scholars of the eighteenth century was a process almost exclusively directed to their appropriation for educational purposes. The apparent aim was to modernize the school curricula, but this did not mean a neutral attitude as to the possible ideological uses of these new ideas–especially the need to establish contact with the ancient heritage and to conform with the doctrines of Orthodoxy. As a result, the assimilation of the scientific ideas involved the production of a new discourse which reflected the *network of local constraints and priorities*. As we tried to show, the process of appropriation refers to the ways devised to overcome cultural resistance and make the new ideas compatible with the local intellectual traditions. As a matter of fact, understanding *the character* of this resistance becomes of paramount importance. And in the case of Greek intellectual life the issue of resistance cannot be discussed independently of the issue of breaking with ancient tradition. The specific ideological and political contingencies of Christian societies under Ottoman rule during the Enlightenment, together with the dominance of the Greek scholars in the Balkans, called for an emphasis not on the break with the ancient modes of thought, but rather on *establishing* the continuity with them. The Greek scholars tended to see the development of modern sciences as a triumph of the programmatic declarations of the ancient Greek thought, with its emphasis on the supremacy of mathematics and rationality, rather than a break with it and the legitimization of a new way of dealing with nature. On the other hand, the absence of a national state and of the relevant intellectual institutions did not allow the Greek society to form those conditions which would favour the exploitation and the respective social assessment of the sciences. Lacking such a corroborative framework, ideological and, in fact, philosophical considerations became the dominant preoccupation of the scholars and comprised the context within which the appropriation of the contemporary natural philosophy took place.

University of Athens

ACKNOWLEDGEMENTS

We would like to thank Jed Buchwald, Jean Christianidis, Dimitri Dialetis, Jürgen Renn and Barbara Spyropoulou for their useful comments.

NOTES

* John Stachel has not only been a very astute observer of what is happening in the uneasy and, perhaps, dangerous times we are living. John is in fact the kind of intellectual Karl Marx must have had in mind when he wrote the 11th thesis on Feuerbach: The point is not (only?) to understand the world but to change it. Though such changes implied by the 11th thesis have proven to be excruciatingly difficult, it is an optimistic sign to know that there are still people like John, with such a committed agenda to what has been envisioned more than 150 years ago.

1. See, indicatively, (Iltis 1977, Schaffer 1980 [esp. sections II. "Natural philosophy as Newtonian matter-theory" and III. "Natural philosophy as the negation of science"], Guerrini 1985, Force 1987, Casini 1988 [esp. sections 2. "The early critics" and 3. "Toland and Berkeley"], Thijssen 1992).
2. On these issues see the articles "Newtonianisme ou Philosophie Newtonienne", "Philosophie Experimentale" and "Mechanique" in *Encyclopédie ou dictionnaire raisonné des sciences, des arts et des métiers*. All of them were written by d'Alembert.
3. An analytical discussion of a number of cases can be found in (Dialetis et al. 1999).
4. There are very few studies on neoaristotelianism. C.B. Schmitt's works reprinted in the collections (Schmitt 1983 and 1984) offer a good overview of the subject. See especially the paper titled "Cesare Cremonini: un aristotelico al tempo di Galilei," contained in (Schmitt 1984, part XI), originally published in 1980.
5. On Theophilos Korydaleas' life and philosophy see (Tsourkas 1967).
6. On this subject see (Klein 1985).
7. Newton's views on space and time have been extensively discussed among historians and philosophers of science. This is not the place to review the respective bibliography, since, from the historical point of view, what we are interested in is the way Newton's contemporary philosophers perceived his ideas on space and time. See, however, J.E. McGuire's elaboration on the subject in (McGuire 1978). According to McGuire, Newton believes that space and time are general conditions of being which attach to a thing's existence. As we shall see below, this interpretation displays significant similarity to the way Eugenios Voulgaris handles the notion of time. For a further elaboration as well as for a criticism of McGuire's thesis see (McGuire 1990 and Carriero 1990), respectively.
8. A closer examination of Leibniz's philosophy, however, indicates that three different notions of this principle appear in his works. The first is almost identical with the principle of causality and states the dependence of every effect from the respective causes. This is the notion, which also Clarke ascribes to the principle of sufficient reason and that is why sometimes he claims that the only reason for an event is the will of God. The second option is contradictory to the previous one. According to it, the principle of sufficient reason functions as a motive. In this respect, even God should have a motive for his actions. The third option is connected to Leibniz's idea that God's actions aim always at the best possible world. Leibniz embraces this option when he argues against the existence of vacuum: God could not have allowed vacuum since the more matter there is in the universe, the more perfect it is (Alexander 1956, xxii-xxiii).
9. Cited by W.P. Carvin in (Carvin 1972), whence the ensuing argument.
10. We note the expressions which summarize the difference between the Newtonian and the Leibnizian perception of time according to Madame du Châtelet: The former states that time exists "independently of the things that exist in time" while the latter that time consists of "successive beings." In what follows we shall see how Voulgaris elaborates *between* the two.
11. Οὐδέν ἐστι πραγματιῶδες ὂν ὁ χρόνος, οὐδὲ οὐσία τις τῶν ἐν χρόνῳ διῃρημένη.
12. Like in many other contemporary treatises of Metaphysics, the two other books are *Ontology* (the first) and *Psychology* (the third).
13. The (ancient) Greek word used here by Voulgaris is "Διαμονή". Its original meaning is "continuance" or "persistence." See (Liddell and Scott). Voulgaris was always very sensitive about the use of terms; thus in this case he was very careful not to confuse the meaning of the word with other terms signifying "duration," "continuity," or "permanence." Unfortunately, in vernacular English such a clear distinction is difficult to keep, at least for the case of continuance-continuity.
14. Διαμονή ἐστὶν ἡ κατὰ τὸ συνεχὲς παράτασις τῆς τοῦ ὄντος ὑπάρξεως.

15. The Greek word (ἄσχετος) also means "irrelevant."
16. Τὰ ἀμιγῆ ὕλης εἴδη. Although Voulgaris unreservedly adopts Newtonian atomism, he is quite reluctant to expel the Aristotelian dipole of matter and form from his philosophy. A few pages below he gives an even more eloquent evidence of his belief, referring to the "immaterial, perpetual forms" that would keep existing even though the material world had ceased to exist (*Cosmology*, 122).
17. As mentioned above, Voulgaris does not feel obliged to juxtapose Aristotelian tradition to the philosophy of his time. In fact, like most Greek-speaking scholars of the eighteenth century, he tends to understand the attainments of his contemporary philosophy as mature fruits of the ancient intellectual heritage. Hence in the introduction of *Placita* he declares that he is going to pay special attention to the origins of the various theories, so as not to underrate the contribution of the "real finders." (*Placita*, *2 [without page-numbering]). Along the same line here, he is careful to avert any misconception: The "real finder" of the theory about the perception of time was Aristotle and not Locke. And, most importantly, he doesn't seem to see any inconsistency in this ascertainment.
18. Although he just mentioned Locke (whose Essay, it must be noted, he himself had translated into Greek, probably in the late 1740s), Voulgaris prefers to quote Aristotle from the fourth book of *Physics*: (Συνάπτομεν γὰρ τὸ πρότερον νῦν, τῷ ὑστέρῳ νῦν καὶ ἓν ποιοῦμεν ἐξαιροῦντες, διὰ τὴν ἀναισθησίαν τὸ μεταξύ).
19. Οὐδὲν ἐστὶ ἡ διαμονὴ πράγματι διακεκριμένον τοῦ διαμένοντος.
20. We shall not discuss here Newton's idea, that God, by existing always and everywhere, constitutes [absolute] duration and space (which occurs in the "General Scholium"), neither his perception of absolute space as God's *sensorium* (which appears in *Opticks* and in *Leibniz-Clarke correspondence*). Although, at first glance, Voulgaris' perception of time displays some similarity to these views, his "entity-oriented" definition of the various aspects of "continuance" points to a quite different direction than Newton's contemplations do.
21. Ἄριστα ὁ Ἀριστοτέλης [...] μηδὲν εἶναι πραγματιῶδες τὸν χρόνον δείκνυσι.
22. Χρόνος γάρ ἐστιν ἡ τῶν ὑφεστώτων διαμονὴ καὶ διάρκεια, εἴτοὖν πρόοδος τῆς τούτων ὑπάρξεως, καὶ προαγωγὴ, καὶ προέκτανσις.
23. According to theorem XXX, "Neither eternity, nor perpetuity are measurable" (*Cosmology*, 124).

REFERENCES

Abattouy, M., J. Renn, and P. Weinig. 2001. "Transmission as Transformation: The translation movements in the Medieval East and West in a comparative perspective." *Science in Context* 14:1-12.

Alexander, H.G. ed. 1956. *The Leibniz-Clarke Correspondence.* Manchester: Manchester University Press.

Carriero, J. 1990. "Newton on Space and Time: Comments on J.E. McGuire" Pp. 109-133 in *Philosophical Perspectives on Newtonian Science,* eds. P. Bricker and R.I.G. Hughes. Cambridge, MA-London: The MIT Press.

Carvin, W.P. 1972. "Leibniz on Motion and Creation." *Journal of the History of Ideas* 33:425-438.

Casini, P. 1988. "Newton's Principia and the Philosophers of the Enlightenment." Pp. 35-52 in *Newton's Principia and its Legacy. Proceedings of a Royal Society discussion meeting, held on 30 June 1987,* eds. D.G. King-Hele and A.R. Hall. London: The Royal Society.

Cassirer, E. 1943. "Newton and Leibniz." *The Philosophical Review* 52:366-391.

Cohen, I.B. and A. Koyré. 1962. "Newton and the Leibniz-Clarke Correspondence." *Archives Internationales d'Histoire des Sciences* 15:63-126.

Cohen, I. B. and A. Whitman. 1999. *Isaac Newton, The Principia. Mathematical Principles of Natural Philosophy. A new translation.* Berkeley, Los Angeles and London: University of California Press.

Dialetis, D., K. Gavroglu, and M. Patiniotis. 1999. "The Sciences in the Greek-speaking Regions during the 17th and 18th Centuries." Pp. 41-72 in (Gavroglu 1999).

Force, J.E. 1987. "Science, Deism and William Whiston's 'Third Way'." *Ideas and Production: A Journal in the History of Ideas* 7:18-33.

Gavroglu, Kostas. ed. 1999. "The Sciences in the European periphery During the Enlightenment." In *Archimedes,* v. 2, series ed. Jed Buchwald. Dordrecht: Kluwer Academic Publishers.

Gay, Peter. 1972. "Why was Enlightenment?" Pp. 61-71 in *18th Century Studies*, idem ed. New Hampshire: University Press of New England.

Guerrini, A. 1985. "James Keill, George Cheyne, and Newtonian physiology, 1690-1740." *Journal of the History of Biology* 18:247-266.

Iltis, C. 1977. "Madame du Châtelet's metaphysics and mechanics." *Studies in the History and Philosophy of Science* 8:29-48.

Klein, Jacob. 1985. "The World of Physics and the 'Natural' World." Pp. 1-34 in idem, *Lectures and Essays*, eds. Robert Williamson and Elliott Zuckerman. Annapolis, MD: St. John's College Press.

Liddell, H.G. and R. Scott. *A Greek-English Lexicon*, (electronic version at www.perseus.tufts.edu).

Madame du Châtelet (Gabrielle-Émilie le Tonnelier de Breteuil, Marquise du Châtelet). 1742. *Institutions Physiques adressées à Mr. son Fils*. Amsterdam (first edition 1740).

McGuire, J.E. 1978. "Existence, Actuality and Necessity: Newton on Space and Time." *Annals of Science* 35:463-508.

———. 1990. "Predicates of Pure Existence: Newton on God's Space and Time." Pp. 91-108 in *Philosophical Perspectives on Newtonian Science*, eds. P. Bricker and R.I.G. Hughes. Cambridge, MA-London: The MIT Press.

Schaffer, S. 1980. "Natural Philosophy." Pp. 55-91 in *The Ferment of Knowledge: Studies in the historiography of eighteenth-century science*, eds. G.S. Rousseau and R. Porter. Cambridge: Cambridge University Press.

Schmitt, C. B. 1983. *Aristotle and the Renaissance*. Cambridge, MA & London: Harvard University Press.

———. 1984.*The Aristotelian Tradition and Renaissance Universities*. London: Variorum reprints.

Thijssen, J.M.M.H. 1992. "David Hume and John Keill and the Structure of Continua." *Journal of the History of Ideas* 53:271-286.

Tsourkas, Cl. 1967. *Les débuts de l'enseignement philosophique et la libre pensée dans les Balkans. La vie et l'oeuvre de Théophile Corydalée* (1570-1646), 2nd revised edition. Thessaloniki (originally published in 1948).

Ventouri, F. 1972. "The European Enlightenment." Pp. 1-33 in idem, *Italy and the Enlightenment: Studies in a Cosmopolitan Century*, ed. Stuart Woolf. London: Longman.

Voulgaris, E. 1805. *Metaphysics*. Venice (in Greek).

———. 1805. *Placita Philosophorum*. Vienna (in Greek).

WOLFGANG LEFÈVRE

DARWIN, MARX, AND WARRANTED PROGRESS:

Materialism and Views of Development in Nineteenth-Century Germany*

> While Marx clearly saw no historical inevitability forcing every society to pass through a capitalist stage, he did see capitalism as a necessary condition for some societies if his vision of a post-capitalist society was to be realised. I should perhaps emphasise that such 'visions' were significant to Marx only if based on concrete analysis of the internal dynamics of development of capitalist society, and of the social forces set in motion in the course of this development. This is all he meant when contrasting 'utopian' with 'scientific' socialism (Stachel 1984, 85.)

INTRODUCTION—DARWIN AS CHIEF WITNESS OF PROGRESS

> Set-backs in political and social, moral and scientific life, which the unified and selfish efforts of priests and tyrants have sought to provoke in all periods of history, might hinder or seem to suppress general progress, but the more unnatural and anachronistic these reactionary pursuits are, the faster and more energetically they induce the progress which inevitably follows them. For, this progress is a law of nature which cannot persistently be suppressed by human power, through either weapons of tyrants or curses of priests.[1]

No doubt, it was a sign of not little civil courage if one dared to voice such thoughts in the Germany of 1863. More remarkable, however, is the identity of the person who uttered them and the occasion he chose to deliver such a statement. The sentences quoted are from the opening address to the 38th Conference of German Natural Researchers and Physicians (*Versammlung deutscher Naturforscher und Ärzte*), given by the morphologist and embryologist Ernst Haeckel (1834-1919). The topic of his speech was, of course, not the political situation of the German states still shaped by the restoration following the failed revolution of 1848. Rather, he was talking about the theory of evolution that Charles Darwin (1809-1882) had published in his *Origin of Species by Means of Natural Selection* from 1859. Haeckel was at that time the most active and influential German propagandist of Darwin's theory, which had already stirred many disputes in Germany, among scientists as well as within the broader public, by the early 1860s.[2] But, as the quotation shows, what Haeckel also propagated was the belief that Darwin had established and proved a "law of progress" for the realm of living beings and furthermore, that this law ruled the history of mankind.

A. Ashtekar et al. (eds.), Revisiting the Foundations of Relativistic Physics, 593-613.
© 2003 *Kluwer Academic Publishers. Printed in the Netherlands.*

This speech, as well as other talks and articles of Haeckel on this topic, should be seen and assessed in the context of the German Scientific Materialism of the time, also known as Vulgar Materialism and associated mostly with the names of Carl Vogt (1817-1895), Jakob Moleschott (1822-1893) and Ludwig Büchner (1824-1899). This materialism played an important and today often underestimated role in German cultural life after the failure of the revolution of 1848. Especially in the 1850s and 1860s, it was accorded a significance which basically overburdened it.[3] Being almost the only voice left under the practice of censorship in the restoration period which could afford to argue against a Christian creed that then dominated public opinion, it was immediately (if indirectly) assigned a critical political character since the alliance of throne and altar had been firmly re-established. The *raison d'être* of this materialism was to confront the Christian creed with results of the sciences—for example, to face the myth of creation with the "conservation laws" of energy and matter (hence the title of the most famous book of Ludwig Büchner, *Kraft und Stoff*), or, to give another example, to confront the doctrine of the immortality of the soul with physiological claims according to which all psychic functions were mere effects of organic activities (Carl Vogt).

Whereas physiology served in the 1850s as the most important weapon in the materialistic side's arsenal in this ideological battle, this role switched over to the Darwinian theory during the 1860s. On the one hand, the biological theory of evolution was employed as an ideological weapon in the same way as the physiological arguments before it—for example, as a refutation of the myth of creation or of the still widespread and popular physico-theology. But, on the other hand, by using the theory of evolution in this way, the ideological criticism of the materialists was transformed into an openly political statement. As the quotation of Haeckel showed, this new openly political character was closely connected with the view that the biological theory of evolution confirmed a "law of progress." If it was a fact established by science that "progress" cannot be held up, then the inevitability of the overthrow of those trying to hold it up was also a fact established by science. And, in view of the clear-cut political situation then given in the German states, the materialists did not even need to name the political and social forces whose fate was thought to be sealed by science.

It is probably unnecessary to stress that Darwin's theory was invoked in this way as chief witness for a conviction which arguably cannot be derived from it. It was a belief which initially existed entirely independently of it and which later hailed Darwin's theory as its supposedly scientific confirmation but whose connection was contingent rather than logical. Even if Haeckel were right, and if Darwin actually had established a "law of progress" for the realm of living beings, it would of course not decide the question whether the history of mankind is necessarily a history of progress. The same is true with respect to the fact that Darwin himself believed that cultivation, competition, and selection of individuals, races, and nations brings about continuous progress within the history of mankind.[4] This belief was not a necessary outcome of his theory of evolution, but shows rather how widespread the conviction then was that the historical development of men constitutes a progressive history, a development through which men obtain ever higher degrees of excellence with respect to technology and science, regarding the forms of their social and political life, and ultimately as moral beings.

In discussions on the use or abuse of the biological theory of evolution for the justification of economical, social, and political convictions, Social Darwinism[5] occupies the centre of interest, and rightly so, if we think about the murderous racism of the Nazi movement. But it should be remembered that this strain of Social Darwinism first materialised in Germany in the two last decades of the nineteenth century, whereas in the 1860s and 1870s, Darwin's theory was used as evidence by those who believed in a progress brought about by solidarity rather than competition or selection. I refer, of course, to the German labour movement taking shape at that time. For Ludwig Büchner, the veteran of the German Scientific Materialism, the "struggle for life" was a unshakable law of nature that also ruled the history of men but had to develop a more humane form in the course of the historical advancement of human reason (*Vernunft*).[6] In contrast with this, the German Social Democrats paid almost no attention to the "struggle for life" or the "survival of the fittest" when claiming Darwin as a witness for their cause.[7] For them, his theory primarily confirmed the principal of development itself, implying a two-fold meaning of this principle. On the one hand, they saw confirmation of the general principle that nothing is permanent and, hence, that the existing social and political order would not last forever. On the other hand, they saw confirmation of their belief that historical development is an objectively progressive development, independent of our wills and wishes, like a law of nature.

In this contribution, I will not go into the political implications of a view which considers the collapse of the criticized social and political state, as well as the emergence of the desired state, inevitable and warranted by certain laws of development. I will not deal explicitly with the problem of reductionism either, that is, with the question of if and how ideas that consider progressive development as a physical necessity miss the peculiarity of the human mode of life and its historical dynamic by ascribing it to alleged (or well established) laws of nature. Rather, I will at first try to outline roughly the ideas of development that referred to a natural law of progress (section 1.1). In a next step, I will briefly examine the relationship of these ideas to much more general ideas of development from that period which defy any classification under the labels "materialism" or "idealism" (section 1.2). I will then confront the ideas of development that considered themselves supported by Darwin's theory with the mechanism of development inherent in the theory of the latter (section 2). Against this background, I will finally discuss some aspects of Marx' theory of history (section 3).

1. VIEWS OF DEVELOPMENT

1.1. *Progress as a Physical Necessity*

Long before the proclamation of the end of history came into fashion, it was already a mark of good breeding to look down on the nineteenth century's unquestioned optimism about progress. It cannot be doubted that the belief in historical progress was an essential component of the political and historical convictions in that century. But, as is well known, this belief was neither peculiar to the nineteenth century nor did it originate with it. It was particularly the eighteenth century that developed the idea that the history of man is a development ruled by laws and following a determined direction—mankind overcoming its barbaric beginnings and achieving ever "higher" forms of civilization. In contrast with this, the historians of the nineteenth century became more and more hesitant to subscribe to laws of history and progressive development.

However, my subject matter is neither the question of whose confidence in historical progress was less deliberate, that of the eighteenth or that of the nineteenth century, nor the attempt to delineate the peculiarity of the nineteenth century's ideas on progress. Rather, my theme is a certain strand of these ideas which apparently occurred first in the nineteenth century—the conviction that the progression of history is governed by a *law of nature*. As the quotation from Haeckel's opening address shows, the use of the term "law of nature" in this context was neither merely metaphorical, nor a rhetorical analogy, nor a reference to the "nature" of man. Rather, it was literally about a law of nature, that is, a law also functioning in nature and which can—at least in principle—be established scientifically.

The fact that convictions of that kind occur within the views of history in the nineteenth century marks a sharp dividing line with respect to earlier ideas on progress. For, by assuming such laws of nature, all the familiar sources and guarantees of progress are dismissed—the perfectibility peculiar to man, man's creativity and inventiveness, and, above all, his ability to reason (*Vernunft*). It was precisely by abandoning the dominant belief that the historical development of man is ultimately determined by the mind—either the human or the divine—that the idea of a progress as a "physical necessity" distinguished itself from the known views of progress and why it deserves attention as an interesting element of nineteenth century's materialism.

The assumption of progress as a physical necessity was as little the outcome of Darwin's theory of evolution as the general idea of a progressive development in history. That becomes immediately obvious if we remember that Herbert Spencer's (1820-1903) theory of universal laws of development, probably the most prominent instance of a theory of progress as a physical necessity at that time, was shaped in its essential features before the appearance of Darwin's *Origin*. Spencer's programmatic essay *Progress: its Law and Cause* was published in 1857.[8] Furthermore, such theories of progress as a physical necessity were not reflections of a historization of the image of nature (*Naturbild*) dawning within the sciences. Notwithstanding evolutionary speculations of nature around the turn of the eighteenth century,[9] the sciences of the first half of the nineteenth century did not actually understand nature as something which evolves historically. The approaches to a historical cosmology or, at least, to a history of the solar system, which Immanuel Kant (1724-1804), Johann Heinrich Lambert (1728-1777) and Pierre Simon Laplace (1749-1827) had attempted in the eighteenth century were all but forgotten by the middle of the nineteenth century, dismissed as a speculative ballast by the scientific community of astronomers. It is true that the geologists of the time did agree, in light of the then known palaeontologic facts, that the state of the planet Earth must have undergone considerable changes in its past. But whereas these antiquities of the Earth testified to a history of repeated catastrophes intelligible only as God's interventions for those in the tradition of George Cuvier (1769-1832), for those in the tradition of Charles Lyell (1797-

1875), they proved the periodical oscillations of states that show the Earth in a state of dynamic equilibrium if a sufficient long period of time is taken into account. It seems to be clear that neither can be called a historical theory of nature. Darwin's theory of evolution was so desirable and attractive for those who believed in progress as a physical necessity precisely because they had not been able, up to that point, to refer to a scientific theory that dealt with historical developments in nature to support their conviction. For instance, Spencer's starting-point within the realm of science was neither cosmology nor geology, but the morphological theories of the 1830s and 1840s, which claimed that the arrangement of the morphological forms within taxonomic classes conformed to a principle of increasing differentiation of function, to the principle of increasing "physiological division of labour" as Henri Milne Edwards (1800-1886) put it.

Given this state of affairs, one may ask what explanatory models of development did exist then that could legitimately be invoked as models of scientific explanations? To my knowledge, there are only two such explanatory models of development within the realm of sciences of modern times preceding Darwin. One is a pure mechanistic model, in which development results from the mechanical interactions between the entities involved. Rene Descartes' (1596-1650) cosmology, or Kant's and Laplace' hypotheses on the emergence of the solar system, are good examples for this first model. The other model supposes a special cause of development in the developing object. Good examples for this second model are theories of ontogenesis supposing an inner mechanism of development, either a specific force, a formative drive, or the working of a specific structure.

1.1.1. Accumulation of Effects
The first model, which explains development as result of merely mechanical interactions between material entities, allows a variety of forms in which states can succeed each other—oscillating successions like the alternation of day and night, or successions periodically moving through a cycle of states like the sequence of the seasons, or irreversible successions following a determined direction like the sequential order of geological strata. Furthermore, it is possible to connect this last form of succession with the assumption of progress. This would be the case, for instance, if the succession of states is not understood merely as a transition from one state to another but as a process of cumulative transformation, that is, as a process in which the structures achieved at a certain moment serve as new points of departure for further advances or further differentiation. Herbert Spencer's law of "advance from homogeneity of structure to heterogeneity of structure" (Spencer 1891, 19) seems to exemplify to this model, as does Jean Baptiste Lamarck's (1744-1829) theory of transformation of species from the beginning of the nineteenth century.[10] Needless to say that it is impossible to achieve neat results, like a sequence of animal forms in the case of the latter theory, from purely mechanical interactions without explicitly or implicitly assuming that certain stable constraints and necessary prerequisites are always given.

However, because of its indefinite nature, this mechanistic model was not very suitable as an explanatory model of progress in history. It is true that this model is strictly deterministic, and supposes that every given state of the natural world is completely determined by a former state. This meant that if the state of the world at its beginning were known, it would be possible to calculate every state of the world, the present as well as any of the future—possible, that is, at least in principle, or for the *Laplacean Demon*. But, as Friedrich Engels (1820-1895) has observed,[11] it makes no real difference whether you call an event completely determined in this sense or a coincidental result. One should recall that this mechanistic model was originally framed by philosophers like Descartes and Kant for demonstrating the possibility to comprehend a *given* structure of high complexity, namely the solar system, as the mere result of mechanical interactions between particles of matter. No wonder, therefore, that this explanatory model had to yield more than it could if applied as a framework of laws that have to guarantee the progress of historical development.

1.1.2. Programmed Development
The second explanatory model, developed in the context of biological theories and assuming an inner mechanism of development, was as strictly non-teleological and strictly deterministic as the first one. It was, however, at the same time, an attempt to overcome the limits of the mechanistic model. In the framework of the latter, it was not possible to explain plausibly why a germ covers all the stages of ontogenesis peculiar to its species and matures to the specific adult form of its species purely by a blind mechanistic interaction with its environment. The supposition of an inner mechanism seemed to be much more adequate for an account of the ontogenetic development regardless of how the biologists conceived this inner mechanism of development—as a specific force, a *vis vitalis* (*Lebenskraft*) inherent in all organic matter as assumed by Caspar Friedrich Wolff (1738-1794), as a formative drive (*Bildungstrieb*) peculiar to all organisms as supposed by Johann Friedrich Blumenbach (1752-1840), or as the outcome of the specific structure of organisms (Lamarck).

It was the epigenetic view of the processes of ontogenesis that forced biologists to grant the germ the ability to develop, under normal conditions, in a way typical to its species from an almost unstructured stage at the beginning to the articulated and differentiated form of the adult organism. It was not a valid scientific assumption to take this ability as a faculty which anticipated the mature stage in the manner of subjective teleology. Natural processes of development therefore had to be conceived which were non-teleological but nevertheless followed a determined direction. If it is permitted to call the mature stage of an organism a "higher" stage of development than that of the fertilized egg, then they would constitute in addition a form of "progress." While it is true that my expressions become more and more tautological as I progress, there seems to be an inevitability to it at this point, because the process of ontogenesis, if conceived of as an epigenetic process, is the proper paradigm for the idea of a progressive development towards a "higher" stage of perfection—a paradigm with precarious consequences.

1.2. Progress as Destination and Justification

The ontogenetic process of maturation apparently served as *the* natural model-process of development, at least for the nineteenth century's ideas of progress. It did so entirely independently of the explanatory efforts of biologists and the philosophical orientations of those convinced of a progressive development, that is, regardless of whether they subscribed to a materialistic or an idealistic philosophical system.

This paradigm of maturation points back to primeval forms in which history was conceived of according to the paradigm of the life-cycle. This paradigm was not silent about decline and death that comes after the mature stage, considering (and rationalizing) it as a necessary stage in the reproductive process by which life renews itself. This cyclic view of history, rooted in a relationship with nature peculiar to dominantly agrarian societies, imagined the life of the collective as lasting eternally. Taking all stages of the life-cycle to be present at the same time, this view of history comprised neither past nor future nor the idea of a destination which had to be realised in the course of history.

By confining its view to the rising half of the life-cycle, the paradigm of maturation makes ideas of development conceivable in which the subject of history has a future, a destination. It is advancing towards the state that corresponds to its essential nature. Oversimplifying matters tremendously, we can distinguish two basic views of the form in which the subject of history accomplishes its course.

According to the first of these views, history is a process in which several obstacles have to be overcome and in which pains and problems connected with birth and growth are an inevitable part of a ceaselessly advancing process—*per aspera ad astra*. Good examples for this view are Auguste Comte's (1798-1857) three-stages theory of the development of the human mind, the ideas of progress as a physical necessity nourished within and by Haeckel's *Monistenbund*, or the confidence in future developments widespread among German Social Democrats following Eduard Bernstein (1850-1932).

According to the second view, the development towards the essential state of mankind is a highly dramatic process in which phases of steady advance are interrupted by upheavals, revolutions, and crises in which the entire success of the process is at stake. With respect to nineteenth-century Germany, it is inevitable to mention Georg Wilhelm Friedrich Hegel (1770-1831) in this context. His conception of development is also shaped by the paradigm of ontogenetic maturation,[12] but it is not the conception of a steady and gradual process of development from the stage of "in itself (*An sich*)" to that of "for itself (*Für sich*)." Hegel conceives of this process rather as a contradictory one, comprising struggles for life and death in which the essential state can only be achieved through stages of "self-externalization (*Selbstentäußerung*)" and "self-alienation (*Selbstentfremdung*)." The conception of loss-of-oneself (*Selbstverlust*) as an indispensable prerequisite for gaining one's proper nature goes back not only to Jean Jacques Rousseau (1712-1778), but also to the above-mentioned mythological rationalizations of death. It is, however, essential to notice a decisive difference here. In the old paradigm of the life-cycle, death was rationalized as a necessary mediating step in the reproduction of life. In contrast with this, within the frame of the ontogenetic paradigm, death is rationalized as a ferment of maturation.

Alienation from oneself as the essential intermediate stage of the course from undeveloped beginnings to the fully evolved essential nature remained a fundamental pattern in the conceptions of development among the disciples of Hegel, and even of those who had publicly renounced the doctrines of their former master. This pattern occurs in Ludwig Feuerbach's (1804-1872) philosophy as well as in the writings of the young Karl Marx (1818-1883)—the "complete loss of man" is thought of as a prerequisite of its "complete re-winning" in the famous *Foreword* to his critique of Hegel's *Philosophy of Right*[13]—and of the old Friedrich Engels who distinguished three principal steps—primeval communism / private property / modern communism—through which world history accomplishes its course.[14]

It is not possible to discuss here the precarious implications of the ideas of development following the ontogenetic paradigm as elaborately as they deserve. To mention just a few: by assuming an inner force of development—regardless of whether in a teleological or in a non-teleological sense—these ideas suppose a subject of development which is, in principle, the unit of development in the same way an organism is the unit of the ontogenetic process. The mind of historical mankind (Comte), or the Human species (Feuerbach), or communism (Engels), was then embryonic in its beginning stage, like an organism, and developed in a programmed way towards its mature form. Such an elevation of abstractions to the rank of autonomous entities which, in this case, are moving as maturing spirits, does certainly not favour soberminded studies of historical processes and their forms, but might deserve respect as an indication of an understandable belief in history having a sense, even one justified by nothing more reliable than a *horror vacui*. Much more precarious than this is probably the fact that these ideas are not less deterministic than the mechanistic model framed within the context of cosmology, as discussed above. Views of development that follow the ontogenetic paradigm not only understand the future fatalistically as destiny. They also consider the mutilations and devastation of human lives produced by real history vindicated within this model of development. They are justified as pains of martyrs, as sacrifices, or merely as inevitable costs of progress.

In this contribution, I will content myself with the remark that those models of development in the second half of the nineteenth century that claimed progress according to laws of nature did not provide alternatives to this precarious paradigm. Rather, they must be seen as mere variants of this archaic scheme. The theoretical naiveté of these models can probably be best made clear by a confrontation with Darwin's theory of evolution or, at least, with those features of his theory which are of special interest in this respect.

2. DARWIN'S THEORY OF EVOLUTION[15]

"Variation" and "natural selection" are key notions for the mechanism by which Darwin tried to explain the alterations of species. This is well-known. It might be less known that the assumption of an alteration of species became originally attractive for Darwin as a possible scientific explanation of the phenomenon that species are generally well adapted to their natural environment.

The often stupefying phenomena of adaptation had long served as an inexhaustible reservoir of suitable objects for physico-theological contemplation. As long as one could assume a by-and-large stable natural environment, these phenomena could be admired as the "wise pre-ordination" of nature by its "author." If, however, the natural environment proved to be not stable, such a "wise pre-ordination" did not any longer suffice. Rather, one was in need of an account not only for the phenomenon of adaptation as such but also for its continuous adjustment. And this was exactly the situation in the first half of the nineteenth century when geology and paleontology showed historical changes in the conditions of life—changes of the "places," in Darwin's words—which could not be ignored. On the background of these geological changes, a scientific explanation of adaptation, that is, one without recourse to supernatural causes, was an entirely new challenge. This was Darwin's starting-point from which the decisive assumptions that shaped his initial investigations become clear.

From the outset, Darwin ruled out the possibility that species may undergo alteration because of an internal developmental cause like that of the ontogenesis of an individual organism. In view of the problem of adaptation, a conception of the phylogenetic development of species as an internally determined process would lead to the monstrous assumption of a "pre-stabilized harmony" between the geological processes of the Earth and the supposed developmental determination innate to species. This is the background to why Darwin consistently did not use the term "evolution." This starting point shaped furthermore his initial view of the alterations of species which considered them merely reactions to the changing geological states of the Earth. Geological changes played the active part in this process, the living beings the passive role. Moreover, in connection with that view, he took the historical development of species to be a process of replacement of poorly adapted variants of a species by fitter ones. As he later grasped, this may account for the phenomenon of adaptation, but not for the historical development of living beings on the Earth. Their development is in general characterised by an increase of the number of species, in spite of massive extinction and, above all, the differentiation of species into descendent species with different directions of development ("development of divergence").

Darwin achieved his final solution not through deeper consideration of the general problems of development but by gaining new insights into the peculiarity of his object—the *teleonomic*[16] nature of the relationships among organisms and with their environment. In order to clarify the point it might be helpful to look at some of the crucial revisions of certain of his initial assumptions which helped Darwin finally understand the development of divergence as the key for the historical development of the life forms.

Firstly, he modified his understanding of adaptation. Whereas his initial conviction had been that the pressure of selection leads to a "perfect" adaptation to a certain "place," he later came to assume that it only caused adaptations sufficient for a given constellation of competition among certain species. That means Darwin no longer believed that the "place" completely determined the structure of the organisms that utilized it as a resource of their life. Rather, the "place" determined the parameters of performances in connection with its utilization only in the rough sense of a frame, but did not determine the structures of organisms that proved to be suitable for such performances.

Linked with that, Darwin revised secondly his initial conception of an one-way dependency of the organism's evolution on the "place." As Darwin came to realize, the perspective that the "place" determined the development of organic structures (in the aforementioned rough sense) had to be completed by the opposite perspective that the organic structures developed at a certain time also determined what counts as a "place." To give an example, without structures that enable organisms to live outside of the sea, land is only potentially a "place." It was only with the development of the first organisms with structures suitable for a life on shore that it was realised as a "place."

Based on that realization, Darwin developed thirdly a view which was crucial for his understanding of the development of divergence: the pressure of selection favours not only those mutants better able to use an already realised "place," but also those able to use new, not yet realised "places" as resources. In addition, this "creation" of new "places" was two-sided: new organic structures did not only make new resources of life accessible, but constituted themselves new resources of life, if other new structures used them as such resources.

The historical process of development and differentiation of the organic forms thus appeared as a process induced and propelled by this differentiation. With respect to the ideas and models of development discussed in this contribution, I would like to summaries this process as follows.[17]

Alteration is part of the reproduction of living beings. These alterations are not caused primarily by external factors, i.e. changes in inorganic conditions. The prevailing development of divergence in the realm of living beings, the increasing differentiation of function and the organic structures' increasing capacity for performance are therefore not side effects caused by geological changes. Thus, evolution is not the result of an accumulation of effects originating in periodic adaptations to different geological states. Based on such adaptations to the respective inorganic conditions, the historical development of organic forms is an internal process to the extent that it results from the 'self-relation' of the biosphere. That must not be mistaken as a 'self-relation' of "life." We are not dealing with relationships that abstract concepts might have in this article. Rather, our topic is the relationships between species, or to be more precise, among the individuals of such species, which are external and independent beings to each other.

It is true, this internal development of organic forms does not rest on a developmental urge innate to organisms. It is a process, however, which is initiated and maintained independently of external impulses and incentives. In addition, it is a process with the general tendency to multiply types of organisms through a development of divergence, to create structures with more differentiation of function, and to make them more suitable for fulfilling increasingly difficult tasks.

This tendency cannot be comprehended adequately by the concept of an accumulation of effects. It is true that within this process, the result reached at a certain time serves as the starting point for further development, and it may be possible to plausibly explain differentiation in this way. But it does not explain the correlation among the types of organisms, which makes up the peculiarity of this differentiation along with its tendency. This correlation has a teleonomic character. Within this correlation, each organic structure constitutes a means of appropriating resources and is, at the same time, itself a potential resource. Furthermore, the development of structures suitable for the appropriation of such potential resources receives its impetus from the "struggle for life." The process then continues on a higher level. It is this teleonomic peculiarity of the correlation among the types of organisms that introduces what can be characterised as the tendency to develop and optimize "Nature's technology" (Marx and Engels 1991, XXXV 375 fn 4; 1956 XXIII 392 fn 89).

The historical development of the types of organisms thus has a "logic"; it is subject to the laws governing the development of "Nature's technology." That does not mean, however, that the process involved can be calculated. It is impossible to predict how the potential within a certain stage will be realised, because the factors which work together in this process are not coordinated by a law. Each tiny accidental step, which partly releases such potentials, generally has irreversible consequences and cuts off the realization of others. The accidental event thus irrevocably determines the further development, and becomes a necessary condition of it. The historical development of organisms is thus not only an irreversible process, but also unique and unrepeatable. It is an open process, in spite of its "logic" of development, to the extent that its results cannot be derived from origins that predetermined them. The historical development of the organic forms on our planet is thus, according to Darwin's theory, a self-activating, irreversible, and unrepeatable process which is subject to certain laws and tendencies. It does not, however, have a goal. Despite uninterrupted causal determination, the outcome of this process is not "inevitable," but necessary and accidental at the same time.

Naturally, it was not the intention of this short depiction of some basic features of Darwin's theory of evolution to recommend the replacement of certain old yet still influential paradigms of development by this theory. Darwin's theory is one[18] theory of the historical development within a special sphere of nature, that is, within the realm of living beings, which deserves attention - not because it can be generalized but because it comprises possible ways to frame theories on development which are not yet well-known. The purpose of this depiction was, rather, to show that in the second half of the nineteenth century, the peculiarity of development Darwin had conceived of was not grasped fully by the advocates of ideas of development thought to be supported by Darwin. Although one has to admit that Darwin can be misunderstood easily, since the basic features of his theory emphasized here remain rather implicit in *The Origin,* these faulty interpretations of his model of development were probably not due to misunderstandings alone.

In any case, with respect to one crucial point, Darwin was not misunderstood by his adherents. They understood that he did not offer any explanation of the direction of development; they realised that his theory did not grant a permit to conceive of, for instance, the origin of *homo sapiens* as a necessary result of the biological evolution. "I believe [...] in no law of necessary development" Darwin wrote in *The Origin* (Darwin 1859, 351). Even though he used the term "progress" in *The Descent of Man* (1871), he never distanced himself from this position. That was soon generally considered a decisive fault of his theory of evolution and ultimately led to the first neo-Lamarckistic reaction. Thus, his adherents tended to "amend" his theory in the sense of the old paradigms of development rather than searching for new insights from it.

3. A DARWIN OF THE HISTORY OF MANKIND?

A contribution on materialism and views of development in the nineteenth century that mentioned the name Karl Marx only casually would probably look strange even now when his theories have lost much of their appeal. On the other hand, it would not be reasonable to discuss his theory of history briefly in the final section of such a contribution, since the interpretation of this theory is still highly controversial.[19] It might be fitting instead to discuss some of Marx' remarks on Darwin's theory, which can lead to insights into his understanding of history.

3.1. "Natural-historical Basis"

About one year after the appearance of Darwin's *Origin,* Marx observed in a letter to Engels that Darwin's book provided the "natural-historical basis for our opinion."[20] A short time later, he wrote to Ferdinand Lassalle (1825-1864) that Darwin's book "serves me as a basis in natural science for the class struggle in history."[21] This latter phrase has often impelled Marxists to stress hastily that Marx never used this formulation again[22] defending Marx against the suspicion of Social Darwinism. Marx doesn't need this protection.

To my knowledge, Marx and Engels were among the first who opposed the opinion that the "struggle for life" was an universal law which provided a key for the understanding of the history of man as well as for that of the biological species. This issue was their main objection to Friedrich Albert Lange's (1828-1875) essay Die Arbeiterfrage.[23] This critique is by no means only a tactical one,[24] that is, it cannot be derived from the fact that Marx and Engels considered Lange a rival in their struggle for influence upon the German working class. Rather this critique pertained to the core of Marx' argument against classical economical theories' conception of economical laws originating with and peculiar to capitalistic economies as universal laws valid in all historical eras. This critique also constituted the background of Marx' well known statement that, in Darwin, "the animal kingdom figures as civil society."[25] Marx considered a rendering of economical laws of the modern capitalistic societies as historically universal laws not only confusing, but an ideological deception that created the illusion that these laws were the "natural" laws of every economy. Thus, it does not need much fantasy to imagine how he assessed essays in which modern social relations were portrayed as relations that are basically not different from that between animals and plants.

For an understanding of what Marx might have meant when he wrote that Darwin's work provided the "natural-historical basis for our opinion," it is important to remember that Marx and Engels had believed since the 1840s that "positive science" (*positive Wissenschaft*), which they thought should replace philosophical speculations, had to be a *historical* science which addressed nature as much as the world of men. To quote a famous phrase from the *Deutsche Ideologie*: "We recognize only one kind of science, namely the science of history. Seen from two different points of view, history can be classified as history of nature and as history of men. The two sides are inseparable, however."[26] But, as stated above, the scientists of the nineteenth century before Darwin did not understand nature as something that evolved historically. Hence the enthusiastic reaction with which Engels hailed Darwin's *Origin*: "Never before has so grandiose an attempt been made to demonstrate historical evolution in Nature, and certainly never to such good effect."[27]

It was not for the sake of coherence of his theoretical system why Marx was particularly interested in the connection between history of men and history of nature (Arndt 1985, 96). For Marx, both histories were "inseparable" in practice. In the 1840s, his attention may have centred mainly on the historical changes of nature caused by the historical development of agriculture and industry, but in the 1850s, he was more interested in the question whether nature might constitute a limit for the historical development of man (Arndt 1985, 84). In 1851, he wrote in a letter to Engels: "But the more I go into the stuff, the more I become convinced that the reform of agriculture, and hence the question of property based on it, is the alpha and omega of the coming upheaval. Without that, Father Malthus will turn out to be right."[28] It is not by chance that Marx was among the first (if not the first) who recognized the implications of Darwin's theory for Malthus' "law of population." In economic drafts of the early 1860s, he noted with evident satisfaction: "In its excellent book, Darwin did not see that he toppled Malthus' theory by discovering the 'geometric' progression within the kingdoms of animals and plants. [...] In Darwin's work [...] the natural-historical refutation of Malthus' theory can be found in detail as well as in fundamental principle."[29]

For Marx, the history of man possessed a "natural-historical basis." According to his own understanding, the materialistic character of his view of history depended not at least on the insight that the history of the relation between man and nature constituted the basis of the history of man.

3.2. "The Law of Development of Human History"

In the address delivered at Marx' burial, Engels stated: "Just as Darwin discovered the law of development of organic nature, so Marx discovered the law of development of human history."[30]

Was it not Marx and Engels who emphasized that the laws which rule social life possess a historically specific character, and that it was decisive not to be content with establishing laws valid in all historical ages but to investigate in detail how they act within specific historical forms of society? Thus, Marx reproached F.A. Lange's Social Darwinistic essay *Die Arbeiterfrage* for "subsuming the whole history [...] under a single law of nature," namely under the "struggle for life." "Instead of analyzing how the 'struggle for life' turns out within different forms of society, one has merely to insert the phrase 'struggle for life' for any real struggle and to put this phrase in the Malthusian 'population-fantasy'."[31]

Nevertheless, Marx did state very general principles concerning history. There is, for example, the well known *Preface* to his *Zur Kritik der politischen Ökonomie* from 1859 with its famous phrase: "The general conclusion at which I arrived and which, once reached, became the guiding principle of my studies, can be summarized as follows: In the social production of their existence, men inevitably enter into definite relations, which are independent of their will, namely relations of production appropriate to a given stage in the development of their material forces of production," and so on.[32] These propositions contain not only the "principal features" of the "materialist conception of history," as Engels understood it,[33] but can also be read as the formulation of a law of human history applicable to all historical epochs. There are other general statements of this kind in Marx' writings which can be understood as laws of history. For instance, comments on the mode of transmission and continuity peculiar to the history of man, in contrast with genetic heredity in biology and thus with the specific mechanism of transmission and continuity in biological evolution: "The simple fact that every succeeding generation finds productive forces acquired by the preceding generation and which serve it as the raw material of further production, engenders a relatedness in the history of man, engenders a history of mankind."[34]

With arguments like these, which had to be completed by further ones,[35] Marx tried to frame mechanisms of development peculiar to the history of man. It is precisely the formulations of such general mechanisms by Marx that can be compared with Darwin's findings and formulations regarding the general mechanisms peculiar to the biological evolution of living beings. Whereas hardly a counterpart can be found in Darwin's work to Marx' attempts to detect laws of development peculiar to historically specific types of society, it is possible, in my opinion, to draw a parallel between Marx' arguments about *general* forms in which human societies develop historically and Darwin's arguments about *general* forms of the evolution of species. No matter how different their contents, these arguments seem to be situated on the same level.

It is not only the categorical level that these arguments have in common, there is further correspondence. A predictable course of history follows neither from the general forms of historical development that rule the human history, according to Marx, nor from the forms detected by Darwin for biological evolution. It is just as difficult to derive from Darwin's theory of evolution that reptiles had to arise from fish, and birds and mammals from reptiles, as it is to derive the scheme of history known as *Historical Materialism* from the general laws of human history stated by Marx. According to this scheme, mankind's historical development, beginning with primal communistic collectives, covers a fixed sequence of specific societies with private property and a class-structure—societies based on the exploitation of slaves, feudal societies, bourgeois societies—and will necessarily reach a final form of society without class differences: the modern communism. Rather, according to Marx' general laws, history is not only an irreversible and unique process which cannot be repeated but also a process with open end despite its conformity to laws. In these respects, there is a striking congruity between Marx' theory of the history of mankind with Darwin's theory of the historical development of the realm of living beings—with possibly one exception: Darwin's general mechanisms of biological evolution include a tendency; it seems unclear to me if the same is true with respect to the general mechanisms of human history formulated by Marx.

3.3. "Necessary Progress"

Except for the so called "Chapter on Feuerbach" of *Deutsche Ideologie,* written conjointly with Engels in 1844/45 but not published during his lifetime, Marx never set down his general theory of history in an essay or book. His main interest was the specific historical dynamics of modern bourgeois society, and he produced well known, controversial, extremely far-reaching statements on the tendencies inherent in this type of society.

It was the goal of his life's work to prove not only that the dynamics of the capitalistic economic system is a *circulus vitiosus* but that this system also necessarily produces precisely the results which, according to Marx, constitute the decisive prerequisites for a new form of society—namely, means of production that could only be applied by society, not by private owners, and a class of workers who, not being private proprietors of these means, were able to apply them co-operatively. He was convinced that the capitalist society created these prerequisites independent of the awareness and will of its acting subjects. At least until the 1870s, he believed, additionally, that the capitalist system was a necessary, inevitable form of society that could not be omitted by revolutionary will because it provides the starting-points for a socialist society. Thus, as he wrote in the *Foreword* of *Das Kapital*: "And even when a society has got upon the right track for the discovery of the natural laws of its movement—and it is the ultimate aim of this work, to lay bare the economic law of motion of modern society—it can neither clear by bold leaps, nor remove by legal enactments, the obstacles offered by the successive phases of its normal development. But it can shorten and lessen the birth-pangs."[36] In view of the social developments in Russia, Marx later modified his conviction that the capitalism constituted an inevitable stage of development, even for those societies which were not yet affected by it.[37]

What is important in the context of this article is the following point. Marx risked statements about crucial prerequisites for the historical emergence of a certain form of society and about whether these prerequisites are produced by the preceding form of society by chance or by necessity. But such statements must not be mistaken for the claim that the historical succession of different types of societies results from a law of development. One has to be the more careful in this respect since Marx, indeed, attempted repeatedly to design schemes of the historical succession of types of society. In this contribution, it is neither possible to discuss elaborately which schemes Marx tried out at different times nor to show how their increasing complexity reflected his expanding historical knowledge, especially his studies on historical as well as contemporary economic systems outside of Europe.[38] I will only try to outline their principle.

Marx saw the coexistence of different types of economy in history as well as the different ways in which these forms had developed in the past—partly independent of each other, partly in interaction. He realised, furthermore, that a given constellation of coexisting forms of society is just as decisive for the further development of each of the forms involved as their inner potential of development. In order to grasp this complicated relationship, he made use of the geological concepts of "stratum" and "formation." He distinguished, on the one hand, the different forms of society chronologically in correspondence to different "strata" with which their type originated and, on the other, marked the complex patterns shaped by coexisting forms as different "formations" and tried to classify these latter as "secondary," "tertiary" etc. formations (following the example of the geological distinction between "secondary," "tertiary" etc. transformations).

The most precarious point of this procedure was, of course, classifying the "formations." The geological analogy suggested two different possibilities. First, a morphological classification, in which the degree of transformations the aggregation of originally independent strata has undergone provides the criterion, and second, a diagnostic classification, in which the absence or presence of certain strata is decisive, similar to paleontology where the absence or presence of certain fossils, the so called "characteristic fossils" (*Leitfossilien*), is the key for the diagnosis of strata. Interestingly, in his several attempts to schematize, Marx did not classify the "formations" morphologically in correspondence to degrees of entanglement but by whether or not certain "characteristic strata" occur in them. The key was whether the "formations" consisted of forms of society that all belong to the type of "primal collective" (*Urgemeinschaft*), whether they contained as well forms "based on slavery, serfdom," or, finally, whether they comprised even "bourgeois" forms.[39] However, these "characteristic strata" were not actually of the same diagnostic significance for Marx that "characteristic fossils" are for the paleontologist. Rather, the types of society which characterize these strata are singled out by Marx because of the significance he accredited to them with respect to the emergence of prerequisites necessary for the genesis of the modern form of society.[40]

Marx was not only completely conscious of the methodical procedure applied but also explained it with his famous aphorism that "human anatomy contains a key to the anatomy of the ape:"[41] The significance of certain historical developments for a subsequent development only becomes visible through the latter. Starting with the already developed matter in question, more precisely with the insight in the preconditions of its existence, the historian can single out those factors which can or must be understood as prerequisites for the emergence of these preconditions. This methodical procedure is entirely normal and inevitable, and is, either with consciousness or not, applied in every historical reconstruction of a given result of history. However, as Marx warned and stressed,[42] this procedure of retrospective reconstruction of historical prerequisites can easily cause the illusion that the identified historical prerequisites of a historical matter came into being because of that matter.

It is an extremely important warning, indeed, since these retrospective historical reconstructions are not only necessarily one-sided, but suggest a teleological understanding of history if the historian looses sight of their heuristic function. In addition, by singling out historical developments as significant, since some components of the interesting historical result originated with them under certain circumstances, the historian inevitably presents these developments as necessary stages of the historical emergence of the result. It is, however, obviously that this necessity arises from the theoretical insight into the framework which these components form within the *already developed* historical matter in question. Thus, it may seem as if this framework itself secretly directs the historical process. And this illusion might become the source of the *quid pro quo* of mistaking the theoretical development of a subject, i.e. the theoretical unfolding of the interrelations among its components, for the reconstruction of the intelligible core of its historical development that is ruled by laws and can be understood rationally.

Such a lack of a clear distinction between the "logical," that is, the conceptual unfolding, and the "historical," the historical development in time, was not only the secret core of Hegel's conception of development, as Marx rightly saw.[43] It was also a characteristic feature of the interpretations of Marx' writings since their beginning. We could start with Engels, who took Marx' theoretical development of the form of value (*Wertform*) in *Zur Kritik der Politischen Ökonomie* (1859) to be an account of the crucial stages through which this form developed historically—"only stripped of the historical form and diverting chance occurrences."[44] But understood in this way, the development of this form follows the model of development shaped by the paradigm of the ontogenetic process of maturation, with all its precarious implications for the doctrinal *Historical Materialism*.

Coming back to Marx' "formations"-schemes of the historical forms of society, we had seen that he combined in these schemes the sober-minded mechanical model of development-as-accumulation-of-effects applied by geologists with the retrospective methodical procedure of reconstruction applied by historians. This complex combination misled even Marx now and then to take for historically necessary what seemed to be so only because of this kind of reconstruction. Two remarks with references to Darwin may substantiate this point. In 1866, in a letter to Engels, Marx mentioned a book almost unknown today, Pierre Trémaux's *Origine et transformations de l'homme et des autres êtres,* and praised it because "progress, purely accidental according to Darwin, is here [sc. according to Trémaux] necessary."[45] Necessary progress? One year later, again in a letter to Engels in which Marx drafted topics for a planned review of *Das Kapital* by Engels, we find the following formulation, which shows that the above-mentioned parallel between Darwin and Marx seems to be an invention of the latter: "By proving that the present society, regarded economically, bears the seeds of a new higher form, he [sc. Marx] is demonstrating only the same gradual process of revolution within the field of society that Darwin has demonstrated for the field of natural history."[46] What had Marx proved, according to his own understanding? Had he proved that the present society is necessarily producing prerequisites which permit the realization of a socialist form of society based on the collectivization of the means of production, or had he proved that "the present society bears the seeds of a new higher form"? In any case, it is possible that such formulations indicate only that Marx, in spite of his theory, was not less influenced by the belief in progress of his time than Darwin, in spite of his theory as well.

Max Planck Institute for the History of Science

NOTES

*　　This is the translation of a slightly revised article with the title "Darwin, Marx und der garantierte Fortschritt: Materialismus und Entwicklungsdenken im 19. Jahrhundert" published in: Arndt, Andreas and Jaeschke, Walter (eds.): Materialismus und Spiritualismus. Hamburg 2000. I wish to thank the editors for granting permission to this republication.

1. My translation. R ckschritte im staatlichen und sozialen, im sittlichen und wissenschaftlichen Leben, wie sie die vereinten selbsts chtigen Anstrengungen von Priestern und Despoten in allen Perioden der Weltgeschichte herbeizuf hren bem ht gewesen sind, k nnen wohl diesen allgemeinen Fortschritt hemmen oder scheinbar unterdr cken; je unnat rlicher, je anachronistischer aber diese r ckw rts gerichteten Bestrebungen sind, desto schneller und energischer wird durch sie der Fortschritt herbeigef hrt, der ihnen unfehlbar auf dem Fu§e folgt. Denn dieser Fortschritt ist ein Naturgesetz, welcher keine menschliche Gewalt, weder Tyrannenwaffen noch Priesterfl che, jemals dauernd zu unterdr cken verm gen. (Haeckel 1924, V 28.)
2. The first German edition of Darwin s *On the Origin of Species,* translated by the palaeontologist Heinrich Georg Bronn, appeared in 1860.
3. See (Lef vre 1992).
4. See, for example, (Darwin 1871, ch. 5; Greene 1977).
5. In accordance with (Zmarzlik 1963), I will apply the term Social Darwinism only to those social or political views referring to biological theories that particularly emphasise the topic of selection.
6. See (B chner 1872). Very similar arguments can be found in (Lange 1865).
7. See, for example, (Bayertz 1982).
8. Published again in (Spencer 1891, I 8-62).
9. One has to be careful at this point. Many of these speculations for instance, those of Johann Gottfried Herder (1744-1803) or of Jean-Baptiste Robinet (1735-1820) were embedded in the Great-Chain-of-Being idea which is basically a-historical. The same holds more or less for the *Deutsche Naturphilosophie.* Its conception of nature as evolving through different levels of complexity meant generally a conceptual unfolding of these levels rather than their historical evolution in time. The confusions that are easily prompted by this conception with regard to views of history will be addressed in the last section in the context of Marx revisions of Hegel s conception of development.
10. See, for example, (Lef vre 2001).
11. See (Marx and Engels 1956, ff., XX 486ff.; 1991, XXV 497ff).
12. Ironically, Friedrich Engels later tried, in a letter written at the end of the 1850s, to clarify the embryology of his time by returning to Hegel: The cell is Hegelian being in itself and its development follows the Hegelian process step by step right up to the final emergence of the idea i.e. each completed organism — (Marx and Engels 1991, XL 325). Die Zelle ist das Hegelsche Ansichsein und geht in ihrer Entwicklung genau den Hegelschen Proze§ durch, bis sich schlie§lich die Idee , der jedesmalige vollendete Organismus daraus entwickelt. — (Marx and Engels 1956, XXIX 338).
13. (Marx and Engels (1991, III 13). Der v llige Verlust des Menschen als Bedingung seiner v lligen Wiedergewinnung - (Marx and Engels 1956, I 390).
14. See, for instance, (Engels 1884; Marx and Engels 1956, XXI; 1991, XXVI).
15. The following interpretation of certain features of Darwin s theory of evolution is not always adapted to the prevailing accounts of this theory and needed therefore many quotations and detailed argumentation for its support. However, since that is not possible in this contribution, I want to refer generally to the detailed argumentation in (Lef vre, 1984, ch. 6).
16. See for this notion (Mayr 1974, 91-117).
17. The following four paragraphs are taken from (Lef vre 1984, 260-62).
18. The partly far-reaching modifications of the theory of evolution during the twentieth century are not of interest in the context of this contribution.
19. To my knowledge, the most differentiated interpretation can be found in (Arndt 1985).
20. Marx and Engels (1991, XLI 231) [ ] die naturhistorische Grundlage f r unsere Ansicht (1956, XXX 131).
21. (Marx and Engels (1991, XLI 245) [ ] pa§t mir als naturwissenschaftliche Unterlage des Klassenkampfs (1956, XXX 578).

22. See, for instance, (Bayertz 1982, 108).
23. See above all Marx letter to Kugelmann from June 27, 1870, - (Marx and Engels 1956, XXXII 685f.; 1991, XLIII 527f).
24. See, for instance, (Groh 1981, 233f).
25. Marx and Engels (1991, XLI 380) [ ] figuriert das Tierreich als b rgerliche Gesellschaft (1956, XXX 249).
26. My translation. Wir kennen nur eine Wissenschaft, die Wissenschaft der Geschichte. Die Geschichte kann von zwei Seiten aus betrachtet, in die Geschichte der Natur und die Geschichte der Menschen abgeteilt werden. Beide Seiten sind indes nicht zu trennen. (Marx and Engels 1956, III 18.)
27. Marx and Engels (1991, XL 550) [ ] ist bisher noch nie ein so gro§artiger Versuch gemacht worden, historische Entwicklung in der Natur nachzuweisen, und am wenigsten mit solchem Gl ck. (Marx and Engels 1956, XXIX 524.)
28. Marx and Engels (1991, XXXVIII 422) Je mehr ich aber den Dreck treibe, um so mehr berzeuge ich mich, da§ die Reform der Agrikultur [ ] das A und O der kommenden Umw lzung ist. Ohne das beh lt Vater Malthus recht. (Marx and Engels 1956, XXVII 314.)
    According to Thomas R. Malthus (1766-1834), a figure of significance for Darwin as well as Social Darwinistic views of history and society, a fundamental discrepancy exists between the ratio in which the human population tends to grow, namely in geometric proportion, and the arithmetical proportion of the actual growth ratio of the animal and plant population which serve as food for mankind.
29. My translation. Darwin in seiner vortrefflichen Schrift sah nicht, da§ er Malthus Theorie umstie§, indem er die geometrische Progression im Tier- und Pflanzenreich entdeckte. [ ] In Darwins Werk [ ] findet sich auch im Detail (abgesehen von seinem Grundprinzip) die naturhistorische Widerlegung der Malthusschen Theorie. (Marx and Engels 1956, XXVI.2 114.)
30. Marx and Engels (1991, XXIV 467) Wie Darwin das Gesetz der Entwicklung der organischen Natur, so entdeckte Marx das Entwicklungsgesetz der menschlichen Geschichte. (Marx and Engels 1956, XIX 335.)
31. Marx and Engels (1991, XLIII 527f) [ ] die ganze Geschichte [ ] unter ein einziges gro§es Naturgesetz zu subsumieren. Statt also den struggle for life , wie er sich geschichtlich mit verschiedenen bestimmten Gesellschaftsformen darstellt, zu analysieren, hat man nichts zu tun, als jeden konkreten Kampf in die Phrase struggle for life und diese Phrase in die Malthussche Bev lkerungsphantasie einzusetzen. (Marx and Engels 1956, XXXII 685f.)
32. Marx and Engels (1991, XXIX 262) Das allgemeinen Resultat, das sich mir ergab, und einmal gewonnen, meinen Studien zum Leitfaden diente, kann kurz so formuliert werden: In der gesellschaftlichen Produktion ihres Lebens gehen die Menschen bestimmte, notwendige, von ihrem Willen unabh ngige Verh ltnisse ein, Produktionsverh ltnisse, die einer bestimmten Entwicklungsstufe ihrer materiellen Produktivkr fte entsprechen. [ ] (Marx and Engels 1956, XIII 8f.)
33. Marx and Engels (1991, XVI 466). Grundz ge der materialistischen Auffassung der Geschichte (1956, XIII 469). Compare also the argument in (Arndt 1985, 88ff).
34. Marx and Engels (1991, XXXVIII 96) Dank der einfachen Tatsache, da§ jede neue Generation die von der alten Generation erworbenen Produktivkr fte vorfindet, die ihr als Rohmaterial f r eine neue Produktion dienen, entsteht ein Zusammenhang in der Geschichte der Menschen, entsteht die Geschichte der Menschheit. (Marx and Engels 1956, XXVII 552.)
35. Of special interest in this context seems to me a distinction by Marx with respect to the prerequisites for the existence of any special type of society, namely whether or not these prerequisites, which were historic in the first place, could be reproduced by the respective type of society once it had come into being. See (Arndt 1985, 153ff.).
36. Marx and Engels (1991, XXXV 10) Auch wenn eine Gesellschaft dem Naturgesetz ihrer Bewegung auf die Spur gekommen ist - und es ist der letzte Endzweck dieses Werks, das konomische Bewegungsgesetz der modernen Gesellschaft zu enth llen -, kann sie naturgem §e Entwicklungsphasen weder berspringen noch wegdekretieren. Aber sie kann die Geburtswehen abk rzen und mildern. (Marx and Engels 1956, XXIII 15f.)
    It is probably superfluous to state that Marx theory, in spite of expressions like law of nature or natural stages of development, does not belong to the nineteenth-century ideas of development depicted above that claimed that progress was a physical necessity. See, for instance, (Groh 1981, esp. 236f.).

37. See (Arndt 1985, 103ff.).
38. See (Arndt 1985 104f.).
39. Above all, see his drafts for the letter to Vera Zasulic from 1881 in (Marx and Engels 1956, XIX 384ff.; 1991, XXIV 346ff.).
40. See the passage Formen, die der kapitalistischen Produktion vorhergehen in (Marx 1939, 375-413; Marx and Engels 1991, XXVIII 399ff.; see also Arndt 1985, 81f.).
41. See (Marx 1939, 26; Marx and Engels 1991, XXVIII 42). This aphorism was written 1858; thus, for chronological reasons, it did not allude to Darwin s theory and its implication that the man is a descendant of the ape. It seems more probable that Marx had knowledge of the lines of development of the contemporary morphology, which constituted, as already mentioned, a decisive starting-point for Herbert Spencer s theory of development.
42. Ibid.
43. See (Arndt 1985, 140f. and 184f.).
44. Marx and Engels (1991, XVI 476) [ ] nur entkleidet der historischen Form und der st renden Zuf lligkeiten (1956, XIII 475). The Hegelian manner in which Marx discussed the form of value in the first edition of *Das Kapital* was certainly as ill-suitable to prevent such misunderstandings as was the use he made of I.I. Kaufmans review of the *Kapital* in the afterword of its second edition (1956) XXIII 25 ff.; 1991, XXXV 30f.).
45. See (Marx and Engels 1956, XXI 248; 1991, XLII 304).
46. Wenn er [sc. Marx] nachweist, da§ die jetzige Gesellschaft, konomisch betrachtet, mit einer neuen h heren Form schwanger gehe, so zeigt er nur sozial denselben allm hlichen Umw lzungsproze§ nach, den Darwin naturgeschichtlich nachgewiesen hat. (Marx and Engels 1956, XXXI 403ff.; 1991, XLII 496f.).

REFERENCES

Arndt, Andreas. 1985. *Karl Marx. Versuch über den Zusammenhang seiner Theorie*. Bochum: Germinal.
Bayertz, Kurt. 1982. "Darwinismus und Ideologie." In *Darwin und die Evolutionstheorie* (= Dialektik 5), eds. K. Bayertz, B. Heidtmann and H.-J. Rheinberger. Köln: Pahl-Rugenstein.
Büchner, Ludwig. 1872. *Der Mensch und seine Stellung in der Natur in Vergangenheit, Gegenwart und Zukunft*. Leipzig: Thomas.
Darwin, Charles. 1859. *On the Origin of Species by means of Natural Selection, or the Preservation of Favoured Races in the Struggle for Life*. London: Murray.
Darwin, Charles. 1871. *The Descent of Man, and Selection in Relation to Sex*. London: Murray.
Engels, Friedrich. 1884. *Der Ursprung der Familie, des Privateigentums und des Staats*. Hottingen-Zürich.
Greene, John C. 1977. "Darwin as a Social Evolutionist." In *Journal of the History of Biology*, I.
Groh, Dieter. 1981. "Marx, Engels und Darwin: Naturgesetzliche Entwicklung oder Revolution." In *Der Darwinismus. Die Geschichte einer Theorie*, ed. Günter Altner. Darmstadt: Wissenschaftliche Buchgesellschaft.
Haeckel, Ernst. 1924. " ber die Entwicklungstheorie Darwins. In *Gemeinverständliche Werke*, ed. H. Schmidt. Leipzig and Berlin: Kröner.
Lange, Friedrich Albert. 1865. *Die Arbeiterfrage*. Duisburg: Valk & Volmer.
Lefèvre, Wolfgang. 1984. *Die Entstehung der biologischen Evolutionstheorie*. Frankfurt/M., Berlin, Wien: Ullstein.
Lefèvre, Wolfgang. 1992. "Wissenschaft und Philosophie bei Feuerbach." In *Sinnlichkeit und Rationalität - Der Umbruch in der Philosophie des 19. Jahrhunderts*, ed. W. Jaeschke. Berlin: Akademie Verlag.
Lefèvre, Wolfgang. 2001. "Jean Baptiste Lamarck." In *Klassiker der Biologie* Vol. 1, eds. Ilse Jahn and Michael Schmitt. München: C.H. Beck.
Marx, Karl. 1939. *Grundrisse der politischen Ökonomie (Rohentwurf)*. Moscow: Verlag für fremdsprachige Literatur.
Marx, Karl and Friedrich Engels. 1956. *Werke*. Berlin: Dietz.
Marx, Karl and Friedrich Engels. 1991. *Collected Works*. London: Lawrence and Wishart.

Mayr, Ernst. 1974. "Teleological and Teleonomic: A New Analysis." In *Boston Studies in the Philosophy of Science,* XIV. Dordrecht: Kluwer.

Spencer, Herbert. 1891. *Essays*. London: Williams and Norgate.

Stachel, John. 1994. *Marx's Critical Concept of Science*. Preprint 10. Berlin: Max Planck Institute for the History of Science.

Zmarzlik, Hans Günter. 1963. "Der Sozialdarwinismus in Deutschland als geschichtliches Problem." In *Vierteljahreshefte für Zeitgeschichte,* XI.

SILVAN S. SCHWEBER

ALBERT EINSTEIN AND THE FOUNDING OF BRANDEIS UNIVERSITY*

In an erudite, sensitive article, John Stachel explored Einstein's Jewish identity. In it he related the events that helped mold Einstein's attitude towards Judaism and shaped his identification as a Jew. He also explored Einstein's mature views on Jews, Judaism, and Zionism but he "did not attempt to recount [Einstein's] many activities on behalf of the Jewish people, individually or collectively.[1] The present note narrates one such episode. It is primarily intended as a vehicle to convey my admiration and respect for John: as a physicist, as a scholar, and above all as a courageous, exceptional human being who has been a model for the rest of us.

1. ISRAEL GOLDSTEIN

The demise of Middlesex University, a small veterinary and medical school located in Waltham, Massachusetts, on the western outskirts of Boston, was the event that triggered the founding of Brandeis University. By the end of World War II Middlesex was no longer a viable institution. Its medical school, the only one in the country that did not have a quota system for the admission of Jews and other minorities, had closed its doors in 1943 after it lost its accreditation, and it was clear that a similar fate would soon befall its veterinary school. In the fall of 1945 Joseph Cheskis, the dean of Humanities at Middlesex, realizing that the collapse of Middlesex was imminent, informed Ruggles Smith, the step-son of the founder of Middlesex University and its current president, of the long standing aspiration of the American Jewish community to establish a non-sectarian university. Cheskis suggested to Smith that he get in touch with Joseph Schlossberg, a member of the New York Board of Higher Education,[2] who in turn advised Smith to write to Rabbi Israel Goldstein, the energetic spiritual leader of Temple B'nai Yeshurun, a large, well-to-do Conservative congregation on West 88th St. in Manhattan. During the 1930s Goldstein had explored the feasibility of establishing a secular university sponsored by the American Jewish community. This idea had been advanced on a number of occasions during the first third of the century, most notably by Rabbi Louis Newman in 1923.[3] The canonical model for such a university was that of Johns Hopkins and Harvard: a liberal arts college to which was attached a graduate school and some professional schools.

Within a week of receiving letters from Smith and Cheskis in early January 1946, Goldstein visited the Waltham campus and became convinced that it could be the site of the Jewish sponsored liberal arts college he had envisaged. Moreover, the 100 acres campus could easily accommodate the reopening of a medical and veterinary school later on. Indeed, part of the attraction of Middlesex was that its charter allowed the university to grant BA and BS degrees, as well as Doctorates of Medicine and of Veterinary Medicine. Thus in time, the university would be able to alleviate the difficult conditions faced by Jewish students stemming from the discriminatory practices of medical and veterinary schools.

The first person Goldstein turned to after his visit to Waltham was his friend Julius Silver, a prominent New York attorney, who at the time was vice-president and general counsel of the Polaroid Corporation of Cambridge MA.[4] A second visit to the campus with Silver, and further discussions with Smith—who had the legal right to dispose of Middlesex University—, led to an agreement to have Goldstein and his associates become a majority on the Board of Trustees of Middlesex University. Goldstein, aware that the project to be successful must acquire national prominence and have the backing of a broad spectrum of the Jewish community, then called on Albert Einstein in Princeton to tell him of his plans and to obtain his support. Both Einstein and Goldstein were ardent Zionists and the two had shared the podium at several rallies for the establishment of a Jewish national state in Palestine.[5]

Einstein immediately agreed with Goldstein that his objective was important. Einstein at the time was deeply distressed by the plight of Jewish scholars and scientists who, because of discriminatory practices, were finding it extremely difficult to obtain faculty appointments in American colleges and universities. Einstein's primary concern was that the university Goldstein was planning be "first-class and free from non-academic control". On January 21, 1946 Einstein wrote Goldstein:

> I would approve very much the creation of a Jewish College or University provided that it is sufficiently made sure that the Board and administration will remain permanently in reliable Jewish hands. I am convinced that such an institution will attract our best young Jewish people and not less our young scientists and learned men in all fields. Such an institution, provided it is of a high standard, will improve our situation a good deal and will satisfy a real need. As is well known, under present circumstances, many of our gifted youth see themselves denied the cultural and professional education they are longing for.
> I would do anything in my power to help in the creation and guidance of such an institute. It would always be near my heart.
>
> Very sincerely yours
> A. Einstein[6]

During a subsequent visit Goldstein inquired whether Einstein would allow the university to be named after him. Einstein gracefully refused because he was of the opinion that the university should be named for "a great Jew who was also a great American,"[7] but consented to have the fund raising vehicle for the project named the Albert Einstein Foundation for Higher Learning, Inc.[8] As he himself was not going to be actively involved with the running of the Foundation, Einstein wrote Goldstein in early March 1946 asking him to get in touch with "one of his nearest friends,"

Dr. Otto Nathan,[9] who, he was convinced "can in many respect be of valuable help in the upbuilding of the institution."[10] Nathan was an economist with a non-tenured appointment at New York University. He came to the United States as a refugee during the 1930s and became very close to Einstein.[11]

Since the institution would not be called Einstein University, Goldstein then recommended that the university be named after Justice Brandeis and proceeded to obtain the permission of Brandeis' daughter to do so. Goldstein also recognized that as the campus was located in the Boston area it was important to involve the Greater Boston Jewish community. At Silver's suggestion, George Alpert, a prominent Boston lawyer active in Jewish philanthropy,[12] was invited to join the Board of Trustees of the new institution, and he readily accepted.

By the beginning of April 1946 enough had been accomplished by Goldstein for the *Boston Traveler* to comment in an editorial that one of the genuine deficiencies in local education will be remedied when Dr. Israel Goldstein and his associates take over the physical plant of Middlesex University and make it an institution of the first rank.

> At one step Middlesex has shaken off the shackles of the past and entered upon a period of high promise. It is incumbent on the general public as well as the world of scholarship to know and evaluate fully the fact that Middlesex hereafter will be in the main stream of the world's intellectual tradition and that its future graduates will be full-fledged and fully honored members of the ancient company of scholars.[13]

Einstein became deeply committed to the success of the contemplated university for which he had "assumed responsibility". But he also became frustrated by the "somewhat scanty information" that he was receiving from Goldstein about "the new college projects" and complained to him in mid April about the fact that he had not as yet gotten together with Otto Nathan.[14] Nathan saw the school as a place where he might obtain a permanent position.

By late April 1946 Goldstein could write Rabbi Stephen S. Wise[15] that at a recent meeting of the New York Board of Jewish Ministers:

> the project was endorsed and the reception given to my presentation was warm and cordial. A number of Rabbis asked for how they can be helpful; and one of them, who is in a position to do a good deal, undertook to raise a substantial sum.[16]

Wise—the spiritual leader of the Free Synagogue in New York and arguably the best known and most influential Jewish clergyman in the United States—was friends with both Franklin Delano Roosevelt and Felix Frankfurter and had access to the corridors of power in Washington, DC. He was also a good friend of Albert Einstein[17] and of Otto Nathan. He had made their acquaintance shortly after Einstein joined the Institute of Advanced Study in Princeton. Wise was deeply committed to the view that all Jews were members of *Klal Yisroel,* the community of Israel, and that each one of them was responsible for the well being of the others. Ever since hearing and meeting Theodor Herzl at the second Zionist Congress in Basel in 1898 he had been a committed Zionist and had played an important role in organizing American support for the establishment of a Jewish national state in Palestine. Wise's views regarding Judaism and Zionism thus resonated with those of Einstein (Stachel 2002), though he was clearly more ritualistically and religiously inclined. He was liberal in his political outlook, and had supported the labor movement in its efforts to have the right to unionize and obtain better working conditions. He thus had earned the respect of both Einstein and Nathan for his courageous political stand, and in particular that of Nathan who was a Socialist. Nathan, like many other liberals during the 30s and 40s, sympathized with the Soviet experiment, and admired the Soviet Union for its opposition to Nazi Germany and its crucial contributions to the Allied victory. In the immediate postwar years he believed that a peaceful accommodation could be reached between Communist Russia and capitalist United States.

Since in a recent conversation between Goldstein and Wise in which the university project was discussed, Wise had been friendly and helpful, Goldstein in his letter to Wise also asked him whether he would consent to have his name added to the list of sponsors on the letter he was sending out to elicit financial contributions.[18] At the time the list included the following dignitaries: Senators Joseph M. Ball, H.M. Kilgore, Brien (sic) McMahon, Wayne Morse, Albert T. Thomas, Robert F. Wagner; Representative John W. McCormack; Governor Tobin (Massachusetts); Mayors Fiorello LaGuardia and O'Dwyer (NYC); Archbishop of Boston Cushing; President William Green (A.F. of L.); University presidents Karl T. Compton (MIT), Paul F. Douglass (The American University), Bryn J. Hovde (New School for Social Research), Daniel Marsh (Boston University), J.E. Newcomb (U. of Virginia), Eduard C. Lindeman (N.Y. School for Social Work; Albert Einstein and Alvin Johnson.[19]

Although Goldstein was clearly keen on enlisting Wise in the project Wise was reluctant to accept his invitation. In the middle of May he wrote Otto Nathan:

> I am writing specifically to ask whether you have gotten in touch with Israel Goldstein. I myself have decided not to touch the thing until I have your judgment. In confidence, I may say that Goldstein is a tremendous public relations person. He conceives a newspaper heading to be the surest title to immortality. I think he is in earnest about this. The question is, what is "this" to be? Has he invited you to see him?... I want your help, and I think the Great Man in Princeton may need your protection.[20]

Goldstein and Wise had been on warm personal relations until fairly recently.[21] But Zionist politics had made Wise change his opinion of Israel Goldstein.[22] In the negotiations during 1946 that eventually led to the establishment of a Jewish national homeland in Palestine Goldstein had sided with Abba Hillel Silver and Ben Gurion—contra Wise—and had supported their more militant position as against Weizmann's more moderate and patient stand. Weizmann's defeat at the 22nd Zionist Congress in December 1946 triggered the resignation of Stephen Wise from the American Section of the Jewish Agency Executive Committee and his retirement from leadership in the Zionist movement. He thereafter branded the Zionist Organization of America, in which Goldstein remained active, a "collection of personal hatreds, rancours and private ambitions."[23]

It is likely that Wise's talks with Otto Nathan had exacerbated matters further. Thus on the heels of a discussion with Nathan, Wise became convinced "that the Great Man cannot afford to permit his name to be used unless he gets certain guarantees..." These he spelled out in a letter to Nathan in early June 1946: "the most worthwhile guarantee would be the delegating of power to you with a little help from I.G.

[Israel Goldstein] and myself to name an organizing academic committee. If I.G. is not willing to do that,—and you will find him rather difficult and dilatory,—than I warn both you and A.E. against going along with him."[24]

Goldstein did agree and an Academic Committee was set up, but Nathan was not on it. On July 1, 1946 Einstein wrote Goldstein that he was "seriously perturbed about the preparation of the academic institutions of the College." Evidently, after discussions with Einstein, Nathan had prepared a fairly detailed plan of procedure for the academic organization of the College and had given it to Mr. Ralph Lazrus,[25] a member of the Board. Lazrus had given him to understand that the plan would be discussed at the next meeting of the Board. However, in a letter to Wise on June 25 Goldstein informed him that an Advisory Committee had already been appointed to look into these matters and "is beginning to give some thought to the selection of the faculty."[26] When Einstein became aware of this he wrote Goldstein and reminded him that when he allowed his name to be used he took it for granted that "no important step would be taken concerning organization without my consent." He therefore asked that the Board "promptly decide that a body of outstanding, independent, and objective men [be] charged with the selection of an Acting Academic Head - the real organizer of the University" and of an Advisory Board to advise the Acting Academic Head. And he - Einstein - "of course" expected to be consulted about the composition of that body. Furthermore, since he was unable to attend the business meetings he informed Goldstein that "I have asked my friend, Dr. Nathan to act as my representative and I should appreciate if you and the Board would act accordingly. Only under such circumstances shall I be able to continue lending my name to the project."[27]

Einstein's July 1 letter to Goldstein was written against a background of exchanges between Einstein and Wise. On June 26th Einstein had written Wise that he was concerned about the fact that Nathan has not been able to find a suitable position since he returned from France where he had taught in the Armed Forces University. He indicated that Nathan "must be quite worried since he has his parent to take care of too" and inquired of Wise whether he had any ideas regarding "what could be done to help him in the present situation? [Nathan] has not spoken to me about this matter and he would probably not like it if he knew of this letter. But having myself no connections and informations I have to ask you as a man of experience and of good heart."[28]

Wise understood Einstein's letter as a plea to have Nathan given a position in the nascent Brandeis University. In his reply to Einstein on June 28 Wise stated:

> You wrote to me in confidence. I answer you in the same spirit. As a friend, I say to you ought not tie yourself up with the Foundation bearing your name and the Jewish University unless there be some completely trustworthy person, like our friend Otto Nathan, standing at the side of Dr. Israel Goldstein, to give him the benefit of his own wise judgment and your judgment *and thus ensure for him at once a place in relation to the proposed university.* [italics added]
>
> If you would write me a line to that effect - that you wish to associate yourself more closely with the university, but you must have someone at the side of Dr. Goldstein whom you can trust, I think I would be able to do the rest. Will you not be good enough to send me word as promptly as you can.[29]

Einstein answered Wise the next day and indicated to him that he had "completely" misunderstood him. "I am at the present, not worried about Otto Nathan's relationship to the future Jewish University, but about his immediate future." In that same letter he sketched his thoughts regarding "the question of organization and selection of the teaching-staff of the university and the general plan for the initial period of the institute." He agreed with Nathan that the solution of this problem was to have at the helm a man who would give the "cause" his "whole time". This man should fulfil the following conditions:

1. He must be a reliable Jew,
2. He must be acquainted with American University-institutions and must have understanding for educational problems and scholarship,
3. He should have experience in organization and 'Menschenkenntnis', and
4. He must be willing to deal and consult with us."

Einstein added that in discussions with Nathan neither of them could come up with a suitable candidate. Although he had thought of Nathan himself in this connection, Einstein doubted "that he will have enough authority in the eyes of the people who would have to consent to the choice." Nor did he believe that Goldstein had any ambition to undertake this job himself. Moreover, "I would not give my consent and cooperation if he would try to take over the mission and I feel pretty sure that also the very reasonable Mr. Lazrus would feel the same way."[30]

Although Goldstein was aware of the existing tensions and of Nathan's pressure to be involved in the academic planning he clearly hoped that in time these issues would resolve themselves. At the end of July 1946 Goldstein informed Einstein that at the recent "small dinner" in June that was attended by some fifty people at the Waldorf-Astoria in New York over $250,000 had been raised. Thus "a good beginning had been made toward the realization of your and our dream" and he felt that a point had been reached "when we can begin to think about organizing a faculty."[31] The time had therefore come to consider the composition of the Academic Advisory Committee that would be responsible for the selection of "an outstanding man who will be able to give all his time to the organization of the University and who would have to prepare the bylaws and regulations for the School and to appoint the original faculty." The suggestion from the Board of Directors of the Foundation was that Ralph Lazrus, Otto Nathan as Einstein's representative, Paul Klapper, the president of Queens College, and David Lilienthal, the chairman of the Tennessee Valley Authority, be members of the Academic Committee. Einstein drafted the letter to Lilienthal inviting him to be on the Committee. Although he had not met him personally Einstein informed him that he was writing to him "with regard to a cause which should be of interest to every American Jew, [namely] the establishment of a new University which would be under the influence of reliable Jewish personalities." He stressed that only:

> self help alone will enable us to free ourselves slowly from a situation which is truly painful for us - to be compelled to knock, often unsuccessfully, at doors which are being opened to us only reluctantly and conditionally. It should be also be borne in mind that, under existing conditions, our young scientific talents have frequently no access to scholarly professions which means that our proudest tradition—the appreciation of productive work— would be faced with slow extinction if we remain as inactive as we have been in the past.[32]

In view of the success of the "small" June dinner at the Waldorf, the Board made plans for a major fund raising dinner, "one that will be a deciding factor in the entire ultimate success of the project."[33] Its date was set for October 27, 1946 and it was to be held at the Hotel Pierre in New York. Given the name of the contemplated university the Board decided that the main speakers at the dinner should be eminent persons who had known Brandeis, and they recommended that Supreme Court Justices Frankfurter and Jackson be asked. But Einstein balked at the suggestion of Jackson:

> I must confess that I cannot approve the choice of a Gentile as main speaker in behalf of our project. This is a Jewish cause and we have to advocate it ourselves. It is therefore, out of the question that I can sign a letter to Justice Jackson, even if a majority of us is in favor of the action.[34]

Einstein however relented and did write to Jackson indicating that "Your own presence at this function and - if I may presume - an address by you in eulogy of the late Justice Brandeis, would contribute immeasurably to the success of the meeting, and assure the opening of Brandeis University under auspices worthy of the name of that great American."[35] In an identical letter to Justice Frankfurter the conclusion of this last sentence read "under auspices worthy of the name of that great American and outstanding Jew."[36]

Although on the surface Einstein appeared as deeply committed to the project as ever, swayed by Wise's resentment of Goldstein, by Goldstein's reluctance to allow Nathan have too great an influence at this stage of the developments, and by Nathan's eagerness to have a hand in the affairs of the new university,[37] relations between Einstein and Goldstein deteriorated further and became more strained. An invitation by Goldstein to Cardinal Spellman[38] to deliver the invocation at the October 27 fund raising dinner was one of the two straws that broke the camel's back. Einstein also took exception to the fact that Goldstein had discussed with Dr. Abram L. Sachar "the possibility of his appointment as chancellor and organizer of the University-faculty without the authorization or even knowledge of the Advisory-Committee." In an angry letter to Goldstein in early September 1946, Einstein noted that:

> Those two facts represent two new breaches of confidence from your side. I have decided, therefore, not to cooperate any longer with you and I will have to make it clear that from now on I cannot take any more responsibility for any of your acts concerning the planned university.
>
> I also cannot permit that my name is used for fund-raising in behalf of an enterprise in which you play an important part. Finally I must request that my name be removed entirely from the name of your foundation and I expect to be notified as soon as this has been done.

And he added that copies of his letter would be mailed to the members of the board and to the advisory-committee.[39]

After it became clear that Einstein would not alter his position as long as he was associated with the project, Goldstein resigned as chairman of the Board of Directors of the Albert Einstein Foundation and as Chairman of the Board of Trustees of Brandeis University. He appreciated the fact that Einstein's participation in the project was indispensable to its success, but that his own role could be assumed by others. But in a lengthy letter to Einstein Goldstein explained and defended his actions. He had, in fact, done nothing wrong as far as his contact with Dr. Sachar was concerned, nor anything unusual in the matter of the fund raising dinner.[40] At its meeting on 16 September 1946 the Board of Directors of the Albert Einstein Foundation accepted Goldstein's resignation and sent Lazrus, Alpert and Abraham Wexler to meet with Einstein "for the purpose of conveying to him the action of the Board."[41]

When informed of Goldstein's pending resignation Nathan contacted Alpert, the likely new chairman of the Board of Trustees of the University. After their meeting Nathan wrote Einstein that he had found Alpert "Ein sehr angenehmer Mann." (A very agreeable gentleman). Einstein therefore gave his blessing to have Alpert become chairman of the Board of Trustees.[42] Interestingly, Stephen Wise did not agree with Einstein and Nathan's assessment of the man. In November 1946 after having attended some of the Board meetings Wise wrote Nathan:

> It is now too late to talk about it, for the deed has been done. I wonder whether you and Professor Einstein have wisely suggested Alpert for a key place. He is not big enough a man to be President of the Board of Trustees. I think Lazarus [sic] is both finer and wiser. Remember my prediction: you will not be able to work with him long. He will become odious in time, due—or undue—to Professor Einstein.[43]

The vacancy in the chairmanship of the Board of Directors of the Albert Einstein Foundation created by Goldstein's resignation was filled on 30 September 1946 when Ralph Lazrus became the new chairman.

2. THE LASKI EPISODE

With Goldstein's resignation matters seemed to have been resolved to Einstein's satisfaction. He therefore wrote to the various people he had notified that he was severing his ties with the Brandeis Board that "work for the university-project ... had made considerable progress" and that he was now "convinced that we shall be able to overcome the difficulties involved in such a new enterprise"[44] and asked them to rejoin the enterprise. Writing to Sachar in late October 1946 in connection with a Hillel scholarship for a student, Einstein informed him that "Goldstein is not anymore connected with the University project; he resigned on my initiative... I will do whatever I can to help to realize the project and I trust that it will be possible."[45]

One of the things Einstein "did" was to make Otto Nathan a member of the Board of the Albert Einstein Foundation.[46] With Goldstein's departure Nathan was able to assume a more active role. He and Lazrus took charge of the academic component of the project, and later Lazrus was blamed for devoting too much of his time to these activities and not enough of his efforts to fund raising, the primary responsibility of the chairman of the Foundation. But actually, financially matters were going well. At the end of October 1946, Boris Young, the director of the Albert Einstein Foundation Inc., informed Einstein that as of September 1 over $350,000 had been pledged and that the October 27 dinner was likewise a great financial success. Concrete plans for the academic mission of the University were therefore in order.

On November 9, 1946 Otto Nathan submitted to the Board "AN OUTLINE OF

POLICY FOR BRANDEIS UNIVERSITY" which was to serve as a basis for discussion. Nathan envisaged that initially the university would consist of a Liberal Arts College with an enrollment of at least 1000 students. As far as the curriculum was concerned, he favored "a minimum of compulsory courses... Emphasis should be placed on interdepartmental courses to break down the artificial rigidities of departments and fields of teaching. Independent work of students should be encouraged in every way possible... Wherever possible, the seminar method of teaching should be encouraged... The question of the honor system (Swarthmore) should be studied.... In the appointments to the faculty, great emphasis should be placed on the teaching ability of the applicants (and not on the number of their publications)." Nathan wanted to make the College a "living democracy," and recognized that this "would require very different relationships among students, faculty and college administration very different from those in most existing colleges." He wanted the students to have a voice in the administration of the College and to be treated as "free and adult human beings. No compulsion to attend classes. There should be fewer examinations than is customary. ...No supervision of their private lives... No 'permits' for weekend absences or for 'late hours.'"[47]

Nathan was aware of the experiments being carried out at Swarthmore, Antioch, Sarah Lawrence, Black Mountain College and the newly founded Roosevelt College in Chicago, as well as the curricular changes being introduced at Harvard and Columbia. But probably the model that most closely approximated what he had in mind was Hutchins' University of Chicago with its College and separate teaching faculty for the College. In order to make recommendations for both the College and the Graduate School component of the university Nathan visited some of the leading American universities and in December 1946 went on an overseas trip to Great Britain "to investigate the methods of instruction and administration policies. He conferred with many important educators, including Professor Laski, and encountered friendly reactions to the plan for Brandeis University wherever he went."[48]

Nathan's friend Harold Laski was a prominent, outspoken, and articulate Socialist and a distinguished political scientist, who taught at the London School of Economics.[49] In July 1945 Laski had been elected Chairman of the national executive committee of the Labour Party, and his position became even more prominent in the summer of 1945 after Labour won an overwhelming majority in the British General Election. In the middle of June 1945, during the election campaign Laski made a speech in Newark in support of one of the Labour candidates. At the end of his presentation he was questioned by a fairly well-known Conservative journalist who had probably been encouraged by Tory interests to ask Laski provocative questions. On the day after Laski's speech a letter appeared in the *Nottingham Journal* signed by one H.C.C. Carlton, a Conservative member of the local county council, in which Carlton alleged that during his speech in Newark when enumerating the reforms he would like to see enacted, Laski had declared that "If we cannot have [these reforms] by fair means we shall use violence to obtain them." The letter went on to say that when challenged by a member of the audience who claimed he was "inviting revolution from the platform" Laski had replied: "If we cannot get reforms we desire we shall not hesitate to use violence, even if it means revolution" (Eastwood 1977, 140). Upon reading the letter Laski immediately issued the following statement:

> I am going to take out a writ for libel against the man who wrote it and against anybody else who reproduces this letter. My answer at the meeting was entirely different. What I said was: it was very much better to make changes in time of war when men were ready for great changes than to wait for the urgency to disappear through victory, and then to find that there was no consent to change what the workers felt an intolerable burden. That was the way a society drifted to violence. We had it in our power to do by consent what other nations have done by violence" (Eastwood 1977, 141).

The incident got further prominence when the *Daily Express,* a Beaverbrook, Tory national newspaper, featured the story under the headline; "Laski unleashes another general but as yet unpublished election broadside: socialism even if it means violence." Then on June 20 the *Newark Advertiser* carried the story of Laski's Newark market place speech as recorded in shorthand by one of their reporters. According to the *Newark Advertiser,* Laski had been asked why he had openly advocated "revolution by violence" in earlier speeches. His answer, according to the *Newark Advertiser,* was:

> If Labour could not obtain what it needed by general consent, 'we shall have to use violence even if it means revolution.'... Great changes are so urgent in this country, that if they were not made by consent they would have to be made by violence.... When a situation in any society became intolerable ... it did not become possible to prevent what was not given by generosity being taken by the organized will of the people.

On June 20 Laski brought a libel suit against the *Nottingham Guardian,* the *Daily Express* and the *Evening Standard* all based on the publication and reproduction of Carlton's letter, and on the 22nd he sued the *Newark Advertiser* and its editor, C.E. Palby.

The trial lasted five days, and was presided over by a very conservative judge, Lord Goddard. During the trial which opened in late November 1946 it emerged that Laski had in fact never made the statement 'we shall have to use violence even if it means revolution.' That statement had been introduced into the *Newark Advertiser* article after the reporter had seen Carlton's letter. And in answers to questions posed to him by his lawyer Laski made clear that he could not be a member of the Labour Party if he advocated revolution by violence, since by its constitution any member of the Labour party is committed to the acceptance of Constitutional democracy. Furthermore, he had been a critic of Communism ever since 1920, had frequently and extensively criticized in his writings both communist theory and communist strategy, and had recently opposed the admittance of the Communist Party into membership of the Labour Party.[50] The defense, however, pointed to Laski's repeated advocacy of revolution in his writings and his belief that "the time is ripe for revolution". Laski explained that what he had meant by the latter statement was that "The time is ripe for great changes." In his charge to the special jury Goddard put the matter thus: You have to decide "whether these speeches..., remembering the audience to which they are addressed, would be an incitement to violence or revolution, whether, it is preaching revolution and violence as part of a political creed and urging the people to adopt it. [Or is Laski] putting forward views which you or other people may abominate and hate, but which he is at liberty to express and has the right to express [if he is stating them not as a matter of incitement and not as a matter of advocacy but as a matter of argument] (Laski 1947).

The jury, after deliberating for 40 minutes, found the article in the *Newark Advertiser* a fair and accurate report of a public meeting and Laski thus lost his suit. He incurred expenses amounting to some $40,000 in connection with the trial and he became greatly distressed by the amount of his obligations. However Laski, with the help of friends and contributions from the Labour Party was able to cover three/fourth of the costs. Max Lerner, Otto Nathan, and others committed themselves to raise the balance in the United States.[51]

His visit with Laski made a deep impression on Nathan. Ever since Goldstein's resignation he had thought a great deal about Brandeis' academic future.[52] As the financial support of the institution by the American Jewish community seemed assured, the crucial question had become: Who would assume the academic leadership of the institution? After his visit with Laski, Nathan came to believe that Laski would be an excellent choice to be the president of Brandeis. An agreement had existed between Einstein and Goldstein stipulating that he would have a major say in recommending and selecting the president. It was also understood that one of the principal functions of the Academic Advisory Committee—of which Nathan and Lazrus were members—was to make recommendations of suitable candidates to the Board. But evidently in the spring of 1947 Nathan convinced Einstein that he should ask the Board that he—Einstein—rather than the Academic Committee make the recommendations.

On March 30 1947 Alpert visited Einstein in Princeton. In a letter he later wrote to Ralph Lazrus, Einstein described that meeting:

> I discussed with him at great length the difficult problem of selecting a president for Brandeis University. I told him that you [S.R. Lazrus] and Mr. Nathan had suggested that the Board of Trustees be asked to delegate the authority for selection of the President to me. I mentioned to Mr. Alpert that, in case the Board would take such action, I was considering to inquire of Professor Laski whether he might be willing to come over here to help us in organizing the University. Mr. Alpert not only did not object to this suggestion, but approved of it.[53]

Alpert also told Einstein that he would bring the matter up at the next Board meeting for its approval. What happened next is somewhat uncertain for there is no explicit record of the events. It is not clear whether Alpert, given his own conservative inclinations and the tenor of the times, upon learning more about Laski had second thoughts about his candidacy, or whether he became apprehensive about delegating to Einstein—and therefore to Nathan and Lazrus—the authority to pick Brandeis' first president. Most likely, he had strong reservations about both issues. As seemingly there was no quorum[54] at the April 14 meeting of the Board of Directors of the Foundation where Einstein's request was to come up, *de facto* there was no meeting of the Board, "and no resolution appeared in any minutes authorizing or requesting [Einstein] to select a president for the University."[55] What is known is that at the April 14 meeting Nathan reported that "several difficulties had been encountered in the attempt to form such a committee and since it was deemed imperative to insure that the eventual President's qualification be such as to guarantee the implementation of our purposes, it was proposed to authorize Professor Einstein to make the selection of President." Bluestein, one of the directors, then pointed out that, while approval of Einstein's choice was the prerogative of the Board, such authorization would be equivalent to agreement, in advance, to accept whatever choice Einstein made. Nathan then stated that "Prof. Einstein had been reluctant to assume that responsibility but that he had finally agreed to this procedure and was currently engaged in the search for a suitable President."[56]

However, given Alpert's positive response at their March 30 meeting, Einstein had written Laski to ask whether he would be prepared *to consider an invitation* to become Brandeis University's first president—possibly for only two or three years were he reluctant to leave Great Britain permanently. In his letter he had added:

> The University will be in Jewish hands, but we are determined to develop it into an institution which is enlivened by a free, modern spirit, which emphasizes, above all, independent scholarship and research and which does not know of discrimination for or against anybody because of sex, color, creed, national origin or political opinion. All decisions about educational policies, about the organization of teaching and research will be in the hands of the family.
>
> The Board of Trustees has delegated to me the authority of selecting the first President of the University. This man would have the challenging task to help us in determining the basic foundations of the University and to select and organize the initial faculty upon whom so much depends. We all feel that among all living Jews you are the one man who, accepting the great challenge, would be most likely to succeed....
>
> I am writing, therefore, to ask you whether you would be prepared to consider such an invitation. ... You would oblige me by treating this inquiry confidentially.[57]

Laski answered him promptly. On 25 April he wrote Einstein in his diminutive handwriting:

> Few things in my eyes have done more honour or given me greater pleasure, than your most generous suggestion. But, with my deep respect, I fear that I must decline it. First, I am sure that I am not the right person for the post. I have no financial capacity. I like teaching, and not administration. My roots are firmly fixed, after 27 years in the University of London, in this country. To these reasons I must add that my wife would not now wish to live in the United States, and apart from being morally bound to fight here, as best I can, for socialism, I want to spend such leisure as I can find in writing a book it has been my ambition to write since I was a student. I am sorry to decline any offer which comes from you, and from my good friend Otto Nathan. But I am confident this is the right decision. I am not fit for a post which demands the special qualities of a University-President.
>
> I hope very much that I may, nevertheless, have the privilege to being connected, in some loosely continuous way, with the new institution in a teaching capacity. I hope I need not say that in that realm I would seek, very gladly, to serve it with all the energy in my power.[58]

Thus by early May Einstein and Nathan knew that Laski would not accept an offer to be president of Brandeis. When this information was made available to the Board is not clear.[59] But it was quite clear to both Einstein and Nathan that the Laski matter was closed.[60]

Einstein's letter to Laski, however, upset some members of the Board since he had not been authorized to write it. On May 11 Alpert and Silver went to Princeton to see Einstein. In his letter to Lazrus reporting on the meeting, Silver indicated that Einstein had expressed surprise when he learned that because of the absence of a quorum *de facto* no meeting of the Board of Directors of the Foundation had been held, and therefore that no resolution appeared in any minutes authorizing or requesting him to pick a president for the University. Nor had he been made aware of the fact that those Directors who had been informally consulted by Alpert had not been informed that Harold Laski was receiving consideration as an active candidate. He expressed the opinion that it was definitely "unfair" not to have communicated this information to him. But Silver went on to say that:

> Professor Einstein readily agreed that selection of the President should be made either at the recommendation of an educational committee or on the nomination of other qualified persons, but subject to full disclosure and approval by the Board of Directors... This removes the chief obstacle to continued progress.[61]

However, no further progress was to be made. Alpert had evidently determined to use the Laski invitation to marginalize Lazrus and Nathan, and thereby also minimize Einstein's say in the shaping of the University. What is clear from the available written documentation is that by mid April Nathan was playing an ever more active role in shaping academic policy. At the time, consideration was being given to hire a part time provost to coordinate academic matters until a president was selected and Max Grossman, a professor of Journalism at Boston University was being considered for this post. Nathan interviewed him and after their meeting Grossman wrote him to counsel that he: "abandon all other undertakings temporarily and have yourself appointed as 'convener' of the university. This will enable you to do whatever organizational work which is necessary until others are selected to perform tasks specifically." More specifically, Grossman proposed to Nathan that he "ought to be dean of faculty and professor and head of the department of economics." Should his deanship not be compatible with whoever is appointed as President he certainly would be happy as a professor of economics.[62]

That relations between Alpert and Nathan were strained is evident from the fact that during March 1947 Alpert kept pressing Nathan to set up a Committee on Education made up of prominent educators whose responsibility would be to make recommendations to the Board of suitable candidates for the presidency—*evidently to no avail.*[63]

That relations between Alpert and Nathan and Lazrus had deteriorated to the breaking point by early May is indicated by the fact that on May 16 Einstein wrote Lazrus that he had asked Professor Otto Nathan to act as his representative at the meeting of the Board of Trustees that was to be held on Monday May 19.[64] The details of that meeting are scant. Its minutes relate that:

> The President ... reported that he was in receipt of a letter from Professor Einstein in which the latter indicated that Dr. Nathan was authorized to act as Professor Einstein's representative at the meeting. The President then called upon Dr. Nathan to make a statement on behalf of Professor Einstein and himself., and added that such statement also represent the position of the President.
>
> Dr. Nathan stated in substance that disagreements had arisen with regard to the selection of a President of the University and to the policies of the University. Accordingly, Professor Einstein desired to withdraw his name from the Foundation; that the Foundation be liquidated and all of its funds be transferred to the University. Dr. Nathan stated that Professor Einstein, Mr. Lazrus and he would like to help the University in every way.[65]

Thereafter Nathan and Lazrus offered to tender their resignation as soon as the name of the Foundation was changed and an understanding was reached that no public statement would be made.[66] However, relations between Alpert and Lazrus became so rancorous that on June 22nd Lazrus did release a statement stating that Einstein was resigning and withdrawing the use of his name because Alpert and the Board were trying "to bring down the educational standards of the university". Having been privy to the increasingly contentious exchanges between Alpert and Lazrus, Einstein took pen to hand and on June 20 had written Susan Brandeis Gilbert that controversies over the educational policies and academic organization of the university had developed between him and his friends on the one hand, and the Board on the other. As these appeared unbridgeable, he informed her that "I have decided to withdraw my support from the project and therefore to have my name eliminated from the fundraising Foundation."[67]

Lazrus' public statement was of course denied by Alpert.[68] He got Susan Brandeis to issue a statement asserting that Lazrus and Nathan had "arrogated to themselves the shaping of academic policy, the selection of a president and other important educational functions which could rightfully be performed only by the Board of Trustees." Alpert in turn charged that Lazrus and Nathan had "surreptitiously" made overtures to a "thoroughly unacceptable choice" as president and thus were trying to give the school a "radical, political orientation."

> To establish a Jewish-sponsored University and to place at its head a man utterly alien to American principles of democracy, tarred with the Communist brush, would have condemned the University from the start. I made it perfectly clear to Mr. Lazrus then and later that on the issue of Americanism I cannot compromise."[69]

Alpert was using Iron Curtain rhetoric and McCarthyite tactics to justify his assuming control of the affairs of the University. An article by William Zukerman in the July 4, 1947 issue of a small Jewish magazine, *The American Hebrew,* stated the case forcefully:

> [Mr. Alpert's] statement is not only utterly tactless and irrelevant to the issue, but also untrue and vicious. It is the statement of a narrow partisan reactionary politician behooving a member of the un-American Committee, not a President of a University named after the late Justice Brandeis.

Zukerman identified the man whom Mr. Alpert had described in the "entire" American Press as "alien to the principles of American democracy" as Harold Laski, "one of the greatest teachers of our age, a man who brought up a generation of youth [from all over the world that] has been the vanguard of humanity's struggle for democracy and social justice." He stated that Laski has been the target of reactionary forces in England and in the United States "chiefly because he was a pioneer of the newer and wider interpretation of democracy from the political to the economic field." And Zukerman pointed to John Dewey, Roosevelt, Wendell Wilkie, and Brandeis as people who shared Laski's "alien principle of democracy"..."Professor Laski is the living incarnation in England of the spirit of the New Deal movement in the United States. He can be said to be its intellectual father even as the late President Roosevelt was its political founder." Moreover, for Zukerman:

> It would have been a beautiful gesture on the part of a Jewish institution of learning to register its faith in the Rooseveltian interpretation of American democracy in a period of reaction by naming Professor Laski its President."

Zukerman was willing to concede that the trustees of the University had "good and weighty" reasons for opposing the nomination of Professor Laski. From the point of view of fund raising a philosopher of New Dealism might not be the ideal choice in Truman's United States and the trustees certainly had the right to oppose the nomination on ground of expediency. But why, Zukerman asked,

> should the trustees have gone out of their way to denounce Laski's principles as "alien to American way of life and democracy? Who has made this obscure Board the judges of what is native and alien to American democracy? ... And why was it necessary to raise altogether the question of Americanism in this connection and to waive the patriotic American flag? Did they not realize that branding Dr. Einstein's choice as "alien to American way of life, and tarnished with the Communist brush," they ... practically accused Professor Einstein and his colleagues of un-Americanism?

Zukerman's conclusion was that "the tactless and reactionary manner" in which the trustees chose to justify their opposition, demonstrated they were unsuited to be trustees of a Jewish sponsored University, "and especially one named after Justice Brandeis."[70]

Upon reading Alpert's comments Einstein drafted the following statement which he sent to Stephen Wise:

> The press statement, which Mr. George Alpert and another member of the Board of Trustees of the Brandeis University released on the occasion of the withdrawal of myself and of my friends, Professor Otto Nathan and S. Ralph Lazrus, have surprised, even shocked me.
>
> It was I who suggested that the name of an eminent British scholar and educator, Professor Harold J. Laski of the University of London, be considered in connection with the Presidency of the Brandeis University. Mr. Alpert, who now makes an untruthful charge of radicalism against my two associates, in no wise objected to this suggestion, made to him by me in my own home.
>
> The press releases have convinced me anew that it was none too early for us to sever a connection from which no good was to be expected for the community. I should like to state specifically that my associates, Professor Nathan and Mr. Lazrus, and I have always been and have always acted in complete harmony. I feel deep gratitude particularly to my old friend, Professor Otto Nathan, who with great devotion and complete selflessness gave time and effort to a cause which the three of us alike considered good and most urgent.[71]

The edited version signed by Einstein on 24 June was somewhat more restrained. It read:

> The press statements which Mr. George Alpert and another member of the Board of Trustees of Brandeis University released on the occasion of the withdrawal of myself and of my friends, Professor Otto Nathan and Mr. S. Ralph Lazrus, have convinced me that it was none too early for us to sever a connection from which no good was to be expected from the community. My associates and myself had very reluctantly come to the conclusion that the type of academic institution in which we have been interested could not be accomplished under the existing circumstances and the present leadership.

Its second paragraph was the same as the last paragraph in the statement sent to Wise except that in the last sentence, "worthy and most urgent" had been substituted for "good and most urgent."[72]

Alpert's version of the "facts" however became the official history.[73] Most newspaper accounts reported the events as follows:[74] Nathan was to form a Committee to advise the Board of the Einstein Foundation on the selection of a president. However, Nathan and Lazrus informed Alpert that there would be no Educational Advisory Committee and suggested instead that the Board of Trustees designate Prof. Einstein to select a president. Alpert then informed them that this was contrary to the original agreement and contrary to all academic procedures but that he was willing to hear their proposal through. When they recommended Harold Laski as president, Alpert was ready to concede that Laski was brilliant and that he might even be a great educator. But Laski was controversial in his political views since he was "an international socialist of record" who had lost a libel suit against a British newspaper that had called him a communist. Alpert was unwavering in his view that the person heading the institution be an American. The matter came to the Board which backed Alpert. Whereupon Einstein, Nathan and Lazrus resigned "with good will" and assured Alpert that they would make no statements that would jeopardize the project. However Lazrus did subsequently release a statement stating that Einstein was resigning and withdrawing the use of his name because Alpert and the Board were trying "to bring down the educational standards of the university," which statement in turn was denied by Alpert.

In addition to other inaccuracies, there is a glaring one in this version of the story: the issue of Laski never came in front of the Board! *When the Board met on May 19, Laski was no longer a candidate*! One can only conclude that the real issue had been who was to control the affairs of the University.[75]

The subsequent course of events had an ironic twist: in 1948 Abram Sachar was installed as the first president of Brandeis University.[76] Sachar had been Goldstein's choice for the presidency, and Goldstein's informal inquiry of him in the summer of 1946 whether he might be interested in the position had led to the first rift between Einstein and the Brandeis Board of Directors. Also history was to repeat itself: A struggle between Sachar and Alpert over the issue who was to control the future of the University eventually led to Alpert's removal as Chairman of the Board of Trustees.

Relations between Einstein and the University remained frigid after the 1947 break. Einstein's bearing in the matter gives a revealing account of one facet of his character. An attempt by Sachar to mend fences in 1948 was rebuffed with Einstein's refusal to see him.[77]

In March 1952 Sachar again wrote him in order to be able to visit him and tell him, "without any obligation to you, something about the development of the school."[78] Sachar had hoped that Einstein would draw a distinction between "the one or two people who are, after all, a temporary part of the life of an institution and the institution itself". But Einstein answered him as follows:

> I was somewhat astonished by your letter of March 25th. The most concise answer to it has been formulated hundred years ago by Schopenhauer who said: "Erlittene Unbill vergessen, heisst müsahm erworbenes Geld zum Fenster hinauswerfen".[79]
>
>> With kind regards,
>> Sincerely,
>> Albert Einstein.[80]

When later that year a young Indian scholar asked him for his assistance in obtaining a position at Brandeis he answered that events of the past had made it impossible for him to get in touch with Brandeis University, "directly or indirectly."

> I was connected with Brandeis University at the time of its foundation. It happened, however, that a few of the Trustees behaved quite dishonestly against me and my nearest friends. Therefore I had to sever completely my connections with the institution. This happened before Dr. Sachar was connected with the enterprise so that he is not directly involved. But after he has, so to speak, inherited the "Tabu" it is impossible for me to approach him.[81]

And in 1953 when Sachar wanted Brandeis to give Einstein an honorary degree Einstein rejected the offer and sent him the following letter:

> Dear Dr. Sacher [sic]
>
> It is embarrassing not to be able to repay friendly behavior in kind. However in this case I cannot help it. What transpired in the preparatory phase of Brandeis University certainly was not based on misunderstandings and can no longer be compensated for. Thus I cannot accept your offer of an honorary doctorate.
>
> I would not want to have this matter discussed, for this would be harmful to the University, and I shall tell only my closest friend, who has the right to be informed, about it.
>
>> With kind regards,
>> A. Einstein[82]

Einstein's last exchange with Sachar occurred in January 1954. Sachar had once again tried to see him. Einstein answered him that:

> If you would be simply a private person who has written delightful books, I would gladly accept your kind offer to visit me. Under the prevailing circumstances, however, it is not possible for me to do so. As you are informed about the relevant past events you will easily understand.[83]

It should be noted that Einstein's interaction with Brandeis and Sachar after 1947 was passive, whereas his animosity towards Alpert was active. Upon learning that Alpert had been appointed Honorary Chairman of the Albert Einstein College of Medicine campaign Einstein wrote Nathaniel Goldstein that his confidence in Alpert had been completely destroyed when he dealt with him in the early days of Brandeis and that this had led to his withdrawal from that institution. And he concluded his letter by asserting:

> I must tell you that I would never have permitted the use of my name in connection with the College of Medicine had I known that Mr. George Alpert would be asked to play an important role in its development.[84]

3. EPILOGUE

The episode I have related gives further proof that after he became fully aware "of our precarious situation among nations" Einstein's "relationship with the Jewish people had become [his] strongest human bond."[85] Already in 1920 he had described himself as "a person without roots anywhere...[who has] journeyed to and fro continuously - a stranger everywhere."[86] Even though he no longer journeyed continuously after coming to the United States, he never overcame the language barrier and never felt fully at home there. He even came to consider the Institute of Advanced Study as an institution not particularly hospitable for Jews.[87] Ever since coming to Berlin in 1913 he was of the opinion "that Jews ought not to press theirs claim in an attempt to obtain the more desirable positions, particularly academic ones, but should create jobs [and by implication, institutions] to be filled from their own rank" (Born 1971, 17).[88] Hence his strong support for establishing the Hebrew University of Jerusalem and his enthusiastic commitment to the establishment of Brandeis as a Jewish sponsored University in Waltham and his insistence that it be and remain in Jewish hands.

When reviewing this early history of Brandeis University I was struck by the fact that Israel Goldstein's role seems largely to have been forgotten, overwhelmed by Sachar's subsequent accomplishments in building the institution. But as Sachar himself acknowledged in his letter to Goldstein upon being installed as the first president of Brandeis in 1948:

> You are really the "father" of Brandeis University. You put endless energy and devotion into the building of the concept and the corralling of its first support.[89]

Another point: The sentiments that motivated Einstein's strong endorsement of Goldstein's project were the same as those which inspired the Jewish members of the faculty to join the University when it opened its doors to its first class of a 100 freshmen in the fall of 1948. A parallel can be drawn between Einstein's commitment to help create a Jewish national home in Palestine and the commitment of the many non-religious Jews who comprised the bulk of Brandeis' initial faculty to help build a Jewish sponsored institution of higher learning.

The response of Einstein—that cosmopolitan, secular Jew who abhorred nationalism—to the hatred and the xenophobia he had encountered in German academic circles during the Weimar era and to the unmistakable vulnerability of the Jewish community even in "civilized" Germany, was to assert that:

> The best in man can flourish only when he loses himself in a community. Hence the moral danger of the Jew who has lost touch with his own people and is regarded as a foreigner by the people of his adoption.
>
> The tragedy of the Jews is that they ... lack the support of a community to keep them together. The result is a want of solid foundations in the individual which in its extreme form amounts to moral instability.

And in 1933 after he had left Germany, Einstein declared:

> It is not enough for us to play a part as individuals in the cultural development of the human race, we must also attempt tasks which only nations as a whole can perform. Only so can Jews regain social health.

Consequently:

> Palestine is not primarily a place of refuge for the Jews of Eastern Europe, but the embodiment of the re-awakening of the corporate spirit of the entire Jewish nation.

Einstein also stressed that the "Jewish" state he advocated would be a place where three archetypical Jewish ideals, would flourish:

> the pursuit of knowledge for its own sake;
> an almost fanatical love of justice;
> [and] a desire for personal independence.[90]

Einstein's justification for the establishment of a Jewish state can be recast to make manifest why the vision of a Jewish sponsored university resonated so deeply with the secular Jews who joined the rank of its initial faculty. The sanction would then read as follows:

> It is not enough for us to play a part as individuals in the cultural development of the United States, we must also attempt tasks which only its entire Jewish community can perform. Only so can Jews regain Social health.

Consequently:

> Brandeis will not primarily be a place of refuge for Jewish scholars who have been discriminated against in the elite American universities, but —now after Auschwitz— it will be the embodiment of the re-awakening of the corporate spirit of the entire American Jewish community.[91]

And, the three ideals that Einstein hoped the Jewish state would nurture were precisely the ones that made a Jewish sponsored, secular university also attractive to non Jewish scholars.

For the most part these ideals have been cultivated at Brandeis. And Einstein's demand that the university be "first class" and meet the highest academic standards were heeded from the very beginning. The faculty that taught the first class to graduate from Brandeis in 1952 was indeed outstanding.[92] Among them were:
In the fine arts: Leonard Bernstein, Erwin Bodky, Arthur Fine, Harold Shapero, Mitch Siporin, Louis Kronenberger, and Lee Strasberg; in the humanities: Ludwig Lewinsohn, Albert Guerard, Milton Hindus, Nahum Glatzer, and Simon Rawidowicz; in history, the Social sciences and psychology: Max Lerner, Frank Manuel, Lewis Coser, Merril Peterson, Leonard Levy, Marie Boas, and Abraham Maslow; and in the sciences: Saul Cohen, Sidney Golden, Oscar Goldman, and Albert Kellner.

Brandeis University

ACKNOWLEDGMENTS

I would like to thank Saul Cohen for allowing me to read the ms of his autobiography, *At Brandeis;* Gerald Holton for an informative conversation; Art Reis for sharing with me his researches on the early history of Brandeis[93] and for allowing me to consult the materials in the Farber University Archives. I am indebted to the archivists at Hebrew University for their help with the Einstein papers and to Zeev Rozenkranz, the Bern Dibner curator of the Albert Einstein Archives for permission to quote from them; to Abigail A. Schoolman, the archivist at the American Jewish Historical Society for permission to quote from the Wise papers; and to Eliot Wilzcek and Lisa C. Long, the archivists in the Farber University Library for their courteous and helpful assistance and for permission to quote from the Alpert and Sachar papers and from the minutes of the Board of Trustees of Brandeis University and those of the Albert Einstein and Brandeis Foundations.

NOTES

* For John with admiration and affection.
1. J. Stachel (2002). "Einstein's Jewish identity."
2. Schlossberg was the secretary-treasurer of the Amalgamated Clothing Workers union.
3. Newman (1923). For reactions see the Stephen S. Wise Papers (01/21/1999) American Jewish Historical Society, Waltham, Mass. and New York, N.Y.; in particular, Wise to Seiman January 11, 1924 and Gross to Wise March 25, 1930. At the recommendation of Ralph Lazrus, the chairman of the Board of Directors of the Albert Einstein Foundation, the fund raising instrument for the embryonic Brandeis University, Sidney Hertzberg was commissioned in late 1946 to write a brochure to be sent to potential donors. It contains a brief history of the early efforts to found a Jewish sponsored University. His 23 page ms. can be found in the George Alpert papers in the Robert D. Tauber University Archives, Brandeis University. See also chapter 1 of (Sachar 1973).
4. Silver had been associate-counsel of the Senate's committee on Banking and Currency.
5. Goldstein had been president of the Jewish national Fund and of Zionist Organization of America. He played a significant role in the deliberations and negotiations that brought about a Jewish National state in Palestine. See (Goldstein 1984; Lacqueur 1972).
6. (Goldstein 1951). Einstein Papers 40 378. The Einstein Papers are located in the Albert Einstein Archives of the Jewish National & University Library, The Hebrew University of Jerusalem. The Einstein Papers will hereafter be referred to as EP. The original copy of the letter is in the George Alpert papers. Einstein Folder. Brandeis University, Robert D. Farber University Archives, Brandeis University.
7. As quoted in a letter from Boris Young, the director of the Einstein Foundation for Higher Learning, to Dr. Joshua Loth Liebman. 11 March 1947. George Alpert papers. Einstein Folder. Robert D. Farber University Archives. Brandeis University.
8. The latter was organized and incorporated by late February 1946 with Goldstein as president. See Goldstein to Einstein. March 7, 1946. EP 40 375.
9. Nathan was an economist, a socialist in politics. He had come to the United States as a refugee during the 1930s and became very close to Einstein. He became an American citizen in 1939. He and Helen Dukas, Einstein's long time secretary, became the executors of Einstein's estate. In 1955, after Nathan had become the executor of the Einstein estate, he was refused a passport under the McCarran Act, which justified such action by the State department if the trip abroad "might be for the purpose of advancing the Communist movement." Nathan filed suit for to get his passport and at the same time filed an affidavit that he never was a member of the Communist party. In May 1957 was ruled in contempt of Congress and indicted for refusing to answer questions posed by the House un-American

Activities Committee whether he was a member of the Communist party. He refused to answer, pleading the first amendment, and challenged the Committee's jurisdiction.
10. Einstein to Goldstein. March 4, 1946. In his letter he described Nathan as being "one of the most excellent persons I know concerning unselfishness of character and devotion to his work as scholar and teacher and to the Jewish cause. He is also experienced in question of organization and administration."
11. EP 40 374.
12. Alpert was also the president of the Boston and Maine Railroad.
13. Boston Traveler, April 5, 1946.
14. Einstein to Goldstein April 16, 1946. EP 40 376. Nathan was interested in becoming involved in the selection of the faculty and possibly in obtaining either a faculty or an administrative position. He had contacted Goldstein and had spoken to him by telephone. See Goldstein's response to Einstein's letter in which he indicates to him that it is somewhat premature to worry about faculty appointments. Goldstein to Einstein April 19, 1946. Einstein Papers. 40 377.
15. Stephen S. Wise was the spiritual leader of the Free Synagogue in New York whose creed melded a liberal and progressive religious outlook with a traditionally oriented Judaism. He was at that time one of the most influential Reform rabbis in the United States.
16. In that same letter Goldstein informed Wise that he had lunch with Louis I. Newman on the day after the Board of Ministers meeting because Newman could not be present at the meeting, and that he was "helpful, recognizing that there is a tangible opportunity to do something that he had been thinking and dreaming of a good many years." At the meeting Goldstein had paid tribute to Newman as "the man who had thought earlier and more consistently about this idea than any of us." Israel Goldstein to Wise. April 29, 1946. Box 49, folder 9, Stephen S. Wise Papers, American Jewish Historical Society, Waltham, Mass. and New York, N.Y. and Brandeis University, Waltham, Mass.
17. There is an extensive correspondence between Wise and Einstein in the Wise papers; see Boxes 36 and 108. Similarly there is a significant correspondence between Wise and Nathan dating back to the 30s. See Box 78, Stephen S. Wise Papers, American Jewish Historical Society, Waltham, Mass. and New York, N.Y. and Brandeis University, Waltham, Mass. Nathan, as Einstein's protégè and confidant, saw Wise socially on a fairly regular basis.
18. The letter to potential donors that Goldstein had drafted stated that "Our purpose is to make a contribution as a Jewish group to American education by supporting a university which in student body and faculty shall be open to all races and creeds, with merit as the only criterion for admission." In that draft Goldstein also outlined his plans for the university. His aim was to admit students to a College of Liberal Arts in October 1947 and to strengthen and improve the School of Veterinary Medicine that was still operating on a limited scale. He intended to re-open the Medical School only later on when adequate resources for a "first-class Medical School" had been secured. A copy of the draft was included in the April 29 letter of Goldstein to Wise. Goldstein to Wise. April 29, 1946. Box 49, folder 9, Stephen S. Wise Papers, American Jewish Historical Society, Waltham, Mass. and New York, N.Y. and Brandeis University, Waltham, Mass.
19. At Otto Nathan's suggestion the list of "endorsers and sponsors" was enlarged to include "prominent Scientists and Scholars," in particular Paul Klapper, president of Queens College, Karl Compton, president of M.I.T. and Alvin Johnson. See Einstein to Goldstein May 14, 1946. EP 40 386 and Goldstein to Einstein May 16, 1946. EP 40 387.
20. Wise to Nathan, May 16, 1946. Box 78, folder 18, Stephen S. Wise Papers, American Jewish Historical Society, Waltham, Mass. and New York, N.Y. and Brandeis University, Waltham, Mass.
21. Thus in late December 1945 Wise had written Goldstein to let him know "how fully I appreciate your understanding and wisdom, as well as your personal friendship which saved, perhaps averted, a really nasty situation [when dealing with State department]." And he concluded his letter "With warmest regards, dear Israel." Wise to Goldstein, 29 December 1944. Box 49, folder 9, Stephen S. Wise Papers, American Jewish Historical Society, Waltham, Mass. and New York, N.Y. and Brandeis University, Waltham, Mass.
22. See the Goldstein-Wise correspondence in the Wise Papers, in particular Goldstein to Wise 25 May, 1925 and 30 September 1936. Box 49, folder 9, Stephen S. Wise Papers, American Jewish Historical Society, Waltham, Mass. and New York, N.Y. and Brandeis University, Waltham, Mass.

23. Quoted in (Laqueur 1972, 577). Laqueur gives a succinct account of this momentous Congress which came on the heels of the bombing of the King David Hotel by the Irgun. Weizmann viewed such acts of terrorism as a "cancer in the body politic of the yishuv." See also (Goldstein 1984, v.1, 192-210), and p. 201 in particular.
24. Wise to Nathan June 6, 1946. EP 40 388.
25. Lazrus was a wealthy industrialist who had ties with the Benrus Watch Company and with the Allied Department Stores. He was liberal in his political outlook.
26. Wise to Goldstein 25 June, 1946. Box 49, folder 9, Stephen S. Wise Papers, American Jewish Historical Society, Waltham, Mass. and New York, N.Y. and Brandeis University, Waltham, Mass.
27. Einstein to Goldstein, 1 July 1946. Box 36, folder 6, Stephen S. Wise Papers, American Jewish Historical Society, Waltham, Mass. and New York, N.Y. and Brandeis University, Waltham, Mass. Behind the scene Nathan had made his position clear to Einstein. See Nathan to Einstein 28 June 1946. EP 40 390-1.
28. Einstein to Wise. 26 June 1946. EP 35-263.
29. Wise to Einstein, 28 June 1946. Box 36, folder 6. Stephen S. Wise Papers, American Jewish Historical Society, Waltham, Mass. and New York, N.Y. and Brandeis University, Waltham, Mass.
30. Einstein to Wise. 29 June 1946. EP 35-265.
31. Goldstein to Einstein, July 20, 1946. EP 40 395.
32. Einstein to Lilienthal, 9 July 1946. EP 40 398.
33. Einstein to Jackson, 19 August 1946. EP 40 402.
34. Einstein to Boris Young, 17 August 1946. EP 40 403.
35. Einstein to Jackson, 19 August 1946. EP 40 402.
36. Einstein to Frankfurter, 20 August 1946. EP 40 406. In a handwritten addendum to the letter Einstein indicated that this was an "official" letter which he had signed but not written—but that it is "wirklich richtig."
37. See Nathan to Einstein, 28 June 1946. EP 40 390.
38. See Boris Young to Helen Dukas, August 29, 1946. EP 40 409.
39. Einstein to Goldstein, September 2nd, 1946. EP 40. Einstein sent a copy of the letter to Stephen Wise on September 4, 1946 and a copy of it can be found in the Wise papers. AJHS. Einstein sent copies of his letter to Goldstein to the members of the Board and the advisory-committee. See the letter Einstein sent to Frankfurter September 2, 1946 (EP 40 407) in which he informs him that "behind my back, Cardinal Spellman has been invited to deliver the invocation (benediction) at the dinner and since some other irregularities which seem important to me occurred." See Frankfurter to Einstein September 6, 1946 for Frankfurter's reply. EP 40 412; also Paul Klapper to Einstein September 7, 1946. EP 40 414.
40. Goldstein's 12 September 1946 letter to Einstein is reprinted in (Goldstein 1951). Not "the slightest tinge of a commitment" had been made in Goldstein's conversation with Sachar and had been so reported by Goldstein to Nathan. Goldstein's retelling of the story in his memoirs is a fair and even-handed account. See (Goldstein 1984, 172-185).
41. Minutes of the meetings of the Board of Directors of the Albert Einstein Foundation for Higher Learning, Inc. Farber University Archives. Brandeis University.
42. Alpert had also indicated to Nathan that a new academic advisory committee would be set up, and that the original agreements would be adhered to. Nathan to Einstein EP 40.
43. Wise to Nathan. 8 November 1946. Box 78, Folder 18. Stephen S. Wise Papers, American Jewish Historical Society, Waltham, Mass. and New York, N.Y. and Brandeis University, Waltham, Mass.
44. Einstein to Lilienthal, 10 October 1946. EP 40 417.
45. Einstein to Sachar, 23 October 1946. EP 40 425. Einstein and Sachar knew each other since the 1930s. As head of the Hillel foundations Sachar had been active in creating scholarships for refugee students to attend American colleges and universities and had corresponded with Einstein regarding these matters. See the Einstein-Sachar correspondence in the Einstein Papers.
46. On 11 November 1946 Nathan became a member of the Board. Minutes of the Board. Albert Einstein Foundation. Farber University Archives. Brandeis University.
47. Otto Nathan 9 November 1946. EP 40 427.
48. Minutes of the meeting of the Board of Directors of the Albert Einstein Foundation 6 January 1947. The entry in the minutes continued: "In his conference with these educators he discussed many facets

of academic administration, such as life tenure of office, the tutorial system, the size of the student body, the age of the faculty and related subjects. A more complete report of the trip will be submitted to the Board in due course." Nathan's expenses were paid by the Albert Einstein Foundation. His involvement and expenses intensified during the first four months of 1947. See the Otto Nathan 1946-1947 folder in the Alpert Papers. Robert D. Farber University Archives. Brandeis.

49. See (Kramnick and Sheerman 1993) for a thorough account of Laski's life and works. Also (Deane 1955) for an analysis of Laski's writing.
50. See the *Manchester Guardian* and the *New York Times* for 3 December 1946. Both newspapers carried extensive coverage of the trial. A full record of the trial is given in (Laski 1947). Kramnick and Sheerman devote the entire chapter twenty to it (Kramnick and Sheerman 1993, 516-543).
51. Nathan wrote to his friends, among them Ralph Lazrus and Stephen Wise, asking for contributions to help Laski. Wise sent him a cheque for $25 and Lazrus also contributed.
52. Thus in the spring of 1947 Nathan had become involved in finding a new dean for the veterinary school. Whether he had made a certain moral commitment on behalf of the University to a particular candidate became a contentious matter at the Board meeting of June 4, 1947.
53. Einstein to Lazrus. July 19, 1947. Alpert Papers. Einstein Folder. Farber University Archives. Brandeis University. In a letter to Judge Steinbrick written on 2 February 1949 Alpert gives a differing account: "In 1947 at a meeting in Professor Einstein's home at which, in addition to Professor Einstein, Mr. Lazrus, Professor Nathan and I were present, it was suggested by Professor Nathan that authority be given to Professor Einstein to select and engage a President of the University. My reaction to the suggestion was that this was a function of the Board which it could not delegate. Professor Nathan, however, insisted that the Board should have sufficient confidence in Professor Einstein to leave the selection entirely up to him. I then asked whether Professor Einstein had anyone in particular in mind and was informed that it was Professor Laski. I pointed out that I had heard of the libel suit brought by Professor Laski in England and of the outcome of the suit, and that I did not feel that a figure as controversial as he, even though he be a learned person, should be selected as president of Brandeis University. Furthermore, while the University was Jewish sponsored, it was to be an American institution,—a contribution by the Jews of America to American education and that I believed it to be very desirable that its first president be an American. With this point of view, Mr. Lazrus and Professor Nathan did not agree....It was left that a meeting of the Board be held the following week and the matter placed before the Board." Alpert to the Hon. Meier Steienbrink, 2 February 1949. Alpert Papers. Farber Archives. Brandeis University.
54. Boris Young, the director of the Albert Einstein Foundation, attended the April 14 meeting "by invitation". On June 5, on the heels of the resignation of Lazrus and Nathan from the Board, he submitted a report of that meeting to Julius Silver. It stated: Present: S. Ralph Lazrus, chairman, Otto Nathan, Milton Bluestein, Israel Rogosin. Boris Young by invitation. After Chairman inquired if a quorum was present, and upon being informed by Mr. Young that there was no quorum, he decided to hold informal discussion of matters pending for Board consideration. Major Abraham F. Wechsler joined the meeting shortly after this point. Throughout the discussion no resolutions or motions were offered or acted upon, nor was an acting secretary appointed to substitute for Dr. Dushkin who was absent. Nathan who was shown that report by Young, crossed out the last sentence and wrote in pencil: "From this point on a quorum existed." Boris Young to Julius Silver. 5 June 1947. A report submitted relating to the meeting of April 14, 1947 at 4:00 p.m. in suite 4 of Hotel Pierre. Minutes of the Meetings of the Board of Directors of the Albert Einstein Foundation. Page 104. Robert D. Farber University Archives. Brandeis University.
55. Silver to Lazrus. May 12, 1947. Alpert Papers. Einstein Folder. Robert D. Farber University Archives. Brandeis University. At the 4 June 1947 meeting of the Board, after the rift between Einstein, Lazrus, Nathan and the Board had occurred "Dr. Nathan desired the minutes to disclose that it was his understanding that at the meeting held on April 14, 1947 Professor Albert Einstein was authorized to choose a President for Brandeis University despite the contention by others that no quorum was present."
56. Minutes of the Meetings of the Board of Directors of the Albert Einstein Foundation. Page 104. Robert D. Farber University Archives. Brandeis University.
57. Einstein to Laski. 16 April 1947. EP 40 432.
58. Laski to Einstein. 25 April 1947. EP 40 435.

59. There is a further somewhat curious aspect to the story: In 1975 Granville Eastwood, a retired English trade union man, was preparing his biography of Harold Laski. He contacted Nathan, who allowed him to read his correspondence with Laski. Among the letters was a copy of Einstein's letter to Laski offering him the presidency of Brandeis. Eastwood wanted to make use of this material and asked Nathan permission to use its content. In an October 1975 letter to Helen Dukas asking for her approval of the material Eastwood wanted to include in his book, see (Eastwood 1977, 85-6.), Nathan added that for reasons unknown to me I carried Laski's reply to Einstein's letters in my wallet for many years. When I gave you Einstein's letters to me to have them photostated I do not recall whether I gave you also Laski's letter which was addressed to Einstein or whether that letter was lost when I lost my wallet last year. I say all that because Eastwood's draft might lead to inquiries from other people about your and my own files in regard to Brandeis University. There is nothing in it, from our point of view, could not see the light of day but I do not quite know what we ought to do and would like to know how you feel.
60. Nathan to Helen Dukas 28 October 1975 EP 40 436. Nathan had clearly given Laski's reply to Dukas as it is to be found in the Einstein papers.
61. Silver to Lazrus. 12 May 1947. Alpert Papers. Farber Archives. Brandeis University. Again, in his letter of 2 February 1949 to Steinbrick Alpert gives a differing account. In the letter Alpert claimed that immediately after the 14 April 1947 meeting a conference was arranged in New York with Lazrus, Nathan and Silver. "At this conference there was a lengthy and at times, spirited discussion. To my objection that nothing should be done in the selection of a president until the Board authorized, Professor Nathan repeatedly stated that Professor Einstein did not propose to be a "rubber stamp for any Board of Directors." The conference ended in a tone that was anything but amiable. Immediately after this conference, Mr. Silver and I drove to Princeton. We invited Mr. Lazrus and Professor Nathan to go along but they refused, Professor Nathan stating that he didn't understand what right I had to make an appointment to call on Dr. Einstein without his, Nathan's, prior consent." The tenor of the 11 May 1947 letter of Silver to Lazrus does not jibe with Alpert's account. The already strained relations between Alpert and Nathan must have turned acrimonious in late May 1947.
62. Max Grossman to Nathan. 20 April 1947. Alpert Papers. Otto Nathan Folder. Robert D. Farber University Archives. Brandeis.
63. Alpert to Nathan. 24 March 1947. Alpert Papers. Otto Nathan Folder. Robert D. Farber University Archives. Brandeis.
64. Einstein to Lazrus. May 16, 1947. Alpert Papers. Einstein Folder. Farber University Archives. Brandeis University.
65. In the typewritten minutes in the Farber Archives the last part of the last sentence was crossed out and amended by Nathan to read "that Professor Einstein and Mr. Lazrus and he do not want to injure the University and wished the project well". Minutes of the Meetings of the Board of Directors of the Albert Einstein Foundation of 19 May 19 1947. Page 114-5. Robert D. Farber University Archives. Brandeis University.
66. The name of the Foundation was changed to the Brandeis Foundation. The press statement which was released in early June stated that the change was made after conferring with Professor Einstein "because the Albert Einstein had accomplished its purpose, namely the actual establishment of the school." It was also explained that maintaining the Albert Einstein Foundation designation would confuse contributors, and that the new name was descriptive of the its present function: the support of Brandeis University.
67. Einstein to Susan Brandeis Gilbert. 20 June 1947. Alpert Papers. Einstein Folder. Farber University Archives. Brandeis University.
68. See Boston Sunday Herald June 29, 1947.
69. *New York Times*. June 23, 1947.
70. Zukerman, W. "March of Jewish Events." *The American Hebrew*. 4 July, 1947. p. 6.
71. Wise Papers.Box, Folder: Brandeis University. Stephen S. Wise Papers, American Jewish Historical Society, Waltham, Mass. and New York, N.Y. and Brandeis University, Waltham, Mass.
72. EP 40 447. *P.M.* printed the first paragraph on Sunday June 29,1947.
73. Thus Boris Young, the executive director of the Brandeis Foundation, (the new name of the Albert Einstein Foundation after July 1947) wrote to Evelyn Van Gelder on October 14, 1947 stating that Albert

Einstein had opposed Spellman's for political reasons. Also that "In the Spring of this year, 1947, Mr. Lazrus and Dr. Nathan decided (although this was outside their own sphere, for neither Lazrus nor Nathan served on the Board of Trustees of the University) to install as first president of the University one of the most controversial political and educational figures of our times." In fact, earlier that year, on 7 January 1947, Young in his letter to E. Saveth had listed the membership of the Board of Trustees of the University. The list names S. Ralph Lazrus as a member! Alpert Papers. Einstein Folder.

74. See for example the *Boston Sunday Herald* June 29, 1947 from which article the following is abstracted.
75. It should be noted that three other "New York" Board members resigned at the meeting at which Lazrus and Nathan tendered their resignation.
76. Incidentally, for a while Stephen Wise remained very hostile to the Brandeis enterprise. He strongly advised Sachar not to take its presidency. See Wise to Sachar. 26 March 1948. Box, Folder. Stephen S. Wise Papers, American Jewish Historical Society, Waltham, Mass. and New York, N.Y. and Brandeis University, Waltham, Mass. Parts of this letter were reprinted in (Sachar 1976, 39). Wise eventually changed his mind and one of his last public appearances was a visit to the Brandeis campus.
77. On May 17, 1949 Sachar wrote Einstein to inform him of developments at the newly founded university: "We are admitting a second class of freshmen numbering approximately 150, and curriculum and faculty will naturally expand. In developing the area of Biology we have had the cordial cooperation of Dr. Selman Waksman with whom we are clearing personnel, and of course we are delighted with his interest in this first Jewish sponsored university. We must now bring in a fine Physicist. Quite a number of applications and recommendations have come to us. One of the applicants is Dr. Nathan Rosen, Professor of Physics at the University of North Carolina." Rosen was a friend and former associate of Einstein. Einstein had earlier indicated that given past events he would not suggest the names of young physicists who might be approached to join Brandeis, but would give his evaluation of candidates who had applied on their own. In his answer Einstein indicated that he did not believe that Rosen was of the stature requisite for the job.
78. Sachar to Einstein. 5 April 1952. EP 61 213.
79. To forget injuries suffered, is like throwing well earned money out the window.
80. Einstein to Sachar 30 March 1952. EP 61 212.
81. Einstein to Chakravarty. 2 December 1952. EP 40 1952.
82. Einstein to Sachar February 22, 1953. Sachar Papers. Robert D. Farber University Archives. Brandeis University.
83. Einstein to Sachar 12 January 1954. EP 61 214.
84. Einstein to Nathaniel L. Goldstein 9 May 1953. EP 40 450.
85. Einstein to Abba Eban, 18 November 1952. Quoted in Gerald E. Tauber, "Einstein and Zionism," in (French 1979, 207).
86. Einstein to Born, 3 March 1920. In (Born 1971, 26).
87. See the letter from Boris Young to Dr.Joshua Loth Liebman. 11 March 1947 where Einstein is quoted as holding this view. Alpert papers. Einstein Folder. Robert D. Farber University Archives. Brandeis University.
88. See (Stachel 2002) for Einstein's role and stand in the founding of the Hebrew University of Jerusalem in 1923.
89. Sachar to Goldstein. June 2, 1948. In (Goldstein 1951, 113). But see the somewhat less generous account in (Sachar 1973).
90. These quotations from Einstein can be found in Isaiah Berlin's (1982) essay "Einstein and Israel".
91. The speech that Einstein wrote for the fund raising dinner of March 20, 1947 can be interpreted as making this point. EP 40431. The speech was not delivered but its text was published in the *Jewish Advocate* in April 1947.
92. That high academic standard were likewise imposed on the students is evidenced by the fact that of the 100 freshmen admitted in 1948 many did not graduate in 1952. There were 104 students graduating in 1952—but this reflected the fact that after 1948 a sizable number of students were being admitted with advanced standing. Of the 104 BA and BS degrees conferred in 1952, 23 were cum laude, 10 magna and 1 summa cum laude.
93. See his article "The Founding" in the *Brandeis Review* 19 (1990): 42-3.

REFERENCES

Berlin, I. 1982. "Einstein and Israel." Pp 144-155 in *Personal Impressions*. New York: Penguin Books.
Born, M. 1971. *The Born-Einstein Letters 1916-1955*. New York: Walker and Company.
Deane, H. A. 1972. *The political ideas of Harold J. Laski*. Hamden, Conn.: Archon Books.
Eastwood, G. G. 1977. *Harold Laski*. London: Mowbrays.
Frankfurter, F. 1950. "On Harold Laski". *Clare Market Review* -Michaelmas 1950. London: The Students' Union of the London School of Economics.
French, A. P. ed. 1979. *Einstein. A Centenary Volume*. Cambridge MA: Harvard University Press.
Goldstein, I. 1951. *Brandeis University. Chapter of its Founding*. New York: Bloch Publishing Co.
———. 1984. *My World as a Jew. The Memoirs of Israel Goldstein* Vol. 1. New York: Herzl Press.
Kramnick, I. and B. Sheerman. 1993. *Harold Laski: a life on the left*. New York: Allen Lane, Penguin Press.
Laqueur, W. 1972. *A History of Zionism*. New York: Holt, Rinehart and Winston.
Laski, H. J. 1947. *Laski v. Newark Advertiser Co., Ltd. & Parlby: before Lord Goddard, lord chief justice of England and a special jury*. London: Daily Express.
Newman, L. I. 1923. *A Jewish university in America?* New York: Bloch Pub. Co.
Sachar, A. 1976. *A Host at Last*. Special Edition. Waltham: Copygraph Inc.
Stachel, J. 2002. "Einstein's Jewish identity." Pp. 57-83 in *Einstein from 'B' to 'Z.'* Einstein Studies Vol. 9. Boston: Birkhäuser. [Dedicated to the memory of Gerald Tauber and prepared for the Symposium "Einstein in Context."]

APPENDIX

JOHN STACHEL'S PUBLICATIONS

THESES

1959. "Energy Flow in Cylindrical Gravitational Waves." M. S. Thesis, Stevens Institute of Technology.

1962. "The Lie Derivative and the Cauchy Problem in the General Theory of Relativity." Ph. D. Thesis, Stevens Institute of Technology.

ARTICLES

1966. "Cylindrical Gravitational News." *Journal of Mathematical Physics* 7:1321-1331.

1968. "Behavior of Weyl-Levi Civita Coordinates for a Class of Solutions Approximating the Schwarzschild Metric." *Nature* 219:1346-1347.

1968. "Einstein Tensor and Spherical Symmetry." *Journal of Mathematical Physics* 9:269-283 (with Jerzy Plebanski).

1968. "Perturbations of an Arbitrary Spherically Symmetric Metric." *Nature* 220:779-780.

1968. "Structure of the Curzon Metric." *Physics Letters* 27A:60-61.

1969. "Comments on 'Causality Requirements and the Theory of Relativity.'" Pp. 179-197 in *Boston Studies in the Philosophy of Science* V, eds. R. S. Cohen and M. W. Wartofsky. Boston: Reidel. To appear in *Going Critical* (in press).

1969. "Covariant Formulation of the Cauchy Problem in Generalized Electrodynamics and General Relativity." *Acta Physica Polonica* 35:689-709.

1969. "Invariances of Approximately Relativistic Lagrangians and the Center-of-Mass Theorem I." *Physical Review* 185:1636-1647 (with Peter Havas).

1969. "Specifying Sources in General Relativity." *Physical Review* 180:1256-1261.

1969. "The Pure Radiation News Function in General Relativity." *Physical Review* 179:1251-1252.

1970. "Einstein Tensor and 3-Parameter Groups of Isometries with 2-Dimensional Orbits." *Journal of Mathematical Physics* 11:3358-3370 (with Hubert Gonner).

1972. "External Sources in General Relativity." *GRG Journal: General Relativity and Gravitation* 3:257-267.

1973. "Comments on 'Ontic Commitments of Quantum Mechanics.'" Pp. 309-317 in *Logical and Epistemological Studies in Contemporary Physics. Boston Studies in the Philosophy of Science XIII*, eds. R. S.Cohen and M. W. Wartofsky. Boston: Reidel.

1973. "Comments on 'The Formal Representation of Physical Quantities.'" Pp. 214-223 in *Logical and Epistemological Studies in Contemporary Physics, Boston Studies in the Philosophy of Science XIII,* eds. R. S. Cohen and M. W. Wartofsky. Boston: Reidel.

1974. "A Note on Scientific Practice." Pp. 417-433 in *For Dirk Struik. Boston Studies in the Philosophy of Science* XV, eds. R. S. Cohen, J. Stachel, M. W. Wartofsky. Boston: Reidel. Reprinted 1985, pp. 160-176 in *A Portrait: Twenty-five Years - Boston Colloquium for the Philosophy of Science 1960-1985*, R. S. Cohen and M. V. Wartofsky. Boston: Reidel. To appear in *Going Critical* (in press).

1974. "Introduction to the Symposium on 'Current Problems in Cosmology.'" Pp. 179-180 in *Philosophical Foundations of Science. Proceedings of Section L. 1969, AAAS. Boston Studies in the Philosophy of Science*, eds. R. S. Cohen and R. J. Seeger. Boston: Reidel.

1974. "The Rise and Fall of Geometrodynamics." Pp. 31-54 in *PSA 1972 - Proceedings of the 1972 Biennial Meeting of the Philosophy of Science Association. Boston Studies in the Philosophy of Science* XX, eds. K. F. Schaffner and R. S. Cohen. Boston: Reidel. To appear in *Going Critical* (in press).

1976. "Center of Mass Theorem in Post-Newtonian Hydrodynamics." *Physical Review* D14:917-921 (with Thomas Pascoe and Peter Havas).

1976. "Invariances of Approximately Relativistic Hamiltonians and the Center-of-Mass Theorem." *Physical Review* D13:1598-1613 (with Peter Havas).

1976. "The 'Logic' of Quantum Logic." Pp. 515-526 in *PSA 1974 - Proceedings of the 1974 Biennial Meeting of the Philosophy of Science Association. Boston Studies in Philosophy of Science* XXXII, eds. R. S. Cohen et al. Boston: Reidel. To appear in *Going Critical* (in press).

1977. "Achille Papapetrou." *GRG Journal: General Relativity and Gravitation* 8:541-543.

1977. "A Variational Principle Giving Gravitational 'Superpotentials,' The Affine Connection, Riemann Tensor and Gravitational Field Equations." *GRG Journal: General Relativity and Gravitation* 8:705-712.

1977. "Classical Particles with Spin. I: The WKBJ Approximation." *Journal of Mathematical Physics* 18:2368-2374 (with Jerzy Plebanski).

1977. "Notes on the Andover Conference." Pp. vii-xiii in Foundations of Space-Time Theories. Minnesota Studies in the Philosophy of Science VIII, eds. John Earman, Clark Glymour, and John Stachel. Minneapolis: University of Minnesota Press. To appear in Going Critical (in press).

1978. "A New Lagrangian for the Vacuum Einstein Equations and its Tetrad Form." *GRG Journal: General Relativity and Gravitation* 9:1075-1087 (with Achille Papapetrou).

1978. "Conformal Two-Structure as the Gravitational Degrees of Freedom in General Relativity." *Journal of Mathematical Physics* 19:2447-2460 (with R. A. D'Inverno).

1979. "Einstein on Civil Liberty." *Rights* 25:6. Reprinted 2002 in *Einstein from 'B' to 'Z'*.

1979. "Einstein's Odyssey." *The Sciences* 19:14-15, 32-34. Reprinted 2002 in *Einstein from 'B' to 'Z'*.

1980. "Einstein and the Rigidly Rotating Disc." Pp. 1-15 in *General Relativity and Gravitation One Hundred Years After the Birth of Albert Einstein*, ed. A. Held. New York: Plenum. Reprinted 1989, pp. 48-62 in *Einstein and The History of General Relativity: based on the proceedings of the 1986 Osgood Hill Conference*, eds. D. Howard and J.Stachel. Boston/Basel/Stuttgart: Birkhäuser, and 2002 in *Einstein from 'B' to 'Z'*.

1980. "Guide to the Duplicate Einstein Archive and Control Index." 48 pp. Princeton: n.p. Reprinted 2002 (in part) in *Einstein from 'B' to 'Z'*.

1980. "If Maxwell Had Worked Between Ampère and Faraday." *American Journal of Physics* 48:5-7. To appear in *Going Critical* (in press).

1980. "The Anholonomic Cauchy Problem in General Relativity." *Journal of Mathematical Physics* 21:1776-1782.

1980. "The Genesis of General Relativity." Pp. 428-442 in *Einstein Symposion Berlin. Lecture Notes in Physics 100*, eds. H. Nelkowski et al. Berlin/Heidelberg/New York: Springer-Verlag. Reprinted 2002 in *Einstein from 'B' to 'Z'*.

1980. "Thickening the String I: String Dusts." *Physical Review* D21:2171-2181.

1980. "Thickening the String II: The Null String." *Physical Review* D21:2182-2184.

1982. "Albert Einstein: The Man Beyond the Myth." *Bostonia Magazine* 56:8-17. Reprinted 2002 in *Einstein from 'B' to 'Z'*.

1982. "Comments on 'Some Logical Problems Suggested by Empirical Theories,' by Professor Dalla Chiara." Pp. 91-102 in *Logic, Language and Method. Boston Studies in the Philosophy of Science XXXI*, eds. R. S. Cohen and M. W. Wartofsky. Boston: Reidel.

1982. "Einstein and Michelson: The Context of Discovery and the Context of Justification." *Astronomische Nachrichten* 303:47-53. Reprinted 2002 in *Einstein from 'B' to 'Z'*.

1982. "Einstein's First Derivation of Mass-Energy Equivalence." *American Journal of Physics* 50:760-763 (with Roberto Torretti). Reprinted 2002 in *Einstein from 'B' to 'Z'*.

1982. "Globally stationary but locally static space-times: A gravitational analogue of the Aharonov-Bohm effect." *Physical Review* D26:1281-1290.

1982. "String Dusts, Fluids, and Subspaces." Pp. 270-283 in *Physics as Natural Philosophy. Essays in Honor of Laszlo Tisza on His Seventy-Fifth Birthday*, eds. Abner Shimony and Herman Feshbach. Cambridge: MIT Press.

1983. "Special Relativity From Measuring Rods." Pp. 255-272 in *Physics. Philosophy and Pyschoanalysis*, eds. R. S. Cohen and L. Laudan. Boston: Reidel. To appear in *Going Critical* (in press).

1984. "The Dynamical Equations of Black-Body Radiation." *Foundations of Physics* 14:1163-1168.

1984. "The Generally Covariant Form of Maxwell's Equations." Pp. 23-37 in *J.C. Maxwell the Sesquicentennial Symposium, New Vistas in Mathematics, Science and Technology*, ed. M. S. Berger. Amsterdam/New York/Oxford: North-Holland.

1984. "The Gravitational Field of Some Rotating and Non-Rotating Cylindrical Shells of Matter." *Journal of Mathematical Physics* 25:338-341.

1986. "Do Quanta Need a New Logic?" Pp. 229-347 in *From Quarks to Quasars: Philosophical Problems of Modern Physics*, ed. R. Colodny. Pittsburgh: University of Pittsburgh Press. To appear in *Going Critical* (in press).

1986. "Eddington and Einstein." Pp. 225-250 in *The Prism of Science. The Israel Colloquium: Studies in History, Philosophy and Sociology of Science* 2, ed. E. Ullmann-Margalit. Boston: Reidel. Reprinted 2002 in *Einstein from 'B' to 'Z'*.

1986. "Einstein and the Quantum." Pp. 349-385 in *From Quarks to Quasars: Philosophical Problems of Modern Physics*, ed. R. Colodny. Pittsburgh: University of Pittsburgh Press. Reprinted 2002 in *Einstein from 'B' to 'Z'*.

1986. "What a Physicist Can Learn From the Discovery of General Relativity." Pp. 1857-1862 in *Proceedings of the Fourth Marcel Grossmann Meeting on General Relativity*, ed. R. Ruffini. Elsevier. To appear in *Going Critical* (in press).

1987. "'A Man of My Type' - Editing the Einstein Papers." *British Journal for the History of Science* 20:57-66. Revised version printed 2002 in *Einstein from 'B' to 'Z'*, pp. 97-111.

1987. "Congruences of Subspaces." Pp. 447-464 in *Gravitation and Geometry*, eds. W. Rindler and A. Trautman. Naples: Bibliopolis.

1987. "Einstein and Ether Drift Experiments." *Physics Today* 40:45-47. Reprinted 2002 in *Einstein from 'B' to 'Z'*.

1987. "How Einstein Discovered General Relativity: a historical tale with some contemporary morals." Pp. 200-208 in *General Relativity and Gravitation: Proceedings of the 11th International Conference on General Relativity and Gravitation*, ed. M. A. H. MacCallum. Cambridge: Cambridge University Press. Reprinted 2002 in *Einstein from 'B' to 'Z'*.

1989. "Albert Einstein (1879-1955)." Pp. 198-199 in *The Blackwell Companion to Jewish Culture*, ed. Glenda Abramson. Oxford: Blackwell. Reprinted 2002 in *Einstein from 'B' to 'Z'*.

1989. "Editorial Note: Einstein on the Theory of Relativity." Pp. 253-274 in *The Collected Papers of Albert Einstein, vol 2, The Swiss Years: Writings 1900-1909*, eds. John Stachel et al. Princeton: Princeton University Press. Reprinted 2002 in *Einstein from 'B' to 'Z'*.

1989. "Einstein's Search for General Covariance, 1912-1915." Pp. 63-100 in *Einstein and The History of General Relativity: based on the proceedings of the 1986 Osgood Hill Conference*, eds. D. Howard and J.Stachel. Boston/Basel/Stuttgart: Birkhäuser. Reprinted 2002 in *Einstein from 'B' to 'Z'*.

1989. "'Quale canzone cantarono le sirene': Come scopri Einstein la teoria speciale della relatività" "'What song the sirens sang': How did Einstein discover special relativity." Pp. 21-37 in *L'Opera di Einstein*, ed. Umberto Curi. Ferrara: Gabriele Corbino & Co.

1990. "The Theory of Relativity." Pp. 442-457 in *Companion to the History of Modern Science*, eds. R. C. Olby et al. London/New York: Routledge. To appear in *Going Critical* (in press).

1991. "Einstein and Quantum Mechanics." Pp. 13-42 in *Conceptual Problems of Quantum Gravity: Proceedings of the 1988 Osgood Hill Conference*, eds. A. Ashtekar and J. Stachel. Boston/Basel/Stuttgart: Birkhäuser. Reprinted 2002 in *Einstein from 'B' to 'Z'*.

1991. "The Cauchy Problem in General Relativity: The Early Years." Pp. 405-416 in *Proceedings of the Second International Conference on the History of General Relativity*, eds. J. Eisenstaedt and A. Kox. Boston/Basel/Stuttgart: Birkhäuser. To appear in *Going Critical* (in press).

1992. Greek translation of "Marx on Science and Capitalism" (1995). *Outopia* No. 3:*93-105*.

1993. "The Meaning of General Covariance: The Hole Story." Pp. 129-160 in *Philosophical Problems of the Internal and External World: Essays on the Philosophy of Adolf Grünbaum*, eds. John Earman et al. Constance: Universitätsverlag/ Pittsburgh: University of Pittsburgh Press. To appear in *Going Critical* (in press).

1993 "The Other Einstein." *Science in Context* 6:275-290. Prepared for the 1989 Jerusalem Symposium "Einstein in Context" and reprinted 2002 in *Einstein from 'B' to 'Z'*.

1994. "Changes in the Concept of Space and Time Brought About by Relativity." Pp. 141-162 in *Artifacts, Representations and Social Practice: Essays for Marx Wartofsky*, eds. Carol Gould and Robert S. Cohen. Dordrecht/Boston/London: Kluwer Academic. Revised and expanded version of "Mutamenti nel concetto di tempo operati dalla relatività" (1995). To appear in *Going Critical* (in press).

1994. "Lanczos's Early Contributions to Relativity and His Relationship with Einstein." Pp. 201-221 in *Proceedings of the Cornelius Lanczo's International Centenary Conference*, eds. J. David Brown et al. Philadelphia: SIAM. Reprinted 2002 in *Einstein from 'B' to 'Z'*.

1994. "Marx's Critical Concept of Science." Max Planck Institute for the History of Science: *Preprint* 10. Written in 1981 based on a talk to Boston Colloquium for Philosophy and History of Science. To appear in *Going Critical* (in press).

1994. "Scientific Discoveries as Historical Artifacts." Pp. 139-148 in *Trends in the Historiography of Science*, eds. Kostas Gavroglu, Jean Christianidis, Efthymios Nicolaidis. Dordrecht/Boston/London: Kluwer Academic. To appear in *Going Critical* (in press).

1995. "Einstein and Bose." Max Planck Institute for the History of Science: *Preprint 4*. Based on a talk prepared for the Bose Centennial. Reprinted 2002, pp. 519-538 in *Einstein from 'B' to 'Z.'* Boston: Birkhäuser.

1995. "History of Relativity." Pp. 249-356 in *Twentieth Century Physics*, vol. 1, eds. Laurie M. Brown, Abraham Pais, Brian Pippard. Bristol: Institute of Physics Pub./Philadelphia: American Institute of Physics Press. To appear in *Going Critical* (in press).

1995. "Marx on Science and Capitalism." Pp. 69-85 in *Science, Politics and Social Practice*, eds. Kostas Gavroglu, John Stachel, Marx Wartofsky. Dordrecht/Boston/London: Kluwer Academic. To appear in *Going Critical* (in press).

1995. "Mutamenti nel concetto di tempo operati dalla relatività." Pp. *509-543* in *Il Tempo nella Scienza e nella Filosofia*, ed. Evandro Agazzi. Naples: Guida Editori.

1995. "The Manifold of Possibilities: Comments on Norton." Pp. 71-88 in *The Creation of Ideas in Physics: Studies for a Methodology of Theory Construction*, ed. Jarett Leplin. Dordrecht/Boston/London: Kluwer Academic. To appear in *Going Critical* (in press).

1996. "Albert Einstein." Pp. 393-397 in *Macmillan Encyclopedia of Physics*, vol. 2, ed. John S. Rigden. Simon and Schuster Macmillan. Reprinted 2002 in *Einstein from 'B' to 'Z'*.

1996. "Albert Einstein and Mileva Maric: A Collaboration That Failed to Develop." Pp. 207-219, 330-335 in *Creative Couples in Science*, eds. Helena M. Pycior, Nancy G. Slack and Pnina G. Abir-Am. New Brunswick: Rutgers. Reprinted 2002 in *Einstein from 'B' to 'Z'*.

1997. "Belated Decision in the Hilbert-Einstein Priority Dispute." *Science* 278:1270-1273 (with Leo Corry and Jürgen Renn). Reprinted 2002 in *Einstein from 'B' to 'Z'*.

1997. "Feynman Paths and Quantum Entanglement: Is There any More to the Mystery?" Pp. 245-256 in *Potentiality, Entanglement and Passion-at-a-Distance/Quantum-Mechanical Studies for Abner Shimony*, vol. 2, eds. Robert S. Cohen, Michael Horne and John Stachel. Dordrecht/Boston/London: Kluwer Academic. To appear in *Going Critical* (in press).

1997. "The Origin of Gravitational Lensing: A Postscript to Einstein's 1936 *Science* Paper." *Science* 275:184-186 (with Jürgen Renn and Tilman Sauer). Reprinted 2002 in *Einstein from 'B' to 'Z'*.

1998. "Introduction." Pp. 3-27 in *Einstein's Miraculous Year: Five Papers That Changed the Face of Physics*, ed. John Stachel. Princeton: Princeton University Press. Reprinted (in part) 2002 in *Einstein from 'B' to 'Z'*.

1998. "On the Interpretation of the Einstein-Cartan Formalism." Pp. 475-485 in *On Einstein's Path*, ed. Alex Harvey. New York/Berlin/Heidelberg: Springer.

1999. "Did Malament Prove the Non-Conventionality of Simultaneity in the Special Theory of Relativity?" In *Philosophy of Science* 66:208-220 (with Sahotra Sarkar). To appear in *Going Critical* (in press).

1999. "Einstein and Infeld, Seen Through Their Correspondence." *Acta Physica Polonica* 30B:2879-2908. Reprinted 2002 in *Einstein from 'B' to 'Z'*.

1999. "Equivalence (Principe d')," "Espace-temps," "Gravitation," "Mécanique quantique et relativité (Compatibilité entre)," "Relativité (Physique)." Pp. 373-374, 376-379, 464-466, 624-628, 828-830 in *Dictionnaire d'Histoire et Philosophie des Sciences*, ed. Dominique Lecourt. Paris: Presses Universitaires de France.

1999. "Hilbert's Foundation of Physics: From a Theory of Everything to a Constituent of General Relativity" (with Jurgen Renn). Max Planck Institute for the History of Science: *Preprint* 118. To appear in *Alternative Approaches to General Relativity (The Genesis of General Relativity: Sources and Interpretations*, vol. 3, ed. Jürgen Renn). Dordrecht/Boston/London: Kluwer Academic. Also to appear in *Going Critical* (in press).

1999. "Introduction" to, and "Comments" on the papers on "Quantum field theory and spacetime." Pp. *166-175* and pp. 233-240 in *Conceptual Problems of Quantum Field Theory*, ed. Tian Yu Cao. Cambridge: Cambridge University Press. To appear in *Going Critical* (in press).

1999. "New Light on the Einstein-Hilbert Priority Question." *Journal of Astrophysics and Astronomy* 20:91-101. Reprinted 2002 in *Einstein from 'B' to 'Z'*.

1999. "The Early History of Quantum Gravity." Pp. 525-534 in *Black Holes, Gravitational Radiation and the Universe*, eds. Bala Iyer and Biplap Bhawal. Dordrecht/Boston/London: Kluwer Academic. To appear in *Going Critical* (in press).

2000. "Einstein's Light-Quantum Hypothesis, or Why Didn't Einstein Propose a Quantum Gas a Decade-and-a-Half Earlier?" Pp. 231-251 in *Einstein: The Formative Years, 1879-1909*, eds. Don Howard and John Stachel. Boston/Basel/Berlin: Birkhäuser. Reprinted 2002 in *Einstein from 'B' to 'Z'*.

2000. "Introduction." Pp. 1-22 in *Einstein: The Formative Years, 1879-1909*, eds. Don Howard and John Stachel. Boston/Basel/Berlin: Birkhäuser. Reprinted 2002 in *Einstein from 'B' to 'Z'*.

2002. *Einstein from 'B' to 'Z'*. (*Einstein Studies*, vol. 9.) Boston/Basel/Berlin: Birkhäuser.

2002. "Einstein's Jewish Identity." Pp. 57-83 in *Einstein from 'B' to 'Z.'* Boston: Birkhäuser Prepared for the 1989 Jerusalem Symposium "Einstein in Context.".

2002. "The First Two Acts." Pp. 261-292 in *Einstein from 'B' to 'Z'*. Boston: Birkhäuser. Prepared for *General Relativity in the Making: Einstein's Zurich Notebook (The Genesis of General Relativity*, vol. 1), Jürgen Renn et al. Dordrecht/Boston/London: Kluwer Academic.

2002. "'The Relations Between Thing' versus 'The Things Between Relations': The Deeper Meaning of the Hole Argument." Pp. 231-266 in *Reading Natural Philosophy/Essays in the History and Philosophy of Science and Mathematics*, ed. David Malament. LaSalle/Chicago: Open Court. To appear in *Going Critical* (in press).

2002. "The Young Einstein: Poetry and Truth." Pp. 21-38 in *Einstein from 'B' to 'Z.'* Boston: Birkhäuser. Presented at 1990 AAAS Symposium on "The Young Einstein."

2002. "'What Song the Syrens Sang': How Did Einstein Discover Special Relativity?" English text of "'Quale canzone cantarono le sirene': come scopro Einstein la teoria speciale della relatività?" (1989). Pp. 157-169 in *Einstein from 'B' to 'Z.'* Boston: Birkhäuser.

2003. "Critical Realism: Bhaskar and Wartofsky." Talk at the session: In Memoriam: Marx Wartofsky of the Boston Colloquium for the Philosophy of Science, Sept. 22, 1997. In *Constructivism and Practice: Toward a Historical Epistemology*, ed. Carol Gould. Lanham, MD: Rowman and Littlefield. To be reprinted in *Going Critical* (in press).

In press. "A Brief History of Space-Time." In 2001: *A Relativistic Spacetime Odyssey. Experiments and Theoretical Viewpoints on General Relativity and Quantum Gravity*, ed. Luca Lusanna. World Scientific. To appear in *Going Critical* (in press).

In press. "Bohr and the Photon." Prepared for the Bohr Centennial Meeting, Boston Colloquium for Philosophy and History of Science 1987. To appear in *Going Critical* (in press).

In press. "Contradiction, Contrariety and Colletti." To appear in *Going Critical* (in press).

In press. "Einstein's Intuition and the Post-Newtonian Approximation." A talk given at the Symposium "Topics in Mathematical Physics, General Relativity, and Cosmology on the Occasion of the 75th Birthday of Jerzy F. Plebanski," September 17-20, 2002; CINVESTAV, Mexico City. To appear in the Proceedings of the Symposium.

In press. "Fibered Manifolds, Geometric Objects, Structured Sets, G-Spaces and All That: The Hole Story from Space-Time to Elementary Particles." To appear in *Going Critical* (in press).

In press. "Five Not-so-Easy Pieces: revised and translated versions of the five articles in *Dictionnaire d'Histoire et Philosophie des Sciences* (1999) on "Compatibility between Quantum Mechanics and Relativity," "Equivalence Principle," "Gravity," "Relativity," and "Space-Time." To appear in *Going Critical* (in press).

In press. "Fresnel's Dragging Coefficient as a Challenge to 19th Century Optics of Moving Bodies." A talk given at "the Sixth International Conference on the History of General Relativity," Amsterdam, The Netherlands, 26–29 June 2002. To appear in the Proceedings of the Conference.

In press. *Going Critical: Selected Essays*. Dordrecht/Boston/London: Kluwer Academic.

In press. "Structure and Individuality." To appear in *Going Critical* (in press).

In press. "The Optics and Electrodynamics of Moving Bodies." In *History of 19th Century Science*, ed. Jed Buchwald. Rome: Istituto della Enciclopedia Italiana (with Michel Janssen). To appear in *Going Critical* (in press).

In press. "The Story of Newstein, or Is Gravity Just Another Pretty Force?" In *Alternative Approaches to General Relativity (The Genesis of General Relativity*, vol. 4), eds. Jürgen Renn and Matthias Schemmel. Dordrecht/Boston/London: Kluwer Academic. Also to appear in *Going Critical* (in press).

EDITORIAL WORK

n.d. Editor, with L. Pande, of Christian Møller, "Selected Problems in General Relativity." Pp. 7-122 in *Brandeis University 1960 Summer Institute in Theoretical Physics. Lecture Notes*. Waltham: Brandeis University.

1974. 1964. Editor, with L. Infeld, *Proceedings of the International Conference on Relativistic Theories of Gravitation, Warsaw, 30-31 July 1962*. Paris and Warsaw: Gauthier-Villars and PWN.

1977. Editor, with John Earman and Clark Glymour, *Foundations of Space-Time Theories. Minnesota Studies in the Philosophy of Science*, vol. 8. Minneapolis: University of Minnesota Press.

1978. Editor, with R. S. Cohen and M. W. Wartofsky, *For Dirk Struik. Boston Studies in the Philosophy of Science* XV. Boston: Reidel.

1978. Editor, with R. S. Cohen, of *Leon Rosenfeld, Selected Papers on History and Philosophy of Science*. Boston: Reidel.

1987. Editor, *The Collected Papers of Albert Einstein, vol. 1, The Early Years. 1879-1902*. Princeton: Princeton University Press.

1987. Editor, with Robert S. Cohen and Michael Home, *Quantum Mechanical Studies for Abner Shimony*, vol. 1, *Experimental Metaphysics*, vol. 2, *Potentiality, Entanglement and Passion-at-a-Distance*. Dordrecht/Boston/London: Kluwer Academic.

1989. Editor, with Don Howard, *The History of General Relativity: Proceedings of the 1986 Osgood Hill Conference*. Boston/Basel/Stuttgart: Birkhäuser.

1990. Editor, *The Collected Papers of Albert Einstein*, vol. 2, *Writings 1901-1909*. Princeton: Princeton University Press.

1991. Editor, with Abhay Ashtekar, *Conceptual Problems of Quantum Gravity: Proceedings of the 1988 Osgood Hill Conference*. Boston/Basel/Stuttgart: Birkhäuser.

1993. Collaborating Editor, *Relativitées I, II*, vols. 2 and 3 of *Albert Einstein Oeuvres Choisis*, ed. François Balibar. Paris: Editions du Seuil/Editions du CNRS.

1995. Editor, with Kostas Gavroglu and Marx W. Wartofsky, *Essays in Honor of Robert S. Cohen*, vol. 1, *Physics, Philosophy, and the Scientific Community*, vol. 2, *Science, Politics and Social Practice*, vol. 3, *Science, Mind and Art*. Dordrecht/Boston/London: Kluwer Academic.

1998. Editor, *Einstein's Miraculous Year: Five Papers That Changed the Face of Physics*, with an Introduction by John Stachel and a Foreword by Roger Penrose. Princeton: Princeton University Press. Editions in Greek, Hungarian, German, Chinese, Spanish, Portuguese, and Italian, and an Indian edition have been published.

2000. Editor, with Don Howard, of *Einstein: The Formative Years, 1879-1909, (Einstein Studies*, vol. 8). Boston/Basel/Berlin: Birkhäuser.

ABSTRACTS, REVIEWS, ETC.

1961. "New Solutions to the Einstein Field Equations." *Bulletin of the American Physical Society* 6:305.

1969. "Bohm-Aharonov Effect and its Gravitational Analogue." *Bulletin of the American Physical Society* 14:16.

1969. "Prefatory note" to "Geodesic Killing Orbits and Bifurcate Killing Horizons," R. H. Boyer. In *Proceedings of the Royal Society* (London) A311:245 (with Jürgen Ehlers).

1969. "Variational Principle and Conservation Laws in Post-Newtonian Hydrodynamics." *Bulletin of the American Physical Society* 14:69 (with Thomas Pascoe).

1970. "Quasi-Newtonian Approximation Method in General Relativity." *Bulletin of the American Physical Society* 15:881 (with Gustavo Gonzalez-Martin).

1970. "Variational Principles as a Basis for Approximation Methods in General Relativistic Hydrodynamics." *Bulletin of the American Physical Society* 15:882.

1973. "Space-time Problems." Review of *General Relativity. Papers in Honor of J. L. Synge*, ed. L. O'Raifeartaigh,. In *Science* 180:292. To appear in *Going Critical* (in press).

1977. "Marxist Critique?" Review of *Ideology of/in the Natural Sciences*, eds. Hilary Rose and Steven Rose. In *The Sciences* 17:25-26. To appear in *Going Critical* (in press).

1977. "String Matter: Perfect Dust and Hydrodynamics." P. 324 in *Abstracts of Contributed Papers, 8th International Conference on General Relativity and Gravitation, August 7-12*. Waterloo, Ontario: University of Waterloo.

1980. Review of *Quantum Logic*, Peter Mittelstaedt. In *Isis* 71:162.

1980. Review of *Wolfgang Pauli: Scientific Correspondence with Bohr, Einstein, Heisenberg*, vol. 1, eds. A. Hermann, K. von Meyenn, V. F. Weisskopf. In *Nature* 285:515. To appear in *Going Critical* (in press).

1981. Review of *The Greatest Power on Earth: The Story of Nuclear Fission*, Ronald Clark. In *Nature* 290:655-656. To appear in *Going Critical* (in press).

1982. Review of *General Relativity: An Einstein Centenary Survey*, eds. S. W. Hawking and W. Israel. In *GRG Journal: General Relativity and Gravitation* 14:107.

1982. Review of *'Subtle is the Lord...' The Science and the Life of Albert Einstein*, Abraham Pais. In *Science* 218:989-990. Reprinted 2002 in *Einstein from 'B' to 'Z'*.

1983. "Special Relativity from Measuring Rods." Pp. 228-231, vol. 4, in *Abstracts of the 7th International Congress of Logic. Methodology and Philosophy of Science, Salzburg. July 11th-16th. 1983.* Salzburg: J. Hutteger OHG.

1985. "Author and augur in theoretical physics." Review of *Wolfgang Pauli: Scientific Correspondence with Bohr Einstein Heisenberg a. o.,* vol. 2, ed. Karl von Meyenn. In *Nature* 318:24. To appear in *Going Critical* (in press).

1986. Review of *Topology and Geometry for Physicists,* Charles Nash and Siddhartha Sen. In *American Journal of Physics* 54:476.

1987. Review of *On Manifolds with an Affine Connection and the Theory of Relativity,* Eli Cartan. In *Archives Internationales d'Histoire des Sciences* 37:376.

1988. "Inside a Physicist." Review of *H. A. Kramers: Between Tradition and Revolution,* Max Dresden. In *Nature* 332:744-745. To appear in *Going Critical* (in press).

1998. Review of *Albert Einstein: A Biography,* Albrecht Fölsing. In *Physics Today*, January 1998. Reprinted 2002 in *Einstein from 'B' to 'Z'*.

1990. "One man and his lab." Review of *Lawrence and his Laboratory,* J. L. Heilbron and Robert W. Seidel. In *Nature* 346:25-26. To appear in *Going Critical* (in press).

1999. Review of *The Dawning of Gauge Theory,"* Lochlainn O'Raifeartaigh. In *Studies in the History and Philosophy of Modern Physics* 30:453-455.